Survival Guide to General Chemistry

Survival Guide to General Chemistry

Patrick E. McMahon

Rosemary F. McMahon

Bohdan B. Khomtchouk

CRC Press is an imprint of the
Taylor & Francis Group, an **informa** business

CRC Press
Taylor & Francis Group
6000 Broken Sound Parkway NW, Suite 300
Boca Raton, FL 33487-2742

Printed on acid-free paper

International Standard Book Number-13: 978-1-138-33362-8 (Paperback)
978-1-138-33372-7 (Hardback)

Library of Congress Cataloging-in-Publication Data

Names: McMahon, Patrick E., author. | McMahon, Rosemary Fischer, author. | Khomtchouk, Bohdan B., author.
Title: Survival guide to general chemistry / Patrick E. McMahon, Rosemary F. McMahon, Bohdan Khomtchouk.
Other titles: General chemistry
Description: Boca Raton, Florida : CRC Press, 2019. | Includes bibliographical references and index.
Identifiers: LCCN 2018039429| ISBN 9781138333727 (hardback : alk. paper) | ISBN 9780429445828 (ebook) | ISBN 9781138333628 (pbk. : alk. paper)
Subjects: LCSH: Chemistry--Textbooks.
Classification: LCC QD33.2 .M3545 2019 | DDC 540--dc23
LC record available at https://lccn.loc.gov/2018039429

Visit the Taylor & Francis Web site at
http://www.taylorandfrancis.com

and the CRC Press Web site at
http://www.crcpress.com

Dedication

To our twins, Patrick and Grace, who are now ten, and to Professor Paul Robert Young of the University of Illinois, who was a mentor and a friend to both of us.

Patrick E. McMahon and
Rosemary F. McMahon

To my daughter, Bogdana Bohdanovna Khomtchouk.
In loving memory of my dear uncle, Taras Khomtchouk.

Bohdan B. Khomtchouk

Contents

Preface

Similar to our companion book, *Survival Guide to Organic Chemistry*, this work evolved over 30 years of teaching both introductory and major level general chemistry to a variety of student demographics at Benedictine University, Elmhurst College, Dominican University, and Triton College in Illinois. The topics and descriptions in this book offer detailed step-by-step methods and procedures for solving the major types of problems in general chemistry, whether mathematical or conceptual. Included are every major topic in the first semester of general chemistry and most of the major topics from the second semester. The approach to the format in this book is based on a viewpoint that answers the question, "If I were first learning this material myself, how would I want it explained to me and how would I want examples and problem solutions presented?" The explanations, instructional process sequences, solved examples, and completely solved practice problems in this book are greatly expanded with special emphasis placed on overcoming deficiencies and correcting problems that my experience suggests as the most common or most insidious.

Chapter 1 introduces the process of problem solving using density and unit conversion as a central theme. Chapters 4, 7, 9, and 10 provide a solid foundation for writing correct compound formulas, correctly naming compounds, writing balanced equations, as well as understanding the basic concepts for determining and analyzing acid/base, precipitation, and redox equations. Chapters 5, 6, 8, and 11 form the corresponding foundation for working with moles and stoichiometric calculations. Thermochemistry and thermodynamic problems are analyzed in Chapters 19 and 25.

Many chapters also provide alternative viewpoints as a way to help students better understand certain chemical concepts. For example, Chapter 3 includes problems of nuclear mass loss and nuclear energy as a way to grasp the ideas of mass number, atomic mass, and the definition of the atomic mass unit. Chapters 14 and 15 describe alternative methods for constructing Lewis structures, including large covalent molecules. These viewpoints help explain the role of bonding electrons and lone electron pairs for atoms bonding in typical patterns and include ways of viewing electron roles in polyatomic ions, unusual ions, and resonance structures. Included in Chapter 15 is a description of methods for structure notation for large (organic) molecules (condensed structures and line drawings), which are necessary for working with more complicated structures. Chapter 16 includes a description of the skill for drawing geometrical shapes as an aid to determining and picturing electron region and molecular geometry, a technique critical for branching out into organic chemistry. Chapters 22 and 24 provide a much expanded discussion of kinetic processes, reaction mechanisms, and their relationship to reaction rate laws. Chapter 23 deals with all equilibrium calculation types; in this case, unified by problem-solving techniques rather than reaction concepts.

Patrick E. McMahon

Authors

Patrick E. McMahon, PhD, holds a PhD in organic chemistry from the University of Illinois and a Masters of Arts for Teachers from Indiana University. For the past 25 years he has taught both general chemistry and organic chemistry at several Chicago area colleges and universities. Prior to that, he was a research scientist for Amoco Chemical Company at their Naperville, Illinois campus. He has over 30 years of experience teaching both introductory and science major level general chemistry to a variety of student demographics at Benedictine University, Elmhurst College, Dominican University, and Triton College. He is a member of the American Legion and served in the United States Army from 1970 to 1972. Awards that can be accredited to his name include the B.J. Babler award for outstanding contribution to undergraduate instruction at the University of Illinois, the Dean's Award for Teaching Excellence at Benedictine University, and first recipient of the Shining Star award given by the student senate for outstanding contribution to students at Benedictine University.

Rosemary F. McMahon, PhD, earned a BS with highest distinction in chemistry and an MS and PhD in organic chemistry at the University of Illinois. Her industrial career at Amoco Chemical and British Petroleum (BP) spanned chemicals process research and development, chemicals manufacturing, technical service, and customer support. She worked in oxidation chemistry, environmental catalyst chemistry, catalyst recovery processes, and manufacturing for product lines which included purified terephthalic acid, dimethyl 2,6-naphthalenedicarboxylate, and acrylonitrile. Her career later branched out to information technology as a patent and technical literature specialist supporting all technologies at BP worldwide. She is the co-author of four United States patents.

Bohdan B. Khomtchouk, PhD, is an American Heart Association (AHA) Postdoctoral Fellow in the Department of Biology and Department of Medicine at Stanford University. Previously, he was an NIH/NIA Postdoctoral Research Scholar, National Institute on Aging of the National Institutes of Health (Stanford Training Program in Aging Research) at Stanford University and a National Defense Science and Engineering Graduate (NDSEG) Fellow at the University of Miami Miller School of Medicine.

1 Unit Conversion and Density

An Introduction to Problem-Solving Methods

I GENERAL TECHNIQUES FOR PERFORMING UNIT CONVERSIONS

Principle #1: Adjust any conversion factor ratio so that the unit to be calculated is on top of the ratio fraction (numerator), and the unit to be cancelled out is on the bottom of the ratio fraction (denominator).

Principle #2: Volumes in certain problem types may be calculated from the corresponding equations for a rectangular object ($V = l \times w \times h$), a cylinder ($V = \pi r^2 h$), or a sphere ($V = 4/3\ \pi r^3$). It is easier to convert all linear dimensions to the required units **before** multiplication to generate volume in the required units. Alternatively, a unit conversion comparing two different volume units must be generated.

Principle #3: A unit conversion for different volume units can be derived from linear dimensions; any linear dimension cubed produces the corresponding volume unit.

Example: Convert cubic inches to cubic centimeters; use 1 inch (in) = 2.54 centimeters (cm). 1 cubic inch (in^3) is the volume of a cube, which is 1 inch on all three sides based on

$$\text{V(rectangular object)} = \text{length (l)} \times \text{width (w)} \times \text{height (h)}$$

$$\text{V(1 inch-sided cube)} = 1\text{ inch} \times 1\text{ inch} \times 1\text{ inch} = 1\text{ in}^3$$

Since 1 inch = 2.54 cm; the volume of the **identical** cube can be expressed as V (1 inch-sided cube) = $2.54\text{ cm} \times 2.54\text{ cm} \times 2.54\text{ cm} = 16.39\text{ cm}^3$

Therefore, **$1\text{ in}^3 = 16.39\text{ cm}^3$**

Principle #4: To perform a complete unit conversion for a **ratio**, it is easiest to convert the numerator and denominator **separately**, then reform the ratio and divide through.

Example: A common highway speed limit is 70 miles per hour; convert 70.0 miles/hour to meters per second. 1 kilometer = 0.6214 miles; 1 mile = 1.609 kilometers

Step (1): Convert the **numerator:**

$$70.0\text{ miles} \times \frac{1\text{ km}}{0.6214\text{ miles}} \times \frac{1000\text{ m}}{1\text{ km}} = 1.1265 \times 10^5\text{ m}$$

Principle #1: Regardless of which conversion factor is used to convert miles to km, km is on top of the ratio fraction, and miles is on the bottom of the ratio fraction. Therefore, the following conversion setup is equivalent.

$$70.0\text{ miles} \times \frac{1.6093\text{ km}}{1\text{ mile}} \times \frac{1000\text{ m}}{1\text{ km}} = 1.1265 \times 10^5\text{m}$$

Step (**2**): Convert the **denominator**:

$$1\text{ hour} \times \frac{60\text{ minutes}}{1\text{ hour}} \times \frac{60\text{ sec}}{1\text{ min}} = 3.600 \times 10^3\text{ sec}$$

Step (**3**): Reform the ratio for distance divided by time:

$$\frac{70\text{ miles}}{1\text{ hour}} = \frac{1.1265 \times 10^5\text{ m}}{3.600 \times 10^3\text{ sec}} = \mathbf{31.3\text{ m/sec}}$$

Example: The density of visible matter in the universe (i.e., ignoring dark matter) is estimated to be equivalent to 3 hydrogen atoms per cubic meter (m^3) of space. Calculate the density of the universe in units of **grams/cm³**. Although the estimate is very general, express the final answer to three significant figures (extra significant figures can and, often, should be carried through a calculation until the final answer).

mass of 1 hydrogen (H) atom = 1.674×10^{-27} kilogram

1 kilogram (kg) = 1000 grams (g); 1 meter (m) = 100 centimeters (cm)

The ratio is expressed as $\frac{\textbf{(mass)}}{\textbf{(Volume)}}$

The starting complete unit to be converted is $\frac{3\text{ hydrogen atoms}}{1\text{ m}^3} = \frac{\text{(numerator)}}{\text{(denominator)}}$

The complete unit to be calculated is $\frac{\text{grams}}{1\text{ cm}^3} = \frac{\text{(numerator)}}{\text{(denominator)}}$

Use **Principle #4:** Convert the numerator and denominator separately, then reform the ratio.

Step (**1**): Convert the **numerator** from the mass of 3 H atoms to the mass in grams:

$$(3\text{ H atoms}) \times \frac{(1.674 \times 10^{-27}\text{ kg})}{1\text{ H atom}} \times \frac{(1000\text{ g})}{1\text{ kg}} = \mathbf{5.02 \times 10^{-24}\text{ grams}}$$

Step (**2**): Convert the **denominator** from 1 m^3 to 1 cm^3:

No direct conversion unit for m^3 to cm^3 is provided.

Use **Principle #3:**

1 m = 100 cm is used as the **linear** conversion;

1 cubic meter (m^3) is the volume a cube, which is 1 meter on all three sides.

V (1 meter-sided cube) = 1 m × 1 m × 1 m = 1 m^3

Since 1 m = 100 cm; the volume of the **identical** cube can be expressed as

V (1 meter-sided cube) = 100 cm × 100 cm × 100 cm = 1×10^6 cm^3

Therefore **$1\ m^3 = 1 \times 10^6\ cm^3$**

Step (3): Reform the $\frac{(\text{mass})}{(\text{Volume})}$ ratio.

The converted units from steps (1) and (2) are used; the bottom number of the ratio must be 1.

$$\textbf{density of universe } (\mathbf{g/cm^3}) = \frac{(5.02 \times 10^{-24}\ g)}{(1 \times 10^6\ cm^3)} = \mathbf{5.02 \times 10^{-30}\ g/cm^3}$$

II GENERAL PROCEDURES FOR SOLVING DENSITY PROBLEMS

Density (**d**) measures the mass (**m**) of an object relative to its volume (**V**):

$$\text{density } (\mathbf{d}) = \frac{\text{mass } (\mathbf{m})}{\text{volume } (\mathbf{V})} \quad \text{or} \quad \mathbf{d} = \frac{\mathbf{m}}{\mathbf{V}}$$

Density **units of mass**: usually grams (**g**);

Density **units of volume**: cubic centimeters (**cm^3**) for solids; milliliters (**mL**) for liquids; Liters (**L**) for gases.

The density equation can be solved for each possible variable.

$\mathbf{d = \frac{m}{V}}$**:** solves for **density** as the **unknown** variable when mass and volume are known variables.

$\mathbf{m = V \times d}$**:** solves for **mass** as the **unknown** variable when density and volume are known variables.

$\mathbf{V = \frac{m}{d}}$**:** solves for **volume** as the **unknown** variable when density and mass are known variables.

Process for Density Problems

Step (**1**): Identify the correct form of the density equation that is needed to solve for the desired unknown variable.

Step (**2**): Identify or calculate the required values for the known variables in the correct units based on the density units given or required.

Step (**3**): Complete the calculation for the unknown variable based on the results from step (**1**) and step (**2**).

Step (**4**): When necessary, use the calculated variable from step (**3**) to solve for additional information related to a more complex problem.

III GENERAL EXAMPLES FOR DENSITY WITH UNIT CONVERSIONS

Example: A 1.75-liter jug is filled with liquid mercury; what is the mass of the mercury in **pounds**? Density of Hg = 13.6 g/mL; 1 pound = 454 grams.

Step (**1**): **Mass** is the unknown variable; equation is **$m = V \times d$**

Step (**2**): **d** is given as **13.6 g/mL**;

volume must be converted to units of mL:

$$V = (1.75\ L) \times \frac{(1000\ mL)}{1\ L} = \mathbf{1,750\ mL}$$

Principle #1: mL is the unit to be calculated and is on top of the ratio fraction; L is the unit to be cancelled out and is on the bottom of the ratio fraction.

Step (**3**): $m = V \times d : m\ (grams) = (1.750\ mL) \times (13.6\ g/mL) = \mathbf{23,800\ grams}$

$$= \mathbf{2.38 \times 10^4\ g}$$

Step (**4**): Mass in **grams** = 23, 800 g; convert to find the mass in **pounds**:

mass (lbs) = (23, 800 g) × (1 lb/454 g) = **52.4 lbs**

Example: Calculate the mass in **kilograms** of a mahogany wooden block with dimensions of 1.05 **meters** × 0.650 **meters** × 0.110 **meters**; density of mahogany = 0.770 g/cm^3;

1 kilogram (kg) = 1000 grams (g); 1 meter (m) = 100 centimeters (cm).

V(rectangular object) = length (l) × width (w) × height (h).

Step (**1**): **Mass** is the unknown variable; equation is $\mathbf{m = V \times d}$

Step (**2**): The density, **d**, is given as **0.770 g/cm^3**;

volume in **cm^3** must be calculated from the dimensions.

Principle #2: Convert each meter unit in the problem to centimeters using (100 cm/1 m) before multiplying through using each dimension unit to find volume. If the volume is calculated in cubic meters, the conversion factor for m^3 to cm^3 must be calculated.

V in **cm^3** = (105 **cm**) × (65.0 **cm**) × (11.0 **cm**) = **75075 cm^3**

Step (**3**): m = V × d: **m (g)** = (75075 cm^3) × (0.770 g/cm^3) = **5.78 × 10^4 g**

Step (**4**): $mass\ (\mathbf{kg}) = (5.78 \times 10^4\ g) \times \frac{(1\ kg)}{1000\ g} = \mathbf{57.8\ kg}$

Example: Titanium metal rods are often used in human joint replacement due to their low density but high strength and tissue unreactivity. Calculate the length (height) in centimeters (**cm**) of a **cylindrical** rod of titanium, which has a diameter of 3.20 **cm** and a mass of 2.44 **pounds**; density of titanium (Ti) = 4.51 grams/cm^3; 1 pound (lb) = 454 grams (g); V (cylinder) = $\pi r^2 h$ (r = radius; h = height); radius = (½) diameter.

Steps 1 through 3 are used to find the volume of the rod using the density; step 4 then calculates the length (height).

Step (**1**): **Volume** is the unknown variable; equation is $\mathbf{V = \frac{m}{d}}$

Step (**2**): The density, d, is given as 4.51 g/cm^3;

$$\text{mass (g)} = 2.44 \text{ lbs} \times 454 \text{ grams/lb} = \mathbf{1106\ grams}$$

Step (3): $V = \frac{m}{d}$; **V in cm³** $= \frac{1106\text{ g}}{4.51\text{ g/cm}^3} = \mathbf{245.1\ cm^3}$

Step (4): V (cylinder) $= \pi r^2 h$; solve for **h:** $\mathbf{h} = \frac{\mathbf{V}}{\pi \mathbf{r}^2}$

$$r = (½)(3.20\text{ cm}) = 1.60\text{ cm};\ h = \frac{245.1\text{ cm}^3}{(\pi)(1.60\text{ cm})^2} = \mathbf{30.5\ cm}$$

Example: The center of the galaxy M-87 is thought to have conditions that can produce the formation of a black hole. Five **billion** stars the size of the sun occupy a cylindrical (disk) region of space in the galactic center with a diameter of 10 light-years and a height of 2 light-years. Follow parts (a)–(e) to calculate the density of this galactic region in **grams/cm³**.

a. Calculate the mass in **grams** of 5 billion stars the size of the sun; 1 billion = 10^9; mass of the sun is 2.19×10^{27} tons; 1 ton = 2000 pounds; 1 pound (lb) = 454 grams.
b. The velocity (speed) of light (**c**) = 6.70×10^8 **miles/hour**; convert this value to units of **cm/sec**; 1 mile = 1.6093 km; 1 hour = 60 minutes; 1 minute = 60 seconds.
c. A light year is a (linear) measure of distance equal to the distance that light can travel through space in one year. Use the value from part (b) and the equation **distance = velocity × time** to calculate the length of 1 light-year in units of **centimeters**. 1 year = 365.25 days; 1 day = 24 hours.
d. Calculate the volume in **cubic centimeters** (cm³) of the disk (cylinder) of 10 light-year diameter and 2 light-year height using the value from part (c).

$$V \text{ (cylinder)} = \pi r^2 h \text{ (r = radius; h = height); radius} = (½) \text{ diameter.}$$

e. Complete the calculation of the density of the galactic center in grams/cm³ by using the values from previous parts of the problem.

a. **mass in grams** =

$$(5 \times 10^9 \text{ stars}) \times \frac{(2.19 \times 10^{27} \text{ tons})}{1 \text{ star}} \times \frac{(2000 \text{ lbs})}{1 \text{ ton}} \times \frac{(454 \text{ g})}{1 \text{ lb}} = \mathbf{9.95 \times 10^{42}\ grams}$$

b. Convert the numerator:

$$(6.70 \times 10^8 \text{ miles}) \times \frac{(1.6093 \text{ km})}{1 \text{ mile}} \times \frac{(1000 \text{ m})}{1 \text{ km}} \times \frac{(100 \text{ cm})}{1 \text{ m}} = \mathbf{1.078 \times 10^{14}\ cm}$$

Convert the denominator:

$$(1 \text{ hour}) \times \frac{(60 \text{ minutes})}{1 \text{ hour}} \times \frac{(60 \text{ seconds})}{1 \text{ minute}} = \mathbf{3600\ sec}$$

Reform the ratio using the converted units and divide through:

$$\text{Ratio } \frac{\mathbf{(distance)}}{\mathbf{time}} = \frac{(1.078 \times 10^{14} \text{ cm})}{3600 \text{ sec}} = \mathbf{3.00 \times 10^{10}\ cm/sec}$$

c. 1 year in **seconds** =

$$\frac{(365.25\text{ days})}{1\text{ year}} \times \frac{(24\text{ hours})}{1\text{ day}} \times \frac{(60\text{ minutes})}{1\text{ hour}} \times \frac{(60\text{ sec})}{1\text{ min}} = \mathbf{3.156 \times 10^{7}\ sec}$$

$$\mathbf{1\ light\text{-}year} = \frac{(3.00 \times 10^{10}\text{ cm})}{1\text{ sec}} \times (3.156 \times 10^{7}\text{ sec}) = \mathbf{9.47 \times 10^{17}\ cm}$$

d. $$r = (1/2)(10\text{ light-years}) \times \frac{(9.47 \times 10^{17}\text{ cm})}{1\text{ light year}} = 4.735 \times 10^{18}\text{ cm}$$

$$h = (2\text{ light-years}) \times \frac{(9.47 \times 10^{17}\text{ cm})}{1\text{ light-year}} = 1.894 \times 10^{18}\text{ cm}$$

$$V = \pi r^2 h = (\pi)(4.735 \times 10^{18}\text{ cm})^2 (1.894 \times 10^{18}\text{ cm}) = \mathbf{1.33 \times 10^{56}\ cm^3}$$

e. Step (1): **Density** is the unknown variable; equation is $\mathbf{d = \frac{m}{V}}$

Step (2): Mass in grams was calculated in part (a): = **9.95×10^{42} grams**
Volume in cm^3 was calculated in part (d): = **1.33×10^{56} cm^3**

Step (3): $d = \frac{m}{V}$; $$\mathbf{d}(\text{galactic center}) = \frac{(9.95 \times 10^{42}\text{ grams})}{(1.33 \times 10^{55}\text{ cm}^3)} = \mathbf{7.48 \times 10^{-13}\ g/cm^3}$$

IV PRACTICE PROBLEMS

1. The density of $CHCl_3$ = 1.48 g/mL; what is the volume in mL of 55.5 grams of $CHCl_3$?
2. Calculate the volume in gallons of 100 pounds of water. 1 pound (lb) = 454 grams; density of water = 1.00 grams/mL; 1 gallon = 3.79 liters; 1 liter = 1000 milliliters (mL)
3. The density of gasoline = 0.709 g/mL; what is the mass in kilograms (kg) of the amount of gasoline that will fill a 20.0 gallon automobile tank? (1 gallon = 3.79 liters)
4. Calculate the mass of a gold bar in units of grams, kilograms, and pounds. The gold bar measures 30.0 cm × 8.00 cm × 4.40 cm. Density of Au = 19.3 g/cm³; volume of a rectangular object = length (l) × width (w) × height (h).
5. Calculate the mass in grams of a solid copper ball, which has a diameter of 22.6 millimeters. Density of Cu = 8.96 g/cm³; V (sphere) = $(4/3)\ \pi\ r^3$; r = (½) diameter; be certain to convert the radius measurement to cm from mm.
6. An unknown metal alloy is formed into a cylindrical bar, which is 1.25 meters long (length in this case = height of a cylinder) and has a radius of 42.5 millimeters (mm). The mass of the bar was found to be 103 kilograms (kg). Calculate the density of the metal alloy in the required units of g/cm³. V (cylinder) = $\pi\ r^2\ h$; be certain to use the correct units in the calculation of mass and volume.
7. Volume can also be measured by displacement, as demonstrated by the following experiment. A metal ring was found to have a mass of 17.8 grams. It is then placed into a graduated measured container (such as a chemistry lab graduated cylinder) filled with water. The volume mark for the water in the container without the ring was 15.0 mL; after the ring was added to the container, the water level mark showed 16.7 mL. The difference between the two measured volumes must be due to the actual volume of the ring. Calculate the density of the metal in the ring. Determine whether the ring was gold or silver; density of gold = 19.3 g/cm³; density of silver = 10.5 g/cm³.

8. A cylindrical tube is filled with a liquid that has a density of 1.30 g/mL (=1.30 g/cm^3). The inside diameter of the tube, which measures the actual diameter of the cylindrical column of the liquid, is 0.870 millimeters (mm). The mass of the empty tube (no liquid) is 0.7850 grams; the mass of the tube filled with the liquid is 0.8160 grams. Calculate the height of the liquid in the tube; recall that V (cylinder) = $\pi r^2 h$.
   Follow the two parts below:
   a) Calculate the volume of the liquid in the tube from the density and mass information.
   b) Use the volume calculated from part (a) and the equation for the volume of a cylinder to calculate the height of the liquid.

9. Use the techniques similar to those described in problem 8 to calculate the average diameter of a lead bead if 25 beads have a mass of 2.31 grams; density of Pb = 11.3 g/cm^3; V (bead/sphere) = $(4/3)\pi r^3$.

10. A quasar is a distant object that shines with 100 times the light of an entire galaxy but is contained in a volume 1 millionth the size of a galaxy. Matter falling into a black hole is suspected as the ultimate power source. A certain quasar contains the mass equivalent of 100 million stars the size of the sun and is estimated to have a density of 6.52×10^{-5} g/cm^3. Assuming that the quasar is spherical, calculate the quasar diameter in cm and light-years. **Hint**: First find the volume of the sphere using the density, then use the equation for a sphere to calculate the radius then diameter. Use any relevant information from previous examples or problems.

$$V \text{ (sphere)} = (4/3)\,\pi\, r^3 \text{ (r = radius)};\ 1 \text{ million} = 10^6$$

V ANSWERS TO PRACTICE PROBLEMS

1. The density of $CHCl_3$ = 1.48 g/mL; what is the volume in mL of 55.5 grams of $CHCl_3$?

   (1) **Volume** is the unknown variable; equation is $\mathbf{V = \frac{m}{d}}$

   (2) The density, **d**, is given as **1.48 g/mL**; **mass** is given directly as **55.5 grams**

   (3) $V = \frac{m}{d}$: $\mathbf{V\ in\ mL} = \frac{55.5\ g}{1.48\ g/mL} = \mathbf{37.5\ mL}$

2. Calculate the volume in **gallons** of 100. **pounds** of water. 1 pound (lb) = 454 grams; density of water = 1.00 grams/mL; 1 gallon = 3.79 liters; 1 liter = 1000 milliliters (mL)

   (1) **Volume** is the unknown variable; equation is $\mathbf{V = \frac{m}{d}}$

   (2) The density, **d**, is given as **1.00 g/mL**;

   $$\textbf{mass (g)} = 100.\ \text{lbs} \times 454\ \text{grams/lb} = \mathbf{4.54 \times 10^4\ grams}$$

   (3) $V = \frac{m}{d}$: $\mathbf{V\ in\ mL} = \frac{4.54 \times 10^4\ g}{1.00\ g/mL} = \mathbf{4.54 \times 10^4\ mL}$

   (4) V in **gallons** $= (4.54 \times 10^4\ mL) \times \frac{(1\ L)}{1000\ mL} \times \frac{(1\ gal)}{3.70\ L} = \mathbf{12.0\ gallons}$

3. The density of gasoline = 0.709 g/mL; what is the mass in **kilograms** (kg) of the amount of gasoline that will fill a 20.0-gallon automobile tank? (1 gallon = 3.79 liters)
   (1) **Mass** is the unknown variable; equation is $\mathbf{m = V \times d}$
   (2) The density, **d**, is given as **0.709 g/mL; volume** must be converted to units of mL:

   $$V = (20.0\ gal) \times (3.79\ L/gal) \times (1000\ mL/L) = \mathbf{75{,}800\ mL = 7.58 \times 10^4\ mL}$$

(3) m = V × d: **m (grams)** = (7.58 × 10^4 mL) × (0.709 g/mL) = **5.37 × 10^4 grams**
(4) Mass in **grams** = 5.37 × 10^4 g; convert mass to **kilograms:**

mass (kg) = (5.37 × 10^4 g) × (1 kg/1000 g) = **53.7 kg**

4. Calculate the mass of a gold bar in units of **grams, kilograms,** and **pounds**. The gold bar measures 30.0 cm × 8.00 cm × 4.40 cm. Density of Au = 19.3 g/cm^3;

Volume of a rectangular object = length (l) × width (w) × height (h)

(1) **Mass** is the unknown variable; equation is **m = V × d**
(2) The density, **d**, is given as **19.3 g/cm^3; volume** must be calculated from the dimensions:

V in cm^3 = (30.0 cm) × (8.00 cm) × (4.40 cm) = **1056 cm ^{3} = 1.056 × 10^3 cm^3**

(3) m = V × d: **m (g)** = (1.056 × 10^3 cm^3) × (19.3 g/cm^3) = **2.04 × 10^4 g**
(4) Mass in **grams** = 2.04 × 10^4 g; convert mass to **kilograms and pounds:**

$$\text{mass } (\textbf{kg}) = (2.04 \times 10^4 \text{ g}) \times \frac{(1\text{ kg})}{1000\text{ g}} = \textbf{20.4 kg}$$

$$\text{mass } (\textbf{lbs}) = (2.04 \times 10^4 \text{ g}) \times \frac{(1\text{ lb})}{454\text{ g}} = \textbf{44.9 lbs}$$

5. Calculate the mass in grams of a solid copper ball, which has a diameter of 22.6 millimeters. Density of Cu = 8.96 g/cm^3; V (sphere) = (4/3) π r^3; r = (½) diameter.
(1) **Mass** is the unknown variable; equation is **m = V × d**
(2) The density, **d**, is given as **8.96 g/cm^3; volume** must be calculated from the dimesions: $\textbf{r}$ in $\textbf{cm} = (1/2) \times (22.6\text{ mm}) \times \frac{1\text{ cm}}{10\text{ mm}} = 1.13\text{ cm}$

V in **cm^3** = (4/3) × (π) × (1.13 cm)^3 = **6.04 cm^3**

(3) m = V × d: **m (g)** = (6.04 cm^3) × (8.96 g/cm^3) = **54.1 g**

6. An unknown metal alloy is formed into a cylindrical bar, which is 1.25 meters long (length in this case = height of a cylinder) and has a radius of 42.5 millimeters (mm). The mass of the bar was found to be 103 kilograms (kg). Calculate the density of the metal alloy in the required units of g/cm^3. V (cylinder) = π r^2 h.

(1) **Density** is the unknown variable; equation is $\textbf{d} = \frac{\textbf{m}}{\textbf{V}}$ with required units of **g/cm^3**.

(2) The **mass** is given in **kilograms**; it must be converted to grams:

m(g) = (103 kg) × 1000 g/kg) = **1.03 × 10^5 grams**

Volume must be calculated from the dimensions and equation V (cylinder) = π r^2 h:

r in **cm** = (42.5 mm) × (1 cm/10 mm) = 4.25 cm;

h in **cm** = (1.25 m) × (100 cm/m) = 125 cm

V in **cm^3** = (π) × (4.25 cm)^2 × (125 cm) = 7093 cm^3

= **7.10 × 10^3 cm^3** (to 3 significant figures)

(3) $d = \frac{m}{V}$; $\mathbf{d} = \frac{(1.03 \times 10^5 \text{ g})}{(7.10 \times 10^3 \text{ cm}^3)} = \mathbf{14.5\ g/cm^3}$

7. Calculate the density of the metal in the ring. Determine whether the ring was gold or silver; density of gold = 19.3 g/cm^3; density of silver = 10.5 g/cm^3.

(1) **Density** is the unknown variable; equation is $\mathbf{d} = \frac{\mathbf{m}}{\mathbf{d}}$ (required units = **g/cm^3**)

(2) The **mass** is given as **17.8 grams**; volume must be calculated from the water displacement in the graduated container:

$$V(\text{ring}) = V(H_2O \text{ with ring}) - V(H_2O \text{ without ring})$$

V(ring) = (16.7 mL) – (15.0 mL) = **1.7 mL**; convert mL to cm^3 using: 1 mL = 1 cm^3:

$$\mathbf{V(ring)} = (1.7 \text{ mL}) \times (1 \text{ cm}^3/1 \text{ mL}) = \mathbf{1.7\ cm^3}$$

(3) $d = \frac{m}{v}$; $\mathbf{d} = \frac{(17.8 \text{ g})}{(1.7 \text{ cm}^3)} = \mathbf{10.5\ g/cm^3}$

(4) 10.5 g/cm^3 matches the density of **silver.**

8. a) Calculate the volume of the liquid in the tube from the density and mass information.

(1) **Volume** is the unknown variable; equation is $\mathbf{V} = \frac{\mathbf{m}}{\boldsymbol{\pi}\mathbf{d}}$

(2) The density, **d,** is given as **1.30 g/mL; mass** of the liquid is found by difference:

mass (liquid) = (mass of tube full) – (mass of tube empty)

= (0.8160 g) – (0.7850 g) = **0.0310 g**

(3) $V = \frac{m}{d}$; $\mathbf{V\ in\ mL} = \frac{0.0310 \text{ g}}{1.30 \text{ g/mL}} = \mathbf{0.0238\ mL = 0.0238\ cm^3}$

b) Use the volume calculated from part (a) and the equation for the volume of a cylinder to calculate the height of the liquid.

(4) V (cylinder) = $\pi r^2 h$; solve for h: $h = \frac{V \text{ (cylinder)}}{r^2}$

r in **cm** = (½) × (0.870 mm) × (1 cm/10 mm) = 0.0435 cm

$$\mathbf{h\ (cm)} = \frac{0.0238 \text{ cm}^3}{(3.1416)(0.0435 \text{ cm})^2} = \mathbf{4.00\ cm}$$

9. a) Calculate the volume of one average bead from the density and mass information.

(1) **Volume** is the unknown variable; equation is $\mathbf{V} = \frac{\mathbf{m}}{\mathbf{d}}$

(2) The density, **d,** is given as **11.30 g/cm^3**; calculate **mass** of **one** bead:

$$\textbf{mass of one bead} = \frac{(\text{mass of 25 beads})}{25} = \frac{(2.31 \text{ g})}{25} = \mathbf{0.0924\ g}$$

(3) $V = \frac{m}{d}$; **V in cm³** $= \frac{0.0924\text{ g}}{11.3\text{ g/cm}^3} = \mathbf{0.00818\ cm^3}$

b) Use the volume calculated from part (a) and the equation for the volume of a sphere to calculate the radius and then diameter of one bead.

(4) V (sphere) = $(4/3)\pi r^3$; solve for r: $r = \left\{\frac{\text{V (sphere)}}{(4/3)}\right\}^{1/3}$

$$r = \left(\frac{(0.00818\text{ cm}^3)}{(4/3)(3.1416)}\right)^{1/3} = \mathbf{0.125\ cm;\ diameter} = 2 \times r = 2 \times (0.125\text{ cm})$$

$$= \mathbf{0.250\ cm}$$

10. Find the volume of the sphere using the density, then calculate the radius and diameter.

(1) **Volume** is the unknown variable; equation is $\mathbf{V = \frac{m}{d}}$

(2) The density, **d**, is given as $\mathbf{6.52 \times 10^{-5}\ g/cm^3}$; mass (g) is calculated from the number of stars and the mass of a star in grams.

$$100\text{ million} = 100 \times 10^6 = 1.00 \times 10^8$$

mass in grams =

$$(1.00 \times 10^8\text{ stars}) \times \frac{(2.19 \times 10^{27}\text{ tons})}{1\text{ star}} \times \frac{(2000\text{ lbs})}{1\text{ ton}} \times \frac{(454\text{ g})}{1\text{ lb}} = \mathbf{1.99 \times 10^{41}\ grams}$$

(3) $V = \frac{m}{d}$; **V in cm³** $= \frac{(1.99 \times 10^{41}\text{ g})}{(6.52 \times 10^{-5}\text{ g/cm}^3)} = \mathbf{3.05 \times 10^{45}\ cm^3}$

(4) V(sphere) = $(4/3)\ \pi\ r^3$; solve for r: $r = \left(\frac{\text{V(sphere)}}{(4/3)}\right)^{1/3}$

$$r = \left(\frac{3.05 \times 10^{45}\text{ cm}^3}{(4/3)\ (3.1416)}\right)^{1/3} = (7.28 \times 10^{44}\text{ cm}^3)^{1/3}$$

To take a cube root of a number in exponential notation, express the exponential portion in a value evenly divisible by 3 and perform the exponential cube root by dividing the exponent by 3; then take the cube root of the non-exponential number.

$$r = (7.28 \times 10^{44}\text{ cm}^3)^{1/3} = (728 \times 10^{42}\text{ cm}^3)^{1/3} = (728)^{1/3} \times (10^{42})^{1/3}$$

$$= (728)^{1/3} \times 10^{14} = \mathbf{9.00 \times 10^{14}\ cm}$$

Diameter of the quasar (cm) = $2 \times r = 2 \times (9.00 \times 10^{14}\text{ cm}) = \mathbf{1.80 \times 10^{15}\ cm}$

Diameter of the quasar (light-years) = $(1.80 \times 10^{15}\text{ cm}) \times \frac{(1\text{ light-year})}{(9.47 \times 10^{17}\text{ cm})} = \mathbf{1.90 \times 10^{-3}\ light\text{-}year}$

2 Atomic Particles, Isotopes, and Ions

An Initial Look at Atomic Structure

I GENERAL CONCEPTS

For the purposes of chemistry problems, the atom is composed of three major particles; the arbitrary symbols shown in the table are used throughout the chapter. The mass unit abbreviated as "amu," the atomic mass unit, provides a convenient measurement of mass and is explained in detail in Chapter 3; 1 amu = 1.66054×10^{-27} kg

Particle	Symbol	Unit Charge	Mass (kg)	Mass (amu)
Proton	p^+	+1	1.6726×10^{-27} kg	1.007276 amu
Electron	e^-	−1	9.11×10^{-31} kg	0.0005486 amu
Neutron	n^0	0	1.6749×10^{-27} kg	1.008665 amu

A **specific** neutral atom of a **specific** element is characterized by four values:

(1) the number of protons (p^+) in the nucleus; (2) the number of electrons (e^-) outside the nucleus; (3) the number of neutrons (n^0) in the nucleus; and (4) the number of nuclear particles (nucleons), the addition of proton number plus neutron number.

(1) **Protons** (p^+) are particles in the **nucleus**; the number of protons is specified by the **atomic number** of an element, which has the symbol **Z. The atomic number (number of protons) is the defining characteristic of an element**. Each different element has a different number of protons in the nucleus, and every atom of the same element must have the same identical number of protons in the nucleus.

(2) **Electrons** (e^-) are particles that reside **outside** the nucleus; electrons occupy the remaining volume of the atom. For **neutral** atoms, the **number of electrons must equal the number of protons** to achieve electrical neutrality.

(3) **Neutrons** (n^0) are particles in the **nucleus**. Atoms of the same element (same number of protons) may have differing numbers of neutrons.

(4) The total number of nucleons (protons plus neutrons) in the nucleus is termed the **mass number**, which has the symbol **A**. The mass number is **proportional** to the actual mass of the atom. The mass values shown in the above table confirm that the total mass of any atom is determined by the masses of the nuclear particles; the mass of the electrons is generally insignificant, as seen in the following example. Mass determination of atoms is covered in Chapter 3.

The mass number (A) is the sum of the # of protons plus the # of neutrons.

A specific element (atom type) is characterized **only** by the number of protons in the nucleus. Atoms of a specific element can have **variable** numbers of **neutrons** in the nucleus or variable numbers of electrons outside the nucleus.

1. Atoms of the same element (same # of protons), which have different numbers of **neutrons,** are said to be related as **isotopes.**
2. Atoms may lose some of their electrons or gain extra electrons. In these cases, the atoms will **not** be electrically neutral. Atoms that have a surplus or deficiency of **electrons** (not electrically neutral) are termed **ions.**

Example for mass number (A): The mass of an atom is not exactly equal to the sum of the independent particles comprising the nucleus plus electrons (Chapter 3).

However, for this example:

(a) First calculate the mass in **kg** of the nucleus of a neutral atom of bismuth with **Z = 83** and **A = 209** as represented by independent particles.
(b) Then calculate the mass of the electrons and determine whether this mass can be added to the nuclear mass to affect the atomic mass to within four significant figures.

Use the mass value for each particle in kg from the table.

a) Bismuth (Z = 83) must therefore have 83 **protons = 83 p^+**
The number of **neutrons** must be = **A − Z =** (209) − (83) = **126 n^0**

mass of protons in bismuth = (83 p^+) × (1.6726 × 10^{-27} kg/p^+) = **1.3883 × 10^{-25} kg**

mass of neutrons in bismuth = (126 n^0) × (1.6749 × 10^{-27} kg/n^0) = **2.1104 × 10^{-25} kg**

total mass (ignoring electrons) = (1.3883 × 10^{-25} kg) + (2.1104 × 10^{-25} kg) = **3.4987 × 10^{-25} kg = 3.499 × 10^{-25} kg** to four significant figures

b) The number of electrons in neutral bismuth must equal the number of protons; number of electrons = 83 electrons

mass of electrons in bismuth = (83 e^-) × (9.11 × 10^{-31} kg/e^-) = **7.56 × 10^{-29} kg** to three significant figures.

The concept of this example is to add the mass of the electrons to the mass of the nucleus: first convert 7.56 × 10^{-29} kg to a value that has the same exponent as the mass expressed for the nucleus: 7.56 × 10^{-29} kg = 0.000756 × 10^{-25} kg

total mass including electrons = (3.4987 × 10^{-25} kg) + (0.000756 × 10^{-25} kg) = **3.499 × 10^{-25} kg** to four significant figures.

This value is the same as the mass found for the sum of protons plus neutrons to four significant figures. The mass number A (p^+ + n^0) α mass of the atom.

II ATOMIC SYMBOLS AND ISOTOPES

A general symbol for a **neutral** atom uses the following format:

$$^{A}_{Z}X$$

X represents the specific symbol for the element that matches the correct atomic number.
Z is the atomic number = # of protons in the nucleus.
A is the mass number = # of nucleons = # of protons plus # of neutrons.

The number of electrons for a **neutral** atom is always specified by the atomic number (number of protons), since all neutral atoms must have # of electrons = # of protons.

Whenever an atom is neutral, no specific additional information is shown in the symbol; **unless a specific ion charge is shown, the atom is assumed to be neutral**. Ion notation is presented in the next section.

Example: Write the symbol for the specific atom that has 37 protons, 48 neutrons, and 37 electrons; name this element; use the periodic table as necessary.

a) The number of protons is 37; thus, the atomic number **(Z)** = **37** = # of protons. The atomic number, Z, is placed as a subscript in front of the correct symbol.
b) The element with Z = 37 protons is rubidium; symbol = **Rb**
c) Mass number **(A)** = # of nucleons (nuclear particles) = 37 protons + 48 neutrons = **85**
d) The mass number, A, is placed as a superscript in front of the correct symbol.
e) The # of electrons = # of protons = 37; the atom is **neutral**: no ion charge is indicated.

$$^{85}_{37}Rb$$

Example: Write the symbol for neutral lead with 121 neutrons; use the periodic table as necessary.

a) The element lead has the symbol = **Pb** (from the Latin: *plumbum*)
b) The periodic table indicates that Pb has an atomic number **(Z)** = **82**; the # of protons = 82
c) Mass number **(A)** = # of nucleons = 82 protons + 121 neutrons = **207**
d) The atom is **neutral**: no ion charge is indicated.
e) The # of electrons = # of protons = 82

$$^{207}_{82}Pb$$

Example: Write the symbol for neutral tungsten with 115 neutrons; use the periodic table as necessary.

a) The element tungsten has the symbol = **W** (from the German: *wolfram*)
b) The periodic table indicates that W has an atomic number **(Z)** = **74**; the # of protons = 74
c) Mass number **(A)** = # of nucleons = 74 protons + 115 neutrons = **189**
d) The atom is **neutral**: no ion charge is indicated
e) The # of electrons = # of protons = 74

$$^{189}_{74}W$$

Example: Determine the element name and particle (p^+, e^-, n^0) numbers for: $^{108}_{47}\mathbf{Ag}$; use the periodic table as necessary.

a) **Ag** is the symbol for **silver** (from the Latin: *argentum*)
b) The number of protons = **Z = 47**
c) **A** = 108 = # of nucleons = (# of protons) + (# of neutrons); solve for # of neutrons: **# of neutrons** = [# of nucleons (A)] – [# of protons] = [108] – [47] = **61**
d) No charge is indicated; atom must be neutral: # of e^- = # of p^+ = **47**
p^+ = 47; e^- = 47; n^0 = 61

Example: Determine the element name and particle (p^+, e^-, n^0) numbers for: $^{200}_{80}\mathbf{Hg}$; use the periodic table as necessary.

a) **Hg** is the symbol for **mercury** (from the Latin: *hydroargentum*)
b) The number of protons = **Z = 80**
c) **A** = 200 = # of nucleons = (# of protons) + (# of neutrons); solve for # of neutrons: **# of neutrons** = [# of nucleons (A)] – [# of protons] = [200] – [80] = **120**
d) No charge is indicated; atom must be neutral: # of e^- = # of p^+ = **80**
p^+ = 80; e^- = 80; n^0 = 120

Example: Determine the element name and particle (p^+, e^-, n^0) numbers for: $^{197}_{79}\mathbf{Au}$; use the periodic table as necessary.

a) **Au** is the symbol for **gold** (from the Latin: *aurum*)
b) The number of protons = **Z = 79**
c) **A** = 197 = # of nucleons = (# of protons) + (# of neutrons); solve for # of neutrons: **# of neutrons** = [# of nucleons (A)] – [# of protons] = [197] – [79] = **118**
d) No charge is indicated; atom must be neutral: # of e^- = # of p^+ = **79**
p^+ = 79; e^- = 79; n^0 = 118

Atoms of a specific element (atoms with a specific number of protons) may exist in different varieties based on the number of neutrons in the nucleus. These varieties of the same element are related as **isotopes**.

Isotopes: Atoms with the Same Value of Z but with a Different Value for A

A useful notation to distinguish isotopes can use the name of the element (which automatically specifies the atomic number, **Z**), followed by a hyphen and the mass number **A,** which specifies the number of neutrons (by subtraction). The names of all isotopes of an element generally have the same name. Notable exceptions are the isotopes of hydrogen that are most often referred to by individual names.

Examples:

$^{235}_{92}\mathbf{U}$ (92 p^+ and 143 n^0): uranium-235

$^{238}_{92}\mathbf{U}$ (92 p^+ and 146 n^0): uranium-238

$^{12}_{6}\mathbf{C}$ (6 p^+ and 12 n^0): carbon-12

$^{13}_{6}\mathbf{C}$ (6 p^+ and 13 n^0): carbon-13

$^{1}_{1}\mathbf{H}$ (1 p^+ and 0 n^0): hydrogen-1 or hydrogen

$^{2}_{1}\mathbf{H}$ (1 p^+ and 1 n^0): hydrogen-2 or deuterium

$^{3}_{1}\mathbf{H}$ (1 p^+ and 2 n^0): hydrogen-3 or tritium

III ATOMIC SYMBOLS AND ION SYMBOLS

Ions are **always** formed by loss or gain of **electrons** for any element in chemistry; the number of protons in the nucleus can never change during a chemical reaction.

A **positive** ion is formed when an element **loses electrons** such that the number of positive charges (protons) exceeds the number of negative charges (electrons).

A **negative** ion is formed when an element **gains electrons** such that the number of negative charges (electrons) exceeds the number of positive charges (protons).

The symbol for a non-neutral **ion** uses the same format described in the previous part, plus the additional information indicating the surplus of positive or negative charge. This is shown by the symbol (**Q**) = size of charge, along with the + or – sign.

+ Q for a surplus of (+) charge and – Q for a surplus of (–) charge. If the specific size of the charge (value of **Q**) is **one**, the number **1** is not required to be shown.

$^{A}_{Z}\mathbf{X}^{+Q}$ for a surplus of + charge $^{A}_{Z}\mathbf{X}^{-Q}$ for a surplus of – charge

Example: Write the symbol for the atom or ion with 11 protons, 12 neutrons, and 10 electrons; use the periodic table as necessary.

a) # of protons = atomic number (**Z**) = **11**
b) The element with 11 protons is sodium, symbol = **Na** (German from Greek: *natrium*)
c) Mass number (**A**) = # of nucleons = 11 protons + 12 neutrons = **23**
d) The # of electrons = 10; the # of protons = 11; the atom is **not** neutral

The actual charge on the atom is required to be shown; follow the correct format.
Note that **Q** is not a direct count of **electrons; Q is a measure of charge difference**.

To calculate: # of positive charges (# of p^+) = 11 (+)
# of negative charges (# of e^-) = 10 (–)
total charge difference = 11 (+) plus 10 (–) = **1** (+)

$$^{23}_{11}\mathbf{Na}^{+1} \quad \textbf{or} \quad ^{23}_{11}\mathbf{Na}^{+}$$

Example: Write the symbol for the atom or ion with 15 protons, 16 neutrons, and 18 electrons; use the periodic table as necessary.

a) # of protons = atomic number (**Z**) = **15**
b) The element with 15 protons is phosphorous, symbol = P
c) Mass number (A) = # of nucleons = 15 protons + 16 neutrons = 31
d) The # of electrons = 18; the # of protons = 15; the atom is **not** neutral

The actual charge on the atom is required to be shown.
# of positive charges (# of p^+) = 15 (+)
# of negative charges (# of e^-) = 18 (–)
total charge difference = 15 (+) plus 18 (–) = **3 (–)**

$${}^{31}_{15}\mathbf{P}^{-3}$$

Additional Practice Examples

Example: Determine the element name and particle (p^+, e^-, n^0) numbers for: ${}^{56}_{26}\mathbf{Fe}^{+3}$; use the periodic table as necessary.

a) **Fe** is the symbol for **iron** (from the Latin: *ferrum*)
b) The number of protons = **Z = 26**
c) **A** = 56 = # of nucleons = (# of protons) + (# of neutrons);
# of neutrons = [# of nucleons (A)] – [# of protons] = [56] – [26] = **30**
d) A charge of **+3** is indicated; atom must have an excess of **three** positive charges:
of e^- = 26 (p^+) – 3 = **23; p^+ = 26; e^- = 23; n^0 = 61**

Example: Determine the element name and particle (p^+, e^-, n^0) numbers for: ${}^{120}_{50}\mathbf{Sn}^{+4}$; use the periodic table as necessary.

a) **Sn** is the symbol for **tin** (from the Latin: *stannum*)
b) The number of protons = **Z = 50**
c) **A** = 120 = # of nucleons = (# of protons) + (# of neutrons);
# of neutrons = [# of nucleons (A)] – [# of protons) = [120] – [50] = **70**
d) A charge of **+4** is indicated; atom must have an excess of **four** positive charges: # of e^- = 50 (p^+) – 4 = **46; p^+ = 50; e^- = 46; n^0 = 70**

Example: Write the complete symbol for the selenium ion with 44 neutrons and 36 electrons; use the periodic table as necessary.

a) The element selenium has the symbol = **Se**
b) The periodic table indicates that Se has an atomic number (**Z**) = **34**; the # of protons = 34
c) Mass number (**A**) = # of nucleons = 34 protons + 44 neutrons = **78**
d) The # of electrons = 36; the # of protons = 34; the atom is **not** neutral

# of positive charges (# of p^+)	=	34 (+)
# of negative charges (# of e^-)	=	36 (–)
total charge difference = 34 (+) plus 36 (–)	=	**2 (–)**

$${}^{78}_{34}\mathbf{Se}^{-2}$$

Example: Write the complete symbol for the antimony ion with 70 neutrons and 48 electrons; use the periodic table as necessary.

a) The element antimony has the symbol = **Sb** (from the Latin: *stibium*)
b) The periodic table indicates that Sb has an atomic number (**Z**) = **51**; the # of protons = 51
c) Mass number (**A**) = # of nucleons = 51 protons + 70 neutrons = **121**
d) The # of electrons = 48; the # of protons = 51; the atom is **not** neutral

# of positive charges (# of p^+) = 51 (+)
# of negative charges (# of e^-) = 48 (–)
total charge difference = 51 (+) plus 48 (–) = 3 (+)

$$^{121}_{51}Sb^{+3}$$

IV PRACTICE PROBLEMS

1. Complete each row in the following table. Each row provides sufficient information to identify one specific atom; all atoms are **neutral**.

ELEMENT	ATOMIC # (Z)	MASS # (A)	ISOTOPE SYMBOL	# OF PROTONS	# OF ELECTRONS	# OF NEUTRONS
______	______	208	______	______	______	126
______	______	______	______	38	______	50
______	7	14	______	______	______	______
______	______	52	______	24	______	______
______	48	______	______	______	______	64
______	______	______	______	______	32	42

2. Identify each atom described by the letter **X** below and determine the relationship (isotopes, different element) between each pair shown:

a) $^{70}_{34}X$ and $^{70}_{33}X$

b) $^{58}_{28}X$ and $^{64}_{28}X$

3. Write complete symbols for the following ions:
   a) 45 protons, 60 neutrons, 41 electrons
   b) 29 protons, 34 neutrons, 28 electrons
   c) 15 protons, 24 neutrons, 18 electrons
   d) The vanadium ion with 29 neutrons and 20 electrons

V ANSWERS TO PRACTICE PROBLEMS

1. Complete each row in the following table. Each row provides sufficient information to identify one specific atom; all atoms are **neutral**.

ELEMENT	ATOMIC # (Z)	MASS # (A)	ISOTOPE SYMBOL	# OF PROTONS	# OF ELECTRONS	# OF NEUTRONS
lead	**82**	208	$^{208}_{82}Pb$	**82**	**82**	126
strontium	**38**	**88**	$^{88}_{38}Sr$	**38**	**38**	50
nitrogen	7	14	$^{14}_{7}N$	**7**	**7**	**7**
chromium	**24**	52	$^{52}_{24}Cr$	**24**	**24**	**28**
cadmium	**48**	**112**	$^{112}_{48}Cd$	**48**	**48**	**64**
arsenic	**33**	**75**	$^{75}_{33}As$	**33**	**33**	**42**

2. Identify each atom described by the letter **X** below and determine the relationship (isotopes, identical, etc.) between each pair shown:
   a) $^{70}_{34}\mathbf{X}$ and $^{70}_{33}\mathbf{X} = {}^{70}_{34}\mathbf{Se}$ **and** $^{70}_{33}\mathbf{As}$ = **different elements**
   b) $^{58}_{28}\mathbf{X}$ and $^{64}_{28}\mathbf{X} = {}^{58}_{28}\mathbf{Ni}$ **and** $^{64}_{28}\mathbf{Ni}$ = **isotopes**

3. Write complete symbols for the following ions:
   a) 26 protons, 30 neutrons, 23 electrons = $^{105}_{45}\mathbf{Rh}^{+4}$
   b) 29 protons, 34 neutrons, 28 electrons = $^{63}_{29}\mathbf{Cu}^{+}$
   c) 15 protons, 24 neutrons, 18 electrons = $^{39}_{15}\mathbf{P}^{-3}$
   d) The vanadium ion with 29 neutrons and 20 electrons = $^{52}_{23}\mathbf{V}^{+3}$

3 Working with Atomic Mass and Nuclear Mass

I GENERAL CONCEPTS

The mass of any specific atom of any element is proportional to the mass number (A), which is the number of nuclear particles (nucleons) in the nucleus: protons plus neutrons; the much lighter electrons do not contribute measurable mass.

The mass of the atom is not exactly equal to the arithmetic sum of the independent nuclear particles comprising the nucleus. For example, in Chapter 2, the mass of the sum of independent nuclear particles for bismuth-209 (83 protons and 126 neutrons) was calculated as:

$$\text{Mass of } \mathbf{protons} = (83\ p^+) \times (1.6726 \times 10^{-27}\ kg/p^+) = 1.3883 \times 10^{-25}\ kg$$

$$\text{Mass of } \mathbf{neutrons} = (126\ n^0) \times (1.6749 \times 10^{-27}\ kg/n^0) = 2.1104 \times 10^{-25}\ kg$$

$$\mathbf{Total}\ \text{mass} = (1.3883 \times 10^{-25}\ kg) + (2.1104 \times 10^{-25}\ kg) = 3.499 \times 10^{-25}\ kg$$

The actual mass of bismuth-209 to three significant figures = 3.470×10^{-25} kg.

Nuclear particles lose some mass when a nucleus is formed from independent protons plus neutrons; this mass is converted to energy released when the particles are bound together in the nucleus. The formation of a nucleus for each element results in a different amount of mass lost as energy. A simple summation of the masses of all nuclear protons plus neutrons in an atom as if they were independent particles will always give an approximate value for the mass of the nucleus that is **greater** than the actual mass of the atom.

To compensate for variable mass loss, a unit can be defined for atomic mass that is based on nuclear particles, which are already bound into a nucleus. This unit is termed the **atomic mass unit** (amu).

II POTENTIAL ENERGY, KINETIC ENERGY, AND FORCES

The general concepts of energy relationships in chemistry are more completely covered in Chapters 18, 19, and 25. However, for the purposes of this general discussion of energy and nucleus formation, **kinetic energy (kE)** can be defined as the energy of heat or motion, and **potential energy (PE)** can be defined as energy stored based on the position or arrangement of matter. Potential energy can be released, for example, as kinetic energy, in response to a force.

A boulder on the top of a high hill overlooking a valley possesses a (relative) high potential energy (based on the gravitational force) due to its height. If the boulder rolls down the hill, its potential energy decreases; potential energy is converted to kinetic energy (motion and heat).

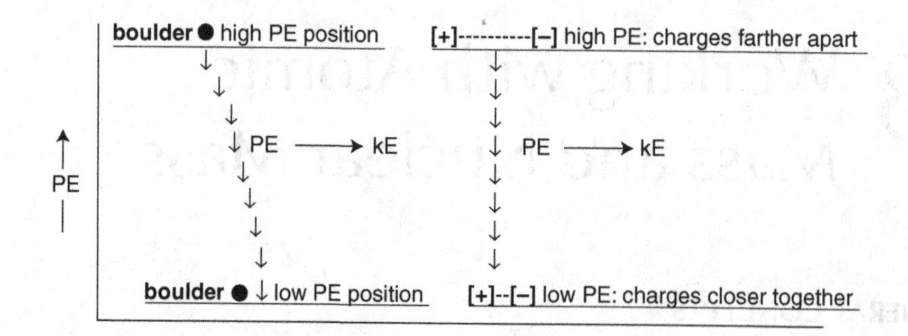

The electromagnetic force is responsible for the energy of chemical bonding and the source of energy conversion in chemical reactions. For the electromagnetic force, opposite charges farther apart represents the higher potential energy position. As opposite charges approach, the decrease in potential energy can be observed as an increase in kinetic energy, such as heat.

Formation of the nucleus from the equivalent of independent protons and neutrons involves the strong nuclear force. Referring only to this specific force, the higher potential energy position is represented by independent nuclear particles that are not part of a nucleus. Protons and neutrons bound together by the strong nuclear force in a nucleus (nuclear fusion) is relatively much lower in potential energy. The release of this "nuclear" energy produces the heat and light of the stars.

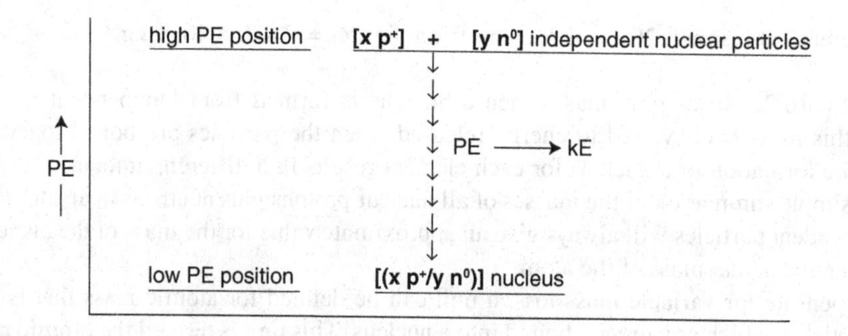

III ENERGY, MASS LOSS, AND THE STRONG NUCLEAR FORCE

An important consideration concerning the interaction of mass with the nuclear force is the amount of energy change relative to the size of the mass involved. As energy is released in the process of nucleus formation, mass is converted energy according to Einstein's equation: $E = mc^2$. Mass conversion to energy also occurs in response to the electromagnetic force, but the energies involved produce a mass loss too small to be readily observable.

The amount of energy released is proportional to the mass lost during nucleus formation. The total energy released by conversion of mass to energy is given by Einstein's equation:

$$\mathbf{E = mc^2}$$

For application of the equation to calculations in the examples:

E = Energy released upon nucleus formation from independent particles; energy is in units of Joules (**J**) = kg × meters²/sec²

m = Mass lost in the process of nucleus formation; units are **kg** for this equation.

c = Speed of light (from the Latin *celeritas*) = 3.00×10^8 meters/sec

There are a large number of particle reaction sequences (in stars) that produce the variety of elemental nuclei; these are not considered here. The purpose of the examples is to demonstrate the general concept of mass loss and energy as part of nucleus formation.

For the calculations, the mass loss is measured by the **difference** between the total mass of the sum of all **independent** nuclear particles that form the nucleus vs. the actual measured mass of the nucleus after formation and energy release:

Mass loss = [mass sum of nucleons as independent particles] − [actual mass of nucleus]

The mass sum of all nucleons as independent nuclear particles:

$$= [(\# \text{ of protons}) \times (1.6726 \times 10^{-27} \text{ kg/proton})]$$

$$+ [(\# \text{ of neutrons}) \times (1.6749 \times 10^{-27} \text{ kg/neutron})]$$

Example: Helium-4 (2 p^+; 2 n^0) is produced in the sun by (as a net result) combining 2 protons and 2 neutrons to form a nucleus; the actual mass of one helium-4 nucleus **in kg** = 6.6463×10^{-27} kg. Calculate the (a) mass loss and (b) energy released in Joules for the formation of one helium nucleus.

(a) **Mass** in **kg** of the sum of **all** independent particles, protons plus neutrons

$$= [(2\ p^+) \times (1.6726 \times 10^{-27} \text{ kg/p}^+)] + [(2\ n^0) \times (1.6749 \times 10^{-27} \text{ kg/n}^0)]$$

$$= [3.3452 \times 10^{-27}] + [3.3498 \times 10^{-27}] = \mathbf{6.6950 \times 10^{-27}\ kg}$$

$$\textbf{Mass loss} = (6.6950 \times 10^{-27} \text{ kg}) - (6.6463 \times 10^{-27} \text{ kg})$$

$$= \mathbf{0.0487 \times 10^{-27}\ kg = 4.87 \times 10^{-29}\ kg}$$

(b) Energy released = **E**; substitute mass lost into **E = mc²**

$$E = (4.87 \times 10^{-29} \text{ kg})(3.00 \times 10^{8} \text{ m/sec})^2 = \mathbf{4.38 \times 10^{-12}\ kg\text{-}m/sec^2 = 4.38 \times 10^{-12}\ J}$$

The mass loss in this case is a little less than 1%, and the amount of energy released seems very small; this value, however, applies to **one atom** of helium. **One gram** of helium contains **1.51×10^{23} atoms** per one gram.

Total energy released per one gram of helium-4 (for this example)
$$= (1.51 \times 10^{23} \text{ atoms}) \times (4.38 \times 10^{-12} \text{ J/atom}) = \mathbf{6.61 \times 10^{11}\ J}$$

Combustion of one gallon of gasoline produces about 2.4×10^8 Joules of energy.
To calculate the energy equivalency:
Number of **equivalent gallons** = total energy released/energy per gallon

$$= \frac{(6.61 \times 10^{11} \text{ J total})}{(2.4 \times 10^{8} \text{ J/gallon})} = \mathbf{2{,}750\ gallons}$$

The formation of one gram of helium-4 by nuclear fusion (for this example) would be equivalent of the energy released by the burning of about 2750 gallons of gasoline.

Example: Assume fluorine-19 (9 p^+; 10 n^0) could be directly produced by combining 9 protons and 10 neutrons to form a nucleus; the actual mass of one fluorine-19 nucleus is 3.154×10^{-26} kg. Repeat the calculation for (a) mass loss and (b) energy released in Joules for the formation of one fluorine nucleus.

(a) **Mass** in **kg** of the sum of **all** independent particles, protons plus neutrons

$$= [(9\ p^+) \times (1.6726 \times 10^{-27}\ kg/p^+)] + [(10\ n^0) \times (1.6749 \times 10^{-27}\ kg/n^0)$$

$$= [1.505 \times 10^{-26}] + [1.675 \times 10^{-26}] = \mathbf{3.180 \times 10^{-26}\ kg}$$

Mass loss = $(3.180 \times 10^{-26}\ kg) - (3.154 \times 10^{-26}\ kg) = \mathbf{0.026 \times 10^{-26}\ kg}$

$$= \mathbf{2.6 \times 10^{-28}\ kg}$$

(b) Energy released = **E**; substitute mass lost into $\mathbf{E = mc^2}$

$$\mathbf{E} = (2.6 \times 10^{-28}\ kg)(3.00 \times 10^8\ m/sec)^2 = \mathbf{2.3 \times 10^{-11}\ kg\text{-}m/sec^2} = \mathbf{2.3 \times 10^{-11}\ J}$$

One gram of fluorine contains **3.17×10^{22} atoms** per one gram.

Total energy released per one gram of fluorine-9 (for this example)

$$= (3.17 \times 10^{22}\ atoms) \times (2.3 \times 10^{-11}\ J/atom) = \mathbf{7.3 \times 10^{11}\ J}$$

IV CALCULATING MASS USING THE ATOMIC MASS UNIT

The definition of atomic mass unit is based on the mass of a chosen standard atom. Since the nuclear particles are already part of a nucleus, the mass loss has already been incorporated into the standard unit.

One **atomic mass unit** (symbol = **amu)** is defined as equal to **exactly** one-twelfth of the mass of the carbon-12 atom (**A** = 12); this means that the mass of carbon-12 is **exactly 12 amu.**

The **metric** value for an **amu** is: 1 amu = 1.66054×10^{-27} kg/amu

or: 1 amu = 1.66054×10^{-24} g/amu

Since carbon-12 has 12 nucleons (6 protons plus 6 neutrons), this means (ignoring the mass of the electrons) that the **average** mass of a nuclear particle **in a carbon-12 nucleus is exactly** 1 amu (1.66054×10^{-27} kg). Note that this value is less than the values for an independent proton (1.6726×10^{-27} kg) or an independent neutron (1.6749×10^{-27} kg). Note also that calculations based on the amu and the mass number (A) count only nuclear particles and do not distinguish between the proton and neutron.

The nuclei of all other elements and isotopes will show slightly different amounts of mass loss in their formation; only the mass of carbon-12 will be an exact whole number in units of amu (12 amu exact). However, **specific** atoms of all elements and isotopes will have a mass **in units of amu** very close to a whole number, usually to at least three significant figures. This whole number is equal to the value of the mass number for that atom **(A).**

# of nucleons (A) = # of protons plus neutrons

= mass of atom in amu usually to three significant figures.

Helium-4 (A = **4**; 2 p^+ + 2 n^0) has a mass of 4.0026 amu
Flourine-19 (A = **19**; 9 p^+ + 10 n^0) has a mass of 18.998 amu
Sodium-23 (A = **23**; 11 p^+ + 12 n^0) has a mass of 22.990 amu

In some cases, for very stable nuclei, the value is not quite accurate to three significant figures.

Iron-58 (A = **58**; 26 p^+ + 32 n^0) has a mass of 57.933 amu
copper-63 (A = **63**; 29 p^+ + 34 n^0) has a mass of 62.94 amu
Nickel-62 (A = **62**; 28 p^+ + 34 n^0) has a mass of 61.9 amu

V CALCULATING THE APPROXIMATE MASS OF AN ATOM IN AMU, KG, OR G

The use of the atomic mass unit provides a direct method to calculate the mass of a **specific** atom of any element and isotope; the calculation is usually accurate to at least **three** significant figures.

Process for Calculating Atomic Masses

(1) Count the **number** of **nucleons** in the atom; this = **A**, (the # of p^+ + # of n^0).
(2) To find the **mass** of the atom in **amu** units, multiply the number of nucleons in the atom times 1.00 amu/nucleon. The **average** mass of a nucleon for almost any nucleus is 1.00 amu/nucleon to **two** or **three** significant figures.

The mass of the specific atom in units of amu =

(# of nucleons) × (1.00 amu/nucleon)

(3) To find the **mass** of the atom in **metric** units of **grams** or **kilograms**, use the metric conversion units to convert amu to kilograms or grams:

$$1 \text{ amu} = 1.66054 \times 10^{-27} \text{ kg/amu}; \; 1 \text{ amu} = 1.66054 \times 10^{-24} \text{ g/amu}$$

The mass of the specific atom in units of kg

$$= (\text{mass in amu}) \times (1.66054 \times 10^{-27} \text{ kg/amu})$$

The mass of the specific atom in units of g

$$= (\text{mass in amu}) \times (1.66054 \times 10^{-24} \text{ g/amu})$$

(4) To use the complete equations in one step:
Combining steps (1), (2), and (3):

$$\textbf{mass(kg)} = (\text{\# of nucleons}) \times (1.00 \text{ amu/nucleon}) \times (1.66054 \times 10^{-27} \text{ kg/amu})$$

$$\textbf{mass(g)} = (\text{\# of nucleons}) \times (1.00 \text{ amu/nucleon}) \times (1.66054 \times 10^{-24} \text{ g/amu})$$

Example: Calculate the mass in amu and **kilograms** of a sodium-23 atom $\left({}^{23}_{11}\mathbf{Na}\right)$ to three significant figures:

(1) **Number** of **nucleons** in the atom: **A** for sodium-23 = **23** (# of p^+ + # of n^0).
(2) **Mass of the** atom in **amu** units:

Mass of sodium-23 in units of amu = (# of nucleons) × (1.00 amu/nucleon)

= (23 nucleons) × (1.00 amu/nucleon) = **23.0 amu**

(The value is correct to at least three significant figures)

(3) **Mass** of the atom in **kilograms** units:

Mass of sodium-23 in units of kg = (mass in amu) × (1.66054×10^{-27} kg/amu)

= (23.0 amu) × (1.66054×10^{-27} kg/amu) = $\mathbf{3.82 \times 10^{-26}}$ **kg**

(4) **Mass** using the **complete** equation:

Mass of sodium-23 in units of kg =

(**23** nucleons) × (**1.00** amu/nucleon) × (1.66054×10^{-27} kg/amu) = $\mathbf{3.82 \times 10^{-26}}$ **kg**

Example: Calculate the mass in amu and **grams** of a silver-107 atom $\left({}^{107}_{47}\mathrm{Ag}\right)$ to three significant figures:

(1) **Number of nucleons** in the atom: **A** for silver-107 = **107** (# of p^+ + # of n^0).
(2) **Mass** of the atom in **amu** units:

Mass of silver-107 in units of amu = (# of nucleons) × (1.00 amu/nucleon)

= (107 nucleons) × (1.00 amu/nucleon) = **107 amu**

(The value is correct to at least three significant figures)

(3) **Mass** of the atom in **grams** units:

Mass of silver-107 in units of g = (mass in amu) × (1.66054×10^{-24} g/amu)

= (107 amu) × (1.66054×10^{-27} kg/amu) = $\mathbf{3.82 \times 10^{-26}}$ **kg**

(4) **Mass** using the **complete** equation:

Mass of silver-107 in units of g =

(**107** nucleons) × (**1.00** amu/nucleon) × (1.66054×10^{-24} g/amu) = $\mathbf{1.78 \times 10^{-22}}$ **g**

VI AVERAGE ATOMIC MASS FOR AN ELEMENT

Atoms of most elements exist in two or more isotope forms; that is, the same number of protons but different numbers of neutrons.

$$\text{Element carbon: } {}^{12}_{6}C \text{ or } {}^{13}_{6}C \text{ or } {}^{14}_{6}C$$

$$\text{Element hydrogen: } {}^{1}_{1}H \text{ or } {}^{2}_{1}H \text{ or } {}^{3}_{1}H$$

$$\text{Element oxygen: } {}^{16}_{8}O \text{ or } {}^{18}_{8}O$$

The periodic table lists two important numerical values along with the symbol for each element. The atomic number (Z) is indicated by the whole number identifying the number of protons in the nucleus. The **average atomic mass** is the other value shown; as one key measurement it specifies the **mass** in **amu** of one average atom of each element.

The **average atomic mass** represents the average of all possible naturally occurring isotopes for an element. The average is found by taking a large sample of the naturally occurring element, containing all isotopes in their natural abundance (percentages), and dividing the total mass by the total number of atoms. **The average depends on the masses of each isotope and the relative percentage of each isotope.**

If an element has **one** predominant isotope, the **average** atomic mass will be close to the mass of this isotope (and will be close to a whole number in units of amu):

$${}^{1}_{1}H = 99.98\%;\ {}^{2}_{1}H = 0.02\%;\ {}^{3}_{1}H = \text{not naturally occurring}$$

The average atomic mass of the element hydrogen is 1.0079 amu, very close to the approximate 1.00 amu mass of the predominant hydrogen-1 isotope.

$${}^{16}_{8}O = 99.76\%;\ {}^{18}_{8}O = 0.24\%$$

The average atomic mass of the element oxygen = 15.999 amu, very close to the approximate 16.0 amu mass of the predominant oxygen-16 isotope.

The average atomic mass can be calculated from the fractional abundance of each isotope and the corresponding masses of each isotope.

Fractional abundance = percentage of the specific isotope written in decimal form.

Average atomic mass of an element in units of amu = Sum of (fractional abundance of each isotope) × (**mass** of each isotope in amu)

Example: Calculate the average atomic mass **in amu** of the element carbon; use the approximation method for determining atomic mass to three significant figures. The natural abundances are shown.

$${}^{12}_{6}C = 98.890\%;\qquad {}^{13}_{6}C = 1.110\%\qquad {}^{14}_{6}C = \text{not naturally occuring}$$

Mass carbon-**12** = (12 nucleons) × (1.00 amu/nucleon) = **12.0 amu** (actually exactly 12 in this case)

Mass carbon-**13** = (13 nucleons) × (1.00 amu/nucleon) = **13.0 amu**

carbon-**14**: does not contribute to the total

Average atomic mass of carbon (extending significant figures)

= (0.98890)(12.000 amu) + (0.01110)(13.000 amu) = 12.011 amu

Example: Calculate the average atomic mass **in amu** of the element nickel; use the approximation method for determining atomic mass. Assume for this example that the calculation is accurate to three significant figures. The natural abundances are shown.

$^{58}_{28}\text{Ni} = 68.3\%$ $^{60}_{28}\text{Ni} = 26.1\%$ $^{61}_{28}\text{Ni} = 1.1\%$ $^{62}_{28}\text{Ni} = 3.6\%$ $^{64}_{28}\text{Ni} = 0.9\%$

mass nickel-**58** = (58 nucleons) × (1.00 amu/nucleon) = **58.0 amu**
mass nickel-**60** = (60 nucleons) × (1.00 amu/nucleon) = **60.0 amu**
mass nickel-**61** = (61 nucleons) × (1.00 amu/nucleon) = **61.0 amu**
mass nickel-**62** = (62 nucleons) × (1.00 amu/nucleon) = **62.0 amu**
mass nickel-**64** = (64 nucleons) × (1.00 amu/nucleon) = **64.0 amu**

Average atomic mass of nickel =

(0.683)(58.0 amu) + (0.261)(60.0 amu) + (0.011)(61.0 amu) + (0.036)(62.0 amu) + (0.009)(64.0 amu) = **58.8 amu** (assuming three significant figures)

Example: In some cases, the atomic mass approximation method is accurate to only two significant figures. Repeat the calculation for the average atomic mass of the element nickel using the actual mass values shown.

$^{58}_{28}\text{Ni} = 68.3\%$ mass = 57.9 amu $^{62}_{28}\text{Ni} = 3.6\%$ mass = 61.9 amu

$^{60}_{28}\text{NI} = 26.1\%$ mass = 59.9 amu $^{64}_{28}\text{Ni} = 0.9\%$ mass = 63.9 amu

$^{61}_{28}\text{NI} = 1.1\%$ mass = 60.9 amu

Average atomic mass of nickel =

(0.683)(57.9 amu) + (0.261)(59.9 amu) + (0.011)(60.9 amu) + (0.036)(61.9 amu) + (0.009)(63.9 amu) = **58.7 amu** (to three significant figures)

VII PRACTICE PROBLEMS

1. Assume that the **actual measured** mass of bismuth-209 can be calculated as 3.470×10^{-25} kg to four significant figures.
   a) As a practice exercise, calculate the mass loss when the nuclear particles of bismuth-209 bind together to form **one nucleus**; that is, calculate the mass difference

between the calculated sum of the independent nuclear particles for bismuth-209 and the actual measured mass given. The sum of independent particles for bismuth-209 was calculated previously at the beginning of the chapter.

b) Use your results from part (a) to calculate the energy released when **one nucleus** of bismuth-209 is formed; use Einstein's equation.

c) Use your results from part (b) to calculate the energy released when **one gram** of bismuth-209 nuclei is formed; first calculate the number of atoms in one gram.

d) Use the value that combustion of one gallon of gasoline produces about 2.4×10^8 Joules of energy. Determine the number of gallons of gasoline that must be burned to equal the amount of energy released by formation of **one gram** of bismuth-209 nuclei; use your answer from part (c).

2. Calculate the **density** of a gold-197 nucleus in **g/cm³** by completing the individual parts labeled (a), (b), and (c):

a) Calculate the **mass** of the gold-197 nucleus in units of **grams** by using the atomic mass approximation method.

b) Calculate the volume of the gold nucleus in cm³; V(sphere) = (4/3) π r^3; radius of the spherical nucleus is 1.00×10^{-13} cm.

c) Complete the calculation for the density.

3. Calculate the average atomic mass in amu of the element silver; use the approximation method for determining atomic mass to three significant figures. The natural abundances are shown.

$$^{107}_{47}Ag = 52.0\% \quad ^{109}_{47}Ag = 48.0\%$$

4. Calculate the average atomic mass in amu of the element chromium; use the approximation method for determining atomic mass to three significant figures; assume three significant figures for the calculation. The natural abundances are shown.

$$^{50}_{24}Cr = 4.35\% \quad ^{52}_{24}Cr = 83.8\% \quad ^{53}_{24}Cr = 9.50\% \quad ^{54}_{24}Cr = 2.36\%$$

VIII ANSWERS TO PRACTICE PROBLEMS

1. a) Mass of protons in bismuth-209 =

$$(83\ p^+) \times (1.6726 \times 10^{-27}\ kg/p^+) = 1.3883 \times 10^{-25}\ kg$$

mass of neutrons in bismuth-209 =

$$(126\ n^0) \times (1.6749 \times 10^{-27}\ kg/n^0) = 2.1104 \times 10^{-25}\ kg$$

$$\text{total mass} = (1.3883 \times 10^{-25}\ kg) + (2.1104 \times 10^{-25}\ kg)$$

$$= 3.4987 \times 10^{-25}\ kg = 3.499 \times 10^{-25}\ kg \text{ to four significant figures}$$

The **actual measured** mass of bismuth-209 = 3.470×10^{-25} kg.

mass loss = mass difference =

$$(3.499 \times 10^{-25}\ kg) - (3.470 \times 10^{-25}\ kg) = 0.029 \times 10^{-25}\ kg = 2.9 \times 10^{-27}\ kg$$

b) Energy released when one nucleus of bismuth-209 is formed.

$$\mathbf{E = mc^2};\ \mathbf{E} = (2.9 \times 10^{-27}\ \text{kg})(3.00 \times 10^{8}\ \text{m/sec})^2$$

$$= 2.61 \times 10^{-10}\ \text{kg-m/sec}^2 = 2.61 \times 10^{-10}\ \text{J}$$

c) Energy released when **one gram** of bismuth-209 nuclei is formed.

mass of **one atom** of bismuth-209 in **grams** =

$$(3.470 \times 10^{-25}\ \text{kg}) \times (10^{3}\ \text{g/kg}) = 3.470 \times 10^{-22}\ \text{g}$$

The **number** of **atoms** in **one gram** = 1 gram/# of grams per one atom

$$= (1\ \text{gram})/(3.470 \times 10^{-22}\ \text{g/atom}) = 2.88 \times 10^{21}\ \text{atoms per one gram}$$

total energy released per one gram of bismuth-209

$$= (2.88 \times 10^{21}\ \text{atoms}) \times (2.61 \times 10^{-10}\ \text{J/atom}) = 7.52 \times 10^{11}\ \text{J}$$

d) **Number of equivalent gallons** = total energy released/energy per gallon

$$= (7.52 \times 10^{11}\ \text{J total})/(2.4 \times 10^{8}\ \text{J/gallon}) = 3{,}130\ \text{gallons}$$

2. Calculate the **density** of a gold-197 nucleus in **g/cm^3**:

a) **Mass** of the gold-197 nucleus in units of **grams**:
(**4**) Use complete equation

mass of gold-197 in units of **g** =

$$(197\ \text{nucleons}) \times (1.00\ \text{amu/nucleon}) \times (1.66055 \times 10^{-24}\ \text{g/amu}) = 3.27 \times 10^{-22}\ \text{g}$$

b) The volume of the spherical gold nucleus in cm^3:

$$\text{V(sphere)} = (4/3)\ \pi\ r^3;\ r = 1.00 \times 10^{-13}\ \text{cm}$$

$$\textbf{V(nucleus)} = (4/3)(3.1416)(1.00 \times 10^{-13}\ \text{cm})^3 = 4.19 \times 10^{-39}\ \text{cm}^3$$

c) Density of the gold-197 nucleus.
Use the process from Chapter 1.
(**1**) Density is the unknown variable; $d = m(g)/V(cm^3)$
(**2**) Mass was found to be 3.27×10^{-22} g;
volume was found to be 4.19×10^{-39} cm^3
(**3**) $d = (3.27 \times 10^{-22}\ g)/(4.19 \times 10^{-39}\ cm^3) = \mathbf{7.80 \times 10^{16}\ g/cm^3}$

3. Calculate the average atomic mass of silver in amu units to three significant figures.

$${}^{107}_{47}\text{Ag} = 52.0\% \qquad {}^{109}_{47}\text{Ag} = 48.0\%$$

mass of silver-107 in amu = (107 nucleons) × (1.00 amu/nucleon) = **107 amu**

mass of silver-109 in amu = (109 nucleons) × (1.00 amu/nucleon) = **109 amu**

Average atomic mass of silver =

(0.520)(107amu) + (0.480)(109 amu) = 107.96 amu

= **108 amu** to three significant figures

4. Calculate the average atomic mass of chromium in amu units to three significant figures.

$$^{50}_{24}Cr = 4.35\% \quad ^{52}_{24}Cr = 83.8\% \quad ^{53}_{24}Cr = 9.50\% \quad ^{54}_{24}Cr = 2.36\%$$

mass of chromium-50 in amu = (50 nucleons) × (1.00 amu/nucleon) = **50.0 amu**
mass of chromium-52 in amu = (52 nucleons) × (1.00 amu/nucleon) = **52.0 amu**
mass of chromium-53 in amu = (53 nucleons) × (1.00 amu/nucleon) = **53.0 amu**
mass of chromium-54 in amu = (54 nucleons) × (1.00 amu/nucleon) = **54.0 amu**

Average atomic mass of nickel = (0.0435)(50.0 amu) + (0.838)(52.0 amu) + (0.0950)(53.0 amu) + (0.0236)(54.0 amu) = **52.1 amu** to three significant figures

4 Procedures for Writing Formulas and Naming Compounds

I GENERAL CONCEPTS

The atoms in chemical compounds are bonded together through either ionic or covalent bonding.

An **ionic bond** is an attractive electromagnetic (electrostatic) force between ions of opposite charge. Ions are produced by an exchange of electrons between atoms such that one atom has a surplus of electrons (negative charge) and one atom has a deficiency of electrons (positive charge).

A **covalent bond** is an attractive force produced by the sharing of electrons between two atoms; the bonding force is based on the favorable energy of the shared electrons.

For purposes of naming and writing formulas, compounds can be generally classified into **four** groups:

(1) **Binary ionic** is a compound composed of **only two** different elements (binary = two) in which one element (a metal) forms a positive ion and one element (a non-metal or certain metalloids) forms a negative ion. The ions combine in the correct combining ratio to produce a **neutral** compound; the bonding type is an ionic bond. These compounds usually exist as large three-dimensional arrays of ions.
Binary ionic compounds can be identified as consisting of **one metal** combined with **one non-metal** (or metalloid).

(2) **Binary covalent** is a compound composed of **only two** different elements in which the bonding type is the covalent bond (sharing of electrons). The elements in this compound-type do **not** form ions. This compound usually exists in the form of molecules.
Binary covalent compounds can be identified as consisting of two elements, ***both*** of which are **non-metals** (or metalloids).

(3) **Multi-atom covalent** is a compound composed of **three** or more different elements in which the bonding type is the covalent bond; they usually exist in the form of molecules. Most of these compounds fall under the classification of organic compounds.
All elements in a multi-atom covalent compound are usually non-metals (or metalloids).

(4) **Ionic compounds containing polyatomic ions**:
1. A **polyatomic ion** is a species that contains **two or more covalently** bonded atoms that, when bonded together, form an ion unit (positive or negative).
2. A complete neutral ionic compound is formed when the polyatomic ion unit is further combined with **any** additional ion of opposite charge in the correct combining ratio to produce net neutrality.

II ELEMENTAL IONS FOR IONIC COMPOUNDS

Binary ionic compounds combine **one metal** atom as a **positive** ion with **one non-metal/ metalloid** atom as the **negative** ion.

1. All non-metals plus the metalloids As and Te will produce **negative** ions of **only one specific charge**; they are always "**fixed**-charged" negative ions.

   Without a detailed discussion of electron configuration of atoms (Chapter 13), a general rule can be stated for finding the specific charge adopted by the non-metals/metalloids in binary ionic compounds:

   The non-metals (H, C, N, O, F, P, S, CI, Se, Br, I) plus As and Te always **gain** the exact number of electrons to achieve a noble gas electron configuration, equal to the number of electrons found in the neutral noble gases (He, Ne, Ar, Kr, Xe). Thus, these numbers are: 2 e^-, 10 e^-, 18 e^-, 36 e^-, or 54 e^-.

   To find the value of the **negative charge** formed on a non-metal/metalloid in an ionic compound:

   (1) Determine the number of electrons in the **neutral** atom.

   (2) For the formation of a ***negative*** ion, ***add*** the correct number of electrons required to reach the closest electron number indicated by the closest noble gas. Alternately, count the number of "steps" (columns) in the periodic table that the specific element must "jump" **forward** to reach the end (noble gas) column.

Examples:

flourine: $F\ (9\ p+; 9\ e^-) + 1\ e^- \longrightarrow F^-\ (9\ p+; \mathbf{10\ e^-})$: matches # e^- in Ne

sulfur: $S\ (16\ p+; 16\ e^-) + 2\ e^- \longrightarrow S^{-2}\ (16\ p+; \mathbf{18\ e^-})$: matches # e^- in Ar

arsenic: $As\ (33\ p+; 33\ e^-) + 3\ e^- \longrightarrow As^{-3}\ (33\ p+; \mathbf{36\ e^-})$: matches # e^- in Kr

Application of this rule to the non-metals/metalloids produces the following specific negative ions when occurring in ionic compounds:

$$\mathbf{H^-, C^{-4}, N^{-3}, O^{-2}, F^-, P^{-3}, S^{-2}, Cl^-, Se^{-2}, Br^-, I^-, As^{-3}, Te^{-2}}$$

The value of the specific negative charge for the fixed-charged non-metals or metalloids must be known or determined in order to write correct formulas.

2. Many metals will produce **positive** ions of **only one specific charge**; these are termed **"fixed-**charged" **metals.** The corresponding guideline for finding the specific charge adopted by certain metals in Groups I and II of the periodic table plus Al can be stated: The metals of **Group I** (Li, Na, K, Rb, Cs), **Group II** (Mg, Ca, Sr, Ba), plus Al **lose** the exact number of electrons required to achieve a noble gas electron configuration, equal to the number of electrons found in the neutral noble gases (He, Ne, Ar, Kr, Xe). Thus, these numbers are: 2 e^-, 10 e^-, 18 e^-, 36 e^-, or 54 e^-.

   To find the value of the **positive charge** formed on metals in Groups I and II or Al in an ionic compound:

   (1) Determine the number of electrons in the **neutral** metal atom.

   (2) For the formation of a ***positive*** ion, ***subtract*** the correct number of electrons required to reach the closest electron number indicated by the closest noble gas. Alternately, count the number of "steps" (columns) in the periodic table that the specific element must "jump" **backward** to reach the end (noble gas) column (i.e., jump backward and then up one row to the closest noble gas). **Examples:**

lithium: Li (3 p+ ; 3 e^-) : lose 1 e^- ⟶ Li^+(3 p+ ; 2 e^-) : matches # e^- in He

aluminum: Al (13 p+ ; 13 e^-) : lose 3 e^- ⟶ Al^{+3}(13 p+ ; **10** e^-) : matches # e^- in Ne

strontium: Sr (38 p+ ; 38 e^-) : lose 2 e^- ⟶ Sr^{+2}(38 p+ ; **36** e^-) : matches # e^- in Kr

Application of this rule to the metals in Groups I and II or Al produces the following specific positive ions when occurring in ionic compounds:

$$Li^+,\ Na^+,\ K^+,\ Rb^+,\ Cs^+,\ Mg^{+2},\ Ca^{+2},\ Sr^{+2},\ Ba^{+2},\ Al^{+3}$$

3. The metals Ag, Zn, Cd, Bi also produce **positive** ions of **only one specific charge** ("**fixed**-charged" metals) but do not follow the noble gas electron configuration rule. The ions formed are: Ag^+, Zn^{+2}, Cd^{+2}, Bi^{+3}.

   The value of the specific positive charge for the fixed-charged metals must be known or determined in order to write correct formulas.
4. **All other metals listed in the periodic table will produce more than one value of positive ion**; these are termed **"variable-**charged" metals. Variable-charged metals include most of the transition metals (excluding Ag, Zn, and Cd listed previously), plus Pb, Sn, and Sb.

Examples:

Iron can form either Fe^{+2} or Fe^{+3}
Copper can form either Cu^+ or Cu^{+2}
Lead can form either Pb^{+2} or Pb^{+4}

Variable-charged metals do not produce one specific positive ion. **Information concerning the value of the positive charge for variable-charged metals must be available for writing formulas and for naming compounds.**

III WRITING FORMULAS FOR BINARY IONIC COMPOUNDS

Ionic compounds must exist as stable **electronically neutral** collections of positive and negative ions.

A **valid** binary ionic compound consists of **one metal** atom, which has formed a **positive ion** by loss of electron(s), plus **one non-metal/metalloid** atom, which has formed a **negative ion** by gain of electron(s). The positive and negative ions **must** be combined in the specific combining ratio to produce **neutrality** (a **neutral** compound).

To write a correct formula for a binary ionic compound:

(1) Identify the specific metal, symbol, and the correct specific metal positive charge; identify the specific non-metal/metalloid, symbol, and the correct specific negative charge. Find all necessary information in the periodic table.
(2) Write the **positive** ion (metal) symbol **first**; write the negative ion (non-metal/metalloid) symbol **second**. Charges can be shown while completing steps (2) and (3); however, do **not** show the charges in the final formula.

(3) Determine the correct **ratio** between the positive ion and the negative ion that leads to the **exact balancing of opposite charges**. This means that the total number of positive charges in the formula plus the total number of negative charges in the formula must be exactly equal. Use the simplest whole number ratio that will balance the charges; indicate the number of each ion with a subscript, with the exception that if the required number is 1 no subscript is shown.

(4) To **verify** that the selected ratio has balanced charges, check to confirm that the total number of positive charges in the formula (as a positive number) plus the total number of negative charges in the formula (as a negative number) **add up to zero.**

Example: Determine the correct formula for a combination of calcium and bromine: use the periodic table as necessary.

(1) Calcium is a metal, the symbol is **Ca**

$$Ca(20p+; 20e^-) : lose\,2e^- \longrightarrow Ca^{+2}(20p+; \mathbf{18}e^-)$$

Bromine is a non-metal, the symbol is **Br**

$$Br(35p+; 35e^-) + 1e^- \longrightarrow Br^- \ (35p+; \mathbf{36}e^-)$$

(2) $\mathbf{Ca^{+2}}$ **plus** $\mathbf{Br^-}$ $\longrightarrow$ $\mathbf{(Ca^{+2})_x\,(Br^-)_y}$

(3) **Each** Ca^{+2} with 2 (+) charges requires **two** Br^-, which each have only 1 (–) charge.

$$Ca^{+2} + 2Br^- \longrightarrow \mathbf{(Ca^{+2})_1\,(Br^-)_2 = CaBr_2;}$$

charges are not shown in final formula.

(4) Check the formula by addition of (+) and (–) values:

$$\text{Formula} = (Ca^{+2})_1\ (F^-)_2$$

$$\text{Total (+) charges} = 1\ (Ca^{+2})\ \text{ion} \times (+2\ \text{per ion}) = +2$$

$$\text{Total (–) charges} = 2\ (Br^-)\ \text{ions} \times (-1\ \text{per ion}) = \underline{-2}$$

$$= 0;\ \text{the total charges balance}$$

Example: Determine the correct formula for a combination of magnesium and phosphorous; use the periodic table as necessary.

(1) Magnesium is a metal, the symbol is **Mg**

$$Mg(12\,p+; 12\,e^-) : lose\ 2e^- \longrightarrow Mg^{+2}(12\,p+; \mathbf{10}\,e^-)$$

Phosphorous is a non-metal, the symbol is **P**

$$P(15p+; 15e^-) + 3e^- \longrightarrow P^{-3}(15p+; \mathbf{18}e^-)$$

(2) $\mathbf{Mg^{+2}}$ **plus** $\mathbf{P^{-3}}$ $\longrightarrow$ $\mathbf{(Mg^{+2})_x(P^{-3})_y}$

(3) **Three** Mg^{+2} with 2 (+) charges each require **two** P^{-3}, which each have 3 (–) charges. (The least common multiple of 2 and 3 = 6.)

$$3\,Mg^{+2} + 2\,P^{-3} \longrightarrow \mathbf{(Mg^{+2})_3(P^{-3})_2 = Mg_3P_2};$$

charges are not shown in final formula.

(4) Check the formula by addition of (+) and (–) values:

$$\text{Formula} = (Mg^{+2})_3\,(P^{-3})_2$$

$$\text{Total (+) charges} = 3\,(Mg^{+2}) \text{ ions} \times (+2 \text{ per ion}) = +6$$

$$\text{Total (–) charges} = 2\,(P^{-3}) \text{ ions} \times (-3 \text{ per ion}) \quad = \underline{-6}$$

$$= 0; \text{ the total charges balance}$$

Example: Determine the correct formula for a combination of tin with a +4 charge and selenium; use the periodic table as necessary.

(1) Tin is a metal, the symbol is **Sn**
Sn is a **variable-**charged **not** a fixed-charged metal. For this type of problem, the value of the positive charge on variable-charged metals must be provided:
The Sn charge is given as **+4** in this problem.
Selenium is a non-metal, the symbol is **Se**

$$Se\ (34\ p+; 34e^-)\ + 2\ e^- \longrightarrow Se^{-2}(34\ p+; 36\ e^-)$$

(2) $\mathbf{Sn^{+4}}$ plus $\mathbf{Se^{-2}} \longrightarrow \mathbf{(Sn^{+4})_x(Se^{-2})_y}$

(3) **Each** Sn^{+4} with 4 (+) charges requires **two** Se^{-2}, which each have 2 (–) charges.

$$Sn^{+4} + 2Se^{-2} \longrightarrow \mathbf{(Sn^{+4})_1(Se^{-2})_2 = SnSe_2};$$

charges are not shown in final formula.

(4) Check the formula by addition of (+) and (–) values:

$$\text{Formula} = (Sn^{+4})(Se^{-2})_2$$

$$\text{Total (+) charges} = 1\,(Sn^{+4}) \text{ ion} \times (+4 \text{ per ion}) \ = +4$$

$$\text{Total (–) charges} = 2\,(Se^{-2}) \text{ ions} \times (-2 \text{ per ion}) = \underline{-4}$$

$$= 0; \text{ the total charges balance}$$

Example: Determine the correct formula for a combination of bismuth and chlorine: use the periodic table as necessary.

(1) Bismuth is a metal, the symbol is **Bi**
Bi is a **fixed-charged** metal but does **not** follow the noble gas electron configuration rule. For this type of problem, the value of the positive charge on Bi (or Ag, Zn, Cd) must come from memory: the memorized charge on Bi is +3.

Chlorine is a non-metal, the symbol is **Cl**

$$Cl(17p+;17e^-)+1e^- \longrightarrow Cl^-\ (17p+;\mathbf{18}e^-)$$

(2) $\mathbf{Bi^{+3}}$ plus Cl^- $\longrightarrow$ $\mathbf{(Bi^{+3})_x\,(Cl^-)_y}$

(3) **Each** Bi^{+3} with 3 (+) charges requires **three** Cl^-, which each have only 1 (–) charge.

$$Bi^{+3}+3Cl^- \longrightarrow \mathbf{(Bi^{+3})_1\,(Cl^-)_3 = BiCl_3};$$

charges are not shown in final formula.

(4) Check the formula by addition of (+) and (–) values:

$$\text{Formula} = (Bi^{+3})_1\ (Cl^-)_3$$

$$\text{Total (+) charges} = 1\ (Bi^{+3})\ \text{ion} \times (+3\ \text{per ion}) = +3$$

$$\text{Total (–) charges} = 3\ (Cl^-)\ \text{ions} \times (-1\ \text{per ion}) = \underline{-3}$$

$$= 0;\ \text{the total charges balance}$$

IV NAMING BINARY IONIC COMPOUNDS

The names of binary ionic compounds are derived from the (adapted) names of the elements contained in the compound and the final resulting formula.

Procedure for Naming a Binary Ionic Compound

(1) Name the **metal** element that forms the **positive** ion **first**; use the name of the metal element directly with no change.

(2) Determine if the metal forms only one specifically charged positive ion (a **fixed-**charged metal) or is a **variable-**charged metal.
If the metal is a fixed-charged metal nothing further is required for the metal portion of the name.
If the metal is a variable-charged metal, the value of the positive charge must be indicated in the name:
  a) Determine the specific numerical value of the metal positive charge in the specific compound to be named; these values will differ depending on the compound.
  b) Place this value, written as a **Roman numeral**, in parentheses directly after the metal name.

(3) Name the **non-metal** (or metalloid) **negative** ion **last**; use the ion form of the elemental name:

The ion form of the non-metal or metalloid is formed by changing the element name ending to **"-ide"**. In some cases an additional syllable is dropped.
The exact names used in compounds are:

$\mathbf{H^-}$ **= hydride**	$\mathbf{C^{-4}}$ **= carbide**	$\mathbf{N^{-3}}$ **= nitride**	$\mathbf{O^{-2}}$ **= oxide**
$\mathbf{F^-}$ **= fluoride**	$\mathbf{P^{-3}}$ **= phosphide**	$\mathbf{S^{-2}}$ **= sulfide**	$\mathbf{Cl^-}$ **= chloride**
$\mathbf{Se^{-2}}$ **= selenide**	$\mathbf{Br^-}$ **= bromide**	$\mathbf{I^-}$ **= iodide**	$\mathbf{As^{-3}}$ **= arsenide**
$\mathbf{Te^{-2}}$ **= telluride**			

Additional Concepts

1. The procedure described applies to binary **ionic** compounds **not** binary covalent compounds. For naming purposes, binary ionic compounds must be identified as a combination of one **metal** combined with one **non-metal** (or metalloid).
2. Variable-charged metals can produce **more than one** possible combining formula when forming ionic compounds with negative ions; two (or more) compound formulas cannot have the exact same name. The inclusion of a Roman numeral to specify the exact charge, therefore, matches one specific name to one specific compound.

It is not necessary to memorize all variable-charged metals. **A metal is variable charged If it is *not* one of the fixed-charged metals** in **Group I** (Li, Na, K, Rb, Cs), **Group II** (Mg, Ca, Sr, Ba), **or** in the **set of five additional** (Al, Ag, Zn, Cd, or Bi).

Examples: Name the compounds from the previous section.

1. Name $CaBr_2$
   (1) **Calcium**: Direct (unchanged) name of the metal that forms the positive ion.
   (2) Calcium is in Group II and is a fixed-charged metal; **no** Roman numeral is required.
   (3) Bromine forms the negative ion; use "bromide," the ion form of the name and state this name last in the complete name of the compound. Name: **calcium bromide**
2. Name Mg_3P_2
   (1) **Magnesium**: Direct (unchanged) name of the metal that forms the positive ion.
   (2) Magnesium is in Group II and is a fixed-charged metal; **no** Roman numeral is required.
   (3) Phosphorous (P) forms the negative ion; use "phosphide," the ion form of the name and state this name last in the complete name of the compound. Name: **magnesium phosphide**
3. Name $SnSe_2$
   (1) **Tin**: Direct (unchanged) name of the metal that forms the positive ion.
   (2) Tin is ***not*** in Groups I or II and is ***not*** one of the additional five fixed-charged metals. Tin is a variable-charged metal; a Roman numeral must be used to indicate the numerical value of the positive charge. The **Roman numeral** in the final name represents the ***charge*** on the positive metal ion, ***not*** the subscript in the formula.

   To determine this value, work backwards from the actual formula of the neutral compound and the *known* charge on the negative ion:

   a) Calculate the total negative charge for the **known** negative portion of the formula: Selenium (Se) can be determined to form a (−2) ion.

(2 Se^{-2} ions in the formula) × (−2 per ion) = **−4** for the negative portion of the formula

   b) For the given formula to be valid, the total of all charges in the formula must add to zero (a neutral compound). The total positive charge for the positive portion of the formula must be equal to **+4**.

   (1 $Sn^{+?}$ ions in the formula) × (+? per ion) must = **+4** for the positive portion of the formula; i.e.,

   1 × (+?) = **+4**; the unknown charge (+?) must be **+4**.

   Each Sn must be **+4**; the Roman numeral is (IV); **tin (IV)**

   (3) Selenium forms the negative ion; "selenide" (ion form of the name) is stated last in the complete name of the compound. Name: **tin (IV) selenide**

4. Name $BiCl_3$
   (1) **Bismuth**: Direct (unchanged) name of the metal that forms the positive ion.
   (2) Bismuth is one of the set of additional five not in Groups I or II and is a fixed-charged metal; **no** Roman numeral is required.
   (3) Chlorine (Cl) forms the negative ion; "chloride" (ion form of the name) is stated last in the complete name of the compound. Name: **bismuth chloride**

Additional Practice Examples

5. Name CoAs
   (1) **Cobalt**: Direct (unchanged) name of the metal that forms the positive ion.
   (2) Cobalt is ***not*** in Groups I or II and is ***not*** one of the additional five fixed-charged metals. Cobalt is a variable-charged metal; a Roman numeral must be used to indicate the numerical value of the positive charge. The **Roman numeral** in the final name represents the ***charge*** on the positive metal ion, ***not*** the subscript in the formula.
      a) Calculate the total negative charge for the **known** negative portion of the formula: Arsenic (As) can be determined to form a (–3) ion.

         (1 As^{-3} ions in the formula) × (–3 per ion) = **–3** for the negative portion of the formula

      b) The total positive charge for the positive portion of the **neutral** formula must be equal to **+3**.

         (1 $Co^{+?}$ ions in the formula) × (+? per ion) must = **+3**; i.e., 1 × (+?) = **+3**;

         the unknown charge (+?) must be **+3**.
         Each Co must be **+3**; the Roman numeral is (III); **cobalt (III)**
   (3) Arsenic forms the negative ion; "arsenide" (ion form of the name) is stated last in the complete name of the compound. Name: **cobalt (III) arsenide**
6. Name Ti_3P_2
   (1) **Titanium**: Direct (unchanged) name of the metal that forms the positive ion.
   (2) Titanium is ***not*** in Groups I or II and is ***not*** one of the additional five fixed-charged metals. Titanium is a variable-charged metal; a Roman numeral must be used to indicate the numerical value of the positive charge. The **Roman numeral** in the final name represents the ***charge*** on the positive metal ion, ***not*** the subscript in the formula.
      a) Calculate the total negative charge for the **known** negative portion of the formula: Phosphorous (P) can be determined to form a (–3) ion.

         (2 P^{-3} ions in the formula) × (–3 per ion) = **–6** for the negative portion of the formula

      b) The total positive charge for the positive portion of the **neutral** formula must be equal to **+6**.

         (3 $Ti^{+?}$ ions in the formula) × (+? per ion) must = **+6**; i.e., 3 × (+?) = **+6**

         Since there are **three** Ti ions in the formula, the unknown charge (+?) must be **+2**.
         Each Ti must be **+2**; the Roman numeral is (II); **titanium (II)**
   (3) Phosphorous forms the negative ion; "phosphide" (ion form of the name) is stated last in the complete name of the compound. Name: **titanium (II) phosphide**

V WRITING FORMULAS FOR IONIC COMPOUNDS WITH POLYATOMIC IONS

A **polyatomic ion** contains two or more **covalently** bonded atoms, which together form an ion unit (positive or negative). Many of the polyatomic ions commonly found in complete neutral compounds are shown below with the formula, net charge, and ion name. **Polyatomic ions are written in compound formulas exactly as shown in the table.**

Polyatomic Ions

NH_4^+ = ammonium	**NO_3^- = nitrate**
OH^- = hydroxide	**NO_2^- = nitrite**
CN^- = cyanide	**PO_4^{-3} = phosphate**
CO_3^{-2} = carbonate	**HPO_4^{-2} = hydrogen phosphate**
HCO_3^- = hydrogen carbonate	**$H_2PO_4^-$ = dihydrogen phosphate**
CrO_4^{-2} = chromate	
MnO_4^- = permanganate	**ClO_4^- = perchlorate**
SO_4^{-2} = sulfate	**ClO_3^- = chlorate**
HSO_4^- = hydrogen sulfate	**ClO_2^- = chlorite**
SO_3^{-2} = sulfite	**ClO^- = hypochlorite**
CH_3COO^- or $CH_3CO_2^-$ or $C_2H_3O_2^-$ = acetate	

The complete neutral ionic compound is formed when the polyatomic ion unit is further combined with **any** additional ion of opposite charge in the correct combining ratio to produce net neutrality. The procedure for writing formulas for ionic compounds containing polyatomic ions is similar to the one used for binary ionic compounds, modified by additional information.

Procedure for Writing Formulas for Ionic Compounds with Polyatomic Ions

(1) Identify the specific metal, symbol, and the correct specific metal positive charge as applicable; the polyatomic ion NH_4^+ is a possible positive ion. Use the periodic table as necessary. Identify the specific negatively charged ion (usually the polyatomic ion unless the compound contains NH_4^+). Always keep the atom formula of the polyatomic ion together and write it exactly as it appears in the table.

(2) Write the **positive** ion symbol (or formula) **first**; write the **negative** ion formula or symbol **second**. Charges can be shown while completing steps number (2) and (3); however do **not** show the charges in the final formula.

(3) Determine the correct **ratio** between the positive ion and the negative ion that leads to the **exact balancing of opposite charges**. Follow the guidelines and methods previously applied to binary ionic compounds. If a subscript is required for the polyatomic ion, enclose the polyatomic ion formula in parentheses.

(4) Verify that the selected ratio has balanced charges. Follow the calculation process previously applied to binary ionic compounds.

Example: Determine the correct formula for a combination of barium and the phosphate polyatomic ion. Use the periodic table as necessary; **structures, charges, and names of polyatomic ions must come from memory**.

(1) Barium is a metal, the symbol is **Ba**

$$Ba\ (56\ p+;56\ e^-) : lose\ 2e^- \longrightarrow Ba^{+2}(56\ p+;\ \mathbf{54\ e^-})$$

Phosphate is a polyatomic ion; formula and charge: **PO_4^{-3}**

(2) Ba^{+2} plus PO_4^{-3} $\longrightarrow$ $(Ba^{+2})_x(PO_4^{-3})_y$

(3) **Three** Ba^{+2} with 2 (+) charges require **two** PO_4^{-3}, which each have 3 (−) charges. (The least common multiple of 2 and 3 = 6.)

$$3Ba^{+2} + 2PO_4^{-3} \longrightarrow (Ba^{+2})_3(PO_4^{-3})_2 = Ba_3(PO_4)_2$$

The entire formula for the phosphate polyatomic ion is enclosed in parentheses to eliminate confusion with the consecutive subscripts: the formula Ba_3PO_{42} would appear to have 42 oxygens!

(4) Check the formula by addition of (+) and (−) values:

$$\text{Formula} = (Ba^{+2})_3\,(PO_4^{-3})_2$$

$$\text{Total (+) charges} = 3\ (Ba^{+2})\ \text{ions} \times (+2\ \text{per ion}) = +6$$

$$\text{Total (−) charges} = 2\ (PO_4^{-3})\ \text{ions} \times (-3\ \text{per ion}) = \underline{-6}$$

$$= 0;\ \text{the total charges balance}$$

Example: Determine the correct formula for a combination of potassium and the carbonate polyatomic ion. Use the periodic table as necessary; **structures, charges, and names of polyatomic ions must come from memory**.

(1) Potassium is a metal, the symbol is **K**

$$K\,(19p+; 19e^-) : \text{lose } 1e^- \longrightarrow K^+\,(19p+; \mathbf{18e^-})$$

Carbonate is a polyatomic ion; formula and charge: CO_3^{-2}

(2) K^+ plus CO_3^{-2} $\longrightarrow$ $(K^+)_x(CO_3^{-2})_y$

(3) **Two** K^+ each with 1 (+) charge are required for **each** CO_3^{-2}, which has 2 (−) charges.

$$2K^+ + CO_3^{-2} \longrightarrow (K^+)_2(CO_3^{-2})_1 = K_2CO_3$$

Since there is no subscript for the polyatomic ion, parentheses are not required.

(4) Check the formula by addition of (+) and (−) values:

$$\text{Formula} = (K^+)_2\,(CO_3^{-2})$$

$$\text{Total (+) charges} = 2\ (K^+)\ \text{ions} \times (+1\ \text{per ion}) = +2$$

$$\text{Total (−) charges} = 1\ (CO_3^{-2})\ \text{ion} \times (-2\ \text{per ion}) = \underline{-2}$$

$$= 0;\ \text{the total charges balance}$$

VI NAMING IONIC COMPOUNDS WITH POLYATOMIC IONS

The procedure for naming ionic compounds containing polyatomic ions is similar to that used for binary ionic compounds, modified by additional information.

Procedure for Naming Ionic Compounds Containing Polyatomic Ions

(1) Name the **metal** element that forms the **positive** ion **first**; use the name of the metal element directly with no change. If the compound contains the polyatomic ion NH_4^+, use the name **ammonia** directly.

(2) Determine if the metal forms only one specifically charged positive ion (a **fixed**-charged metal) or is a **variable**-charged metal. The ammonium polyatomic ion, NH_4^+, has only one specific (+1) positive charge.
If the metal is a fixed-charged metal nothing further is required for the metal portion of the name; ammonia also requires nothing further.
If the metal is a variable-charged metal, the value of the positive charge must be indicated in the name:
   a) Determine the specific numerical value of the metal positive charge in the specific compound to be named; these values will differ depending on the compound.
   b) Place this value, written as a **Roman numeral**, in parentheses directly after the metal name.

(3) Name the **negative ion last**; most often the negative ion will be the polyatomic ion. The names of all polyatomic ions shown in the table are already in the ion form for naming the complete compound; no changes are required. Complete the name of the compound by naming the negative ion; **use the exact name of the polyatomic ion** for the specific compound.

Example: Name $Ba_3(PO_4)_2$

(1) **Barium**: Direct (unchanged) name of the metal that forms the positive ion.
(2) Barium is in Group II and is a fixed-charged metal; **no** Roman numeral is required.
(3) The name **phosphate** matches the formula (PO_4) as the negative ion (charges are not shown); the name is used directly with no change and is stated last in the complete name of the compound. Name: **barium phosphate**

Example: Name K_2CO_3

(1) **Potassium**: Direct (unchanged) name of the metal that forms the positive ion.
(2) Potassium is in Group I and is a fixed-charged metal; **no** Roman numeral is required.
(3) The name **carbonate** matches the formula (CO_3) as the negative ion (charges are not shown); the name is used directly with no change and is stated last in the complete name of the compound. Name: **potassium carbonate**

VII NAMING BINARY COVALENT COMPOUNDS

The method for constructing structures and determining correct formulas for covalent compounds is based on a completely different set of concepts and are not covered here (see Chapter 14). This section describes only **naming** of **binary covalent** compounds; names for **multi-atom covalent** compounds generally fall under the classification of organic compounds and are not covered in this chapter. (However, see Chapter 5 in *Survival Guide to Organic Chemistry*, CRC Press.)

Binary covalent compounds **usually** exist in the form of molecules; formulas most often represent the actual composition of the molecule, not the simplest whole number ratio. Based on the molecular formula, binary covalent compounds must be named according to the **covalent system. Binary covalent compounds** are identified as consisting of **two** different **non-metal** (or metalloid) elements covalently bonded and, thus, are identified as being combinations of **only** the elements: **H, C, N, O, F, P, S, Cl, Se, Br, I, B, Si, As, and Te.**

Since methods for constructing covalent formulas and applying rules for the correct order of elements are not covered here, a procedure for naming binary covalent compounds from a given specific formula is described.

Covalent System for Naming Binary Covalent Compounds

(1) Name the first element listed in the formula with no change; this name is first in the final name of the binary compound.

(2) Name the second element listed in the formula by using the "-ide" ending form of the name; this name is last in the final compound name. Note that **ions** are **not** found in covalent compounds; however, the "ide" ending form of the last element is identical to the ion form of the name used for binary ionic compounds.

(3) The combining ratio of elements in the covalent compound is identified by the use of ***prefixes*** to indicate the number of atoms of each element in the formula:

mono- = **1**	di- = **2**	tri- = **3**	tetra- = **4**
penta- = **5**	hexa- = **6**	hepta- = **7**	octa- = **8**

a) To complete the name, start with the partial name based on steps (1) and (2), then add the appropriate **prefix** to each element that matches the number of each element listed in the formula.

b) **Exception**: If the **first element** in the formula has the number **1**, do **not** use the prefix **mono-**; absence of a prefix for the first element automatically indicates that the number of atoms in the formula is equal to 1.

c) **Exception**: The last vowel on a prefix name **(excluding di- or tri-)** is dropped (for pronunciation) if the element name that it is attached to also begins with a vowel.

d) **Exception** (sometimes optional): **All** prefixes for compounds containing the element **hydrogen** can be dropped; (often either name is acceptable).

Examples: Name the nitrogen/oxygen molecules: N_2O; NO; NO_2; N_2O_4; N_2O_5

N_2O

(1) Nitrogen
(2) ( )nitrogen ( )oxide
(3) N_2 = **di**nitrogen O_1 = **mon**oxide; note that since oxide begins with a vowel, the last "o" on mono is dropped. Name: **dinitrogen monoxide**

NO

(1) Nitrogen
(2) ( )nitrogen ( )oxide
(3) N_1 = nitrogen; note that if the first element in the name has the number "1," the prefix mono is dropped O_1 = **mon**oxide; the prefix mono (with the last "o" dropped) is required for the second element in the formula. Name: **nitrogen monoxide**

NO_2

(1) Nitrogen
(2) ( )nitrogen ( )oxide
(3) N_1 = nitrogen O_2 = **di**oxide; note that the last "i" on the prefix **di** cannot be dropped. Name: **nitrogen dioxide**

N_2O_4

(1) **Nitrogen**
(2) ( )nitrogen ( )oxide
(3) N_2 = **dinitrogen** O_4 = **tetr**oxide; note that the last "a" of tetra is dropped. Name: **dinitrogen tetroxide**

N_2O_5

(1) **Nitrogen**
(2) ( )nitrogen ( )oxide
(3) N_2 = **dinitrogen** O_5 = **pent**oxide; note that the last "a" of penta is dropped. Name: **dinitrogen pentoxide**

Example: Name P_2S_5

(1) **Phosphorous**
(2) ( )phosphorous ( )sulfide
(3) P_2 = **diphosphorous** S_5 = **penta**sulfide; note that the last "a" of penta is retained since sulfide does not begin with a vowel. Name: **diphosphorous pentasulfide**

Example: Name H_2S:

(1) **Hydrogen**
(2) ( )hydrogen ( )sulfide
(3) H_2 = **di**hydrogen S = **mono**sulfide; following the rules but not applying the exception to names containing hydrogen: **dihydrogen monosulfide**

Generally accepted names: **hydrogen disulfide** or simply **hydrogen sulfide**

VIII ADDITIONAL COMBINATION PRACTICE EXAMPLES

Show the correct formulas for the following named compounds. Use the periodic table as necessary; structures, charges, and names of polyatomic ions must come from memory.

1. Cadmium nitride
   (1) Cadmium (Cd) is one of the additional five fixed-charged metals; forms Cd^{+2} ion. Nitride is the negative ion of nitrogen; forms N^{-3}
   (2) $\mathbf{Cd^{+2}}$ plus $\mathbf{N^{-3}} \longrightarrow \mathbf{(Cd^{+2})_x(N^{-3})_y}$
   (3) $3Cd^{+2} + 2N^{-3} \longrightarrow \mathbf{(Cd^{+2})_3(N^{-3})_2 = Cd_3N_2}$
   (4) Total (+) charges = 3 (Cd^{+2}) ions × (+2 per ion) = +6
   Total (−) charges = 2 (N^{-3}) ions × (−3 per ion) = −6 total charges balance

2. Gold (III) cyanide
   (1) Gold (Au) is given as a charge of +3; forms Au^{+3} ion. Cyanide is a polyatomic ion: CN^-
   (2) $\mathbf{Au^{+3}}$ plus $\mathbf{CN^-} \longrightarrow \mathbf{(Au^{+3})_x\,(CN^-)_y}$
   (3) $1Au^{+3} + 3CN^- \longrightarrow \mathbf{(Au^{+3})_1\,(CN^-)_3 = Au(CN)_3}$
   (4) Total (+) charges = 1 (Au^{+3}) ion × (+3 per ion) = +3
   Total (−) charges = 3 (CN^-) ions × (−1 per ion) = −3 total charges balance
3. Vanadium (V) oxide
   (1) Vanadium (V) is given as a charge of +5; forms V^{+5} ion.
   Oxide is the negative ion of oxygen; forms O^{-2}

(2) V^{+5} plus O^{-2} ⟶ $(V^{+5})_x (O^{-2})_y$
(3) $2V^{+5} + 5O^{-2}$ ⟶ $(V^{+5})_2(O^{-2})_5 = V_2O_5$
(4) Total (+) charges = 2 (V^{+5}) ions × (+5 per ion) = +10
Total (−) charges = 5 (O^{-2}) ions × (−2 per ion) = −10 total charges balance

4. Selenium tetraflouride
Selenium (non-metal) **Se**; no prefix for the first element, the subscript must be **1**. Flouride is the compound form of flourine (non-metal) **F**; **tetra** = subscript **4** formula from the covalent system: SeF_4

5. Aluminum perchlorate
(1) Aluminum (Al) follows the noble gas electron rules; forms Al^{+3} ion. Perchlorate is a polyatomic ion: ClO_4^-
(2) Al^{+3} plus ClO_4^- ⟶ $(Al^{+3})_x (ClO_4^-)_y$
(3) $1Al^{+3} + 3ClO_4^-$ ⟶ $(Al^{+3})_1 (ClO_4^-)_3 = Al(ClO_4)_3$
(4) Total (+) charges = 1 (Al^{+3}) ion × (+3 per ion) = +3
Total (−) charges = 3 (ClO_4^-) ions × (−1 per ion) = −3 total charges balance

6. Carbon disulfide
Carbon (non-metal) **C**; no prefix for the first element, subscript must be **1**. Sulfide is the compound form of sulfur (non-metal) **S**; **di** = subscript **2** formula from the covalent system: CS_2

7. Ammonium acetate
(1) Ammonium is a positive polyatomic ion; NH_4^+ ion.
acetate is a polyatomic ion: CH_3COO^-
(2) NH_4^+ plus CH_3COO^- ⟶ $(NH_4^+)_x (CH_3COO^-)_y$
(3) $1NH_4^+ + 1\,CH_3COO^-$ ⟶ $(NH_4^+)_1(CH_3COO^-)_1 = NH_4CH_3COO$ (or NH_4OOCCH_3)
(4) Total (+) charges = 1 (NH_4^+) ion × (+1 per ion) = +1
Total (−) charges = 1 (CH_3COO^-) ion × (− per ion) = −1 total charges balance

8. Strontium nitrate
(1) Strontium (Sr) follows the noble gas electron rules; forms Sr^{+2} ion.
Nitrate is a polyatomic ion: NO_3^-
(2) Sr^{+2} plus NO_3^- ⟶ $(Sr^{+2})_x (NO_3^-)_y$
(3) $1\,Sr^{+2} + 2\,NO_3^-$ ⟶ $(Sr^{+2})_1 (NO_3^-)_2 = Sr(NO_3)_2$
(4) Total (+) charges = 1 (Sr^{+2}) ion × (+2 per ion) = +2
Total (−) charges = 2 (NO_3^-) ions × (−1 per ion) = −2 total charges balance

Name the following compounds; the compound must first be identified as ionic or covalent; use the ionic naming system or the covalent naming system as appropriate.

1. Ni_2CO_3
(1) **Nickel**: Direct (unchanged) name of the metal that forms the positive ion.
(2) Nickel is a variable-charged metal; a Roman numeral is required.
CO_3 (carbonate) has a charge of −2 (CO_3^{-2})
(1 CO_3^{-2} ions in the formula) × (−2 per ion) = −2
(2 $Ni^{+?}$ ions in the formula) × (+? per ion) must = **+2**
Each Ni must be **+1**; the Roman numeral is (I): **nickel (I)**
(3) The name **carbonate** matches the formula (CO_3^{-2}) as the negative ion. Name: **nickel (I) carbonate**

2. PtS_2
   (1) **Platinum**: Direct (unchanged) name of the metal that forms the positive ion.
   (2) Platinum is a variable-charged metal; a Roman numeral is required. Sulfur (S) forms an ion charge of –2: (S^{-2}).
   (2 S^{-2} ions in the formula) × (–2 per ion) = **–4**
   (1 $Pt^{+?}$ ion in the formula) × (+? per ion) must = **+4**
   Each Pt must be **+4**; the Roman numeral is (IV): **platinum (IV)**
   (3) Sulfur forms the negative ion; the ion form of the name is sulfide. Name: **platinum (IV) sulfide**

3. As_2Cl_5
   The compound is composed of two non-metals; the covalent system is required.
   (1) **Arsenic**
   (2) ( )arsenic( )chloride
   (3) As_2 = **di**arsenic Cl_5 = **penta**chloride. Name: **diarsenic pentachloride**

4. $CrCl_3$
   (1) **Chromium**: Direct (unchanged) name of the metal that forms the positive ion.
   (2) Chromium is a variable-charged metal; a Roman numeral is required. Chlorine (Cl) forms an ion of charge of –1: (Cl^-)
   (3 Cl^- ions in the formula) × (–1 per ion) = **–3**
   (1 $Cr^{+?}$ ion in the formula) × (+? per ion) must = **+3**
   Each Cr must be +3; the Roman numeral is (III): **chromium (III)**
   (3) Chlorine forms the negative ion; the ion form of the name is chloride. Name: **chromium (III) chloride**

5. $Zn_3(SO_3)_2$
   (1) **Zinc**: Direct (unchanged) name of the metal that forms the positive ion.
   (2) Zinc is a fixed-charged metal; **no** Roman numeral is required.
   (3) The name **sulfite** matches the formula (SO_3^{-2}) as the negative ion. Name: **zinc sulfite**

6. Mn_2Te_3
   (1) **Manganese**: Direct (unchanged) name of the metal that forms the positive ion.
   (2) Manganese is a variable-charged metal; a Roman numeral is required.
   Tellerium (Te) forms an ion of charge of –2: (Te^{-2})
   (3 Te^{-2} ions in the formula) × (–2 per ion) = **–6**
   (2 Mn +? ion in the formula) × (+? per ion) must = **+6**
   Each Mn must be +3; the Roman numeral is (III): **manganese (III)**
   (3) Tellurium forms the negative ion; the ion form of the name is telluride. Name: **manganese (III) telluride**

7. BrF_3
   The compound is composed of two non-metals; the covalent system is required.
   (1) **Bromine**
   (2) ( )bromine ( )fluoride
   (3) Br_1 = bromine ("mono" prefix not used) F_3 = **tri**chloride. Name: **bromine trifluoride**

8. Li_2CrO_4
   (1) **Lithium**: Direct (unchanged) name of the metal that forms the positive ion.
   (2) Lithium is a fixed-charged metal; **no** Roman numeral is required.
   (3) The name **chromate** matches the formula (CrO_4^{-2}) as the negative ion. Name: **lithium chromate**

IX PRACTICE PROBLEMS

1. **(i)** Write correct formulas for the following combination of elements to form ionic compounds. **(ii)** Write the names of the compounds that were formed in part **(i)**; be certain to use Roman numerals where required.

a) Cesium plus fluorine
b) Lithium plus bromine
c) Rubidium plus selenium
d) Strontium plus bromine
e) Magnesium plus sulfur
f) Magnesium plus arsenic
g) Aluminum plus sulfur
h) Sodium plus sulfur
i) Barium plus chlorine
j) Zinc plus phosphorous
k) Bismuth plus oxygen
l) Potassium plus iodine
m) Iron (+3) plus tellurium
n) Copper (+2) plus phosphorous
o) Chromium (+3) plus selenium
p) Mercury (+1) plus oxygen
q) Zirconium (+2) plus nitrogen
r) Platinum (+2) plus chlorine
s) Platinum (+4) plus sulfur
t) Cobalt (+4) plus bromine

2. **(i)** Write correct formulas for the following combination of elements and/or polyatomic ions to form ionic compounds. **(ii)** Write the names of the compounds that were formed in part **(i)**.

a) Nickel (+1) plus carbonate ion
b) Barium plus nitrite ion
c) Chromium (+2) plus nitrate ion
d) Iron (+2) plus phosphate ion
e) Lead (+4) plus hydroxide ion
f) Tin (+4) plus hydrogen phosphate ion
g) Vanadium (+3) plus hydrogen sulfate ion
h) Silver plus sulfite ion

3. Write formulas for the following compounds based on the name given.

a) Platinum (II) nitrate
b) Aluminum hydroxide
c) Potassium phosphate
d) Manganese (II) hydrogen carbonate
e) Calcium permanganate
f) Ammonium chromate
g) Copper (I) nitrate
h) Titanium (IV) perchlorate
i) Chlorine monofluoride
j) Silicon tetrachloride
k) Culfur trioxide
l) Iodine pentabromide

4. Name the following compounds; use the ionic or the covalent naming system as appropriate.

a) LiH
b) $PdCl_4$
c) Ag_2S
d) Ni_2Se
e) $Cu(H_2PO_4)_2$
f) $(NH_4)_2SO_3$
g) $Zn(CN)_2$
h) $Ba(ClO_3)_2$
i) ClF_3
i) H_2Se
k) SF_4
l) PbO
m) PbO_2
n) PCl_5
o) N_2O_5
p) Cu_3N
q) SnF_2
r) SCl_2
s) BF_3
t) SeO_2

X ANSWERS TO PRACTICE PROBLEMS

1. **(i)** Write correct formulas for the following combination of elements to form ionic compounds. **(ii)** Write the names of the compounds that were formed in part **(i)**; be certain to use Roman numerals where required.

a) Cesium plus fluorine = **CsF = cesium fluoride**
b) Lithium plus bromine = **$LiBr$ = lithium bromide**
c) Rubidium plus selenium = **Rb_2Se = rubidium selenide**
d) Strontium plus bromine = **$SrBr_2$ = strontium bromide**
e) Magnesium plus sulfur = **MgS = magnesium sulfide**

f) Magnesium plus arsenic = Mg_3As_2 = **magnesium arsenide**
g) Aluminum plus sulfur = Al_2S_3 = **aluminum sulfide**
h) Sodium plus sulfur = Na_2S = **sodium sulfide**
i) Barium plus chlorine = $BaCl_2$ = **barium chloride**
j) Zinc plus phosphorous = Zn_3P_2 = **zinc phosphide**
k) Bismuth plus oxygen = Bi_2O_3 = **bismuth oxide**
l) Potassium plus iodine = **KI** = **potassium iodide**
m) m) Iron (+3) plus tellurium = Fe_2Te_3 = **iron (III) telluride**
n) Copper (+2) plus phosphorous = Cu_3P_2 = **copper (II) phosphide**
o) Chromium (+3) plus selenium = Cr_2Se_3 = **chromium (III) selenide**
p) Mercury (+1) plus oxygen = Hg_2O = **mercury (I) oxide**
q) Zirconium (+2) plus nitrogen = Zr_3N_2 = **zirconium (II) nitride**
r) Platinum (+2) plus chlorine = $PtCl_2$ = **platinum (II) chloride**
s) Platinum (+4) plus sulfur = PtS_2 = **platinum (IV) sulfide**
t) Cobalt (+4) plus bromine = $CoBr_4$ = **cobalt (IV) bromide**

2. **(i)** Write correct formulas for the following combination of elements and/or polyatomic ions to form ionic compounds. **(ii)** Write the names of the compounds that were formed in part **(i)**.
   a) Nickel (+1) plus carbonate ion = Ni_2CO_3 = **nickel (I) carbonate**
   b) Barium plus nitrite ion = $Ba(NO_2)_2$ = **barium nitrite**
   c) Chromium (+2) plus nitrate ion = $Cr(NO_3)_2$ = **chromium (II) nitrate**
   d) Iron (+2) plus phosphate ion = $Fe_3(PO_4)_2$ = **iron (II) phosphate**
   e) Lead (+4) plus hydroxide ion = $Pb(OH)_4$ = **lead (IV) hydroxide**
   f) Tin (+4) plus hydrogen phosphate ion = $Sn(HPO_4)_2$ = **tin (IV) hydrogen phosphate**
   g) Vanadium (+3) plus hydrogen sulfate ion = $V(HSO_4)_3$ = **vanadium (III) hydrogen sulfate**
   h) Silver plus sulfite ion = Ag_2SO_3 = **silver sulfite**

3. Write formulas for the following compounds based on the name given.
   a) Platinum (II) nitrate = $Pt(NO_3)_2$
   b) Aluminum hydroxide = $Al(OH)_3$
   c) Potassium phosphate = K_3PO_4
   d) Manganese (II) hydrogen carbonate = $Mn(HCO_3)_2$
   e) Calcium permanganate = $Ca(MnO_4)_2$
   f) Ammonium chromate = $(NH_4)_2CrO_4$
   g) Copper (I) nitrate = $CuNO_3$
   h) Titanium (IV) perchlorate = $Ti(ClO_4)_4$
   i) Chlorine monofluoride = **ClF**
   j) Silicon tetrachloride = $SiCl_4$
   k) Sulfur trioxide = SO_3
   l) Iodine pentabromide = IBr_5

4. Name the following compounds; use the ionic or the covalent naming system as appropriate.
   a) LiH = **lithium hydride**
   b) $PdCl_4$ = **palladium (IV) chloride**
   c) Ag_2S = **siver sulfide**
   d) Ni_2Se = **nickel (I) selenide**
   e) $Cu(H_2PO_4)_2$ = **copper (II) dihydrogen phosphate**
   f) $(NH_4)_2SO_3$ = **ammonium sulfite**
   g) $Zn(CN)_2$ = **zinc cyanide**
   h) $Ba(ClO_3)_2$ = **barium chlorate**

i) ClF_3 = **chlorine trifluoride**
j) H_2Se = **hydrogen selenide (or dihydrogen selenide)**
k) SF_4 = **sulfur tetrafluoride**
l) PbO = **lead (II) oxide**
m) PbO_2 = **lead (IV) oxide**
n) PCl_5 = **phosphorous pentachloride**
o) N_2O_5 = **dinitrogen pentoxide**
p) Cu_3N = **copper (I) nitride**
q) SnF_2 = **tin (II) fluoride**
r) SCl_2 = **sulfur dichloride**
s) BF_3 = **boron trifluoride**
t) SeO_2 = **selenium dioxide**

5 An Introduction to Moles and Molar Mass

I GENERAL CONCEPTS

A mole is a **counting** number; the value of **one mole = 6.02214 × 10^{23}** objects. In principle, this counting number can be applied to any countable object; however, in practice, the mole is used only for objects as small as atoms or molecules. One mole of pennies would cover the area of the state of Illinois to a height of about 10 miles.

COUNTING NUMBERS

pair = 2 dozen = **12** score = **20** gross = **144**

mole = **6.02214 × 10^{23}**

The mole counting number is based on a specific standard. **One mole** is defined as equal to a number represented by the exact number of atoms present in **exactly 12 grams** of the isotope **carbon-12.** The carbon-12 standard used for the definition of one mole is the same standard used for the definition of an amu. This results in the following relationships:

The mass of **one atom** of carbon-12 is exactly **12 amu** (by definition).
The mass of **one mole** (6.02214 × 10^{23}) of carbon-12 atoms is exactly **12 grams** (by definition).

Note that the conversion factor 1.66054 × 10^{-24}g/amu interconverts grams and amu's. Calculate the number of amu's per gram:

$$\textbf{\# amu in 1 gram} = \frac{1\text{ gram}}{1.66054\times10^{-24}\text{ g/amu}} = \mathbf{6.02214\times10^{23}\,amu/gram}$$

The number of atoms (or ions or molecules) in one mole is identical to the number of amu in one gram.

The **molar mass (MM)** of an element is defined as the **mass of one mole of average atoms** of this element.

The **molar mass** (MM = the mass of one mole of atoms) of **any element is numerically equal to the average atomic mass of the element expressed in gram units;** the units of MM = grams/moles.

Recall that the mass of one average atom of an element is defined as the atomic mass (sometimes called atomic weight) and is listed in the periodic table. The definitions of amu and mole based on carbon-12, as described, produce the result that:

The atomic mass listed in the periodic table is numerically identical to the molar mass of that element.

Mass of **one atom** of carbon-12 = 12 **amu**

Mass of **one mole** (6.02214 × 10^{23}) of carbon-12 **atoms** = 12 **grams**

Mass of **one average atom** of carbon = 12.011 **amu**

Mass of **one mole** (6.02214×10^{23}) of **average** carbon **atoms** = 12.011 **grams**

MM of carbon = 12.011 grams

Mass of **one average atom** of copper = 63.546 **amu**

Mass of **one mole** (6.02214×10^{23}) of **average** copper **atoms** = 63.546 **grams**

MM of copper = 63.546 grams

Example:

a) Calculate the number of Cu atoms in 10.0 grams by calculating through amu units.
b) Repeat the calculation by using moles.

The total mass of copper in the problem is given in gram units. Calculate the mass of one (average) copper atom in grams; use the periodic table to find the mass in amu.

a) Mass 1 Cu atom = $(63.55 \text{ amu}) \times (1.66054 \times 10^{-24} \text{g/amu}) = 1.055 \times 10^{-22}$ g/atom

$$\#\text{ of Cu atoms in } 10.0 \text{ g} = \frac{10.0 \text{ gram}}{1.055 \times 10^{-22} \text{ g/atom}} = \mathbf{9.48 \times 10^{22}} \text{ atoms}$$

Alternatively: total mass of copper in amu

$$= (10.0 \text{ g}) \times (6.02214 \times 10^{23} \text{ amu/gram}) = 6.02214 \times 10^{24} \text{ amu}$$

$$\#\text{ of Cu atoms in } 10.0 \text{ g} = \frac{6.02214 \times 10^{24} \text{ amu}}{63.55 \text{ amu/atom}} = \mathbf{9.48 \times 10^{22}} \text{ atoms}$$

b) MM = m(g)/moles; solve for moles: moles = m(g)/MM

MM is read from the periodic table = 63.55 g/mole

$$\#\text{ of moles of Cu atoms} = \frac{10.0 \text{ grams}}{63.55 \text{ g/mole}} = \mathbf{0.157} \text{ moles}$$

# of Cu atoms in 10.0 g

$$= (0.157 \text{ moles}) \times (6.022 \times 10^{23} \text{ atoms/mole}) = \mathbf{9.48 \times 10^{22}} \text{ atoms}$$

Example:

a) Calculate the mass in **grams** of 3.011×10^{23} atoms of Fe by calculating through amu units; use the periodic table to find the mass in amu.
b) Repeat the calculation by using moles.

a) Mass 1 Fe atom in **grams**

$$= (55.85 \text{ amu}) \times (1.66054 \times 10^{-24} \text{g/amu}) = 9.274 \times 10^{-23} \text{ g/atom}$$

mass of 3.011×10^{23} atoms of Fe

$$= (3.011 \times 10^{23} \text{ atoms}) \times (9.274 \times 10^{-23} \text{ g/atom}) = \mathbf{27.9} \text{ grams}$$

b) MM = m(g)/moles; solve for m(g): m(g) = moles × MM

MM is read from periodic table = 55.85 g/mole

$$\text{moles of Fe} = \frac{3.011 \times 10^{23} \text{ atoms}}{6.022 \times 10^{23} \text{ atoms/mole}} = 0.500 \text{ mole}$$

$$\text{total mass of Fe} = (0.500 \text{ mole}) \times (55.85 \text{ g/mole}) = \mathbf{27.9} \text{ grams}$$

The use of the mole as a counting number allows the mass to be related to the number of atoms and to be used with atom ratios in formulas.

Molar mass = mass(grams)/moles; the abbreviated equational form of the relationship is:

$$\mathbf{MM} = \frac{\mathbf{m(g)}}{\mathbf{moles}}$$

The MM for an **unknown** element can be determined if the mass and moles for a specific sample are known variables:

$$\mathbf{MM} = \frac{\textbf{mass (grams) of a specific sample}}{\textbf{moles of the same specific sample}}$$

To solve for mass (g) for a known element when # of moles is given and the MM can be read from the periodic table:

$$\mathbf{m(g) = moles \times MM}$$

To solve for # of moles for a known element when mass (g) is given and the MM can be read from the periodic table:

$$\mathbf{moles} = \frac{\mathbf{m(g)}}{\mathbf{MM}}$$

To convert # of moles to # of atoms:

$$\textbf{\# of atoms} = \textbf{(\# of moles)} \times \mathbf{(6.022 \times 10^{23}} \textbf{ atoms/mole)}$$

II MASS/MOLE/ATOM CONVERSIONS FOR ELEMENTS

Procedure for Solving Mole Problems (Elements)

(1) Identify the correct form of the **[MM = mass(g)/moles]** equation, which is required to solve for the unknown variable.

(2) Identify or calculate the values of the required known variables in the correct units:

a) Perform unit conversions when required.

b) If necessary, **solve** for required known variables from other chemical concepts, such as density.

c) If the **molar mass** of the **element** is a required known variable, determine this value directly from the **periodic table.**

(3) Complete the final calculation for the desired known variable.

(4) Use the value from **step (3)** to solve a more complex chemical problem, when necessary.

Example: Calculate mass in pounds of moles of a gold statue, which contains 12.7 moles of gold; 1 lb = 454 grams.

(1) m(g) = moles × MM

(2) moles given = 12.7 moles

MM is read from the periodic table = 196.97 g/mole

(3) m(g) = (2.7 moles) × 196.97 g/mole = **2.50 × 10³** grams

(4) $\text{mass (lbs)} = 2.50\times10^3\,\text{grams}\times\dfrac{1\,\text{lb}}{454\,\text{grams}} = 5.50\text{ lbs}$

Example: Calculate # of atoms of silver in a silver plate with a volume of 145 cm³; density of silver = 10.50 g/cm³.

(1) Moles = m(g)/MM

(2) MM is read from the periodic table = 107.87 g/mole

mass (g) must be calculated from density: mass (g) = 145 cm³ × 10.50 g/cm³
= 1522.5 grams

(3) $\#\text{ of moles} = \dfrac{1522\,\text{grams}}{107.87\,\text{g/mole}} = \mathbf{14.1}\text{ moles}$

(4) # of Ag atoms = (14.1 moles) × (6.022 × 10^{23} atoms/mole) = **8.49 × 10^{24}** atoms

Example: Calculate # of moles of copper in a copper wire with a 2.00 millimeter radius and a 7.60 meter length. Density of Cu = 8.92 g/cm³; V of cylinder = $\pi r^2 h$

(1) Moles = m(g)/MM

(2) MM is read from the periodic table = 63.55 g/mole

mass (g) must be calculated from density;

volume must be calculated from linear dimensions:

convert all dimensions to cm before inserting into the equation (see Chapter 1):

$$V(\text{cm}^3) = \pi r^2 h = \pi(0.200\text{ cm})^2(760.\text{ cm}) = 95.5\,\text{cm}^3$$

$$\text{mass (g)} = V \times d = (145\text{ cm}^3) \times (8.92\text{ g/cm}^3) = 852\text{ g}$$

(3) $\#\text{ of moles} = \dfrac{852\text{ grams}}{63.55\text{ g/mole}} = \mathbf{13.4}\,\text{moles}$

Example: A titanium ball contains 222 moles of Ti. Calculate the radius of this ball. Density of Ti = 4.51 g/cm³; V of sphere = $(4/3)\pi r^3$

(1) The radius can be determined from the volume; the volume can be determined from the density if mass (g) is calculated.

$$m(g) = \text{moles} \times \text{MM}$$

(2) Moles given = 222 moles
MM is read from the periodic table = 47.88 g/mole
(3) m(g) = (222 moles) × 47.88 g/mole = **1.063 × 10⁴** grams
(4) Solve a density problem: d = m/V; V = m/d

$$V = \frac{1.063\times10^4\,\text{grams}}{4.51\,\text{g/cm}^3} = 2357\ \text{cm}^3$$

$$V = (4/3)\pi r^3 \quad ; \quad r^3 = \frac{V(\text{cm}^3)}{(4/3)\pi} \qquad r^3 = \frac{2357\,\text{cm}^3}{(4/3)\pi} = 562.7\,\text{cm}^3$$

$$r = (562.7\ \text{cm}^3)^{1/3} = \mathbf{8.26\,cm}$$

III MASS/MOLE/MOLECULE CONVERSIONS FOR COMPOUNDS

The **formula unit mass** applies to compounds that come in the form of a large three-dimensional array (usually ionic compounds), and it is equal to the mass of one formula written as the simplest whole number ratio. The formula unit mass (amu) = sum of the masses of each atom (amu) in the simplest whole number ratio formula.

Molecular mass applies to compounds that come in the form of molecules, and it is equal to the mass of one molecule. The molecular mass (amu) = sum of the masses of each atom (amu) in the molecular formula.

In both cases, the calculation involves the summation of the masses (amu) of each atom in the specific formula type used; the masses of elements are taken directly from the periodic table.

Example:

Formula unit mass of $CuCl_2$ (ionic compound): 1 Cu × 63.54 amu = 63.54 amu

+ 2 Cl × 35.45 amu = 70.90 amu

total = **134.44 amu**

Example:

Molecular mass of CO_2 (molecule): 1 C × 12.01 amu = 12.01 amu

+ 2 O × 16.00 amu = 32.00 amu

total = **44.01 amu**

The use of the mole as a counting number can be applied to compounds as well as elements; the relationship is based on the **molar mass of the compound**.

The term molar mass (MM) of a compound has a similar definition to that for an element: The **molar mass** (MM) of a **compound** is defined as the mass of one mole of molecules or formula units.

The molar mass (MM) of any compound is numerically equal to the molecular mass or formula unit mass of the compound expressed in gram units; the units of MM = grams/moles.

Molar mass is most often used directly in calculations. The molar mass (MM) = sum of the molar masses of each atom (in gram units) in the molecule or each atom (in gram units) in the simplest whole number ratio formula.

Example:

MM of $CuCl_2$ (ionic compound): 1 Cu × 63.54 g/mole = 63.54 grams/mole

+ 2 CI × 35.45 g/mole = 70.90 grams/mole

total = **134.44 g/mole**

Example:

MM of CO_2 (molecule): 1 C × 12.01 g/mole = 12.01 grams/mole

+ 2 0 × 16.00 g/mole = 32.00 grams/mole

total = **44.01 g/mole**

Counting all the atoms in formulas containing polyatomic ions requires care.

Example:

MM of $Co_3 (PO_4)_2$ (ionic compound): 3 Co × 63.54 g/mole = 176.70 grams/mole

+ 2 P × 30.97 g/mole = 61.94 grams/mole

+ 8 O × 16.00 g/mole = 128.00 grams/mole

total = **366.73 g/mole**

Certain elements come in the form of **molecules.** The molar mass of elements, which come in the form of molecules, is based on the formula of the complete molecule.

MM of O_2 (element comes as a molecule): 2 O × 16.00 g/mole = 32.00 g/mole

Procedure for Solving Mole Problems (Compounds)

(1) Identify the correct form of the **[MM = mass(g)/moles]** equation, which is required to solve for the unknown variable.
(2) Identify or calculate the values of the required known variables in the correct units:
   a) Perform unit conversions when necessary.
   b) If necessary, **solve** for required known variables from other chemical concepts, such as density.
   c) If molar mass of the compound is a required known variable, calculate this value by **summation** of the **MM** of each atom in the compound formula.
(3) Complete the final calculation for the desired known variable.
(4) Use the value from **step** (3) to solve a more complex chemical problem.

Example: Calculate the # of moles of acetone (C_3H_6O) in a 1.00-liter bottle; density of acetone = 0.788 g/mL.

(1) Moles = m(g)/MM
(2) MM must be calculated from MM of the elements in the formula; the process does **not** require distinguishing the type of compound:

$$\textbf{MM} \text{ of } C_3H_6O = 3\ C \times 12.01 \text{ g/mole} = 36.03 \text{ g/mole}$$

$$+ 6\ H \times 1.01 \text{ g/mole} = 6.06 \text{ g/mole}$$

$$+ 1\ O \times 16.00 \text{ g/mole} = \underline{16.00 \text{ g/mole}}$$

$$\text{total MM} = 58.09 \text{ g/mole}$$

(3) Mass must be calculated from density:

$$V(\text{mL}) = 1.00 \text{ L} \times 1000 \text{ mL/L} = 1000 \text{ mL}$$

$$\text{mass (g)} = V \times d;\ V = 1000 \text{ mL} \times 0.788 \text{ g/mL} = 788 \text{ grams}$$

(4) # of moles acetone $= \dfrac{788 \text{ grams}}{58.09 \text{ g/mole}} = \mathbf{13.6}$ moles

Example: Calculate the # of molecules of propane gas (C_3H_8) in a 10.0-liter tank; density of propane gas = 80.0 g/L (Gas densities are stated in grams/liter)

(1) Moles = m(g)/MM

(2) MM must be calculated from MM of the elements in the formula:

$$\text{MM of } C_3H_8 = 3\ C \times 12.01 \text{ g/mole} = 36.03 \text{ g/mole}$$

$$+ 8\ H \times 1.01 \text{ g/mole} = \underline{8.08 \text{ g/mole}}$$

$$\text{total MM} = 44.11 \text{ g/mole}$$

$$\text{mass (g)} = 10.0 \text{ L} \times 80.0 \text{ g/L} = 800. \text{ g}$$

(3) moles propane $= \dfrac{800 \text{ grams}}{44.11 \text{ g/mole}} = \mathbf{18.2}$ moles

(4) molecules = (18.2 moles) × (6.022×10^{23} molecules/mole) = $\mathbf{1.10 \times 10^{25}}$ molecules

IV CONCEPTS FOR USING RATIOS IN FORMULAS

The formula of a compound provides the ratio of elements in the compound. For example, the formula $\mathbf{C_3H_6O}$ indicates that one molecule of this compound has **3** carbons, **6** hydrogens, and **1** oxygen; this is equivalent to a **ratio** of **6** carbons to **3** hydrogens to **1** oxygen.

The ratio of atoms of an element to one formula must be exactly equal to the ratio of moles of each element to one mole of formulas (molecules or formula units). Thus, the formula $\mathbf{C_3H_6O}$ also specifies that one **mole** of molecules has **3 moles** of carbon, **6 moles** of hydrogen, and **1 mole** of oxygen; this is equivalent to a **ratio** of **6** moles carbon to **3** moles hydrogen to **1** mole oxygen.

This concept is the basis for relating the amount of a specific element to an amount of the compound in which it is contained:

moles of element contained in a sample of compound =

$$\text{moles of the compound} \times \frac{\text{\# of moles of the element}}{\text{(in) one mole of compound}}$$

The ratio listed in this equation is found directly from the compound formula:

$$\text{moles of element contained in a sample of compound} =$$

$$\text{moles of compound} \times \frac{\text{\# of atoms of the element}}{\text{(in) 1 total formula}}$$

Analogy: The "formula" of a bicycle might be:

$$\textbf{(Handle Bar)}_1\ \textbf{(Seat)}_1\ \textbf{(Pedals)}_2\ \textbf{(Wheels)}_2$$

How many wheels are contained in 41 bikes?

Of course, this question can be answered by "inspection," but the process used sets up the ratio of a specific part to one complete unit (using the "formula").

$$\text{\# of wheels} = \text{\# of bikes} \times \frac{2\,\text{wheels}}{1\,\text{bike}};\quad \text{\# of wheels} = 41\ \text{bikes} \times \frac{2\ \text{wheels}}{1\,\text{bike}} = 82\ \text{wheels}$$

Example: Calculate the number of moles of carbon and the number of moles of hydrogen in 625 grams of acetone, C_3H_6O, by first finding the number of moles of acetone.

(1) Moles = m(g)/MM

(2) MM is calculated from MM of the elements in the formula; the result from a previous problem was MM = 58.09 g/mole

(3) $\text{\# of moles acetone} = \dfrac{625\,\text{grams}}{58.09\,\text{g/mole}} = \mathbf{10.8}\ \text{moles of } C_3H_6O$

(4) The number of moles of specific elements in the formula uses the specific ratios in the formula.

$$\text{moles of carbon} = 10.8\ \text{moles } C_3H_6O \times \frac{3\ \text{carbons}}{1\ C_3H_6O} = \mathbf{32.4\ mole}\ \text{carbon}$$

$$\text{moles of hydrogen} = 10.8\ \text{moles } C_3H_6O \times \frac{6\ \text{hydrogens}}{1\ C_3H_6O} = \mathbf{64.8\ mole}\ \text{hydrogen}$$

The inverse of the ratio of elements in a formula can be used to solve for an unknown amount of compound that contains a specific amount of a certain element.

$$\text{moles of a compound which contains a specific amount of an element} =$$

$$\text{moles of the element} \times \frac{\text{one mole of compound (to)}}{\text{\# of moles of each element}}$$

This ratio is equivalent to:

$$\text{moles of a compound which contains a specific amount of an element} =$$

$$\text{moles of the element} \times \frac{\text{1 total formula (to)}}{\text{of atoms of the element (in formula)}}$$

Analogy: Use the bicycle formula to calculate the number of bikes, which can be made from 30 wheels.

$$\textbf{(Handle Bar)}_1\ \textbf{(Seat)}_1\ \textbf{(Pedals)}_2\textbf{(Wheels)}_2$$

In this case, the ratio developed indicates 2 wheels are required for 1 bike.

$$\#\text{ of bikes} = \#\text{ of wheels} \times \frac{1\text{ bike}}{2\text{ wheels}}; \quad \#\text{ of bikes} = 30\text{ wheels} \times \frac{1\text{ bike}}{2\text{ wheels}} = 15\text{ bikes}$$

Example: Calculate the **moles** of C_3H_6O, which can be prepared from 155 **grams** of hydrogen. This is the same as calculating the **moles** of C_3H_6O, which contains 155 grams of hydrogen. First calculate the number of moles of hydrogen atoms represented by 155 grams of hydrogen; note for this problem the MM of diatomic H_2 is not used.

(1) moles = m(g)/MM

(2) MM of the element hydrogen = 1.01 g/mole
m(g) of H = 155 grams

(3) # of moles hydrogen **atoms** = 155 grams/1.01 g/mole = 153 moles of H atoms

(4) moles of C_3H_6O which contain 153 moles of hydrogen atoms =

$$(153\text{ moles H atoms}) \times \frac{1\ C_3H_6O}{6\text{ H atoms}} = 25.5\text{ mole } C_3H_6O$$

Example: Lithium has been used as a pharmaceutical. A bottle contains 29.6 grams of lithium carbonate, Li_2CO_3.

a) Calculate the number of moles of lithium ions and the number of moles of oxygen atoms in the bottle.
b) Calculate the number of moles of lithium carbonate, which can be formed from 1.00 gram of elemental lithium.

a) **(1)** moles = m(g)/MM

(2) m(g) = 29.6 grams
MM is calculated from MM of the elements in the formula:

$$\text{MM of } Li_2CO_3 = 2\text{ Li} \times 6.94\text{ g/mole} = 13.88\text{ g/mole}$$

$$+ 1\text{ C} \times 12.01\text{ g/mole} = 12.01\text{ g/mole}$$

$$+ 3\text{ O} \times 16.00\text{ g/mole} = 48.00\text{ g/mole}$$

$$\text{total MM} = 73.89\text{ g/mole}$$

(3) $\text{moles } Li_2CO_3 = \dfrac{625\text{ grams}}{73.89\text{ g/mole}} = \mathbf{0.400}\text{ moles } Li_2CO_3$

(4) $\text{moles of lithium} = 0.400\text{ moles } Li_2CO_3 \times \dfrac{2\ Li^+}{1\ Li_2CO_3} = \mathbf{0.800}\text{ mole lithium ions}$

$$\text{moles of oxygen} = 0.400 \text{ moles } Li_2CO_3 \times \frac{3\,O}{1\,Li_2CO_3} = \mathbf{1.20} \text{ mole oxygen atoms}$$

b) **(1)** moles = m(g)/MM
(2) MM of the element lithium = 6.94 g/mole

$$m(g) \text{ of Li} = 1.00 \text{ grams}$$

(3) moles lithium atoms = 155 grams/6.94 g/mole = 0.144 moles of Li atoms
(4) moles of Li_2CO_3 which contain 0.144 moles of Li atoms =

$$(0.144 \text{ moles Li atoms}) \times \frac{1\,Li_2CO_3}{2 \text{ Li atoms}} = 0.072 \text{ mole } Li_2CO_3$$

V MASS PERCENT OF AN ELEMENT IN A COMPOUND

The percent mass (% mass) and fractional mass (% mass expressed in decimal form) of an element in a compound can be found from the ratio of elements and the mass of each element stated in the formula of the compound:

$$\text{fractional mass} = \frac{\text{mass of a specific element in a compound sample}}{\text{total mass of the compound in the sample}}$$

$$\%\text{ mass} = \text{fractional mass} \times 100\%$$

If the sample compound has a **known** formula, the calculation of fractional mass or % mass is derived directly from the calculation of the molar mass; in this case, the sample selected is exactly one mole:

$$\text{fractional mass of an element in a compound} = \frac{\text{mass of the specific element in the MM calculation}}{\text{total molar mass of the compound from MM calculation}}$$

Example: Calculate the fractional mass and percent mass of sulfur and oxygen in H_2SO_4.
Set up the method for calculating MM of H_2SO_4:

$$2\,H \times 1.01\,g/mole = 2.02\,g/mole$$

$$1\,S \times 32.07\,g/mole = 32.07\,g/mole$$

$$4\,O \times 16.00\,g/mole = \underline{64.00\,g/mole}$$

$$\mathbf{total = 98.09\,g/mole}$$

$$\text{fractional mass S} = \frac{32.07 \text{ g S contributing to the MM calculation}}{98.09 \text{ g total mass of the compound } (= MM)} = 0.3269$$

$$\%\text{ mass S} = 0.3269 \times 100\% = 32.69\%$$

$$\text{fractional mass O} = \frac{64.00 \text{ g O contributing to the MM calculation}}{98.09 \text{ g total mass of the compound } (= MM)} = 0.6525$$

$$\% \text{ mass O} = 0.6525 \times 100\% = 65.25\%$$

Fractional mass or percent mass can then be used for further calculations; for example, to calculate the mass of one element in a specific sample of compound:

$$\text{mass of an element in a specific sample} = (\text{mass of the sample}) \times (\text{fractional mass of the element})$$

Example: Use the concept of percent mass to calculate the total mass of sulfur and the total mass of oxygen contained in 650 grams of H_2SO_4.

The mass of sulfur and oxygen in **any** sample of H_2SO_4 can be found from fractional masses in the compound:

$$\text{mass oxygen in 650 g sample} = 650 \text{ grams } H_2SO_4 \cdot \frac{0.6525 \text{ g O}}{1\text{g } H_2SO_4} = \mathbf{424\ g} \text{ oxygen}$$

The units for the calculation can be seen by showing the units of fractional mass:

$$\text{mass sulfur in 650 g sample} = 650 \text{ grams } H_2SO_4 \times \frac{0.3269 \text{ g S}}{1 \text{ g } H_2SO_4} = \mathbf{213\ g} \text{ sulfur}$$

$$\text{mass oxygen in 650 g sample} = 650 \text{ grams } H_2SO_4 \times \frac{0.6525 \text{ g O}}{1\text{g } H_2SO_4} = \mathbf{424\ g} \text{ oxygen}$$

Example: Calculate the mass in grams of chlorine contained in 500 grams of CF_2Cl_2 by using the concept of percent mass.

Set up the method for calculating MM of CF_2Cl_2:

$$1 \text{ C} \times 12.01 \text{ g/mole} = 12.01 \text{ g/mole}$$

$$2 \text{ F} \times 19.00 \text{ g/mole} = 38.00 \text{ g/mole}$$

$$2 \text{ Cl} \times 35.45 \text{ g/mole} = \underline{70.90 \text{ g/mole}}$$

$$\mathbf{total = 120.91\ g/mole}$$

$$\text{fractional mass Cl} = \frac{70.90 \text{ g Cl contributing to the MM calculation}}{120.91 \text{ g total mass of the compound } (= \text{MM})} = 0.5864$$

$$\% \text{ mass Cl} = 0.5864 \times 100\% = 58.64\%$$

$$\text{mass Cl in 500. g sample} = 500. \text{ g } CF_2Cl_2 \times (0.5864) = \mathbf{293\ g} \text{ chlorine}$$

$$\text{or mass Cl in 500. g sample} = 500. \text{ } g \text{ } CF_2Cl_2 \times \frac{0.5864 \text{ g Cl}}{1 \text{ g } CF_2Cl_2} = \mathbf{293 g} \text{ Chlorine}$$

VI PRACTICE PROBLEMS

1. Calculate the number of moles represented by 50.0 grams of the following elements:
   a) Ag
   b) V
   c) Se
   d) Rb

2. Calculate the mass of 12.5 moles of each of the elements in problem #1.

3. a) Calculate the mass of an aluminum can if it contains 0.778 moles of Al.
   b) Calculate the mass of 1.00×10^{24} atoms of Fe; first convert the number of atoms to number of moles.
   c) Calculate the number of moles of carbon in a diamond (pure carbon) of 2.50 carats; 1 gram = 4.861 carats.
   d) Calculate the number of moles of tin in a sheet of metal, which has a volume of 136 cm^3; density of Sn = 7.30 g/cm^3

4. a) A 14-carat gold ring has a mass of 32.5 grams. Pure gold is 24 carats; therefore, 14-carat gold is 14/24 or 58.3% gold by mass. How many **moles** of gold atoms and how many gold **atoms** are in the gold ring?
   b) Platinum (Pt) is more expensive and more dense than gold. A platinum bar has a volume of 1562 cm^3; the density of Pt is 21.5 g/cm^3. How many moles of Pt are in the bar?

5. A gold cylinder has a height of 0.500 **meters** and a radius of 3.30 cm.

$$\text{Density of gold} = 19.3\ g/cm^3;\ V(\text{cylinder}) = \pi\, r^2\, h$$

   a) Calculate the mass of the gold in the cylinder by solving a density equation.
   b) Calculate the # of moles of gold in the cylinder.
   c) Calculate the number of gold atoms in the cylinder.
   d) The population of the United States is approximately 300,000,000. How many gold atoms would each person get if the gold in this cylinder was equally distributed among the population?

6. Calculate the number of moles represented by 50.0 grams of the following compounds; first calculate the MM.
   a) $MgCO_3$
   b) C_2H_6O
   c) $FeBr_2$
   d) $C_{14}H_{20}N_2$

7. Calculate the mass of 12.5 moles of each of the compounds in problem #6.

8. a) Calculate the mass of 5.15×10^{24} molecules of CH_4 by first converting this number to moles of CH_4.
   b) Calculate the MM of an **unknown** compound if 4.75 moles of the compound has a mass of 522 grams.

9. a) A person takes 12 tablets of acetaminophen in one day; each tablet is 500 mg of pure **acetaminophen, $C_8H_9O_2N$**. Calculate the number of moles of acetaminophen consumed.
   b) **Aspirin is $C_8H_9O_4$**. On the next day, this same person decides to consume the exact same number of **moles** of aspirin; that is, the # of moles of acetaminophen the first

day = # of moles of aspirin the second day. Use your answer from part (a) to calculate the **mass** of aspirin that must be consumed to equal the same number of moles of acetaminophen calculated from part (a).

c) These pain relievers are needed because the wedding party consumed a 1.75-liter bottle of gin over the weekend. Calculate the number of moles of alcohol, C_2H_6O, in a 1.75-liter bottle of 80-proof gin. 80 proof = 40.0% alcohol by **volume;** density of C_2H_6O = 0.890 g/mL.

10. a) Glucose has a formula of $C_6H_{12}O_6$. How many moles of glucose are contained in a 1.75 liter bottle of a sports drink, if the drink is 5.50% glucose by **mass**? The density of the solution in the bottle is 1.10 g/mL.
    b) How many **moles** of carbon atoms from the glucose are in the bottle? Use your answer from part (a).
    c) How many hydrogen **atoms** from the glucose are in the bottle? Use your answer from part (a) then calculate the number of moles of hydrogen atoms.
    d) How many grams of pure glucose can be formed from 75.0 grams of oxygen? That is, how many grams of glucose contain 75.0 grams of oxygen atoms?

11. Lactic acid is $C_3H_6O_3$ and is formed by glucose metabolism.
    a) How many moles of oxygen atoms are in 2.30 grams of lactic acid?
    b) Based on the formula for lactic acid, how many moles of lactic acid could be formed from 150.0 grams of carbon?

12. a) Calculate the percent nitrogen in the amino acid histidine, $C_6H_9N_3O_2$.
    b) Calculate the mass of nitrogen that would be excreted by the complete metabolism of 250.0 grams of the amino acid (i.e., calculate the mass of nitrogen in 250.0 grams of the amino acid).

13. a) Calculate the percent nitrogen in the amino acid cysteine, $C_3H_7NO_2S$.
    b) Calculate the mass of nitrogen that would be excreted by the complete metabolism of 250.0 grams of this amino acid.

VII ANSWERS TO PRACTICE PROBLEMS

1. Calculate the number of moles represented by 50.0 grams of the following elements:
   a) Ag
      (1) Moles = m(g)/MM
      (2) Mass given = 50.0 grams; MM is read from the periodic table = 107.87 g/mole
      (3) # of moles = 50.0 grams/107.87 g/mole = **0.464 moles**

   Parts (b)–(d) are set up using the exact same process; the condensed answers:
   b) V: # of moles = 50.0 grams/50.94 g/mole = **0.982 moles**
   c) Se: # of moles = 50.0 grams/78.96 g/mole = **0.633 moles**
   d) Rb: # of moles = 50.0 grams/85.47 g/mole = **0.585 moles**

2. Calculate the mass of 12.5 moles of each of the elements in problem #1.
   a) Ag
      (1) m(g) = moles × MM
      (2) Moles given = 12.5 moles; MM is read from the periodic table = 107.87 g/mole
      (3) Mass = (12.5 mole) × (107.87 g/mole) = **1348 grams** (1350 to 3 sig. fig.)

   Parts (b)–(d) are set up using the exact same process; the condensed answers:
   b) V: mass = (12.5 mole) × (50.94 g/mole) = **637 grams**
   c) Se: mass = (12.5 mole) × (78.96 g/mole) = **987 grams**
   d) Rb: mass = (12.5 mole) × (85.47 g/mole) = **1068 grams** (1070 to 3 sig. fig.)

3. a) Calculate the mass of an aluminum can if it contains 0.778 moles of Al.
   (1) m(g) = moles × MM
   (2) Moles given = 0.778 moles; MM is read from the periodic table = 26.98 g/mole
   (3) Mass = (0.778 mole) × (26.98 g/mole) = **20.99 grams** (21.0 to 3 sig. fig.)
   b) Calculate the mass of 1.00×10^{24} atoms of Fe; first convert the number of atoms to number of moles.
   (1) m(g) = moles × MM
   (2) MM is read from the periodic table = 55.85 g/mole; moles must be calculated from the number of atoms using the conversion: 6.022×10^{23} atoms per mole

$$\text{moles} = 1.00 \times 10^{24}\,\text{atoms} \times \frac{1\,\text{mole}}{6.022 \times 10^{23}\,\text{atoms}} = 1.66\,\text{moles}$$

   (3) mass = (1.66 mole) × (55.85 g/mole) = **92.7 grams**
   c) Calculate the number of moles of carbon in a diamond (pure carbon) of 2.50 carats; 1 gram = 4.861 carats.
   (1) moles = m(g)/MM
   (2) MM is read from the periodic table = 12.01 g/mole; mass in carats must be converted to mass in grams: $2.50\,\text{carats} \times \frac{1\,\text{gram}}{4.861\,\text{carats}} = 0.514\,\text{grams}$
   (3) # of moles = 0.514 grams/12.01 g/mole = **0.0428 moles**
   d) Calculate the number of moles of tin in a sheet of metal, which has a volume of 136 cm^3; density of Sn = 7.30 g/cm^3
   (1) moles = m(g)/MM
   (2) MM is read from the periodic table = 118.71 g/mole; mass must be calculated by solving a density problem: m(g) = V × d = (136 cm^3) × (7.30 g/cm^3) = 992.8 grams
   (3) # of moles = 992.8 grams/118.71 g/mole = **8.36 moles**

4. a) A 14-carat gold ring has a mass of 32.5 grams. Pure gold is 24 carats; therefore, 14-carat gold is 14/24 or 58.3% gold by mass. How many **moles** of gold atoms and how many gold **atoms** are in the gold ring?
   (1) moles = m(g)/MM
   (2) MM is read from the periodic table = 196.97 g/mole; mass ***Au*** must be calculated from percent gold: mass Au = (mass ring) × (% gold) = (32.5 g) × (0.583) = 18.95 g
   (3) # of moles = 18.95 grams/196.97 g/mole = **0.0962 moles**
   (4) # of atoms = (# of moles) × (6.022×10^{23} atoms/mole)
   # of gold atoms = 0.0962 moles × (6.022×10^{23} atoms/mole) = **5.79×10^{22} atoms**
   b) Platinum (Pt) is more expensive and more dense than gold. A platinum bar has a volume of 1562 cm^3; the density of Pt is 21.5 g/cm^3. How many moles of Pt are in the bar?
   (1) Moles = m(g)/MM
   (2) MM is read from the periodic table = 195.08 g/mole; mass must be calculated by solving a density problem: m(g) = V × d = (1562 cm^3) × (21.5 g/cm^3) = 33583 grams
   (3) # of moles = 33583 grams/195.08 g/mole = **172 moles** (3 sig. fig.)

5. A gold cylinder has a height of 0.500 **meters** and a radius of 3.30 cm.
   Density of gold = 19.3 g/cm^3; V(cylinder) = $\pi \mathbf{r^2} h$
   a) Calculate the mass of the gold in the cylinder by solving a density equation.

$$h = 50.0\ \textit{cm};\ V = \pi r^2 h = (3.1416)(3.30\ \text{cm})^2(50.0\ \text{cm}) = 1711\ \text{cm}^3$$

$$m(g) = V \times d = (1711\ \text{cm}^3) \times (19.3\ \text{g/cm}^3) = \mathbf{33000\ grams}\ (3\ \text{sig. fig.})$$

b) Calculate the # of moles of gold in the cylinder.
c) Calculate the number of gold atoms in the cylinder.
(1) Moles = m(g)/MM
(2) MM is read from the periodic table = 196.97 g/mole; mass part (a) = 33000 g
(3) # of moles = 33000 grams/196.97 g/mole = **168 moles**
(4) # of gold atoms = 168 moles × (6.022 × 10^{23} atoms/mole) = **1.01 × 10^{26} atoms**
d) The population of the United States is approximately 300,000,000. How many gold atoms would each person get if the gold in this cylinder was equally distributed among the population?

$$\#\text{of gold atoms per person} = \frac{1.01\times10^{26}\text{ atoms}}{3.00\times10^{8}\text{ people}} = \mathbf{3.37\times10^{17}\ atoms/person}$$

6. Calculate the number of moles represented by 50.0 grams of the following compounds; first calculate the MM.
a) $MgCO_3$
(1) Moles = m(g)/MM
(2) Mass given = 50.0 grams; MM must be calculated from the periodic table:

MM of $MgCO_3$ = 1 Mg × 24.31 g/mole = 24.31 g/mole

+ 1 C × 12.01 g/mole = 12.01 g/mole

+ 3 O × 16.00 g/mole = 48.00 g/mole

total MM = 84.32 g/mole

(3) # of moles = 50.0 grams/84.32 g/mole = **0.593 moles**
Parts (b)–(d) are set up using the exact same process; the condensed answers:
b) C_2H_6O: # of moles = 50.0 grams/46.08 g/mole = **1.09 moles**
c) $FeBr_2$: # of moles = 50.0 grams/215.66 g/mole = **0.232 moles**
d) $C_{14}H_{20}N_2$: # of moles = 50.0 grams/216.35 g/mole = **0.231 moles**

7. Calculate the mass of 12.5 moles of each of the compounds in problem #6.
a) $MgCO_3$: mass = (12.5 mole) × (84.32 g/mole) = **1054 grams**
b) C_2H_6O: mass = (12.5 mole) × (46.08 g/mole) = **576 grams**
c) $FeBr_2$: mass = (12.5 mole) × (215.66 g/mole) = **2696 grams**
d) $C_{14}H_{20}N_2$: mass = (12.5 mole) × (216.35 g/mole) = **2704 grams**

8. a) Calculate the mass of 5.15 × 10^{24} molecules of CH_4 by first converting this number to moles of CH_4.
(1) m(g) = moles × MM
(2) MM is ***calculated*** from the periodic table = 16.05 g/mole; moles must be calculated from the number of atoms using the conversion: 6.022 × 10^{23} atoms per mole

$$\text{moles} = 5.15\times10^{24}\text{ atoms}\times\frac{1\text{ mole}}{6.022\times10^{23}\text{ atoms}} = 8.55\text{ moles}$$

(3) Mass = (8.55 mole) × (16.05 g/mole) = **137 grams**
b) Calculate the MM of an **unknown** compound if 4.75 moles of the compound has a mass of 522 grams.
(1) MM = m(g) (of a sample)/moles of the sample
(2) m(g) = 522 grams; moles = 4.75 moles
(3) MM = (522 g)/(4.75 moles) = **109.0 g/mole**

9. a) A person takes 12 tablets of acetaminophen in one day; each tablet is 500 mg of pure **acetaminophen, $C_8H_9O_2N$**. Calculate the number of moles of acetaminophen consumed.
   (1) Moles = m(g)/MM
   (2) Total mass(g) of acetaminophen = (500 mg/tablet)(12 tablets)(1 g/1000 mg) = 6.00 g
   MM must be calculated from the periodic table:

$$\text{MM of } C_8H_9O_2N = 8\ \text{C} \times 12.01\ \text{g/mole} = 96.08\ \text{g/mole}$$

$$+\ 9\ \text{H} \times 1.01\ \text{g/mole} = 9.09\ \text{g/mole}$$

$$+\ 2\ \text{O} \times 16.00\ \text{g/mole} = 32.00\ \text{g/mole}$$

$$+\ 1\ \text{N} \times 14.01\ \text{g/mole} = \underline{14.01\ \text{g/mole}}$$

$$\text{total } \mathbf{MM} = 151.18\ \text{g/mole}$$

   (3) # of moles = 6.00 grams/151.18 g/mole = **0.0397 moles**

b) **Aspirin is $C_8H_9O_4$**. Use your answer from part (**a**) to calculate the **mass** of aspirin that must be consumed to equal the same number of moles of acetaminophen calculated.
   (1) m(g) = moles × MM
   (2) # of moles comes from the answer to part (a) = 0.0397 moles
   MM is ***calculated*** from the periodic table:

$$\text{MM of } C_8H_9O_4 = 8\ \text{C} \times 12.01\ \text{g/mole} = 96.08\ \text{g/mole}$$

$$+\ 9\ \text{H} \times 1.01\ \text{g/mole} = 9.09\ \text{g/mole}$$

$$+\ 4\ \text{O} \times 16.00\ \text{g/mole} = 64.00\ \text{g/mole}$$

$$\text{total MM} = 169.17\ \text{g/mole}$$

   (3) Mass = (0.0397 mole) × (169.17 g/mole) = **6.72 grams**

c) Calculate the number of moles of alcohol, C_2H_6O, in a 1.75-liter bottle of 80-proof gin. 80 proof = 40.0% alcohol by **volume;** density of C_2H_6O = 0.890 g/mL.
   (1) Moles = m(g)/MM
   (2) Total mass(g) of alcohol is solved from a density equation: m(g) = V(L) × d
   V(L) of alcohol = (1.75 L)(0.40) = 0.700 L;
   m(g) alcohol = (0.700 L)(1000 mL/L)(0.890 g/mL) = 623 grams
   MM is calculated from the periodic table = 46.08 g/mole:
   (3) # of moles = 623 grams/46.08 g/mole = **13.5 moles**

10. a) How many moles of glucose ($C_6H_{12}O_6$) are contained in a 1.75-liter bottle of a sports drink, if the drink is 5.50% glucose by **mass**? Density of the **drink** is 1.10 g/mL
   (1) Moles = m(g)/MM
   (2) Total mass(g) of glucose is solved first from a density equation: m(g) = V(L) × d
   total mass of drink = (1.75 L)(1000 mL/L)(1.10 g/mL) = 1925 grams;
   m(g) glucose = (1925 g)(0.0550) = 105.9 grams
   MM is ***calculated*** from the periodic table:

$$\text{MM of } C_6H_{12}O_6 = 6\ \text{C} \times 12.01\ \text{g/mole} = 72.06\ \text{g/mole}$$

$$+\ 12\ \text{H} \times 1.01\ \text{g/mole} = 12.12\ \text{g/mole}$$

$$+ 6\ O \times 16.00\ g/mole = 96.00\ g/mole$$

$$\text{total MM} = 180.18\ g/mole$$

(3) # of moles = 105.9 grams/180.18 g/mole = **0.588 moles**

b) How many **moles** of carbon atoms from the glucose are in the bottle? Use your answer from part (a)

(4) moles of carbon $= (0.588 \text{ moles } C_6H_{12}O_6) \times \frac{6C}{1C_6H_{12}O_6} =$ **3.53 mole carbon**

c) How many hydrogen **atoms** from the glucose are in the bottle? Use your answer from part (a) and first calculate the number of moles of hydrogen atoms.

(4) moles of hydrogen $= (0.588 \text{ moles } C_6H_{12}O_6) \times \frac{12H}{1C_6H_{12}O_6} =$ **7.056 mole hydrogen**

# of H atoms = 7.056 moles × (6.022×10^{23} atoms/mole) = **4.25×10^{24} atoms**

d) How many grams of pure glucose can be formed from 75.0 grams of oxygen? That is, how many grams of glucose contain 75.0 grams of oxygen atoms? For this problem, first find # of moles of oxygen atoms and then # of moles of glucose that contain this amount of oxygen

(1) Moles = m(g)/MM

(2) MM of the element oxygen = 16.00 g/mole

(3) # of moles oxygen atoms = 75.0 grams/16.00 g/mole = 4.688 moles of O atoms

(4a) Moles of $C_6H_{12}O_6$ which would contain this amount of oxygen atoms =

$$(4.688 \text{ moles O atoms}) \times \frac{1C_6H_{12}O_6}{6O \text{ atoms}} = \mathbf{0.7813\ mole\ C_6H_{12}O_6}$$

(4b) Mass of $C_6H_{12}O_6$ = (0.7813 mole) (180.18 g/mole) = **141 grams**

11. Lactic acid is $C_3H_6O_3$ and is formed by glucose metabolism.

a) How many moles of oxygen atoms are in 2.30 grams of lactic acid?

(1) Moles = m(g)/MM

(2) m(g) lactic acid = 2.30 grams

MM is ***calculated*** from the periodic table:

$$\text{MM of } C_3H_60_3 = 3\ C \times 12.01\ g/mole = 36.03\ g/mole$$

$$+ 6\ H \times 1.01\ g/mole = 6.06\ g/mole$$

$$+ 3\ O \times 16.00\ g/mole = 48.00\ g/mole$$

$$\text{total MM} = 90.09\ g/mole$$

(3) # of moles = 2.30 grams/90.09 g/mole = **0.02553 moles**

(4) moles of oxygen $= (0.02553 \text{ moles } C_3H_6O_3) \times \frac{3O}{1C_3H_6O_3} =$ **0.0766 mole oxygen**

b) Based on the formula for lactic acid, how many moles of lactic acid could be formed from 150.0 grams of carbon? Use the procedure described for problem10(d).

(1) Moles = m(g)/MM

(2) MM of the element carbon = 12.01 g/mole

(3) # of moles carbon atoms = 150.0 grams/12.01 g/mole = 12.49 moles of C atoms

(4) Moles of $C_3H_60_3$ which would contain this amount of carbon atoms =

$$(12.49 \text{ moles C atoms}) \times \frac{1C_3H_6O_3}{3C \text{ atoms}} = \mathbf{4.16\ mole\ C_3H_6O_3}$$

12. a) Calculate the percent nitrogen in the amino acid histidine, $C_6H_9N_3O_2$
    b) Calculate the mass of nitrogen which would be excreted by complete metabolism of 250.0 grams of the amino acid (i.e., calculate the mass of nitrogen in 250.0 grams of the amino acid).
    a) Set up the method for calculating MM of histidine, $C_6H_9N_3O_2$

$$\text{MM of } C_6H_9N_3O_2 = 6\text{ C} \times 12.01 \text{ g/mole} = 72.06 \text{ g/mole}$$

$$+\,9\text{ H} \times 1.01 \text{ g/mole} = 9.09 \text{ g/mole}$$

$$+\,3\text{ N} \times 14.01 \text{ g/mole} = 42.03 \text{ g/mole}$$

$$+\,2\text{ O} \times 16.00 \text{ g/mole} = 32.00 \text{ g/mole}$$

$$\text{total MM} = 155.18 \text{ g/mole}$$

$$\text{fractional mass N} = \frac{42.03 \text{ g N contributing to the MM calculation}}{155.18\text{g total mass of the compound } (= \text{MM})} = 0.2708$$

$$\%\text{ mass} = 0.2708 \times 100\% = 27.08\%$$

b) mass N in 250 g sample = (250 grams $C_6H_9N_3O_2$) × (0.2708) = **67.7 g N**

Units for this calculation can be seen by showing the units of fractional mass:

$$\text{mass N in 250 g sample} = 250 \text{ grams } C_6H_9N_3O_2 \times \frac{0.2708\text{g N}}{1\text{g } C_6H_9N_3O_2} = \mathbf{67.7\,g\,N}$$

13. a) Calculate the percent nitrogen in the amino acid cysteine, $C_3H_7NO_2S$.
    b) Calculate the mass of nitrogen that would be excreted by the complete metabolism of 250.0 grams of this amino acid.
    a) Set up the method for calculating MM of cysteine, $C_3H_7NO_2S$.

$$\text{MM of } C_3H_7NO_2S = 3\text{ C} \times 12.01 \text{ g/mole} = 36.03 \text{ g/mole}$$

$$+\,7\text{ H} \times 1.01 \text{ g/mole} = 7.07 \text{ g/mole}$$

$$+\,1\text{ N} \times 14.01 \text{ g/mole} = 14.01 \text{ g/mole}$$

$$+\,2\text{ O} \times 16.00 \text{ g/mole} = 32.00 \text{ g/mole}$$

$$+\,1\text{ S} \times 32.07 \text{ g/mole} = 32.07 \text{ g/mole}$$

$$\text{total MM} = 121.18 \text{ g/mole}$$

$$\text{fractional mass N} = \frac{14.01\,\text{g N contributing to the MM calculation}}{121.18\,\text{g total mass of the compound (= MM)}} = 0.116$$

$$\%\ \text{mass} = 0.006 \times 100\% = 11.6\%$$

b) mass N in 250 g sample = (250 grams $C_3H_7NO_2S$)(0.116) = **28.9 g N**
Units for this calculation can be seen by showing the units of fractional mass:

$$\text{mass N in 250 g sample} = 250\,\text{grams } C_3H_7NO_2S \times \frac{0.116\,\text{g N}}{1\,\text{g } C_3H_7NO_2S} = \mathbf{28.9\,g\,N}$$

6 Procedures for Calculating Empirical and Molecular Formulas

I GENERAL CONCEPT

The **simplest formula** (or **empirical formula**) is the simplest whole number ratio of elements in a compound. This type of formula is the **only** formula commonly used for ionic compounds or for those compounds that come as a large three-dimensional (3-D) array.

The **molecular formula** (or **"true" molecular formula**) represents the numbers of each element in one molecule; these numbers need **not** represent the simplest whole number ratio. Molecular formulas apply to covalent molecules that come in the form of molecules.

The simplest formula (simplest whole number ratio of elements) can be calculated for **any** compound, regardless of whether the compound comes as a large 3-D array or as molecules.

Concepts of Empirical or Molecular Formulas

(1) The formula of a compound states the number of atoms of each element in one formula unit or one molecule. This also means that the formula of a compound states **the number of moles of each element in one mole of formula units or one mole of molecules.**

(2) The number ratio of elements in any formula must be equal to the mole ratios of these elements in a specific sample size. **Calculation of the mole ratio of all elements in a specific sample, expressed in whole numbers, provides the simplest formula of any compound**. For compounds that come as molecules, the mole ratio may **not** provide the actual molecular formula.

(3) Mole ratio calculations are based on the determination of the **mass ratio of elements in a compound**. Mass ratios are converted to mole ratios through mass/mole/MM calculations.

(4) The mass ratio is found by experimentation, hence the name **empirical** formula. In practice, the calculations that determine the masses of elements in a compound require varying degrees of complexity; mass values are subject to experimental error.

II PROCEDURE FOR CALCULATION OF SIMPLEST (EMPIRICAL) FORMULA OF ANY COMPOUND

(1) Determine the masses of each element in a **specific sample** of the compound; this produces the **mass ratio of elements**. A variety of experimental techniques are used for this step, producing values of varying degrees of experimental accuracy.

(2) Convert the mass ratio of elements to the **mole ratio of elements** for the specific sample by solving a mass/mole/MM calculation for each element mass determined from step **(1)**:

$$\textbf{moles of each element} = \frac{\textbf{mass of each element in sample}}{\textbf{MM of each element}}$$

(3) Convert the mole ratio of elements to the **mole ratio of elements expressed as simplest whole numbers**; this is the simplest (empirical) formula. To do this:

a) The element mole ratios calculated from step **(2)** represent the number of moles of each element in the specific sample. Identify the **smallest** mole value found in step **(2)** and divide all mole values by this smallest value. This step is often the only mathematical operation required to achieve a whole number ratio for all elements.

b) In some cases, step **(3a)** will not achieve all whole numbers; some mole values will still be fractional. For these cases, multiply all mole values by the required number to clear all fractions. Generally, convert the decimal form of the fractional number to a fraction; then multiply all mole numbers by the denominator of this fraction.

Example: A certain compound of cobalt and chlorine (Co_xCl_y) was analyzed; it was found that a 100.0 gram sample of this compound contained 64.35 grams of chlorine. Calculate the simplest formula.

(1) Mass **Cl = 64.35 g**; mass **Co** = (total mass of sample) – (mass of chlorine)

= (100.0 g – 64.35 g) = **35.65 g**

(2) Moles Cl = 64.35 g/35.45 g/mole = **1.815 moles Cl**

moles Co = 35.65 g/58.93 g/mole = **0.605 moles Co**

Note that step **(2)** actually completes the "chemistry" portion of the problem; the mole ratio of elements is a form of the formula (essentially can be thought of as $\mathbf{Co_{0.605}Cl_{1.815}}$) The purpose of step **(3)** is to convert the step **(2)** ratio to whole numbers:

(3a) **0.605** is the smallest mole value found; divide all mole numbers by this value:

1.815 mole Cl / 0.605 = **3.00 mole Cl**
0.605 mole Co / 0.605 = **1 mole Co**
} Formula is $\mathbf{CoCl_3}$

Step (3b) is not necessary for this problem.

Example: A certain compound of cobalt and oxygen (Co_xO_y) was analyzed; an 8.91 gram sample was found to contain 6.54 grams of cobalt. Calculate the simplest formula.

(1) Mass **Co = 6.54 g**; mass **O** = (total mass of sample) – (mass of cobalt)

= (8.91 g – 6.54 g) = **2.37 g**

(2) Moles Co = 6.54 g/58.93 g/mole = **0.111 moles Co**

moles O = 2.37 g/16.00 g/mole = **0.148 moles O**

(3a) **0.111** is the smallest mole value found; divide all mole numbers by this value:

0.111 mole Co/0.111 = **1 mole Co**

0.148 mole O/0.111 = **1.33 mole O**

(3b) The ratio 1 to 1.33 is **not** a simplest **whole number** ratio. The decimal form of the fractional number of moles of oxygen is 0.33; convert this decimal to the fraction 1/3. The denominator of this fraction is 3; multiply ***all*** mole numbers through by **3** to clear the fraction:

$$3 \times 1 \text{ mole Co} = \mathbf{3 \text{ mole Co}}$$

$$3 \times 1.33 \text{ mole O} = \mathbf{4 \text{ mole O}}$$

Note that all mole numbers, not just the fractional mole value, must be multiplied by the same number to keep the mole ratio correct.
The whole number mole ratio is equivalent to the simplest formula: $\mathbf{Co_3O_4}$

Example: A certain compound of titanium, selenium, and oxygen ($Ti_xSe_yO_z$) was analyzed; a 7.22 gram sample showed the following masses: 1.45 grams of titanium; 3.59 grams of selenium; and 2.18 grams of oxygen. Calculate the simplest formula.

(1) All masses are given directly in the problem:

Ti = 1.45 g; Se = 3.59 g; O = 2.18 g

(2) Moles Ti = 1.45 g/58.93 g/mole = **0.0303 moles Ti**

moles Se = 3.59 g/78.96 g/mole = **0.0455 moles Se**

moles O = 2.18 g/16.00 g/mole = **0.1363 moles O**

(3a) **0.303** is the smallest mole value found; divide all mole numbers by this value:

0.0303 moles Ti/0.0303 = **1 mole Ti**

0.0455 moles Se/0.0303 = **1.50 mole Se**

0.1363 moles O/0.0303 = **4.50 mole O**

(3b) The decimal 0.50 is converted to the fraction 1/2. The denominator of this fraction is 2; multiply **all** mole numbers through by **2** to clear the fraction:

$$2 \times 1 \text{ mole Ti} = \mathbf{2 \text{ mole Ti}}$$

$$2 \times 1.50 \text{ mole Se} = \mathbf{3 \text{ mole Se}}$$ Simplest formula: $\mathbf{Ti_2Se_3O_9}$

$$2 \times 4.50 \text{ mole O} = \mathbf{9 \text{ mole O}}$$

III DETERMINATION OF THE SIMPLEST FORMULA FROM ELEMENT MASS PERCENT

Mass ratio data is often stated in the form of mass percent (mass %) of each element in a compound. Convert mass percent to the mass ratio of elements required for step (**1**). Although any sample size could be used as a basis for multiplying through by the given percentages, it is simplest to assume a specific sample size of exactly 100 grams. In this case the numerical values for percent mass can be used directly as mass values in grams (for an exact 100 gram sample).

Example: A certain compound of potassium, manganese, and oxygen ($K_xMn_yO_z$) was analyzed; a sample showed the following percent mass results: 24.7% mass K; 34.8% mass Mn; 40.5% mass O. Calculate the simplest formula.

(1) Assume that a sample size is selected to be exactly 100 grams:

mass **K** = (100 g × 24.7%) = (100 g × 0.247) = **24.7 g**

mass **Mn** = (100 g × 34.8%) = (100 g × 0.348) = **34.8 g**

mass **O** = (100 g × 40.5%) = (100 g × 0.405) = **40.5 g**

(2) moles K = 24.7 g/39.10 g/mole = **0.632 moles K**

moles Mn = 34.8 g/54.94 g/mole = **0.633 moles Mn**

moles O = 40.5 g/16.00 g/mole = **2.531 moles O**

(3a) 0.632 is the smallest mole value found; divide all mole numbers by this value:

0.632 moles K/0.632 = **1 mole K**

0.633 moles Mn/0.632 = **1.00 mole Mn** simplest formula: $\mathbf{KMnO_4}$

2.531 moles O/0.632 = **4.00 mole O**

Step (3b) Is not necessary for this problem.

IV DETERMINATION OF THE TRUE MOLECULAR FORMULA FOR MOLECULES

The simplest (empirical) formula (simplest whole number ratio) is the only formula necessary for ionic compounds. Molecular formulas, which apply to most covalent compounds, can be found from the simplest formula if the actual molar mass of the compound is known.

A true molecular formula must be either identical to, or a multiple of, the simplest whole number ratio of elements in one molecule. For example, if a simplest whole number ratio for one molecule of a compound were CH_3O, the actual molecular formula must be either CH_3O or a multiple such that the ratio of all atoms remains the same: $C_2H_6O_2$ or $C_3H_9O_3$, and so on. A certain structure of glucose has a molecular formula of $C_6H_{12}O_6$; however, a determination of the simplest (empirical) formula would produce a formula of CH_2O.

Determining and Using the Multiple

The **multiple** relates the elemental simplest ratio to the actual numbers of atoms in a molecular formula and can be determined if the actual molar mass (MM) of the compound is known from other types of analysis. The multiple is calculated as equal to the true molar mass divided by the calculated molar mass of the simplest formula:

Multiple = actual MM of compound/MM of simplest formula

PROCEDURE FOR FINDING THE SIMPLEST FORMULA AND TRUE MOLECULAR FORMULA

(1) Determine the masses of each element in a specific sample of the compound; this produces the **mass ratio of elements.**
(2) Convert the mass ratio of elements to the **mole ratio of elements** for the specific sample by solving a mass/mole calculation for each element mass.
(3a) and **(3b)** Convert the mole ratio of elements to the mole ratio of elements expressed as **simplest whole numbers**; this is the simplest (empirical) formula.
(4) Determine the multiple; this is found by the calculation:

[multiple = actual MM of compound/MM of simplest formula].

Multiply the simplest formula through **by this multiple** to find the true molecular formula.

Example: A compound of carbon and hydrogen was found to contain 85.60% carbon and 14.40% hydrogen by mass. Calculate the simplest formula and the **true molecular formula** if the true molar mass (MM) of the compound was found to be 56.1 g/mole.

(1) Assume a sample size of exactly 100 grams; therefore, out of a 100-gram sample, 85.60 grams must be carbon (100 g × 0.8560) and 14.40 grams must be hydrogen (100 g × 0.1440). **C = 85.60 g; H = 14.40 g**
(2) Mole C = 85.60 g/12.01 g/mole = **7.127 mole C**

mole H = 14.40 g/1.01 g/mole = **14.257 mole H**

(3a) **7.127** mole is the smallest mole value found; divide all mole numbers by this value:

7.127 mole C/7.127 = **1 mole C**

14.257 mole H/7.127 = **2.00 mole H** ***simplest*** formula: $\mathbf{CH_2}$

(4) **MM of** $\mathbf{CH_2}$ **= 14.03** g/mole; **true MM of compound = 56.1** g/mole

$$\textbf{multiple} = \frac{56.1 \text{ (true MM of compound)}}{14.03 \text{ (MM of simplest formula)}} = \mathbf{4}$$

true **molecular** formula = $\{\mathbf{CH_2}\}$ (simplest formula) × {4} (multiple) = $\mathbf{C_4H_8}$

V EXPERIMENTAL DETERMINATION OF COMPOUND FORMULAS

A variety of techniques can be employed to determine the mass ratio of elements to satisfy step **(1)** for formula calculation. Simple **decomposition** separates the elements in a compound or separates two or more compounds that are combined in a grouping termed a complex.

Example: The compound $CuSO_4$ is found in combination with H_2O molecules in several possible ratios as a complex; this can be written as $CuSO_4(H_2O)_x$ with varying values for x. The following decomposition experiment was performed: a 5.000 gram sample of $CuSO_4(H_2O)_x$ was heated to drive off the water; after all the water in the complex was removed as steam, 2.981 grams of pure $CuSO_4$ remained. From this experiment, determine the value of x in the complex $CuSO_4(H_2O)_x$.

$CuSO_4(H_2O)_x$	$\longrightarrow$	$CuSO_4$ +	$XH_2O_{(g)}$
complex with unknown ratio		pure compound	pure compound
5.000 grams		**2.981 grams**	

(1) Note that this problem involves the whole number mole ratio between two **compounds,** rather than two or more elements. That is, the formulas for $CuSO_4$ and H_2O are fixed; only the value of x in the complex is to be determined. The individual masses of pure $CuSO_4$ and pure H_2O are required.

mass pure $\mathbf{CuSO_4}$ = **2.981 g**;

mass pure $\mathbf{H_2O}$ = (total mass of sample) – (mass of $CuSO_4$)

= (5.000 g – 2.981 g) = **2.019 g**

(2) The molar masses of the **compounds** $CuSO_4$ and H_2O must be used:

mole $CuSO_4$ = 2.981 g/159.62 g/mole = **0.01868 mole $\mathbf{CuSO_4}$**

mole H_2O = 2.019 g/18.02 g/mole = **0.1120 mole $\mathbf{H_2O}$**

(3a) 0.01868 mole is the smallest mole value found; divide all mole numbers by this value:

0.01868 mole $CuSO_4$/0.01868 mole = **1 mole $\mathbf{CuSO_4}$**

0.1120 mole H_2O/0.01868 mole = **6.00 mole $\mathbf{H_2O}$** complex: $\mathbf{CuSO_4(H_2O)_6}$

Most compounds cannot be decomposed into elements by simple heating; chemical reactions are used to separate individual elements for analysis. Determination of the masses of elements in a specific sample of a compound is often done by combustion with oxygen; this results in the separated elements combined with oxygen. Once the masses of the independent oxygen-containing products are measured, the actual mass of each element in the original compound can be calculated by other techniques.

Example: Assume a certain compound of aluminum and sulfur (Al_xS_y) can be analyzed by combustion analysis. A 17.50 gram sample was burned in air to form 22.40 grams of sulfur dioxide (SO_2). All of the sulfur in the original sample is found in the SO_2. Determine the simplest formula.

(1) Step (1) first requires the determination of the mass of **elemental S** in the original sample. The mass S in the molecule SO_2 can be calculated using the concept of percent mass and, thus, fractional mass (see Chapter 5).

Fractional mass using a molar mass calculation:

$$\begin{pmatrix}\text{fractional mass of}\\ \text{an element in a compound}\end{pmatrix} = \left(\frac{\text{mass of the specific element in the MM calculation}}{\text{total molar mass of the compound from MM calculation}}\right)$$

fractional mass **S** in $\mathbf{SO_2}$ = 32.07 g S/64.07 g SO_2 = **0.5005**

Using fractional mass to find the mass of any element in a specific sample:

mass of an element in a specific sample = (mass of the sample) × (fractional mass of the element)

mass of **S** in original sample = mass of SO_2 formed × fractional mass of S in SO_2

= 22.40 g SO_2 × 0.5005 = **11.21 g sulfur**

mass Al = (total mass of sample) – (mass of sulfur)

= (17.50 g – 11.21 g) = **6.29 g aluminum**

Step **(1)** has produced the mass ratio; steps **(2)** and **(3)** are performed as previously. As an alternative to fractional mass conversion through moles can be used:

moles SO_2 from combustion of original sample = 22.40 g/64.07 g/mole

= **0.3496 mole SO_2**

$$\textbf{moles of sulfur in sample} = 0.3496 \text{ moles } SO_2 \times \frac{1\text{ S}}{1\text{ }SO_2} = \mathbf{0.3496 \text{ mole S}}$$

mass of **S** in original sample = 0.3496 mole S × 32.07 g/mole = **11.21 g sulfur**

(2) moles Al = 6.29 g/26.98 g/mole = **0.233 moles Al**

moles S = 11.21 g/32.07 g/mole = **0.350 moles S**

(3a) **0.233** is the smallest mole value found; divide all mole numbers by this value:

0.233 mole Al/0.233 = **1 mole Al**

0.350 mole S/0.233 = **1.50 mole O**

(3b) 2 × 1 mole Al = **2 mole Al**

2 × 1.50 mole S = **3 mole** S simplest formula: $\mathbf{Al_2S_3}$

Example: The organic compound Vitamin C contains only C, H, and O ($C_xH_yO_z$). A 7.750 sample of vitamin C was combusted in air to form 11.620 g of CO_2 and 3.170 g of H_2O. All of the carbon in the sample was found in the CO_2 and all of the hydrogen in the sample was found in the H_2O. Determine the simplest formula of vitamin C. The true MM of vitamin C is 176 g/mole; determine the molecular formula from the simplest formula calculated.

(1) The element oxygen cannot be analyzed through combustion since an unknown mass of oxygen is added to the original elements through combustion. For a compound originally containing oxygen, the mass of oxygen in the sample must be found by difference.
To determine the mass of C and H in the sample, first calculate the fractional mass of carbon in CO_2 and the fractional mass of hydrogen in H_2O; these values are used to calculate the masses of carbon and hydrogen in the original sample:

fractional mass **C** in $\mathbf{CO_2}$ = 12.01 g/44.01 g = **0.2729**

fractional mass **H** in $\mathbf{H_2O}$ = 2.016 g/18.016 g = **0.1119**

Mass of **C** in original sample = mass of CO_2 formed × fractional mass of C in CO_2

= 11.620 g × 0.2729 = **3.171 g carbon**

Mass of **H** in original sample = mass of H_2O formed × fractional mass of H in H_2O

= 3.170 g × 0.1119 = **0.3547 g hydrogen**

Mass of **O** in original sample = (mass of original sample) – (masses of other elements)

= (7.750 g) – (3.171 g C) – (0.3547 g H) = **4.224 g oxygen**

(2) **Mole C** = 3.171 g/12.01 g/mole = **0.2640 mole C**

mole H = 0.3547/1.008 g/mole = **0.3519 mole H**

mole O = 4.224 g/16.00 g/mole = **0.2640 mole O**

(3a) 0.2640 mole C/0.2640 = **1 mole C**

0.3519 mole H/0.2640 = **1.33 mole H**

0.2640 mole O/0.2640 = **1.00 mole O**

(3b) The decimal 0.33 is converted to the fraction 1/3; multiply each value by **3**.

3 × 1 mole C = **3 mole C**

3 × 1.33 mole H = **4 mole H** simplest formula: $\mathbf{C_3H_4O_3}$

3 × 1.00 mole O = **3 mole O**

(4) **MM of $\mathbf{C_3H_4O_3}$ = 88.07** g/mole; **true MM of compound** = **176** g/mole

$$\textbf{multiple} = \frac{\text{176 (true MM of Vitamin C)}}{\text{88.07 (MM of simplest formula)}} = 2$$

true **molecular** formula = {$\mathbf{C_3H_4O_3}$} (simplest formula) × {**2**} (multiple) = $\mathbf{C_6H_8O_6}$

VI PRACTICE PROBLEMS

1. Determine the simplest formula for the following compounds:
   a) A compound containing carbon and fluorine (C_xF_y): a 60.00 gram sample contains 51.81 grams of fluorine.
   b) A sample of a compound that contains 351.9 grams of selenium and 142.6 grams of oxygen (Se_xO_y).

c) A compound containing copper and phosphorous (Cu_xP_y): a 50.00 gram sample contains 6.99 grams of phosphorous.
d) A sample of compound that contains 10.07 grams of lithium, 20.32 grams of nitrogen, and 69.62 grams of oxygen ($Li_xN_yO_z$).

2. A compound contains 49.5% carbon, 3.20% hydrogen, 22.0% oxygen, and 25.2% manganese ($C_xH_yO_zMn_w$); the true MM of the compound is 436 g/mole. (The percent mass values have some experimental error; round off mole numbers to the nearest whole number in this case.) Determine the simplest formula and true molecular formula.

3. The organic compound deoxyribose contains only C, H, and O ($C_xH_yO_z$). A 5.000 g sample of deoxyribose was combusted in air to form 8.202 g of CO_2 and 3.358 g of H_2O. All of the carbon in the sample was found in the CO_2, and all of the hydrogen in the sample was found in the H_2O. The true MM of deoxyribose is 134 g/mole. Determine the simplest formula and true molecular formula.

4. A compound contains carbon, hydrogen, oxygen, and iron ($C_xH_yO_zFe_w$).

A 1.5173 gram sample was analyzed by combustion: 2.838 grams of CO_2 was formed and 0.8122 grams of H_2O was formed. A separate experiment showed that the original sample contained 0.2398 grams of iron. The true MM of the compound is 353 g/mole. Determine the simplest formula and true molecular formula.

VII ANSWERS TO PRACTICE PROBLEMS

1. Determine the simplest formula for the following compounds:

a) Compound (C_xF_y): a 60.00 gram sample contains 51.81 grams of fluorine.
(1) mass **F = 51.81 g**; mass **C** = (60.0 g – 51.81 g) = **8.19 g**
(2) moles F = 51.81 g/19.00 g/mole = **2.727 moles F**

moles C = 8.19 g/12.01 g/mole = **0.6819 moles C**

(3a) 0.6819 is the smallest mole value found; divide all mole numbers by this value:

2.727 mole F/0.6819 = **4.00 mole F**

0.6819 mole C/0.6819 = **1 mole C**

The whole number mole ratio is equivalent to the simplest formula: $\mathbf{CF_4}$

b) A sample of a compound that contains 351.9 grams of selenium and 142.6 grams of oxygen (Se_xO_y).
(1) Mass **O = 142.6 g**; mass **Se = 351.9 g**
(2) Moles O = 142.6 g/16.00 g/mole = **8.913 moles O**

moles Se = 351.9 g/78.96 g/mole = **4.457 moles Se**

(3a) 4.457 is the smallest mole value found; divide all mole numbers by this value:

8.913 mole O/4.457 = **2.00 mole O**

4.457 mole Se/4.457 = **1 mole Se**

The whole number mole ratio is equivalent to the simplest formula: $\mathbf{SeO_2}$

c) A compound containing copper and phosphorous (Cu_xP_y): a 50.00 gram sample contains 6.99 grams of phosphorous.
   (1) Mass **P = 6.99 g**; mass **Cu** = (50.00 g – 6.99 g) = **43.01 g**
   (2) Moles P = 6.99 g/30.97 g/mole = **0.2257 moles P**

moles Cu = 43.01 g/63.55 g/mole = **0.6768 moles Cu**

(3a) 0.2257 is the smallest mole value found; divide all mole numbers by this value:

0.2257 mole P/0.2257 = **1 mole P**

0.6768 mole Cu/0.2257 = **3.00 mole Cu**

The whole number mole ratio is equivalent to the simplest formula: **Cu_3P**

d) A sample of compound that contains 10.07 grams of lithium, 20.32 grams of nitrogen, and 69.62 grams of oxygen ($Li_xN_yO_z$).
   (1) Mass **Li = 10.07 g**; mass **N = 20.32 g**; mass **O = 69.62 g**
   (2) Moles Li = 10.07 g/6.941 g/mole = **1.451 moles Li**

moles N = 20.32 g/14.01 g/mole = **1.450 moles N**

moles O = 69.62 g/16.00 g/mole = **4.351 moles O**

(3a) 1.450 is the smallest mole value found; divide all mole numbers by this value:

1.451 mole Li/1.450 = **1.00 mole Li**

1.450 mole N/1.450 = **1 mole N**

4.351 mole O/1.450 = **3.00 mole O**

The whole number mole ratio is equivalent to the simplest formula: **$LiNO_3$**

2. Compound contains: 49.5% carbon, 3.20% hydrogen, 22.0% oxygen, and 25.2% manganese ($C_xH_yO_zMn_w$); the true MM of the compound is 436 g/mole.
   (1) Select a sample size of exactly 100 grams; therefore, out of a 100-gram sample:

mass **C = 49.5 g** (100 g × 0.495); mass **H = 3.20 g** (100 g × 0.0320).

mass **O = 22.0 g** (100 g × 0.220); mass **Mn = 25.2 g** (100 g × 0.252).

(2) Mole C = 49.5 g/12.01 g/mole = **4.122 mole C**

mole H = 3.20 g/1.01 g/mole = **3.168 mole H**

mole O = 22.0 g/16.00 g/mole = **1.375 mole O**

mole Mn = 25.2 g/54.94 g/mole = **0.4587 mole Mn**

(3a) **0.4587** is the smallest mole value found; divide all mole numbers by this value:

4.122 mole C/0.4587 = **8.99 mole C**

3.168 mole H/0.4587 = **6.91 mole H**

1.375 mole O/0.4587 = **3.00 mole O** simplest formula: $\mathbf{C_9H_7O_3Mn}$

0.4587 mole Mn/0.4587 = **1 mole Mn**

(4) **MM of $\mathbf{C_9H_7O_3Mn}$ = 218.1** g/mole; **true MM of compound = 436** g/mole

$$\textbf{multiple} = \frac{436\ (\text{true MM of compound})}{218.1\ (\text{MM of simplest formula})} = \mathbf{2}$$

true **molecular** formula = $\mathbf{\{C_9H_7O_3Mn\}}$ (simplest formula) × **{2}** (multiple) = $\mathbf{C_{18}H_{14}O_6Mn_2}$

3. Deoxyribose contains only C, H, and O ($C_xH_yO_z$). A 5.000 g sample was combusted in air to form 8.202 g of CO_2 and 3.358 g of H_2O. The true MM is 134 g/mole.
(1) Fractional mass **C** in $\mathbf{CO_2}$ = 12.01 g/44.01 g = **0.2729**

fractional mass **H** in $\mathbf{H_2O}$ = 2.016 g/18.016 g = **0.1119**

mass of **C** in original sample = mass of CO_2 formed × fractional mass of C in CO_2

= 8.202 g × 0.2729 = **2.238 g carbon**

mass of **H** in original sample = mass of H_2O formed × fractional mass of H in H_2O

= 3.358 g × 0.1119 = **0.376 g hydrogen**

Mass of **O** in original sample = (mass of original sample) − (masses of other elements)

= (5.000 g) − (2.238 g C) − (0.376 g H) = **2.386 g oxygen**

(2) **Mole C** = 2.238 g/12.01 g/mole = **0.186 mole C**

mole H = 0.376/1.008 g/mole = **0.373 mole H**

mole O = 2.386 g/16.00 g/mole = **0.149 mole O**

(3a) **0.149** is the smallest mole value found; divide all mole numbers by this value:

0.186 mole C/0.149 = **1.25 mole C**

0.373 mole H/0.149 = **2.50 mole H**

0.149 mole O/0.149 = **1 mole O**

(3b) The smallest decimal 0.25 is converted to 1/4; multiply each value by **4.**

4×1.25 mole C = **5 mole C**

4×2.50 mole H = **10 mole H** simplest formula: $\mathbf{C_5H_{10}O_4}$

4×1 mole O = **4 mole O**

(4) MM of $C_5H_{10}O_4$ = 134.15 g/mole; **true MM of compound = 134** g/mole

$$\textbf{multiple} = \frac{134 \text{ (true MM of deoxyribose)}}{134.15 \text{ (MM of simplest formula)}} = \mathbf{1}$$

true **molecular** formula = $\{\mathbf{C_5H_{10}O_4}\}$ (simplest formula) × **{1}** (multiple) = $\mathbf{C_5H_{10}O_4}$

4. A compound of C, H, O, Fe ($C_xH_yO_zFe_w$): a 1.5173 gram sample produced 2.838 grams of CO_2 and 0.8122 grams of H_2O; the original sample contained 0.2398 grams of iron. The true MM of the compound is 353.2 g/mole.

(1) Fractional mass **C** in $\mathbf{CO_2}$ = 12.01 g/44.01 g = **0.2729**

fractional mass **H** in $\mathbf{H_2O}$ = 2.016 g/18.016 g = **0.1119**

mass of **C** in original sample = mass of CO_2 formed × fractional mass of C in CO_2

= (2.838 g) × (0.2729) = **0.7745 g carbon**

mass of **H** in original sample = mass of H_2O formed × fractional mass of H in H_2O

= (0.8122 g) × (0.1119) = **0.09105 g hydrogen**

mass Fe is given in the problem as: = **0.2398 g iron**

Mass of **O** in original sample = (mass of original sample) – (masses of other elements)

= (1.5173 g) – (0.7745 g C) – (0.09105 g H) – (0.2398 g Fe) = **0.4120 g oxygen**

(2) Mole C = 0.7745 g/12.01 g/mole = **0.06449 mole C**

mole H = 0.09105/1.008 g/mole = **0.09015 mole H**

mole O = 0.4120 g/16.00 g/mole = **0.02575 mole O**

mole Fe = 0.2398 g/55.85 g/mole = **0.004294 mole Fe**

(3a) 0.004294 is the smallest mole value found; divide all mole numbers by this value:

0.06449 mole C/0.004294 = **15.0 mole C**

0.09015 mole H/0.004294 = **21.0 mole H**

0.02575 mole O/0.004294 = **6.00 mole O** simplest formula: $\mathbf{C_{15}H_{21}O_6Fe}$

0.004294 mole Fe/0.004294 = **1 mole Fe**

(4) **MM of** $\mathbf{C_{15}H_{21}O_6Fe}$ **= 253.21** g/mole; **true MM of compound = 253** g/mole

$$\textbf{multiple} = \frac{253\ \text{(true MM of compound)}}{253.21\ \text{(MM of simplest formula)}} = \mathbf{1}$$

molecular formula = $\mathbf{\{C_{15}H_{21}O_6Fe\}}$ (simplest formula) × **{1}** (multiple) = $\mathbf{C_{15}H_{21}O_6Fe}$

7 Writing Chemical Equations

I GENERAL CONCEPTS

A chemical reaction involves the exchange of atoms between an original set of compounds (or elements) to form a new set of compounds (or elements). A chemical equation is used to describe these atom exchanges.

The **reactants** are the original set of elements or compounds; these are listed on the left side of the equation. The **products** are the newly formed set of elements or compounds; these are listed on the right side of the equation. The left and right side of the equation are connected by a left-to-right arrow ($\longrightarrow$).

The **state of matter** of each reactant or product is often also listed as a subscript in parentheses after the formula: **(s)** = solid; **(l)** = liquid; **(g)** = gas; **(aq)** = aqueous (water) solution. Additional information, such as temperature or the presence of a catalyst, can be listed above or below the connecting arrow.

Conservation Laws

The law of conservation of atoms states that atoms cannot be created or destroyed in chemical reactions. This requires that the total number of atoms of each element on each side of the equation be equal.

The law of conservation of electrons states that electrons are not created or destroyed in chemical reactions. Electrons can be exchanged or transferred, but net neutrality must remain.

The law of conservation of mass derives from the law of conservation of atoms. Mass conversion to energy in chemical reactions is generally too small to measure.

The requirement that atoms of each element on each side of an equation be equal describes a **balanced equation**. Numbers written directly in front of an element symbol or compound formula are termed **stoichiometric coefficients**. The stoichiometric coefficient states the number of formulas (molecules or formula units) of each reactant or product required to balance all elements in the equation.

All reactants and products must be specifically included in the number ratios described by the stoichiometric coefficients. However, a coefficient of **1** is not written; absence of a coefficient indicates that only **one** of a particular formula is required for a balanced equation.

II WRITING BALANCED EQUATIONS

A number of techniques can be used as an aid to balancing an equation such as balancing an acid to a base (Chapter 9) or balancing oxidation numbers (Chapter 10). However, stoichiometric coefficients are often determined simply through trial-and-error by counting total atoms of each element on both sides of an equation and then adjusting stoichiometric coefficients accordingly. **Formulas for compounds can never be altered to achieve a balanced equation.**

Process for Writing Balanced Equations from Descriptions

(1) Write the correct formulas for all products and reactants; reactants are written to the left of the arrow; products are written to the right.
 a) Compound formulas are determined from names or other descriptions.
 b) Formulas for elements are the symbol for the element, with the following general exception. Some elements come as diatomic molecules: two atoms of the same element per one molecule. These elements are $\mathbf{H_2}$, $\mathbf{N_2}$, $\mathbf{O_2}$, $\mathbf{F_2}$, $\mathbf{Cl_2}$, $\mathbf{Br_2}$, $\mathbf{I_2}$, and are written in their diatomic molecule form. Certain other elements may sometimes appear as multi-atom such as S_8.

(2) Include additional information such as states of matter, temperature, catalyst or other data as required.

(3) Balance the equation by adjusting the stoichiometric coefficients of each reactant and product; the number of atoms of each element on each side of the equation should be equal. **Formulas for compounds can never be altered to achieve a balanced equation.**
 a) The total number of atoms of each element is found by multiplying the coefficient of a compound by the number of atoms of the element in the formula; perform this calculation for each compound containing the element being counted.
 b) To balance by trial-and-error, count the total atoms of one element at a time; change the coefficients of the compounds that contain this element such that the numbers on each side of the equation match. Continue this process in turn for each element.
 c) Adjustment of coefficients for some compounds performed later in the process may alter an element that had been previously balanced. As a general sequence, start by balancing elements in the more complicated molecules first; balance elements that exist by themselves last.

Example: Write a balanced equation based on the description: nitrogen gas plus hydrogen gas react to form ammonia gas at a temperature of 400°C in the presence of an iron catalyst.

(1) $\mathbf{N_2 + H_2 \longrightarrow NH_3}$ Nitrogen and hydrogen are written as diatomic molecules.
Extra information is added: all three elements/compounds are gases, indicated with a subscript **(g)** after the formula; catalyst and temperature are shown above and below the arrow.

(2) $N_{2(g)} + H_{2(g)} \xrightarrow[400\,°C]{Fe} NH_{3(g)}$
Nitrogen does not balance. The **two** nitrogens on the left must be balanced by a coefficient of **2** in front of the NH_3.

(3) $N_{2(g)} + H_{2(g)} \longrightarrow \mathbf{2}\,NH_{3(g)}$
Hydrogen does not balance. There are now **six** hydrogens on the right; balance with a coefficient of **3** in front of the H_2:

(4) $N_{2(g)} + \mathbf{3}H_{2(g)} \longrightarrow \mathbf{2}\,NH_{3(g)}$
Each side has 2 nitrogens and 6 hydrogens; the complete equation:

(5) $N_{2(g)} + \mathbf{3}H_{2(g)} \xrightarrow[400\,°C]{Fe} \mathbf{2}\,NH_{3(g)}$

Example: Write a balanced equation based on the description: phosphorous solid plus chlorine gas reacts to form phosphorous pentachloride solid.

(1) $\mathbf{P + Cl_2 \longrightarrow PCl_5}$ chlorine is written as a diatomic molecule.
Extra information is added: chlorine is a gas **(g)**; phosphorous and phosphorous pentachloride are solids **(s)**.

(2) $P_{(s)} + Cl_{2(g)} \longrightarrow PCl_{5(s)}$

Since phosphorous is by itself, start with chlorine:chlorine does not balance. To balance diatomic Cl_2, an even number of chlorines from PCl_5 on the right is required; the smallest number to achieve that is **2**. Cl is then balanced by a **5** for Cl_2.

(3) $P_{(s)} + \mathbf{5}Cl_{2(g)} \longrightarrow \mathbf{2}PCl_{5(s)}$

Phosphorous can now be balanced by adding the required **2** coefficient; since P is by itself any added coefficient does not change the balance for chlorine.

(4) $\mathbf{2}P_{(s)} + \mathbf{5}Cl_{2(g)} \longrightarrow \mathbf{2}PCl_{5(s)}$

Example: Write a balanced equation based on the description: aqueous glucose ($C_6H_{12}O_6$) reacts with oxygen gas to produce carbon dioxide gas plus water liquid at 37°C in the presence of enzyme catalysts.

(1) $\mathbf{C_6H_{12}O_6 + O_2 \longrightarrow CO_2 + H_2O}$ oxygen is diatomic.

Extra information is added: oxygen and carbon dioxide are gases (**g**); water is liquid (**l**); glucose is in aqueous solution (**aq**). The temperature and catalysts are included.

(2) $C_6H_{12}O_{6(aq)} + O_{2(g)} \xrightarrow[37\,°C]{enzymes} CO_{2(g)} + H_2O_{(l)}$

Although oxygen is diatomic it is by itself; start with carbon (or hydrogen). To balance the six carbons on the left, a coefficient of **6** is required for carbon dioxide.

(3) $C_6H_{12}O_{6(aq)} + O_{2(g)} \longrightarrow \mathbf{6}CO_{2(g)} + H_2O_{(l)}$

To balance the twelve hydrogens on the left, a coefficient of **6** is required for water.

(4) $C_6H_{12}O_{6(aq)} + O_{2(g)} \longrightarrow \mathbf{6}CO_{2(g)} + \mathbf{6}H_2O_{(l)}$

There are 18 oxygens on the right; oxygen can now be balanced by adding the required **6** coefficient; since O is by itself any added coefficient does not change the balance for carbon or hydrogen. The complete equation:

(5) $C_6H_{12}O_{6(aq)} + \mathbf{6}O_{2(g)} \xrightarrow[37\,°C]{enzymes} \mathbf{6}CO_{2(g)} + \mathbf{6}H_2O_{(l)}$

Example: Balance the equation shown:

$$Fe_2O_{3(s)} + CO_{(g)} \longrightarrow Fe_{(s)} + CO_{2(g)}$$

Often balancing an equation requires readjustment of coefficients as balancing one element upsets the balance of another. Fe (by itself) is left to last; at one carbon on each side, the carbons are balanced in the given equation. Balance oxygen with a coefficient of **2** for CO_2, showing four O per side.

$$Fe_2O_{3(s)} + CO_{(g)} \longrightarrow Fe_{(s)} + \mathbf{2}CO_{2(g)}$$

The oxygens are now balanced, but the carbons no longer are. To balance carbons, use of coefficient of **2** for CO.

$$Fe_2O_{3(s)} + \mathbf{2}CO_{(g)} \longrightarrow Fe_{(s)} + \mathbf{2}CO_{2(g)}$$

The carbons are now balanced, but the balance for oxygen has been lost. Finally, coefficients of **3** for CO and CO_2 will balance both the oxygen and the carbon.

$$Fe_2O_{3(s)} + \mathbf{3}CO_{(g)} \longrightarrow Fe_{(s)} + \mathbf{3}CO_{2(g)}$$

Iron can now be balanced by adding the required **2** coefficient for the element Fe, which does not change the balance for carbon or oxygen. The balanced equation:

$$Fe_2O_{3\,(s)} + \mathbf{3}CO_{(g)} \longrightarrow \mathbf{2}Fe_{(s)} + \mathbf{3}CO_{2\,(g)}$$

III PRACTICE PROBLEMS

1. Write balanced equations based on the following descriptions:
   a) Sulfur dioxide gas reacts with oxygen gas to form sulfur trioxide gas.
   b) Calcium oxide solid plus water forms calcium hydroxide solid.
   c) Calcium carbonate solid decomposes to form calcium oxide solid plus carbon dioxide gas.
   d) Chromium (II) oxide solid is converted to chromium metal plus oxygen gas.
   e) Cobalt (II) nitrate in an aqueous solution plus sodium phosphate in an aqueous solution forms cobalt (II) phosphate solid plus sodium nitrate in an aqueous solution.
   f) Gold (III) chloride solid reacts with iron metal to form gold metal plus iron (III) chloride solid.
   g) Aluminum hydroxide solid decomposes to form aluminum oxide solid plus water.
   h) Lead (II) acetate in an aqueous solution reacts with potassium sulfate in an aqueous solution to form lead (II) sulfate solid plus potassium acetate in an aqueous solution.

2. Balance the equations.
   a) $C_3H_6 + NH_3 + O_2 \longrightarrow C_3H_3N + H_2O$
   b) $NH_3 + HNO_3 \longrightarrow N_2 + H_2O$

IV ANSWERS TO PRACTICE PROBLEMS

1. Write balanced equations based on the following descriptions:
   a) Sulfur dioxide gas reacts with oxygen gas to form sulfur trioxide gas.

$$2SO_{2\,(g)} + O_{2\,(g)} \longrightarrow 2SO_{3\,(g)}$$

   b) Calcium oxide solid plus water forms calcium hydroxide solid.

$$CaO_{(s)} + H_2O_{(l)} \longrightarrow Ca(OH)_{2\,(s)}$$

   c) Calcium carbonate solid decomposes to form calcium oxide solid plus carbon dioxide gas.

$$CaCO_{3\,(s)} \longrightarrow CaO_{(s)} + CO_{2\,(g)}$$

   d) Chromium (II) oxide solid is converted to chromium metal plus oxygen gas.

$$2CrO_{(s)} \longrightarrow 2Cr_{(s)} + O_{2\,(g)}$$

   e) Cobalt (II) nitrate in an aqueous solution plus sodium phosphate in an aqueous solution forms cobalt (II) phosphate solid plus sodium nitrate in an aqueous solution.

$$3Co(NO_3)_{2\,(aq)} + 2Na_3PO_{4\,(aq)} \longrightarrow Co_3(PO_4)_{2\,(s)} + 6NaNO_{3\,(aq)}$$

f) Gold (III) chloride solid reacts with iron metal to form gold metal plus iron (III) chloride solid.

$$AuCl_{3\,(s)} + Fe_{(s)} \longrightarrow Au_{(s)} + FeCl_{3\,(s)}$$

g) Aluminum hydroxide solid decomposes to form aluminum oxide solid plus water.

$$2Al(OH)_{3\,(s)} \longrightarrow Al_2O_{3\,(s)} + 3H_2O_{(l)}$$

h) Lead (II) acetate in an aqueous solution reacts with potassium sulfate in an aqueous solution to form lead (II) sulfate solid plus potassium acetate in an aqueous solution.

$$Pb(CH_3COO)_{2\,(aq)} + K_2SO_{4\,(aq)} \longrightarrow PbSO_{4\,(s)} + 2CH_3COOK_{(aq)}$$

2. Balance the equations.
   a) $\mathbf{2}\ C_3H_6 + \mathbf{2}\ NH_3 + \mathbf{3}O_2 \longrightarrow \mathbf{2}C_3H_3N + \mathbf{6}H_2O$
   b) $\mathbf{5}NH_3 + \mathbf{3}HNO_3 \longrightarrow \mathbf{4}N_2 + \mathbf{9}H_2O$

8 Techniques for Performing Stoichiometric Calculations

I GENERAL CONCEPTS

Stoichiometry ("measuring elements") and stoichiometric calculations are methods for comparing and calculating specific amounts of elements or compounds used (as reactants) or formed (as products) in a chemical reaction based on the ratios in the balanced equation.

The **balanced equation** represents the specific molar ratio of reactants that must be combined to form a specific molar ratio of corresponding products. The molar ratios of all reactants and products for a specific reaction are represented by the **stoichiometric coefficients**.

The balanced equation shows the recipe, or blueprint, for converting fixed mole values of reactants to the corresponding fixed mole values of products. Initial given amounts and final calculated amounts of either reactants or products may be stated in a variety of units, such as mass, moles, and concentrations. However, since stoichiometric calculations depend on balanced equation mole ratio comparisons, all calculations **must** always proceed through **mole** values.

Stoichiometric calculations provide a method for comparing **any** given amount of reactant or product to the specific amount of another reactant required in the process or the specific amount of a product that can be formed by the process.

1. The calculation can determine the **maximum** specific amount of a selected product that can be formed from a specific given amount of reactant. This value is called the **theoretical yield (T.Y.)** of that product, expressed in moles, grams, or other units.
2. The calculation can determine the specific amount of a selected reactant that is required to form a specific given amount of product based on theoretical yield.
3. The calculation can determine the specific amount of a one selected reactant that must be reacted in the proper ratio with another given amount of reactant, assuming no reactant is in excess and the theoretical yield applies.
4. The actual (experimental) yield of a specific product in a reaction can be compared to the maximum possible yield (theoretical yield) through a determination of **percent yield**.
5. Under conditions where one reactant is present in a limited amount **(limiting reagent)**, compared to other reactants that are in excess, a complete calculation can be performed to determine the amount of all reactants and products present after completion of a reaction based on theoretical yield.

Analogy: A (severely limited) blueprint/balanced equation for producing a car is shown below. Plant inventory reports 1200 pounds of wheels in the factory.

1. Calculate the maximum number of cars (theoretical yield) that can be produced from the wheels, assuming sufficient numbers of all other parts.

$$\textbf{1 Engine + 1 Car Body + 2 Bumpers + 4 Wheels} \longrightarrow \textbf{1 Car}$$

The information provided for the amount of wheels is insufficient. The blueprint is based on **counting number** ratios, not mass ratios. Before a calculation for the number of cars can be performed, a calculation for the **number** of wheels is required. Assume it is determined that the mass in pounds per wheel is 25 pounds/wheel.

$$\#\text{ of wheels} = \frac{1200 \text{ pounds}}{25 \text{ pounds/wheel}} = 48 \text{ wheels}$$

The number of cars can now be found from the ratio in the blueprint/balanced equation:

$$\#\text{ cars} = 48 \text{ wheels} \times \frac{1 \text{ car}}{4 \text{ wheels}} = 12 \text{ cars}$$

2. Plant inventory finds 6 dozen bumpers in supply. Calculate the theoretical yield of cars from this value for bumpers.
   Note that "dozen" is a counting number representing 12 objects; it can be converted to a number for use in the balanced equation:

$$\#\text{ of bumpers} = 6 \text{ dozen} \times 12 \text{ bumpers/dozen} = 72 \text{ bumpers}$$

$$\#\text{ cars} = 72 \text{ bumpers} \times \frac{1 \text{ car}}{2 \text{ bumpers}} = 36 \text{ cars}$$

   This problem can also be solved directly from the value of the dozen unit. Since dozen represents a specific counting number value, the ratios in the balanced equation apply equally to this unit:

$$\#\text{ dozen cars} = 6 \text{ dozen bumpers} \times \frac{1 \text{ dozen cars}}{2 \text{ dozen bumpers}} = 3 \text{ dozen cars}$$

Example: Chemical inventory shows 6 moles of iron (III) oxide available for reaction; calculate the theoretical yield of carbon dioxide from this value for iron (III) oxide based on the equation.

$$\mathbf{Fe_2O_3 + 3CO \longrightarrow 2Fe + 3CO_2}$$

The mole also represents a specific (very large) counting number value, the ratios in the balanced equation apply equally to moles:

$$\#\text{ moles } CO_2 = 6 \text{ mole } Fe_2O_3 \times \frac{3 \text{ moles } CO_2}{1 \text{ mole } Fe_2O_3} = 18 \text{ moles } CO_2$$

II GENERAL PROCESS FOR STOICHIOMETRIC CALCULATIONS BASED ON BALANCED EQUATIONS

(1) If necessary, first write the balanced equation from a word description (Chapter 7). Then identify the compound or element for which an **initial** starting amount is given; the initial compound (or element) given may be a reactant or product. If the given starting amount is not in moles, **convert this amount to moles by solving a mass/MM problem**. Use the periodic table as necessary to determine molar masses. (Other possible conversions are examined in Chapter 11)

$$\textbf{Moles of Compound} = \frac{\text{mass (g) of compound}}{\text{Molar Mass of compound}}$$

(2) Identify the compound (or element) for which a specific amount must be calculated. Then set up a **mole** ratio between this compound (or element), and the one identified in step (1) based on the **balanced** equation. This ratio is expressed such that the moles of the **compound to be calculated** is on **top**, and the moles of the **compound with the amount given** (to be cancelled out) is on the **bottom**:

$$\textbf{ratio based on balanced equation} = \frac{\text{moles compound to be calculated}}{\text{moles compound given}}$$

The mole ratio of these two compounds (or elements) is determined from the corresponding **stoichiometric coefficients** from the balanced equation:

$$\textbf{ratio based on balanced equation} = \frac{\text{coefficient of compound to be calculated}}{\text{coefficient of compound given}}$$

(3) Use the moles of the given compound (or element) found in step (1) and the ratio established in step (2) to calculate the **moles** of compound (or element) to be calculated.

$$\textbf{moles of compound to be calculated} = (\text{moles of compound given}) \times \frac{\text{coefficient of compound to be calculated}}{\text{coefficient of compound given}}$$

(4) Use the value for the moles of the compound to be calculated from step (3) to calculate the amount of this compound in mass (g), (or other units), if necessary. Solve a mass/molar mass calculation (or a concentration (Molarity) problem) as required.

$$\textbf{Mass of Compound} = (\text{moles of compound}) \times (\text{Molar Mass of compound})$$

Example: Calculate the theoretical yield (T.Y.) in grams of CO_2 which can be prepared from 2.00 kilograms of Fe_2O_3 based on the balanced equation. The process can be written in a diagram form using the balanced equation to emphasize that the calculation proceeds through **moles**. The steps are identified with the step numbers from the general process.

The mass in grams of the starting Fe_2O_3 compound = 2.00 kg × 1000 g/kg = 2000 grams (three significant figures). The MM of Fe_2O_3 is calculated from values in the periodic table as 159.6 g/mole.

Diagram Form

Fe_2O_3 + 3CO ⟶ 2 Fe + $3CO_2$

mass (**start**) = 2000 g ↓ ... mass = **1650** g ↑

↓ (1) moles Fe_2O_3 = 2000 g / 159.6 g/mole = 12.5 mole

(4) ↑ mass CO_2 = 37.5 mole × 44.01 g/mole = **1650 grams**

↓ 12.5 mole ⟶ (2) $\frac{3\ CO_2}{\times\ 1\ Fe_2O_3}$ (3) ⟶ = 37.5 mole CO_2 ↑

VIEWING EACH STEP

(1) Mole Fe_2O_3 = 2000 g/159.6 g/mole = 12.5 mole Fe_2O_3

(2) Ratio in balanced equation is $\frac{3CO_2}{1Fe_2O_3}$

(3) Moles of CO_2 which can be formed from the Fe_2O_3

$$= (12.5 \text{ mole } Fe_2O_3) \times \frac{3CO_2}{1Fe_2O_3} = 37.5 \text{ mole } CO_2$$

(4) Theoretical yield (g) = (37.5 mole) × (44.01 g/mole) = **1650** grams CO_2

Example: Calculate the mass in grams of CO required to **exactly react** with 2.00 kilograms of Fe_2O_3 based on the balanced equation.

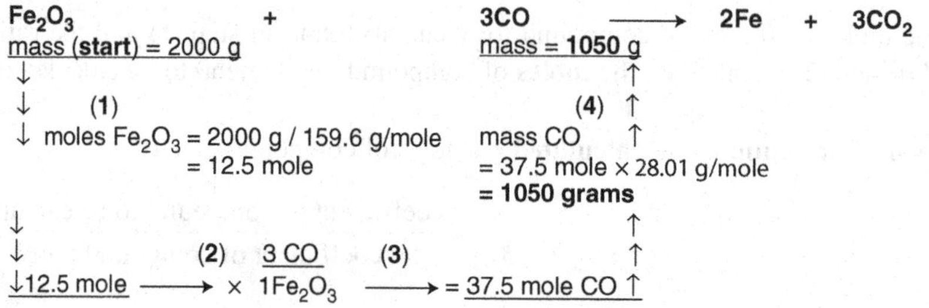

Fe_2O_3 + 3CO ⟶ 2Fe + $3CO_2$

Fe_2O_3: mass (**start**) = 2000 g ↓

↓ (1) ↓ moles Fe_2O_3 = 2000 g / 159.6 g/mole = 12.5 mole

↓ 12.5 mole ⟶ (2) × $\frac{3\ CO}{1Fe_2O_3}$ (3) ⟶ = 37.5 mole CO ↑

(4) ↑ mass CO = 37.5 mole × 28.01 g/mole = **1050 grams**

3CO: mass = **1050** g

The first example compared a desired product amount to a given reactant amount. The second example compared another reactant amount to a given reactant amount. It is also possible to calculate "backwards," comparing an unknown required reactant amount to a given product amount. The calculation process is identical, even while viewing the direction of calculation in the diagram as product to reactant.

Example: Calculate the mass(g) of H_2SO_4 required to produce 1.00 kg of $Al_2(SO_4)_3$ according to the balanced equation:

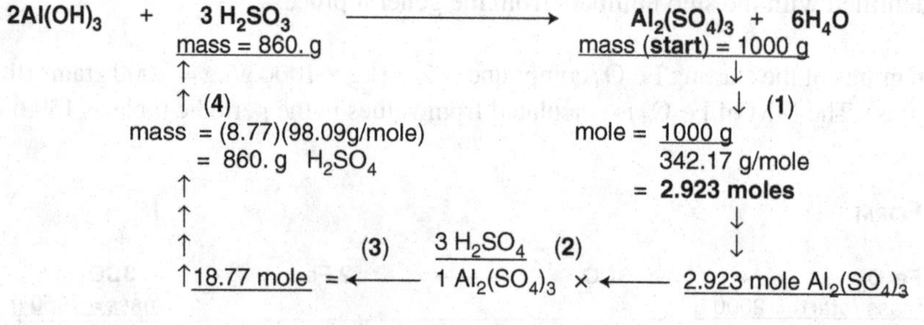

$2Al(OH)_3$ + $3\ H_2SO_3$ ⟶ $Al_2(SO_4)_3$ + $6H_4O$

$Al_2(SO_4)_3$: mass (**start**) = 1000 g ↓

↓ (1) mole = $\frac{1000\text{ g}}{342.17\text{ g/mole}}$ = **2.923 moles**

↓ 2.923 mole $Al_2(SO_4)_3$ ⟵ × (2) $\frac{3\ H_2SO_4}{1\ Al_2(SO_4)_3}$ (3) ⟵ = ↑ 18.77 mole

↑ (4) mass = (8.77)(98.09g/mole) = 860. g H_2SO_4

$3\ H_2SO_3$: mass = 860. g

Complete Example: Calculate the theoretical yield (T.Y.) in grams of (a) NO and (b) H_2O, which can be formed from 150 grams of NH_3 according to the equation given. (c) Calculate the mass of O_2, which is required to exactly react with 150 grams of NH_3.

$4\,NH_3$ + $5\,O_2$ ⟶ $4NO$ + $6H_2O$

mass (**start**) = 150 g (c) mass = **352 g** mass = **264 g** mass = **238 g**

(1) moles NH_3 = $\frac{150.\ g}{17.03\ g/mole}$ = 8.81 mole

(4) mass NO = 8.81 mole × 30.01 g/mole = **264 grams**

mass H_2O = 13.2 mol × 18.02 g/mol = **238 grams**

(a) 8.81 mole ⟶ (2) $\frac{4\ NO}{\times\ 4\ NH_3}$ ⟶ (3) = 8.81 mole NO

(b) 8.81 mole ⟶ (2) $\frac{6\ H_2O}{\times\ 4\ NH_3}$ ⟶ (3) = 13.2 mole H_2O

(c) 8.81 mole ⟶ (2) $\frac{5\ O_2}{\times\ 4\ NH_3}$ ⟶ (3) = 11.0 mole O_2

(c) (4) mass O_2 = (11.0 mole) × (32.00 g/mole) = **352 grams O_2**

III PERCENT YIELD

Chemical reactions do not usually produce 100% of the expected amount (the full theoretical yield) of a specific product. Reactants may not have complete efficiency for combining in the reaction, undesirable side products may be formed, or recovery of the product may result in losses of certain compounds.

Measurements termed **fractional yield** and **percent yield** are used to describe the efficiency of a product forming reaction; yields can take any unit since the measurement is a ratio:

fractional yield = actual measured yield/theoretical (calculated) yield

percent yield = actual measured yield/theoretical (calculated) yield × 100%

Most chemical synthetic processes combine many chemical reactions in sequence; optimization of percent yield is especially important for commercial products. The percent yield of a complete process based on multiple reactions is found by multiplication of the fractional yield of each reaction in the sequence. **Example:**

$$\mathbf{A} \xrightarrow{\text{90\% yield}} \mathbf{B} \xrightarrow{\text{85\% yield}} \mathbf{C} \xrightarrow{\text{97\% yield}} \mathbf{D}$$

The net yield for the process A ⟶ D is (0.90) × (0.85) × (0.97) = 0.74 = 74%

General Example: 10.0 grams of C_2H_6O reacts in the reaction shown. The actual measured yield of $C_4H_8O_2$ is 14.8 grams. (a) Calculate the % yield of the reaction.

a) First calculate the **theoretical** yield of product starting with the given reactant amount of 10.0 grams. Then determine the percent yield based on the actual yield of 14.8 grams.

C_2H_6O + CH_3COOH ⟶ H_2O + $C_4H_8O_2$

C_2H_6O: mass = 10.0 g ↓

↓ (1)

↓ moles C_2H_6O = 10.0 g / 46.08 g/mole = 0.217 mole

↓

↓ (2)

↓ 0.217 mole ⟶ $\times \frac{1\ C_4H_8O_2}{1\ C_2H_6O}$ ⟶ (3) = 0.217 mole $C_4H_8O_2$ ↑

(4) ↑ mass $C_4H_8O_2$ = 0.217 mole × 88.1 g/mole = **19.1 grams** ↑

$C_4H_8O_2$: mass = **19.1** g

The theoretical yield (T.Y.) is 19.1 grams; % **yield** = 14.8 g/19.1 g × 100% = **77.4%**

b) In a separate problem, an **actual** measured yield of 200.0 grams of $C_4H_8O_2$ is required. If the % yield is 77.4%, what must the **theoretical** yield (T.Y.) of $C_4H_8O_2$ be to ensure that the actual yield be 200.0 grams?

Rearrange the fractional yield equation: fractional yield = actual yield/T.Y.;

T.Y. = actual yield/fractional yield

= 200.0 g/0.774 = **258.4 grams**

IV LIMITING REAGENT

By design or by circumstance, reactants are often not mixed in the exact stoichiometric proportions dictated by the balanced equation. In syntheses, one reactant is often expensive or difficult to obtain or produce. In these cases, the yield based on this reactant can be maximized by using an excess of the other cheaper, easier to get reactants. The most important reactant is then in the shortest supply, while the other reactants have an excess amount.

For these reactions, the reactant in shortest supply will run out first and will limit the total amount of product that can be formed. The reactants in excess supply will have some amount leftover after the reaction is completed.

The **limiting reagent** is the reactant in shortest supply that limits the formation of product and, thus, determines the true theoretical yield. The **excess reagents** are those reagents, which are not completely consumed in the reaction (some amount leftover); these do **not** determine the theoretical yield.

Analogy: The blueprint/balanced equation for producing a car is shown.

For the day, the plant inventory reports 10 engines, 12 car bodies, 18 bumpers, and 32 wheels in the factory. Calculate the maximum number of cars (theoretical yield), which can be produced from the parts present in the factory.

The given counting numbers can be used directly. The technique to find the limiting part in this analogy is to **first** calculate the maximum number of cars, which can be produced from **each** of the parts **separately**. The ratio in the "balanced equation" is used in each case. The results indicate that, based on all the parts stated, the **32 wheels** in inventory limit the total number of cars that can be produced to **8**.

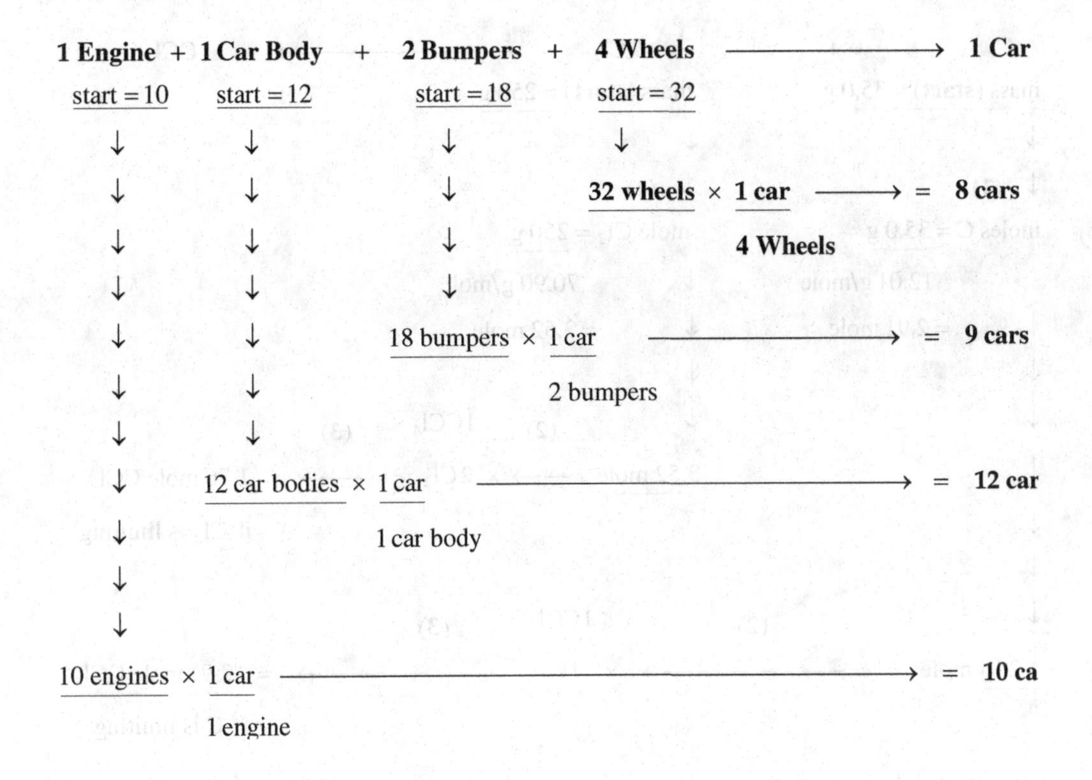

The limiting "reagent" in the example is the 32 wheels. Note that the limiting reagent cannot simply be found by selecting the part with the lowest numerical value. The lowest numerical value of 10 engines does not limit the final number of cars; the balanced equation was required to convert the starting number of parts to a specific number of cars produced. With the production of the limited 8 cars, 2 engines, 4 car bodies, and 2 bumpers will be left over; these are termed to be in excess.

A limiting reagent problem for chemical stoichiometric calculations can be identified as one in which **more than one** reactant has a **given amount** stated. The **limiting reagent** in this type of problem is the **one that will produce the *fewest* number of moles for any product** in a standard theoretical yield calculation.

The technique is to perform a theoretical yield calculation for **each** of the reactants for which a specific given amount is stated. The actual correct theoretical yield is the **smallest** product mole value calculated based on the separate calculations for each of the given reactants. The reactant, which produces this smallest product mole value, is the **limiting reagent**; all other given reactants are said to be in **excess**. The theoretical yield calculation can be performed for **any** of the possible products; a limiting reagent for any one product must be limiting for all products in the equation.

Example: 35.0 grams of carbon is reacted with 250 grams of diatomic chlorine to form carbon tetrachloride. Find the limiting reagent and calculate the true theoretical yield of CCl_4 in grams.

C + $2Cl_2$ ⟶ CCl_4

mass (**start**) = 35.0 g mass (**start**) = 250 g

↓ ↓

↓ (1) ↓ (1)

moles C = $\frac{35.0\text{ g}}{12.01\text{ g/mole}}$ = 2.91 mole

mole Cl_2 = $\frac{250\text{ g}}{70.90\text{ g/mole}}$ = 3.52 mole

↓ 3.52 mole ⟶(2) $\times \frac{1CCl_4}{2Cl_2}$ ⟶(3) = 1.76 mole CCl_4

if Cl_2 is limiting

↓ 2.91 mole ⟶(2) $\times \frac{1CCl_4}{1C}$ ⟶(3) = 2.91 mole CCl_4

if C is limiting

2.91 mole CCl_4, the theoretical yield based only on the given amount of the C reactant, would be the correct theoretical yield **only** if **carbon** is the **limiting** reagent. 1.76 mole CCl_4, the theoretical yield based only on the given amount of the Cl_2 reactant, would be the correct theoretical yield **only** if **chlorine** is the **limiting** reagent.

The **smaller** product mole value from the two separate calculations performed is **1.76** mole of CCl_4 based on the given amount of Cl_2. This means that **1.76 mole of CCl_4** is the true **correct theoretical yield**; therefore, **Cl_2** is identified as the **limiting reagent**, even though the starting mass of Cl_2 was much larger than that for carbon.

The theoretical yield value calculated based on carbon has no meaning for the completion of this problem. It merely indicates that **carbon is the excess reagent**, and that some carbon will be leftover after the reaction has consumed all of the chlorine. To complete the problem, the theoretical yield in **grams** from the true theoretical yield in moles can be calculated by following step **(4)**. It was not necessary to calculate the theoretical yield in grams for the carbon reactant to determine the limiting reagent.

Theoretical yield (**g**) = (1.76 mole) × (153.8 g/mole) = **271 grams CCl_4**

V CALCULATION OF ALL PRODUCTS AND REACTANTS IN A LIMITING REAGENT PROBLEM

The total amount of all products and all reactants after the completion of a reaction can be determined by the extension of stoichiometry calculations. The theoretical yield of **all products** in a reaction can be calculated from the limiting reagent; the calculation process is simply extended to each possible product. The limiting reagent, determined for any one product in the complete reaction, must apply to **all** products.

The total amount of **all unused reactants** (excess amount leftover) after completion of a reaction depends on determination of the limiting reagent. By definition, the limiting reagent is assumed to be 100% reacted (i.e., none leftover); the reagents supplied in excess will **not** be completely consumed. The actual amount of an excess reagent **consumed** in the reaction is found by calculating the amount required to exactly react with the completely consumed limiting reagent. The amount of unused (leftover) excess reagent is then found by subtraction from the original given amount.

A summary table of all reactants and products in the reaction is often used to keep track of all the required information. These tables are especially valuable for equilibrium calculations.

Example:

Reaction Data	**Reactant 1 +**	**Reactant 2 ⟶**	**Product 1 +**	**Product 2**
Starting moles	Given	Given		
Mole (Δ) change	Must be − (negative)	Must be − (negative)	Must be + (positive)	Must be + (positive)
Finishing moles	From subtraction	From subtraction	From addition	From addition
Mass (g)	From MM	From MM	From MM	From MM

Example: Use the reaction shown to calculate the masses of all products and reactants after the completion of the reaction of 200 grams of C_3H_8 with 200 grams of O_2; use the molar masses shown.

MM : 44.11 g/mol   MM : 32.00 g/mol   MM : 82.16 g/mol   MM : 18.02 g/mol

$$\mathbf{4C_3H_8 + 3O_2 \longrightarrow 2C_6H_{10} + 6H_2O}$$

mass (**start**) = 200 g   mass (**start**) = 200 g

↓ (1)   ↓ (1)

$$\text{mole } C_3H_8 = \frac{200\ \text{g}}{44.11\ \text{g/mole}} = 4.53\ \text{mole} \qquad \text{mole } O_2 = \frac{200\ \text{g}}{32.00\ \text{g/mole}} = 6.25\ \text{mole}$$

$$6.25\ \text{mole} \xrightarrow{(2)} \times \frac{2C_6H_{10}}{3O_2} \xrightarrow{(3)} = 4.17\ \text{mole } C_6H_{10}$$

if O_2 is limiting

$$4.53\ \text{mole} \xrightarrow{(2)} \times \frac{2C_6H_{10}}{4C_3H_8} \xrightarrow{(3)} = 2.27\ \text{mole } C_6H_{10}$$

if C_3H_8 is limiting

The selection of C_6H_{10} for these first calculations was arbitrary; the same limiting reagent would have been indicated if H_2O were the selected product. The **smaller** product mole value from the two separate calculations indicates that **2.27** mole of C_6H_{10} based on the reactant C_3H_8 is the correct theoretical yield; therefore, C_3H_8 is the limiting reagent. The requirement for the complete calculation is the **complete** consumption of the limiting reagent; the entire starting amount of 4.53 mole C_3H_8 must be shown as a change of minus 4.53 mole. This is the first new entry into the table after placing the given starting amounts. The C_3H_8 column for moles is completed by subtraction.

Reaction Data	**4 C_3H_8+**	**3 O_2 ⟶**	**2 C_6H_{10}+**	**6 H_2O**
Starting moles	4.53 moles	6.25 moles	0 mole	0 mole
Mole (Δ) change	−4.53 moles			
Finishing moles	0 mole			
Mass (g)				

The C_6H_{10} mole column can also be completed using the correct 2.27 mole theoretical yield calculated from consuming the C_3H_8 limiting reagent. Since the C_6H_{10} product is formed, the mole change must be plus 2.27 moles; addition (to zero moles in this case) then completes this part of the column.

Reaction Data	**4 C_3H_8+**	**3O_2 ⟶**	**2 C_6H_{10}+**	**6 H_2O**
Starting moles	4.53 moles	6.25 moles	0 mole	0 mole
Mole (Δ) change	−4.53 moles		+2.27 moles	+6.80 moles
Finishing moles	0 mole		2.27 moles	6.80 moles
Mass (g)				

The **same** limiting reagent must be used to calculate the theoretical yield of the other product. Complete the steps to find the theoretical yield of H_2O **based on the reactant C_3H_8**. The product H_2O must show a positive change; addition completes this part of the column.

$$\text{moles of } H_2O = \underset{(1)}{(4.53 \text{ mole } C_3H_8)} \times \underset{(2)}{\frac{6H_2O}{4C_3H_8}} \underset{(3)}{=} \mathbf{6.80} \text{ mole } H_2O \text{ produced}$$

The 4.17 mole theoretical yield of C_6H_{10} calculated based on O_2 is not used in the problem; O_2 **is the excess reagent**. To determine the **actual** number of moles of O_2 consumed, the calculation must be based on the complete consumption of the C_3H_8 limiting reagent. This calculation is identical to several of the previous examples: calculate the number of moles of O_2 required to **exactly** react with the 4.53 moles of limiting C_3H_8.

$$O_2 \text{ consumed} = \underset{(1)}{(4.53 \text{ mole } C_3H_8)} \times \underset{(2)}{\frac{3O_2}{4C_3H_8}} \underset{(3)}{=} \mathbf{3.40} \text{ mole } O_2 \text{ consumed}$$

Note that the result of this calculation does **not** show the amount of oxygen leftover, but the moles consumed; subtraction is required to complete this part of the column. Since oxygen is consumed as a reactant, the mole change must be minus 3.40 moles.

The final **row** in the table (step **4**) is completed by:

$$\text{mass (g)} = \text{(moles)} \times \text{(MM in g/mole)}$$

Reaction Data	$4\ C_3H_8+$	$3\ O_2 \longrightarrow$	$2\ C_6H_{10}+$	$6\ H_2O$
Starting moles	4.53 moles	6.25 moles	0 mole	0 mole
Mole (Δ) change	−4.53 moles	−3.40 moles	+2.27 moles	+6.80 moles
Finishing moles	0 mole	2.85 moles	2.27 moles	6.80 moles
Mass (g)	0 gram	91.2 grams	187 grams	123 grams

As a final check on the accuracy of the complete calculations in this example, all masses of the compounds, after the reaction is completed, can be added and compared to the starting masses given for the reactants. The law of conservation of mass requires that they match (approximately, depending on rounding results) to within the accepted number of significant figures:

$$\text{Starting mass} = 200.\text{ g of } C_3H_8 + 200.\text{ g of } O_2 = 400.\text{ g starting mass.}$$

$$\text{Final mass} = 91.2\text{ g } O_2 + 187\text{ g } C_6H_{10} + 123\text{ g } H_2O = 401.2\text{ g}$$

Example: Use the reaction shown to calculate the masses of all products and reactants after the completion of the reaction of 175 grams of NO with 145 grams of O_2; use the molar masses provided.

$$\begin{array}{llllll} \text{MM} = & 30.01\ \text{g/mole} & 32.00\ \text{g/mole} & & 108.02\ \text{g/mole} & 46.01\ \text{g/mole} \\ & 8\,NO \quad + & 5\,O_2 & \longrightarrow & 2\,N_2O_5 \quad + & 4\,NO_2 \end{array}$$

This can be identified as a limiting reagent problem since there are **two** given starting amounts for the **two reactants**. Select **any** product to determine the limiting reagent; in this case, N_2O_5 is selected.

Theoretical yield of N_2O_5 **based on the reactant NO**:

(1) Mole NO = 175 g/30.01 g/mole = 5.83 mole

(2) Ratio in balanced equation is $\frac{2\,N_2O_5}{8\,NO}$

(3) Moles of N_2O_5 that can be formed from the NO

$$= (5.83\text{ mole } N_2O_5) \times \frac{2\,N_2O_5}{8\,NO} = \mathbf{1.46\ mole}\ N_2O_5$$

This mole value is the theoretical yield based on the given amount of N_2O_5. It is the correct theoretical yield only if N_2O_5 is the limiting reagent.

Theoretical yield of N_2O_5 **based on the reactant O_2**: (separate calculation)

(1) Mole O_2 = 145 g/32.00 g/mole = 4.53 mole O_2

(2) Ratio in balanced equation is $\frac{2\,N_2O_5}{5\,O_2}$

(3) Moles of N_2O_5 which can be formed from the O_2

$$= \left(4.53\text{ mole } O_2\right) \times \frac{2\,N_2O_5}{5\,O_2} = \mathbf{1.81\ mole}\ N_2O_5$$

This mole value is the theoretical yield based on the given amount of O_2. It is the correct theoretical yield only if O_2 is the limiting reagent.

The **smaller** product mole value from the two separate calculations is **1.46 mole** of N_2O_5 based on the given amount of NO. Thus, 1.46 mole of N_2O_5 is the correct theoretical yield, and NO is identified as the limiting reagent.

The calculations for the other NO_2 product are based on the **same** limiting reagent:

$$\text{moles } NO_2 = 5.83 \text{ mole NO} \times \frac{4\,NO_2}{8\,NO} = \mathbf{2.92} \text{ mole } NO_2$$

Reaction Data	8 NO +	$5\,O_2 \longrightarrow$	$2\,N_2O_5$+	$4\,NO_2$
Starting moles	5.83 moles	4.53 moles	0 mole	0 mole
Mole (Δ) change	−5.83 moles		+1.46 moles	+2.92 moles
Finishing moles	0 mole		1.46 moles	2.92 moles
Mass (g)				

In this case, the O_2 is the **excess** reagent. Calculate the number of moles of O_2 required to **exactly** react with the 5.83 moles of limiting NO. As a reactant, this value is a minus change of moles when placed in the table.

$$O_2 \text{ consumed} = 5.83 \text{ mole NO} \times \frac{5O_2}{8\,NO} = \mathbf{3.64} \text{ mole } O_2 \text{ consumed}$$

Perform step **(4)** to complete the table: mass **(g)** = (moles) × (MM in g/mole)

Reaction Data	8 NO +	$5\,O_2 \longrightarrow$	$2\,N_2O_5$+	$4\,NO_2$
Starting moles	5.83 moles	4.53 moles	0 mole	0 mole
Mole (Δ) change	−5.83 moles	−3.64 moles	+1.46 mole	+2.92 mole
Finishing moles	0 mole	0.89 moles	1.46 moles	2.92 moles
Mass (g)	0 gram	28.5 grams	158 grams	134 grams

$$\text{Starting mass} = 175 \text{ g of NO} + 145 \text{ g of } O_2 = 320 \text{ g starting mass}$$

$$\text{Final mass} = 28.5 \text{ g } O_2 + 158 \text{ g } N_2O_5 + 134 \text{ g } NO_2 = 320.5 \text{ g}$$

VI PRACTICE PROBLEMS

1. 146 grams of Fe_2O_3 is reacted in the equation shown. Calculate the mass of CO required to exactly react with the 146 grams of Fe_2O_3 and calculate the mass of Fe produced in the reaction of 146 grams of Fe_2O_3.

$$\begin{array}{lllll} MM = 159.6\,g/mol & MM = 28.01\,g/mol & & MM = 55.85\,g/mol & \\ Fe_2O_3 & + \; 3CO & \longrightarrow & 2Fe \; + & 3CO_2 \end{array}$$

2. 2.43 grams of Al is reacted in the equation shown. Calculate the mass of H_2SO_4 required to exactly react with the 2.43 grams of Al and calculate the mass of H_2 produced in the reaction of 2.43 grams of Al.

$$\begin{array}{llllll} MM = 27.0\,g/mol & MM = 98.09\,g/mol & & & & MM = 2.02\,g/mol \\ 2Al \; + & 3H_2SO_4 & \longrightarrow & Al_2(SO_4)_3 & + & 3H_2 \end{array}$$

3. The following reaction shows the synthesis of the explosive trinitrotoluene (TNT):

$$\begin{array}{lllll} MM = 92.15\,g/mol & MM = 63.02\,g/mol & & MM = 227.15\,g/mol & \\ C_7H_8 \; + & 3HNO_3 & \longrightarrow & C_7H_5N_3O_6\,(TNT) & + \; 3H_2O \end{array}$$

a) How many grams of HNO_3 are required to exactly react with 454 grams of C_7H_8?
b) How many grams of TNT (theoretical yield) can be formed from 829 grams of C_7H_8?
c) How many grams of C_7H_8 are required to form 2.00 kilograms of TNT?

4. The following reaction shows a possible synthesis of aspirin:

$$\begin{array}{lllll} MM = 138.1\,g/mol & MM = 102.1\,g/mol & & MM = 180.2\,g/mol & \\ 2C_7H_6O_3 \; + & C_4H_6O_3 & \longrightarrow & 2C_9H_8O_4\,(aspirin) & + \; H_2O \end{array}$$

a) The reaction has a percent yield of 87.0%. A synthetic chemist requires an actual yield of 75.0 grams of aspirin. What **theoretical** yield of aspirin is required if the **actual** yield must be 75.0 grams?
b) The chemist now prepares to perform the reaction. Based on the theoretical yield calculated in part (a), perform separate calculations to determine the required masses of $C_7H_6O_3$ and $C_4H_6O_3$ that must be reacted to produce the theoretical yield from part (a).
c) 200.0 grams of $C_7H_6O_3$ and 100.0 grams of $C_4H_6O_3$ are reacted together. Calculate the correct theoretical yield in grams of aspirin. This is a limiting reagent problem.

5. Pyromelletic tetracarboxylic acid (PMTA) is a compound used to synthesize high strength materials used in aerospace and marine applications. It can be synthesized through the reaction shown. Calculate the masses of all products and reactants after the completion of the reaction of 100.0 grams of $C_{10}H_{14}$ with 125.0 grams of O_2.

$$\begin{array}{lllll} MM = 134.2\,g/mole & 32.00\,g/mole & & 254.2\,g/mole & 18.02\,g/mole \\ C_{10}H_{14} & + \; 6O_2 & \longrightarrow & C_{10}H_6O_8\,(PMTA) \; + & 4H_2O \end{array}$$

6. Using the equation for the synthesis of TNT shown in problem #4, calculate the masses of all reactants and products after completion of the reaction between 1000 grams of C_7H_8 and 1800 grams of HNO_3. This is a limiting reagent problem; be sure to calculate the mass of the excess reagent, which is leftover after the reaction is complete. Use a reaction table to organize the data.

VII ANSWERS TO PRACTICE PROBLEMS

1. Calculate the mass of CO required to exactly react with the 146 grams of Fe_2O_3 and calculate the mass of Fe produced in the reaction of 146 grams of Fe_2O_3.

Fe_2O_3 + **3CO** $\longrightarrow$ **2Fe** + **$3CO_2$**

mass = 146 g　　mass = 79.6 g　　mass = **102 g**

↓　　↑　　↑

(1)　　(4)

moles Fe_2O_3
= 146 g/159.6 g/mole
= 0.915 mole

mass **CO_2**
= 2.75 mole × 28.01 g/mole
= **76.9 grams**

↑

(4)

mass **Fe**
= 1.83 mol × 55.85 g/mol
= **102 grams**

↓　　↑

↓ 0.915 mole $\longrightarrow$ (2) $\times \frac{3CO_2}{1\,Fe_2O_3}$ (3) $\longrightarrow$ = 2.75 mole CO ↑

↑

↓ 0.915 mole $\longrightarrow$ (2) $\times \frac{2\,Fe}{1\,Fe_2O_3}$ (3) $\longrightarrow$ = 1.83 mole Fe ↑

2. Calculate the mass of H_2SO_4 required to exactly react with the 2.43 grams of Al and calculate the mass of H_2 produced in the reaction of 2.43 grams of Al.

2 Al + **3 H_2SO_4** $\longrightarrow$ **$Al_2(SO_4)_3$** + **3 H_2**

mass = 2.43 g　　mass = **13.2 g**　　mass = **0.273 g**

↓　　↑　　↑

(1)　　(4)　　(4)

moles Al = 2.43 g / 27.0 g/mole
= 0.0900 mole

mass H_2SO_4
= 0.135 mole × 98.09 g/mole
= **13.2 grams**

mass H2
= 0.135 mol × 2.02 g/mol
= **0.273 grams**

↓ 0.0900 mole → (2) $\times \frac{3\,H_2SO_3}{2\,Al}$ (3) → = 0.135 mole H_2SO_4 ↑

↓ 0.0900 mole $\longrightarrow$ (2) $\times \frac{3\,H_2}{2\,Al}$ (3) $\longrightarrow$ = 0.135 mole H_2 ↑

3. a) How many grams of HNO_3 are required to exactly react with 454 grams of C_7H_8?

$$C_7H_8 + 3HNO_3 \longrightarrow C_7H_5N_3O_6 + 3H_2O$$

mass = 454 g ↓   mass = **933 g** ↑

(1)

moles C_7H_8

= 454 g/92.15 g/mole

= **4.93 mole**

(4)

mass HNO_3

= 14.8 mole × 63.02 g/mole

= **933 grams**

(2) $\frac{3HNO_3}{1C_7H_8}$ (3)

↓ 4.93 mole ⟶ × $\frac{3HNO_3}{1C_7H_8}$ ⟶ = 14.8 mole HNO_3 ↑

b) How many grams of TNT (theoretical yield) can be formed from 829 grams of C_7H_8?

$$C_7H_8 + 3HNO_3 \longrightarrow C_7H_5N_3O_6 + 3H_2O$$

mass = 829 g ↓   mass = **2044 g** ↑

(1)

moles C_7H_8

= 829 g / 92.15 g/mole

= **9.00 mole**

(4)

mass $C_7H_5N_3O_6$

= 9.00 mole × 227.15 g/mole

= **2044 grams**

(2) $\frac{1C_7H_5N_3O_6}{1C_7H_8}$ (3)

↓ 9.00 mole ⟶ × $\frac{1C_7H_5N_3O_6}{1C_7H_8}$ ⟶ = 9.00 mole $C_7H_5N_3O_6$ ↑

c) How many grams of C_7H_8 are required to form 2.00 kilograms of TNT?

C_7H_8 + $3HNO_3$ ⟶ $C_7H_5N_3O_6$ + $3H_2O$

mass = **811** g   mass = **2000** g

↑ (4)   ↓ (1)

mass C_7H_8
= 8.805 mole × 92.15 g/mol
= **811 grams**

moles $C_7H_5N_3O_6$
= 2000 g/227.15 g/mole
= **8.805 moles**

(3) $1C_7H_8$ (2)

↑ 8.805 mole C_7H_8 = ⟵ $1C_7H_5N_3O_6$ × ⟵ 8.805 mole $C_7H_5N_3O_6$ ↓

4. a) The reaction has a percent yield of 87.0%. To calculate the **theoretical** yield of aspirin required to produce an **actual** yield of 75.0 grams:

% yield = 87.0%; **fractional yield = 0.870**

fractional yield = (actual yield)/(theoretical yield [T.Y.])

0.870 = (75.0 grams)/(T.Y.); solve for theoretical yield:

T.Y. = (75.0 grams)/0.87 = **86.2 grams**

b) The required masses of $C_7H_6O_3$ and $C_4H_6O_3$ that must be reacted to produce the theoretical yield from part (a):

$2C_7H_6O_3$ + $C_4H_6O_6$ ⟶ $2C_9H_8O_4$ + H_2O

mass = **66.1** g   mass = **24.4** g   mass = 86.2 g

↑ (4)   ↑ (4)   ↓ (1)

mass $C_4H_6O_3$
= 0.2392 mole × 102.1 g/mol
= **24.4 grams**

moles $C_9H_8O_4$
= 86.2 g/180.2 g/mol
= **0.4783 moles**

(3) $1C_4H_6O_3$ (2)

↑ 0.2392 mol $C_4H_6O_3$ = ⟵ $2C_9H_8O_4$ × ⟵ 0.4783 mole $C_9H_8O_4$ ↓

mass $C_7H_6O_3$
= 0.4783 mol × 138.1 g/mol
= **66.1 g**

(3) $2C_7H_6O_3$ (2)

↑ 0.4783 mole $C_7H_6O_3$ = ⟵ $2C_9H_8O_4$ × ⟵ 0.4783 mole $C_9H_8O_4$ ↓

c) 200.0 grams of $C_7H_6O_3$ and 100.0 grams of $C_4H_6O_3$ are reacted together; the correct theoretical yield in grams of aspirin.

To find the theoretical yield of aspirin **based on the reactant $C_7H_6O_3$**:

(1) Mole $C_7H_6O_3$ = 200.0 g/138.1 g/mole = 1.448 mole

(2) Ratio in balanced equation is $\frac{2\,C_9H_8O_3\ \text{(aspirin)}}{2\,C_7H_6O_3}$

(3) Moles of aspirin which can be formed from the $C_7H_6O_3$

$$= (1.448\text{ mole } C_7H_6O_3) \times \frac{2\,C_9H_8O_3}{2\,C_7H_6O_3} = \mathbf{1.448\ mole\ C_9H_8O_3(aspirin)}$$

This mole value is the theoretical yield based on the given amount of $C_7H_6O_3$. It is the correct theoretical yield only if $C_7H_6O_3$ is the limiting reagent.

To find the theoretical yield of aspirin **based on the reactant $C_4H_6O_3$**:

(1) Mole $C_4H_6O_3$ = 100.0 g/102.1 g/mole = 0.9794 mole

(2) Ratio in balanced equation is $\frac{2\,C_9H_8O_3\ \text{(aspirin)}}{1\,C_4H_6O_3}$

(3) Moles of aspirin which can be formed from the $C_4H_6O_3$

$$= (0.9794\text{ mole } C_4H_6O_3) \times \frac{2\,C_9H_8O_3}{1\,C_4H_6O_3} = \mathbf{1.959\ mole\ C_9H_8O_3(aspirin)}$$

This mole value is the theoretical yield based on the given amount of $C_4H_6O_3$. It is the correct theoretical yield only if $C_4H_6O_3$ is the limiting reagent.

Completion of 4c: The **smaller** product mole value from the two separate calculations is **1.448 mole** of $C_9H_8O_3$ (aspirin) based on the given amount of $C_7H_6O_3$. 1.448 mole of $C_9H_8O_3$ (aspirin) is the true correct theoretical yield; **$C_7H_6O_3$** is the **limiting reagent**. $C_4H_6O_3$ is the **excess reagent**; some $C_4H_6O_3$ will be leftover after the reaction has consumed all of the limiting reagent.

(4) Theoretical yield (g) of **$C_9H_8O_3$** = (1.448 mole) × (180.2 g/mole) = **260.9 g**

5. Calculate the masses of all products and reactants after the completion of the reaction of 100.0 grams of $C_{10}H_{14}$ with 125.0 grams of O_2.

MM = 134.2 g/mole		32.00 g/mole		254.2 g/mole		18.02 g/mole
$C_{10}H_{14}$	+	$6O_2$	$\longrightarrow$	$C_{10}H_6O_8$ (PMTA)	+	$4H_2O$

Theoretical yield of PMTA based on the reactant $C_{10}H_{14}$:

(1) Mole $C_{10}H_{14}$ = 100.0 g/134.2 g/mole = 0.7452 mole

(2) Ratio in balanced equation is $\frac{1\,\text{PMTA}}{1\,C_{10}H_{14}}$

(3) Moles of PMTA which can be formed from the $C_{10}H_{14}$

$$= (0.7452\text{ mole } C_{10}H_{14}) \times \frac{1\,\text{PMTA}}{1\,C_{10}H_{14}} = \mathbf{0.7452\ mole}\ \text{PMTA}$$

Theoretical yield of PMTA based on the reactant O_2:

(1) Mole O_2 = 125.0 g/32.00 g/mole = 3.906 mole O_2

(2) Ratio in balanced equation is $\frac{1PMTA}{6O_2}$

(3) Moles of PMTA which can be formed from the O_2

$$= (3.906 \text{ mole } O_2) \times \frac{1PMTA}{6O_2} = \mathbf{0.6510} \text{ mole PMTA}$$

The smaller product mole value is 0.6510 mole of PMTA based on the given amount of O_2. This is the correct theoretical yield; O_2 is the limiting reagent.

Complete the mole calculations by determining the moles of the product H_2O formed based on the **same** limiting reagent:

$$\text{moles } H_2O = 3.906 \text{ mole } O_2 \times \frac{4H_2O}{6O_2} = \mathbf{2.604} \text{ mole } H_2O$$

$$C_{10}H_{14} \textbf{ consumed} = 3.906\ O_2 \times \frac{1C_{10}H_{14}}{6O_2} = \mathbf{0.6510} \text{ mole } C_{10}H_{14} \text{ consumed}$$

The following table displays all the required values; all masses in grams follow step **(4)**: mass **(g)** = (moles) × (MM in g/mole)

Reaction Data	$C_{10}H_{14}$+	$6\ O_2$ ⟶	$C_{10}H_6O_8$+ (PMTA)	$4\ H_2O$
Starting moles	0.7452 mole	3.906 moles	0 mole	0 mole
Mole (Δ) change	−0.6510 mole	−3.906 moles	+0.6510 moles	+2.604 mole
Finishing moles	0.0942 mole	0 mole	0.6510 mole	2.604 moles
Mass (g)	12.6 g	0 g	165.5 g	46.92 g

6. The **masses of all reactants and products** after completion of the reaction between 1000 grams of C_7H_8 and 1800 grams of HNO_3.

Reaction Data	C_7H8 +	$3\ HNO_3$ ⟶	$C_7H_5N_3O_6$+(TNT)	$3\ H_2O$
Starting moles	10.86 moles	28.56 moles	0 mole	0 mole
Mole (Δ) change	−9.521 moles	−28.56 moles	+9.521 moles	+28.56 moles
Finishing moles	1.34 mole	0 mole	9.521 moles	28.56 moles
Mass (g)	123 g	0 g	2162 g	515 g

Theoretical yield of TNT based on the reactant $\mathbf{C_7H_8}$:

$$= (10.86 \text{ mole } C_7H_8) \times \frac{1C_7H_5N_3O_6}{1C_7H_8} = \mathbf{10.86} \text{ mole } C_7H_5N_3O_6\text{(TNT)}$$

Theoretical yield of TNT based on the reactant $\mathbf{HNO_3}$:

$$= (28.56 \text{ mole } HNO_3) \times \frac{1C_7H_5N_3O_6}{3HNO_3} = \mathbf{9.521} \text{ mole } C_7H_5N_3O_6(\text{TNT}) \ (\textbf{limiting})$$

Moles of the product H_2O formed based on the limiting reagent:

$$= (28.56 \text{ mole } HNO_3) \times \frac{3H_2O}{3HNO_3} = \mathbf{28.56} \text{ mole } H_2O \text{ formed}$$

The **actual** number of moles of C_7H_8 consumed

$$= (28.56 \text{ mole } HNO_3 \text{ consumed}) \times \frac{1C_7H_8}{3HNO_3} = \mathbf{9.521} \text{ mole } C_7H_8 \text{ consumed}$$

9 Precipitation and Acid/Base Aqueous Reactions

Concepts and Methods to Design Complete Balanced Equations

I GENERAL CONCEPTS

The description of a reaction as precipitation or acid/base involves explaining how electron exchange or sharing alters during conversion of reactants to products, that is, how ionic or covalent bonds are changed during a reaction. Precipitation reactions and most acid/base reactions occur in solution with water as the solvent; these are termed **aqueous solutions**.

Precipitation reaction: An exchange of ions and ionic bonding in an aqueous solution to produce new ionic compounds such that one of the new compounds is no longer soluble in water.

Acid/base reaction: Process by which an H atom usually in the form of an $H^+_{(aq)}$ ion (i.e., in water) is transferred between two compounds.

II SOLUBILITY EQUATIONS AND AQUEOUS SOLUTION FORMATS

Compounds dissolve at least to some measurable amount in a particular solvent due to attractive forces between the compound being dissolved (solute) and the compound doing the dissolving (solvent). These attractive forces are termed intermolecular (see Chapter 21) and ion-dipole forces and are combined with the entropy of solvation.

For water as a solvent, the form of a compound in aqueous solution affects the reaction and the method of writing an equation. Ionic and covalent compounds generally behave differently when dissolved in water.

Covalent compounds often contain polar covalent bonds; **if** they dissolve in water this is due to attractive forces between the polar ends of the neutral covalent molecule and the polar ends of water molecules. Most solution applicable covalent compounds come in the form of molecules; solution from bulk matter occurs by separation of individual molecules, surrounded by water molecules, into the water solvent.

Covalent compounds generally remain as neutral intact molecules during solvation; individual molecular covalent bonds are not broken in the solution forming process. Important exceptions to this, however, are certain covalent strong acids described later.

Ionic compounds, **if soluble in water**, dissolve through the attractive forces between polar water molecules and the positive or negative charges on the ions.

Ionic compounds do not come as molecules but as large 3-dimensional arrays of ions. Thus, in contrast to most covalent compounds, ***soluble*** **ionic compounds must break up into individual ions** during solvation; individual **ions** are surrounded by water molecules during the solution process.

Solution Formation Equations

A process of solution can be described by an equation: formation of an aqueous solution shows the change from the pure compound, solid (s), liquid (l), or gas (g) to the aqueous solution form denoted by the **(aq)** subscript. A ***formula* format** for the equation shows a complete intact formula for the compound being dissolved.

Formula Format

$$\text{(covalent/non-acid) } CH_3OH_{(l)} \xrightarrow{H_2O \text{ solvent}} CH_3OH_{(aq)} \text{ (methanol)}$$

$$\text{(covalent/non-acid) } CH_3CHO_{(l)} \xrightarrow{H_2O \text{ solvent}} CH_3CHO_{(aq)} \text{ (ethanal)}$$

$$\text{(ionic/soluble) } NaCl_{(s)} \xrightarrow{H_2O \text{ solvent}} NaCl_{(aq)}$$

$$\text{(ionic/soluble) } Fe(NO_3)_{3(s)} \xrightarrow{H_2O \text{ solvent}} Fe(NO_3)_{3(aq)}$$

The ***ionic* format** requires that soluble **ionic** compounds be shown as individual **separate ions** in aqueous solution, depicting their actual form in water. The format requires writing **all** ions separately, showing the **charge** and the designation of **(aq)**.

$$KBr_{(s)} \xrightarrow{H_2O \text{ solvent}} K^+_{(aq)} + Br^-_{(aq)}$$

If the compound contains more than one of the same ion type, these ions are also separated as individual ions in solution. To indicate this requirement, a **coefficient** must be used to show the number of each dissolved ion type in a balanced equation:

$$CaCl_{2\,(s)} \xrightarrow{H_2O \text{ solvent}} Ca^{+2}_{(aq)} + 2\ Cl^-_{(aq)}$$

$$\textbf{It is \textit{not} correct to write: } CaCl_{2\,(s)} \xrightarrow{H_2O \text{ solvent}} Ca^{+2}_{(aq)} + Cl_2^-{}_{(aq)}$$

This would indicate that the two Cl ions are attached in solution, which is ***not*** the case.

The atoms that comprise a polyatomic ion are covalently bonded; the formula for each **specific** polyatomic ion is kept together when indicating formation of an aqueous solution. For the **ionic format**, soluble ionic compounds, which contain polyatomic ions, show all ions separately as required:

$$(NH_4)_2SO_{4(s)} \xrightarrow{H_2O \text{ solvent}} 2\ NH_4^+{}_{(aq)} + SO_4^{-2}{}_{(aq)}$$

It is critical to note that the ionic format requires ***only* ionic** compounds (and some strong covalent acids) to be written as separate aqueous ions. Covalent/non-acid molecules are ***not*** composed of ions and are, therefore, not shown as separated molecular fragments.

Ionic Format

$$\text{(covalent/non-acid)}\quad CH_3OH_{(l)} \xrightarrow{H_2O\text{ solvent}} CH_3OH_{(aq)}\quad \text{(methanol)}$$

$$\text{(covalent/non-acid)}\quad CH_3CHO_{(l)} \xrightarrow{H_2O\text{ solvent}} CH_3CHO_{(aq)}\quad \text{(ethanal)}$$

$$\text{(ionic/soluble)}\quad NaCl_{(s)} \xrightarrow{H_2O\text{ solvent}} Na^{+}_{(aq)} + Cl^{-}_{(aq)}$$

$$\text{(ionic/soluble)}\quad Fe(NO_3)_{3\,(s)} \xrightarrow{H_2O\text{ solvent}} Fe^{+3}_{(aq)} + 3\,NO_3^{-}{}_{(aq)}$$

Equation Formats for Complete Aqueous Solution Reactions

A complete reaction in aqueous solution can be shown in three different ways: a **formula** equation, a **total ionic** equation, or a **net ionic** equation.

A **formula equation** shows all compounds as intact, complete neutral formulas. States of matter and aqueous solutions are indicated by the general equation subscripts:

Formula Format Equation

$$Fe(NO_3)_{3(aq)} + 3\,NaOH_{(aq)} \longrightarrow Fe(OH)_{3\,(S)} + 3\,NaNO_{3(aq)}$$

$$CH_3OH_{(aq)} + NaCl_{(aq)} \longrightarrow CH_3Cl_{(aq)} + NaOH_{(aq)}$$

A **total ionic** equation shows all aqueous soluble compounds in the form they take in an aqueous solution; i.e., the product side of the solution formation equations. This means that **all soluble ionic compounds (and certain covalent acids) must be written in the separated, individual ion form** described previously. No changes are necessary for non-acid covalent compounds.

Compounds that are listed as solid (or liquid) and not aqueous are by definition **not** soluble and are not written as separate ions. A **balanced** equation requires that ion coefficients are calculated as:

coefficient of each ion = (coefficient of the original compound) × (subscript of each ion in formula)

Total Ionic Format Equation

Example: $Fe(NO_3)_{3(aq)} + 3\,NaOH_{(aq)} \longrightarrow Fe(OH)_{3(s)} + 3\,NaNO_{3(aq)}$

$Fe(NO_3)_{3\,(aq)}$ = aqueous (soluble) ionic compound: write as separate ions

$$\text{Coefficient of } Fe^{+3}_{(aq)} = 1\,Fe(NO_3)_3 \times \frac{1\,Fe^{+3}_{(aq)}}{1\,Fe(NO_3)_3} = \mathbf{1\,Fe^{+3}_{(aq)}}$$

$$\text{Coefficient of } NO_3^{-}{}_{(aq)} = 1\,Fe(NO_3)_3 \times \frac{3\,NO_3^{-}{}_{(aq)}}{1\,Fe(NO_3)_3} = \mathbf{3\,NO_3^{-}{}_{(aq)}}$$

$NaOH_{(aq)}$ = aqueous (soluble) ionic compound: write as separate ions

$$\text{Coefficient of } Na^+_{(aq)} = 3\text{ NaOH} \times \frac{1\text{ Na}^+_{(aq)}}{1\text{ NaOH}} = \mathbf{3\ Na^+_{(aq)}}$$

$$\text{Coefficient of } OH^-_{(aq)} = 3\text{ NaOH} \times \frac{1\text{ OH}^-_{(aq)}}{1\text{ NaOH}} = \mathbf{3\ OH^-_{(aq)}}$$

$Fe(OH)_{3(s)}$ = shown as a ***non***-soluble **solid**: write as a complete formula
$NaNO_{3(aq)}$ = aqueous (soluble) ionic compound: write as separate ions
Coefficient of $\mathbf{Na^+_{(aq)}}$ **= 3**; Coefficient of $\mathbf{NO_3^-{}_{(aq)}}$ **= 3**

$$Fe^{+3}_{(aq)} + 3\ NO_3^-{}_{(aq)} + 3\ Na^+_{(aq)} + 3\ OH^-_{(aq)} \longrightarrow Fe(OH)_{3(s)} + 3\ Na^+_{(aq)} + 3\ NO_3^-{}_{(aq)}$$

Example: $CH_3OH_{(aq)} + NaCl_{(aq)} \longrightarrow CH_3Cl_{(aq)} + NaOH_{(aq)}$

$CH_3OH_{(aq)}$ = aqueous covalent compound: write as intact molecule
$NaCl_{(aq)}$ = aqueous (soluble) ionic compound: write as separate ions
$CH_3Cl_{(aq)}$ = aqueous covalent compound: write as intact molecule
$NaOH_{(aq)}$ = aqueous (soluble) ionic compound: write as separate ions

NET IONIC EQUATION

$$CH_3OH_{(aq)} + Na^+_{(aq)} + Cl^-_{(aq)} \longrightarrow CH_3Cl_{(aq)} + Na^+_{(aq)} + OH^-_{(aq)}$$

A ***net* ionic** equation is derived from a total ionic equation by canceling out all **exactly identical** species (usually ions) on each side of the equation. The exactly identical species can be termed **spectator ions** because they remain completely unchanged during the reaction. The effect is to focus the ionic equation down to the species that actually undergo reaction.

Example:

$$Fe^{+3}_{(aq)} + \mathbf{3\ NO_3^-{}_{(aq)}} + \mathbf{3\ Na^+_{(aq)}} + 3\ OH^-_{(aq)} \longrightarrow Fe(OH)_{3(s)} + \mathbf{3\ Na^+_{(aq)}} + \mathbf{3\ NO_3^-{}_{(aq)}}$$

$\mathbf{3\ NO_3^-{}_{(aq)}}$ is identical on both sides of the equation and is cancelled out.
$\mathbf{3\ Na^+_{(aq)}}$ is identical on both sides of the equation and is cancelled out.
Note that $Fe^{+3}_{(aq)}$, as an aqueous ion is ***not*** identical to Fe in the solid compound as $Fe(OH)_{3\ (s)}$;
$3\ OH^-_{(aq)}$ as aqueous ions are ***not*** identical to OH in the solid compound as $Fe(OH)_{3\ (s)}$.

net ionic equation: $Fe^{+3}_{(aq)} + 3\ OH^-_{(aq)} \longrightarrow Fe(OH)_{3\ (s)}$

Example:

$$CH_3OH_{(aq)} + \mathbf{Na^+_{(aq)}} + Cl^-_{(aq)} \longrightarrow CH_3Cl_{(aq)} + \mathbf{Na^+_{(aq)}} + OH^-_{(aq)}$$

$\mathbf{Na^+_{(aq)}}$ is identical on both sides of the equation and is cancelled out.
No other set of molecules or ions are exactly identical on both sides of the equation.
Net ionic equation: $CH_3OH_{(aq)} + Cl^-_{(aq)} \longrightarrow CH_3Cl_{(aq)} + OH^-_{(aq)}$

III PRECIPITATION REACTIONS

One of the previous examples is a precipitation reaction:

$$Fe(NO_3)_{3(aq)} + 3\ NaOH_{(aq)} \longrightarrow \mathbf{Fe(OH)_{3(s)}} + 3\ NaNO_{3(aq)}$$

The general definition of a **precipitation reaction** requires that:

1. The reactants must start as soluble aqueous ions in solution.
2. The mixing of the reactant aqueous solutions produces a set of possible ion combinations, which form new potential ionic compounds.
3. One of the new ionic compounds formed is no longer soluble in water. It becomes an insoluble solid, which can be removed from the mixed aqueous solution. The insoluble ionic compound, which separates from the solution, is called the **precipitate**.

If all possible new ionic compounds remain soluble, the product result is the same solution of soluble ions as formed by the reactants: no reaction can be detected.

The mixing of aqueous solutions of two different **binary** ionic compounds or ionic compounds with polyatomic ions, will always produce the possibility for forming two new ionic compounds by a direct exchange of ions. Considering precipitation as the only reaction, **only two new ionic combinations** can form.

Each solution of a single ionic compound contains only one type of positive ion and one type of negative ion; the total combination for the two solutions is **two** positive ions plus **two** negative ions. Two positive ions cannot combine to form a (neutral) compound: two negative ions cannot combine to form a (neutral) compound. **An exchange of ions between two ionic compounds will produce only two new possible ionic compounds.**

The two new possible ionic compounds may be soluble or insoluble in water. If both potentially new ionic compounds are soluble, the ion components of each compound will remain separated in aqueous solution: no detectable change in the solution is termed **no reaction**. If one of the new potential compounds is **not** soluble in water, the component ions will join together to form the insoluble solid **precipitate**.

Example:

Formula: $AgNO_{3(aq)} + NaCl_{(aq)} \longrightarrow \mathbf{AgCl_{(s)}} + NaNO_{3(aq)}$

Total ionic: $Ag^+_{(aq)} + NO_3^-{}_{(aq)} + Na^+_{(aq)} + Cl^-_{(aq)} \longrightarrow \mathbf{AgCl_{(s)}} + Na^+_{(aq)} + NO_3^-{}_{(aq)}$

Net ionic: $Ag^+_{(aq)} + Cl^-_{(aq)} \longrightarrow \mathbf{AgCl_{(s)}}$

$AgNO_{3(aq)}$ and $NaCl_{(aq)}$ are stated as ionic aqueous soluble reactants: the total ionic equation depicts the aqueous ions. The products show the precipitate, $\mathbf{AgCl_{(s)}}$, as solid, non-soluble and $NaNO_{3(aq)}$ as soluble aqueous ions, emphasized in the total ionic equation. The net ionic equation displays the specific precipitate forming process; spectator ions that remain unchanged in the complete reaction are not seen.

DETERMINING IONIC COMPOUND SOLUBILITY

Writing complete balanced equations for precipitation reactions requires determination of the aqueous solubility of potential reactant and product ionic compounds. A list of solubility rules for ionic compounds can be used. These rules can take many forms; a simple version is shown.

The identification of solubility is not absolute (see Chapters 21 and 23): the most soluble ionic compound will reach a maximum soluble amount and the least soluble will have some minimal solubility. The solubility rules employed will distinguish between "mostly" soluble vs. "mostly" insoluble for the purposes of completing precipitation reaction equations.

General Solubility Guideline Rules

An ionic compound will show general solubility in water if ***either*** the positive ion ***or*** the negative ion (can be both) is listed under the classification of **soluble ions**; certain exceptions to the general rules are then considered.

1. **Soluble Cations** (positive ions): Ionic compounds containing: Li^+, Na^+, K^+, Rb^+, Cs^+, and NH_4^+ with **any** negative ion will be soluble.
2. **Soluble Anions** (negative ions): Ionic compounds containing: NO_3^-, NO_2^-, Cl^-, Br^-, I^- SO_4^{-2}, HSO_4^-, SO_3^{-2}, ClO_4^-, CH_3COO^- with **any** positive ion will be soluble.
3. **Insoluble Exceptions**: Ag^+, Pb^{+2}, and Hg^+ with Cl^-, Br^-, and I^- are **not** soluble even though the negative ions are listed in rule #2.
   $BaSO_4$, Ag_2SO_4, and $PbSO_4$ are **not** soluble even though SO_4^{-2} is listed in rule #2.
4. **Soluble Exceptions**: $Ba(OH)_2$, $Ca(OH)_2$, $Sr(OH)_2$, MgS, CaS, BaS are **soluble** even though neither ion is listed in rule #1 or rule #2.

Examples from the Previous Reactions

$Fe(NO_3)_3$: NO_3^- is listed as a soluble anion; therefore, $Fe(NO_3)_3$ is aqueous/soluble.

NaOH: Na^+ is listed as a soluble cation; OH^- is not listed as a soluble anion. Only one soluble ion in a compound is required to designate aqueous/soluble; NaOH is aqueous/soluble.

$Fe(OH)_3$: Fe^{+3} is not listed as a soluble cation; OH^- is not listed as a soluble anion. Since neither ion in the compound is listed as a soluble, and $Fe(OH)_3$ is not listed as an exception, this compound is not soluble.

$NaNO_3$: NO_3^- is listed as a soluble anion; Na^+ is also listed as a soluble cation. Based on either of these designations $NaNO_3$ is aqueous/soluble.

$AgNO_3$: NO_3^- is listed as a soluble anion; therefore, $AgNO_3$ is aqueous/soluble.

NaCl: Cl^- is listed as a soluble anion; Na^+ is also listed as a soluble cation. Based on either of these designations NaCl is aqueous/soluble.

AgCl: Cl^- is listed as a soluble anion; Ag^+ is not listed as a soluble cation. Although only one soluble ion in a compound is required to designate aqueous/soluble; in this case, AgCl is listed in **rule #3** as an **insoluble exception**.

Predicting Products for Precipitation Reactions and Writing Equations

Analysis of a precipitation reaction problem requires the determination of the correct products formed upon mixing aqueous solutions of two reactant ionic solutions, the identification of the new compound, which produces the insoluble precipitate, and the description of the complete reaction through the three forms of the reaction equation.

Process for Completing and Writing Precipitation Reaction Equations

(1) The problem starts with identification of two aqueous ionic compound reactants. Write correct formulas for the compounds to start the **formula** format equation. Then write the **reactant side** of a **total ionic equation** for these compounds (i.e., write both aqueous ionic compounds in the separated ion form); this simplifies the exchange process for step (2).

Use correct coefficients for each ion based on the starting formulas identified, even though the complete reaction is not balanced at this point. This will simplify the final balance in step (4).

(2) Complete the **unbalanced formula** format equation by directly exchanging the positive and negative ion combinations to form the two new potential ionic compounds; there is only one way the exchange can be performed.

a) Write the **formula format** for each potential ionic **product** in the formula format equation; that is, write the ions together as a neutral compound, not as separated ions.

b) Be certain to **write the correct ionic formula** for the product based on the charges of each component ion and the resulting correct ion ratio. **Do not simply combine the number of ions resulting from the description in step (1).** The equation is not balanced at this point.

(3) Complete the **unbalanced formula** equation by identifying the correct states of matter for the two new potential ionic product compounds formed in step (2). Use the solubility rules to determine which compound, if either, is insoluble in water. The insoluble precipitate will be a solid **(s)** and the soluble compound(s) will be aqueous **(aq)**.

(4) Complete the **balanced formula** format of the reaction equation. The reactants were determined in step (1); the products and states of matter were determined from steps (2) and (3). **Balance the formula equation**; this eliminates the discrepancies between the number of ions listed in the original reactant formulas and the number of ions required for correct product formulas.

(5) Complete the **total ionic** format of the reaction equation. The reactants were written in the ionic format in step (1). Write the solid insoluble precipitate product as the neutral ionic compound; write the aqueous soluble ionic product as separate ions. **Use the balanced formula equation from step (4) to correct all ion coefficients so that the total ionic equation is balanced.**

(6) Write the **net ionic** equation by canceling exactly identical species from both sides of the total ionic equation.

Example: Predict the products and write all three equation formats for the aqueous solution combination of lead (II) nitrate (aqueous) and potassium iodide (aqueous).

(1) $Pb(NO_3)_{2\,(aq)} + KI_{(aq)} \longrightarrow$

(1) $Pb^{+2}{}_{(aq)} + 2\,NO_3^-{}_{(aq)} + K^+{}_{(aq)} + I^-{}_{(aq)} \longrightarrow$

(2) $Pb^{+2}{}_{(aq)} + 2\,NO_3^-{}_{(aq)} + K^+{}_{(aq)} + I^-{}_{(aq)} \longrightarrow PbI_2 + KNO_3$ **(not balanced)**

PbI_2 and KNO_3 are the correct formulas for these ion combinations. Do **not** simply combine the number of ions shown in the partial equation. **One** Pb^{+2} is shown on the reactant side and **one** I^- is shown on the product side; however, PbI is **not** the correct formula. Similarly, $K(NO_3)_2$ is not the correct formula despite **one** K^+ and **two** NO_3^- being shown. The equation, as written at this point, is **not** balanced; the **numbers of each ion in the complete equation have not yet been specified**.

(3) Check for solubility using the solubility table. Although I^- is listed as a soluble anion, the combination of Pb^{+2} and I^- is listed as one of the **insoluble** exceptions in rule #3. PbI_2 is insoluble in water and is the precipitate. Both K^+ and NO_3^- are listed as soluble ions; based on either of these ions, KNO_3 is aqueous soluble.

unbalanced formula equation: $Pb(NO_3)_{2(aq)} + KI_{(aq)} \longrightarrow PbI_{2\,(s)} + KNO_{3(aq)}$

(4) The NO_3^- ions are balanced by a coefficient of (2) for KNO_3
The I^- ions are balanced by a coefficient of (2) for KI

balanced formula equation: $Pb(NO_3)_{2\,(aq)} + \mathbf{2\,KI}_{(aq)} \longrightarrow PbI_{2\,(s)} + \mathbf{2KNO_3}_{(aq)}$

(5) The coefficients needed to balance the formula format of the equation must be included to balance the **total ionic** form of the equation.

$$Pb^{+2}_{(aq)} + 2NO_3^-{}_{(aq)} + 2K^+_{(aq)} + 2I^-_{(aq)} \longrightarrow PbI_{2\,(s)} + 2K^+_{(aq)} + 2NO_3^-{}_{(aq)}$$

(6) **2** $\mathbf{NO_3^-}_{(aq)}$ is identical on both sides of the equation and is cancelled out.
2 $\mathbf{K^+}_{(aq)}$ is identical on both sides of the equation and is cancelled out.

***net* ionic equation:** $Pb^{+2}_{(aq)} + 2I^-_{(aq)} \longrightarrow PbI_{2\,(s)}$

Example: Predict the products and write all three equation formats for the combination of platinum (IV) perchlorate (aqueous) and sodium phosphate (aqueous).

(1) $Pt(ClO_4)_{4\,(aq)} + Na_3PO_{4\,(aq)} \longrightarrow$

(1) $Pt^{+4}_{(aq)} + 4\,ClO_4^-{}_{(aq)} + 3\,Na^+_{(aq)} + PO_4^{-3}{}_{(aq)} \longrightarrow$

(2) $Pt^{+4}_{(aq)} + 4\,ClO_4^-{}_{(aq)} + 3\,Na^+_{(aq)} + PO_4^{-3}{}_{(aq)} \longrightarrow Pt_3(PO_4)_4 + NaClO_4$
(**not** balanced)

$\mathbf{Pt_3(PO_4)_4}$ and $\mathbf{NaClO_4}$ are the correct formulas for these ion combinations. **Three** Pt^{+4} (with a +4) charge must be combined with **four** PO_4^{-3} (with a −3 charge) to produce a neutral formula (12 positive charges and 12 negative charges). The formula $NaClO_4$ is neutral with 1 positive charge and 1 negative charge. The combinations based on the unbalanced equation, $PtPO_4$ and $Na_3(ClO_4)_4$, are not correct.

(3) The negative ion PO_4^{-3} is not listed as a soluble anion, Pt^{+4} is not listed as a soluble cation and the compound $Pt_3(PO_4)_4$ is not listed as an exception. The insoluble precipitate is $Pt_3(PO_4)_4$.
Both Na^+ and ClO_4^- are listed as soluble ions; based on either of these ions, $NaClO_4$ is aqueous soluble.
Unbalanced formula equation:

$$Pt(ClO_4)_{4\,(aq)} + Na_3PO_{4\,(aq)} \longrightarrow Pt_3(PO_4)_{4\,(s)} + NaClO_{4(aq)}$$

(4) Pt is balanced by a coefficient of (3) for $Pt(ClO_4)_4$
The ClO_4^- ions are balanced by a coefficient of (12) for $NaClO_4$
The PO_4^{-3} ions are balanced by a coefficient of (4) for Na_3PO_4
The Na^+ ions are already balanced by the coefficient of (12) for $NaClO_4$
Balanced formula equation:

$$\mathbf{3}\,Pt(ClO_4)_{4(aq)} + \mathbf{4}\,Na_3PO_{4(aq)} \longrightarrow Pt_3(PO_4)_{4(s)} + \mathbf{12}\,NaClO_{4\,(aq)}$$

(5) The balance for the **total ionic** form of the equation is taken directly from the coefficients needed to balance the formula format of the equation.

$$\mathbf{3}\ Pt^{+4}_{(aq)} + \mathbf{12}\ ClO_4{}^{-}_{(aq)} + \mathbf{12}\ Na^{+}_{(aq)} + \mathbf{4}\ PO_4{}^{-3}_{(aq)} \longrightarrow Pt_3(PO_4)_{4\,(s)} + \mathbf{12}Na^{+}_{(aq)} + \mathbf{12}\ ClO_4{}^{-}_{(aq)}$$

(6) **12** $\mathbf{ClO_4{}^{-}_{(aq)}}$ is identical on both sides of the equation and is cancelled out.
12 $\mathbf{Na^{+}_{(aq)}}$ is identical on both sides of the equation and is cancelled out.

***net* ionic equation:** $3\ Pt^{+4}_{(aq)} + \mathbf{4}\ PO_4{}^{-3}_{(aq)} \longrightarrow Pt_3(PO_4)_{4\,(s)}$

Example: Predict the products and write all three equation formats for the combination of vanadium (V) nitrite (aqueous) and lithium chromate (aqueous).

(1) $V(NO_2)_{5\,(aq)} + Li_2CrO_{4\,(aq)} \longrightarrow$

(1) $V^{+5}{}_{(aq)} + 5\ NO_2{}^{-}{}_{(aq)} + 2\ Li^{+}{}_{(aq)} + CrO_4{}^{-2}{}_{(aq)} \longrightarrow$

(2) $V^{+5}{}_{(aq)} + 5\ NO_2{}^{-}{}_{(aq)} + 2\ Li^{+}{}_{(aq)} + CrO_4{}^{-2}{}_{(aq)} \longrightarrow V_2(CrO_4)_5 + LiNO_2$ **(not balanced)**

$\mathbf{V_2(CrO_4)_5}$ and $\mathbf{LiNO_2}$ are the correct formulas for these ion combinations. **Two** V^{+5} (with a +5) charge must be combined with **five** $CrO_4{}^{-2}$ (with a −2 charge) to produce a neutral formula (10 positive charges and 10 negative charges). The formula $LiNO_2$ is neutral with 1 positive charge and 1 negative charge. The combinations based on the unbalanced equation, $VCrO_4$ and $Li_2(NO_2)_5$, are not correct.

(3) The negative ion $CrO_4{}^{-2}$ is not listed as a soluble anion, V^{+5} is not listed as a soluble cation and the compound $V_2(CrO_4)_5$ is not listed as an exception. The insoluble precipitate is $V_2(CrO_4)_5$. Both Li^+ and $NO_2{}^-$ are listed as soluble ions; based on either of these ions, $LiNO_2$ is aqueous soluble. **Unbalanced** formula equation:

$$V(NO_2)_{5(aq)} + Li_2CrO_{4(aq)} \longrightarrow V_2(CrO_4)_{5(s)} + LiNO_{2(aq)}$$

(4) V is balanced by a coefficient of (2) for $V(NO_2)_5$
The $NO_2{}^-$ ions are balanced by a coefficient of (10) for $LiNO_2$
The $CrO_4{}^{-2}$ ions are balanced by a coefficient of (5) for Li_2CrO_4
The Li^+ ions are already balanced by the coefficient of (10) for $LiNO_2$.
Balanced formula equation:

$$\mathbf{2}\ V(NO_2)_{5\,(aq)} + \mathbf{5}\ Li_2CrO_{4\,(aq)} \longrightarrow V_2(CrO_4)_{5\,(s)} + \mathbf{10}\ LiNO_{2\,(aq)}$$

(5) The balance for the **total ionic** form of the equation is taken directly from the coefficients needed to balance the formula format of the equation.

$$\mathbf{2}\ V^{+5}_{(aq)} + \mathbf{10}\ NO_2{}^{-}_{(aq)} + \mathbf{10}\ Li^{+}_{(aq)} + \mathbf{5}\ CrO_4{}^{-2}_{(aq)} \longrightarrow V_2(CrO_4)_{5(S)} + \mathbf{10}\ Li^{+}_{(aq)} + \mathbf{10}\ NO_2{}^{-}_{(aq)}$$

(6) **10** $\mathbf{NO_2{}^{-}_{(aq)}}$ is identical on both sides of the equation and is cancelled out.
10 $\mathbf{Li^{+}_{(aq)}}$ is identical on both sides of the equation and is cancelled out.

net ionic equation: $2\ V^{+5}_{(aq)} + 5\ CrO_4{}^{-2}_{(aq)} \longrightarrow V_2(CrO_4)_{5(S)}$

IV ACID/BASE REACTIONS: GENERAL CONCEPTS

A first discussion of acid/base equations will employ somewhat restricted definitions of acid and base. An **acid** is any molecule that **transfers**, in a chemical reaction, a hydrogen atom as an **H^+ ion** to another atom or molecule called a **base**. Prior to a detailed look at covalent structures for molecules, the notation shown will be used; the pair of dots indicate two electrons: the electron pair representing a covalent bond, or a free (originally unused for bonding) electron pair, usually termed a lone pair.

Notation: H—••Acid. = acid molecule with an H for transfer, which shows the electron pair in the H—Acid covalent bond

: **Acid⁻** or **••Acid⁻** = portion of the acid molecule left over after the H^+ was transferred, showing the remaining pair of electrons

A **base** is any atom or molecule, which picks up (**accepts**) the **H^+ ion** from the acid in a chemical reaction. Acid and base are complementary definitions.

Notation: **:Base** or **••Base** = neutral base with an electron pair

(H—••Base)⁺ = base molecule after the H^+ was accepted by a neutral base; shows the electron pair being used to form a covalent bond

or

:Base⁻ or **••Base⁻** = negative ion base portion of a neutral ionic base with an electron pair

H—••Base = base molecule after the H^+ was accepted by a negative ion base

An **acid/base reaction** is the complete description of the transfer of one or more H^+ ions from the acid to the base. An H^+ ion does not have an independent existence. The reaction involves the exchange of one covalent bond involving the hydrogen on the acid for another covalent bond involving the hydrogen on the base. The hydrogen is transferred without its electron; thus, the species **transferred** is an **H^+ ion.**

The most common isotope of hydrogen is ^1H with 1 proton, 0 neutrons, and 1 electron. An H^+ ion has 1 proton, 0 neutrons, and 0 electrons and is, thus, a proton. Alternatively, this reaction is termed a proton transfer reaction.

General Example: For a neutral starting acid and base, the process can be described by following the roles of the electron pairs shown. (These reactions are most often in aqueous solution, not indicated in the example.)

as H^+

H—••Acid + ••Base ⟶ ••Acid⁻ + (H—••Base)⁺

Viewing the role of the acid molecule:

as H^+

H ⟨ —••Acid + ••Base ⟶ ••Acid⁻ + (H —••Base)⁺

The H—••Acid covalent bond breaks such that both electrons in the original bond remain with the atom(s) of the acid molecule left over after bond breaking. Since hydrogen's bonding electron is left behind, the H is transferred as H^+ to the base. The free pair of electrons on the base are then used to form a new covalent bond with the transferred H^+ ion. The result indicates that the remaining acid molecule acquires an extra negative charge, and the base molecule acquires an extra positive charge.

If the base is a negative ion (anion) derived from a neutral ionic compound, the general equation shows that the H—Base is neutral:

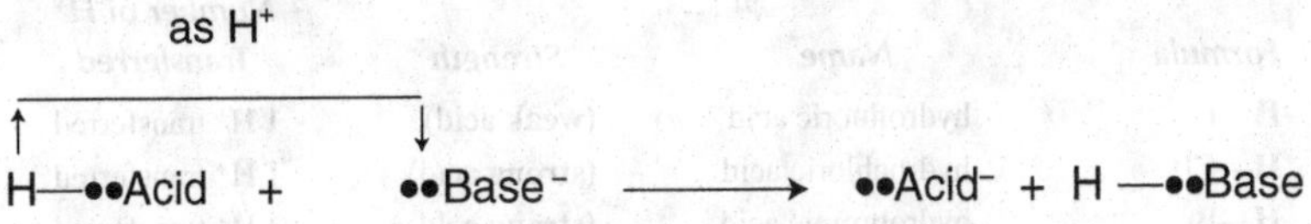

as H^+

H—••Acid + ••Base⁻ ⟶ ••Acid⁻ + H—••Base

The complete neutral ionic base compound can be included in the general equation: (Metal⁺) (Base⁻). The positively charged ion is often a spectator ion with no role in the reaction, but can be shown in a formula format as changing negative ion partners:

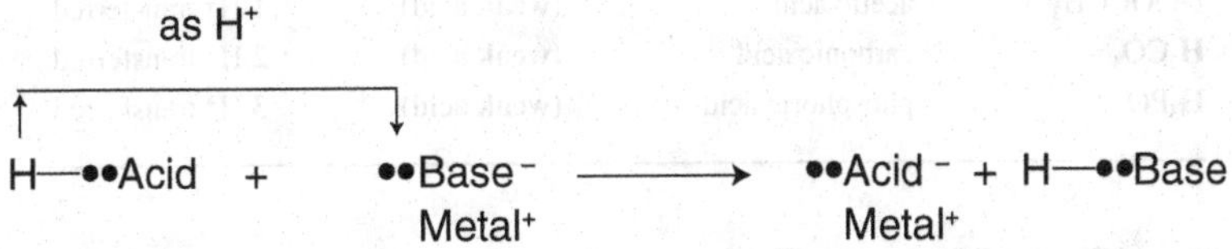

as H^+

H—••Acid + ••Base⁻ Metal⁺ ⟶ ••Acid⁻ Metal⁺ + H—••Base

Properties of Acids

Acids are covalent molecules; common acids show one or more hydrogen atoms covalently bonded to another non-metal atom (X) such that the **hydrogen represents the relatively positive portion (δ+) of a very polar covalent bond.**

$$\overset{\delta+}{H}—\overset{\delta-}{X} \quad \text{and} \quad \overset{\delta+}{H}—\overset{\delta-}{X}—R$$

X = usually O, F, Cl, Br, I; R = other bonded atoms

One group of common acids has hydrogen bonded to one of the Group 7A halogens: H—F, H—Cl, H—Br, H—I. Another major group of common acids has hydrogen(s) covalently bonded to one or more oxygen atoms in polyatomic combinations with structures related to certain oxygen containing polyatomic ions. The resulting acid molecules are **not** ionic, but have a covalent structure based on structures of the polyatomic ions.

Examples:

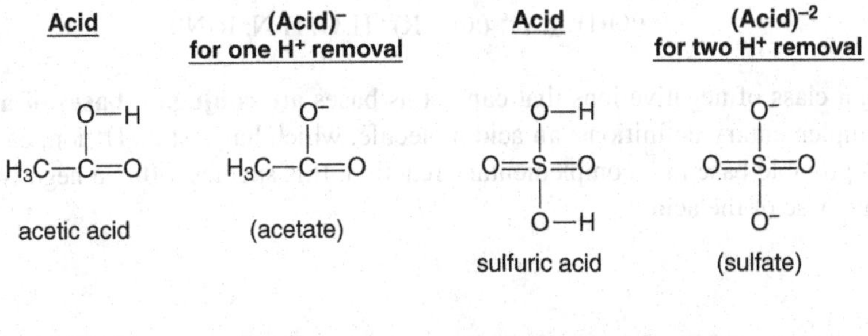

Acid	(Acid)⁻ for one H⁺ removal	Acid	(Acid)⁻² for two H⁺ removal
$H_3C—C(=O)—O—H$ acetic acid	$H_3C—C(=O)—O^-$ (acetate)	$H—O—S(=O)_2—O—H$ sulfuric acid	$^-O—S(=O)_2—O^-$ (sulfate)

Some acids used for reaction examples are shown in the table. The names of the acids are derived from the compound or related polyatomic ion names by changing the endings to "-ic acid." The designation of strong acid vs. weak acid is described in the next section. Some acids can transfer more than one H^+ ion; these are termed **polyprotic** acids. For the purposes of the compounds in this table and the reactions generated, a general rule is that all H's shown in the compound formula can be transferred as H^+ for all acids shown except for acetic acid. Only the one hydrogen shown as being bonded to the oxygen in the full structure indicated for acetic acid can be transferred as an H^+ ion.

Table of Common Acids

Formula	*Name*	*Strength*	*Number of H^+ Transferred*
H—F	hydrofluoric acid	(**weak** acid)	1 H^+ transferred
H—Cl	hydrochloric acid	(**strong** acid)	1 H^+ transferred
H—Br	hydrobromic acid	(**strong** acid)	1 H^+ transferred
H—I	hydroiodic acid	(**strong** acid)	1 H^+ transferred
H_2SO_4	sulfuric acid	(**strong** acid)	2 H^+ transferred
$HClO_4$	perchloric acid	(**strong** acid)	1 H^+ transferred
H_2CrO_4	chromic acid	(**strong** acid)	2 H^+ transferred
HNO_3	nitric acid	(**strong** acid)	1 H^+ transferred
$HOOCCH_3$	acetic acid	(**weak** acid)	1 H^+ transferred
H_2CO_3	carbonic acid	(**weak** acid)	2 H^+ transferred
H_3PO_4	phosphoric acid	(**weak** acid)	3 H^+ transferred

Example of a Specific Acid from the Table:

as H^+

H—••Cl + ••Base⁻ ⟶ ••Cl⁻ + H—••Base

Properties of Bases

A **base** is any atom or molecule that covalently bonds to the **H^+ ion**, which was transferred from the acid during a chemical reaction. The atom of the base that bonds to the H^+ ion must, therefore, have a free pair of electrons available for sharing, commonly, a lone pair.

Common bases are ionic compounds containing negative ions, especially those ions that contain negatively charged oxygen(s). The combining positive ion in these ionic compounds usually has no role in the acid/base reaction itself; the negative ion is referred to as the base portion of the compound. Other bases are neutral compounds containing neutral oxygen or nitrogen; these atoms have a lone electron pair to accept an H^+ ion.

Common Base or Base Portions: R = other bonded atoms

$$(:OH)^- \ (:X)^- \ (:O{-}R)^- \ H_2O: \ H_3N: \ R_3N:$$

An important class of negative ions that can act as bases are **conjugate bases of acids**. Acid and base are complementary definitions: an acid molecule, which has lost its H^+ ion, can often be considered as a possible base in a complementary reaction; this species, often a negative ion, is called the conjugate base of the acid.

as H^+

$$H{-}{\bullet\bullet}Acid + {\bullet\bullet}Base^- \longrightarrow {\bullet\bullet}Acid^- + H{-}{\bullet\bullet}Base$$

acid **conjugate <u>base</u>**

as H^+

$$H{-}{\bullet\bullet}Cl + {\bullet\bullet}Acid^- \longrightarrow {\bullet\bullet}Cl^- + H{-}{\bullet\bullet}Acid$$

Another acid **conjugate <u>base</u> of an acid**

V AQUEOUS ACID/BASE REACTIONS

Acid Ionization Reaction and Acid Strength

Acid molecules react with water both in a process of solvation to form an aqueous solution and in an acid/base reaction with water molecules acting as the base. Recall that the neutral oxygen in water has a lone electron pair to act as a base.

$$H{-}{\bullet\bullet}Acid_{(aq)} + H_2O{\bullet\bullet} \longrightarrow {\bullet\bullet}Acid^-_{(aq)} + (H_2O{\bullet\bullet}{-}H)^+_{(aq)}$$

The protonated water molecule formed, $(H_2O{\bullet\bullet}{-}H)^+_{(aq)}$, is written as H_3O^+. Using the simpler notation:

$$H{-}Acid_{(aq)} + H_2O \longrightarrow (Acid)^-_{(aq)} + H_3O^+_{(aq)}$$

H_3O^+ and $H^+_{(aq)}$ can be considered as identical descriptions of the same species.

An alternative notation is:

$$H{-}Acid_{(aq)} \xrightarrow{H_2O\ solvent} H^+_{(aq)} + (Acid)^-_{(aq)}$$

This reaction is termed an **ionization** reaction because a **covalent** molecule is converted (at least partially) into **ions** by reaction with water. This process is **not** the same as for simple solution of soluble ionic compounds; ions are formed from covalent compounds by a specific acid/base reaction as opposed to simple solution.

The ionization reaction of acids with water as a base is the reference reaction for classification of acid strength. While equilibrium calculations are required for quantification of acid strength (Chapter 26), the analysis in these sections will very generally define an acid as either **weak** or **strong** based on the approximate efficiency of an acid's ionization reaction with water.

A **strong acid** is one that undergoes approximately 100% ionization when dissolved in an aqueous solution; example:

$$H{-}Cl\ (\textbf{strong}\ acid) \xrightarrow{H_2O\ solvent} H^+_{(aq)} + Cl^-_{(aq)} \approx \textbf{100\% reaction}$$

A weak acid is one that ionizes (usually) less than 1% to 5% when dissolved in an aqueous solution; example:

$$CH_3COO\text{—}H\ (\textbf{weak}\ \text{acid}) \xrightarrow{H_2O\ \text{solvent}} H^+_{(aq)} + CH_3COO^-_{(aq)}\ \mathbf{< 0.5\%\ reaction}$$

An alternate aqueous solution definition of an acid is a species that produces $H^+_{(aq)}$ during formation of an aqueous solution.

Aqueous Base Reactions

Many common bases are ionic compounds, which contain the hydroxide (OH^-) polyatomic ion. In these cases, OH^- acts as the base portion of the compound; the **net** ionic equation depicts the formation of water:

$$H^+_{(aq)} + OH^-_{(aq)} \longrightarrow H_2O_{(l)}$$

Base strength can be measured as the concentration of aqueous hydroxide ion, $[OH^-_{(aq)}]$. Examples of strong bases are hydroxide containing ionic compounds, which have high solubility in water and complete separation of dissolved ions, thus, producing high concentrations of hydroxide ion:

$$\text{Metal (OH)} \xrightarrow{H_2O\ \text{solvent}} \mathbf{Metal}^+_{(aq)} + OH^-_{(aq)}\quad \text{(high solubility)}$$

Common strong bases are: NaOH (sodium hydroxide), KOH (potassium hydroxide), $Ba(OH)_2$ (barium hydroxide), and $Ca(OH)_2$. Non-soluble or slightly soluble hydroxide containing bases, such as $Mg(OH)_2$, are considered weak bases; very low solubility in water produces very low concentrations of hydroxide ion.

Most non-hydroxide bases (NH_3 or X^-) to be used in reactions in this chapter are classified as weak. **Aqueous** non-hydroxide bases react with water in a complementary ionization reaction in which water acts as an **acid**. This ionization proceeds to only a small extent and produces a low concentration of OH^-. Note that OH^- is the conjugate base of water as an acid. For an **anion base**:

$$H\text{—}{\bullet\bullet}OH + {\bullet\bullet}Base^-_{(aq)} \longrightarrow {\bullet\bullet}OH^-_{(aq)} + H\text{—}{\bullet\bullet}Base_{(aq)}$$

Neutral NH_3 acts as a base through a free lone electron pair on the nitrogen:

$$H\text{—}{\bullet\bullet}OH + {\bullet\bullet}NH_{3\,(aq)} \longrightarrow {\bullet\bullet}OH^-_{(aq)} + (H\text{—}{\bullet\bullet}NH_3)^+_{(aq)}\ \mathbf{< 0.5\%}$$

The acetate anion, CH_3COO^-, is the **conjugate base** of acetic acid.

$$H\text{—}{\bullet\bullet}OH + {\bullet\bullet}^-OOCCH_{3(aq)} \longrightarrow {\bullet\bullet}OH^-_{(aq)} + H\text{—}{\bullet\bullet}OOCCH_{3(aq)}\ \mathbf{< 0.5\%}$$

An alternate aqueous solution definition of a base is a compound that forms OH^- ions when forming an aqueous solution.

Completing Acid/Base Reactions and Equations

The complete reaction that describes the transfer of an H^+ from the acid compound to the base compound is termed the **acid/base reaction**. The reaction can involve any combination of strong or weak acids or bases. A potential equation can be designed for reaction of any base with any acid, regardless of the degree the forward reaction proceeds. In general, if either the acid or the base is considered strong, the acid/base reaction is assumed to proceed approximately 100% toward formation of the products.

The reaction of a **strong hydroxide** containing base with a **strong** acid is termed a **neutralization** reaction: the resulting products are water plus a neutral ionic compound (termed a salt).

Example: Aqueous hydrochloric acid (strong) reacts with aqueous sodium hydroxide

$$H{-}{\bullet\bullet}Cl_{(aq)} + {\bullet\bullet}OH^-_{(aq)}\ Na^+_{(aq)} \longrightarrow {\bullet\bullet}Cl^-_{(aq)}\ Na^+_{(aq)} + H{-}{\bullet\bullet}OH_{(aq)}$$

The base portion of sodium hydroxide is the hydroxide ion; the sodium ion merely comes along for the ride. However, a more conventional notation for the **formula** equation shows the familiar exchange of partners:

$$H{-}Cl_{(aq)} + NaOH_{(aq)} \longrightarrow NaCl_{(aq)} + H_2O_{(l)}$$

The total ionic equation for acid/base reactions is written following the principles outlined previously. Show all aqueous soluble compounds in the form they take in an aqueous solution: all soluble ionic compounds are written in the separated, individual ion form; non-acid covalent compounds are written as intact molecules. Compounds that are listed as solid (or liquid) and not aqueous are not written as separate ions. The following rules apply to covalent acids:

Write all **strong acids as aqueous ions** based on the approximately 100% ionization reaction of strong acids in water; all **weak acids** are written **as neutral intact covalent aqueous** (aq) molecules based on ionization of less that 1% to 5%.

The total **ionic** equation is:

$$H^+_{(aq)} + Cl^-_{(aq)} + Na^+_{(aq)} + OH^-_{(aq)} \longrightarrow H_2O_{(l)} + Na^+_{(aq)} + Cl^-_{(aq)}$$

In this case, the net ionic equation shows the formation of water; $Cl^-_{(aq)}$ and $Na^+_{(aq)}$ are cancelled from both sides of the total ionic equation:

$$H^+_{(aq)} + OH^-_{(aq)} \longrightarrow H_2O_{(l)}$$

Examination of Reaction Requirements

Example: Aqueous hydrobromic acid (strong) reacts with aqueous barium hydroxide.

Formulas for products in this example are written by transferring the H^+ to OH^- to form water; then combine Ba^{+2} with Br^- to produce the correct formula. Note that the equation is not balanced at this point.

$$H{-}Br_{(aq)} + Ba(OH)_{2(aq)} \longrightarrow BaBr_{2(aq)} + H_2O\ (\textbf{not}\ \text{balanced})$$

For these equations, the combination of H^+ and OH^- must always produce only H_2O; combining the unbalanced acid and base portions to form $H(OH)_2$ is not valid; BaBr is not a correct formula. To balance the complete equation, first balance the acid to the base by matching the correct number of H^+ ions to the correct number of base portions. Each formula of barium hydroxide contains two hydroxide ion base portions; each hydrobromic acid transfers only one H^+ ion: two HBr are required for each $Ba(OH)_2$, which then forms two H_2O molecules. If this is performed correctly, the correct formula for the other product, $BaBr_2$ in this example, will be confirmed.

$$2\,HBr_{(aq)} + Ba(OH)_{2\,(aq)} \longrightarrow BaBr_{2\,(aq)} + 2\,H_2O_{(l)}$$

Based on solubility guidelines and the strong acid, the total **ionic** equation is:

$$2\,H^+_{(aq)} + 2\,Br^-_{(aq)} + Ba^{+2}_{(aq)} + 2\,OH^-_{(aq)} \longrightarrow 2\,H_2O_{(l)} + Ba^{+2}_{(aq)} + 2Br^-_{(aq)}$$

The net ionic equation shows the formation of water; 2 $Br^-_{(aq)}$ and $Ba^{+2}_{(aq)}$ are cancelled from both sides of the total ionic equation:

$$2\,H^+_{(aq)} + 2\,OH^-_{(aq)} \longrightarrow 2\,H_2O_{(l)}$$

Example: Aqueous hydrobromic acid (strong) reacts with aqueous ammonia.

Use of the neutral base NH_3 produces a variation in the exchange process. Transfer of an H^+ forms the ammonium cation (NH_4^+), which then combines with the remainder of the acid molecule to form only one final ionic product when viewed as a **formula** equation. The equation is balanced; NH_3 always accepts only one transferred H^+.

$$H{-}Br_{(aq)} + NH_{3\,(aq)} \longrightarrow NH_4Br_{(aq)}$$

total ionic equation: $H^+_{(aq)} + Br^-_{(aq)} + NH_{3(aq)} \longrightarrow NH_4^+{}_{(aq)} + Br^-_{(aq)}$

net ionic equation: $H^+_{(aq)} + NH_{3\,(aq)} \longrightarrow NH_4^+{}_{(aq)}$

A Specific Process for Completing and Writing Acid/Base Reaction Equations

(1) Start a partial **formula** equation with the acid and base reactants given by writing correct formulas; at least one of the components is usually in an aqueous solution.

(2) Identify the correct number of hydrogen atom(s) in the acid that will be transferred as H^+ ion(s). Follow the guidelines previously described based on the formula of the acid. The polyprotic acids are $\mathbf{H_2CO_3}$ (transfers **two H^+** ions), $\mathbf{H_2SO_4}$ (transfers **two H^+** ions), $\mathbf{H_2CrO_4}$ (transfers **two H^+** ions), $\mathbf{H_3PO_4}$ (transfers **three H^+** ions).

(3) Identify the base portion(s) of the base that will accept the H^+ ion(s).

- **a)** This portion will be OH^- if the base is a compound containing the hydroxide ion; include **all** hydroxide ions as H^+ acceptors.
- **b)** The base portion for compounds containing another anion will be the X^- (charge is variable).
- **c)** If the base is ammonia (NH_3), the **complete** molecule NH_3 is the base portion.

(4) Predict the final products by transferring all required acid H^+ ions from the acid to the correct number of base portions of the base molecules. Write the **correct** formula of the specific product formed by combining the H^+ ion with the base portion: **H^+ plus OH^-** will form water **H_2O; H^+ plus NH_3** will form **NH_4^+; H^+ plus X^-** will form **HX**.

(5) Preliminarily, **balance** the reactants by matching the number of H^+ ions to the corresponding number of base acceptors; complete this initial balance by determining the correct coefficient for the resulting first product.

(6) Write the **correct** formula of the other product by combining the remainder of the acid molecule (Acid⁻ without the H^+) with the remaining part of the base molecule.

a) This product is often the ionic compound produced when the negative ion portion of the acid combines with the positive ion portion of the base.
b) The reaction of an acid with ammonia is a special case in which only one compound from steps (4) and (5) is formed.

(7) Complete the formula format of the equation by finishing any balancing if required and determining the correct states of matter for all species; use the solubility guidelines for aqueous solubility.

(8) Write the **total ionic** format of the reaction equation.
a) Write the aqueous **soluble** ionic products as separate ions with the **(aq)** designation; **non**-soluble ionic compounds are written as intact solid (s) compounds.
b) Write all strong acids as aqueous ions based on the ≈100% ionization reaction of strong acids in water; all weak acids are written as neutral covalent intact aqueous (aq) molecules based on ionization of less that 1% to 5%.

(9) Write the **net ionic** equation by canceling exactly identical species from both sides of the total ionic equation.

Example: Complete the following reaction and write all three equation formats: Aqueous sulfuric acid (strong acid) is reacted with aqueous potassium hydroxide.

(1) $H_2SO_{4(aq)} + KOH_{(aq)} \longrightarrow$
(2) H_2SO_4 transfers **two** H+ ions;
(3) The base portion of KOH is $\mathbf{OH^-}$; each OH^- accepts one H^+

(4) $H_2SO_{4\,(aq)}$ + **KOH** $_{(aq)} \longrightarrow$ $\mathbf{H_2O}$ **unbalanced equation**

(5) ***Two*** KOH are required to accept the two H^+ ions from ***one*** H_2SO_4; two H^+ transferred to two OH^- will produce two water molecules.

$$H_2SO_{4\,(aq)} + 2\,KOH_{(aq)} \longrightarrow 2\,H_2O \text{ “partially” balanced equation}$$

(6) $H_2\mathbf{SO_4}\,_{(aq)} + 2\,\mathbf{K}OH_{(aq)} \longrightarrow 2\,H_2O + \mathbf{K_2SO_4}$

If the acid is correctly balanced to the base, the correct formula for the other product is usually confirmed and no other balancing is required.

(7) $H_2SO_{4\,(aq)} + 2KOH_{(aq)} \longrightarrow 2H_2O_{(l)} + K_2SO_{4\,(aq)}$ **balanced equation**
The equation was completely balanced in step (6); solubility rules indicate that K_2SO_4 is **aqueous** soluble.

(8) H_2SO_4 was stated to be a strong acid; write as separated ions based on the ionization reaction:

$$2\,H^+_{(aq)} + SO_4^{-2}{}_{(aq)} + 2\,K^+_{(aq)} + 2\,OH^-_{(aq)} \longrightarrow 2\,H_2O_{(l)} + 2\,K^+_{(aq)} + SO_4^{-2}{}_{(aq)}$$

(9) $SO_4^{-2}{}_{(aq)}$ and $2\,K^+_{(aq)}$ are cancelled from both sides of the total ionic equation.

$$2\,H^+_{(aq)} + 2\,OH^-_{(aq)} \longrightarrow 2\,H_2O_{(l)}$$

Example: Complete the following reaction and write all three equation formats: Aqueous chromic acid (strong acid) is reacted with aqueous ammonia.

(1) $H_2CrO_{4\,(aq)} + NH_{3\,(aq)} \longrightarrow$
(2) H_2CrO_4 transfers **two** H+ ions;

(3) The base portion of NH_3 is the complete molecule; each NH_3 accepts one H^+

(4) $H_2CrO_{4\,(aq)}$ + $NH_{3\,(aq)}$ $\longrightarrow$ NH_4^+ **unbalanced equation**

The first product, NH_4^+, is temporarily shown as an ion in this format for clarity.

(5) ***Two*** NH_3 are required to accept the two H^+ ions from ***one*** H_2CrO_4; two H^+ transfers to two NH_3 will produce two NH_4^+ ions.

$$H_2CrO_{4\,(aq)} + 2\,NH_{3\,(aq)} \longrightarrow 2\,NH_4^+ \text{ “partially” balanced equation}$$

(6) In this special case, CrO_4^{-2} remains as a free aqueous ion; it is shown as combining with the NH_4^+ product formed in the complete **formula** format equation:

$$H_2CrO_{4\,(aq)} + 2\,NH_{3\,(aq)} \longrightarrow 2\,NH_4^+ + CrO_4^{-2}{}_{(aq)} \text{ (partial ion format)}$$

$$H_2CrO_{4(aq)} + 2\,NH_{3(aq)} \longrightarrow (NH_4)_2CrO_4 \text{ (complete formula format)}$$

(7) $H_2CrO_{4\,(aq)} + 2\,NH_{3(aq)} \longrightarrow (NH_4)_2CrO_{4(aq)}$ **balanced equation**
The equation was completely balanced in step (6); solubility rules indicate that $(NH_4)_2CrO_4$ is **aqueous** soluble.

(8) H_2CrO_4 was stated to be a strong acid; write as separated ions based on the ionization reaction:

$$2\,H^+{}_{(aq)} + CrO_4^{-2}{}_{(aq)} + 2\,NH_{3(aq)} \longrightarrow 2\,NH_4^+{}_{(aq)} + CrO_4^{-2}{}_{(aq)}$$

(9) $CrO_4^{-2}{}_{(aq)}$ is cancelled from both sides of the total ionic equation.

$$2H^+{}_{(aq)} + 2NH_{3\,(aq)} \longrightarrow 2NH_4^+{}_{(aq)}$$

Example: Complete the following reaction and write all three equation formats: Aqueous perchloric acid (strong acid) is reacted with solid, **non**-soluble chromium (III) hydroxide.

(1) $HClO_{4(aq)} + Cr(OH)_{3(s)} \longrightarrow$
(2) $HClO_4$ transfers **one** H+ ion;
(3) The base portion of $Cr(OH)_3$ is $\mathbf{OH^-}$; each OH^- accepts one H^+

(4) $HClO_{4\,(aq)}$ + $Cr(\mathbf{OH})_{3\,(s)}$ $\longrightarrow$ H_2O **unbalanced equation**

(5) ***Three*** $HClO_4$ are required to transfer three H^+ ions to each $Cr(OH)_3$, which contains three OH^-; three H^+ transferred to three OH^- will produce three water molecules.

$$3\,HClO_{4\,(aq)} + Cr(OH)_{3\,(s)} \longrightarrow 3\,H_2O \text{ “partially” balanced equation}$$

(6) $3\,HClO_{4\,(aq)}$ + $Cr(OH)_{3\,(s)}$ $\longrightarrow$ $3\,H_2O$ + $Cr(ClO_4)_3$

The correct formula for the other product confirms the initial balancing; no other balancing is required.

(7) $3HClO_{4\,(aq)} + Cr(OH)_{3\,(s)} \longrightarrow 3H_2O_{(l)} + Cr(ClO_4)_{3\,(aq)}$ **balanced**
Solubility rules indicate that $Cr(ClO_4)_3$ is **aqueous** soluble.

(8) $HClO_4$ was stated to be a strong acid; write as separated ions based on the ionization reaction:

$$3H^+_{(aq)} + 3ClO_4^-{}_{(aq)} + Cr(OH)_{3(s)} \longrightarrow 3H_2O_{(l)} + Cr^{+3}{}_{(aq)} + 3ClO_4^-{}_{(aq)}$$

(9) $3\ ClO_4^-{}_{(aq)}$ is cancelled from both sides of the total ionic equation.

$$3\,H^+_{(aq)} + Cr(OH)_{3\,(s)} \longrightarrow 3\,H_2O_{(l)} + Cr^{+3}{}_{(aq)}$$

Example: Complete the following reaction and write all three equation formats: Aqueous acetic acid (weak acid) is reacted with aqueous calcium hydroxide.

(1) $CH_3COOH_{(aq)} + Ca(OH)_{2(aq)} \longrightarrow$
(2) CH_3COO**H** transfers **one** H^+ ion;
(3) The base portion of $Ca(OH)_2$ is $\mathbf{OH^-}$; each OH^- accepts one H^+

(4) $CH_3COO\mathbf{H}_{(aq)} + Ca(\mathbf{OH})_{2\,(aq)} \longrightarrow \mathbf{H_2O}$ **unbalanced equation**

(5) ***Two*** CH_3COOH are required to transfer two H^+ ions to each $Ca(OH)_2$, which contains two OH^-; two H^+ transferred to two OH^- will produce two water molecules.

$$2\,CH_3COOH_{(aq)} + Ca(OH)_{2\,(aq)} \longrightarrow 2\,H_2O$$ **"partially" balanced equation**

(6) $2\,\mathbf{CH_3COO}H_{(aq)} + \mathbf{Ca}(OH)_{2\,(aq)} \longrightarrow 2\,H_2O + \mathbf{Ca(CH_3COO)_2}$

The correct formula for the other product confirms the initial balancing; no other balancing is required.

(7) $2\,CH_3COOH_{(aq)} + Ca(OH)_{2\,(aq)} \longrightarrow 2\,H_2O_{(l)} + Ca(CH_3COO)_{2\,(aq)}$ **balanced**
The solubility rules indicate that $Ca(CH_3COO)_2$ is **aqueous** soluble.

(8) CH_3COOH was stated as a weak acid; **weak** acids are written as aqueous neutral intact covalent compounds, **not** ions.

$$2\,CH_3COOH_{(aq)} + Ca^{+2}{}_{(aq)} + 2\,OH^-{}_{(aq)} \longrightarrow 2\,H_2O_{(l)} + Ca^{+2}{}_{(aq)} + 2\,CH_3COO^-{}_{(aq)}$$

(9) $Ca^{+2}{}_{(aq)}$ is cancelled from both sides of the total ionic equation.

$$2\,CH_3COOH_{(aq)} + 2\,OH^-{}_{(aq)} \longrightarrow 2\,H_2O_{(l)} + 2\,CH_3COO^-{}_{(aq)}$$

VI PRACTICE PROBLEMS

1. Predict the products and write all three equation formats for the following **precipitation** reactions. Use the solubility guidelines from Section III.
   a) Aqueous silver nitrate reacts with aqueous potassium chromate.
   b) Aqueous manganese (II) bromide reacts with aqueous sodium hydroxide.
   c) Aqueous chromium (III) chloride reacts with aqueous sodium sulfide.

2. Complete the reaction and write all three equation formats for the following partial acid/base reactions. Use the solubility guidelines and refer to Section IV for names, strengths, and properties of acids and bases if necessary.
   a) Aqueous nitric acid reacts with insoluble solid magnesium hydroxide.

b) Aqueous phosphoric acid reacts with aqueous strontium hydroxide.
c) Aqueous acetic acid reacts with aqueous ammonia.
d) Aqueous sulfuric acid reacts with aqueous sodium acetate. Acetate is the conjugate base of acetic acid; acetic acid is a weak acid and is aqueous soluble.

VII ANSWERS TO PRACTICE PROBLEMS

1. a) Aqueous silver nitrate reacts with aqueous potassium chromate.

(1) $AgNO_{3(aq)} + K_2CrO_{4(aq)} \longrightarrow$

(1) $Ag^+{}_{(aq)} + NO_3^-{}_{(aq)} + 2\,K^+{}_{(aq)} + CrO_4^{-2}{}_{(aq)} \longrightarrow$

(2) $Ag^+{}_{(aq)} + NO_3^-{}_{(aq)} + 2\,K^+{}_{(aq)} + CrO_4^{-2}{}_{(aq)} \longrightarrow Ag_2CrO_4 + KNO_3$ (**not** balanced)

(3) $AgNO_{3(aq)} + K_2CrO_{4(aq)} \longrightarrow Ag_2CrO_{4(s)} + KNO_{3(aq)}$ (not balanced)

(4) $2\,AgNO_{3(aq)} + K_2CrO_{4(aq)} \longrightarrow Ag_2CrO_{4(s)} + 2\,KNO_{3(aq)}$

(5) $2\,Ag^+{}_{(aq)} + 2\,NO_3^-{}_{(aq)} + 2\,K^+{}_{(aq)} + CrO_4^{-2}{}_{(aq)} \longrightarrow Ag_2CrO_{4(s)} + 2\,K^+{}_{(aq)} + 2\,NO_3^-{}_{(aq)}$

(6) $2\,Ag^+{}_{(aq)} + CrO_4^{-2}{}_{(aq)} \longrightarrow Ag_2CrO_{4(s)}$

b) Aqueous manganese (II) bromide reacts with aqueous sodium hydroxide.

(1) $MnBr_{2(aq)} + NaOH_{(aq)} \longrightarrow$

(1) $Mn^{+2}{}_{(aq)} + 2\,Br^-{}_{(aq)} + Na^+{}_{(aq)} + OH^-{}_{(aq)} \longrightarrow$

(2) $Mn^{+2}{}_{(aq)} + 2\,Br^-{}_{(aq)} + Na^+{}_{(aq)} + OH^-{}_{(aq)} \longrightarrow Mn(OH)_2 + NaBr$ (**not** balanced)

(3) $MnBr_{2(aq)} + NaOH_{(aq)} \longrightarrow Mn(OH)_{2(s)} + NaBr_{(aq)}$ (not balanced)

(4) $MnBr_{2\,(aq)} + 2NaOH_{(aq)} \longrightarrow Mn(OH)_{2(s)} + 2\,NaBr_{(aq)}$

(5) $Mn^{+2}{}_{(aq)} + 2Br^-{}_{(aq)} + 2Na^+{}_{(aq)} + 2OH^-{}_{(aq)} \longrightarrow$

$Mn(OH)_{2(s)} + 2\,Na^+{}_{(aq)} + 2\,Br^-{}_{(aq)}$

(6) $Mn^{+2}{}_{(aq)} + 2\,OH^-{}_{(aq)} \longrightarrow Mn(OH)_{2(s)}$

c) Aqueous chromium (III) chloride reacts with aqueous sodium sulfide.

(1) $CrCl_{3\,(aq)} + Na_2S_{(aq)} \longrightarrow$

(1) $Cr^{+3}{}_{(aq)} + 3Cl^-{}_{(aq)} + 2\,Na^+{}_{(aq)} + S^{-2}{}_{(aq)} \longrightarrow$

(2) $(Cr^{+3}{}_{(aq)} + 3\,Cl^-{}_{(aq)} + 2\,Na^+{}_{(aq)} + S^{-2}{}_{(aq)} \longrightarrow Cr_2S_3 + NaCl$ (**not** balanced)

(3) $CrCl_{3\,(aq)} + Na_2S_{\,(aq)} \longrightarrow Cr_2S_{3\,(s)} + NaCl_{\,(aq)}$ (not balanced)

(4) $2CrCl_{3(aq)} + 3Na_2S_{(aq)} \longrightarrow Cr_2S_{3(s)} + 6NaCl_{(aq)}$

(5) $2Cr^{+3}{}_{(aq)} + 6Cl^-{}_{(aq)} + 6Na^+{}_{(aq)} + 3S^{-2}{}_{(aq)}$

$\longrightarrow Cr_2S_{3(s)} + 6Na^+{}_{(aq)} + 6Cl^-{}_{(aq)}$

(6) $2Cr^{+3}{}_{(aq)} + 3S^{-2}{}_{(aq)} \longrightarrow Cr_2S_{3(s)}$

2. a) Aqueous nitric acid reacts with insoluble solid magnesium hydroxide.

(1) $HNO_{3(aq)} + Mg(OH)_{2(s)} \longrightarrow$

(2) HNO_3 transfers one H+ ion

(3) The base portion of $Mg(OH)_2$ is OH^-

(4) $HNO_{3(aq)} + Mg(OH)_{2(s)} \longrightarrow H_2O$

(5) $2HNO_{3(aq)} + Mg(OH)_{2(s)} \longrightarrow 2H_2O$

(6) $2HNO_{3(aq)} + Mg(OH)_{2(s)} \longrightarrow 2H_2O + Mg(NO_3)_2$

(7) $2HNO_{3(aq)} + Mg(OH)_{2(s)} \longrightarrow 2H_2O_{(l)} + Mg(NO_3)_{2(aq)}$

(8) $2H^+{}_{(aq)} + 2NO_3^-{}_{(aq)} + Mg(OH)_{2(S)} \longrightarrow 2H_2O_{(l)} + Mg^{+2}{}_{(aq)} + 2NO_3^-{}_{(aq)}$

(9) $2H^+{}_{(aq)} + Mg(OH)_{2(s)} \longrightarrow 2H_2O_{(l)} + Mg^{+2}{}_{(aq)}$

b) Aqueous phosphoric acid reacts with aqueous strontium hydroxide.

(1) $H_3PO_{4(aq)} + Sr(OH)_{2(aq)} \longrightarrow$

(2) H_3PO_4 transfers three H+ ion

(3) The base portion of $Sr(OH)_2$ is OH^-

(4) $H_3PO_{4(aq)} + Sr(OH)_{2(aq)} \longrightarrow H_2O$

(5) $2H_3PO_{4(aq)} + 3Sr(OH)_{2(aq)} \longrightarrow 6H_2O$

(6) $2H_3PO_{4(aq)} + 3Sr(OH)_{2(aq)} \longrightarrow 6H_2O + Sr_3(PO_4)_2$

(7) $2H_3PO_{4(aq)} + 3Sr(OH)_{2(aq)} \longrightarrow 6H_2O_{(l)} + Sr_3(PO_4)_{2(s)}$

(8) $2H_3PO_{4(aq)} + 3Sr^{+2}{}_{(aq)} + 6OH^-{}_{(aq)} \longrightarrow 6H_2O_{(l)} + Sr_3(PO_4)_{2(s)}$

(9) $2H_3PO_{4(aq)} + 3Sr^{+2}{}_{(aq)} + 6OH^-{}_{(aq)} \longrightarrow 6H_2O_{(l)} + Sr_3(PO_4)_{2(s)}$

c) Aqueous acetic acid reacts with aqueous ammonia.

(1) $CH_3COOH_{(aq)} + NH_{3(aq)} \longrightarrow$

(2) CH_3COOH transfers one H+ ion

(3) The base portion of NH_3 is the complete molecule.

(4) $CH_3COOH_{(aq)}$ + $NH_{3\,(aq)} \longrightarrow NH_4(CH_3COO)$

(5)

(6) $CH_3COOH_{(aq)} + NH_{3(aq)} \longrightarrow NH_4(CH_3COO)$ (balanced)

(7) $CH_3COOH_{(aq)} + NH_{3(aq)} \longrightarrow NH_4(CH_3COO)_{(aq)}$

(8) $CH_3COOH_{(aq)} + NH_{3(aq)} \longrightarrow NH_4^+{}_{(aq)} + CH_3COO^-{}_{(aq)}$

(9) $CH_3COOH_{(aq)} + NH_{3(aq)} \longrightarrow NH_4^+{}_{(aq)} + CH_3COO^-{}_{(aq)}$

d) Aqueous sulfuric acid reacts with aqueous sodium acetate. Acetate is the conjugate base of acetic acid; acetic acid is a weak acid and is aqueous soluble.

(1) $H_2SO_{4\,(aq)} + NaOOCCH_{3\,(aq)} \longrightarrow$

(2) H_2SO_4 transfers two H+ ions;

(3) The base portion of $NaOOCCH_3$ is the ion $^-OOCCH_3$ accepting one H^+.

(4) $H_2SO_{4\,(aq)} + NaOOCCH_{3\,(aq)} \longrightarrow HOOCCH_3$

(5) $H_2SO_{4\,(aq)} + 2\,NaOOCCH_{3(aq)} \longrightarrow 2\,HOOCCH_3$

(6) $H_2SO_{4\,(aq)} + 2\,NaOOCCH_{3\,(aq)} \longrightarrow 2\,HOOCCH_3 + Na_2SO_4$

(7) $H_2SO_{4\,(aq)} + 2\,NaOOCCH_{3\,(aq)} \longrightarrow 2\,HOOCCH_{3(aq)} + Na_2SO_{4\,(aq)}$

(8) $2\,H^+{}_{(aq)} + SO_4{}^{-2}{}_{(aq)} + 2\,Na^+{}_{(aq)} + 2\,{}^-OOCCH_{3\,(aq)} \longrightarrow 2HOOCCH_{3\,(aq)} +$

$2Na^+{}_{(aq)} + SO_4{}^{-2}{}_{(aq)}$

(9) $2H^+{}_{(aq)} + 2\,{}^-OOCCH_{3\,(aq)} \longrightarrow 2HOOCCH_{3\,(aq)}$

10 Oxidation Numbers
A First Look at Redox Reactions

I GENERAL CONCEPTS

A redox reaction describes the transfer of electrons between elements or ions to form new or differently charged elements or ions. For covalent molecules, the reaction describes the alteration in the type of polar covalent bond.

The term **redox** is derived from the combination of reactions termed **reduction** plus **oxidation**. A complete redox reaction involves the simultaneous process of the reduction of one (or more) reactant and oxidation of another reactant(s).

1. **Oxidation** is the **removal** of electrons or electron share from a specific atom or atomic ion. The specific atom or ion may be independent or may be part of a larger compound or molecule.

   For individual elements, or component elemental ions as part of ionic compounds, **oxidation is the direct loss of electrons**; the result is a change in element or ion charge to a more positive or less negative value.

Examples

(These are not balanced equations):

Pb	$\longrightarrow$	Pb^{+4}	element Pb loses 4 electrons
Pb^{+2}	$\longrightarrow$	Pb^{+4}	positive ion Pb^{+2} loses 2 electrons
Cl^-	$\longrightarrow$	Cl	negative ion Cl^- loses 1 electron

Elements in **covalent** molecules are not ions and cannot change charge. For a **specific** atom in a covalent compound, **oxidation is a decrease in the relative portion of the electron share** an atom has in its covalent bonds with other atoms. In general, this change is measured by changes in the polarities of the covalent bonds formed.

Example for the specific atom **H** (This is not a balanced equation):

H_2 (= **H**—H)	$\longrightarrow$	HF (= **H**—F), with $\delta+$ on H and $\delta-$ on F
each **H** has an **equal** share of electrons in the H—H bond		**H** has a **decreased** share of electrons in the H—F bond

2. **Reduction** is the **addition** of electrons or electron share to a specific atom or atomic ion. The specific atom or ion may be independent or may be part of a larger compound or molecule.

   For individual elements, or component ions as part of ionic compounds, **reduction is the direct gain of electrons**; the result is a change in element or ion charge to a more negative or less positive value.

Examples

(These are not balanced equations):

$Pb^{+4} \longrightarrow Pb$ positive ion Pb^{+4} gains 4 electrons

$Pb^{+4} \longrightarrow Pb^{+2}$ positive ion Pb^{+4} gains 2 electrons

$Cl \longrightarrow Cl^{-}$ neutral Cl atom gains 1 electron

Elements in **covalent** molecules are not ions and cannot change charge. For a **specific** atom in a covalent compound, **reduction is an increase in the relative portion of the electron share** an atom has in its covalent bonds with other atoms. In general, this change is measured by changes in the polarities of the covalent bonds formed.

Example for the specific atom **F** (This is not a balanced equation.):

F_2 (= **F**—F) ⟶ HF (= H—**F**) with $\delta+$ on H and $\delta-$ on F

each **F** has an **equal** share of electrons in the F—F bond

F has a **increased** share of electrons in the H—F bond

The concepts of oxidation and reduction are **complementary**; electron/electron share loss by one atom/ion must result in an electron/electron share gain by another atom/ion. A complete redox reaction must involve a combination of oxidation and reduction such that electrons **balance** on each side of the equation.

Example: Ionic compounds:

$$CuCl_2 + Zn \longrightarrow Cu + ZnCl_2$$

Cu^{+2} (from $CuCl_2$) $\longrightarrow$ Cu positive ion Cu^{+2} gains 2 electrons

Zn $\longrightarrow$ Zn^{+2}(in $ZnCl_2$) element Zn loses 2 electrons

Electrons balance: The gain of 2 electrons by Cu^{+2} is balanced by the loss of 2 electrons by Zn. Cl^{-} ion does not change from reactant side to product side.

Example: Covalent compounds:

This is a conceptual (partial) example that does not show how electrons are balanced; this is covered in the next section.

$$C_{(s)} + O_{2\,(g)} \longrightarrow \overset{\delta-}{O}=\overset{\delta+}{C}=\overset{\delta-}{O}_{(g)} = (CO_{2\,(g)})$$

All C atoms in $C_{(s)}$ have equal electron shares in covalent bonding.

Each O atom in O_2 has an equal electron share in the covalent bond.

In the molecule $CO_2 = \overset{\delta-}{O}=\overset{\delta+}{C}=\overset{\delta-}{O}$, each oxygen gains a share of covalently bonded electrons to carbon; the carbon loses shares of electrons covalently bonded to oxygen.

Specific atoms that **lose** electrons/electron share are said to be **oxidized**. The term is specific for the actual atom losing the electrons but is sometimes applied to the compound or molecule, which contains this specific atom.

In the reaction: $CuCl_2 + Zn \longrightarrow Cu + ZnCl_2$; **Zn is oxidized**

In the reaction: $C_{(s)} + O_{2\,(g)} \longrightarrow CO_{2\,(g)}$; **carbon is oxidized**

Specific atoms that **gain** electrons/electron share are said to be **reduced.**

In the reaction: $CuCl_2 + Zn \longrightarrow Cu + ZnCl_2$; **Cu is reduced**

In the reaction: $C_{(s)} + O_{2\,(g)} \longrightarrow CO_{2\,(g)}$; **oxygen is reduced**

Terminology for redox reactions can be based on the concept of an oxidizing or reducing **agent**; the elements or their corresponding molecules can be identified based on the **complementary** result they have on another compound.

A **reducing agent** is a species that causes another atom to gain electrons; that is, to be reduced. The reducing **agent** supplies the electrons; therefore, the reducing **agent** is the atom (or corresponding molecule), which itself is oxidized. In the reactions previously described, Zn was oxidized and supplied electrons to Cu; therefore, Zn was the reducing **agent**; carbon was oxidized and supplied electrons to oxygen; carbon was the reducing **agent**.

An **oxidizing agent** is a species that causes another atom to lose electrons; that is, to be oxidized. The oxidizing **agent** removes the electrons; therefore, the oxidizing **agent** is the atom (or corresponding molecule), which itself is reduced. In the reactions previously described, Cu was reduced and removed electrons from Zn; therefore, Cu was the oxidizing **agent**; oxygen was reduced and removed electrons from carbon; oxygen was the oxidizing **agent**.

II OXIDATION NUMBERS

Changes in electron numbers or electron sharing are analyzed through **oxidation numbers**, a method for assigning relative values for absolute ion charge or relative degrees of covalent bond electron share.

A complete redox reaction is analyzed by comparing each element in the reactant molecules to the same element in the product molecules. Changes in oxidation number for each element are identified:

If the oxidation number of an element becomes **more positive** or **less negative** during the reaction, the element is **oxidized**.

If the oxidation number of an element becomes **more negative** or **less positive** during the reaction, the element is **reduced**.

Calculation of oxidation numbers for an element in any compound is related to the technique for determining ion charges and correct formulas covered in chapter 4.

General Rules for Oxidation Numbers

(**1**) All elements that are **not in compounds** (that is, the atoms are by themselves or bonded only to other atoms of the same element) are defined as having an oxidation number (ox #) of zero: **ox # = 0**.

(**2**) Single elemental ions (**not** polyatomic ions) have an oxidation number equal to the charge on the ion; oxidation number and absolute charge are the same value: **ox # = ion charge**.

(3) As required by rule **(2)**, the ox # of all **fixed** charged metals **in ionic compounds** are non-variable:
Ox # for Li, Na, K, Rb, Cs, and Ag = **+1**
Ox # for Mg, Ca, Sr, Ba, Zn, and Cd = **+2**
Ox # for Al and Bi = **+3**

(4) The oxidation number for variable charged metals must be calculated from the compound formulas through the methods described in rules **(7)** and **(8)**.

(5) Oxidation numbers for non-metals in **covalent** compounds must be assigned based on relative polarity of the covalent bonds. However, some general resulting oxidation numbers can be applied to certain non-metals in compounds under specific conditions.

a) Hydrogen **(H)** always has **ox #** of **(–1)** when combined with metals or boron.
b) **H** always has an **ox #** of **(+1)** when combined with any other element.
c) Oxygen **(O)** almost always has an **ox #** of **(–2)** in compounds when combined with any other element; the only exceptions are when it is combined with fluorine or combined with itself in a **compound** structure termed a peroxide.
d) Fluorine **(F)** always has an **ox #** of **(–1)**.

(6) Other Group 7A elements (Cl, Br, I) have an **ox #** of **(–1)**, except when bonded to oxygen or fluorine or another member of the Group 7A, which is above it in the periodic table. In these cases, the ox # is positive and must be calculated from the compound through rules **(7)** and **(8)**.

(7) All atoms not covered by rules **(1)** through **(6)** are considered variable and must be calculated from the compound.

a) The sum of all ox #'s for all atoms in a **neutral** compound must add up to **zero: sum ox # for a neutral compound = 0.**
b) The sum of all ox #'s for all atoms in a **polyatomic ion** must add up to the **charge** on the ion: **sum ox # = charge on polyatomic ion.**

(8) Calculation process: add up all the **known ox #'s**, fixed by rules **(1)** through **(6)**, for each atom of an element in a neutral compound or polyatomic ion; **be certain to add the ox #'s using the correct sign.**

a) To find the **one variable** atom in a **neutral** compound, subtract the total of all known ox #'s from **zero**; this must be the ox # for the **one** variable atom.
b) To find the **one variable** atom in a **polyatomic ion**, subtract the total of all known ox #'s from the polyatomic ion **charge**; this must be the ox # for the **one** variable atom.
c) If more than one variable atom occurs in a compound, each variable atom must be isolated and analyzed by further bonding analysis. A determination of the complete covalent structure of a molecule (Chapter 14) is required for this more advanced topic and is not covered here.

Example: Determine the ox # of **each** element in HNO_3; this molecule is not a peroxide. (For this chapter, any peroxide will be specifically identified.)

ox # of **H** must be **+1** from rule **(5)** since H cannot be bonded to any metal or boron; the H is actually bonded to O in this molecule.

ox # of **O** = **–2** from rule **(5)**; if the molecule is not a peroxide, oxygen cannot be bonded to itself in this molecule (all O are actually bonded to **N**).

molecule is **neutral**; sum of all ox # must = 0.
sum of known ox #'s = [one H × (+1)] + (three O × (–2)] = [+1] + [–6] = **–5**
(ox # for **N**) + (–5) = 0; solve for ox # for **N**:
ox # for **N** = (0) – (–5) = **+5**

Example: Determine the ox # of **each** element in the polyatomic ion, NO_2^-,

ox # of **O** = **−2** from rule **(5)**; (both O are actually bonded to N).

ion is a **polyatomic ion**; sum of all ox # must = charge on polyatomic ion = **(−1)**
sum of known ox #'s = [(two O × (−2)] = **−4**
(ox # for **N**) + (−4) = −1; solve for ox # for **N**:
ox # for **N** = (−1) − (−4) = **+3**

Example: Determine the ox # of **each** element in NH_3

ox # of **H** must be **+1** from rule **(5)** since H cannot be bonded to any metal or boron.

molecule is **neutral**; sum of all ox # must = 0.
sum of known ox #'s = [three H × (+1)] = **+3**
(ox # for **N**) + (+3) = 0; solve for ox # for **N**:
ox # for **N** = (0) − (+3) = **−3**

Example: Determine the ox # of **each** element in C_2H_6O; the actual structure is not required for this problem.

ox # of **H** must be **+1** from rule **(5)** since H cannot be bonded to any metal or boron;
ox # of **O** = **−2** from rule **(5)**;
molecule is **neutral**; sum of all ox # must = 0.
sum of known ox #'s = [six H × (+1)] + (one O × (−2)] = [+6] + [−2] = **+4**
(ox # for **C**) + (+4) = 0; solve for ox # for **C**:
ox # for **C** = (0) − (+4) = −4

III PROCESS FOR THE ANALYSIS OF REDOX REACTIONS USING OXIDATION NUMBERS

(1a) Identify or write the correct balanced equation.

(1b) Determine the ox # for **each** element in **each** reactant and product species (individual element, polyatomic ion, or neutral compound) found in the complete **balanced** equation. The ox # of each element in each reactant or product can be written in parentheses directly above the element symbol in the balanced equation.

(2) For **each element** used in the balanced equation, **compare** the **ox #'s** of this element found in the **reactants** to this **same** element found in the **products**. More than one value of ox # may be determined for either side of the equation if the element occurs in more than one reactant or product compound.

(2a) Identify the element or elements being **oxidized**, this is the element(s) for which the oxidation number becomes **more positive** or **less negative** during the reaction; that is, when comparing the element in the reactants to the element in the products. Determine the numerical value by which the oxidation number **changes** for each atom in the formula.

(2b) Identify the element or elements being **reduced**, this is the element(s) for which the oxidation number becomes **more negative** or **less positive** during the reaction; that is, when comparing the element in the reactants to the element in the products. Determine the numerical value by which the oxidation number **changes** for each atom in the formula.

(3) Complete the display of this analysis by writing **partial** balanced equations showing the **oxidized** element and the **reduced** element **separately**; these are forms of a method termed **half reactions**. Label the half reaction species as oxidized and reduced.

(4) To confirm correct balance and ox #'s, check that electrons balance for the balanced equation. Based on balanced atoms and correct ox #'s, **the total change in oxidation number for a redox reaction** (using correct signs) **must add to zero**. This means that the total positive change for the oxidized element(s) must exactly match the total negative change for the reduced element(s).

Example: Analyze the following redox reaction using steps (1) through (4). Sodium metal reacts with chlorine gas to form sodium chloride solid.

(1a) $2Na_{(s)} + Cl_{2(g)} \longrightarrow 2\,NaCl_{(s)}$

ox # of **Na** (metal solid) = **0** from rule **(1)**

ox # of **Cl** (in diatomic Cl_2) = **0** from rule **(1)**

ox # of Na^+ (Na^+ ion in NaCl) = **+1** from rules **(2, 3)**

ox # of Cl^- (Cl^- in NaCl) = **−1** from rule **(2)**

(1b) **ox #:** $\overset{(0)}{2\,Na_{(s)}} + \overset{(0)}{Cl_{2(g)}} \longrightarrow \overset{(+1)(-1)}{2\,NaCl_{(s)}}$

(2a) $\overset{(0)}{2\,Na_{(s)}} \longrightarrow \overset{(+1)}{2\,Na^+}$ **each Na** changes by +**1**

(2b) $\overset{(0)}{Cl_{2(g)}} \longrightarrow \overset{(-1)}{2\,Cl^-}$ **each Cl** changes by − **1**

(3) $\overset{(0)}{2\,Na_{(s)}} \longrightarrow \overset{(+1)}{2\,Na^+}$ **oxidized; Δ ox # = +1/Na**

$\overset{(0)}{Cl_{2(g)}} \longrightarrow \overset{(-1)}{2Cl^-}$ **reduced; Δ ox # = −1/Cl**

(4) Total oxidation ox # change for Na = 2 Na × (+1) = **+2**

Total reduction ox # change for Cl = 2 Cl × (−1) = **−2**

Total ox # change for the reaction = **0**

Example: Analyze the following redox reaction using steps (1) through (4). Magnesium metal reacts with aqueous hydrochloric acid to form aqueous magnesium chloride plus hydrogen gas.

(1a) $Mg_{(s)} + 2HCl_{(aq)} \longrightarrow MgCl_{2(aq)} + H_{2(g)}$

ox # of **Mg** (metal solid) = **0** from rule **(1)**

ox # of **H** (in HCl) = **+1** from rule **(5)**

ox # of **Cl** (in HCl) = **0** from rule **6** (or by calculation from HCI with H = +1)

ox # of Mg^{+2} (Mg^{+2} ion in $MgCl_2$) = **+2** from rules **(2, 3)**

ox # of Cl^- (Cl^- in MgCl) = **−1** from rule **(2)**

ox # of **H** (in diatomic H_2) = **0** from rule **(1)**

(1b) **ox #** $\overset{(0)}{Mg_{(s)}} + \overset{(+1)(-1)}{2\,HCl_{(aq)}} \longrightarrow \overset{(+2)(-1)}{MgCl_{2(aq)}} + \overset{(0)}{H_{2(g)}}$

(2a) $\overset{(0)}{Mg_{(s)}} \longrightarrow \overset{(+2)}{Mg^{+2}}$ **each Mg** changes by +**2**

(2b) $\overset{(+1)}{2H^+} \longrightarrow \overset{(0)}{H_2}$ **each H** changes by − **1**

The oxidation number of Cl remains **−1** and does not change in the reaction

(3) $\overset{(0)}{2Mg_{(s)}} \longrightarrow \overset{(+2)}{Mg^{+2}}$ **oxidized; Δ ox # = +2/Mg**

$\overset{(+1)}{2H^+} \longrightarrow \overset{(0)}{H_2}$ **reduced; Δ ox # = −1/H**

(4) Total oxidation ox # change for Mg = 1 Mg × (+2) = +2
Total reduction ox # change for H = 2 H × (−1) = −2
Total ox # change for the reaction = **0**

Example: Analyze the following redox reaction using steps (1) through (4). Ethane (C_2H_6) burns in oxygen to form carbon dioxide and water; all molecules are gases; covalent structure of carbon dioxide was shown previously.

(1a) $2C_2H_{6(g)} + 7O_{2(g)} \longrightarrow 4CO_{2(g)} + 6H_2O_{(g)}$
ox # of **H** (in C_2H_6) = **+1** from rule **(5)**
ox # of **H** (in H_2O) = **+1** from rule **(5)**
ox # of **O** (in diatomic O_2) = **0** from rule **(1)**
ox # of **O** (in CO_2) = **−2** from rule **(5)**
ox # of **O** (in H_2O) = **−2** from rule **(5)**
ox # of carbon in C_2H_6 must be calculated:
sum of known ox #'s = [six H × (+1)] = **+6**

There are **2** carbons in the molecule:

$$(\mathbf{2} \times \text{ox \# for } \mathbf{C}) + (+6) = 0; \quad \text{ox \# for } \mathbf{C} \text{ in } C_2H_6 = \frac{(0)-(+6)}{\mathbf{2}} = \mathbf{-3}$$

ox # of carbon in CO_2 must be calculated:
sum of known ox **#'s** = [two O × (−2)] = **−4**
(ox # for **C**) + (−4) = 0; ox # for **C** in CO_2 = (0) − (−4) = **+4**

(1b) $\overset{(-3)(+1)}{2C_2H_{6(g)}} + \overset{(0)}{7\,O_{2(g)}} \longrightarrow \overset{(+4)(-2)}{4CO_{2(g)}} + \overset{(+1)(-2)}{6H_2O_{(g)}}$

(2a) $\overset{(-3)}{4C} \longrightarrow \overset{(+4)}{4C}$ **each C** changes by **+7**

(2b) $\overset{(0)}{7O_{2(g)}} \longrightarrow \overset{(-2)}{14O}$ (in 4 CO_2 and 6 H_2O) **each O** changes by **−2**

The half-reactions are balanced by including correct coefficients and subscripts.

(3) $\overset{(-3)}{4C} \longrightarrow \overset{(+4)}{4C}$ **oxidized**; Δ ox # = **+7/C**

$\overset{(0)}{7O_{2(g)}} \longrightarrow \overset{(-2)}{14O}$ **reduced;** Δ ox # = **−2/O**

(4) Total oxidation ox # change for C = 4 C × (+7) = **+28**
Total reduction ox # change for O = 14 O × (−2) = **−28**
Total ox # change for the reaction = **0**

Example: Analyze the following redox reaction using steps (1) through (4). Iron metal reacts with aqueous gold(III)nitrate to produce iron(II)nitrate aqueous plus gold metal.

(1a) $3Fe_{(s)} + 2Au(NO_3)_{3(aq)} \longrightarrow 3Fe(NO_3)_{2(aq)} + Au_{(s)}$
ox # of **Fe** (metal solid) = **0** from rule **(1)**
ox # of $\mathbf{Au^{+3}}$ (Au^{+3} ion in $Au(NO_3)_3$) = **+3** from rule **(2)**
ox # of $\mathbf{Fe^{+2}}$ (Fe^{+2} ion in $Fe(NO_3)_2$) = **+2** from rule **(2)**
ox # of **Fe** (metal solid) = **0** from rule **(1)**
ox # of **O** (in NO_3^-) = **−2** from rule **(5)**

The nitrate ion does not change during this reaction; to calculate ox # **N** sum of known ox #'s = [(three O × (−2)] = **−6**
(ox # for **N**) + (−6) = −1; ox # for **N** = (−1) − (−6) = **+5**

ox #: (0) (+3)(+5)(−2) (+2)(+5)(−2) (0)

(1b) $3Fe_{(s)} + 2Au(NO_3)_{3(aq)} \longrightarrow 3Fe(NO_3)_{2(aq)} + Au_{(s)}$

(0) (+2)

(2a,3) $3Fe_{(s)} \longrightarrow 3Fe^{+2}$ **oxidized;** Δ ox # = **+2/Fe**

(+3) (0)

(2b,3) $2Au^{+3} \longrightarrow 2Au_{(s)}$ **reduced;** Δ ox # = **−3/Au**

(4) Total oxidation ox # change for Fe = 3 Fe × (+2) = **+6**
Total reduction ox # change for Au = 2 Au × (−3) = **−6**
Total ox # change for the reaction = **0**

IV PRACTICE PROBLEMS

1. Determine the oxidation number of each element in the following neutral compounds or polyatomic ions.
   a) H_2SO_4 b) SO_4^{-2} c) SO_3^{-2} d) SO_2 e) CH_4 f) CO_3^{-2}
   g) ClO^- h) ClO_2^- i) ClO_3^- j) ClO_4^- k) MnO i) CH_2F_2

2. For the following **redox** reactions,
   Use the given equation or write the correct balanced equation; then analyze the redox reaction using steps (1) through (4).
   a) Calcium metal reacts with water to produce calcium hydroxide solid plus hydrogen gas.
   b) Lead metal reacts with aqueous sulfuric acid to form lead (IV) sulfate solid plus hydrogen gas.
   c) Iron metal reacts with oxygen gas to form iron (III) oxide.
   d) Acetylene gas (C_2H_2) burns in oxygen gas to form carbon dioxide plus water.
   e) Iron metal reacts with aqueous silver nitrate to form aqueous iron (II) nitrate plus silver metal.
   f) $2NH_4Cl + K_2Cr_2O_7 + 6HCl \longrightarrow N_2 + 2CrCl_3 + 2KCl + 7H_2O$
   g) $KClO_3 + 6HBr \longrightarrow 3Br_2 + KCl + 3H_2O$

V ANSWERS TO PRACTICE PROBLEMS

1. Determine the oxidation number of each element in the following neutral compounds or polyatomic ions.

a) **(+1) (+6)(−2)** $H_2\ S\ O_4$ b) **(+6)(−2)** $S\ O_4^{-2}$ c) **(+4)(−2)** $S\ O_3^{-2}$

d) **(+4)(−2)** $S\ O_2$ e) **(−4)(+1)** $C\ H_4$ f) **(+4)(−2)** $C\ O_3^{-2}$

g) **(+1)(−2)** $Cl\ O^-$ h) **(+3)(−2)** $Cl\ O_2^-$ i) **(+5)(−2)** $Cl\ O_3^-$

j) **(+7)(−2)** $Cl\ O_4^-$ k) **(+2)(−2)** $Mn\ O$ l) **(0)(+1)(−1)** $C\ H_2\ F_2$

2. a) Calcium metal reacts with water to produce calcium hydroxide solid plus hydrogen gas.

$$\overset{(0)}{Ca_{(s)}} + \overset{(+1)(-2)}{2H_2O_{(l)}} \longrightarrow \overset{(+2)(-2)(+1)}{Ca(OH)_{2(s)}} + \overset{(0)}{H_{2(g)}}$$

$\overset{(0)}{Ca} \longrightarrow \overset{(+2)}{Ca^{+2}}$ = oxidized; **each Ca** changes by **+2**; total change = **+2**

$\overset{(+1)}{2H} \longrightarrow \overset{(0)}{H_2}$ = reduced; **each H** changes by **–1**; total change = $\frac{\mathbf{-2}}{\mathbf{0}}$

b) Lead metal reacts with aqueous sulfuric acid to form lead (IV) sulfate solid plus hydrogen gas.

$$\overset{(0)}{Pb_{(s)}} + \overset{(+1)(+6)(-2)}{2H_2SO_{4(aq)}} \longrightarrow \overset{(+4)(+6)(-2)}{Pb(SO_4)_{2(s)}} + \overset{(0)}{2H_{2(g)}}$$

$\overset{(0)}{Pb} \longrightarrow \overset{(+4)}{Pb^{+4}}$ = oxidized; **each Pb** changes by **+ 4**; total change = **+ 4**

$\overset{(+1)}{4H} \longrightarrow \overset{(0)}{2H_2}$ = reduced; **each H** changes by **– 1**; total change = $\frac{\mathbf{-4}}{\mathbf{0}}$

c) Iron metal reacts with oxygen gas to form iron (III) oxide.

$$\overset{(0)}{4Fe_{(s)}} + \overset{(0)}{3O_{2(g)}} \longrightarrow \overset{(+3)(-2)}{2Fe_2O_{3(s)}}$$

$\overset{(0)}{4Fe} \longrightarrow \overset{(+3)}{4Fe^{+3}}$ = oxidized; **each Fe** changes by **+3**; total changes = **+12**

$\overset{(0)}{3O_2} \longrightarrow \overset{(-2)}{6O}$ = reduced; **each O** changes by **–2**; total change = $\frac{\mathbf{-12}}{\mathbf{0}}$

d) Acetylene gas (C_2H_2) burns in oxygen gas to form carbon dioxide plus water.

$$\overset{(-1)(+1)}{2C_2H_{2(g)}} + \overset{(0)}{5O_{2(g)}} \longrightarrow \overset{(+4)(-2)}{4CO_{2(g)}} + \overset{(+1)(-2)}{2H_2O_{(g)}}$$

$\overset{(-1)}{4C} \longrightarrow \overset{(+4)}{4C}$ = oxidized; **each C** changes by **+5**; total changes = **+ 20**

$\overset{(0)}{5O_{2(g)}} \longrightarrow \overset{(-2)}{10O}$ = reduced; **each O** changes by **–2**; total changes = $\frac{\mathbf{-20}}{\mathbf{0}}$

e) Iron metal reacts with aqueous silver nitrate to form aqueous iron (II) nitrate plus silver metal.

$$\overset{(0)}{Fe_{(s)}} + \overset{(+1)(+5)(-2)}{2AgNO_{3(aq)}} \longrightarrow \overset{(+2)(+5)(-2)}{Fe(NO_3)_{2(aq)}} + \overset{(0)}{2Ag_{(s)}}$$

$$\overset{(0)}{Fe} \longrightarrow \overset{(+2)}{Fe^{+2}} = \text{oxidized; } \textbf{each Fe} \text{ changes by } \mathbf{+2}\text{; total change} = \mathbf{+2}$$

$$\overset{(+1)}{2\,Ag^{+1}} \longrightarrow \overset{(0)}{2\,Ag} = \text{reduced; } \textbf{each Ag} \text{ changes by } \mathbf{-1}\text{; total change} = \frac{\mathbf{-2}}{\mathbf{0}}$$

f) $$\overset{(-3)(+1)(-1)}{2NH_4Cl} + \overset{(+1)(+6)(-2)}{K_2Cr_2O_7} + \overset{(+1)(-1)}{6HCl} \longrightarrow \overset{(0)}{N_2} + \overset{(+3)(-1)}{2CrCl_3} + \overset{(+1)(-1)}{2KCl} + \overset{(+1)(-2)}{7H_2O}$$

$$\overset{(-3)}{2N} \longrightarrow \overset{(0)}{N_2} = \text{oxidized; } \textbf{each N} \text{ changes by } +3\text{; total change} = \mathbf{+6}$$

$$\overset{(+6)}{2Cr} \longrightarrow \overset{(+3)}{2Cr^{+3}} = \text{reduced; } \textbf{each Cr} \text{ changes by } \mathbf{-3}\text{; total change} = \frac{\mathbf{-6}}{\mathbf{0}}$$

g) $$\overset{(+1)(+5)(-2)}{K\ ClO_3} + \overset{(+1)(-1)}{6HBr} \longrightarrow \overset{(0)}{3Br_2} + \overset{(+1)(-1)}{KCl} + \overset{(+1)(-2)}{3H_2O}$$

$$\overset{(-1)}{6Br} \longrightarrow \overset{(0)}{3Br_2} = \text{oxidized; } \textbf{each Br} \text{ changes by } \mathbf{+1}\text{; total change} = \mathbf{+6}$$

$$\overset{(+5)}{Cl} \longrightarrow \overset{(-1)}{Cl^{-1}} = \text{reduced; } \textbf{each Cl} \text{ changes by } \mathbf{-6}\text{; total change} = \frac{\mathbf{-6}}{\mathbf{0}}$$

11 Solution Concentration, Molarity, and Solution Stoichiometry

I GENERAL CONCEPTS

A **solution** is a **homogeneous mixture** of two or more pure compounds or elements; **homogeneous** is defined as mixing at the atomic, ionic, or molecular level. For general practice:

The **solvent** is the compound doing the dissolving.
The **solute** is the compound or element being dissolved.
The **solution** is the mixture of solute plus solvent; the solution can be almost any combination of gases, liquids, or solids.

A solution **concentration** is a measurement of the amount of solute present per specific amount of solvent, or per specific amount of solution.

$$\textbf{concentration of solute} = \frac{\text{amount of solute}}{\text{amount of solvent}} \quad \textbf{or} \quad \frac{\text{amount of solute}}{\text{amount of solution}}$$

A number of different units can be used to express either of these ratios:

For general solutions:

weight percent (% weight) = mass (grams) solute/mass (grams) solution × 100%

For liquid solutions:

volume percent (% volume) = V (mL) liquid solute/V (mL) liquid solution × 100%

mass per volume in grams/milliliter = mass (g) solute/V (mL) of solution

The concentration of **liquid** solutions is very often expressed in units of **Molarity (M)**. Enclosing the identity of the solute in brackets (**[solute]**) is a symbol for "concentration of" the specific solute usually in molarity.

$$\textbf{[solute] (M)} = \frac{\text{moles of solute}}{\text{Liters of solution}} = \frac{\text{moles (solute)}}{\text{V (L) (solution)}} \qquad \text{(units are } \textbf{moles / liter}\text{)}$$

The equational form of the definition can be used to solve for any of the three variables given the other two:

$$\text{moles (solute)} = \text{V(L) (solution)} \times \text{M}; \quad \text{V(L) solution} = \frac{\text{moles (solute)}}{\text{M}}$$

Molarity can apply to any liquid solution; the solute can be gas (e.g., oxygen in water), liquid (e.g., alcohol in water), or solid (e.g., salt or sugar in water).

II CONCENTRATION CALCULATIONS BASED ON MOLARITY

General Procedure for Solving Molarity Problems

1. Identify the correct form of the molarity equation which is needed to solve for the desired unknown variable.
2. Identify or calculate the required values for the known variables in the correct units.
3. Complete the calculation for the unknown variable based on steps **1** and **2**.
4. Use the calculated value from step **3** to solve for additional information related to a more complex problem such as solution stoichiometry.

Example: Calculate the molarity (M) of 12.5 grams of H_2SO_4 dissolved in 1.75 L of an aqueous solution; use the periodic table as necessary.

(1) $\text{Molarity } (\mathbf{M}) = \dfrac{\text{moles (solute)}}{\text{V (L) (solution)}}$

(2) $\text{V(L)} = \mathbf{1.75\ L}; \quad \text{moles solute} = \dfrac{12.5\text{ g } H_2SO_4}{98.1\text{ g/mole}} = \mathbf{0.127\ mole}$

(3) $\mathbf{M} = \dfrac{0.127\text{ mole}}{1.75\text{ L}} = \mathbf{0.0726\ M}$

Example: Calculate the number of moles and the total mass of $FeCl_2$ contained in 3.00 L of a 0.250 M (molar) solution; use the periodic table as necessary.

(1) moles (solute) = V(L) (solution) × M
(2) V(L) = **3.00 L; M = 0.250 M**
(3) moles $FeCl_2$ = (3.00 L) × (0.250 M) = **0.750 moles**
Units of M = moles/L; thus, moles = (3.00 L) × (0.250 moles/L) = 0.750 moles
(4) mass of $FeCl_2$ = (0.750 moles) × (126.8 g/mole) = **95.1 grams**

Example: Calculate the volume of an aqueous $FeCl_2$ solution required to equal 100 grams of $FeCl_2$; the solution is 0.250 M.

(1) $\text{V(L) solution} = \dfrac{\text{moles (solute)}}{\mathbf{M}}$

(2) $\mathbf{M = 0.250\ M}; \quad \text{moles solute} = \dfrac{100.\text{ g } FeCl_2}{126.8\text{ g/mole}} = \mathbf{0.789\ mole}$

(3) $\text{V(L) solution} = \dfrac{0.789\text{ moles}}{0.250\text{ M}} = \mathbf{3.16\ L}$

Ionic compounds break up into individual ions upon formation of an aqueous solution. The molarity (**M**) of ionic compounds can also be expressed as molarities of the individual ions: **[ion]**.

[M] of a specific ion = ([M] of the ionic compound) × (# of specific ions per formula)

Example: Calculate the molar concentration of Cl^- ions and Ca^{+2} ions (abbreviated as $[Cl^-]$ and $[Ca^{+2}]$) if 225 grams of $CaCl_2$ is dissolved into 4.75 L of aqueous solution.

(1) First calculate **M** of $CaCl_2$; Molarity **(M)** $= \dfrac{\text{moles (solute)}}{\text{V(L) solution}}$

(2) V(L) = **4.75 L**; moles solute $= \dfrac{225 \text{ g } CaCl_2}{111.0 \text{ g/mole}} = \mathbf{2.027 \text{ mole}}$

(3) $M = \dfrac{2.027 \text{ mole}}{4.75 \text{ L}} = \mathbf{0.427 \text{ M}}$

(4) $[Ca^{+2}] = (0.427 \text{ M}) \times \dfrac{1 \text{ } Ca^{+2}}{1 \text{ } CaCl_2} = \mathbf{0.427 \text{ M } Ca^{+2}}$

$[Cl^-] = (0.427 \text{ M}) \times \dfrac{2 \text{ } Cl^-}{1 \text{ } CaCl_2} = \mathbf{0.854 \text{ M } Cl^-}$

Example: Calculate the number of moles of Li_3PO_4, which are required to form 12.5 liters of a 0.740 M solution of Li^+.

(1) First calculate the number of moles of Li^+ required; moles solute = V(L) × M
(2) V(L) = **12.5 L; M = 0.740 M**
(3) moles Li^+ = (12.5 L) × (0.740 M) = **9.25 moles**
(4) moles $Li_3PO_4 = (9.25 \text{ moles } Li^+) \times \dfrac{1 \text{ } Li_3PO_4}{3 \text{ } Li^+} = \mathbf{3.08 \text{ mole } Li_3PO_4}$

The molarity equations can be used in calculations for preparation of solutions with specific concentration.

Example: Calculate the mass of $Ca(NO_3)_2$ required to prepare 3.50 liters of a 1.60 M solution of aqueous nitrate ions.

(1) First calculate the number of moles of NO_3^- in the required solution:
moles solute = V(L) × M

(2) V(L) = **3.50 L; M = 1.60 M**
(3) moles NO_3^- = (3.50 L) × (1.60 M) = **5.60 moles**

(4) moles $Ca(NO_3)_2 = (5.60 \text{ moles } NO_3^-) \times \dfrac{1 \text{ } Ca(NO_3)_2}{2 \text{ } NO_3^-} = \mathbf{2.80 \text{ mole}}$

(5) mass $Ca(NO_3)_2$ = (2.80 moles) × (164.1 g/mole) = **460 grams**

Dilution involves changing the concentration of a specific solution by addition of extra solvent. A certain amount of the **initial**, more concentrated, solution is diluted with extra solvent to produce a specific amount of a **final**, less concentrated, solution.

The key to solving dilution problems is recognizing that the **number of moles of solute supplied** by the initial concentrated solution must be **equal** to the **number of moles of solute required** to be in the specific amount of the final diluted solution:

moles solute from initial solution = moles solute required in the final solution

Example: Calculate the volume of an 18.0 M solution of H_2SO_4, which must be diluted with water to produce 2.50 L of a 2.40 M aqueous H_2SO_4 solution.

Calculate **moles of solute** required in the **final (diluted)** solution:

(1) moles solute (final) = V(L) (final) × M (final)
(2) V(L) (final) = **2.50 L**; M (final) = **2.40 M**
(3) moles H_2SO_4 (final solution) = (2.50 L) × (2.40 M) = **6.00 moles**

Use the relationship: moles solute initial solution = moles solute final solution; calculate the **volume** of the **initial (concentrated)** solution which supplies the required moles solute:

(1) $\text{V(L) initial solution} = \dfrac{\text{moles solute (initial)}}{\text{M (initial)}}$

(2) M (initial) = **18.0 M**; moles solute initial = moles solute final = **6.00 mole H_2SO_4**

(3) $\text{V(L) initial solution} = \dfrac{6.00 \text{ moles}}{18.0 \text{ M}} = \mathbf{0.333\ L}$

A **general dilution equation** is based on the molarity equation and the mole solute relationship:

$$\text{moles solute initial} = \text{V(L) (initial)} \times \text{M (initial)}$$

$$\text{moles solute final} = \text{V(L) (final)} \times \text{M (final)}$$

Substitute each V(L) × M expression for the specific mole value:

$$\text{moles solute final} = \text{moles solute initial}$$

$$\mathbf{V(L)\ (final) \times M\ (final) = V(L)\ (initial) \times M\ (initial)}$$

$$\mathbf{Vf \times Mf = Vi \times Mi}$$

Example: Solve the previous H_2SO_4 problem by using the dilution equation.

The unknown variable was initial volume; solve Vf × Mf = Vi × Mi for **Vi:**

$$\text{Vi} = \frac{(\text{Vf}) \times (\text{Mf})}{(\text{Mi})} = \frac{(2.50 \text{ L}) \times (2.40 \text{ M})}{18.0 \text{ M}} = \mathbf{0.333\ L}$$

Example: Calculate the final molarity of a solution of H_2SO_4 if 550 mL of an 18.0 M solution is diluted with water to a final volume of 2.00 L.

(1) $\text{Vf} \times \text{Mf} = \text{Vi} \times \text{Mi}$; solve for Mf: $\text{Mf} = \dfrac{(\text{Vi}) \times (\text{Mi})}{\text{Vf}}$

(2) Identification of the known variables requires care:

$$\text{Vi} = 0.550 \text{ L}; \text{Mi} = 18.0 \text{ M}; \text{Vf} = 2.00 \text{ L}$$

(3) $\text{Mf} = \dfrac{(0.550 \text{ L}) \times (18.0 \text{ M})}{(2.00 \text{ L})} = \mathbf{4.95\ M}$

Example: Calculate the final $[Na^+]$ (molarity) of a solution if 525 mL of a 3.48 M solution of Na_3PO_4 is diluted to a final volume of 1.50 L.

(1) $Vf \times Mf = Vi \times Mi$; solve for Mf: $Mf = \dfrac{(Vi)\times(Mi)}{Vf}$

(2) $Vi = 0.525$ L; $Mi = 3.48$ M; $Vf = 1.50$ L

(3) $Mf = \dfrac{(0.525\ L)\times(3.48\ M)}{(1.50\ L)} = \mathbf{1.22\ M}\ Na_3PO_4$

(4) $[Na^+] = (1.22\ M)\times\dfrac{3\ Na^+}{1\ Na_3PO_4} = \mathbf{3.66\ M}\ Na^+_{(aq)}$

III SOLUTION STOICHIOMETRY

Solution stoichiometry problems can be solved using the general techniques described in Chapter 8; reactions in an aqueous solution use the molarity equation to interconvert moles and solution measurements.

PROCESS FOR SOLUTION STOICHIOMETRY

1. Identify the compound (or element) for which an initial starting amount is given; the initial compound (or element) given may be a reactant or product. When required, convert this amount to moles by solving:

   $$\textbf{moles solute} = \mathbf{V(L) \times M}$$

2. Identify the compound (or element) for which a specific amount must be calculated. Set up a **mole** ratio between this compound (or element) and the one identified in step **(1)** based on the balanced equation. This ratio is:

   $$\frac{\text{moles compound to be calculated}}{\text{moles compound given}} = \frac{\text{coefficient of compound to be calculated}}{\text{coefficient of compound given}}$$

3. Use the moles of the given compound (or element) found in step **(1)** and the ratio established in step **(2)** to calculate the **moles** of compound (or element) to be calculated.

   $$\text{moles of compound to be calculated} = \text{moles of compound given} \times \frac{\text{coefficient of compound to be calculated}}{\text{coefficient of compound given}}$$

4. Use the value for the moles of the compound to be calculated from step (3) to calculate the amount of this compound in other units (if necessary). Solve for **volume** or **molarity** in the problem as required.

Example: Calculate the **volume** of a 0.0200 M aqueous solution of $Ba(OH)_2$, which is required to exactly react with 100 mL of a 0.0500 M aqueous solution of HCl based on the balanced equation shown.

2 HCl	+	**Ba(OH)$_2$**	$\longrightarrow$	**BaCl$_2$ + 2 H$_2$O**
V and M known		V = **unknown**; M = known		
100 mL; 0.0500 M		***V = 0.125 L***; 0.0200 M		
↓		↑		
(1)		(4)		
moles HCl		$V(L) = \frac{2.50\times10^{-3}\text{moles}}{0.0200\text{ M}} = \mathbf{0.125\ L}$		
= (0.100 L)(0.0500 M)				
= 5.00×10^{-3} moles				

(2)

$$\downarrow 5.00\times10^{-3}\text{mole} \rightarrow \times \frac{1\,Ba(OH)_2}{2\,HCL} = 2.50\times10^{-3}\text{moles} \uparrow \quad \textbf{(3)}$$

Titration is an experimental technique to **measure** the (unknown) amount of one reactant, which is required to exactly react with a known amount of another reactant. The measurement then allows a calculation for the original amount of moles or mass or concentration of the unknown reactant. The experiment often uses an **indicator**, which identifies the point at which both reactants are present in the exact molar ratio required by the balanced equation, termed the **stoichiometric equivalence point**; this can be the **neutralization point** for certain acid/base reactions.

Example: A 1.25 M solution of aqueous potassium hydroxide is slowly added to 214 mL of an **unknown concentration** aqueous solution of hydrochloric acid. The stoichiometric equivalence point, as measured by an indicator color change, was reached when 42.53 mL of the KOH (aq) was added. Calculate the number of moles of HCl in the original 214 mL solution and calculate the original molarity of the HCL (aq).

KOH	+	**HCl**	$\longrightarrow$	**KCl + H$_2$O**
V and M known		V = known; M = **unknown**		
42.53 mL; 1.25 M		214 mL; **0.248 M**		
↓		↑		
moles KOH		M = moles/V(L)		
= (0.04253 L)(1.25 M)		$= \frac{0.05316\text{ moles}}{0.214\text{ L}} = 0.248\text{ M}$		
= 0.05316 moles				

$$\downarrow 0.05316\text{ mole} \longrightarrow \times \frac{1\,HCl}{1\,KOH} = 0.05316\text{ moles} \uparrow$$

Stoichiometry calculations can include reactants and products in **any combination** of solutions and solids (or liquids, or gases).

Example: Calculate the theoretical yield in **grams** of Ag_2CO_3, which can be prepared from combination of 55.0 mL of 0.0500 M aqueous $AgNO_3$ with excess aqueous K_2CO_3 based on the balanced equation shown.

MM = 275.8 g/mole

2 $AgNO_{3(aq)}$ + **$K_2CO_{3(aq)}$** ⟶ **$Ag_2CO_{3(s)}$** + **2 $KNO_{3(aq)}$**

V and M known (mass = **unknown**)

55.0 mL; 0.0500 M excess **0.381 g**

↓ ↑

(1) **(4)**

moles $AgNO_3$ mass Ag_2CO_3

$= (0.0550 \text{ L})(0.0500 \text{ M})$ $= (1.375\times10^{-3} \text{ moles})\times 275.8 \text{ g/mole}$

$= \mathbf{2.75\times10^{-3}}$ **mole** **= 0.381 grams**

(2)

$$\downarrow 2.75\times10^{-3} \text{ mole} \longrightarrow \times \frac{1\,Ag_2CO_3}{2\,AgNO_3} \longrightarrow = 1.375\times10^{-3} \text{ moles} \uparrow \quad \textbf{(3)}$$

Example: Calculate the concentration of aqueous NaOH ($[NaOH_{(aq)}]$) from the titration information using the balanced equation; **$KHC_8H_4O_4$** is a solid acid that transfers one hydrogen. A mass of 0.7996 grams of solid $KHC_8H_4O_4$ requires 43.6 mL of the solution of unknown molarity NaOH to reach the stoichiometric equivalence point.

MM = 204.2 g/mole

$KHC_8H_4O_4$ + NaOH ⟶ $NaKC_8H_4O_4$ + H_2O

V = known; **M = unknown**

10.7996 g V = 0.0436 L; **0.0898 M**

↓ ↑

(1) **(4)**

moles $KHC_8H_4O_4$

$$= \frac{(0.7996 \text{ g})}{204.2 \text{ g/mole}}$$

$= 3.916\times10^{-3}$ mole

$$M = \frac{3.916\times10^{-3} \text{moles}}{0.0436 \text{ L}}$$

$= 0.0898$ M

$$\downarrow 3.916\times10^{-3} \text{ mole} \rightarrow \times \frac{1\,NAOH}{1\,KHC_8H_4O_4} = 3.916\times10^{-3} \text{ moles} \uparrow$$

(2) (3)

IV PRACTICE PROBLEMS

1. a) Calculate the molarity (M) of 42.0 grams of alcohol (C_2H_6O) dissolved in a 325 mL bottle of aqueous drink.
   b) Calculate the number of moles and total mass of alcohol (C_2H_6O) in a 20.0-liter barrel of a 0.165 M aqueous drink.
   c) Calculate the volume in liters of a 0.0620 M solution of aqueous alcohol (C_2H_6O), which will equal 500 grams of alcohol.

2. Calculate the molar concentration of ions, $[Na^+]$ and $[SO_4^{-2}]$, in a solution of 100 grams of Na_2SO_4 dissolved in 2.50 L of aqueous solution.

3. a) Calculate the molarity of ammonium ions ($[NH_4^+]$) for a solution of 7.50 grams of NH_4NO_3 dissolved in 300 mL of aqueous solution.
   b) Calculate the volume of the solution from part (a), which must be diluted with water to produce 750 mL of a 5.00×10^{-4} M aqueous solution of NH_4^+ ions.

4. Solve the following general molarity problems.
   a) Calculate the molar concentration of all ions in a solution of 6.73 grams of Na_2CO_3 dissolved in 250 mL of aqueous solution.
   b) Calculate the mass of $KMnO_4$ present in 250 mL of a 0.0125 M aqueous solution of potassium permanganate.
   c) Calculate the volume of a 2.06 M aqueous solution of $KMnO_4$, which contains 322 grams of potassium permanganate.
   d) 25.0 mL of a 1.50 M aqueous solution of HCl is diluted to a final volume of 500 mL. Calculate the molarity of the final diluted solution.
   e) Calculate the volume of a 2.56 M solution of HCl, which must be diluted with water to produce 250 mL of a 0.102 M solution of aqueous H^+ ions.

5. Calculate the theoretical yield in grams of Ag_2CO_3, which can be prepared from combination of 55.0 mL of 0.0500 M aqueous $AgNO_3$ with 25.0 mL of 0.0350 M aqueous K_2CO_3. This is a **limiting reagent** problem.

6. Calculate the theoretical yield in grams of solid precipitate, which can be prepared from combination of 75.0 mL of 0.750 M aqueous $Pb(NO_3)_2$ with 125.0 mL of 0.855 M aqueous NH_4Cl. **This is a limiting reagent problem**; first write the correct balanced equation.

7. Calculate the molarity of an H_3PO_4 (aq) solution if 143 mL of this solution is found by titration to require 42.6 mL of 0.495 M aqueous KOH to reach the stoichiometric equivalence point.

8. Titrations can also be performed for **redox** reactions:
   Calculate the mass of Fe^{+2} (aq) present in a solution if 22.25 mL of a 0.0123 M solution of $KMnO_4$ is required to reach the stoichiometric equivalence point; use the partial ionic equation.

$$MnO_4^-{}_{(aq)} + 5\,Fe^{+2}{}_{(aq)} + 8\,H^+{}_{(aq)} \longrightarrow Mn^{+2}{}_{(aq)} + 5\,Fe^{+3}{}_{(aq)} + 4\,H_2O_{(l)}$$

9. A titration experiment is performed to check the purity of a vitamin C tablet; vitamin C is ascorbic acid, $HC_6H_7O_6$:

$$HC_6H_7O_{6\,(aq)} + NaOH_{(aq)} \longrightarrow H_2O_{(l)} + NaC_6H_7O_{6\,(aq)}$$

   Complete the parts below:
   a) One vitamin C tablet, dissolved in water, was analyzed: a titration showed that 24.45 mL of a 0.1045 M solution of NaOH was required to reach the stoichiometric equivalence point. Calculate the number of **moles** of ascorbic acid (vitamin C) present in the original tablet.

b) Calculate the **mass** of ascorbic acid (vitamin C) in the original tablet.
c) The original tablet had a mass of 500 milligrams; calculate the percent mass of ascorbic acid (vitamin C) in the original tablet.

V ANSWERS TO PRACTICE PROBLEMS

1. a) Calculate the molarity (M) of 42.0 grams of alcohol (C_2H_6O) dissolved in a 325 mL bottle of aqueous drink.

(1) $\text{Molarity(M)} = \dfrac{\text{moles(solute)}}{\text{V(L)(solution)}}$

(2) V(L) = 325 mL × 1 L/1000 mL = **0.325 L**;
moles solute = 42.0 g C_2H_6O/46.08 g/mole = **0.9115 mole**

(3) M = 0.9115 mole/0.325 L = **2.80 M**

b) Calculate the number of moles and total mass of alcohol (C_2H_6O) in a 20.0-liter barrel of a 0.165 M aqueous drink.

(1) moles (solute) = V(L) (solution) × **M**
(2) V(L) = **20.0 L**; M = **0.165 M**
(3) moles C_2H_6O = (20.0 L) × (0.165 M) = **3.30 moles**
(4) mass of C_2H_6O = (3.30 moles) × (46.08 g/mole) = **152 grams**

c) Calculate the volume in liters of a 0.0620 M solution of aqueous alcohol (C_2H_6O), which will equal 500 grams of alcohol.

(1) $\text{V(L)solution} = \dfrac{\text{moles(solute)}}{\text{M}}$

(2) M = **0.0620 M**; moles solute = 500 g C_2H_6O/46.08 g/mole = **10.85 mole**

(3) $\text{V(L)solution} = \dfrac{10.85\,\text{moles}}{0.0620\,\text{M}} = \mathbf{175\,L}$

2. Calculate the molar concentration of ions, [Na^+] and [SO_4^{-2}], in a solution of 100 grams of Na_2SO_4 dissolved in 2.50 L of aqueous solution.

(1) First calculate **M** of Na_2SO_4; Molarity (**M**) = moles (solute)/V(L) solution
(2) V(L) = **2.50 L**; moles solute = 100 g Na_2SO_4/142.1 g/mole = **0.704 mole**
(3) M = 0.704 mole/2.50 L = **0.281 M**
(4) [Na^+] = (0.281 M) × (2 Na^+ per formula) = **0.562 M Na^+**
[SO_4^{-2}] = (0.281 M) × (1 SO_4^{-2} per formula) = **0.281 M SO_4^{-2}**

3. a) Calculate the molarity of ammonium ions ([NH_4^+]) for a solution of 7.50 grams of NH_4NO_3 dissolved in 300 mL of aqueous solution.

(1) First calculate **M** of NH_4NO_3; Molarity (**M**) = moles (solute)/V(L) solution
(2) V(L) = **0.300 L**; moles solute = 7.50 NH_4NO_3/80.0 g/mole = **0.0938 mole**
(3) M = 0.0938 mole/0.300 L = **0.3125 M**
(4) [NH_4^+] = (0.3125 M) × (1 NH_4^+ per formula) = **0.3125 M NH_4^+**

b) Calculate the volume of the solution from part (a), which must be diluted with water to produce 750 mL of a 5.00 × 10^{-4} M aqueous solution of NH_4^+ ions.
Use **Vf** × **Mf** = **Vi** × **Mi**; solve for **Vi**:

Vi = (Vf) × **(Mf)/(Mi)** = (0.750 L) × (5.00 × 10^{-4} M)/(0.3125 M) = **0.0012 L**

4. Solve the following general molarity problems.

a) Calculate the molar concentration of all ions in a solution of 6.73 grams of Na_2CO_3 dissolved in 250 mL of aqueous solution.

$$[Na^+] = \mathbf{0.508\,M\,Na^+};\quad [CO_3^{-2}] = \mathbf{0.254\,M\,CO_3^{-2}}$$

b) Calculate the mass of $KMnO_4$ present in 250 mL of a 0.0125 M aqueous solution of potassium permanganate.

$$\text{moles } KMnO_4 = (0.250\text{ L}) \times (0.0125\text{ M}) = \mathbf{0.003125\ moles}$$

$$\text{mass of } KMnO_4 = (0.003125\text{ moles}) \times (158.0\text{ g/mole}) = \mathbf{0.494\ grams}$$

c) Calculate the volume of a 2.06 M aqueous solution of $KMnO_4$, which contains 322 grams of potassium permanganate.

$$\text{moles solute} = 322\text{ g } KMnO_4/158.0\text{ g/mole} = \mathbf{2.04\ mole}$$

$$\text{V(L) solution} = 2.04\text{ moles}/2.06\text{ M} = \mathbf{0.989\ L}$$

d) 25.0 mL of a 1.50 M aqueous solution of HCl is diluted to a final volume of 500 mL. Calculate the molarity of the final diluted solution. Use **Vf × Mf = Vi × Mi**; solve for **Mf**; Mf = (Vi) × (Mi)/(Vf) = (0.0250 L) × (1.50 M)/(0.500 L) = **0.0750 M**

e) Calculate the volume of a 2.56 M solution of HCl, which must be diluted with water to produce 250 mL of a 0.102 M solution of aqueous H^+ ions. Use **Vf × Mf = Vi × Mi**:

$$\mathbf{[H^+]\ initial} = (2.56\text{ M}) \times (1\ H^+ \text{ per HCl } \mathbf{strong\ acid)} = \mathbf{2.56\ M\ H^+}$$

Solve for **Vi**: Vi = (Vf) × (Mf)/(Mi) = (0.250 L) × (0.102 M)/(2.56 M) = **0.00996 L**

5. Calculate the theoretical yield in grams of Ag_2CO_3, which can be prepared from combination of 55.0 mL of 0.0500 M aqueous $AgNO_3$ **with 25.0 mL of 0.0350 M aqueous K_2CO_3** (instead of excess) as related to the demonstration example. This is a limiting reagent problem.

$$\text{mole } AgNO_3 = (0.0550\text{ L}) \times (0.0500\text{ M}) = 0.00275\text{ moles}$$

$$\text{mole } Ag_2CO_3 \text{ from } AgNO_3 = 0.00275\text{ mole} \times \frac{1\ Ag_2CO_3}{2\ AgNO_3} = 0.001375\text{ mole } Ag_2CO_3$$

$$\text{mole } K_2CO_3 = (0.0250\text{ L}) \times (0.0350\text{ M}) = 0.000875\text{ moles}$$

$$\text{mole } Ag_2CO_3 \text{ from } K_2CO_3 = 0.000875\text{ mole} \times \frac{1\ Ag_2CO_3}{1\ K_2CO_3} = 0.000875\text{ mole } Ag_2CO_3$$

K_2CO_3 is limiting reagent; correct mole T.Y. of Ag_2CO_3 = **0.000875 mole Ag_2CO_3** mass Ag_2CO_3 = (0.000875 mole) × (275.8 g/mole) = **0.241 g Ag_2CO_3**

6. Calculate the theoretical yield in grams of solid precipitate, which can be prepared from combination of 75.0 mL of 0.750 M aqueous $Pb(NO_3)_2$ with 125.0 mL of 0.855 M aqueous NH_4Cl. This is a limiting reagent problem; first write the correct balanced equation.

$$\mathbf{Pb(NO_3)_{2(aq)} + 2\ NH_4Cl_{(aq)} \longrightarrow PbCl_{2(s)} + 2\ NH_4NO_{3(aq)}}$$

$$\text{mole } Pb(NO_3)_2 = (0.0750\text{ L}) \times (0.750\text{ M}) = 0.05625\text{ moles}$$

$$\text{mole } PbCl_{2(s)} \text{ from } Pb\ (NO_3)_2 = 0.05625\text{ mole} \times \frac{1\ PbCl_{2(s)}}{1\ Pb\ (NO_3)_2} = 0.05625\text{ mole } PbCl_{2(s)}$$

$$\text{mole } NH_4Cl = (0.125\text{ L}) \times (0.855\text{ M}) = 0.1069\text{ moles}$$

$$\text{mole } PbCl_{2(s)} \text{ from } NH_4Cl = 0.1069 \text{ mole} \times \frac{1\ PbCl_{2(s)}}{2\ NH_4Cl} = 0.05344 \text{ mole } PbCl_{2(s)}$$

NH_4Cl is a limiting reagent; correct mole T.Y. of $PbCl_{2(s)}$ = **0.05344 mole $PbCl_{2(S)}$**

$$\text{mass } PbCl_{2(s)} = (0.05344 \text{ mole}) \times (278.2 \text{ g/mole}) = \mathbf{14.9\ g\ PbCl_{2(s)}}$$

7. Calculate the molarity of an $H_3PO_4(aq)$ solution if 143 mL of this solution is found by titration to require 42.6 mL of 0.495 M aqueous KOH to reach the stoichiometric equivalence point.

First write the correct balanced equation:

$$\mathbf{3\ KOH_{(aq)} + H_3PO_{4(aq)} \longrightarrow 3\ H_2O_{(l)} + K_3PO_{4(aq)}}$$

V and M known — V = known; M = **unknown**

42.6 mL; 0.495 M — 143 mL; **0.0492 M**

↓ — ↑

moles KOH — M = moles/V(L)

= (0.0426 L)(0.495 M) — = 0.007029 moles/0.143 L

= 0.02109 moles — = **0.0492 M**

$$\downarrow 0.02109 \text{ mole} \longrightarrow \times \frac{1\ H_3PO_4}{3\ KOH} = \mathbf{0.007029\ moles} \uparrow$$

8. Titrations can also be performed for **redox** reactions:
Calculate the mass of $Fe^{+2}(aq)$ present in a solution if 22.25 mL of a 0.0123 M solution of $KMnO_4$ is required to reach the stoichiometric equivalence point.

$$MnO_4^-{}_{(aq)} + 8\ E^+{}_{(aq)} + 5\ Fe^{+2}{}_{(aq)} \longrightarrow Mn^{+2}{}_{(aq)} + 5\ Fe^{+3}{}_{(aq)} + 4\ H_2O_{(l)}$$

V and M known — (mass = **unknown**)

22.25 mL; 0.0123 M — **0.0765 g**

↓ — ↑

moles MnO_4^- = moles $KMnO_4$

moles MnO_4^- — mass Fe^{+2}

= (0.02225 L)(0.0123 M) — = $(1.37 \times 10^{-3}$ moles$) \times 55.85$ g/mole

= **2.74×10^{-4} mole** — = **0.0765 grams**

$$\downarrow 2.74 \times 10^{-4} \text{ mole} \longrightarrow \times \frac{5\ Fe^{+2}}{1\ MnO_4^-} \longrightarrow = 1.37 \times 10^{-3} \text{ moles} \uparrow$$

9. A titration experiment is performed to check the purity of a vitamin C tablet; vitamin C is ascorbic acid, $HC_6H_7O_6$:
   a) One vitamin C tablet, dissolved in water, was analyzed: a titration showed that 24.45 mL of a 0.1045 M solution of NaOH was required to reach the stoichiometric equivalence point. Calculate the number of **moles** of ascorbic acid (vitamin C) present in the original tablet.
   b) Calculate the **mass** of ascorbic acid (vitamin C) in the original tablet.

MM = 176.1 g/mole

$NaOH_{(aq)}$ + $HC_6H_7O_{6(aq)} \longrightarrow H_2O_{(l)} + NaC_6H_7O_{6(aq)}$

V and M known (mass = **unknown**)

24.45 mL; 0.1045 M **0.4499 g**

↓ ↑

moles NaOH mass $HC_6H_7O_6$

= (0.02445 L)(0.1045 M) $= (2.555 \times 10^{-3} \text{ moles}) \times 176.1 \text{ g/mole}$

$= \mathbf{2.555 \times 10^{-3}}$ **mole** = **0.4499 grams**

$$\downarrow 2.555 \times 10^{-3} \text{ mole} \longrightarrow \times \frac{1\ HC_6H_7O_6}{1\ NaOH} = 2.555 \times 10^{-3} \text{ moles} \uparrow$$

   c) The original tablet had a mass of 500 milligrams; calculate the percent mass of ascorbic acid (vitamin C) in the original tablet.

500 mg × 1 g/1000 mg = 0.500 g; % mass = 0.4499 g/0.500 g × 100% = **89.98%**

12 Light, Matter, and Spectroscopy

I GENERAL PROPERTIES OF LIGHT

Wave Properties

Energy transmission through wave properties allows transfer of energy (or information) through a distance in the absence of net forward motion of any matter, that is, no net displacement of matter in the direction of wave motion.

The **wave** nature of an energy transmission is based on the **mathematical** cyclical behavior of some variable essential to the particular wave propagation. Cyclical implies a repeating oscillation of this variable above (greater than) and below (less than) a specific average value. Examples are the vertical position of water molecules in a water wave or the pressure of air molecules in a sound wave.

A wave form is characterized by the following variables.

The **amplitude (A)** of a wave is the peak height (or depth) of the wave description above (or below) the average.

The **wavelength** is defined as the distance between any two successive peaks (or valleys) of the wave form; the symbol for wavelength is the Greek letter lambda, λ. Units of wavelength are distance units.

The **frequency** of a wave form is the number of waves that pass a specific point per unit time; the symbol for frequency is the Greek letter nu, ν. Units of frequency can conveniently be expressed as waves or cycles per second (waves/sec = cycles/sec); the unit "**hertz**" is defined as one cycle per second. The terms "wave" or "cycle" are omitted from the mathematical form of frequency; the mathematical (equational) units of frequency are "per second" or, equivalently, reciprocal seconds: **1/sec = sec**$^{-1}$.

The **velocity** of any wave form = (length of each wave) × (# of waves per unit time).
Equivalent to: **velocity** of wave form = (wavelength) × (frequency).
The equation is **v (velocity)** $= \lambda\nu$.

The Wave Nature of Light

The wave nature of light can be defined as a self-propagating oscillation of electric and magnetic fields, termed **electromagnetic radiation.**

Light does not require a "medium"; it does not require matter to move through such as water waves in water or sound waves through air. The **wave form** for light is **self-propagating,** indicating that the electric field generates the magnetic field while the magnetic field generates the electric field.

Light is characterized by typical wave variables of amplitude, wavelength (λ), and frequency (ν). The velocity of light (symbol = **c**) in a vacuum is constant at 3.00×10^8 meters/second.

The equational form of the velocity relationship for light is $\mathbf{c} = \lambda\nu$.

Solving for λ: $\lambda = \mathbf{c}/\nu$; solving for ν: $\nu = \mathbf{c}/\lambda$.

Since the velocity of light (in a vacuum) is constant, therefore, it is true that wavelength and frequency are inversely proportional: $\lambda \propto \mathbf{1/\nu}$ or $\nu \propto \mathbf{1/\lambda}$. As the wavelength decreases, the frequency increases; as the wavelength increases, the frequency must decrease. Velocity of light is measured in meters/second (m/sec), the wavelength is expressed in meters and the frequency is expressed in reciprocal seconds (1/sec or sec^{-1}).

Example: Vitamin A absorbs light with a wavelength of 4.97×10^{-7} meters; what is the frequency of this light?

(1) $c = \lambda\nu$, solve for frequency: $\nu = c/\lambda$

(2) $c = 3.00 \times 10^8$ m/sec; $\lambda = 4.97 \times 10^{-7}$ m

(3) $\nu = \dfrac{(3.00 \times 10^8 \text{ m/sec})}{(4.97 \times 10^{-7} \text{ m})} = 6.04 \times 10^{14}\text{/sec or written as sec}^{-1}$

Example: Strong ultraviolet radiation has a frequency of 5.0×10^{17} sec^{-1}. Calculate the wavelength of this light.

(1) $c = \lambda\nu$, solve for wavelength: $\lambda = c/\nu$

(2) $c = 3.00 \times 10^8$ m/sec; $\nu = 5.0 \times 10^{17}$/sec

(3) $\lambda = \dfrac{(3.00 \times 10^8 \text{ m/sec})}{(5.0 \times 10^{17}\text{/sec})} = 6.0 \times 10^{-10} \text{ m}$

II ENERGY AND THE QUANTUM THEORY OF LIGHT

Light and matter interact during transfer of energy. Electromagnetic radiation can transfer energy to matter, producing heat or electricity or work. Conversely, matter can convert certain types of energy to electromagnetic radiation, producing light from other forms of energy.

The **quantum theory of light** states that transfer of energy between electromagnetic radiation and matter must **only** occur through discrete amounts termed **quanta** (singular = **quantum**). That is, only specific energy amounts, energy "packets," can be transferred to, or released from, specific forms of matter.

A **quantum** is the smallest fixed amount of energy that can be transferred from electromagnetic radiation of one specific frequency; one quantum of light = one energy "packet." One quantum of light is termed a **photon**.

The quantum concept that light acts as discrete packets of energy is termed the **particle nature of light**; the concept that light is a wave form but transfers energy in discrete amounts is an example of **wave-particle duality**.

The energy value (**E**) of one quantum of light (one photon of light) is proportional to the frequency of the light: **E(photon)** $\propto \nu$. The intensity of light is proportional to the number of photons transferred per unit time; total light energy is proportional to the (# of photons per time) × (energy per photon).

The common term **"light"** includes all forms of electromagnetic radiation; the **electromagnetic spectrum** includes all values for frequency and energy. Since **E(photon)** $\propto \nu$, the high energy end of the spectrum will have the highest frequencies and shortest wavelengths; the low energy end will have the lowest frequencies and the longest wavelengths:

highest E ⟷ **lowest E**
highest ν **lowest** ν
shortest λ **longest** λ

gamma rays > x-rays > UV rays > visible light > infrared > radio waves

$\nu = 10^{21}$ sec^{-1} $\nu = 10^{18}$ sec^{-1} $\nu = 10^{16}$ sec^{-1} $\nu = 10^{14}$ sec^{-1} $\nu = 10^{12}$ sec^{-1} $\nu = 10^{7}$ sec^{-1}

$\lambda = 10^{-13}$ m $\lambda = 10^{-10}$ m $\lambda = 10^{-8}$ m $\lambda = 10^{-7}$ m $\lambda = 10^{-4}$ m $\lambda = 10$ m

Transfer of energy can occur between electromagnetic radiation and the electrons in matter. An electron in an atom must interact with a light energy "packet" (photon) in an **all or nothing** process.

The requirements of the "all or nothing" electron/light interaction are:

1. An electron cannot interact with a fraction of a photon (hence the concept of a quantum).
2. An electron can only interact with **one** photon at a time.
3. The energies of photons are not additive for a single specific electron change; photon energies cannot be "stored up" by an electron in the process of undergoing one specific atomic change.

Energy and Light Calculations

The proportionality **E (photon)** $\propto$ ν is converted to an equation through Plank's constant: **E (photon) = hν**. Plank's constant **(h) = 6.63 × 10⁻³⁴ Joule-seconds**

The **Joule (J)** is the metric unit of energy ($kg\text{-}m^2/sec^2$).

The units of the constant **h** are: Joules × seconds **(J-sec)**; this allows the energy value from the equation to equal units of Joules:

$$\mathbf{E\ (photon) = h}\ (\text{Joules} \times \text{seconds}) \times \nu\ (1/\text{seconds}) = \mathbf{Joules}$$

The equation can also be expressed in terms of the wavelength of light rather than the frequency. Since $c = \lambda\nu$, $\nu = c/\lambda$. Substitute **(c/λ)** for (ν) in the equation for energy:

$$\mathbf{E\ (photon) = hc/\lambda}$$

Example: Compact disc players use light of frequency = 3.85×10^{14} sec⁻¹.

Calculate the energy and wavelength one photon of this light.

(1) Since the frequency is given directly, use E (photon) = hν

(2) $h = 6.63 \times 10^{-34}$ J-sec; $\nu = 3.85 \times 10^{14}$ sec⁻¹

(3) E(photon) = $(6.63 \times 10^{-34}\text{ J-sec}) \times (3.85 \times 10^{14}\text{ sec}^{-1}) = \mathbf{2.55 \times 10^{-19}\ J}$

(4) $c = \lambda\nu$; $\lambda = c/\nu$

$$\lambda = \frac{(3.00 \times 10^8\,\text{m/sec})}{(3.85 \times 10^{14}\,\text{sec}^{-1})} = 7.79 \times 10^{-7}\,\text{m}$$

Example: Calculate the energy of blue light, wavelength = 469 nanometers.

(1) λ is directly given; use E (photon) = hc/λ

(2) $h = 6.63 \times 10^{-34}$ J-sec; units of λ must be **meters** = 4.69×10^{-7} m

(3) $\text{E(photon)} = \dfrac{(6.63 \times 10^{-34}\ \text{J-sec}) \times (3.00 \times 10^8\,\text{m/sec})}{4.69 \times 10^{-7}\ \text{m}} = \mathbf{4.24 \times 10^{-19}\ J}$

Example: The energy that holds a certain electron (the electron binding energy) in cesium is 6.22×10^{-19} J. Calculate the wavelength of the photon required to release this electron.

(1) λ is requested; solve E (photon) = hc/λ; $\lambda = \dfrac{hc}{E}$

(2) $h = 6.63 \times 10^{-34}$ J-sec; $E = 6.22 \times 10^{-19}$ J

(3) $\lambda = \dfrac{(6.63 \times 10^{-34}\,\text{J-sec}) \times (3.00 \times 10^8\,\text{m / sec})}{6.22 \times 10^{-19}\,\text{J}} = \mathbf{3.20 \times 10^{-7}\,m}\ (320\ \text{nm})$

III ELECTRONS AND ATOMS

Electrons in atoms are described as having a potential energy based on occupying a specific potential energy level. Potential energy levels are a function, in part, of the average distance of the electron from the nucleus. Recall, as discussed in Chapter 3, that for consideration of the electromagnetic force, the higher potential energy position represents opposite charges farther apart; the lower potential energy position represents charges closer together.

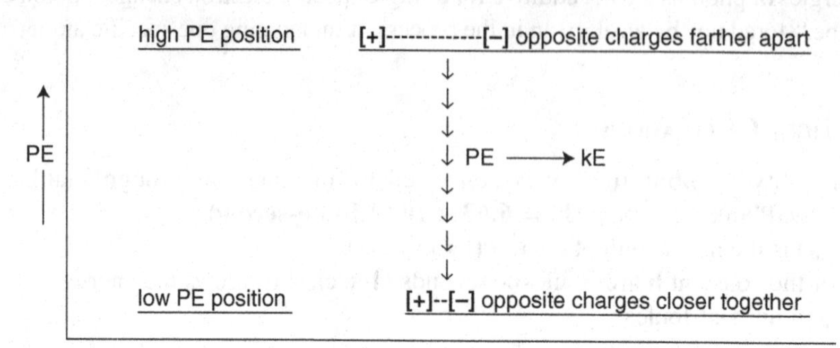

The electromagnetic force of attraction is:

$$\mathbf{F(el) = \frac{k\,(q\ of\ [+])\,(q\ of\ [-])}{r^2}}\text{; } \mathbf{q} = \text{size of charge, } \mathbf{r} = \text{distance between charges}$$

The attractive force is proportional to the negative charge on an electron, the number of positively charged protons in the nucleus and inversely proportional to the distance of the electron from the nucleus: **the closer the electron to the nucleus, the greater the attractive force and, thus, the lower the potential energy.**

The **stability** of an electron in an atom is related to its potential energy such that the most stable electrons have the lowest potential energy and are closest to the nucleus. A summary of these relationships:

An electron in:

lower energy level ←——————→ **higher energy level**
lower potential energy ←——————→ **higher potential energy**
closer to the nucleus ←——————→ **farther from nucleus**
more stable ←——————→ **less stable**

Electrons in an atom **cannot** have every possible potential energy; only certain values are available. This means that the **energy levels** in an atom are spaced apart such that they have only limited specific potential energies, like rungs on a ladder. The energy levels in an atom are said to be **quantitized**, a concept related to the quantum nature of light.

Prior to a complete discussion of atomic orbitals (Chapter 13), one description of the energy levels occupied by electrons in an atom can be used. Electrons can be thought of as occupying a series of **major shells**, a concept that views these energy levels as concentric "shells" around the nucleus distinguished by the average distance from the nucleus; thus, the average distance provides the approximate potential energy. The major shell number is identified by a value termed the **principal quantum number** (symbol = **n**) such that **n = 1** is the first shell, **n = 2** is the second shell, up to a theoretical value of infinity. The distance and energy relationship is:

Energy Level n = 1 ←——————→ **Energy Level n = ∞**
closest to nucleus ←——————→ **farthest from nucleus**
lowest potential energy ←——————→ **highest potential energy**

The distribution of electrons in specific energy levels in an atom produces a specific energy state; **only specific potential energy states are available**. Since each element has a different number of protons (positive charges) in the nucleus and different distribution of electrons, each element has different specific values for the electron energy levels. Thus, each specific element will occupy a unique set of energy states.

IV ENERGY STATES, LIGHT INTERACTION, AND ELECTRON TRANSITIONS

An atom can move from one energy state to another by absorbing or emitting energy. The basis for a discussion of spectroscopy for this chapter is that a **change** in energy state can be produced by interaction of light with an electron in a specific energy level, termed an **electron transition**.

An electron can ***absorb*** **one** photon to move from an **initial** lower electron energy level to a **final** higher energy level. **Absorption spectroscopy** measures the wavelengths of light that are **absorbed** during electron transitions in an atom. The selected wavelengths (colors, if in the visible region of the spectrum) of absorbed photons appear as dark lines against a background of all other unabsorbed wavelengths of light.

An electron can ***emit*** **one** photon to move from an **initial** higher electron energy level to a **final** lower energy level. **Emission spectroscopy** measures the wavelengths of light photons that are **emitted** during electron transitions in an atom. The electron(s) in the atom reach a higher energy level through other added energy such as heat or electricity. The emitted photon wavelengths (colors, if in the visible region of the spectrum) appear as bright (colored) lines against a dark background of all other non-emitted wavelengths of light.

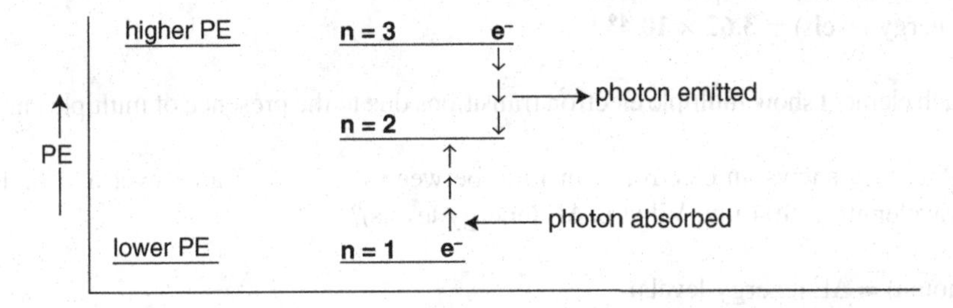

Since an electron must always occupy one of a fixed number of energy levels, **an electron can only gain or lose specific values of energy corresponding to a fixed energy difference between two energy levels**.

The change in energy of an electron in a transition must equal the difference between the final energy level and the initial energy level:

$$\mathbf{\Delta E\ (electron) = \Delta\ E\ (energy\ levels) = [E\ (final\ level) - E\ (initial\ level)]}$$

V SPECTROSCOPY CALCULATIONS

Recall that an electron in an atom must interact with **one** photon at a time in an all or nothing process: the electron cannot interact with a fraction of a photon, nor with more than one photon by "storing up" energy. An electron can undergo a transition through absorption or emission of a photon only when the energy of the photon exactly matches the energy difference between two electron energy levels.

$$\mathbf{\Delta E\ (electron) = \Delta\ E\ (energy\ levels) = E\ (photon)}$$

The specific photon energies responsible for electron transitions for any specific element are measured by the relationship between photon energy and wavelength or frequency: **E (photon) =** $\mathbf{h\nu}$ and **E (photon) =** $\mathbf{hc/\lambda}$. An electron transition through absorption or emission of a photon will correspond to a specific wavelength (or frequency) of light; the wavelength specifies the correct matching photon energy. Since each element has different specific values for the electron energy levels, each element has a unique set of wavelengths in absorption or emission spectroscopy.

The following equations describing spectroscopy are used to solve for wavelengths and energies:

ΔE (electron) = ΔE (energy levels)

ΔE (energy levels) = [E (final level) – E (initial level)]

E (photon) = ΔE (energy levels)

E (photon) = [E (final level) – E (initial level)]

Example: An atom emits light (photons) of wavelength 550 nanometers. What must be the energy difference between the two electron energy levels, ΔE (energy levels), involved in the corresponding electron transition?

(1) ΔE (energy levels) = E (photon); select the equation that solves for E (photon): wavelength of the photon is given, use E (photon) = hc/λ

(2) $h = 6.63 \times 10^{-34}$ J-sec; $\lambda = (550 \text{ nm}) \times \dfrac{10^{-9}\text{m}}{1\text{nm}} = 5.50 \times 10^{-7}$ m

(3) $\text{E(photon)} = \dfrac{(6.63 \times 10^{-34}\text{ J-sec}) \times (3.00 \times 10^{8}\text{ m/sec})}{5.50 \times 10^{-7}\text{ m}} = \mathbf{3.62 \times 10^{-19}\ J}$

(4) ΔE (energy levels) = E (photon)
ΔE (energy levels) = $\mathbf{3.62 \times 10^{-19}}$ J

Atoms of each element show multiple electron transitions due to the presence of multiple energy levels.

Example: Mercury shows an electron transition between level n = 2 and level n = 6. The corresponding wavelength is 404 nm. What is ΔE (energy levels)?

(1) E (photon) = ΔE (energy levels)
E (photon) = hc/λ

(2) $h = 6.63 \times 10^{-34}$ J-sec; $\lambda = (404 \text{ nm}) \times \dfrac{10^{-9}\text{ m}}{1\text{nm}} = 4.04 \times 10^{-7}$ m

(3) $\text{E (photon)} = \dfrac{(6.63 \times 10^{-34}\text{ J-sec}) \times (3.00 \times 10^{8}\text{ m/sec})}{(4.04 \times 10^{-7}\text{ m})} = \mathbf{4.92 \times 10^{-19}\ J}$

(4) ΔE (energy levels) = E (photon)
ΔE (energy levels) = $\mathbf{4.92 \times 10^{-19}\ J}$

Example: The energy difference between level n = 5 and level n = 8 in mercury is 3.44×10^{-19} J. What is the wavelength of an emitted photon for this transition?

(1) ΔE (photon) = E (energy levels); select the equation that solves for wavelength of the photon with E (photon) given. $\lambda = \dfrac{hc}{\text{E (photon)}}$

(2) $h = 6.63 \times 10^{-34}$ J-sec; E(photon) = 3.44×10^{-19} J

(3) $\lambda = \dfrac{(6.63 \times 10^{-34}\text{ J-sec}) \times (3.00 \times 10^{8}\text{ m/sec})}{3.44 \times 10^{-19}\text{ J}} = \mathbf{5.78 \times 10^{-7}\,m}$ (578nm)

VI SPECTRA CALCULATIONS USING [E (FINAL LEVEL) – E (INITIAL LEVEL)]

Exact potential energy values for electron energy levels in atoms **cannot** usually be calculated directly. However, exceptions occur for atoms or ions with only **one electron**, such as H, He^{+}, Li^{+2}, Be^{+3} and any other possible ions with only one electron; these have potential energy levels that can be calculated directly from the following equation:

$$E(n) = \frac{(Z)^2 \times (-2.18 \times 10^{-18}\,J)}{n^2}$$

n = shell number and principal quantum number (**1** to ∞)
E(n) = potential energy of shell number "n"
Z = charge on the nucleus = # of protons in the nucleus

The value (-2.18×10^{-18} J) is a constant generated from a set of more fundamental constants. The value is, by convention, a **negative** number; the potential energy of all energy levels will be stated as a negative number.

A free electron, that is, an electron not in an atom, undergoes no electromagnetic force of attraction to a nucleus; a free electron is also called an **unbound** electron. The system for describing electron potential energy levels is based on the convention that a free (unbound) electron is arbitrarily set at a potential energy of zero:

PE (unbound electron) = 0

The potential energy of the lowest energy level (**n = 1**) will have the largest negative value (the largest absolute value for the negative number). The absolute value of the negative number decreases as the number for **n** increases.

Energy Level n = 1	⟷	**Energy Level n = ∞**
lowest potential energy	⟷	**highest potential energy**
largest numerical value of the negative number (largest absolute value)	⟷	**smallest numerical value of the negative number (smallest absolute value)**

Example: Calculate E(1) through E(4) for the hydrogen atom; **Z** = 1 for $_1H$

$$E(\mathbf{1}) = \frac{(1)^2 \times (-2.18 \times 10^{-18}\,J)}{1^2} = \frac{(1) \times (-2.18 \times 10^{-18}\,J)}{1} = \mathbf{-2.18 \times 10^{-18}\,J}$$

$$E(\mathbf{2}) = \frac{(1)^2 \times (-2.18 \times 10^{-18}\,J)}{2^2} = \frac{(1) \times (-2.18 \times 10^{-18}\,J)}{4} = \mathbf{-0.545 \times 10^{-18}\,J}$$

$$E(\mathbf{3}) = \frac{(1)^2 \times (-2.18 \times 10^{-18}\,J)}{3^2} = \frac{(1) \times (-2.18 \times 10^{-18}\,J)}{9} = \mathbf{-0.242 \times 10^{-18}\,J}$$

$$E(\mathbf{4}) = \frac{(1)^2 \times (-2.18 \times 10^{-18}\,J)}{4^2} = \frac{(1) \times (-2.18 \times 10^{-18}\,J)}{16} = \mathbf{-0.136 \times 10^{-18}\,J}$$

Note that the exponent for each of these energies can be kept the same, since using [E (final level) – E (initial level)] requires subtraction. The energies and wavelengths for electron transitions for one electron atoms or ions can be calculated directly from potential energy levels.

Example: Calculate the wavelength of the absorbed photon required to cause a transition from E(2) to E(4) for hydrogen.

(1) ΔE (energy levels) = E (photon) and λ = hc/E(photon)
(2) h = 6.63 × 10^{-34} J-sec; ΔE (energy levels) must be calculated from ΔE (energy levels) = [E (final level) – E (initial level)]

$$\mathbf{E(2)} = \frac{(1)^2 \times (-2.18 \times 10^{-18}\,\mathrm{J})}{2^2} = \frac{(1) \times (-2.18 \times 10^{-18}\,\mathrm{J})}{4} = \mathbf{-0.545 \times 10^{-18}\,J}$$

$$\mathbf{E(4)} = \frac{(1)^2 \times (-2.18 \times 10^{-18}\,\mathrm{J})}{4^2} = \frac{(1) \times (-2.18 \times 10^{-18}\,\mathrm{J})}{16} = \mathbf{-0.136 \times 10^{-18}\,J}$$

ΔE (energy levels) = [E (final level) – E (initial level)] = [E(4) – E(2)]
= (–0.136 × 10^{-18} J) – (–0.545 × 10^{-18} J) = **0.409 × 10^{-18} J**

E(photon) = 0.409 ×10^{-18} J or 4.09 × 10^{-19} J

$$\mathbf{(3)}\ \lambda = \frac{(6.63 \times 10^{-34}\,\mathrm{J\text{-}sec}) \times (3.00 \times 10^{8}\,\mathrm{m/sec})}{4.09 \times 10^{-19}\,\mathrm{J}} = \mathbf{4.86 \times 10^{-7}\,m}\ (486\ \mathrm{nm})$$

Example: Calculate the wavelength of the emitted photon required for a transition from E(5) to E(2) for the C^{+5} ion. Carbon with a +5 charge has only one electron.

(1) ΔE (energy levels) = E (photon) and λ = hc/E(photon)
(2) h = 6.63 × 10^{-34} J-sec; **Z** for carbon = **6** (Do not use the charge in the equation.)

$$\mathbf{E(2)} = \frac{(6)^2 \times (-2.18 \times 10^{-18}\,\mathrm{J})}{2^2} = \frac{(36) \times (-2.18 \times 10^{-18}\,\mathrm{J})}{4} = \mathbf{-19.62 \times 10^{-18}\,J}$$
(keep exponent the same)

$$\mathbf{E(5)} = \frac{(6)^2 \times (-2.18 \times 10^{-18}\,\mathrm{J})}{5^2} = \frac{(36) \times (-2.18 \times 10^{-18}\,\mathrm{J})}{25} = \mathbf{-3.14 \times 10^{-18}\,J}$$

ΔE (energy levels) = [E(2) – E(5)]
= (–19.62 × 10^{-18} J) – (–3.14 × 10^{-18} J) = **–16.48 × 10^{-18} J**

A negative value for ΔE (energy levels) means the photon was emitted; **E(photon)** must always be a **positive** value = **+16.48 × 10^{-18} J**

$$\mathbf{(3)}\ \lambda = \frac{(6.63 \times 10^{-34}\,\mathrm{J\text{-}Sec}) \times (3.00 \times 10^{8}\,\mathrm{m/sec})}{16.48 \times 10^{-18}\,\mathrm{J}} = \mathbf{1.21 \times 10^{-8}\,m}\ (12.1\mathrm{nm})$$

When the number for n approaches infinity, the potential energy approaches zero; at this point, the electron in the atom is free of attraction to the nuclear charge and is unbound. To achieve an unbound state, an electron must absorb sufficient energy to increase its potential energy from a specific negative value (given by the energy level) to zero (E(∞) = 0). The energy required for this transition, which allows the electron to escape from the atom, is termed the **Ionization Energy (I.E.)**.

Example:

(a) Calculate the **Ionization Energy** for an electron in energy level **n = 3** for hydrogen.

$$\mathbf{E(3)} = \frac{(1)^2 \times (-2.18 \times 10^{-18}\,\mathrm{J})}{3^2} = \frac{(1) \times (-2.18 \times 10^{-18}\,\mathrm{J})}{9} = \mathbf{-0.242 \times 10^{-18}\,J}$$

I.E. = [E (final level) – E (initial level)]
= (0) – (-0.242×10^{-18}J) = **+ 0.242×10^{-18} J**

(b) Calculate the wavelength and frequency of the photon that must be absorbed to cause this ionization.

(1) E (photon) = **I.E.** = ΔE (energy levels); λ = hc/E (photon)

(2) h = 6.63×10^{-34} J-sec; E (photon) = 0.242×10^{-18} J

(3) $$\lambda = \frac{(6.63\times10^{-34}\text{J-sec})\times(3.00\times10^{8}\text{ m/sec})}{(0.242\times10^{-18}\text{J})} = \mathbf{8.22\times10^{-7}\,m\ (822\,nm)}$$

(4) ν: ν = E (photon)/h (The ν = c/λ equation could also be used.)
ν = 0.242×10^{-18} J/(6.63×10^{-34} J-sec) = **3.65×10^{14} sec^{-1}**

Example: Calculate the wavelength and frequency of a photon required to reach the ionization energy for an electron in energy level **n = 3** for the ion C^{+5}

$$E(3) = \frac{(6)^2\times(-2.18\times10^{-18}\text{J})}{3^2} = \frac{(36)\times(-2.18\times10^{-18}\text{J})}{9} = \mathbf{-8.72\times10^{-18}\,J}$$

I.E. = [E (final level) – E (initial level)]
= (0) – (-8.72×10^{-18} J) = **+ 8.72×10^{-18} J**

(1) E (photon) = **I.E.** = ΔE (energy levels); λ = hc/E (photon)

(2) h = 6.63×10^{-34} J-sec; E (photon) = 8.72×10^{-18} J

(3) $$\lambda = \frac{(6.63\times10^{-34}\text{J-sec})\times(3.00\times10^{8}\text{ m/sec})}{(8.72\times10^{-18}\text{J})} = \mathbf{2.28\times10^{-8}\ m}\ (22.8\,\text{nm})$$

(4) ν: ν = E (photon)/h
ν = 0.242×10^{-18} J/(6.63×10^{-34} J-sec) = **3.65×10^{14} sec^{-1}**

VII PRACTICE PROBLEMS

1. A specific metal atom absorbs blue light of wavelength = 4.98×10^{-7} m; absorption of one photon of this light causes one electron to be ejected from the metal.
   a) Calculate the frequency of this light.
   b) Calculate the energy of one photon of this light.
   c) Assuming the ratio is one electron ejected per one photon absorbed, how much energy must the metal absorb to release (eject) one **mole** of electrons?

2. Certain minerals absorb ultraviolet light and then emit visible light after converting some of the light energy to heat.
   a) A certain mineral absorbs one photon of wavelength 366 nanometers. How much energy does the mineral absorb?
   b) The mineral converts some of this energy to heat and then emits a photon of wavelength 540 nanometers. How much energy was emitted?
   c) How much of the photon energy was converted to heat?

3. Mercury shows spectral lines for the following electron transitions; calculate the wavelengths and frequencies for emitted photons corresponding to these transitions:
   a) n = 4 ⟶ n = 1 ΔE (energy levels) = 1.075×10^{-18} J
   b) n = 3 ⟶ n = 1 ΔE (energy levels) = 7.83×10^{-19} J
   c) n = 6 ⟶ n = 4 ΔE (energy levels) = 3.64×10^{-19} J

4. Neon shows spectral lines corresponding to the following wavelengths; calculate the ΔE (energy levels) for each of these electron transitions:
   a) λ = 465 nm (green)
   b) λ = 540 nm (yellow)
   c) λ = 680 nm (red)

5. Calculate the following energies, wavelengths, and frequencies for hydrogen; use the equation for exact energy levels: $E(n) = (Z)^2 \times (-2.18 \times 10^{-18}\ J)/n^2$
   a) Calculate all necessary E(n), ΔE (energy levels), and the wavelength of a photon emitted for the electron transition: $n = 6 \longrightarrow n = 3$
   b) Repeat the calculation for the electron transition: $n = 2 \longrightarrow n = 5$
   c) Calculate E(n), ΔE (energy levels), and the frequency of a photon that must be absorbed for the ionization of an electron in energy level n = 1.

6. Repeat the complete problem #5 for the one-electron ion He^+. The equation for exact energy levels can be used; be certain to use the correct value for Z.

7. a) Calculate the wavelength of a photon emitted for the electron transition $n = 5 \longrightarrow n = 2$ for the fluoride +8 ion (F^{+8}).
   b) Calculate the frequency of a photon that must be absorbed for the ionization of an electron in energy level n = 1 for the fluoride +8 ion (F^{+8}).

VIII ANSWERS TO PRACTICE PROBLEMS

1. A specific metal atom absorbs blue light of wavelength = 4.98×10^{-7} m; absorption of one photon of this light causes one electron to be ejected from the metal.
   a) Calculate the frequency of this light.
   $c = \lambda\nu$, solve for frequency: $\nu = c/\lambda$
   $\nu = (3.00 \times 10^8\ m/sec)/(4.98 \times 10^{-7}\ m) = 6.02 \times 10^{14}/sec$

   b) Calculate the energy of one photon of this light that is absorbed by the one electron.
   E (photon) = hc/λ (or use part (a) and E (photon) = $h\nu$
   E (photon) = $(6.63 \times 10^{-34}\ J\text{-}sec) \times (3.00 \times 10^8\ m/sec)/(4.98 \times 10^{-7}\ m)$
   $= 3.99 \times 10^{-19}\ J$

   c) Assuming the ratio is one electron ejected per one photon absorbed, how much energy must the metal absorb to release (eject) one mole of electrons?
   E per mole = $(3.99 \times 10^{-19}\ J/electron) \times (6.022 \times 10^{23}\ electrons/mole) = 2.40 \times 10^5\ J$

2. Certain minerals absorb ultraviolet light and then emit visible light after converting some of the light energy to heat.
   a) A certain mineral absorbs one photon of wavelength 366 nanometers. How much energy does the mineral absorb?
   E (photon) = hc/λ;;$\lambda = 366\ nm \times 10^{-9}\ m/nm = 3.66 \times 10^{-7}\ m$
   E (photon) = $(6.63 \times 10^{-34}\ J\text{-}sec) \times (3.00 \times 10^8\ m/sec)/(3.66 \times 10^{-7}\ m)$
   $= 5.43 \times 10^{-19}\ J$ (absorbed)

   b) The mineral converts some of this energy to heat and then emits a photon of wavelength 540 nanometers. How much energy was emitted?
   E (photon) = hc/λ;;$\lambda = 540\ nm \times 10^{-9}\ m/nm = 5.40 \times 10^{-7}\ m$
   E (photon) = $(6.63 \times 10^{-34}\ J\text{-}sec) \times (3.00 \times 10^8\ m/sec)/(5.40 \times 10^{-7}\ m)$
   $= 3.68 \times 10^{-19}\ J$ (emitted)

c) How much of the photon energy was converted to heat?
(5.43×10^{-19} J absorbed) – (3.68×10^{-19} J emitted) = 1.75×10^{-19} J (heat)

3. Mercury shows spectral lines for the following electron transitions; calculate the wavelengths and frequencies for emitted photons corresponding to these transitions:

a) $n = 4 \longrightarrow n = 1$ ΔE (energy levels) = 1.075×10^{-18} J
E (photon) = ΔE (energy levels); E (photon) = 1.075×10^{-18} J
E (photon) = hc/λ; solve directly for λ: λ = hc/E (photon)
λ = (6.63×10^{-34} J-sec) × (3.00×10^{8} m/sec)/(1.075×10^{-18} J)
= 1.85×10^{-7}m or 185 nm
c = λν, solve for frequency: ν = c/λ
ν = (3.00×10^{8} m/sec)/(1.85×10^{-7} m) = 1.62×10^{15} sec^{-1}

b) $n = 3 \longrightarrow n = 1$ ΔE (energy levels) = 7.83×10^{-19} J
Solve as for (a): λ = 2.54×10^{-7} m (254 nm); ν = 1.18×10^{15} sec^{-1}

c) $n = 6 \longrightarrow n = 4$ ΔE (energy levels) = 3.64×10^{-19} J
Solve as for (a): λ = 5.46×10^{-7} m (546 nm); ν = 5.49×10^{14} sec^{-1}

4. Neon shows spectral lines corresponding to the following wavelengths; calculate the ΔE (energy levels) for each of these electron transitions:

a) λ = 465 nm (green) E (photon) = Δ E (energy levels)
E (photon) = hc/λ; λ = 465 nm × 10^{-9} m/ nm = 4.65×10^{-7}m
E (photon) = (6.63×10^{-34} J-sec) × (3.00×10^{8} m/sec)/(4.65×10^{-7} m)
= 4.28×10^{-19} J = ΔE (energy levels)

b) λ = 540 nm (yellow): Solve as for (a): ΔE (energy levels) = 3.68×10^{-19} J

c) λ = 680 nm (red): Solve as for (a): ΔE (energy levels) = 2.93×10^{-19} J

5. Calculate the following energies, wavelengths, and frequencies for hydrogen; use the equation for exact energy levels: $E(n) = \frac{(Z)^2 \times (-2.18 \times 10^{-18}\ J)}{n^2}$

a) Calculate all necessary E(n), ΔE (energy levels), and the wavelength of a photon emitted for the electron transition: $n = 6 \longrightarrow n = 3$

$$E(3) = \frac{(1)^2 \times (-2.18 \times 10^{-18} J)}{3^2} = \frac{(1) \times (-2.18 \times 10^{-18} J)}{9} = -0.242 \times 10^{-18}\ J$$

$$E(6) = \frac{(1)^2 \times (-2.18 \times 10^{-18} J)}{6^2} = \frac{(1) \times (-2.18 \times 10^{-18} J)}{36} = -0.0606 \times 10^{-18}\ J$$

Δ E (energy levels) = [E(3) – E(6)] = (– 0.242×10^{-18} J) – (-0.0606×10^{-18} J)
= -0.1814×10^{-18} J
The value is negative because a photon was emitted; the energy of the photon must always be a positive value.
E (photon) = | ΔE (energy levels) |; E (photon) = + 0.1814×10^{-18} J
λ = hc/E (photon);
λ = (6.63×10^{-34} J-sec) × (3.00×10^{8} m/sec)/(0.1814×10^{-18} J)
= 1.096×10^{-6} m or 1096 nm

b) Calculate all necessary E(n), ΔE (energy levels), and the wavelength of a photon absorbed for the electron transition: $n = 2 \longrightarrow n = 5$

$$E(2) = \frac{(1)^2 \times (-2.18 \times 10^{-18} J)}{2^2} = \frac{(1) \times (-2.18 \times 10^{-18} J)}{4} = -0.545 \times 10^{-18} J$$

$$E(5) = \frac{(1)^2 \times (-2.18 \times 10^{-18} J)}{5^2} = \frac{(1) \times (-2.18 \times 10^{-18} J)}{25} = -0.0872 \times 10^{-18} J$$

Δ E (energy levels) = [E(5) – E(2)] = (-0.0872×10^{-18} J) – (-0.545×10^{-18} J)
= $+0.4578 \times 10^{-18}$ J
E (photon) = + 0.4578×10^{-18} J; λ = hc/E (photon) = 4.35×10^{-7} m (435 nm)

c) Calculate E(n), ΔE (energy levels), and the frequency of a photon that must be absorbed for the ionization of an electron in energy level n = 1.

$$E(1) = \frac{(1)^2 \times (-2.18 \times 10^{-18} J)}{1^2} = \frac{(1) \times (-2.18 \times 10^{-18} J)}{1} = -2.18 \times 10^{-18} J$$

ΔE (energy levels) = [E(ionized) – E(1)] = (0 J) – (– 2.18×10^{-18} J)
= $+2.18 \times 10^{-18}$ J
E (photon) = $+2.18 \times 10^{-18}$ J; ν = E (photon)/h = 3.29×10^{15} sec^{-1}

6. Repeat the complete problem #5 for the one-electron ion He^+. The equation for exact energy levels can be used; be certain to use the correct value for Z.

a) Electron transition: $n = 6 \longrightarrow n = 3$

$$E(3) = \frac{(2)^2 \times (-2.18 \times 10^{-18} J)}{3^2} = \frac{(4) \times (-2.18 \times 10^{-18} J)}{9} = -0.969 \times 10^{-18} J$$

$$E(6) = \frac{(2)^2 \times (-2.18 \times 10^{-18} J)}{6^2} = \frac{(4) \times (-2.18 \times 10^{-18} J)}{36} = -0.242 \times 10^{-18} J$$

Δ E (energy levels) = [E(3) – E(6)] = (-0.969×10^{-18} J) – (-0.242×10^{-18} J)
= -0.727×10^{-18} J
E (photon) = |Δ E (energy levels)|; E (photon) = + 0.727×10^{-18} J
λ = hc/E (photon);
λ = (6.63×10^{-34} J-sec) × (3.00×10^{8} m/sec)/(0.727×10^{-18} J)
= 2.74×10^{-7} m or 274 nm

b) Electron transition: $n = 2 \longrightarrow n = 5$

$$E(2) = \frac{(2)^2 \times (-2.18 \times 10^{-18} J)}{2^2} = \frac{(4) \times (-2.18 \times 10^{-18} J)}{4} = -2.18 \times 10^{-18} J$$

$$E(5) = \frac{(2)^2 \times (-2.18 \times 10^{-18} J)}{5^2} = \frac{(4) \times (-2.18 \times 10^{-18} J)}{25} = -0.349 \times 10^{-18} J$$

Δ E (energy levels) = [E(5) – E(2)] = (-0.349×10^{-18} J) – (-2.18×10^{-18} J)
= $+1.83 \times 10^{-18}$ J
E (photon) = + 1.83×10^{-18} J; λ = hc/E (photon) = 1.09×10^{-7} m (109 nm)

c) Ionization of an electron in energy level n = 1.

$$E(1) = \frac{(2)^2 \times (-2.18 \times 10^{-18}\,J)}{1^2} = \frac{(4) \times (-2.18 \times 10^{-18}\,J)}{1} = -8.72 \times 10^{-18}\,J$$

Δ E (energy levels) = [E(ionized) – E(1)] = (0 J) – (-8.72×10^{-18} J)
= $+8.72 \times 10^{-18}$ J
E (photon) = $+8.72 \times 10^{-18}$ J; ν = E (photon)/h = 1.32×10^{16} sec^{-1}

7. a) Calculate the wavelength of a photon emitted for the electron transition $n = 5 \longrightarrow n = 2$ for the fluoride +8 ion (F^{+8}).

b) Calculate the frequency of a photon that must be absorbed for the ionization of an electron in energy level n = 1 for the fluoride +8 ion (F^{+8}).

a) electron transition: $n = 5 \longrightarrow n = 2$

$$E(2) = \frac{(9)^2 \times (-2.18 \times 10^{-18}\,J)}{2^2} = \frac{(81) \times (-2.18 \times 10^{-18}\,J)}{4} = -44.15 \times 10^{-18}\,J$$

$$E(5) = \frac{(9)^2 \times (-2.18 \times 10^{-18}\,J)}{5^2} = \frac{(81) \times (-2.18 \times 10^{-18}\,J)}{25} = -7.06 \times 10^{-18}\,J$$

Δ E (energy levels) = [E(2) – E(5)] = (-44.15×10^{-18} J) – (-7.06×10^{-18} J)
= -37.09×10^{-18} J
E (photon) = + 37.09×10^{-18} J; photon was emitted.
λ = hc/E (photon) = 5.36×10^{-9} m (5.36 nm)

b) Ionization of an electron in energy level n = 1.

$$E(1) = \frac{(9)^2 \times (-2.18 \times 10^{-18}\,J)}{1^2} = \frac{(81) \times (-2.18 \times 10^{-18}\,J)}{1} = -176.6 \times 10^{-18}\,J$$

Δ E (energy levels) = [E(ionized) – E(1)] = (0 J) – (-176.6×10^{-18} J)
= $+176.6 \times 10^{-18}$ J
E (photon) = + 176.6×10^{-18} J or + 1.77×10^{-16} J
ν = E (photon)/h = 2.67×10^{17} sec^{-1}

13 Atomic Orbitals and the Electronic Structure of the Atom

I QUANTUM THEORY AND ELECTRON ORBITALS

The nucleus remains unchanged during chemical reactions. Chemical processes are based on electron behavior, such as exchange or sharing of electrons between atoms. The electronic structure of the atom, thus, provides the basis for the chemical behavior of the elements. The electromagnetic force of attraction between protons in the nucleus and the electrons is proportional to the size of the total positive and negative charges and inversely proportional to the square of the distance between the charges:

$$F(el) = \frac{k(q\ of\ p^+)(q\ of\ e^-)}{r^2}$$

q = total size of the nuclear charge
r = distance between charges

A fundamental requirement of physics (Heisenberg uncertainty principle) states that both the position and velocity of an electron cannot both be known with precision. For a description of the atom, the result is that neither the **exact** position of the electron nor its path of motion can be known. Electrons in an atom cannot have all possible potential energies; only certain values are available (Chapter 12). This means that the energy levels in an atom are spaced apart such that they have only limited specific potential energies, like rungs on a ladder. The energy levels in an atom are said to be quantized. A summary of the relationships described in Chapter 12 are:

An electron in:

lower energy level ⟷ **higher energy level**
lower potential energy ⟷ **higher potential energy**
closer to the nucleus ⟷ **farther from nucleus**
more stable ⟷ **less stable**

Electrons in atoms are described according to the theory of **quantum mechanics**, which uses the concept that electrons in the atom can be described by their wave nature **(wave/particle duality)**. Quantum mechanics states that each electron in an atom can be identified with:

(1) A potential energy based on an electron occupying a specific energy level, and
(2) A probable region of space around the nucleus in which the electron may be located.

Since each element has a different number of protons (positive charges) in the nucleus, each element has different specific values for the electron energy levels. However, all atoms of all elements are structured such that energy levels form the same general patterns and can be described by the same rules of electron distribution.

A complete description of an electron in an atom is based on the concept that all electrons in an atom occupy specific **orbitals**. An **orbital** is an identification of a value for potential energy (a specific energy level) and a description of the probable region of space around the nucleus in which the electron occupying that orbital is expected to be found.

Orbitals require three designators to fully describe the electron information; these are termed **quantum numbers (n, l, m_l**, although letters are also used as labels). This is the "address" of the electron. The **principal** quantum number (symbol = n) designates what is termed the major shell number. The major shell number occupied by the electron provides the approximate distance from the nucleus and, thus, the approximate potential energy. The major shell number (**n**) ranges from **n = 1** (first shell) up to **n** = infinity. Atoms in their normal state require only up to the eighth shell. The distance and energy relationships for the electron configurations are:

n = 1 ⟵⟶ **n = 8**
closest to nucleus ⟵⟶ **farthest from nucleus**
lowest potential energy ⟵⟶ **highest potential energy**

The **angular momentum** quantum number (symbol = l) designates the **orbital shape**; the shape of the region of space in which the electron has a high probability of being found. The possible orbital shapes based on probability are termed subshells. The possible values for (**l**) and, therefore, the possible subshells depend on the principal quantum number (**n**): (**l**) can have all integer values from **0** to **n–1**. Subshell shapes influence the approximate average distance of the electron from the nucleus and, thus, contribute to determining the potential energy of an orbital.

The **magnetic** quantum number (symbol = m_l) is determined from the subshell type and can have all possible integer values from the negative value of (**l**) to the positive value of (**l**). The magnetic quantum number (m_l) can be thought of as designating the specific preferred directions for each of the possible orbital shapes. Thus, they depend on the subshell. The orbital preferred direction derived from the m_l quantum number can be visualized on a set of three-dimensional axes; the numerical quantum numbers are then translated to x-axis, y-axis, z-axis, or combinations.

The general pattern of orbitals is derived from solving certain equations called **wave equations**. The orbital structure of all atoms is defined as a sequence of all possible **valid** combinations of the three quantum numbers (**n, l, m_l**); each orbital is a unique set of these three numbers. A valid combination requires following the numerical rules for quantum number values.

Example: Determine all possible orbitals for shell number **3 (n = 3)** of an atom.

The values of (**l**) are all integer values from **0** to **n–1**. Since n = 3, n–1 = 3 – 1 = **2**.
The values of (**l**), therefore, are **0** or **1** or **2**.

For each possible value of (**l**), the possible integer values of m_l range from [(**–l**) to (**+l**)].

For (**l**) = **0**, m_l can only be **0**;
for (**l**) = **1**, m_l can be **–1** or **0** or +**1**;
for (**l**) = **2**, m_l can be **–2** or **–1** or **0** or **+1** or +**2**.

The atomic orbitals in shell 3 are all possible valid combinations defined as (**n, l, m_l**).

n = 3, (l) = 0, m_l = 0 is the orbital **(3, 0, 0)**
n = 3, (l) = 1, m_l = –1 or **0** or **+1** produce the orbitals **(3, 1, –1) (3, 1, 0)** and **(3, 1, +1)**
n = 3, (l) = 2, m_l = –2 or **–1** or **0** or **+1** or **+2** produce the orbitals:
(3, 2, –2) (3, 2, –1) (3, 2, 0) and **(3, 2, +1) (3, 2, +2)**

The complete list of orbitals for shell number 3 has been generated. All other shells can be generated using the same set of quantum rules.

The subshell quantum number **(l)** is usually designated with identifying letters: **(l)** = **0** is identified as an s-subshell; **(l)** = **1** is identified as a **p**-subshell; **(l)** = **2** is identified as a **d**-subshell; **(l)** = **3** is identified as an **f**-subshell. Higher value atomic subshells are possible but are not normally occupied by electrons. Using the letter designations, the possible shapes of the orbitals (probability region shapes) are: **s**-subshell = spherical shape; **p**-subshell = dumbbell shape; **d**-subshell = (generally) cloverleaf shape; **f**-subshell = complicated. The influence of the orbital shape on the approximate average distance from the nucleus and potential energy of an orbital is:

s - subshell < **p - subshell** < **d - subshell** < **f - subshell**
closest to nucleus ⟷ **farthest from nucleus**
lowest potential energy ⟷ **highest potential energy**

The combination of shell number plus subshell letter specifies the energy of a specific orbital. The values of $\mathbf{m_l}$ (different preferred directions) do not normally influence the average distance from the nucleus and, thus, do not normally affect the energy of the orbital. The preferred directions (values of $\mathbf{m_l}$) are often not indicated in electron descriptions. The importance of the $\mathbf{m_l}$ value is to count the total number of possible orbitals with any given shape; that is, the total number of possible orbitals per each subshell. Based on the subshell dependent range of $\mathbf{m_l}$ values:

s-subshell (spherical) has only one possible direction, **one** orbital per **s**-subshell.
p-subshell (dumbbell) has **three** possible directions, **three** orbitals per **p**-subshell.
d-subshell (cloverleaf) has five possible directions, **five** orbitals per **d**-subshell.
f-subshell (complicated) has seven possible directions, **seven** orbitals per **f**-subshell.

Orbital Architecture and Electron Spin

An electron in an atom is associated with a spin. The spin can be thought of as clockwise or counterclockwise and is measured by its magnetic field ("north pole" or "south pole"). The electron is assigned a fourth quantum number; a spin quantum number with the symbol $\mathbf{m_s}$. The electron can take the values $\mathbf{m_s} = (+\frac{1}{2})$ or $\mathbf{m_2} = (-\frac{1}{2})$; the [$\frac{1}{2}$] value in this context is important only to establish the two possible values.

The distribution of electrons in orbitals is governed by the **Pauli Exclusion Principle**, which states that no two electrons in an atom can have the same set of four quantum numbers (**n, l,** $\mathbf{m_l}$, $\mathbf{m_s}$). Each orbital is a unique set of the first three quantum numbers (**n, l,** $\mathbf{m_l}$). The principle requires that no two electrons in the same orbital can have the same spin. The two possible values for $\mathbf{m_s}$ restricts the maximum number of electrons in any orbital to two; one with each of the two possible spins.

Each element will have different specific electron energies and electron distances based on the relationship between its nucleus and electrons. However, all atoms are constructed using the same pattern of shells, subshells (shapes), and directions (number of orbitals per subshell). The complete set of orbitals generated from valid combinations of the three orbital quantum numbers (**n, l,** $\mathbf{m_l}$) are summarized in the table below. For the purposes of electron configurations, shells #5 through #8 are filled with electrons in the manner of shell #4; subshells above the f-subshell are not normally used for electron description purposes. The total number of electrons for each shell is determined by 2 e^- per orbital × # of orbitals per shell.

Shell Number	Subshell Letters for Each Shell	# of Orbitals per Each Subshell	Total # of Orbitals	Total # of e^- per Shell
1	s	one	1	2
2	s	one	4	8
	p	three		
3	s	one	9	18
	p	three		
	d	five		
4	s	one	16	32
	p	three		
	d	five		
	f	seven		
5–8	s	one	16	32
	p	three		
	d	five		
	f	seven		

II DETERMINING ELECTRON CONFIGURATIONS OF ELEMENTS

The **electron configuration** of an element is the full description of all orbitals occupied by all the electrons in the atom of the specific element. The description identifies each orbital occupied and the number of electrons per orbital; the information can be displayed in the full form or different abbreviated forms.

An orbital is identified by symbols indicating the shell #, the subshell letter, and (optionally) the preferred direction as a subscript:

{shell #}{subshell letter}$_{\textbf{preferred direction}}$

Example:

The one orbital of shell **#1**, subshell **s = 1s**
One of the three orbitals of shell **#2**, subshell **p = 2p**
The orbital of shell **#2**, subshell **p** that lies along the **x**-axis = **$2p_x$**

The combination of shell number plus subshell letter (shape) specifies the energy of the orbital. The preferred direction does not (normally) affect the energy; thus, all orbitals of a specified shell and subshell will have the same energy when determining an electron configuration.

The energy sequence for shell number showed that potential energy increases from shell #1 to shell #8; the sequence for subshells showed energy increasing in the order s < p < d < f. If either the shell or subshell is kept constant for energy comparisons, these relationships always hold true. For example, 1s < 2s < 3s < 4s < 5s < 6s <7s, or 4s < 4p < 4d < 4f. For the complete ("mix and match") combinations of all shells and subshells, the potential energy sequence generally follows the major shell order with, however, the following qualifications. Electrons fill the (n)d subshells only after the (n + 1)s is completely filled, and (with a single electron exception) electrons fill the (n)f subshell directly after the (n + 2)s. The complete potential energy sequence up to atomic number 88 is:

lowest potential energy ⟷ highest potential energy
1s < 2s < 2p < 3s < 3p < 4s < 3d < 4p < 5s < 4d < 5p < 6s < 4f < 5d < 6p < 7s

The restriction for electron spin means that orbitals can have only three different occupancy results: (a) an orbital can have no electrons; (b) an orbital can have one electron of **either** spin; or (c) an orbital can have two electrons of **opposite** spin (one of each spin type); two electrons of opposite spin in an orbital are **spin-paired.**

There are two general methods for displaying an electron configuration.

The **orbital diagram** is a line, box, or circle representing an orbital; the orbital is labeled with the shell # and subshell letter. Each electron in an orbital is represented by an "up" arrow (↑) or "down" (↓) arrow to indicate the two possible electron spins (or magnetic field).

The **orbital notation** shows only the shell # and subshell letter with the number of electrons in each shell and subshell combination indicated by a superscript after the number and letter symbol. The information does not display individual orbitals or electron spins.

Example: Show a possible electron configuration for six electrons (carbon) in the orbital diagram method and the orbital notation method.

The orbital diagram used here is a simple line to represent one orbital; all three orbitals of the p-subshell are shown, even though one of them has no electrons. Each arrow represents an electron in the orbital; completely filled orbitals with two electrons show the spin-pairing of opposite direction arrows.

(6 e⁻) = 1s ↑↓ 2s ↑↓ 2p ↑. ↑. __

The orbital notation is more concise, showing the orbital labels with the superscript indicating only the number of electrons in the subshell instead of each individual orbital.

$$1s^2\ 2s^2\ 2p^2$$

Construction of electron configurations for atoms is based on the requirement that all electrons are in the lowest potential energy orbitals possible. Under these conditions, the atom is said to be in the **ground state**. When one or more electrons are in higher energy orbitals vs. the ground state, the atom is said to be in an **excited state**. The "expected" electron configuration is determined by following the general rules listed. Some atoms may show a one electron exception to the rules; these exceptions do not alter the overall concepts and conclusions.

The rules for producing the ground state electron configuration of an atom are derived from the Pauli exclusion principle and Hund's rule. Hund's rule states that electrons in atoms must maximize electron spin (magnetic field).

Rules for Ground State Electron Configurations

(1) Count the number of electrons in the neutral form of the element; this must be equal to the number of protons in the nucleus, the atomic number.

(2) Each orbital must take a maximum of two electrons per orbital.

(3) Two electrons in one specific orbital must have opposite spins (spin-paired).

(4a) Fill available orbitals in order of increasing potential energy of the orbitals; follow the potential energy sequence:

$$1s < 2s < 2p < 3s < 3p < 4s < 3d < 4p < 5s < 4d < 5p < 6s < 4f < 5d < 6p < 7s$$

(4b) Fill a higher energy orbital only after all lower energy orbitals have been completely filled subject to rule **(2)**. Use all available orbitals in each subshell: **one** orbital per **s**-subshell; **three** orbitals per **p**-subshell; five orbitals per **d**-subshell; **seven** orbitals per **f**-subshell.

(5) Electrons will not pair up in an orbital if another empty orbital of **equal energy** (same subshell) is available; this is due to electron-electron repulsion. Electrons remain unpaired until all equal energy orbitals become half-filled; additional electrons must then pair up based on rule **(4b)**.
(6) Electrons that singly occupy an orbital are said to be unpaired. All unpaired electrons in an atom must have the same spin (Hund's rule). The type of spin ("up" or "down") is **not** specifically known in the absence of an experiment; therefore, either spin type can be selected for an orbital diagram.

Example: Determine the ground state electron configuration of nitrogen; show both the orbital diagram method and the orbital notation method.

Nitrogen has **7** electrons; the **first** electron must be placed in the lowest potential energy orbital, the 1s: **1s ↑.**; the spin arrow can be shown either "up" or "down."

The **second** electron must enter the same orbital based on rule **4**: **1s ↑↓**; rule **3** requires that the two electrons in the 1s orbital are shown as spin-paired.

The next potential energy orbital is the 2s; the **third** and **fourth** electrons enter this orbital following the same rules: **2s ↑↓**. The **fifth** electron enters the 2p-subshell, which contains three equal energy orbitals. It is not important which of where to place the three 2p orbitals is selected for placement of the fifth electron. The arrow indicating spin can be shown in either direction.

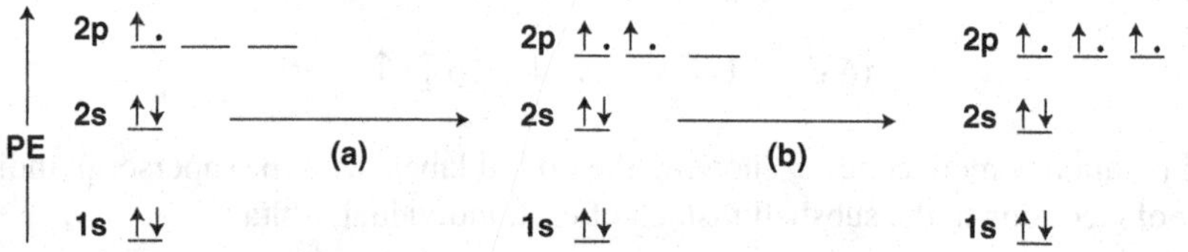

PE		(a)		(b)	
	2p ↑. __ __		2p ↑. ↑. __		2p ↑. ↑. ↑.
	2s ↑↓	⟶	2s ↑↓	⟶	2s ↑↓
	1s ↑↓		1s ↑↓		1s ↑↓

a. The **sixth** electron must enter one of the 2p-orbitals according to rule **(4)**; rule **(5)** requires that the electron enter an empty 2p-orbital and rule **(6)** requires that both electrons in the 2p-subshell have the same spin.
b. The **seventh** electron must also enter one of the 2p-orbitals according to rule **(4)**; applying rules **(5)** and **(6)** produces the complete orbital diagram.

The orbital notation is **$1s^2$ $2s^2$ $2p^3$**; no information concerning the three electrons in the 2p-subshell is indicated.

Example: Determine the ground state electron configuration of oxygen; show both the orbital diagram method and the orbital notation method.

Oxygen has **8** electrons. All atoms follow the same orbital potential energy sequence and follow the same electron filling rules. Thus, the first seven electrons of oxygen must fill orbitals in the exact manner that was described for nitrogen; the electron configuration of oxygen is solved by determining the result for the eighth electron.

⟶ PE ⟶

$_8O$ ($8\ e^-$) = 1s ↑↓ 2s ↑↓ 2p ↑↓ ↑. ↑. ; notation = **$1s^2$ $2s^2$ $2p^4$**

The **eighth** electron must pair up in one of the 2p-orbitals since there are no more equal energy orbitals available in the 2p-subshell. The next available empty orbital is the higher energy 3s-orbital; this cannot be used according to rule **(4)**.

Example: Determine the ground state electron configuration of neon; show both the orbital diagram method and the orbital notation method.

Neon has **10** electrons:

—————— PE ——————→

$_{10}$**Ne (10 e$^-$) = 1s ↑↓ 2s ↑↓ 2p ↑↓ ↑↓ ↑↓ ; notation = $1s^2 2s^2 2p^6$**

All available orbitals in shell #2 are filled; this is termed a closed shell. The noble gas neon represents a completely filled s-subshell and p-subshell of the last shell to be filled.

Example: Show both forms of the electron configuration for the two elements selected from row 3 in the periodic table; silicon and argon.

—————— PE ——————→

$_{14}$**Si (14 e$^-$) = 1s ↑↓ 2s ↑↓ 2p ↑↓ ↑↓ ↑↓ 3s ↑↓ 3p ↑ ↑ __**

= $1s^2 2s^2 2p^6 3s^2 3p^2$

$_{18}$**Ar (18 e$^-$) = 1s ↑↓ 2s ↑↓ 2p ↑↓ ↑↓ ↑↓ 3s ↑↓ 3p ↑↓ ↑↓ ↑↓**

= $1s^2 2s^2 2p^6 3s^2 3p^6$

The noble gas argon represents a completely filled s-subshell and p-subshell of the last shell to be filled (shell #3). Note that each row number in the periodic table matches the shell # that begins to fill. The periodic table always ends a row with a filled s and p-subshell of the shell number that matches the row number.

Example: Show the electron configuration for the row 4 element calcium.

—————— PE ——————→

$_{20}$**Ca (20 e$^-$) = 1s ↑↓ 2s ↑↓ 2p ↑↓ ↑↓ ↑↓ 3s ↑↓ 3p ↑↓ ↑↓ ↑↓ 4s ↑↓**

= $1s^2 2s^2 2p^6 3s^2 3p^6 4s^2$

The potential energy sequence for orbitals shows that the next subshell to fill after the 3p is the 4s-orbital in shell #4, **not** one of the orbitals in the 3d-subshell; the 4s-orbital must fill before electrons can enter the 3d subshell of shell #3. This is shown in the periodic table by the start of row 4, which begins the filling of shell #4.

Closed Shells and the Noble Gas Electron Configuration

As the number of electrons in the atom increases, both the orbital diagram and orbital notation methods begin to become cumbersome; a combined method can be used that abbreviates both methods.

The **outer shell** of an atom is defined as the shell with the highest shell number (highest principal quantum number), which is occupied by at least one electron; this shell is farthest from the nucleus. An inner shell is any shell with a shell number lower than the outer shell. The outer shell of an atom can never be filled past the s-subshell plus p-subshell. This is because a new outer shell must be started before the d-subshell of any shell can begin to acquire electrons: the 4s-orbital fills before the 3d-subshell; the 5s-orbital fills before the 4d-subshell; the 6s-orbital fills before the 5d-subshell.

Any outer shell with a completely filled s-subshell plus p-subshell is termed a **closed outer** shell. This specific electron configuration is represented by the last element in each row of the periodic table, the noble gases: He, Ne, Ar, Kr, Xe.

$$_{2}He = 1s^2$$

$$_{10}Ne = 1s^2\ 2s^2\ 2p^6$$

$$_{18}Ar = 1s^2\ 2s^2\ 2p^6\ 3s^2\ 3p^6$$

$$_{36}Kr = 1s^2\ 2s^2\ 2p^6\ 3s^2\ 3p^6\ 4s^2\ 3d^{10}\ 4p^6$$

$$_{54}Xe = 1s^2\ 2s^2\ 2p^6\ 3s^2\ 3p^6\ 4s^2\ 3d^{10}\ 4p^6\ 5s^2\ 4d^{10}\ 5p^6$$

The noble gas electron configuration can be used to abbreviate the orbital notation for any element. Since electrons fill orbitals in the same sequence for all elements, it is true that the **first 10** electrons for all elements show the same pattern as **neon**, the **first 18** electrons will show the same pattern as **argon**, the **first 36** electrons will show the same pattern as **krypton**, and the **first 54** electrons will show the same pattern as **xeon**.

The noble gas symbol enclosed in brackets can represent the orbital notation for the first 10, 18, 36, or 54 electrons; the atomic number shown with the symbol is optional. To use the abbreviation, replace the orbital notation of the selected number of electrons with the noble gas symbol; select the noble gas abbreviation nearest to the element to be described:

[$_{10}$Ne] or [Ne] represents $1s^2\ 2s^2\ 2p^6$

[$_{18}$Ar] or [Ar] represents $1s^2\ 2s^2\ 2p^6\ 3s^2\ 3p^6$

[$_{36}$Kr] or [Kr] represents $1s^2\ 2s^2\ 2p^6\ 3s^2\ 3p^6\ 4s^2\ 3d^{10}\ 4p^6$

[$_{54}$Xe] or [Xe] represents $1s^2\ 2s^2\ 2p^6\ 3s^2\ 3p^6\ 4s^2\ 3d^{10}\ 4p^6\ 5s^2\ 4d^{10}\ 5p^6$

Example: Show the abbreviated orbital notation for the configuration of calcium.

$$_{20}\mathbf{Ca}\ \mathbf{(20\ e^-)} = 1s^2\ 2s^2\ 2p^6\ 3s^2\ 3p^6\ 4s^2 = \mathbf{[_{18}Ar]\ 4s^2}$$

The orbital diagram and the orbital notation can be combined to show the maximum information within the minimum space. All completely filled subshells must show completely filled orbitals with spin-paired electrons; only the last subshell being filled can provide different information on electron occupancy and spin. The combined method shows the complete electron configuration using the abbreviated orbital notation; an additional orbital **diagram** is then drawn only for the last (usually partially filled) subshell, which is in the process of being filled.

Once the 4s orbital is filled, the **3d**-subshell of shell **#3** will begin to accept electrons; the d-subshells have **five** orbitals per subshell. This is shown in the periodic table by the B-group metals, the ten elements (Sc to Zn) in row 4 inserted between the tall (A) columns. All of the B-group metals fill a d-subshell of an **inner** shell between the s-subshells and p-subshells of an outer shell. They are termed **transition metals** because this section in the periodic table represents a transition between the two outer shell subshells.

Example: Show the combined abbreviated methods for the electron configurations of manganese, zinc and ruthenium.

$_{25}$**Mn = [$_{18}$Ar]4s^2 3d^5 3d** ↑· ↑· ↑· ↑· ↑· The five electrons in the 3d-subshell remain unpaired.
$_{30}$**Zn = [$_{18}$Ar]4s^{2}3d^{10} 3d** ↑↓ ↑↓ ↑↓ ↑↓ ↑↓ The 3d-subshell of shell #3 is filled; further elements will now begin to fill the p-subshell in shell #4.
$_{44}$**Ru = [$_{36}$Kr]5s^2 4d^6 4d** ↑↓· ↑· ↑· ↑· ↑· This transition metal shows electrons filling the 4d-subshell inserted between the 5s-subshell and the 5p-subshell.

Exceptions to the expected electron configuration sometimes occur for transition metals. A completely filled d-subshell or an exactly half-filled d-subshell has extra stability. The small potential energy difference between the (n + 1)s and (n)d subshells can produce a one (or two) electron exception.

Example: Show the expected electron configuration for chromium then compare to the given actual configuration.

Expected $_{24}$Cr=[$_{18}$Ar]4s^{2}3d^{4}3d ↑· ↑· ↑· ↑· ___

Actual $_{24}$Cr=[$_{18}$Ar]4s^{1}3d^{5}3d ↑· ↑· ↑· ↑· ↑·

The actual configuration shows one electron from the 4s "moved" to the empty 3d orbital.

Row **6** begins by filling the **6**s-orbital of shell **#6**. Row 6 represents the first insertion of a sequence of elements that fill an **f**-subshell; the **f**-subshell has **seven** orbitals per subshell (14 total electrons). Based on the orbital potential energy sequence, the 4f-subshell is expected to start filling after the completion of the 6s-orbital and before the filling of the 5d-subshell. In fact, a one electron exception occurs; one electron enters the 5d-subshell before the **4f-subshells** begins to fill.

Example: Show the expected electron configuration for lanthanum (La), then compare to the given actual configuration.

Expected $_{57}$**La = [$_{54}$Xe]6s^{2}4f^1 4f ↑.** ___ ___ ___ ___ ___ ___

Actual $_{57}$**La = [$_{54}$Xe]6s^{2}5d^1 5d ↑.** ___ ___ ___ ___

The actual configuration shows one electron in the 5d before the 4f fills.

The 4f-subshell sequence is represented in the periodic table by the upper 14-element row (Ce to Lu) shown detached from the rest of the table. The electron configurations of these elements will show this one electron exception; upon completion of the 4f-subshell, the remaining electron configurations for the other row 6 elements return to the expected result.

Example: Show the expected electron configurations for halfnium (Hf), iridium (Ir), and bismuth (Bi).

$_{72}$Hf = [$_{54}$Xe] 6s^2 4f^{14} 5d^1 5d ↑· ___ ___ ___ ___

$_{77}$Ir = [$_{54}$Xe] 6s^2 4f^{14} 5d^7 5d ↑↓ ↑↓· ↑· ↑· ↑·

$_{83}$Bi = [$_{54}$Xe] 6s^2 4f^{14} 5d^{10} 6p^3 6p ↑· ↑· ↑·

III ELECTRON CONFIGURATIONS AND ORGANIZATION OF THE PERIODIC TABLE

The periodic table was originally organized to place elements with similar chemical behavior in the same column. Chemical behavior is dominated by the electron configuration of the outer shell of an element; elements with the same outer shell s- and p-subshell (or inner d-subshell) electron configuration fall in the same column.

Example: All noble gases (column 8) have a closed outer shell configuration of **ns^2 np^6**. Each has **2** electrons in the **s**-subshell and **6** electrons in the **p**-subshell of the outer shell; **n** represents the different outer shells. All halogens (column 7), which are one column inward from the noble gases have an outer shell configuration of **ns^2 np^5**.

$[_9F] = [He]\ \mathbf{2s^22p^5}$ $[_{10}Ne] = [He]\ \mathbf{2s^22p^6}$ outer shell = shell **#2**

$[_{17}Cl] = [Ne]\ \mathbf{3s^23p^5}$ $[_{18}Ar] = [Ne]\ \mathbf{3s^23p^6}$ outer shell = shell **#3**

$[_{35}Br] = [Ar]\ \mathbf{4s^23d^{10}4p^5}$ $[_{36}Kr] = [Ar]\ \mathbf{4s^23d^{10}4p^6}$ outer shell = shell **#4**

$[_{53}I] = [Kr]\ \mathbf{5s^24d^{10}5p^5}$ $[_{54}Xe] = [Kr]\ \mathbf{5s^24d^{10}5p^6}$ outer shell = shell **#5**

The shell, subshell, and orbital pattern for the periodic table up to element $_{86}Rn$ is diagrammed below.

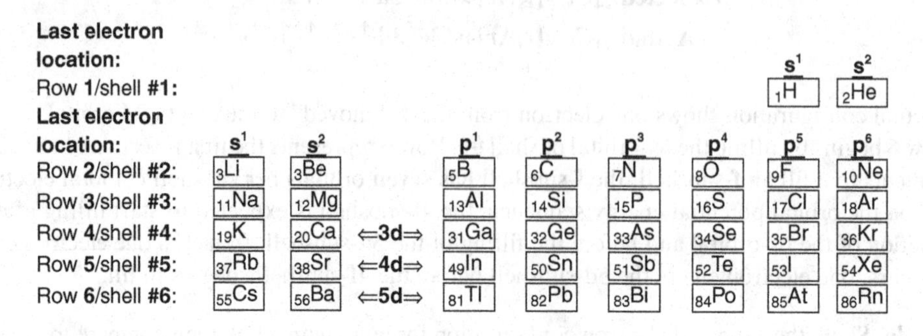

For the A-groups (tall columns), the **row** in the periodic table represents the outer shell number, which is in the process of being filled. The **columns** are sequenced from left-to-right showing the successive filling of the outer shell s-subshell and p-subshell; the specific column in the periodic table represents the subshell position of the last electron to enter atomic orbitals.

	d^1		d^2	d^3	d^4	d^5	d^6	d^7	d^8	d^9	d^{10}
3d:	$_{21}Sc$		$_{22}Ti$	$_{23}V$	$_{24}Cr$	$_{25}Mn$	$_{26}Fe$	$_{27}Co$	$_{28}Ni$	$_{29}Cu$	$_{30}Zn$
4d:	$_{39}Y$		$_{40}Zr$	$_{41}Nb$	$_{42}Mo$	$_{43}Tc$	$_{44}Ru$	$_{45}Rh$	$_{46}Pd$	$_{47}Ag$	$_{48}Cd$
5d:	$_{57}La$	⇐**4f**⇒	$_{72}Hf$	$_{73}Ta$	$_{74}W$	$_{75}Re$	$_{76}Os$	$_{77}Ir$	$_{78}Pt$	$_{79}Au$	$_{80}Hg$

	f^1	f^2	f^3	f^4	f^5	f^6	f^7	f^8	f^9	f^{10}	f^{11}	f^{12}	f^{13}	f^{14}
4f:	$_{58}Ce$	$_{59}Pr$	$_{60}Nd$	$_{61}Pm$	$_{62}Sm$	$_{63}Eu$	$_{64}Gd$	$_{65}Tb$	$_{66}Dy$	$_{67}Ho$	$_{68}Er$	$_{69}Tm$	$_{70}Yb$	$_{71}Lu$

The B-groups (transition metals) represent elements in the process of filling an inner d-subshell. The three rows for the B-groups indicate the shell number of the d-subshell being filled in the sequence 3d, 4d, and 5d. The columns for the transition metals follow the sequential filling of the inner d-subshell from d1 to d10. The detached row at the bottom of the periodic table represents the sequential filling of the 4f-subshell (the 5f-subshell is not shown).

The atomic number sequence 1–88 automatically reveals the sequential pattern of the shells and subshells and the corresponding orbital potential energy sequence. It is also possible to read the electron configuration directly from the periodic table without counting and placing each electron. Since the orbital sequence pattern is the same for all elements (ignoring exceptions), the identification of the orbital configuration of the last electron specifies the complete electron configuration of the atom.

Process to Read the Electron Configuration Directly from the Periodic Table

Step (**1**): Locate the element in the periodic table.

Step (**2**): If it is an **A**-group element, determine the row number directly by counting from the top from row **1** down to row **7**; the row number is the outer shell number into which the last electron is being filled.

Step (**3**): If it is a **B**-group element (transition metal), determine the correct inner d-subshell being filled from the fact that the first d-subshell to fill is 3d, followed by 4d, followed by 5d: first B-row = 3d; second B-row = 4d; third B-row = 5d.

Step (**4**): Count across to determine the position of the last electron; this determines the configuration of the last subshell being filled. Therefore, the complete electron configuration is all filled shells and subshells up to this last subshell configuration.

Example: Determine electron configurations of selenium, antimony, and tungsten by reading the periodic table directly; show the orbital notation.

Step (**1**): Selenium = **Se**

Step (**2**): Se is an **A**-group element located in the **4**th row.

Step (**4**): Se is in the column identified as $\mathbf{p^4}$. The last electron is the fourth electron to enter the 4p subshell, producing the last subshell designation of $\mathbf{4p^4}$.

Se is the complete sequence up to $4p^4$ = $\mathbf{[Ar]\ 4s^2\ 3d^{10}\ 4p^4}$.

Step (**1**): Antimony = **Sb**

Step (**2**): Sb is an **A**-group element located in the **5**th row.

Step (**4**): Sb is in the column identified as $\mathbf{p^3}$. The last electron is the third electron to enter the 5p subshell, producing the last subshell designation of $\mathbf{5p^3}$.

Sb is the complete sequence up to $5p^3$ = $\mathbf{[Kr]\ 5s^2\ 4d^{10}5p^3}$.

Step (**1**): Tungsten = **W**

Step (**3**): W is a **B**-group element.

Step (**4**): W is in the third row of the d-block, which is identified as the **5d**-row. The column is identified as $\mathbf{d^4}$. The last electron is the fourth electron to enter the 5d subshell, producing the last subshell designation of $\mathbf{5d^4}$.

W is the complete sequence up to $5d^4$ = $\mathbf{[Xe]\ 6s^2\ 4f^{14}\ 5d^4}$.

Valence Shell Electrons

For **A-group** elements, only the outer shell electrons affect the chemical properties of an atom. For A-groups, the **outer shell** of the atom is termed the **valence shell**; electrons in the outer shell are termed the **valence electrons**. The valence shell counts the complete outer shell, that is, all subshells of the highest shell number (highest principal quantum number). Valence electrons include all electrons in this complete **outer shell**, not just those in the last **subshell**.

Example: Determine the electron configuration and count the number of valence electrons for carbon, oxygen, phosphorous, and chlorine.

$$\text{Carbon} = 1s^2\ 2s^2\ 2p^2$$

The outer shell and valence shell is shell **#2**; shell #2 has four total electrons: two electrons in the 2s-subshell and two electrons in the 2p-subshell. The number of valence electrons for carbon is (**4**). Note that **all** electrons in shell #2 are counted, not just the 2 electrons in the 2p-subshell.

$$\text{Oxygen} = 1s^2\ 2s^2\ 2p^4$$

The outer shell and valence shell is shell #2; shell #2 has six total electrons: two electrons in the 2s-subshell and four electrons in the 2p-subshell. The number of valence electrons for oxygen is (**6**).

$$\text{Phosphorous} = [\text{Ne}]3s^2\ 3p^3$$

The outer shell and valence shell is shell #**3**; shell #3 has five total electrons: two electrons in the 3s-subshell and three electrons in the 3p-subshell. The number of valence electrons for phosphorous is (**5**).

$$\text{Chlorine} = [\text{Ne}]3s^2\ 3p^5$$

The outer shell and valence shell is shell #**3**; shell #3 has seven total electrons: two electrons in the 3s-subshell and five electrons in the 3p-subshell. The number of valence electrons for chlorine is (**7**).

IV IONIZATION ENERGY AND ELECTRON CONFIGURATION RELATIONSHIP

The **ionization energy (I.E.)** of an electron is the amount of energy required to remove a specific electron from the atom. The I.E. is proportional to the strength of the attractive force between the electron and the positively charged nucleus; this provides a measure of how tightly the nucleus holds the electron in the atom. Under normal conditions, ionization occurs such that the weakest held electrons are removed first; these are the electrons in the highest energy orbitals and/or in the atom's outer shell (farthest from the nucleus).

The force of attraction increases with the size of the positive charge in the nucleus; thus, the force of attraction increases as the number of protons in the nucleus increases. The force of attraction also increases as the distance from the nucleus to the electron decreases. The closer the electron is to the nucleus the greater the attractive force; the farther the electron is (farther outer shell), the weaker the attractive force.

The attractive force between an outer shell electron and the nucleus depends on the relationship between the number of protons in the nucleus, the number of inner shell electrons, and the corresponding distance to the outer shell. Although the diagrams below are highly schematic and not to scale, they can help to visualize these relationships.

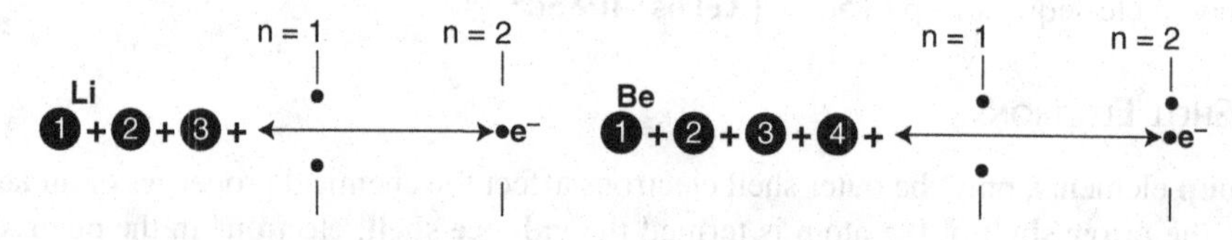

Lithium ($1s^2\ 2s^1$) has three protons in the nucleus; the outer shell is #2 with one electron. The strength of the attractive force holding the outer shell is proportional to the three positive charges. However, the outer shell electron is "screened" by the electron density of the two inner shell (shell #1) electrons; essentially, the outer shell electron must "look through" two electrons of negative charge to see the nuclear positive charges. The net result of these effects shows the ionization energy (I.E.) of the weakest held electron, the 2s electron to be 520 kJ per mole.

Now predict how the I.E. of beryllium should change compared to lithium. The weakest held electron for beryllium ($1s^2\ 2s^2$) is also in outer shell #2. Beryllium, however, has four protons in the nucleus holding the same outer shell. Therefore, the outer shell of beryllium should be held more tightly, and the I.E. should increase. The increased value, however, is not expected to be an even multiple based only on counting protons. The increased nuclear charge pulls the outer shell closer to the nucleus, increasing the attractive force. Partly countering this is the additional electron in the 2s orbital, which produces electron-electron repulsion. The resulting I.E. of beryllium is 900 kJ/mole.

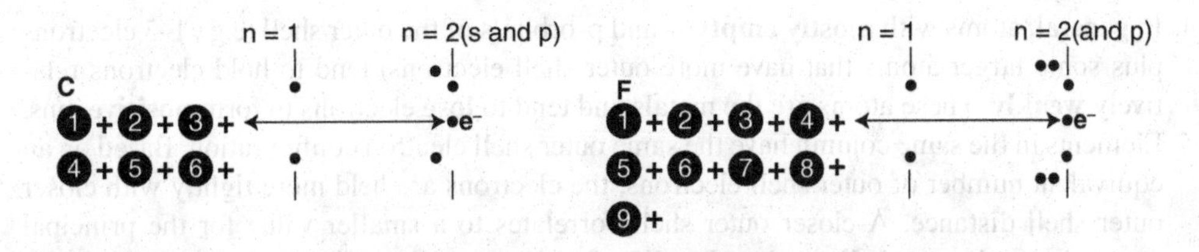

Now consider the trend in the sequence moving from left-to-right across row 2. The elements in the row increase in atomic numbers from 3 to 8. Increasing atomic numbers in the same row represent atoms with increasing nuclear charge holding the same outer shell. The result is that the attractive force increases, and the outer shell is increasingly pulled closer to the nucleus.

The balance of effects for carbon ($1s^2\ 2s^2\ 2p^2$) with six positive charges increasing the attractive force but with four electrons in shell #2 contributing to electron–electron repulsion results in an I.E. of 1, 090 kJ/mole. The I.E. of fluorine ($1s^2\ 2s^2\ 2p^5$) with nine protons holding shell #2 is 1680 kJ mole.

Neon ($1s^2\ 2s^2\ 2p^6$) has a closed shell with the maximum number of outer shell electrons. Neon's ten protons represent the maximum "holding power" for row 2; it has the greatest number of protons, which can attract electrons in outer shell #2. The I.E of Neon is 2080 kJ/mole; predict what occurs for sodium, which begins row 3.

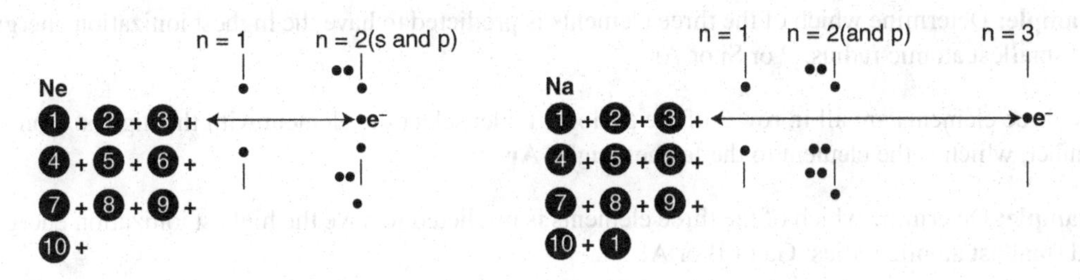

Sodium ($1s^2\ 2s^2\ 2p^6\ 3s^1$) has eleven protons in the nucleus, one more than neon. However, the beginning of a new row completely changes the relative attractive force. The weakest held electron is now in outer shell #3, with a corresponding large increase in the distance from the nucleus. The strength of the attractive force holding the outer shell decreases as the square of the separating distance. In addition, the outer shell electron is screened by the electron density of the ten inner shell electrons; the outer shell electron must "look through" ten electrons of negative charge to see the nuclear positive charges. The net result is that the ionization energy (I.E.) of the 3s electron decreases to 500 kJ per mole.

The balance between the number of protons holding a specific outer shell, the distance to the outer shell, and screening of inner shell electrons produces the following general results.

1. Increasing atomic numbers for elements in any specific row in the periodic table (reading from left-to-right along any row) represent atoms, which hold outer shell electrons more and more tightly. Elements with increasing atomic number along any row represent greater numbers of protons attracting the same outer shell.
2. For any specific outer shell number (row), the attractive force is maximized for those atoms that have a completely filled outer shell s- and p-subshell; this represents the maximum number of protons, which hold a specific outer shell before a new outer shell is started. This situation is represented by the noble gas (He, Ne, Ar, Kr, Xe) electron configurations.
3. In general, atoms (especially smaller atoms) with **mostly filled** s- and p-subshells of the outer shell (e.g., 4–8 electrons) tend to hold electrons relatively **tightly**. These atoms are the non-metals (or metalloids) and tend to **gain** electrons to form **negative** ions. This result does apply to the largest atoms in certain columns due to conclusion (**5**).

4. In general, atoms with **mostly empty** s- and p-orbitals of the outer shell (e.g., 1–3 electrons plus some larger atoms that have more outer shell electrons) tend to hold electrons relatively **weakly**. These atoms are the metals and tend to **lose** electrons to form **positive** ions.
5. Elements in the same column have the same outer shell electron configuration. Based on an equivalent number of outer shell electrons, the electrons are held more tightly with closer outer shell distance. A closer outer shell correlates to a smaller value for the principal quantum number or shell number. Reading from bottom-to-top in any column represents atoms with a decreasing numerical value for the shell number, a corresponding closer outer shell and, thus, more tightly held outer shell electrons.
6. The atomic radius for an element is a measure of average distance from the nucleus to the outer shell electrons. The ionization energy increases and the atomic radius decreases with increasing atomic number along a specific row (reading left-to-right across a row in the periodic table). The ionization energy increases and the atomic radius decreases with a closer outer shell value in any specific column (reading bottom-to-top up a column in the periodic table).
7. The **electron affinity** of an atom is a measure of the potential energy change upon addition of one electron to form a negative ion. Very generally (with some exceptions), elements that hold electrons most tightly have the most favorable potential energy changes for electron affinity (larger decrease or smaller increase in potential energy).

Example: Determine which of the three elements is predicted to have the highest ionization energy and smallest atomic radius: P or Si or Ar

The three elements are all in row 3 of the periodic table; select the element with the highest atomic number, which is the element to the farthest right: **Ar**

Example: Determine which of the three elements is predicted to have the highest ionization energy and smallest atomic radius: Ga or B or Al

The three elements are all in column 3A of the periodic table; select the element with the smallest outer shell number (smallest row number), which is the element closest to the top: **B**

Example: Determine which of the three elements is predicted to have the highest ionization energy and smallest atomic radius: Sb or Rb or Br

The three elements are in different rows or columns. However, one element is both farthest to the right and closer to a column top; select **Br**

V DETERMINATION OF ELECTRON CONFIGURATION FOR IONS

The electron configurations of ions are determined by starting with the electron configuration of the parent atom. Specific rules can then be established based on the concepts of ionization energy and the favorable energetics of achieving a noble gas electron configuration. A noble gas electron configuration is energetically favorable because it represents the maximum "holding power" of an ion's outer shell. Completely filling the s- and p-subshell takes exactly eight electrons; therefore, the rule is also termed the **octet rule**. (Hydrogen requires only two electrons to fill shell n = 1.) The concept that many ions reach a noble gas electron configuration was introduced and used in Chapter 4 without a complete explanation.

Electron Configuration for Negative Ions

Step (**1**): Identify the non-metal (or metalloid) as one of the following elements: **H, C, N, O, F, P, S, Cl, As, Se, Br, Te, I**. These elements hold electrons relatively tightly and have a relatively favorable energy for addition of extra electrons to form negative ions.

Step (2): Write the electron configuration of the neutral atom from which the ion is derived.

Step (3): Most atoms that form negative ions will produce only one specific ion for each element. Follow the corresponding rule to determine the correct number of electrons to add: each element will **gain** the exact number of extra electrons to completely (exactly) fill the outer shell s- and p-subshell; this achieves a noble gas electron configuration.

Step (4): To form the negative ion, **add** the correct number of **extra** electrons determined from step (3) to the proper orbitals based on the orbital energy sequence.

Example: Determine the **one** correct ion charge and the ion electron configuration for ions formed from fluorine, oxygen, and phosphorous.

(1) (2) (3) (4) Noble Gas #

$_9F = 1s^2 2s^2 2p^5$ (needs 1 e^- to fill 2s and 2p) $\xrightarrow{\text{gain } 1\,e^-}$ $_9F^- = 1s^2 2s^2 2p^6 = [_{10}Ne]$

$_8O = 1s^2 2s^2 2p^4$ (needs 2 e^- to fill 2s and 2p) $\xrightarrow{\text{gain } 2\,e^-}$ $_8O^{-2} = 1s^2 2s^2 2p^6 = [_{10}Ne]$

$_{15}P = [Ne]3s^2 3p^3$ (needs 3 e^- to fill 3s and 3p) $\xrightarrow{\text{gain } 3\,e^-}$ $_{15}P^{-3} = [Ne]3s^2 3p^6 = [_{18}Ar]$

Electron Configuration for Most Fixed-Charge Positive Ions

Step (1): Identify the alkali or alkaline earth metal from periodic groups 1A and 2A; include aluminum: **Li, Na, K, Rb, Cs, Mg, Ca, Sr, Ba, Al**. These elements hold electrons relatively weakly and, therefore, have a relatively favorable energy for loss of electrons (ionization) to form positive ions.

Step (2): Write the electron configuration of the neutral atom from which the ion is derived.

Step (3): These atoms that form positive ions will produce only one specific ion for each element. Follow the corresponding rule to determine the correct number of electrons to subtract: each element will **lose** the exact number of outer shell electrons to completely (exactly) empty out the s- and p-subshell of the original outer shell. These elements achieve a noble gas electron configuration because the full next-to-last outer shell becomes the filled outer shell after the correct ion is formed.

Step (4): To form the positive ion, **subtract** the correct number of electrons determined from step (3) from the outer shell of the atom.

Example: Determine the **one** correct ion charge and the ion electron configuration for ions formed from sodium, aluminum, and calcium.

(1) (2) (3) (4) Noble Gas #

$_{11}Na = [Ne]3s^1$ (loses 1 e^- from $3s^1$) $\xrightarrow{\text{lose } 1\,e^-}$ $_{11}Na^+ = (1s^2 2s^2 2p^6) = [_{10}Ne]$

$_{13}Al = [Ne]3s^2 3p^1$ (loses 3e^- from $3s^2 3p^1$) $\xrightarrow{\text{lose } 3\,e^-}$ $_{13}Al^{+3} = (1s^2 2s^2 2p^6) = [_{10}Ne]$

$_{20}Ca = [Ar]4s^2$ (loses 2e^- from $4s^2$) $\xrightarrow{\text{lose } 2\,e^-}$ $_{20}Ca^{+2} = [Ne]3s^2 3p^6 = [_{18}Ar]$

Ion Formation for Metals in Groups 4A and 5A

Step (**1**): Identify the larger atoms in periodic groups 4A and 5A: the metals: **Pb, Bi, Sn, Sb**. Despite having 4 or 5 electrons in the outer shell, these elements near the bottom of the columns hold electrons relatively weakly because the outer shells (n = 5 and n = 6) are relatively farther away. These elements have a relatively favorable energy for loss of electrons (ionization) to form positive ions.

Step (**2**): Write the electron configuration of the neutral atom from which the ion is derived.

Step (**3**): These elements are variable-charged metals (except for Bi) due to two different possibilities. The element may lose the exact number of outer shell **p-subshell** electrons **only** or the element may lose the exact number of outer shell electrons contained in **both** the **p-subshell** and **s-subshell** of the original outer shell. Follow the required option to determine the correct number of electrons to subtract: empty out **only** the electrons in the p-subshell or empty out **all** electrons in the p-subshell and s-subshell.

Step (**4**): To form the positive ion, **subtract** the correct number of electrons determined from step (**3**) from the outer shell of the atom.

Example: Determine the ion electron configuration for the specified ions: Sn^{+2}, Sn^{+4}, Sb^{+3}, Sb^{+5}. (A zero superscript is retained in the notation for empty subshells for clarity.)

(1) (2) (3) (4)

$$_{50}Sn = [Kr]5s^2 4d^{10} 5p^2 (\text{loses } 2\ e^- \text{ from } 5p^2) \xrightarrow{\text{lose } 2e^-} {}_{50}Sn^{+2} = [Kr]5s^2 4d^{10} 5p^0$$

$$_{50}Sn = [Kr]5s^2 4d^{10} 5p^2 (\text{loses } 4\ e^- \text{from } 5s^2 5p^2) \xrightarrow{\text{lose } 4e^-} {}_{50}Sn^{+4} = [Kr]5s^0 4d^{10} 5p^0$$

$$_{51}Sb = [Kr]5s^2 4d^{10} 5p^3 (\text{loses } 3\ e^- \text{ from } 5p^3) \xrightarrow{\text{lose } 3e^-} {}_{51}Sb^{+3} = [Kr]5s^2 4d^{10} 5p^0$$

$$_{51}Sb = [Kr]5s^2 4d^{10} 5p^3 (\text{loses } 5\ e^- \text{ from } 5s^2 5p^3) \xrightarrow{\text{lose } 5e^-} {}_{51}Sb^{+5} = [Kr]5s^0 4d^{10} 5p^0$$

Ion Formation for Transition Metals (B Groups)

Step (**1**): Identify the metal in the periodic B-group block, the transition metals. These metals have a maximum of two outer shell electrons located in the outer shell s-orbital. They hold electrons relatively weakly and have a relatively favorable energy for loss of electrons (ionization) to form positive ions.

Step (**2**): Write the electron configuration of the neutral atom from which the ion is derived.

Step (**3**): Most transition metals form variable-charged positive ions due to the general behavior in which the outer shell s-orbital electrons are lost first. Transition metals may then also lose from 0 to 3 additional electrons from the inner d-subshell orbitals. This number is variable and explains the variable nature of these metals.

Step (**4**): To form the positive ion, **subtract** the correct number of electrons from the outer shell s-orbital electrons **before** the loss of any inner d-subshell electrons; this behavior does **not** simply follow the reverse order of orbital electron filling. Subtract additional electrons, if required, from the next inner d-orbitals only after the outer shell has been completely emptied.

Example: Show the electron configurations for Fe^{+2} and Fe^{+3}; Pt^{+2} and Pt^{+4}.

(1) **(2)** **(3)** **(4)**

$$_{26}Fe = [Ar]4s^2 3d^6 (\text{loses 2 e}^- \text{ from } 4s^2) \xrightarrow{\text{lose } 2e^-} {}_{26}Fe^{+2} = [Ar]4s^0 3d^6$$

$$_{26}Fe = [Ar]4s^2 3d^6 (\text{loses 2 e}^- \text{ from } 4s^2 \text{ plus 1 e}^- \text{ from } 3d^6) \xrightarrow{\text{lose } 3e^-} {}_{26}Fe^{+3} = [Ar]4s^0 3d^5$$

$$_{78}Pt = [Xe]6s^2 4f^{14} 5d^8 (\text{loses 2 e}^- \text{ from } 6s^2) \xrightarrow{\text{lose } 2e^-} {}_{78}Pt^{+2} = [Xe]6s^0 5d^8$$

$$_{78}Pt = [Xe]6s^2 4f^{14} 5d^8 (\text{loses 2 e}^- \text{ from } 6s^2 \text{ plus 2 e}^- \text{ from } 5d^8)$$

$$\xrightarrow{\text{lose } 4e^-} {}_{78}Pt^{+4} = [Xe]6s^0 5d^6$$

VI PRACTICE PROBLEMS

1. Use the combination display system to show the **expected** electron configuration of the following elements: Y, Mo, Rh, Os, Tl
2. Use the orbital notation system to demonstrate that elements in the same column in the periodic table have similar outer shell electron configurations; prove this for Group 5A (N, P, As, Sb, Bi).
3. Use the orbital notation system to determine the number of valence electrons in the following A-group elements: Si, S, K, Mg, Te
4. Determine which of the three elements is predicted to have the highest ionization energy and smallest atomic radius.
   a) Ge or Se or Ca
   b) Br or Kr or Cu
   c) C or Al or Ge
5. Determine the **one** correct ion charge and the ion electron configuration for each of the following metals or non-metals: N, CI, Se, Rb, Sr, Ba
6. a) Determine **both** possible ion charges and the ion electron configurations for Pb.

   b) Show the electron configuration of Bi^{+3}.
7. Show the expected electron configurations for: Co^{+2}, Co^{+3}, Zn^{+2}, Hg^{+2}, Pd^{+2}, Pd^{+4}.

VII ANSWERS TO PRACTICE PROBLEMS

1. Y (39 e^-) = [Kr] $5s^2$ $4d^1$ 4d ↑. ___ ___ ___ ___

   Mo (42 e^-) = [Kr] $5s^2$ $4d^4$ 4d ↑. ↑. ↑. ↑. ___ (expected)

   Mo (42 e^-) = [Kr] $5s^1$ $4d^5$ 4d ↑. ↑. ↑. ↑. ↑. (actual)

   Rh (45 e^-) = [Kr] $5s^2$ $4d^7$ 4d ↑↓ ↑↓ ↑. ↑. ↑.

   Os (76 e^-) = [Xe] $6s^2$ $4f^{14}$ $5d^6$ 5d ↑↓ ↑. ↑. ↑. ↑.

   Tl (81 e^-) = [Xe] $6s^2$ $4f^{14}$ $5d^{10}$ $6p^1$ 6p ↑. ___ ___

2. $N = [He]\mathbf{2s^2\ 2p^3}$; $P = [Ne]\mathbf{3s^2\ 3p^3}$; $As = [Ar]\mathbf{4s^2}\,3d^{10}\,\mathbf{4p^3}$;

   $Sb = [Kr]\mathbf{5s^2}\,4d^{10}\,\mathbf{5p^3}$; $Bi = [Xe]\mathbf{6s^2}\,4f^{14}\,5d^{10}\,\mathbf{6p^3}$;

3. Si = [Ne]**$3s^2 3p^2$** = **4** valence e^-
   S = [Ne]**$3s^2 3p^4$** = **6** valence e^-
   K = [Ar] **$4s^1$** = **1** valence e^-
   Mg = [Ne]**$3s^2$** = **2** valence e^-
   Te = [Kr]**$5s^2 4d^{10}$ $5p^4$** = **6** valence e^-

4. Determine which of the three elements is predicted to have the highest ionization energy and smallest atomic radius.
   a) Ge or **Se** or Ca; **Se** is farthest to the right in the same row 4.
   b) Br or **Kr** or Cu; **Kr** is farthest to the right in the same row 4.
   c) **C** or Al or Ge; **C** is farthest to the right or closer to the top.

5. $N = 1s^2 2s^2 2p^3$ (needs 3 e^- to fill 2s + 2p) $\xrightarrow{\text{gain } 3e^-}$ $N^{-3} = 1s^2 2s^2 2p^6 = [Ne]$

   $Cl = [Ne]3s^2 3p^5$ (needs 1 e^- to fill 3s + 3p) $\xrightarrow{\text{gain } 1e^-}$ $Cl^- = [Ne]3s^2 3p^6 = [Ar]$

   $Se = [Ar]4s^2 3d^{10} 4p^4$ (needs 2 e^- to fill 4s + 4p) $\xrightarrow{\text{gain } 2e^-}$ $Se^{-2} = [Ar]4s^2 3d^{10} 4p^6 = [Kr]$

   $Rb = [Kr]5s^1$ (loses 1 e^- from $5s^1$) $\xrightarrow{\text{loses } 1e^-}$ $Rb^+ = [Kr]$

   $Sr = [Kr]5s^2$ (lose 2 e^- from $5s^2$) $\xrightarrow{\text{loses } 2e^-}$ $Sr^{+2} = [Kr]$

   $Ba = [Xe]6s^2$ (lose 2 e^- from $6s^2$) $\xrightarrow{\text{loses } 2e^-}$ $Ba^{+2} = [Xe]$

6. $Pb = [Xe]6s^2 4f^{14} 5d^{10} 6p^2$ (loses 2 e^- from $6p^2$) $\xrightarrow{\text{lose } 2e-}$ $Pb^{+2} = [Xe]6s^2 4f^{14} 5d^{10} 6p^o$

   $Pb = [Xe]6s^2 4f^{14} 5d^{10} 6p^2$ (loses 4 e^- from $6s^2 6p^2$) $\xrightarrow{\text{lose } 4e-}$ $Pb^{+4} = [Xe]6s^0 4f^{14} 5d^{10} 6p^o$

   $Bi = [Xe]6s^2 4f^{14} 5d^{10} 6p^3$ (loses 3 e^- from $6p^3$) $\xrightarrow{\text{lose } 3e-}$ $Bi^{+3} = [Xe]6s^2 4f^{14} 5d^{10} 6p^o$

7. $Co = [Ar]4s^2 3d^7$ (loses 2 e^- from $4s^2$) $\xrightarrow{\text{lose } 2e-}$ $Co^{+2} = [Ar]4s^0 3d^7$

   $Co = [Ar]4s^2 3d^7$ (loses 2 e^- from $4s^2$ plus 1 e^- from $3d^7$) $\xrightarrow{\text{lose } 3e-}$ $Co^{+3} = [Ar]4s^0 3d^6$

   $Zn = [Ar]4s^2 3d^{10}$ (loses 2e^- from $4s^2$) $\xrightarrow{\text{lose } 2e-}$ $Zn^{+2} = [Ar]4s^0 3d^{10}$

   $Hg = [Xe]6s^2 4f^{14} 5d^{10}$ (loses 2e^- from $6s^2$) $\xrightarrow{\text{lose } 2e-}$ $Hg^{+2} = [Xe]6s^0 4f^{14} 5d^{10}$

   $Pd = [Kr]5s^2 4d^8$ (loses 2e^- from $5s^2$) $\xrightarrow{\text{lose } 2e-}$ $Pd^{+2} = [Kr]5s^0 4d^8$

   $Pd = [Kr]5s^2 4d^8$ (loses 2e^- from $5s^2$ plus 2 e^- from $4d^8$) $\xrightarrow{\text{lose } 4e-}$ $Pd^{+4} = [Kr]5s^0 4d^6$

14 Alternate Methods for Visualizing and Constructing *Lewis Structures of Covalent Molecules*

I INTRODUCTION TO INTERPRETATION OF LEWIS STRUCTURES

Lewis structures are pictorial representations, not exact theoretical analyses, of the role of all valence electrons (outer shell electrons for non-metals) in a covalent molecule. All valence electrons for all atoms in the molecule are accounted for as either covalent bonding electron pairs, shown as a line, or as electrons not used in bonding, termed lone electron pairs. A lone electron pair is usually shown as a pair of dots; occasionally, non-octet atoms will have an unpaired electron, shown as a single dot.

The Lewis structure for an individual atom shows all valence electrons as dots around the symbol of the atom. The generally accepted method is to show the electrons as pairs or to show them symmetrically distributed around the symbol, especially when indicating bonding in molecules.

Example: Acceptable for carbon with four valence electrons: •C• or •C•

Example: Acceptable for oxygen with six valence electrons: •O•

The standard method for constructing Lewis structures involves counting up all valence electrons for all atoms in the molecule, then distributing them into a bonding pattern according to a set of fixed rules. The methods described in this chapter adapts and extends the general bonding pattern information generated from this approach. The purpose is to generate techniques for understanding the electron behaviors responsible for bonding in typical covalent molecules. Corresponding techniques are then produced for determining valid atom connection patterns for larger covalent molecules without counting all valence electrons and in the absence of a provided bonding pattern.

Covalent molecules are described as being constructed from two types of structural atoms.

(**1**) Central atoms are bonded to more than one other atom in any specific molecule.
(**2**) Outside atoms are bonded to only one other atom.

Simple covalent molecules and polyatomic ions often consist of one central atom surrounded by outside atoms bonded only to the central atom. More complex molecules with more than one central atom can be interpreted with a knowledge of normal bonding behavior. Lewis structures for these molecules are put together by considering each central atom and its outside atoms connected in ways to generate a framework for larger molecules.

The techniques in this chapter provide different viewpoints for understanding how both smaller molecules (one to three central atoms) and larger molecules are assembled. The beginning approach employs the most common way in which atoms bond to form molecular structures of all sizes. The additional bonding options for the non-metal atoms are then described as logical alternatives ("exception rules") to the most common ("normal") behavior. In should be noted that the terminology for many of the process steps and rules described in the chapter are specific to this chapter. The methods represent certain ways to solve structure problems and do not generally represent specific fundamental laws; they would not necessarily be described in this specific manner in other references.

II COMMON BONDING BEHAVIOR FOR NON-METALS IN COVALENT MOLECULES

A covalent bond is a shared pair of electrons between two atoms. The shared electron pair can most simply be described as occupying a region of space formed by orbital overlap of one atomic orbital from each of the two atoms forming the bond. This overlap is called a bonding molecular orbital and always takes only two electrons.

Covalent bonding pairs are shown as one line; one line equals one bonding electron pair. Although not strictly proper notation, in some cases, the electron pair is shown as a pair of dots on the bonding line; this is done either as an aid to counting or to demonstrate how the electrons are used in a chemical process. Multiple covalent bonds represent additional bonding electron pairs between two atoms. Each line represents a separate set of orbital overlaps (molecular orbital) and, thus, a separate electron bonding pair:

Two lines equals double bond (X═X); three lines equals a triple bond (X≡X).

General Concept of Normal Bonding in Covalent Molecules

Covalent bonds can most often be considered as being formed by a contribution of one electron from each of the two atoms involved in the bond formation. Under this condition, each covalent bond adds **one additional** electron to each of the bonding atom's outer **bonding** valence shell.

If personalities can be ascribed to the atoms, consider the situation when a hydrogen atom and a fluorine atom come together to make a compound. Hydrogen says to the fluorine, "Give me an electron and we'll make an ionic compound, $(F^+)(H^-)$." Since fluorine has such a high ionization energy (Chapter 13), this is not energetically feasible. Therefore, fluorine says, "No thank you; why don't you give me your electron and we can make the ionic compound $(H^+)(F^-)$?" This is not energetically feasible either since hydrogen also has a relatively high ionization energy. The two atoms compromise by each sharing one electron to form a covalent bond; therefore, each atom adds one electron to its outer bonding valence shell.

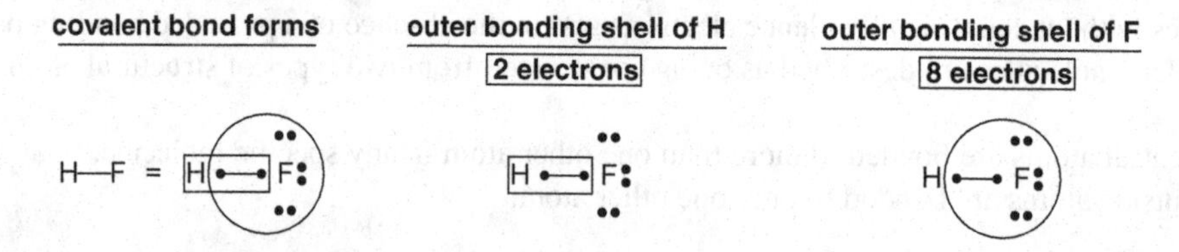

In the process, each atom achieves a full (closed) outer valence shell through bonding. Hydrogen starts with one valence electron; from its perspective it has outsmarted the fluorine by adding one electron to complete its outer shell (n = 1). Fluorine starts with seven valence electrons; from its perspective it has outsmarted hydrogen by adding one electron to complete its octet, a full (n = 2) outer shell.

Oxygen and hydrogen can come to the same agreement in the formation of water. Two hydrogens each share an electron pair with oxygen; thus, all three atoms add one electron to its outer bonding valence shell.

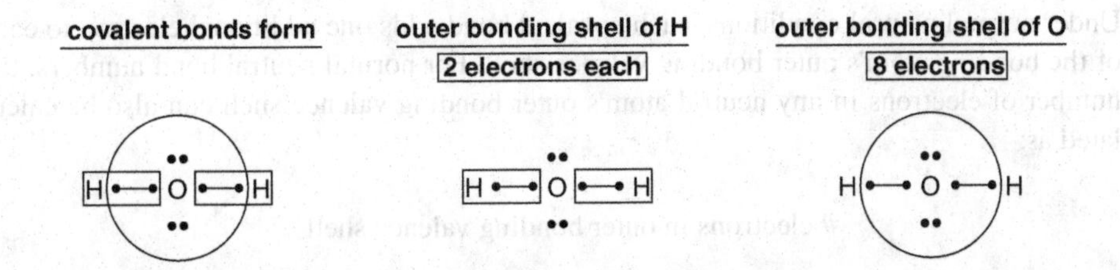

In the process, all atoms achieve a full (closed) outer bonding valence shell through the formation of covalent bonds. Each hydrogen adds one electron to complete its outer shell; oxygen starts with six valence electrons; it has added two electrons, one from each of the hydrogens, to complete its octet, a full (n = 2) outer shell.

Nitrogen, with five valence electrons, can achieve a full (closed) (n = 2) outer bonding valence shell through formation of three bonds and carbon, with four valence electrons, can achieve a full (closed) (n = 2) outer bonding valence shell through formation of four bonds. Single bonds are shown in these examples; double or triple bonds are also possible as long as the bond count remains correct.

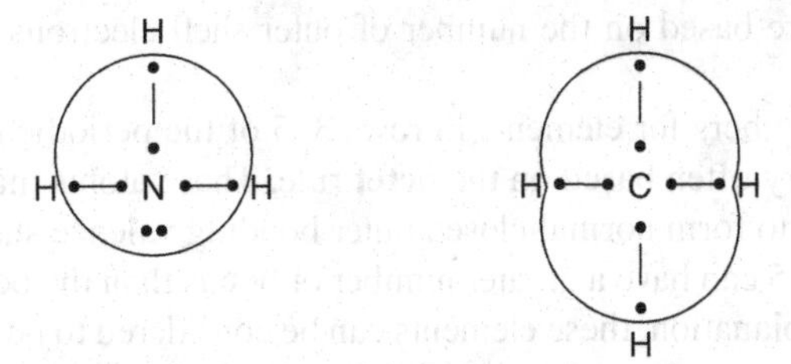

Most molecules can be constructed using (terminology for this chapter) **normal neutral bonding rules** for elements in covalent molecules. Normal neutral bonding rules for hydrogen (H), plus the row-2 elements, (C, N, O, F) are based on the concept that each atom has a tendency to form a sufficient number of covalent bonds to obtain the exact number of electrons required to completely fill the outer bonding valence shell. For H (shell n = 1), this equals two electrons; for the row-2 elements (n = 2), this equals eight electrons, an "octet." The specific bond numbers are determined from the fact that under normal neutral conditions, **each** covalent bond adds **one** additional electron to the atom's outer bonding valence shell.

Normal Neutral Bonding Rules

(**1**) For neutral atoms, which follow the octet rule, the number of covalent bonds an atom will make will be equal to the number of additional valence electrons the atom needs to complete the outer bonding valence shell (total to equal 2 electrons for hydrogen or 8 electrons for C, N, O, and F).

(**2**) The number of electrons in any neutral atom's outer bonding valence shell can be calculated from:

electrons in outer bonding valence shell

= [# of atom's original valence electrons]

+ [# of electrons provided from bonds]

Under normal neutral conditions, each covalent bond adds one additional electron to each of the bonding atom's outer bonding valence shell. For normal neutral bond numbers, the number of electrons in any neutral atom's outer bonding valence shell can also be calculated as:

electrons in outer bonding valence shell

= [# of atom's original valence electrons]

+ [# of normal neutral covalent bonds formed]

(3) If each atom in the molecule follows normal neutral bonding rules, formal charges on all atoms in the molecule will be zero (i.e., the atoms are neutral). Under circumstances where atoms show exceptions to normal neutral bond numbers, a formal charge of (+1) or (−1) will be calculated.

(4) A general **restriction rule** applies to all atoms: the total maximum number of covalent bonds possible for a **neutral** atom is limited to the total number of valence electrons. A neutral atom cannot share valence electrons that it does not have. Bonding numbers for neutral Be and B are based on the number of outer shell electrons available for forming covalent bonds.

(5) Neutral bonding numbers for elements in rows 3–5 of the periodic table (Si, P, S, Cl, As, Se, Br, Te, I) are very often based on the **octet** rule. These atoms may behave similarly to their column-mates to form normal closed outer bonding valence shells.

(6) Elements in rows 3–5 can have a greater number of bonds than the octet number. Although not the complete explanation, these elements can be considered to be covered by an **expansion rule**. These elements have the potential for an expanded valence shell, which can be defined as the use of the empty outer shell (the highest shell number that has electrons in the s and p subshells) d-orbitals to accept extra bonding electrons. This provides the possibility of more than eight electrons in the valence shell and, thus, more bonds than normal for an octet. The maximum number of bonds is still limited to the number of available valence electrons stated in rule (**4**).

(7) Normal neutral bonding rules assign one electron from each of the bonding atoms to form the shared electron pair in the covalent bond. The "leftover" valence electrons, the ones not used in bonding, are shown written in pairs and designated as lone pairs. The number of non-bonding electrons found in lone pairs is equal to the number of original valence electrons minus the number of electrons used in covalent bonding.

unused (leftover) electrons for lone pairs

= [# of atom's original valence electrons]

− [# of electrons used in bonding]

The calculation is presented in diagram form in the examples.

SUMMARY OF NORMAL NEUTRAL BOND NUMBERS FOR H, BE, B, C, N, O, F

Hydrogen (**H**) has **one** valence electron; it needs **one** additional electron to completely fill shell n = 1. H always forms **one** covalent bond; it cannot have any electrons leftover.

Atom Lewis structure: H •; H uses its only electron for bonding: H •—

Beryllium (**Be**) has **two** valence electrons; it needs **six** additional electrons to completely fill shell n = 2 but has only two valence electrons. Be is limited by the **restriction** rule and always forms **two** covalent bonds as a neutral atom; as a neutral atom it has no electrons leftover.

Atom Lewis structure: • Be •; Be uses both electrons for bonding: —• Be •—

Boron (**B**) has **three** valence electrons; it needs **five** additional electrons to completely fill shell n = 2 but has only three valence electrons. B is limited by the **restriction** rule and always forms **three** covalent bonds as a neutral atom; as a neutral atom it has no electrons leftover.

Atom Lewis structure: • B •; B uses all three electrons for bonding: —• B •—

Carbon (**C**) has **four** valence electrons; it needs **four** additional electrons to completely fill shell n = 2. C as a neutral atom forms **four** covalent bonds to reach an **octet**; as a neutral atom, it has no electrons leftover.

Atom Lewis structure: • C •; C uses all four electrons for bonding: —• C •—

Nitrogen (**N**) has **five** valence electrons; it needs **three** additional electrons to completely fill shell n = 2. N forms **three** covalent bonds to reach an **octet**. As a neutral atom, **N** has **two** electrons leftover, shown as **one lone pair**.

Atom Lewis structure: • N •; N uses three electrons for bonding: —• N •—

Oxygen (**O** has **six** valence electrons; it needs **two** additional electrons to completely fill shell n = 2. O forms **two** covalent bonds to reach an **octet**. As a neutral atom, O has **four** electrons leftover, shown as **two lone pairs**.

Atom Lewis structure: • O •; O uses two electrons for bonding: —• O •—

Fluorine (**F**) has **seven** valence electrons; it needs **one** additional electron to completely fill shell n = 2; F forms **one** covalent bond to reach an **octet**. As a neutral atom, F has **six** electrons leftover, shown as **three lone pairs**.

Atom Lewis structure: • F :; F uses one electron for bonding: —• F :

Summary Table

Atom/# of Valence e⁻	# of e⁻ Needed to Fill Shell	# of Covalent Bonds	# e⁻ Leftover/# of Lone Pairs
H (1 valence e^-)	**1** (1 + 1 = 2)	**1** covalent bond (full n = 1)	0 e^- leftover = **0** lone pairs
B (3 valence e^-)	**5** (5 + 3 = 8)	maximum of **3** bonds (restriction rule)	0 e^- leftover = **0** lone pairs
C (4 valence e^-)	**4** (4 + 4 = 8)	**4** covalent bonds (octet)	0 e^- leftover = **0** lone pairs
N (5 valence e^-)	**3** (3 + 5 = 8)	**3** covalent bonds (octet)	2 e^- leftover = **1** lone pair
O (6 valence e^-)	**2** (2 + 6 = 8)	**2** covalent bonds (octet)	4 e^- leftover = **2** lone pairs
F (7 valence e^-)	**1** (1 + 7 = 8)	**1** covalent bond (octet)	6 e^- leftover = **3** lone pairs

Summary of Normal Neutral Bond Numbers for Elements in Rows 3–5

Normal neutral bonding rules (**5**) and (**6**) apply to elements in **rows 3–5** (Si, P, S, Cl, As, Se, Br, Te, I). They very often form an **octet** with the same bonding and lone pair configuration as their column mates. These elements, however, are variable based on formation of additional bonds; the additional valence shell electrons can be thought of as entering empty outer shell d-orbitals. The number of possible bonds can reach a maximum corresponding to the number of valence electrons in the element.

Silicon (**Si**) with **four** valence electrons usually forms **four** covalent bonds to form an octet similar to carbon.

Phosphorous (**P**) and arsenic (**As**) have **five** valence electrons; they often form **three** covalent bonds to reach an octet similar to nitrogen. An expanded valence shell for P and As can produce a bond number of **five** covalent bonds.

Sulfur (S), selenium (Se), and tellurium (Te) have **six** valence electrons; they often form **two** covalent bonds to reach an octet similar to oxygen. An expanded valence shell for S, Se, and Te can produce a bond number of **four** covalent bonds (with two electrons leftover) or **six** covalent bonds (with no electrons leftover).

The remaining halogens, chlorine (Cl), bromine (Br), and iodine (I), each have **seven** valence electrons; they often form **one** covalent bond to reach an octet similar to fluorine. An expanded valence shell for Cl, Br, and I can produce a bond number of **three** covalent bonds (with four electrons leftover), **five** covalent bonds (with two electrons leftover), or **seven** covalent bonds (with no electrons leftover).

Summary Table

Atom/# of Valence e⁻	# of e⁻ Needed to Fill Shell	# of Octet Covalent Bonds	# of Bonds for Expanded Valence Shell
Si (4 valence e^-)	**4** (4 + 4 = 8)	**4** covalent bonds	Not applicable
P (5 valence e^-) As	**3** (3 + 5 = 8)	**3** covalent bonds	**5** covalent bonds (10 e^- in outer shell)
S (6 valence e^-) Se Te	**2** (2 + 6 = 8)	**2** covalent bonds	**4** covalent bonds (10 e^- in outer shell) or **6** covalent bonds (12 e^- in outer shell)
Cl (7 valence e^-) Br I	**1** (1 + 7 = 8)	**1** covalent bond	**3** covalent bonds (10 e^- in outer shell) **5** covalent bonds (12 e^- in outer shell) **7** covalent bonds (14 e^- in outer shell)

III CONSTRUCTING LEWIS STRUCTURES FOR COVALENT MOLECULES FROM H AND THE ROW-2 ELEMENTS: BONDING THAT FOLLOWS NORMAL NEUTRAL BONDING RULES

Lewis Structure Concepts

Lewis structures identify two types of atoms. **Outside atoms** are atoms that are bonded to only **one** other atom. All atoms that take only **one** bond must be outside atoms and can be termed **required** outside atoms. Atoms that can take two or more bonds **can** be outside atoms if all of their bonds are to only one other atom through a double or triple bond. **Central atoms** are those atoms that take more than one bond and are bonded to **two or more** other atoms. Any atom that can take two or more bonds can be a central atom and can be termed a **possible** central atom.

Specific definitions and requirements are used for the purposes of building molecular structures.

(1) Atoms that take only one bond are required outside atoms.
(2) Atoms capable of taking two or more bonds are possible central atoms.

All valence electrons for all atoms must be shown in a Lewis structure.

(3) Each covalent bond, shown by a line, represents one bonding electron pair.
(4) Multiple bonds are shown by a double line for a double bond and a triple line for a triple bond.
(5) Non-bonding electrons are shown by dots as lone-pairs.
(6) As a first requirement, Lewis structures are formed in such a way as to determine at least one valid bonding structure for a particular molecular formula; more specific structures can be constructed if additional information is available. Sometimes only one Lewis structure is possible for a molecular formula. In other cases, especially for larger molecules, there may be many possible bonding patterns. Different bonding patterns (shown as Lewis structures) for the same molecular formula are said to be related as (constitutional) **isomers**.

Process for Molecules Containing H, Be, B, C, N, O, F

H plus the row-2 elements (H, Be, B, C, N, O, F) cannot have expanded valence shells. If no exceptions need to be used, the normal neutral bond numbers for each of these elements can be used directly to form a valid molecule.

Step (**1**): For a neutral covalent molecule, identify the normal neutral bonding numbers for each of the atoms in the molecular formula.

Step (**2**): Identify all **possible** central atoms and **required** outside atoms.

Step (**3**): If there is more than one central atom, connect the central atoms with a single bond (to start with). More than one possible connection order is possible with three or more central atoms. As the number of central atoms in the formula increases, a valid structure can usually be found if they are connected in any order. In these cases, the easiest starting point is to connect the central atoms in a continuous chain; central atoms can, however, "branch" off from any selected chain.

Step (**4**): A valid Lewis structure will be any atom bonding pattern, which will form a molecule with correct bond numbers for **each** atom. Begin connecting all the required outside atoms to the possible central atoms using single bonds. Try to find a connection pattern such that all atoms have the correct number of bonds; this is like putting a puzzle together. Depending on the ratio between outside and central atoms, a useful guideline is to start by distributing outside atoms somewhat symmetrically around the central atoms as opposed to necessarily "filling up" the first ones encountered.

Step (**5**): If there are not enough outside atoms to give all central atoms the correct numbers of bonds, this means that double or triple bonds between central atoms will be required. Use these types of bonds to increase the number of bonds to deficient atoms; rearrangements of the initial pattern may be necessary to accommodate all atoms. Continue by trial-and-error until all bonding requirements are satisfied.

Step (**6**): Add the required lone electron pairs to each atom; these are the number of electrons, arranged in pairs, leftover after normal bonding. For circumstances where hydrogen and the row-2 elements are following normal neutral bond numbers, the number of lone pairs determined by the valence shell analysis in Section II will always apply. Since the elements in rows 3–5 are variable, lone pair analysis for these elements must be performed for each new molecule using the techniques described in Section II.

Example: Construct a valid Lewis structure for the compound C_2H_5F.

Steps (**1**) and (**2**): 2 carbons; each C = 4 bonds — possible central atom
5 hydrogens; each H = 1 bond — required outside atoms
1 fluorine; each F = 1 bond — required outside atom

```
                       H  H                      H  H
           Step (4)    |  |        Step (6)      |  |  ••
Step (3): C—C ——————> H—C——C—F  ——————————> H—C——C—F:
                       |  |                      |  |  ••
                       H  H                      H  H
```

The number of outside atoms was sufficient to satisfy both central atoms; step (**5**) was not required. Note that there is only one way to construct this molecule; there are no possible isomers. Constitutional isomers describe different bonding patterns (different atom connection patterns) for the same molecular formula. An important idea in working with molecular structure is recognition of the difference between different atom connection patterns and different views (different "drawings") of the same atom connection pattern (same constitutional isomer). Consider the different "pictures" of the molecule that could have been generated:

```
                      H  H                                  H  H
      Alt Step (4b)   |  |                Alt Step (4c)     |  |
C—C ——————————————> F—C——C—H ; or C—C ——————————————> H—C——C—H
                      |  |                                  |  |
                      H  H                                  H  F
```

The molecule produced in alternate step (**4b**) is the identical mirror image of the original molecule formed. It is not a different molecule, but a different view of the same molecule with the same bonding pattern. The molecule produced in alternate step (**4c**) also has the exact same bonding pattern as the original molecule formed; It is not a different molecule or isomer. A Lewis structures does not imply any information about molecular shape; in this case, it does not matter which direction any of the bonded atoms are pointing. Atoms do not know if they are marching from left-to-right, right-to-left, up or down, or sideways. The information conveyed by a Lewis structure is "who is bonded to whom." All three of the structures generated show that this molecule is composed of two identical carbons bonded to each other, with three hydrogens bonded to one carbon and two hydrogens and a fluorine bonded to the other carbon.

Example: Construct a valid Lewis structure for the compound CF_3N.

Steps (**1**) and (**2**): 1 carbon; each C = 4 bonds — **possible** central atom
3 fluorines; each F = 1 bond — **required** outside atoms
1 nitrogen; each N = 3 bonds — **possible** central atom

```
                     Step (4)                Step (5)
Step (3):  C—N ——————————————> F—C—N—F ——————————————> F—C═N—F
                                 |                        |
                                 F                        F
```

The bonding is checked at step (**4**). Each fluorine has one bond that is the correct number; carbon has only three bonds but needs four; nitrogen has only two bonds but needs three. Step (**5**) states that if there are not enough outside atoms to give all central atoms the correct numbers of bonds, double or triple bonds must be used. One solution is to give the carbon and nitrogen one extra bond apiece by forming a **double** bond between them. All bond numbers are now correct: each F has 1 bond; the C has four bonds; the N has three bonds. This is a valid bonding pattern.

Note that an initial bonding pattern for the outside and central atoms from step (**4**) may need to be rearranged if double or triple bonds are required. For example, consider the result if the initial bonding pattern were the one shown as **initial pattern (b)**:

F—C—N (with F above and below C) —Step (5)→ F—C=N (with F above and below C); F—C—N (with F above and below C) —Step (5)→ F—C≡N (with F above and below C)

initial pattern (b) **incorrect result** **initial pattern (b)** **incorrect result**

A double bond could not be used between the central atoms: carbon would have five bonds and nitrogen would have two bonds. A triple bond would give nitrogen the correct number of three bonds but would give carbon six bonds. A shift of one of the F atoms from C to N, as shown is required before the double bond can be written correctly.

F—C—N (with F above and below C; upper F boxed, moved to N) —rearrange→ F—C—N—F (with F below C) —Step (5)→ F—C=N—F (with F below C)

initial pattern (b) **final correct pattern** **Lewis structure**

Step (**6**): Add the required lone electron pairs to each atom; the calculation for any row-2 atom need not be performed again assuming valid normal neutral bonding. The valid Lewis structure for this molecule is:

:F̤̈—C=N̈—F̤̈: with :F̤̈: below C

Example: Construct a valid Lewis structure for the compound HNO_2.

Steps (**1**) and (**2**): 1 nitrogen; each N = 3 bonds **possible** central atom
1 hydrogen; each H = 1 bond **required** outside atom
2 oxygens; each O = 2 bonds **possible** central atoms

Step (**3**): There are three central atoms and two possible central atom patterns:

(**a**) N—O—O and (**b**) O—N—O; note that O—O—N is the same as (**a**)

Step (**3a**): N—O—O —Step (4)→ (**4a**) N—O—O—H or (**4b**) H—N—O—O

There is only one outside atom; the nitrogen and one oxygen each require one more bond, therefore, step (**5**) is required. This atom connection pattern for the central atoms cannot lead to a valid Lewis structure. A double bond between nitrogen and oxygen in the (4a) atom pattern, N=O—O—H leaves nitrogen deficient and oxygen with an extra bond. A double bond in the (4b) atom pattern, H—N=O—O leaves one oxygen deficient and the other oxygen with an extra bond. (A possible solution is the formation of a ring compound, which is not considered in this chapter.) This is an example of three central atoms that must be connected in a specific order.

```
                    Step (4)                  Step (5)
Step (3b): O—N—O ------------> O—N—O—H ------------> O=N—O—H

                                           Step (6)    ••  ••  ••
All atoms have the correct numbers of bonds. ------------> O=N—O—H
                                                       ••      ••
```

Example: Construct a valid Lewis structure for the compound C_2H_6O.

Steps (**1**) and (**2**): 2 carbons; each C = 4 bonds — **possible** central atoms
6 hydrogens; each H = 1 bond — **required** outside atoms
1 oxygen; each O = 2 bonds — **possible** central atom

Step (**3**): There are three central atoms and two possible central atom patterns:

(**a**): C—C—O and (**b**) C—O—C ; note that O—C—C is the same as (**a**)

```
                        H  H                       H  H
             Step (4)   |  |          Step (6)     |  |  ••
(3a): C—C—O ---------> H—C—C—O—H  -------------> H—C—C—O—H
                        |  |                       |  |  ••
                        H  H                       H  H

                        H     H                    H     H
             Step (4)   |     |       Step (6)     |  •• |
(3b): C—O—C ---------> H—C—O—C—H  -------------> H—C—O—C—H
                        |     |                    |  •• |
                        H     H                    H     H
```

This is an example demonstrating that the three central atoms can be connected in either of the two possible patterns to yield a valid Lewis structure.

Example: Find **three** constitutional isomers for the formula $C_3H_2F_2O_2$.

Steps (**1**) and (**2**): 3 carbons; each C = 4 bonds — **possible** central atoms
2 oxygens; each O = 2 bonds — **possible** central atoms
2 hydrogens; each H = 1 bond — **required** outside atoms
2 fluorines; each F = 1 bond — **required** outside atoms

```
                                      H     H
                     Step (4)         |     |
Step (3): O—C—C—C—O ---------> F—O—C—C—C—O—F
```

Five possible central atoms for one formula can usually be connected in any order. There are a number of different central atom sequences including patterns in which central atoms branch off a particular atom chain. For a formula with this many atoms, the first goal is to find one valid bonding pattern; a symmetrical pattern is selected for the central atoms and the outside atoms.

The bonding is checked at this point. Each fluorine has one bond (correct number); each hydrogen has one bond (correct number); each oxygen has two bonds (correct number). Two carbons have only three bonds; each needs four. One carbon has only two bonds; it needs four.

```
              H     H                       ••  ••  H     H  ••  ••
              |     |        Step (6)           |     |
Step (5): F—O—C=C=C—O—F  -------------> :F—O—C=C=C—O—F:
                                         ••  ••          ••  ••
```

The technique of placing outside atoms approaching a symmetrical pattern around central atoms aids in recognizing positions for multiple bonds. The solution for the pattern in step (**4**) results in the symmetrical double bond pattern shown; all atoms have the correct number of bonds.

Constitutional isomers are different bonding patterns for the same molecular formula. Another isomer could be formed by simply exchanging a hydrogen on a carbon with a fluorine on an oxygen. A more interesting variation would be the use of one triple bond instead of the two double bonds as shown in constitutional isomer **#2**; note that this is not the only possible structure with a triple bond.

Isomer #2: :F—O—C(H)(H)—C≡C—O—F: Isomer #3: :F—C(F)=C(H)—C(=O)—O—H

The original central atom connection pattern can be rearranged in a variety of ways. Possible central atoms need not be in a continuous sequence, and in fact some need not be central atoms at all as shown by the double bonded oxygen in constitutional isomer **#3**. As will be demonstrated in Chapter 15, all constitutional isomers of the same molecular formula must contain the same number of multiple bonds (pi-bonds) if ring patterns are specifically excluded.

IV CONSTRUCTING LEWIS STRUCTURES FOR COVALENT MOLECULES USING ELEMENTS FROM ROWS 3–5: BONDING PATTERNS THAT FOLLOW NORMAL NEUTRAL BONDING RULES

Since atoms from rows 3–5 can have a variable number of bonds, three variations in Lewis structure processes are described.

Problem Type (**1**): If the bond numbers of all row 3–5 elements are specified as additional information, the identical **steps** (**1**) through (**6**) described for row-2 elements are applicable; adapt step (**1**) by using the additional information concerning bond numbers. This method is useful because the row 3–5 elements very often show octet bonding; these elements are always octet bonded when they act as outside atoms.

Example: Find a valid Lewis structure for C_3HBr_3, using the additional information that bromine shows octet bonding.

Bromine is stated as following octet bonding; it has seven valence electrons and will, therefore, form one bond with three lone pairs similar to its column-mate fluorine.

Steps (**1**) and (**2**): 3 carbons; each C = 4 bonds — **possible** central atoms
1 hydrogen; each H = 1 bond — **required** outside atom
3 bromines; each Br = 1 bond — **required** outside atoms

Step (**3**): C—C—C —Step (**4**)→ H—C(Br)(Br)—C—C—Br —Step (**5**)→ H—C(Br)(Br)—C≡C—Br

The bonding is checked at step (**4**). Each bromine has one bond (correct number); the hydrogen has one bond (correct number). One carbon has four bonds (correct number); two connected carbons each have two bonds but need four. One solution is to give the deficient carbons two extra bond apiece by forming a **triple** bond between them.

Step (**6**): The final Lewis structure for this molecule is shown along with another possible constitutional isomer.

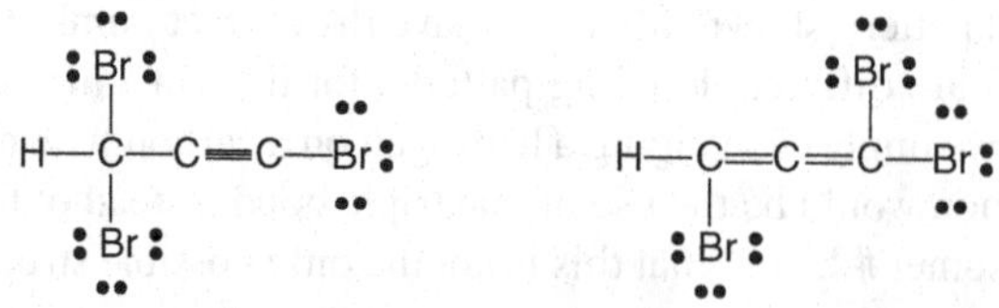

Problem Type (**2**): Many molecules can have one (**only** one) row 3–5 atom as a central atom; all other atoms are outside atoms connected to this central atom. The problem consists of one row 3–5 central atom with an unspecified number of bonds. If the normal neutral bond numbers for all other atoms are known, a modified procedure for these molecules can be applied.

For step (**1**), identify the correct bond numbers for all atoms **except** the row 3–5 central atom; do **not** specify the bond number for this central atom at this time. For step (**2**), identify the row 3–5 atom as the central atom; identify the other atoms as outside atoms For steps (**3**), (**4**), and (**5**), arrange all the outside atoms around the unspecified central atom; use double or triple bonds as necessary (can be applicable to the third problem type) to give all the outside atoms the correct number of bonds determined from steps (**1**) and (**2**).

The key to this process is that if the molecular formula represents a valid molecule, no additional determination for the unspecified central is necessary; the number of bonds to the unspecified row 3–5 central atom must be equal to the total number of bonds given to this atom in order to correctly satisfy all of its outside atoms. The process may result in a normal octet for the central atom or may result in an expanded valence shell. If the count is made correctly, the central atom will not have exceeded its maximum number of bonds for any valid molecule.

Example: Determine the structure of PF_5.

Steps (**1**) and (**2**): 1 phosphorous; each P = **unspecified** central atom
5 fluorines; each F = 1 bond **required** outside atoms

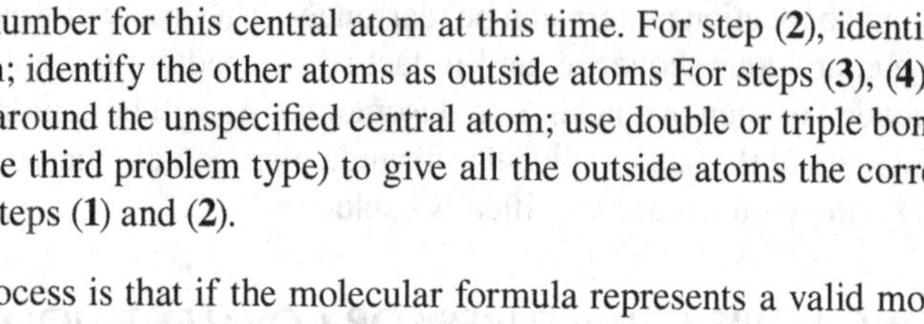

Step (3): P —Step (4)→ F—P—F (with three more F) —Step (6)→ Lewis structure of PF_5

Step (**4**): There is only one possible structure that can be formed; all five fluorines must be bonded to the phosphorous as the only central atom. Assuming PF_5 is a valid molecule, the bonding pattern must be correct. Note that the value of five covalent bonds was listed as a possible normal neutral bond number for phosphorous. In this case, phosphorous has an expanded valence shell with 10 electrons in the outer bonding valence shell, 5 original valence electrons plus 5 electrons from the bonds to fluorine.

Step (**6**): Fluorine with one bond will always have three lone pairs. Lone pairs for the variable row 3–5 central atoms must be determined for each specific case. Phosphorous starts with five valence electrons and uses all five electrons for bonding; it has no electrons leftover (no lone pairs).

Atom Lewis structure: •P•; P uses all five electrons for bonding: —•P•—

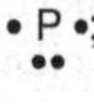

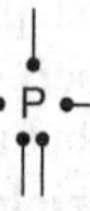

Example: Determine the structure of Cl F_5.

Fluorine cannot be a central atom; chlorine as the row 3–5 atom, even though a halogen, must be the unspecified central atom.

Steps (**1**) and (**2**): 1 chlorine; each Cl = **unspecified** central atom
5 fluorines; each F = 1 bond **required** outside atoms

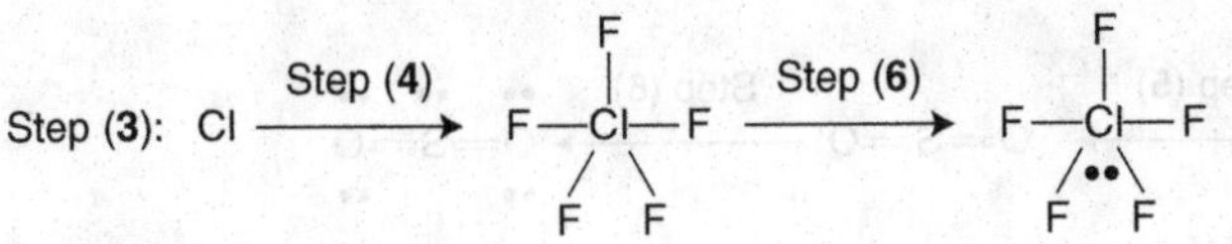

Step (**4**): There is only one possible structure that can be formed; all five fluorines must be bonded to the chlorine as the only central atom. Assuming Cl F_5 is a valid molecule, the bonding pattern must be correct. Note that the value of five covalent bonds was listed as a possible normal neutral bond number for chlorine. In this case, chlorine has seven original valence electrons plus five electrons from the bonds to fluorine for an expanded valence shell with 12 electrons in the outer bonding valence shell.

Step (**6**): F with one bond will always have three lone pairs; for clarity these are left off the final Lewis structure. Cl uses five electrons for bonding and has two electrons left over for a total of one lone pair. The atom Lewis structure is drawn to help match the bonding.

Atom Lewis structure: ; Cl uses five electrons for bonding: 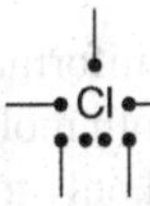

Example: Determine the structure of Cl F_3.

Steps (1) and (2):	1 chlorine;	each Cl =	**unspecified** central atom
	5 fluorines;	each F =	1 bond **required** outside atoms

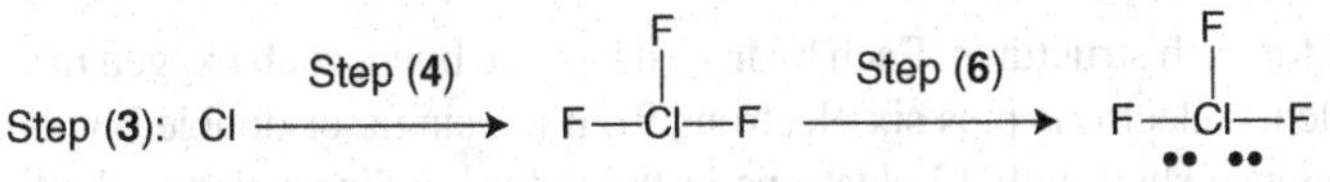

Step (**4**): There is only one possible structure that can be formed with the three fluorines bonded to the chlorine as the only central atom; three covalent bonds is a possible normal neutral bond number for chlorine. In this case, chlorine has seven original valence electrons plus three electrons from the bonds to fluorine for an expanded valence shell with 10 electrons in the outer bonding valence shell.

Step (**6**): F with one bond will always have three lone pairs; for clarity these are left off the final Lewis structure. Cl uses three electrons for bonding and has four electrons leftover for a total of two lone pairs. The atom Lewis structure is drawn to help match the bonding.

Atom Lewis structure: 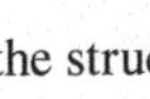; Cl uses three electrons for bonding: 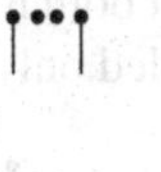

Problem Type (3): Many molecules with one unspecified row 3–5 atom as a central atom may also contain other specified possible central atoms. In these cases, one specific structure is required; additional information is included in the problem to indicate a unique bonding connection pattern.

Example: Determine the structure of SO_2. The extra information is that S is the **only** central atom for SO_2; both oxygens are outside atoms in this structure.

Steps (**1**) and (**2**): 1 sulfur; each S = **unspecified** central atom
2 oxygens; each O = 2 bonds: must be outside atoms

Oxygen, which takes two bonds, is always a **possible** central atom. It may be an outside atom if both bonds are to the same atom; the extra information specifies this result in SO_2.

Step (**3**): S —Step (**4**)→ O—S—O —Step (**5**)→ O=S=O —Step (**6**)→ O=S=O (with lone pairs: two on each O, one on S)

Step (**4**): The bonding pattern was indicated by the additional information.
Step (**5**) was necessary to insert double bonds for both oxygens. Sulfur has six original valence electrons plus four electrons from the double bonds to the oxygens for an expanded valence shell with 10 electrons in the outer bonding valence shell.

Step (**6**): O with two bonds will always have two lone pairs. S uses four electrons for bonding and has two electrons leftover for a total of one lone pair.

Atom Lewis structure: •S̈•; S uses four electrons for bonding: —•S̈•—

Example: Use the additional information to draw the final structures of sulfuric acid, H_2SO_4 and phosphoric acid, H_3PO_4. In both molecules, all oxygens are bonded to the row 3–5 central atom, and all hydrogens are bonded only to oxygens.

:Ö—H / Ö=S=Ö / H—Ö: (sulfuric acid: S bonded to two O=, and two O—H)

H—Ö: / H—Ö—P=Ö / :Ö—H (phosphoric acid: P bonded to one O= and three O—H)

The additional information is satisfied for both structures. Each hydrogen has one bond, each oxygen has two bonds. Sulfur has six original valence electrons plus six electrons from the single or double bonds to the oxygens; it has an expanded valence shell with 12 electrons in the outer bonding valence shell. Sulfur uses all six electrons for bonding and has no lone pairs. Phosphorous has an expanded valence shell with 10 electrons in the outer bonding valence shell, five original valence electrons plus five electrons from bonds to oxygen. Phosphorous uses all five electrons for bonding and has no lone pairs.

V CONSTRUCTING MOLECULES THAT REQUIRE EXCEPTIONS TO NORMAL NEUTRAL BONDING RULES

The elements in rows 3–5 may not always be neutral; these elements can show formal charges in neutral Lewis structures. However, the use of expanded valence shells as a possible solution for certain structures can most often produce one acceptable electron distribution without fixing formal charges in neutral molecules.

Lewis structures of positive or negative ions composed of one unspecified row 3–5 atom will often require a formal charge on the central atom; these cases can be solved through an adaption to the step (**1**). For step (**1**), add or subtract an electron to/from the unspecified central atom; include this result in the calculation of lone pairs.

Example: Determine the structure of $(BrF_4)^-$.

Steps (**1**) and (**2**): 1 bromine; each Br = **unspecified** central atom
4 fluorines; each F = 1 bond **required** outside atoms

Step (3): $\overset{-}{Br}$ —Step (4)→ F—$\overset{-}{Br}$—F (with two F below Br) —Step (6)→ F—$\overset{-}{Br}$—F (with two lone pairs on Br, two F below Br)

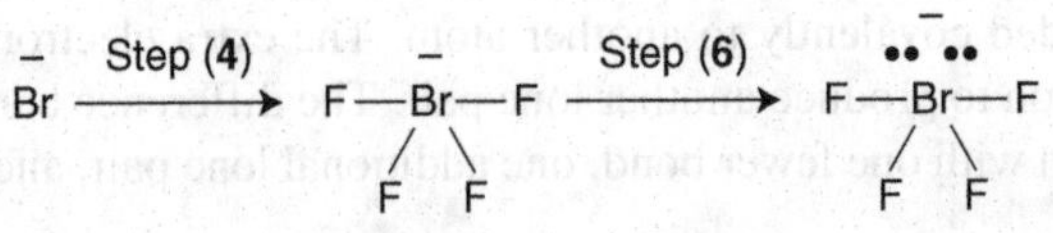

Step (**3**) indicates that bromine has an extra electron to account for the negative charge.

Step (**4**) follows the procedure described for neutral molecules of problem type (**2**).

Step (**6**) is required to calculate the number of lone pairs on bromine; lone pairs on fluorine are not shown. A neutral bromine atom has seven valence electrons; a bromine ion with a charge of (−1) has eight electrons in the outer shell. The bromine ion uses four electrons for bonding resulting in four electrons leftover (two lone pairs). The Br^{-1} ion Lewis structure is shown to match the bonding.

Ion Lewis structure: •Br• (with two lone pairs above and one pair below, charge −); Br uses four electrons for bonding: —•Br•— (with bonds below)

Example: Determine the structure of $(ClF_2)^+$.

Steps (**1**) and (**2**): 1 chlorine; each Cl = **unspecified** central atom
2 fluorines; each F = 1 bond **required** outside atoms

Step (3): $\overset{+}{Cl}$ —Step (4)→ F—$\overset{+}{Cl}$—F —Step (6)→ F—$\overset{+}{Cl}$—F (with two lone pairs on Cl)

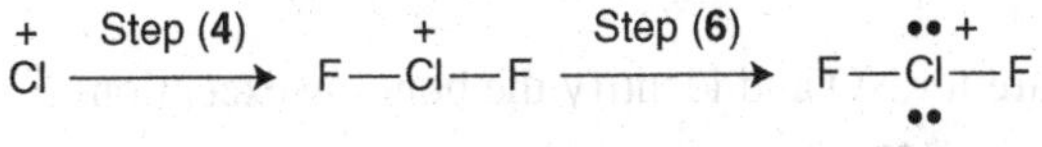

Step (**3**) indicates that chlorine has one fewer electron and a positive charge.

Step (**4**) follows the procedure described for neutral molecules of problem type (**2**).

Step (**6**) is required to calculate the number of lone pairs on chlorine; lone pairs on fluorine are not shown. A neutral chlorine atom has seven valence electrons; a chlorine ion with a charge of (+1) has six electrons in the outer shell. The chlorine ion uses two electrons for bonding resulting in four electrons leftover (two lone pairs). The Cl^{+1} ion Lewis structure is shown to match the bonding.

Ion Lewis structure: •Cl• (with lone pairs above and below, charge +); Cl uses two electrons for bonding: —•Cl•— (with lone pairs above and below, charge +)

Bonding Patterns through Exception Rules

The use of the Exception Rules applies to the row-2 elements, which cannot have outer bonding valence shells with more than eight electrons; they have no d-orbitals in shell number 2. F and H have no exceptions; they must always take only one bond. The common exceptions to normal neutral bonding rules for B, C, N, and O occur whenever these atoms are not neutral. A formal charge of (+1) or (−1) indicates bonding numbers different from the normal neutral values. The term **formal charge** is used to describe charge values generated by specific placement of electrons according to standard concepts of valence shell count and bond formation. In terms of chemical behavior, however, the atoms with formal charges share the electron deficiency or surplus with their bonded atom neighbors. The different bonding patterns exhibited by B, C, N, and O are organized though the three exception rules.

Exception Rule #1

An extra electron can compensate an atom for one fewer bond while allowing the atom to reach a normal octet; the difference is designated by a formal charge of (−1). The bonding difference occurs when an atom acquires an extra electron all to itself; the electron is held completely in its orbital and not shared with another atom through a covalent bond. The atom showing this exception can be thought of as being a (−1) ion bonded covalently to another atom. The extra electron pairs up with the unused normal bonding electron to produce another lone pair. The difference compared to normal neutral bonding shows the atom with one fewer bond, one additional lone pair, and a formal charge of (−1).

Exception Rule #2

In rare cases, an atom that normally shows octet bonding cannot achieve a full eight electrons (octet) in the outer bonding shell. This central atom has one missing bond and only seven bonding shell electrons; one of the electrons is unpaired. This atom is termed a radical or free radical. A radical is usually very reactive; the unpaired electron has a great tendency to form an additional bond.

Exception Rule #3

Exceptions to normal bonding occur during the formation of what is termed a **coordinate** covalent bond. This bonding occurs when one atom **donates both** electrons to the covalent bond and one atom **accepts both** electrons for the covalent bond. The donor atom converts a lone pair to a bonding pair; this changes its formal charge by (+1). The net bonding result for the donor atom is formation of one additional bond with one fewer lone pair while retaining a normal octet. The acceptor of both electrons must have an empty orbital available. Acceptance of a bonding pair results in the formation of an additional bond without any electron contribution; the acceptor atom changes its formal charge by (−1).

Example: Determine the Lewis structure for NO and identify the bonding exception.

Steps (**1**) through (**6**) produce the structure: • N═O (with lone pairs: •• above N; •• above and •• below O)

The nitrogen requires three bonds and oxygen two bonds. The optimum structure satisfies the oxygen but leaves the nitrogen with only two bonds and only seven bonding outer shell valence electrons. Nitrogen is non-octet with an unpaired electron; it is described by Exception Rule **#2**.

Working with Exception Rule #1

Exception Rule **#1** is used to construct most polyatomic ions as well as certain neutral molecules. To understand the bonding resulting from this process, consider the molecule hydrogen peroxide. Assume energy is added to the molecule; the oxygen–oxygen single bond breaks in the process to form two OH fragments.

H—O•—•O—H ⟶ H—O•/•O—H ⟶ H—O• + •O—H (each O with two lone pairs, •• above and •• below)

The species formed are hydroxyl radicals, the oxygen of the hydroxyl radical has only seven outer bonding shell valence electrons and has one unpaired electron. The oxygen can be satisfied in two ways. If it were to meet a hydrogen radical (H•) or could pull a hydrogen atom from another molecule, the oxygen could form another standard covalent bond to match its normal neutral bonding pattern:

H—O• + •H ⟶ H—O•—•H (each O with two lone pairs, •• above and •• below)

To apply Exception Rule **#1**, consider another possibility. Assume another atom is available (for example Na) to transfer an electron directly to the oxygen "free and clear"; that is, not as part of a covalent bond but as a direct electron transfer:

H—O• + o Na ⟶ H—O•o + Na^+

The hydroxide polyatomic ion is formed in the process and represents the restoration of a stable octet around oxygen. The difference in the normal neutral bonding pattern is signified by the formal charge of (−1) derived from the extra electron. The resulting octet oxygen bond pattern is formation of one bond with three lone pairs.

Exception Rule **#1** applies to C, N, and O. Each can achieve an octet through formation of one fewer bond with one additional lone pair and a formal charge of (−1). The acceptable bonding patterns for these atoms according to this rule are:

—O:⁻ —N—⁻ —C—⁻

Lewis Structures for Polyatomic Ions

The concept of a polar covalent bond was described in Chapter 10. The polarity of the bond, as measured by a bond dipole, is caused by unequal sharing of the electron pair in a covalent bond. The strength with which an atom holds a bonding electron pair in a covalent bond is measured by the atom's **electronegativity (EN)**; the higher the value, the more tightly the atom holds the electron pair. EN values generally follow the outer shell electron trends observed in Chapter 13. Atoms that hold electrons most tightly as measured by ionization energy generally hold electrons more tightly (higher EN value) in covalent bonds. For any bond, the atom that holds the greater share of the electron pair is termed relatively negative (δ−); the atom that holds the lesser share of the electron pair is termed relatively positive (δ+).

The following is a list of typical values for electronegativity (**EN**).

F = 4.0	**O** = 3.5	**Cl** = 3.2	**N** = 3.0	**Br** = 3.0
I = 2.7	**S** = 2.6	**Se** = 2.6	**C** = 2.5	**P** = 2.2
H = 2.2	**As** = 2.2	**Te** = 2.1	**B** = 2.0	**Si** = 1.9

Polyatomic ions, by definition, have a charge and, therefore, cannot be neutral; thus, they require use of the exception rules. Anions (ions with a negative charge) can be solved using steps (**1**) through (**6**) for normal neutral bonding after determination of modified bond numbers for the exception atoms through Exception Rule **#1**. The sum of all formal charges on atoms in a polyatomic ion must be equal the actual charge on the polyatomic ion. For anions, the atom(s) that get the extra electron and a formal charge of (−1) should be the more electronegative element (very often oxygen). Never give more than one extra electron to any one atom. If the charge on the polyatomic ion is (−2) or (−3), distribute one extra electron to two or three different atoms.

Example: Determine the structure of the carbonate anion, $(CO_3)^{-2}$. Use the additional information that carbon is the **only** central atom; all three oxygens are bonded only to carbon.

Steps (**1**) and (**2**):	1 carbon;	C = 4 bonds	central atom
	1 neutral oxygen;	O = 2 bond	outside atom
	2 anionic oxygens; each	O = 1 bond	outside atoms

Step (**1**) is modified by the requirement that carbonate must have a charge of (–2). Two separate oxygens must each follow Exception Rule **#1**: they must have one bond, three lone pairs, and a formal charge of (–1). The neutral carbon and neutral oxygen follow normal neutral bonding rules. Step (**2**) indicates the additional information.

Step (4): ⁻O—C(—O)—O⁻ —Step (5)→ ⁻O—C(=O)—O⁻ —Step (6)→ :Ö⁻—C(=Ö:)—Ö⁻:

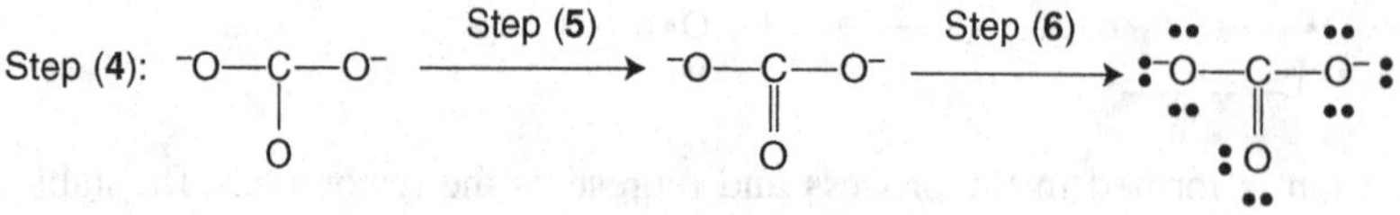

Step (**6**): Neutral carbon has no lone pairs, and neutral oxygen has two lone pairs. Oxygens with one bond and a formal charge of (–1) have three lone pairs.

Example: Determine the structure of the sulfite anion, $(SO_3)^{-2}$. Use the additional information that sulfur is the **only** central atom; all three oxygens are bonded only to sulfur.

Steps (**1**) and (**2**):	1 sulfur	S = unspecified	central atom
	1 neutral oxygen	O = 2 bond	outside atom
	2 anionic oxygens; each	O = 1 bonds	outside atoms

Step (**1**): Sulfite has a charge of (–2). Two separate oxygens each follow Exception Rule **#1** with one bond, three lone pairs, and a formal charge of (–1). The neutral oxygen follows normal neutral bonding rules; sulfur has an expanded valence shell and has an unspecified number of bonds. Step (**2**) indicates the additional information.

Step (4): ⁻O—S(—O)—O⁻ —Step (5)→ ⁻O—S(=O)—O⁻ —Step (6)→ :Ö⁻—S̈(=Ö:)—Ö⁻:

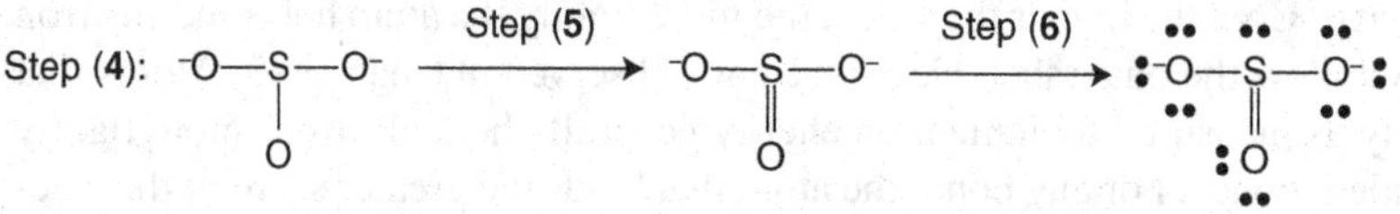

Step (**6**): Lone pairs on the unspecified atom must be calculated.

Atom Lewis structure: •S̈•(with two lone pairs); S uses four electrons for bonding: —•S̈•—

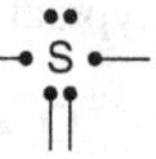

Example: Determine the structure of the amide anion, $(NH_2)^-$.

Steps (**1**) and (**2**):	1 nitrogen;	N = 2 bonds	central atom
	2 hydrogens;	each H = 1 bond	outside atoms

Step (**1**) is modified by the requirement that the amide anion has a charge of (–1). Hydrogen cannot have exceptions and cannot take a negative charge in a covalent compound; the extra electron must go to nitrogen. Nitrogen follows Exception Rule **#1** with two bonds, two lone pairs, and a formal charge of (–1).

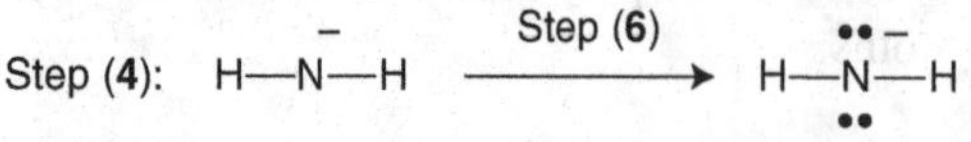

Step (4): H—N⁻—H —Step (6)→ H—N̈⁻—H (with two lone pairs)

Example: Determine the structure of the cyanide anion, $(CN)^-$.
The problem is best visualized by first forming the Lewis structure as if the molecule were neutral.

Steps (**1**) and (**2**):	1 carbon;	C = 4 bonds	possible central atom
	1 nitrogen	N = 3 bonds	possible central atom

Steps (**3**) to (**6**) would result in the neutral structure $\bullet\ddot{N}$ $\bullet C\equiv \ddot{N}$

The solution is now clear; the only way to achieve an octet for both nitrogen and carbon is to place the extra electron from the cyanide anion on carbon, even though carbon is not the more electronegative atom.

Carbon is an uncommon participant in Exception Rule **#1**: the anion is $^{-}\ddot{C}\equiv\ddot{N}$

Using Exception Rule #3

The **donor** atom in a coordinate covalent bond donates both electrons to the covalent bond. A lone pair is converted to a bonding pair, and, thus, the donor atom forms one additional bond at the expense of one fewer lone pairs. This type of bonding does not change the bonding valence shell electron number; atoms following octet bonding will still have eight outer shell bonding valence electrons. The difference in bonding occurs because two lone pair valence electrons must now be shared; the donor atom changes its formal charge by (+1) to indicate that the two electrons originally held by the atom all to itself must now be equally shared with the other bonding atom. For counting purposes, the donor atom appears to be providing the acceptor's contribution of one electron to a normal covalent bond.

For this exception rule, donor atoms must have at least one electron pair, which can be converted to a bonding electron pair. For row-2, the rule is restricted to nitrogen and oxygen, which have lone pairs to donate. Oxygen can form three bonds by donation of one lone pair resulting in one lone pair leftover; it has a formal charge of (+1). Nitrogen can form four bonds by the donation of one lone pair resulting in no lone pairs leftover; it has a formal charge of (+1).

The **acceptor** atom in a coordinate covalent gains two additional electrons to its outer bonding valence shell; an empty orbital accepts two bonding electrons from the donor. The acceptor atom changes its formal charge by (–1) since it gets a share of an electron pair that it did not originally have. (For counting purposes, the acceptor is getting its normal contribution of one electron for a normal covalent bond from the donor atom.) Acceptor atoms must have an empty orbital, which can take two additional electrons; this excludes the row-2 neutral atoms C, N, and O. Boron, with an empty p-orbital, is a common example of an acceptor from row-2; boron can form four bonds and achieve an octet with a formal charge of (–1). Acceptable octet bonding patterns for nitrogen, oxygen and boron are shown below.

$$-N^{+}- \qquad -\underset{\bullet\bullet}{O^{+}}- \qquad -B^{-}-$$

Example: Show the formation of a coordinate covalent bond between ammonia and boron trifluoride.

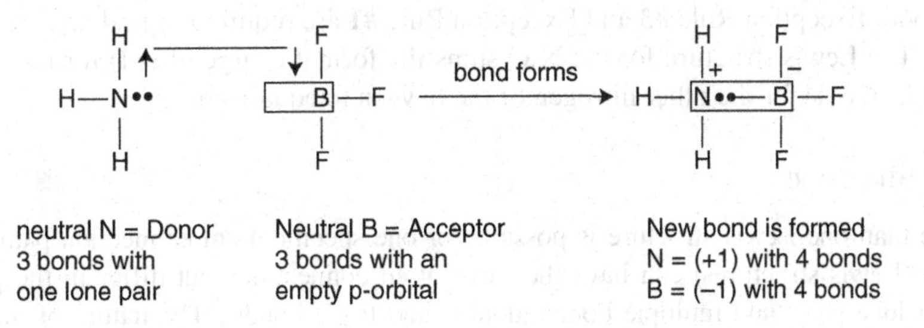

Example: Show the formation of a coordinate covalent bond between H^+ and a water molecule to for the hydronium ion (H_3O^+).

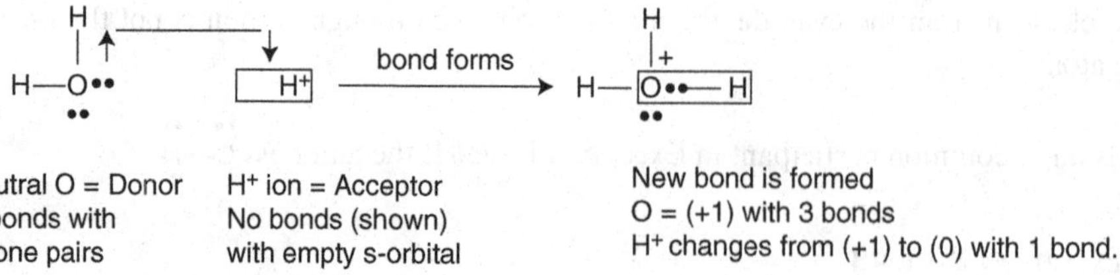

$$H{-}\ddot{O}{:}(H) \quad + \quad H^+ \xrightarrow{\text{bond forms}} H{-}\ddot{O}^{+}{-}H \,(H)$$

neutral O = Donor
2 bonds with
2 lone pairs

H^+ ion = Acceptor
No bonds (shown)
with empty s-orbital

New bond is formed
O = (+1) with 3 bonds
H^+ changes from (+1) to (0) with 1 bond.

Other Uses for Exception Rules

Polyatomic ions are not neutral and will always require the use of an exception rule. Lewis structures for **neutral** compounds may also sometimes require the use of Exception Rules; this occurs whenever no bonding pattern can be found that will satisfy all **normal neutral** bonding requirements. Although this situation is not common, bonding patterns for the row-2 exceptions are used to explain certain neutral compounds.

Example: Determine the structure of ozone, O_3.

Steps (**1**) through (**6**) cannot produce a unique structure that satisfies all normal neutral bonding rules; the central oxygen cannot satisfy the outside oxygens if all oxygens must have only two bonds. Employing the exception rules provides a possible structure allowing all oxygens to achieve an octet at the expense of requiring formal charges in a neutral molecule.

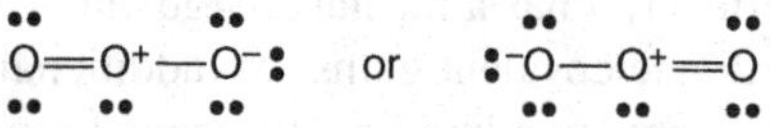

$$O{=}O^{+}{-}O^{-} \quad \text{or} \quad {}^{-}O{-}O^{+}{=}O$$

The central oxygen is given an extra bond with one lone pair and a formal charge of (+1) by Exception Rule **#3** and one outside oxygen is given one bond with three lone pairs and a formal charge of (–1) by Exception Rule **#1**. Two structures are shown at this stage assuming that each end oxygen can be distinguished. If a molecule is **neutral**, the sum of all formal charges for all atoms must add to **zero**. The Lewis structure for the O_3 sums the formal charge (+1) plus the formal charge of (–1) to equal zero.

Example: Determine the structure of dinitrogen oxide.

As for ozone, steps (**1**) through (**6**) for this molecule cannot produce a unique structure that satisfies all normal neutral bonding rules. Employing the exception rules provides possible structures with all atoms achieving an octet at the expense of formal charges.

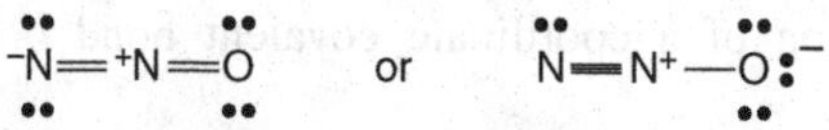

$$^{-}N{=}{}^{+}N{=}O \quad \text{or} \quad N{\equiv}N^{+}{-}O^{-}$$

The use of both Exception Rule **#3** and Exception Rule **#1** are required to find an acceptable bonding pattern. The Lewis structure for the N_2O sums the formal charge of (+1) for nitrogen plus the formal charge of (–1) for the other nitrogen or the oxygen to equal zero.

Resonance Structures

Often, more than one Lewis structure is possible for one specific atom connection pattern. Bonding descriptions (Lewis structures) can have the same atom connections but differ in the placement of electrons for lone pairs and multiple bonds (double and triple bonds). The nature of multiple bonds (pi-bonds) allows potential variable electron distributions within the same bonding connection pattern.

Lewis structures for a specific molecule that show the exact same bonding connection pattern but differ only in the placement of the lone pair and multiple bond electrons are termed **resonance structures**.

Resonance structures do not represent different molecules, but instead represent different electron descriptions of the same molecule. Resonance structures are not isomers; they have the same atom connection pattern (the same description of who is bonded to whom). The best description of the electron distribution in a molecule is usually a composite of the most probable resonance structure descriptions.

The previous examples for O_3 and N_2O illustrate examples of resonance structures. Resonance structures for a molecule are usually indicated by connecting them with a double-headed arrow.

$$O{=}O^{+}{-}O{:}^{-} \longleftrightarrow {}^{-}{:}O{-}O^{+}{=}O \;;\quad {}^{-}N{=}{}^{+}N{=}O \longleftrightarrow N{\equiv}N^{+}{-}O{:}^{-}$$

The two Lewis structures for ozone indicate that either of the end oxygen can have the negative charge. The fact that both end oxygens are found to be identical and share the negative charge confirms the representation of the "true" molecule as a composite of the two Lewis structures. The two Lewis structures for N_2O differ as to the distribution of electrons within the same atom bonding connection pattern; electron distribution in multiple bonds contributes to the differing distribution. The structure showing the negative charge on the more electronegative oxygen would contribute more to the overall picture of the molecule.

For the nitrate polyatomic anion, three resonance structures are necessary to show that each of the three oxygens can have the double bond or the negative charges. In fact, it can be shown that **all** oxygens in this polyatomic ion are identical and show equal sharing of charge and double bond character.

$$NO_3^{-} = {}^{-}{:}O{-}{}^{+}N({-}O{:}^{-}){=}O \longleftrightarrow O{=}N^{+}({-}O{:}^{-}){-}O{:}^{-} \longleftrightarrow {}^{-}{:}O{-}N^{+}({=}O){-}O{:}^{-}$$

VI PRACTICE PROBLEMS

1. Draw Lewis structures for the following two molecules; sulfur shows octet bonding.
   a) N_2H_4 b) C_2S_2
2. Draw two Lewis structures (two isomers) for each of the following formulas; all rows 3–5 elements will follow octet bonding. For formulas (b), (c), and (d), many isomers are possible.
   a) C_2H_4O b) $C_3H_3NO_2$ c) C_4H_5NS d) C_3H_4BrPSO
3. Draw Lewis structures for the following molecules that contain elements from rows 3–5 as unspecified central atoms.
   a) SF_4 b) SF_6 c) AsF_3
4. Draw Lewis structures for the following molecules that contain an element from rows 3–5 as an unspecified central atom. Chlorine is always a central atom; all oxygens are bonded to chlorine and all hydrogens are bonded to oxygen.
   a) $HClO_2$ b) $HClO_3$ c) $HClO_4$
5. Draw Lewis structures for the following polyatomic ions; use Exception Rule #1. Se and As are the only central atoms; all oxygens are bonded to the central atom.
   a) SeO_3^{-2} b) AsO_3^{-3} c) AsO_4^{-3}
6. Draw Lewis structures for the following ions; place the charge on the central atom.
   a) ClF_2^{-} b) BrF_4^{+}

VII ANSWERS TO PRACTICE PROBLEMS

1. (a)

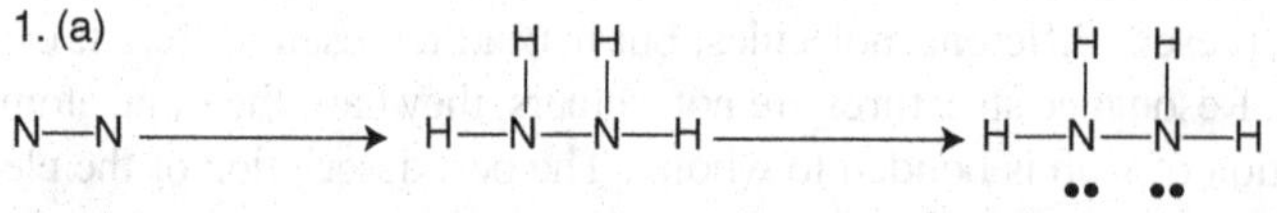

N—N ⟶ H—N(H)—N(H)—H ⟶ H—N(H)—N(H)—H (lone pairs on each N)

(b) S—C—C—S ⟶ S=C=C=S ⟶ S=C=C=S (two lone pairs on each S)

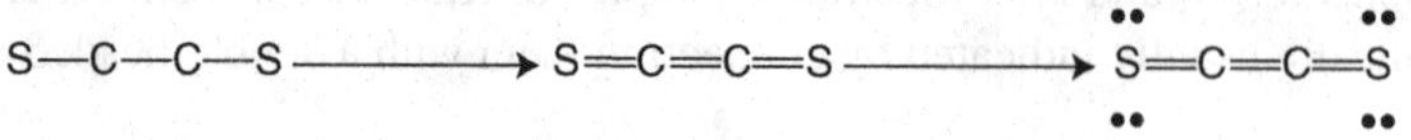

2. (a)

C—C—O ⟶ H—C(H)—C(H)—O—H ⟶ H—C(H)=C(H)—O—H

H—C(H)=C(H)—O—H (two lone pairs on O) or H—C(H)(H)—C(H)=O (two lone pairs on O)

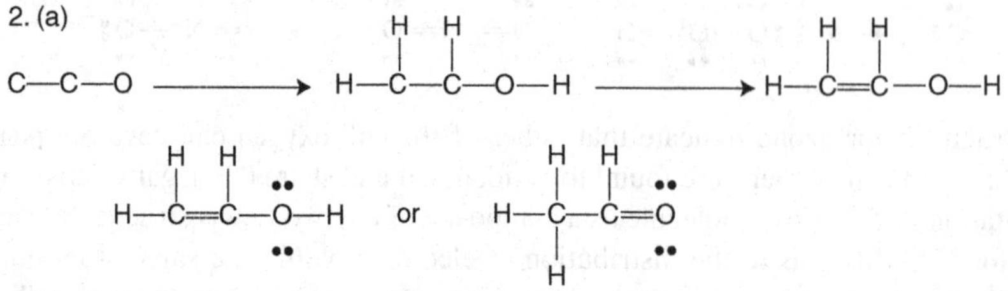

2. (b), (c), (d) Many answers are possible, two are shown. Other answers may be checked by counting the number of multiple bonds (see Chapter 15). A double bond counts as one multiple bond, and a triple bond counts as two multiple bonds. If a structure matches the same number of multiple bonds as the answers here, then the structure is correct assuming no miscounts for bond numbers.

(b)

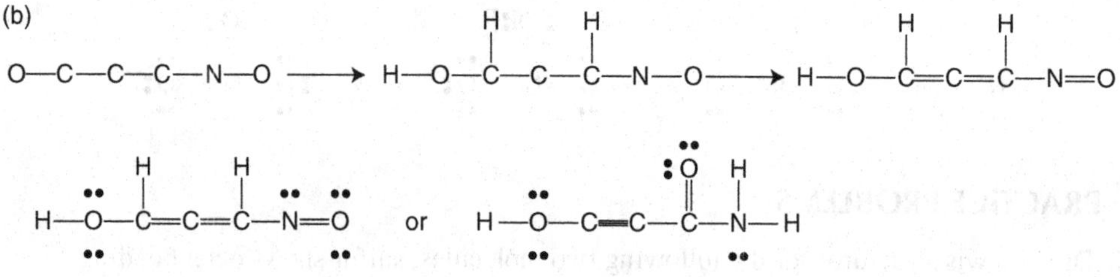

O—C—C—C—N—O ⟶ H—O—C(H)—C—C(H)—N—O ⟶ H—O—C(H)=C=C(H)—N=O

H—O—C(H)=C=C(H)—N=O (with lone pairs) or H—O—C≡C—C(=O)—N(H)—H (with lone pairs)

(c)

N—C—C—C—C—S ⟶ H—N—C(H)—C(H)—C(H)—C(H)—S ⟶ H—N=C(H)—C(H)=C(H)—C(H)=S

H—N=C(H)—C(H)=C(H)—C(H)=S (with lone pairs) or H—C(H)(H)—C≡C—C(=S)—N(H)—H (with lone pairs)

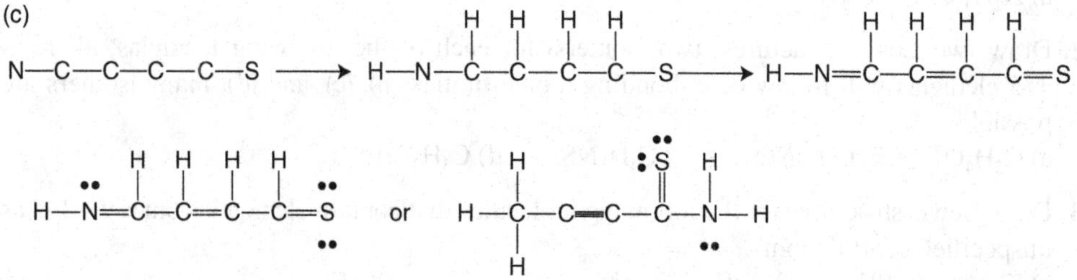

(d)

P—C—C—O—C—S ⟶ Br—P(H)—C(H)—C(H)—O—C(H)—S ⟶ Br—P(H)—C(H)=C(H)—O—C(H)=S

Br—P(H)—C(H)=C(H)—O—C(H)=S (with lone pairs) or P≡C—S—C(H)(H)—O—C(H)(H)—Br (with lone pairs)

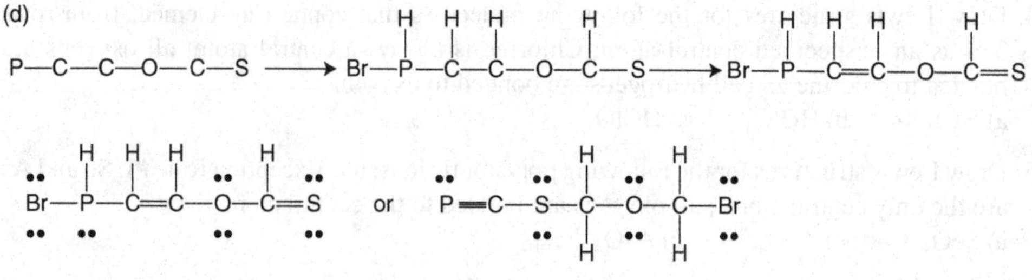

3. The lone pairs on fluorine are not shown.

(a) 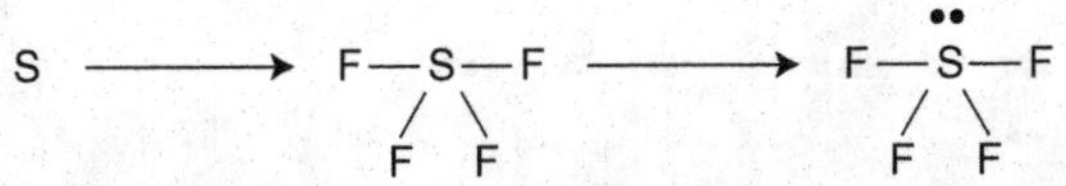 S ⟶ F—S—F (with two F below) ⟶ F—S—F (with two F below, lone pair on S)

Atom Lewis structure: •S• (with two lone pairs); S uses four electrons for bonding: —•S•— (with two bonds below)

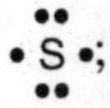

(b) 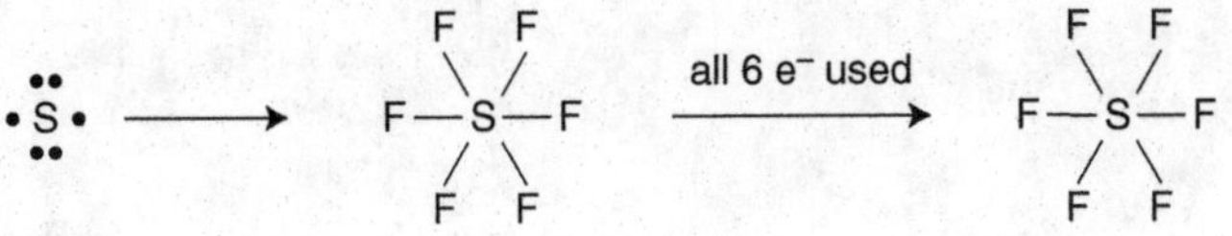 •S• ⟶ F—S—F (with F F above and F F below) $\xrightarrow{\text{all 6 e}^- \text{ used}}$ F—S—F (with F F above and F F below)

(c) 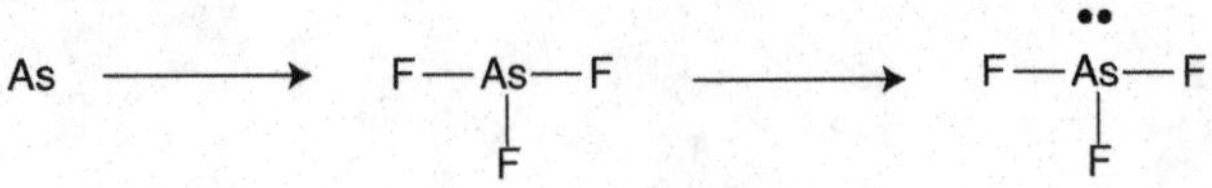 As ⟶ F—As—F (with F below) ⟶ F—As—F (with F below, lone pair on As)

4. Atom Lewis structure: •As• (with lone pair); As uses three electrons for bonding: —•As•— (with bond below)

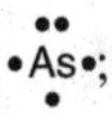

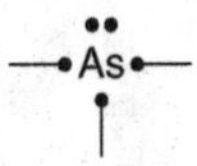

(a) 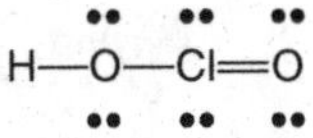 H—O—Cl=O (b) 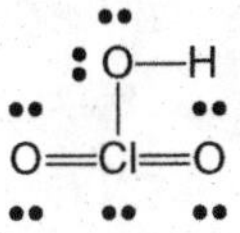 O=Cl=O (with O—H above) (c) 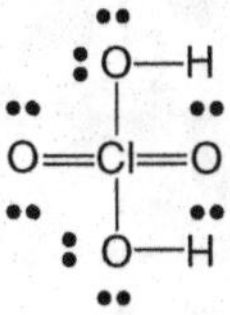 O=Cl=O (with O—H above and O—H below)

5.

(a) 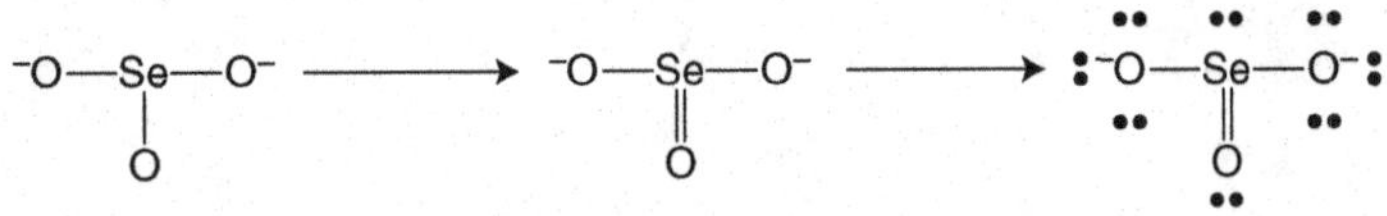 $^-$O—Se—O$^-$ (with O below) ⟶ $^-$O—Se—O$^-$ (with =O below) ⟶ :$^-$O—Se—O$^-$: (with =O below, lone pairs shown)

(b) 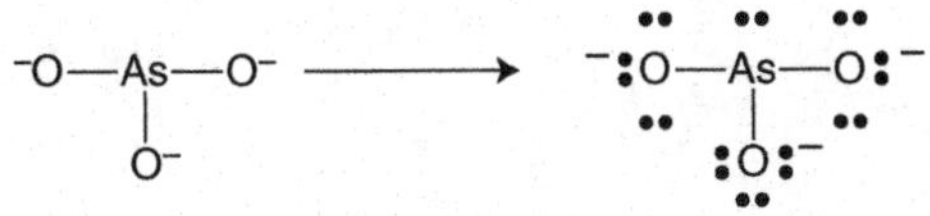 $^-$O—As—O$^-$ (with O$^-$ below) ⟶ $^-$:O—As—O:$^-$ (with :O:$^-$ below, lone pairs shown)

(c) 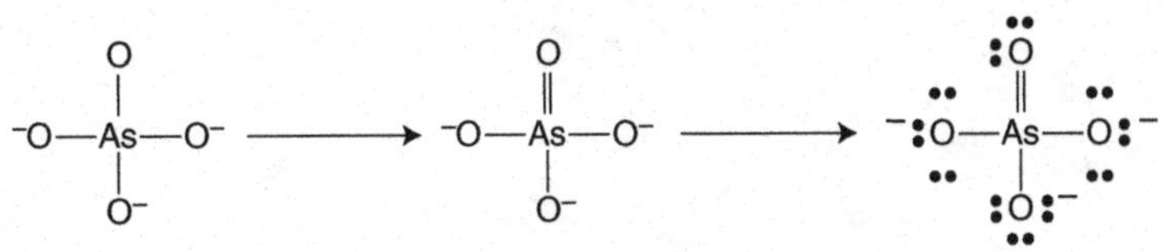 $^-$O—As—O$^-$ (with O above and O$^-$ below) ⟶ $^-$O—As—O$^-$ (with =O above and O$^-$ below) ⟶ $^-$:O—As—O:$^-$ (with :O= above and :O:$^-$ below, lone pairs shown)

6.

(a) 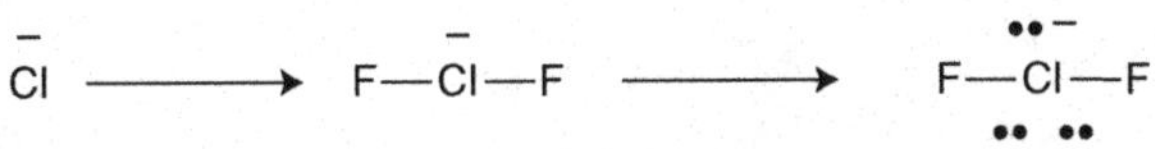 Cl$^-$ ⟶ F—Cl—F ($^-$) ⟶ F—Cl—F ($^-$, three lone pairs on Cl)

(b) 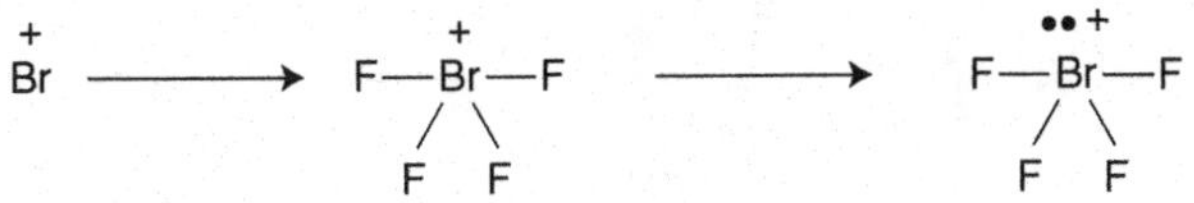 Br$^+$ ⟶ F—Br—F (with two F below, $^+$) ⟶ F—Br—F (with two F below, $^+$, lone pair on Br)

15 Additional Techniques for Designing and Representing Structures of Large Molecules

I TECHNIQUES FOR UNDERSTANDING AND DESIGNING ISOMERS OF LARGE MOLECULES

Larger, more complex molecules generally fall under the classification of organic or biochemical molecules. Most organic or biomolecules contain only C, H, N, O, F, Cl, Br, I, S, and P. Except for the common occurrence of phosphate (PO_4^{-3}) and sulfate (SO_4^{-2}) derivatives, these atoms almost always follow either normal neutral octet bonding rules or follow the row-2 exceptions described by Exception Rule #1 or Exception Rule #3 presented in Chapter 14. The process for forming constitutional isomers (different atom bonding patterns based on the same molecular formula) of larger molecules has added steps to identify and place multiple bonds (double and triple bonds), useful for writing more specific Lewis structures.

Most large (usually organic) molecules will contain carbon atoms and commonly, at least, some hydrogen atoms. The halogen elements Cl, Br, and I are almost always octet bonded (1 bond) and will almost always be **outside** atoms, unless specifically stated otherwise. Hydrogen and the halogens are the required outside atoms; after inserting them in the central atom bonding pattern, the number of vacant bonding positions can be counted. If sufficient outside atoms are available to fill all vacant bonding positions for all central atoms, a valid structure is produced without the requirement for multiple bonds. If some atoms are left with fewer than the required number of bonds, the result is a requirement for at least one **multiple** bond; a multiple bond implies a double or triple bond. Vacant bond positions can also indicate that the atoms are bonded in a "ring" pattern; this type of bonding is discussed in Section III. The following notation will be used for a calculation method. Each **double** bond counts as **one** multiple bond; each **triple** bond counts as **two** multiple bonds. A multiple bond is the same as a pi bond (π-bond) in this notation (Chapter 17).

General Rule #1 for the Number of Multiple Bonds (or Rings)

Steps (**1**) through (**6**) from Chapter 14 are used to produce valid structures of large molecules. A bonding template can be constructed by sequential addition of required outside atoms to any central atom combination. **General Rule #1** states that based on the ratio of required outside atoms to all central atom bonding requirements, the molecule being constructed will contain one multiple bond for every two vacant bonding positions. Or, stated alternatively, the molecule will contain one multiple bond for every two "missing" required outside atoms (hydrogen plus the halogens). General Rule #1 can include a ring pattern to substitute for a multiple bond (Section III).

> **General Rule #1** leads to **General Rule #2** and to a new step (**7**) and a new step (**8**) for generating other isomers of a specific formula.
>
> **General Rule #2:** Once one possible Lewis structure has been generated with the number of multiple bonds found by General Rule #1, all other possible Lewis structures for this same formula must contain this exact same number of total multiple bonds (or rings).

(New) Step (**7**): Use the bonding template method to determine the number of vacant bonding positions generated by a specific central atom pattern. This number can also be used to help verify the validity of the first isomer formed.

(New) Step (**8**): If isomer formation requires changes in multiple bonds, generate the required structures by adapting certain multiple bonds to different combinations; a triple bond can be exchanged for two double bonds or different atoms may be involved in the multiple bond. General Rules #1 and #2 can be used to construct a more specific molecule based on additional information.

Example: Construct isomers for the molecular formula $C_5H_5NOClBr$.

a) Determine the number of multiple bonds required for this specific formula.
b) Determine any valid Lewis structure for this formula.
c) Construct a molecule from the same central atom connection pattern that contains no triple bonds.
d) Construct a molecule from the same central atom connection pattern that contains a different C—C multiple bond pattern.
e) Construct a molecule from a different central atom connection pattern that contains a cyano group; cyano represents an example of a C—N triple bond.
f) Construct a molecule from a different central atom connection pattern that contains an example of a C—O double bond.

Steps (**1**) through (**6**) are described in Chapter 14.

Halogens in large molecules are always considered octet bonded.

Steps (**1**) and (**2**):	5 carbons;	each C = 4 bonds	possible central atom
	5 hydrogens;	each H = 1 bond	required outside atoms
	1 nitrogen;	each N = 3 bonds	possible central atom
	1 oxygen;	each O = 2 bonds	possible central atom
	1 chlorine;	each Cl = 1 bond	required outside atom
	1 bromine;	each Br = 1 bond	required outside atom

Step (**3**):

C—C—C—C—C—N—O

For seven central atoms, the atoms can almost always be connected in any order. A pattern simply following the order in the formula is selected.

Step (**4**):

```
      H   H   H   ?   ?
      |   |   |   |   |
  Br—C—C—C—C—C—N—O—?
      |   |   |   |   |   |
      Cl  H   H   ?   ?   ?
```

Step (**4**) explanation: The bonding template for placement of required outside atoms has been constructed. In this case, atoms were added sequentially from left-to-right on the central atom sequence without consideration of locating possible multiple bonds.

(a) Step (**7**): The new step (7) is applied at this point. The vacant bonding positions, determined from completely matching each central atom's normal neutral bond number, are marked with an empty bond line and a question mark. There are **6** vacant positions (marked as "?"). **General Rule #1** states that the molecule being constructed will contain one multiple bond for every two vacant bonding positions.

$$\text{The \# of multiple bonds} = \frac{\textbf{6 vacant bonding positions (6 "?" marks)}}{\textbf{2 vacant positions per multiple bond}} = \textbf{3 multiple bonds}$$

The notation is **one** double bond = **one** multiple bond; **one** triple bond = **two** multiple bonds. A total of **three** multiple bonds can be formed for this molecular formula. The requirement can be satisfied by three double bonds or one triple bond plus one double bond.

```
Step (5):
     H  H  H                                      H  H  H
     |  |  |            Step (6):                 |  |  |              ••  ••
Br—C—C—C—C≡C—N=O      ————————→     Br—C—C—C—C≡C—N=O
     |  |  |                                      |  |  |                  ••
     Cl H  H                                      Cl H  H
```

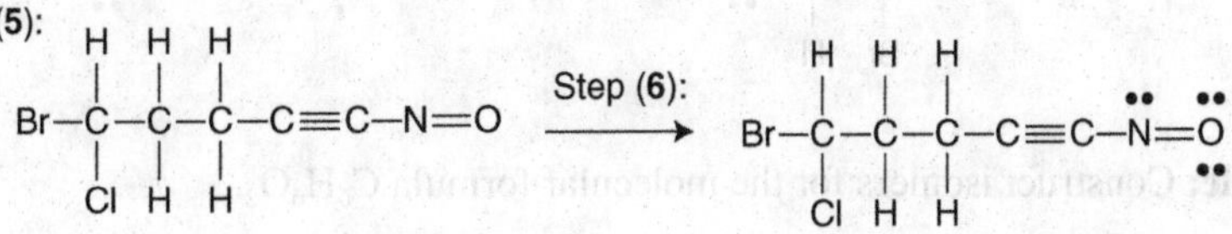

(b) Interestingly, despite simply adding outside atoms in sequence from one side, a valid structure can be found for the original bonding pattern without rearrangement. Counting of vacant bonding positions can be accomplished equally well from a more symmetrical bonding pattern. Lone pairs are not shown for the outside atoms in step (**6**).

(c) Step (**8**): Many possible Lewis structures can be drawn for this molecular formula; all must have three multiple bonds (with the restriction that the ring pattern is not being used). A structure with no triple bonds must, therefore, contain three double bonds. In this case, the original outside atom positions cannot completely accommodate the required set of multiple bond positions; the bonding pattern can be rearranged in a variety of ways to accommodate the required three multiple bonds.

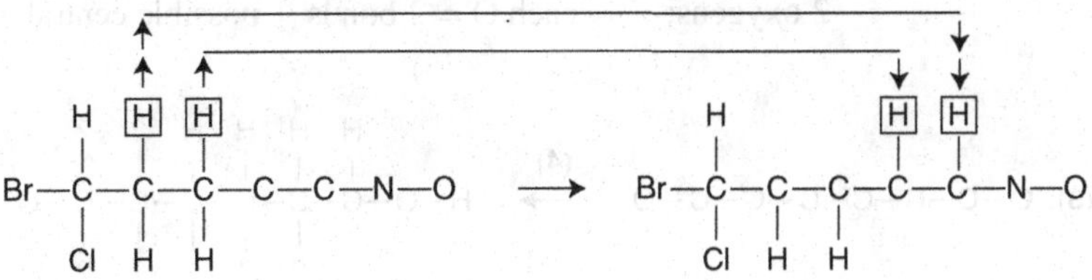

```
     H [H][H]                                  H          [H][H]
     |  |  |                                   |           |  |
Br—C—C—C—C—C—N—O      ——→      Br—C—C—C—C—C—N—O
     |  |  |                                   |  |  |
     Cl H  H                                   Cl H  H
```

The rearranged pattern can accommodate the required double bonds.

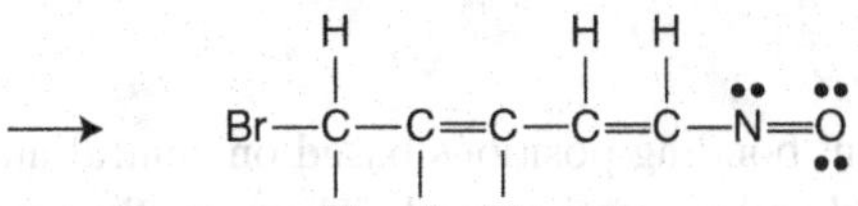

```
              H           H  H
              |           |  |      ••  ••
   ——→  Br—C—C=C—C=C—N=O
              |  |  |                    ••
              Cl H  H
```

(d) Step (**8**): A different triple bond plus double bond structure must contain three multiple bonds.

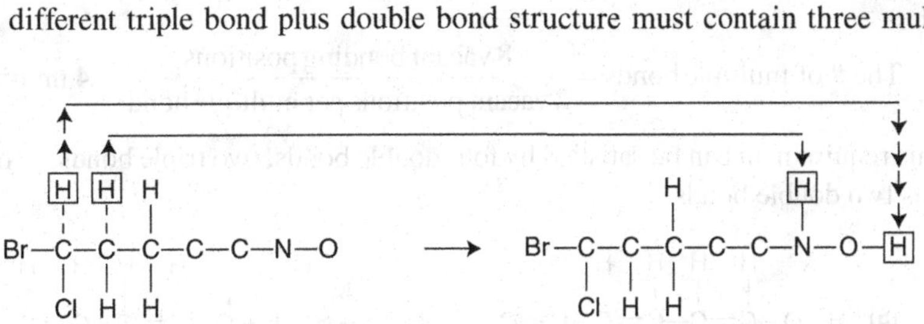

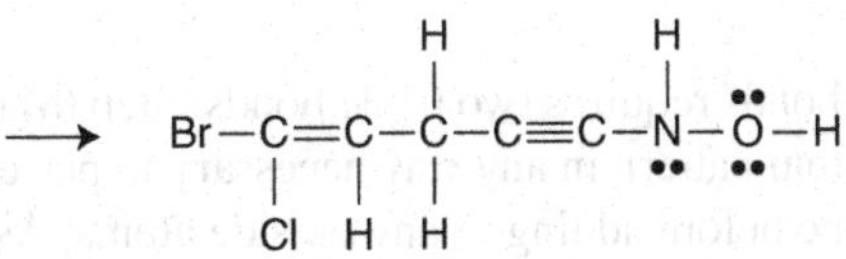

```
    [H][H] H                                   H        [H]
     |  |  |                                   |         |
Br—C—C—C—C—C—N—O      ——→      Br—C—C—C—C—C—N—O—[H]
     |  |  |                                   |  |  |
     Cl H  H                                   Cl H  H

              H           H
              |           |   ••
   ——→  Br—C=C—C—C≡C—N—O—H
                  |  |  |  ••  ••
                  Cl H  H
```

(e) and (f) Once the multiple bond requirement is determined, major changes in the structure, including the original central atom pattern can be easily visualized. Application of step

(**8**) can most easily be accomplished by first inserting the required multiple bonds in the desired position. Once this is done, unless atoms or bond numbers are miscounted, the structure produced must always be correct.

```
                    Cl  H     H  H                      ••
      ••            |   |  •• |  |               Cl  H  O: H   H  Br
e)    N≡C—C—C—O—C=C—H    f) H—C=C—C—N—C=C—H
           |   |  ••                              |  |  ||  |   |  |
           Br  H                                           ••
```

Example: Construct isomers for the molecular formula $C_6H_6O_2$.

a) Determine the number of multiple bonds required for this specific formula.
b) Determine any valid Lewis structure for this formula.
c) Construct a molecule that contains no double bonds.
d) Construct a molecule that contains no triple bonds.
e) Construct a molecule that contains a carboxylic acid group: —C(=O)—O—H

Steps (**1**) and (**2**):	6 carbons;	each C = 4 bonds	possible central atom
	6 hydrogens;	each H = 1 bond	required outside atoms
	2 oxygens;	each O = 2 bonds	possible central atoms

```
                                       H  H  H  H  ?  ?
                          (4)          |  |  |  |  |  |
Step (3): (3): O—C—C—C—C—C—C—O   ⟶   H—O—C—C—C—C—C—C—O—H
                                       |  |  |  |  |  |
                                       ?  ?  ?  ?  ?  ?
```

Step (**4**): Outside atoms were placed more symmetrically to anticipate the requirement for multiple bonds.

(a) Step (**7**): The vacant bonding positions based on central atom bond numbers are marked with an empty bond and a question mark. There are **8** vacant positions; **General Rule #1** is applied.

$$\text{The \# of multiple bonds} = \frac{\textbf{8}\text{ vacant bonding positions}}{\textbf{2}\text{ vacant positions per multiple bond}} = \textbf{4}\text{ multiple bonds}$$

(b) This requirement can be satisfied by four double bonds, two triple bonds, or one triple bond plus two double bonds.

```
            H  H  H  H                                 H  H  H  H
            |  |  |  |                   (6)      ••   |  |  |  |        ••
(5): H—O—C=C—C=C—C≡C—O—H   ⟶   H—O—C=C—C=C—C≡C—O—H
                                                  ••                     ••
```

(c) A structure with no double bonds requires two triple bonds. Step (**8**) is best accomplished by first arranging the central atom pattern in any way necessary to place two triple bonds in the desired locations; this is done before adding in any outside atoms. Assuming no miscalculation of atoms or bond numbers, the outside atoms will automatically be sufficient to complete the structure and can be added sequentially to all remaining empty positions. The example shown keeps the same central atom connection pattern, but many patterns are possible.

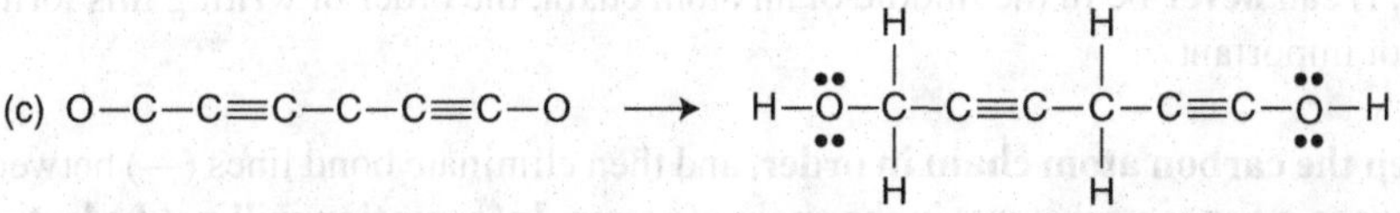

(d) Step (**8**): A structure with no triple bonds requires four double bonds; arrange the central atom pattern in any dlnesired way. After locating the double bond positions, add the outside atoms.

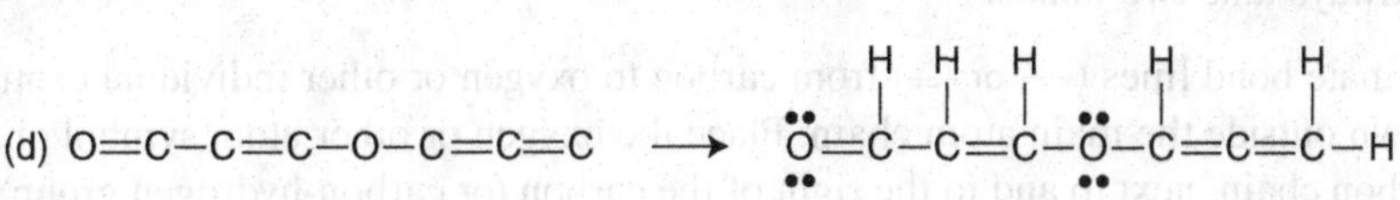

(e) Step (**8**): The carboxylic acid group (atom bonding combination) requires one double bond; three additional multiple bonds are required.

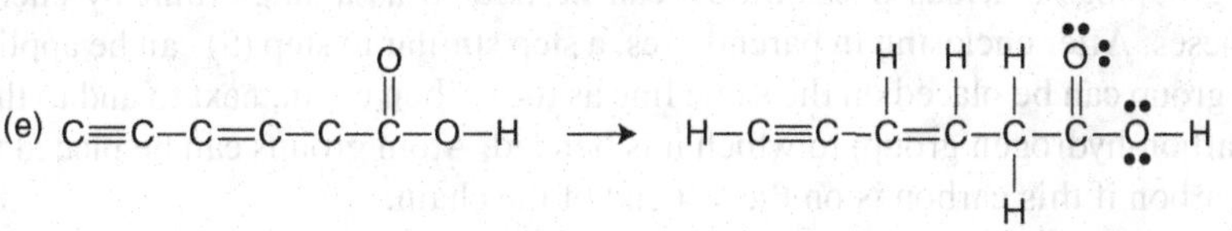

II USING CONDENSED STRUCTURAL FORMULAS

The examples in this chapter demonstrate that as molecules become larger, drawing full structurers becomes more cumbersome. Full Lewis structures can be simplified to a condensed formula by elimination of bond lines and lone pair notation. The key requirement to any simplification is that all information about the correct structure must be retained in the condensed form. Simplification is possible by following well-established normal bonding principles and lone pair counts for the major atoms in the molecule (C, H, O, N, plus halogens)

Guidelines for Converting a Full Structure to a Condensed Structural Formula

Any structural formula can be condensed to a variety of intermediate stages. It is not necessary to use all the steps listed. It is important to realize that **all** stages of condensed or expanded formulas are equally acceptable as long as the complete molecule can be reconstructed from the information. The level to which a Lewis structure is condensed is dictated by writing convenience, not by rigid requirements.

Step (**1**): Start with the correct complete Lewis structure.

Step (**2**): Eliminate the dots for all lone pairs of electrons. **No information is lost**, the explanation in that the number of lone pairs on each atom is specified by the normal neutral bond number combined with any information on lone pair number required by any stated formal charge derived from Exception Rule #1 or Exception Rule #3.

Step (**3**): Eliminate the bond lines between hydrogen and its bonded atoms (i.e., the structural line (—) connecting H—to its bonded atom). Replace these with the standard subscript formula system, where the hydrogens are written next to the atoms they are bonded to, with the number of hydrogens bonded to the same atom indicated by a subscript. **No information is lost**, since hydrogen can take

only one bond, the bonding pattern will be specifically designated. Halogens (F, Cl, Br, I) almost always take **one** bond; step **(3)** can be applied to halogen atoms also. Since required outside atoms (H, F, Cl, Br, I) can **never** be in the middle of an atom chain, the order of writing this formula combination is not important.

Step (**4**): **Keep the carbon atom chain in order**, and then eliminate bond lines (—) between carbon atoms or between other central atoms in the chain of atoms. **Information will not be lost only** if the carbon/central atom sequence remains in the same order as the original Lewis structure. The bonding pattern for carbon or the other central atoms is determined by normal neutral bonding rules. Neutral carbon must always take four bonds; neutral nitrogen must always take three bonds; neutral oxygen must always take two bonds.

Step (**5**): Eliminate bond lines (— or ═) from carbon to oxygen or other individual central atoms, which are drawn outside the main atom chain. Place the oxygen or other atom symbol on the same line as the carbon chain, next to and to the right of the carbon (or carbon-hydrogen group) to which it is bonded. **No information is lost**, since the bonding pattern from carbon to oxygen or the other atom is specified by established bonding rules.

Step (**6**): Atom groupings (various possibilities) can be treated as a single unit by enclosing the group in parentheses. After enclosing in parentheses, a step similar to step (**5**) can be applied to this atom group; the group can be placed on the same line as the carbon chain, next to and to the right of the carbon (or carbon-hydrogen group) to which it is bonded. Atom groups can be placed to the left of the bonding carbon if this carbon is on the left end of the chain.

Step (**7**): More than one **identical** atom grouping (enclosed in parentheses) can take a numerical subscript in a condensed formula. This step also includes identical group combinations that are part of a long carbon chain. After enclosing in parentheses with a subscript, the group combination can be placed on the same line as the carbon chain, next to and usually to the right of the carbon (or carbon-hydrogen group) to which it is bonded. Atom groups can be placed to the left of the bonding carbon if this carbon is on the left end of the chain.

Example: Show the process for condensing the following two Lewis structures.

```
      H  :Ö: H                        H  H  :Ö:
      |  ||  |      ..                |  |  ||   ..
    H—C——C——C—:Br:    Step (1)      H—C——C——C——O—H
      |      |      ..                |  |       ..
      H      H                        H :N—H       (alanine)
                                         |
      (bromoacetone)                     H
```

Step (**2**): Eliminate the dots for all lone pairs of electrons. Neutral octet bromine always has three lone pairs; neutral oxygen always has two lone pairs; neutral nitrogen always has one lone pair.

```
      H  O  H                  H  H  O
      |  ||  |                 |  |  ||
    H—C——C——C—Br             H—C——C——C—O—H
      |      |                 |  |
      H      H                 H  N—H
                                  |
                                  H
```

Step (**3**): Eliminate the bond lines between hydrogen and its bonded atoms and replace with the standard subscript formula system. Note that the format (CH_3—C) cannot mean that the hydrogens are connected sequentially in a chain such as (C—H—H—H—C). Hydrogen always takes only one bond;

this incorrect chain reconstruction shows each with two bonds and, thus, cannot be correct. The condensed format (**CH_3**—C) can only be regenerated to the original molecule; no information is lost.

$$H_3C-\overset{\overset{\large O}{\|}}{C}-CH_2Br \quad \text{or} \quad \mathbf{CH_3}-\overset{\overset{\large O}{\|}}{C}-CH_2Br \qquad CH_3-\underset{\underset{\large NH_2}{|}}{CH}-\overset{\overset{\large O}{\|}}{C}-OH$$

Step (**4**): If the carbon atom chain is kept in order, bond lines between carbon atoms can be eliminated. Neutral carbon must always take four bonds; any double or triple bonds between carbons would be indicated by the requirement for four bonds for all carbons.

$$CH_3\overset{\overset{\large O}{\|}}{C}CH_2Br \qquad CH_3\underset{\underset{\large NH_2}{|}}{C}H\overset{\overset{\large O}{\|}}{C}OH$$

Step (**5**): Condensation through step (**4**) often provides the most useful description of the molecule; further condensation is possible. Bond lines from carbon to oxygen can be eliminated. For double bonded oxygens, which are not drawn in the line of the carbon chain atoms, place the oxygen atom symbol on the same line as the carbons, to the right and next to the carbon (or carbon-hydrogen group) to which it is bonded. Note that the format (CH_3COCH_2Br) cannot mean that the oxygen is part of the carbon chain such as ($H_3C—C—O—CH_2Br$). Although the oxygen has two bonds in this incorrect reconstruction, the second carbon has only two bonds and, thus, cannot be correct. The condensed format can only be regenerated to the original molecule; no information is lost.

$$CH_3COCH_2Br \qquad CH_3\underset{\underset{\large NH_2}{|}}{C}HCOOH \quad \text{or} \quad CH_3\underset{\underset{\large NH_2}{|}}{C}HCO_2H$$

Step (**6**): The atom group (—NH_2) can enclosed in parentheses; a step similar to step (**5**) can then be applied to this group following the procedure for atoms or groups not in the carbon chain.

$$CH_3CH(NH_2)COOH$$

Example: Condense the following molecule.

$$H-\underset{\underset{\large H}{|}}{\overset{\overset{\large H}{|}}{C}}-\underset{\underset{\large Br}{|}}{\overset{\overset{\large Br}{|}}{C}}-\underset{\underset{\large H}{|}}{\overset{\overset{\large H}{|}}{C}}-O-H \xrightarrow{\text{Steps (3) through (5)}} CH_3CBr_2CH_2OH$$

Example: Further condense the following molecule.

In this case, the atom grouping {**CH_3**}, which is bonded to the middle chain carbon, is placed on the same line as the carbon chain next to and to the right of the carbon to which it is bonded. The parentheses indicate that this atom grouping is not part of the carbon chain sequence.

$$CH_3-CH_2-\underset{\underset{\large \mathbf{CH_3}}{|}}{CH}-CH_2-CH_3 \xrightarrow{\text{Steps (4), (5), and (6)}} CH_3CH_2CH(\mathbf{CH_3})CH_2CH_3$$

Example: Condense the following molecule.

More than one **identical** atom grouping (enclosed in parentheses) can take a numerical subscript in a condensed formula. In this case, **two** atom groups {$\mathbf{CH_3}$} are bonded to the same carbon; this is indicated by the subscript (**2**) after the parentheses.

```
       H  H
       |  |                Steps (4) through (7)
CH3—C—C—F   ——————————————————————→   (CH3)2CHCHF2
       |  |
      CH3 F
```

Example: Further condense the following molecule.

$$CH_3\mathbf{CH_2CH_2CH_2CH_2}CH_3 \xrightarrow{\text{Step}(\mathbf{7})} CH_3\mathbf{(CH_2)_4}CH_3$$

Guidelines for Expanding Condensed Formulas: Reconstructing the Complete Molecule

Although the level to which a structural formula is condensed is arbitrary, it is always necessary to be able to reconstruct, or "picture," the complete Lewis structure from any condensed formula. If a condensed formula is valid, it must contain all information necessary to reproduce the original molecule.

To expand a condensed formula to any specific level of detail, simply reverse the process steps used in condensation. Reversing step (**2**) is often not necessary unless one or more lone pairs are used to demonstrate a further concept.

Example: Expand $F_2CH(CH_2)_3CH_2COBr$.

```
                    Reverse Step (7)                              Reverse Step (5)
F2CH(CH2)3CH2COBr ——————————————————→ F2CHCH2CH2CH2CH2COBr ——————————————————→

                      O                                                      O
                      ||   Reverse Step (4)                                  ||
F2CHCH2CH2CH2CH2CBr ——————————————————→ F2CH—CH2—CH2—CH2—CH2—CBr

Reverse step (3)        F  H  H  H  H  O
                        |  |  |  |  |  ||
——————————————→     H—C—C—C—C—C—C—Br
                        |  |  |  |  |
                        F  H  H  H  H
```

Double and triple bonds in a formula can be identified through normal neutral bonding rules for each central atom as a guide.

Example: Expand $(CH_3)_2CBrCHO$.

Expansion of the structure through the higher steps up to reverse step (**4**) shows that the end carbon has only three bonds and the oxygen has only one bond. The complete original structure must have a double bond to the oxygen.

```
                                             Br H                 Br H
             Reverse through Step (4)        |  |                 |  |
(CH3)2CBrCHO ——————————————————————→ H3C—C—C—O  ——→  H3C—C—C═O
                                             |                    |
                                            CH3                  CH3
```

Example: Expand (a) $CH_3CHCHCH_3$ and (b) CH_3CCCH_3.

(a) $CH_3CHCHCH_3 \longrightarrow CH_3{-}CH{-}CH{-}CH_3$; middle carbons are each missing one bond.

Insert a double bond $\longrightarrow CH_3{-}CH{=}CH{-}CH_3 \longrightarrow H{-}C(H)(H){-}C(H){=}C(H){-}C(H)(H){-}H$

(b) $CH_3CCCH_3 \longrightarrow CH_3{-}C{-}C{-}CH_3 \longrightarrow CH_3{-}C{\equiv}C{-}CH_3$

Example: Expand $(CH_2Cl)_2CHCHCHCCCOOCH_3$ through the reverse of step (**4**).

$(CH_2Cl)_2CHCHCHCCCOOCH_3 \xrightarrow{\text{Reverse Steps (6), (7)}} ClCH_2{-}CH(CH_2Cl)CHCHCCCOOCH_3$

$\xrightarrow{\text{Reverse Step (5)}} ClCH_2{-}CH(CH_2Cl)CHCHCCC(=O)OCH_3 \xrightarrow{\text{Reverse Step (4)}}$

$ClCH_2{-}CH(CH_2Cl){-}CH{=}CH{-}C{\equiv}C{-}C(=O){-}O{-}CH_3$

III CONCEPT OF RING STRUCTURES AND LINE DRAWINGS

The method described for counting multiple bonds in Section I specifically excluded ring structures. In fact, the number of vacant bonding positions can be used to count the total number of multiple bonds and/or rings. The concept of a ring formation and its relationship to "missing" outside atoms (vacant bonding positions) is demonstrated through the following examples.

Example: Draw a structure for C_6H_{14} in which all carbons are part of one chain of central atoms; use a partially condensed format.

The formula C_6H_{14} has sufficient hydrogens to satisfy all carbons; no multiple bonds are required. The carbons are numbered and the bond lines for the end hydrogens are shown.

$$H{-}\overset{1}{CH_2}{-}\overset{2}{CH_2}{-}\overset{3}{CH_2}{-}\overset{4}{CH_2}{-}\overset{5}{CH_2}{-}\overset{6}{CH_2}{-}H$$

Example: Draw a similar structure for C_6H_{12}.

C_6H_{12} has two fewer hydrogens than C_6H_{14}. A chain of six carbons with single bonds only will have two vacant bonding positions; these are shown on the end carbons.

$$?{-}\overset{1}{CH_2}{-}\overset{2}{CH_2}{-}\overset{3}{CH_2}{-}\overset{4}{CH_2}{-}\overset{5}{CH_2}{-}\overset{6}{CH_2}{-}?$$

A valid Lewis structure could be formed from this formula by shifting a hydrogen from C#2 to C#6 and inserting a double bond between C#1 and C#2:

$$\overset{1}{CH_2}{=}\overset{2}{CH}{-}\overset{3}{CH_2}{-}\overset{4}{CH_2}{-}\overset{5}{CH_2}{-}\overset{6}{CH_2}{-}H$$

Example: Draw a valid structure for C_6H_{12} that does not have a double bond.

An alternative solution is to wrap the chain around so that C#1 and C#6 are close to each other; in this position, C#1 and C#6 can form an additional bond between them. A chain of central atoms closed up "head-to-tail" is termed a **ring**.

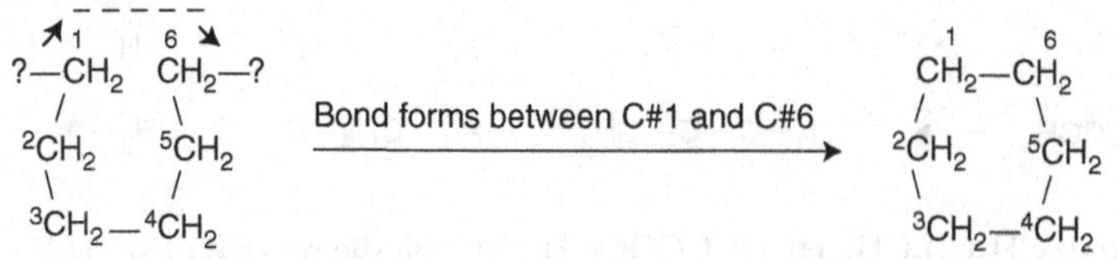

```
  ↗1       6 ↘                                            1      6
?—CH2    CH2—?                                           CH2—CH2
   /        \        Bond forms between C#1 and C#6      /       \
2CH2       5CH2     ─────────────────────────────→    2CH2      5CH2
   \        /                                            \       /
  3CH2—4CH2                                             3CH2—4CH2
```

The ring construction for the formula C_6H_{12} now has all carbons with four bonds and represents a valid Lewis structure without the use of a double bond. The essential result is the exchange of the second bond of a double bond between two carbons for an additional single bond between certain carbons. Assuming a molecule has the capacity to form a ring, any formula with a set of two central atoms each with a vacant bonding position can be satisfied by ring formation or multiple bond formation. A more complete statement for **General Rule #1** is that the molecule being constructed will contain one multiple bond and/or ring for every two vacant bonding positions.

$$\text{The total \# of multiple bonds and/or rings} = \frac{\text{\# of vacant bonding positions}}{\mathbf{2}}$$

Ring formation is very common in nature and especially in biological systems; biochemical rings most often contain carbon along with nitrogen, oxygen, and sulfur as possible central atom ring members. The most common ring sizes range from five to ten central atom members per ring; one molecule may contain multiple numbers of rings.

Structural Notation: Guidelines for Producing Line Drawings

Condensed structures are not suitable for rings; elimination of bond lines for central atoms in a ring renders the structure unrecognizable. Line drawings are an alternate notational system for molecular structures and are standard for ring systems. The key requirement is the same as for condensed structures; all information contained in the Lewis structure must be retained in the line drawing. The simplification employs a somewhat opposite approach to condensed structures; most bond lines are kept but certain atom symbols are dropped.

Step (**1**): Start with a complete structural formula for the molecule; lone pairs are optional.

Step (**2**): Eliminate the atom symbol for **carbon** (**C**) while leaving **all** bond lines in place.

Step (**3**): Specifically and **only** for **hydrogens** bonded to carbon (C—H), eliminate both the atom symbol for hydrogen and the C—H bond line. All other atom symbols must be shown; symbols for all hydrogens bonded to any atom other than carbon must be shown. Bond lines from hydrogen to other atoms are optional; use the rules for condensing formulas for further simplification if desired.

Step (**4**): Line drawings may be used for **non**-ring structures. For carbon atoms in a sequential chain, use "zig-zag" drawings for the bonds in the chain to help visualize the carbons that are located at the vertices (intersection of two bond lines); this mimics the true tetrahedral geometrical shape of carbon (Chapter 16).

Example: Draw a line drawing for the previous ring structure of C_6H_{12}.

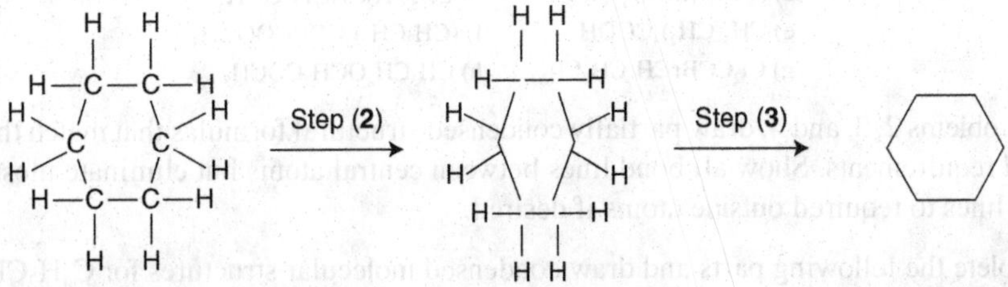

Step (2) → Step (3)

Example: Draw a line drawing for the following molecule.

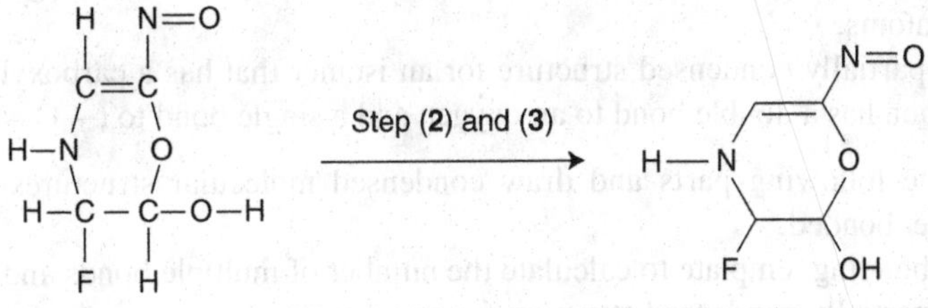

Step (2) and (3)

Example: Draw a line drawing for the following molecule.

$CH_2{=}CH{-}CH_2{-}CH_2{-}CH_2{-}CH_3$ —Step (4)→ $CH_2{=}CH\ CH_2\ CH_3$ (with CH_2 CH_2 above)

—Steps (2) and (3)→

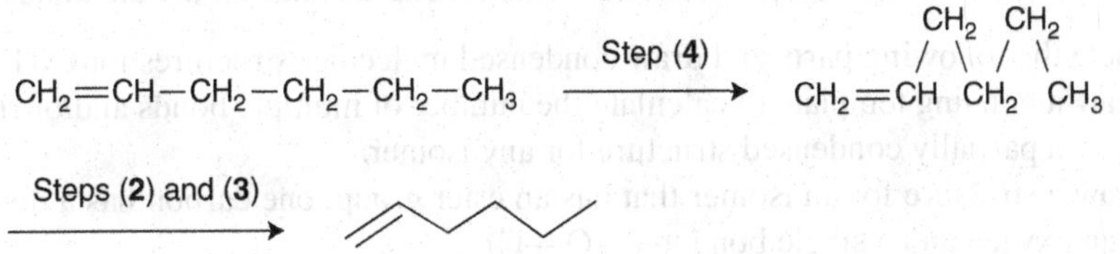

CONVERTING LINE DRAWINGS TO STRUCTURAL FORMULAS

Step (**1**): Start from a line drawing and replace carbons at each line end or line vertex (intersection of two lines) **only** if no other atom is shown.

Step (**2**): Replace the hydrogens (**H** symbols) on carbons. Use the normal neutral bonding rules to determine the correct number of hydrogens on each carbon. Neutral carbon (in normal compounds) **must** have **four** bonds; add sufficient numbers of hydrogens to give each carbon exactly four bonds. It is not necessary to show all C—H bond lines; select a convenient condensed formula format.

Example: Draw a structural formula for the line drawing shown.

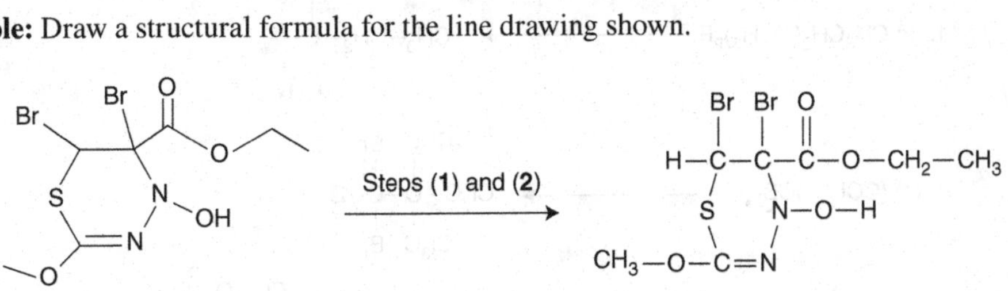

Steps (**1**) and (**2**)

IV PRACTICE PROBLEMS

1. Expand the following condensed formulas through the reverse of Step (**4**). Show all central atom—central atom bond lines for all possible central atoms but eliminate most bond lines between central atoms and required outside atoms; some may be shown for clarity.

a) $(CH_3)_3CCHO$ b) $CH_3CH_2C(CH_3)_2F$
c) $(CCl_3)_3CCBr_3$ d) $CH_3CH_2CHCl\ COCH_3$
e) $CH_3(CH_2)_4COOH$ f) $(CH_3CH_2)_2CCHCOOCH_3$
g) $Cl_2CCBrCH(CH_2Cl)_2$ h) $CH_3CH_2OCH_2COCH_3$

For problems 2, 3, and 4, draw partially condensed structural formulas that match the additional requirements. Show all bond lines between central atoms but eliminate most single bond lines to required outside atoms if desired.

2. Complete the following parts and draw condensed molecular structures for $C_4H_2Cl_2O_3$.
   a) Draw a bonding template to calculate the number of multiple bonds and/or rings.
   b) Draw a partially condensed structure for any isomer.
   c) Draw a partially condensed structure for an isomer that has all oxygens as resulting outside atoms.
   d) Draw a partially condensed structure for an isomer that has a carboxylic acid group: one carbon has a double bond to an oxygen and a single bond to (—O—H)

3. Complete the following parts and draw condensed molecular structures for $C_5H_6N_2S$; sulfur is octet bonded.
   a) Draw a bonding template to calculate the number of multiple bonds and/or rings.
   b) Draw a partially condensed structure for any isomer.
   c) Draw a partially condensed structure for an isomer that has no triple bonds.
   d) Draw a partially condensed structure for an isomer that has no double bonds.

4. Complete the following parts and draw condensed molecular structures for $C_6H_4Br_2O_2$.
   a) Draw a bonding template to calculate the number of multiple bonds and/or rings.
   b) Draw a partially condensed structure for any isomer.
   c) Draw a structure for an isomer that has an ester group: one carbon has a double bond to an oxygen and a single bond to (—O—C)
   d) Draw a partially condensed structure for an isomer that has no double bonds.

V ANSWERS TO PRACTICE PROBLEMS

1. (a) $(CH_3)_3CCHO$ ⟶ H_3C–C(H_3C)(H_3C)–C(=O)–H

1. (b) $CH_3CH_2C(CH_3)_2F$ ⟶ CH_3–CH_2–C(CH_3)(CH_3)–F

1. (c) $(CCl_3)_3CCBr_3$ ⟶ Cl_3C–C(Cl_3C)(Cl_3C)–C(Br)(Br)–Br

1. (d) $CH_3CH_2CHCl\ COCH_3$ ⟶ CH_3–CH_2–CH(Cl)–C(=O)–CH_3

1. (e) $CH_3(CH_2)_4COOH$ ⟶ CH_3–CH_2–CH_2–CH_2–CH_2–C(=O)–OH

```
                                     H3C—CH2     O
                                            |    ||
1. (f) (CH3CH2)2CCHCOOCH3  ———→   CH3—CH2—C═CH—C—O—CH3
```

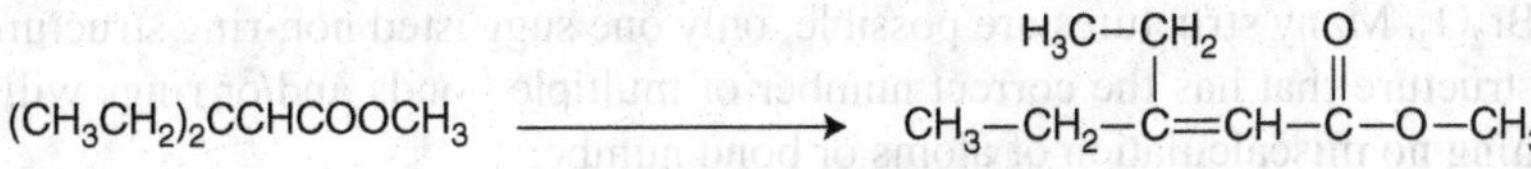

```
                                Br  CH2Cl
                                |   |
1. (g) Cl2CCBrCH(CH2Cl)2  ———→  Cl2C═C—CH—CH2Cl
```

```
                                                      O
                                                      ||
1. (h) CH3CH2OCH2COCH3  ———→  CH3—CH2—O—CH2—C—CH3
```

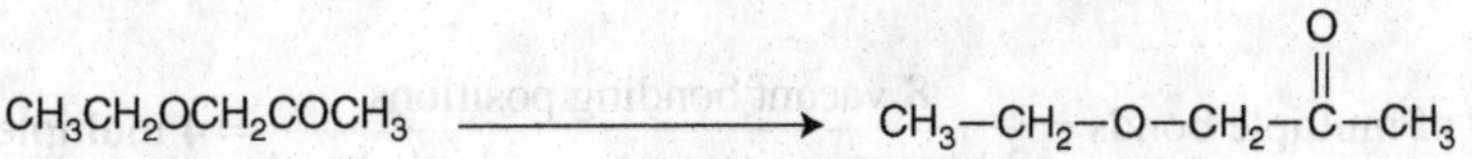

2. $C_4H_2Cl_2O_3$. Many structures are possible; only one suggested non-ring structure is shown. Any structure that has the correct number of multiple bonds and/or rings will be correct assuming no miscalculation of atoms or bond numbers.

```
                                    Cl Cl ?  ?
                                    |  |  |  |
(a) O—C—C—C—C—O—O  ——→  H—O—C—C—C—C—O—O—H
                                    |  |  |  |
                                    ?  ?  ?  ?
```

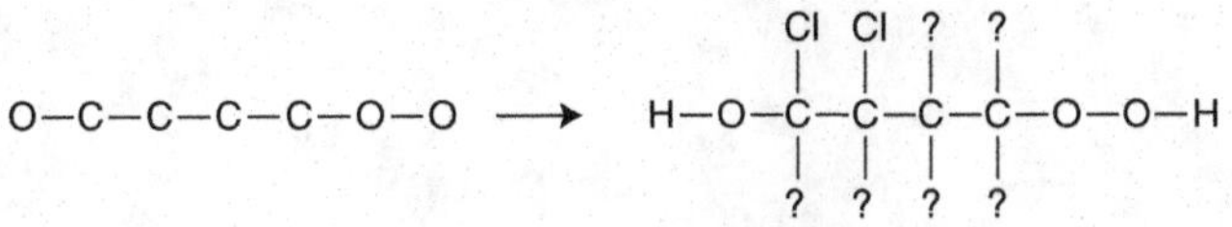

$$\text{The \# of multiple bonds} = \frac{\textbf{6 vacant bonding positions}}{\textbf{2 vacant positions per multiple bond}} = \textbf{3 multiple bonds}$$

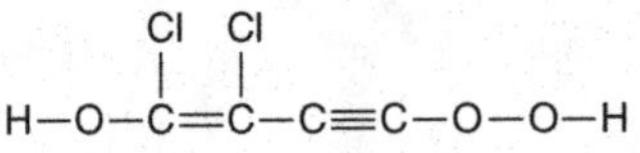

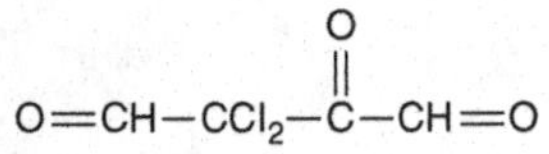

```
          Cl Cl                                        O
          |  |                                         ||
(b) H—O—C═C—C≡C—O—O—H    (c) O═CH—CCl2—C—CH═O
```

```
                        O
                        ||
(d) HO—CCl2—C≡C—C—O—H
```

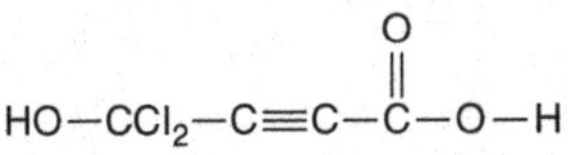

3. $C_5H_6N_2S$; sulfur is octet bonded. Many structures are possible; only one suggested non-ring structure is shown. Any structure that has the correct number of multiple bonds and/or rings will be correct assuming no miscalculation of atoms or bond numbers.

```
                                       H  H  H  ?  ?  H
                                       |  |  |  |  |  |
(a) N—C—C—C—C—C—N—S  ——→  H—N—C—C—C—C—C—N—S—H
                                    |  |  |  |  |  |
                                    ?  ?  ?  ?  ?  ?
```

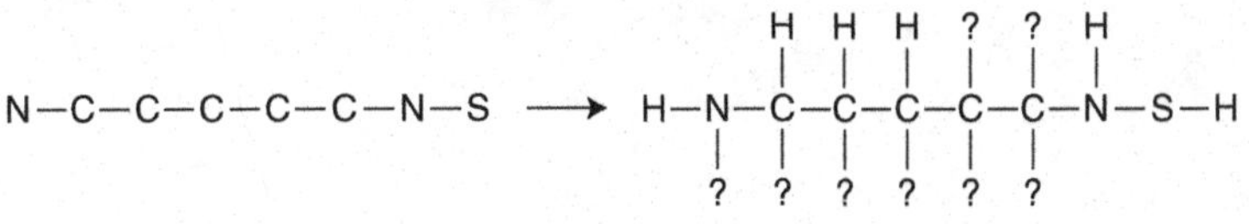

$$\text{The \# of multiple bonds} = \frac{\textbf{8 vacant bonding positions}}{\textbf{2 vacant positions per multiple bond}} = \textbf{4 multiple bonds}$$

```
          H  H  H         H                                      S
          |  |  |         |                                      ||
(b) H—N═C—C═C—C≡C—N—S—H    (c) HN═CH—CH═CH—C—N═CH2
```

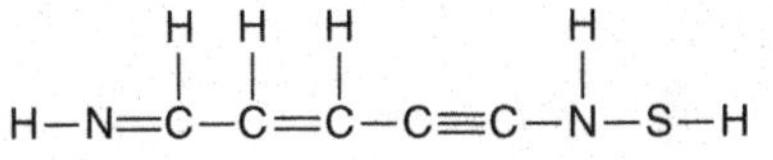

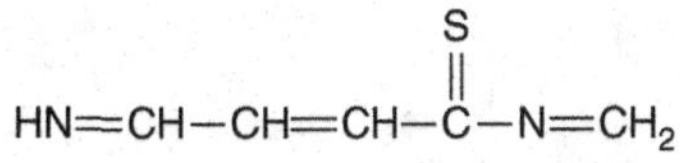

(d) $H_2N—C{\equiv}C—C{\equiv}C—S—CH_2—NH_2$

4. $C_6H_4Br_2O_2$. Many structures are possible; only one suggested non-ring structure is shown. Any structure that has the correct number of multiple bonds and/or rings will be correct assuming no miscalculation of atoms or bond numbers.

(a) O–C–C–C–C–C–C–O ⟶ ?–O–C(H)(?)–C(H)(?)–C(H)(?)–C(H)(?)–C(Br)(?)–C(Br)(?)–O–?

$$\text{The \# of multiple bonds} = \frac{\textbf{8 vacant bonding positions}}{\textbf{2 vacant positions per multiple bond}} = \textbf{4 multiple bonds}$$

(b) O=C(H)–C(H)=C(H)–C(H)=C(Br)–C(Br)=O

(c) $Br_2C{=}CH{-}C{\equiv}C{-}C({=}O){-}O{-}CH_3$

(d) $H{-}O{-}C{\equiv}C{-}CH_2{-}CBr_2{-}C{\equiv}C{-}O{-}H$

16 Determining and Drawing Molecular Geometry and Polarity

I CONCEPTS OF MOLECULAR GEOMETRY

Lewis structures show the bonding atom connection pattern and number of lone pairs around each atom in a covalent molecule; the structures do not represent any direct information concerning the actual three-dimensional (3-D) shape of the molecule. The three-dimensional (3-D) 3-D shape of a complete molecule, termed the **molecular geometry**, is based on the arrangement of all electron pairs around each central atom in a complete molecule.

(**1**) The 3-D geometry around a central atom describes the relative positions of each lone electron pair and each bonding electron pair. By extension, the relative positions of the bonding electron pairs must specify the corresponding positions of all atoms, whether central or outside, attached to a central atom.

(**2**) Analysis of 3-D geometry applies **only** to **central** atoms in a molecule; the geometry around outside (atoms does not contribute to overall molecular geometry. A minimum of two bonded-atoms is necessary to make central atom positional comparisons relevant to molecular properties.

(**3**) The first analysis of the shape of a complete molecule proceeds through determination of the 3-D geometry of one central atom at time. The shape of each central atom, in turn, produces a starting composite view of the shape of the molecule. A first analysis determines the geometry of the central atoms; the positional relationships between central atoms require further levels of analysis.

Types of Geometry Analysis for Central Atoms

Electron Region (E.R.) Geometry describes the arrangement of all electron pairs (bonding electron pairs and lone electron pairs) around a central atom; another term for this description is "electron pair geometry."

Atom Geometry (sometimes called **molecular geometry**, used as a more narrow description) describes the arrangement of only the bonded atoms around a central atom. Bonding electron pairs specify the positions of the bonded atoms; thus, lone electron pairs are considered invisible in this description. Atom geometry must be derived from E.R. geometry.

Counting Electron Regions

Both electron region (**E.R.**) and atom geometry ultimately depend on the distribution of **all** electron regions (bonding and lone pair) around a central atom. The following rules are used to count electron regions around a central atom.

(**1**) Assign **one** electron region to **each** lone electron pair (non-bonding electron pair).
(**2**) Assign **one** electron region to **each** single bond.
(**3**) Assign **one** electron region to **each** double bond.
(**4**) Assign **one** electron region to **each** triple bond.

Although double and triple bonds are formed by separate orbital overlaps (molecular orbitals), the separate overlaps occupy approximately the same region of space. Double and triple bonds are each considered one electron region apiece.

Summary of E.R. Count

One lone pair = one **E.R.**
One single bond = one **E.R.**
One double bond = one **E.R.** (not two)
One triple bond = one **E.R.** (not three)

II PROCESS FOR COMPLETE DETERMINATION OF MOLECULAR GEOMETRY AND POLARITY

Electron regions repel each other; the distribution of electron regions around a central atom is based on the fact that electron regions will occupy positions around a central atom that allow them to be as far away from each other as possible. This concept is termed the **V**alence **S**hell **E**lectron **P**air **R**epulsion (**VSEPR** theory).

Step (**1**): Draw the correct Lewis structure for the complete molecule. This step specifies the number and types of electron regions (E.R.) around each central atom. Follow the procedure for Lewis structures described in Chapter 14.

Step (**2**): Count the number of electron regions for a specified central atom using the electron region count rules:

Each lone electron pair = one E.R.
Each single bond = one E.R.
Each double bond = one E.R.
Each triple bond = one E.R.

Step (**3**): Determine the electron region (E.R.) geometry for the specified central atom; each central atom is analyzed separately.

The electron region (E.R.) geometry is found by matching the E.R. count to the corresponding geometrical figure, which describes the 3-D distribution of electron regions in space. Use the summary table to identify the geometrical figure from the electron region (E.R.) count.

Step (**4**): Draw the geometry of the central atoms by placing bonded atoms and lone pairs at the corners of the correct geometrical figure; the specific central atom is placed at the center of the figure.

(**4a**) For linear, trigonal planar, and tetrahedral geometrical figures, any corner can be selected for lone pairs or atoms.

(**4b**) The trigonal bipyramidal geometrical figure has the two distinct positions of axial and equatorial; place lone pairs first in the equatorial positions.

(**4c**) For the octahedral geometrical figure, any corner can be selected for the first lone pair; place a second lone pair, if applicable, in a position directly opposite the first lone pair. The lone pair occupied corners of the octahedral figure will have an angle of 180°.

Step (**5**): Determine the atom geometry (molecular geometry). The atom geometry describes the 3-D position of only the atoms around the central atom, as if the lone pairs are invisible. Start with the drawing from step (**4**); the atom geometry is found by "covering up" the lone pairs and identifying the figure that remains. Do not rearrange atoms from the E.R. geometry; the lone pairs are still present in the designated corners, even if they are considered invisible for the atom geometry.

Step (6): Complete the Polarity Analysis.

(6a) Use a table of atom electronegativities (EN) to determine the polarity of each bond in the molecule; describe the polarity with a dipole arrow.

(6b) Analyze each central atom separately to determine whether all the bond dipole arrows around a central atom show a net additive direction or whether they all exactly cancel.

Descriptions of Geometric Figures for 2 to 4 Electron Regions

Linear, trigonal planar, and tetrahedral geometrical figures are symmetrical; all corner positions are equivalent and all angles connecting any two corners through the central atom are identical. Any arrangement of atoms and lone pairs will always produce equivalent final shapes.

Summary Table of Geometries for 2 to 4 Electron Regions

Notation: (○) = E.R. (●) = central atom

(ııııı) = figure line going backward (away from you)

(▬) = figure line coming forward (toward you)

# of Electron Regions	Geometrical Figure	Drawn Figure
2	**Linear** all angles = 180°	
3	**Trigonal planar** all angles = 120°	
4	**Tetrahedral** all angles = 109.5°	

Descriptions of Geometric Figures for 5 and 6 Electron Regions

Lone electron pairs occupy a larger volume than bonding electron pairs; this results in differing degrees of electron-electron repulsion as a function of lone-pair positions. The difference does not affect geometrical figures of 2 to 4 electron regions, but does affect geometries of 5 (trigonal bipyramidal) and 6 (octahedral) electron regions. The corner positions for the trigonal bipyramidal shape are marked on the summary table drawing as axial or equatorial. The separation angle between any two equatorial positions is 120° and is larger than the 90° separation between an axial and equatorial position. Lone electron pairs favor equatorial positions, equivalent to greater separation; the three equatorial positions are equivalent.

The octahedral geometrical figure is symmetrical; all angles connecting any two corners through the central atom are identical at 90°. All corner positions therefore are equivalent for the placement of the first lone electron pair. If present, the second lone pair is most stable with maximum separation; the angle between the corners of the figure occupied by the lone pairs should be 180°.

Summary Table of Geometries for 2 to 4 Electron Regions

Notation: (○) = E.R. (●) = central atom

(⋯⋯) = figure line going backward (away from you)

(◀━) = figure line coming forward (toward you)

# of Electron Regions	Geometrical Figure	Drawn Figure
5	**Trigonal bipyramidal** angles = 120° or 90° **A = axial position** **E = equatorial position**	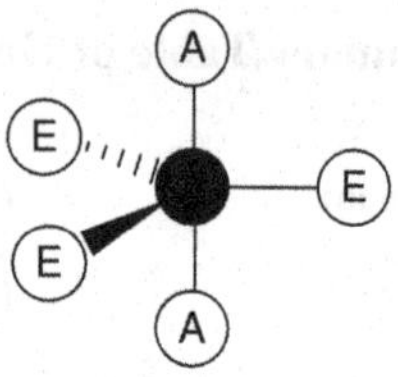
6	**Octahedral** all angles = 90°	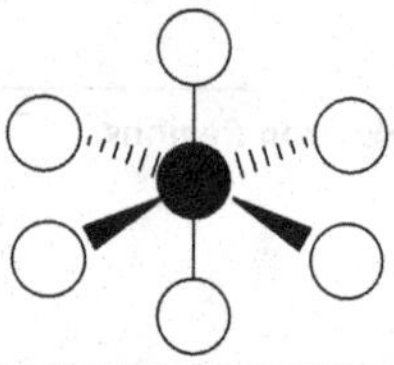

The complete central atom analysis begins with a concentration on steps (**1**) through (**4**).

Example: Complete steps (1) through (4) for the determination of the 3-D shape of methane, CH_4.

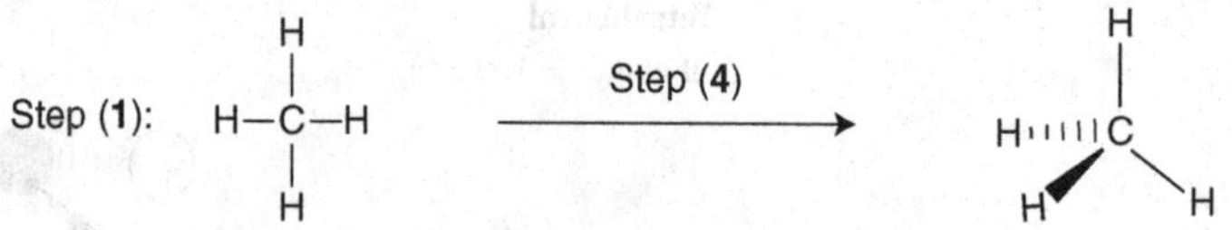

Step (**2**): Carbon is the central atom. Carbon has four single bonds and no lone pairs. Each single bond = 1 E.R.: **4** single bonds with no lone pairs = **4 E.R.**

Step (**3**): The summary table states that 4 E.R. will take a **tetrahedral** shape.

Step (**4**): Adapt the general drawing of a tetrahedron to this specific molecule. The carbon is drawn in the center (dark circle) of the figure and the hydrogens are placed at the corners (light, open circles) of the figure.

Example: Complete steps (1) through (4) for the determination of the 3-D shape of ammonia, NH_3.

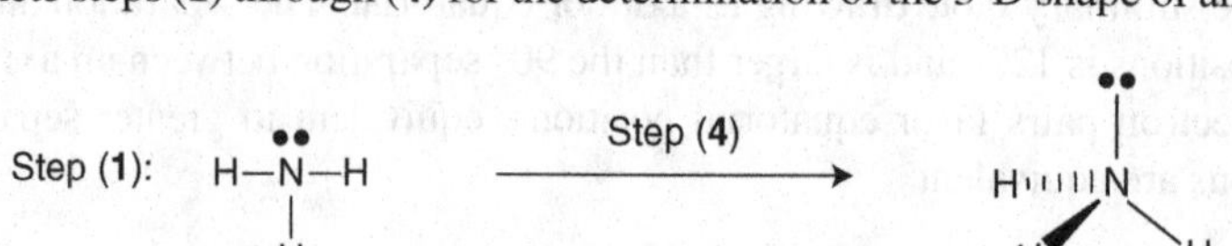

Step (**2**): Nitrogen is the central atom. Nitrogen has three single bonds and one lone pair.

Each single bond = 1 E.R.; each lone pair = 1 E.R.:
3 single bonds plus **1** lone pairs = **4 E.R.**

Step (**3**): The summary table states that 4 E.R. will take a **tetrahedral** shape.

Step (**4**): Adapt the general drawing of a tetrahedron to this specific molecule. The nitrogen is drawn in the center of the figure and the hydrogens are placed at the three corners of the figure; a lone pair is placed at the fourth corner. The lone pair is arbitrarily shown at the top corner; any corner could have been selected to achieve an identical shape.

Example: Complete steps (**1**) through (**4**) for the determination of the 3-D shape of water, H_2O.

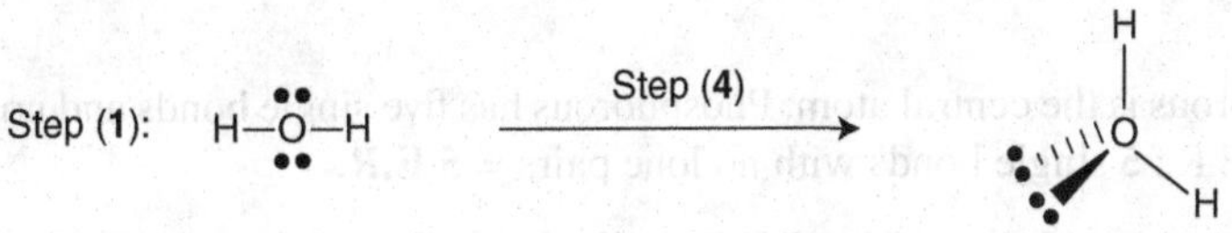

Step (1): H—O—H
Step (4)

Step (**2**): Oxygen is the central atom. Oxygen has two single bonds and two lone pairs. Each single bond = 1 E.R.; each lone pair = 1 E.R.:

2 single bonds plus **2** lone pairs = **4 E.R.**

Step (**3**): The summary table states that 4 E.R. will take a **tetrahedral** shape.

Step (**4**): Adapt the general drawing of a tetrahedron to this specific molecule. The oxygen is drawn in the center of the figure and the two hydrogens are placed at two corners of the figure. Two lone pairs are placed at arbitrarily selected corners; any corners could have been selected to achieve an identical shape.

Example: Complete steps (**1**) through (**4**) for the determination of the 3-D shape of boron trifluoride, BF_3. Lone pairs on fluorine are not shown.

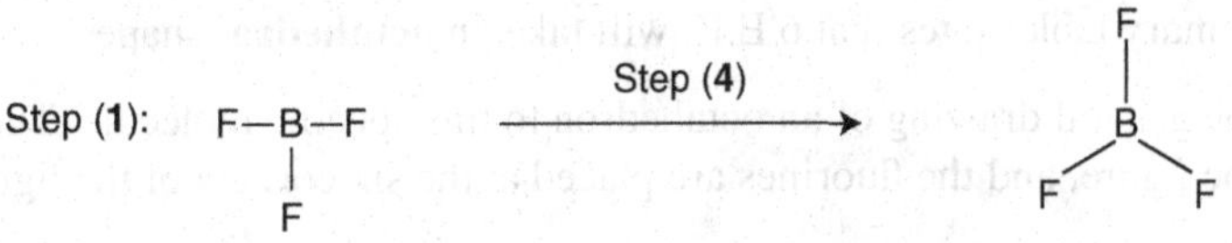

Step (1): F—B—F
Step (4)

Step (**2**): Boron is the central atom. Boron has three single bonds and no lone pairs. Each single bond = 1 E.R.: **3** single bonds and no lone pairs = **3 E.R.**

Step (**3**): The summary table states that 3 E.R. will take a **trigonal planar** shape.

Step (**4**): Adapt the general drawing of trigonal planar to this specific molecule. The boron is drawn in the center of the figure, and the three fluorines are placed at the three corners of the figure.

Example: Complete steps (**1**) through (**4**) for the determination of the 3-D shape of carbon dioxide, CO_2. Lone pairs on the oxygens are not shown.

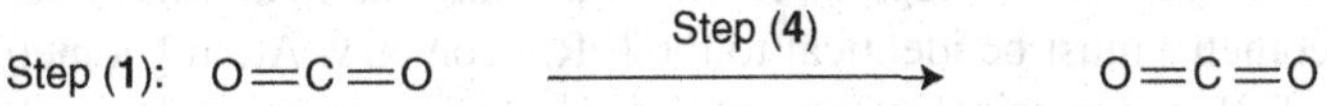

Step (1): O=C=O
Step (4)
O=C=O

Step (**2**): Carbon is the central atom. Carbon has two double bonds and no lone pairs. Each double bond = 1 E.R.: **2** double bonds and no lone pairs = **2 E.R.**

Step (**3**): The summary table states that 2 E.R. will take a **linear** shape.

Step (**4**): Adapt the general drawing of linear to this specific molecule. The carbon is drawn in the center of the figure, and the two oxygens are placed at the two corners of the figure. Note that the oxygens could have been central atoms, but they are outside atoms in this molecule.

Example: Complete steps (**1**) through (**4**) for the determination of the 3-D shape of phosphorous pentafluoride, PF_5. Lone pairs on the fluorines are not shown.

Step (**1**): Step (**4**)

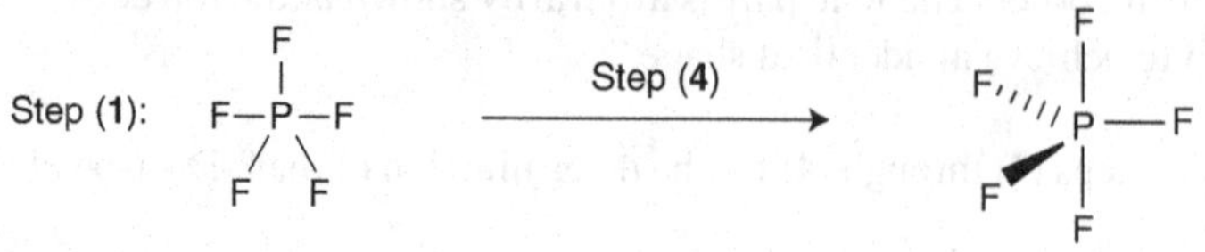

Step (**2**): Phosphorous is the central atom. Phosphorous has five single bonds and no lone pairs. Each single bond = 1 E.R.: **5** single bonds with no lone pairs = **5 E.R.**

Step (**3**): The summary table states that 5 E.R. will take a **trigonal bipyramidal** shape.

Step (**4**): Adapt the general drawing of trigonal bipyramidal to this specific molecule. The phosphorous is drawn in the center of the figure, and the fluorines are placed at the five corners of the figure.

Example: Complete steps (**1**) through (**4**) for the determination of the 3-D shape of sulfur hexafluoride, SF_6. Lone pairs on the fluorines are not shown.

Step (**1**): Step (**4**)

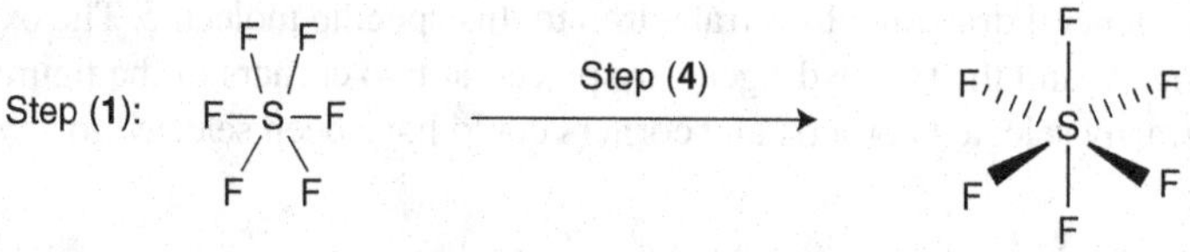

Step (**2**): Sulfur is the central atom. Sulfur has six single bonds and no lone pairs. Each single bond = 1 E.R.: **6** single bonds with no lone pairs = **6 E.R.**

Step (**3**): The summary table states that 6 E.R. will take an **octahedral** shape.

Step (**4**): Adapt the general drawing of an octahedron to this specific molecule. The sulfur is drawn in the center of the figure, and the fluorines are placed at the six corners of the figure.

III ATOM GEOMETRY DETERMINATION FROM ELECTRON REGION GEOMETRY

Atom geometry describes the shape of the bonded atoms around a central atom. For the atom geometry, the lone pairs are not included and are considered "invisible." Atom geometry must be derived from electron region geometry. The specific corners of a geometrical figure occupied by atoms cannot be determined without first locating all electron regions, including the lone pairs. Lone pairs always influence the position of bonded atoms, even if the lone pairs are not specifically designated in the atom geometry shape.

If a central atom does not have lone pairs, then atoms will occupy all corners of the E.R. geometrical figure. Whenever the electron regions (E.R.) of a central atom are all bonded atoms (no lone pairs), the atom geometry must be identical to the E.R. geometry. Atom geometry shapes include all of the possible E.R. geometrical figures plus a number of additional possibilities describing various combinations of bonding corners and lone pair corners. These additional geometries are described in the accompanying two tables; the major combinations are indicated but not all possible shapes are included. Each figure is derived from the correct E.R. geometrical shape, modified by treating the lone pairs as invisible. Lone pairs are placed in arbitrary corners and shown shaded out to indicate the shape of just the atoms. The geometrical figure name applies to the remaining atom-only shape.

Step (**5**): Determine the atom geometry (molecular geometry). The atom geometry describes the 3-D position of only the atoms around the central atom, as if the lone pairs are invisible. Start with the drawing from step (**4**); the atom geometry is found by "covering up" the lone pairs and identifying the figure that remains. Do not rearrange atoms from the E.R. geometry; the lone pairs are still present in the designated corners, even if they are considered invisible for the atom geometry. If the E.R. geometry around the central atom contains no lone pairs, the E.R. geometry and the atom geometry must be identical.

Additional Atom Geometries Derived from Trigonal Planar and Tetrahedral

# of Electron Regions and Distribution	Geometrical Figure	Drawn Figure
3 **2 Atoms/1 lone pair**	**Angular/bent** angles ≅ 120°	
4 **2 Atoms/2 lone pair**	**Angular/bent** angles ≅ 109.5°	
4 **3 Atoms/1 lone pair**	**Trigonal pyramidal** angles ≅ 109.5°	

Example: Complete step (**5**); determine the atom geometry of methane, CH_4. Start with step (**4**) from the corresponding previous example.

Step (4): Cover up lone pairs → [CH₄ structure] Step (5) → no change

Step (**5**): The E.R. geometry around the central carbon atom contains no lone pairs. If the E.R. geometry around the central atom contains no lone pairs, the E.R. geometry and the atom geometry must be identical; both the E.R. geometry and the atom geometry are **tetrahedral**.

Example: Complete step (**5**); determine the atom geometry of ammonia, NH_3. Start with step (**4**) from the corresponding previous example.

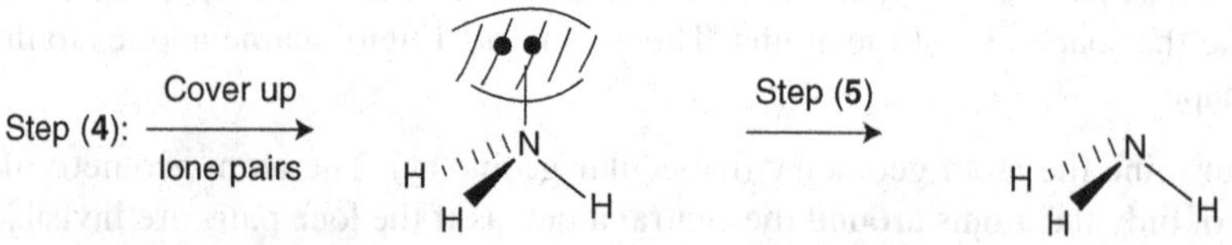

Step (4): Cover up lone pairs → Step (5) →

Step (**5**): The E.R. geometry contains 3 single bonds and one lone pair. After covering up the lone pair, the resulting atom geometry (shape of just the atoms) is termed **trigonal pyramidal**. The bond angles will be very close to those of the tetrahedral shape; the larger size of the lone pair alters the bond angles slightly.

Example: Complete step (**5**); determine the atom geometry of water, H_2O. Start with step (**4**) from the corresponding previous example.

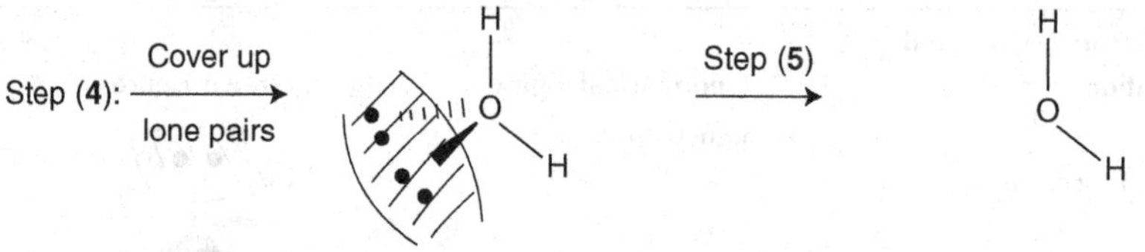

Step (4): Cover up lone pairs → Step (5) →

Step (**5**): The E.R. geometry contains 2 single bonds and two lone pairs. After covering up the lone pairs, the resulting atom geometry (shape of just the atoms) is termed **bent or angular**. The bond angles will be very close to those of the tetrahedral shape; the larger size of the lone pairs alters the bond angles slightly.

Example: Complete step (**5**); determine the atom geometry of boron trifluoride (BF_3), phosphorous pentafluoride (PF_5), and sulfur hexafluoride (SF_6). Start with step (**4**) from the corresponding previous examples.

Step (**5**): The E.R. geometries around the central atom for all three molecules contain no lone pairs; in this case, the E.R. geometry and the atom geometry must be identical.

For BF_3, both the E.R. geometry and the atom geometry are **trigonal planar**.
For PF_5, both the E.R. geometry and the atom geometry are **trigonal bipyramidal**.
For SF_6, both the E.R. geometry and the atom geometry are **octahedral**.

Additional Atom Geometries Derived from Trigonal Bipyramidal and Octahedral

# of Electron Regions and Distribution	Geometrical Figure	Drawn Figure
5 4 Atoms/1 lone pair	**See-saw** lone pairs equatorial	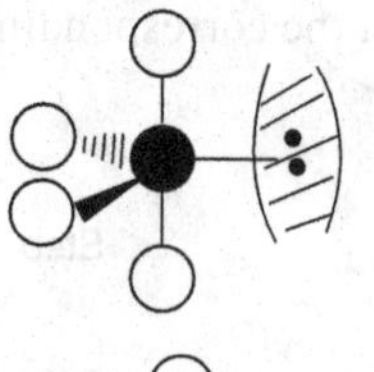
5 3 Atoms/2 lone pair	**Tee-shape** lone pairs equatorial	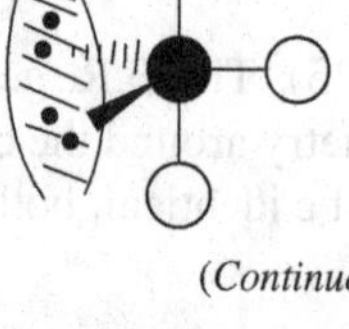

(Continued)

# of Electron Regions and Distribution	Geometrical Figure	Drawn Figure
6 5 Atoms/1 lone pair	**Square pyramidal** angles ≅ 90°	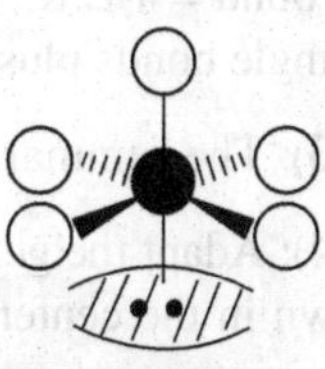
6 4 Atoms/2 lone pair	**Square planar** lone pairs 180° apart	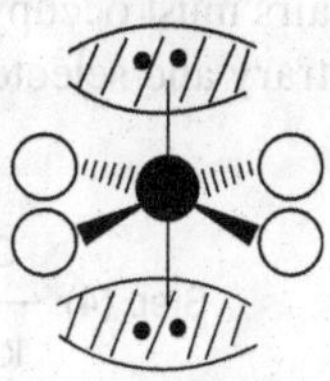

Example: Complete steps (**1**) through (**5**); determine the electron region geometry and the atom geometry for sulfur tetrafluoride (SF_4). Show lone pairs only for the central atom.

Step (1): H—S—F (with F, F) → Step (4)

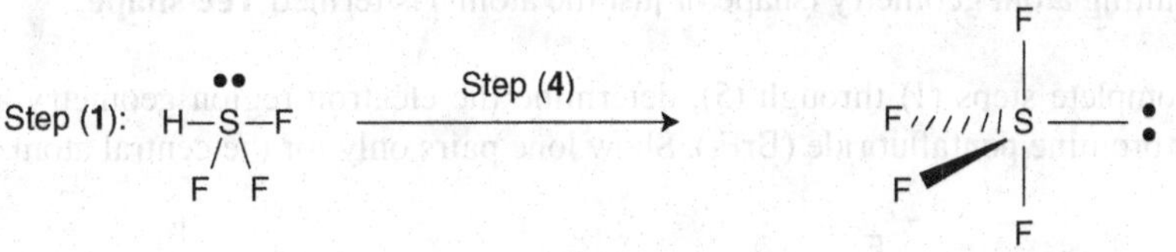

Step (**2**): Sulfur is the central atom. Sulfur has four single bonds and one lone pair. Each single bond = 1 E.R.; each lone pair = 1 E.R.:

4 single bonds plus **1** lone pair = **5 E.R.**

Step (**3**): The summary table states that 5 E.R. will take a **trigonal bipyramidal** shape.

Step (**4**): Adapt the general drawing of trigonal bipyramidal to this specific molecule. The sulfur is drawn in the center of the figure. Four fluorines are placed at four corners of the figure. The lone pair must occupy one of the equatorial positions; the choice of the specific equatorial corner is arbitrary and selected for easier visualization of the atom geometry.

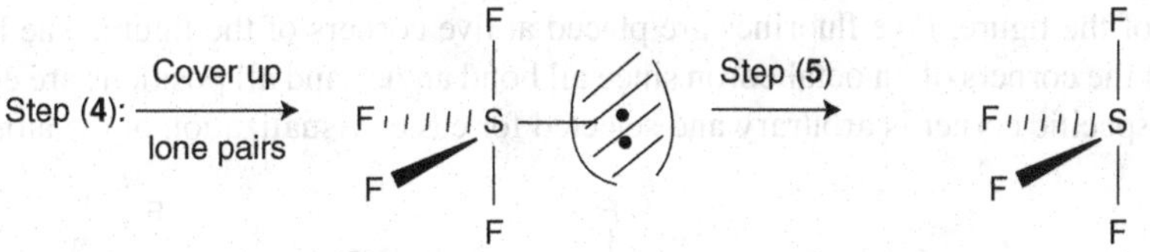

Step (4): Cover up lone pairs → Step (5)

Step (**5**): The E.R. geometry contains 4 single bonds and one lone pair. After covering up the lone pair, the resulting atom geometry (shape of just the atoms) is termed **see-saw**.

Example: Complete steps (**1**) through (**5**); determine the electron region geometry and the atom geometry for bromine trifluoride (BrF_3). Show lone pairs only for the central atom.

Step (1): F—Br—F (with F) → Step (4)

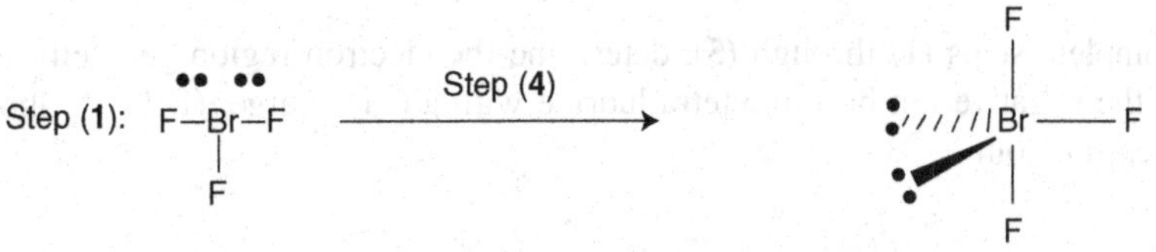

Step (**2**): Bromine is the central atom. Bromine has three single bonds and two lone pairs. Each single bond = 1 E.R.; each lone pair = 1 E.R.:

3 single bonds plus 2 lone pairs = 5 E.R.

Step (**3**): The summary table states that 5 E.R. will take a **trigonal bipyramidal** shape.

Step (**4**): Adapt the general drawing of trigonal bipyramidal to this specific molecule. The bromine is drawn in the center of the figure. Three fluorines are placed at three corners of the figure. The lone pairs must occupy two of the equatorial positions; the choice of the specific equatorial corners is arbitrary and selected for easier visualization of the atom geometry.

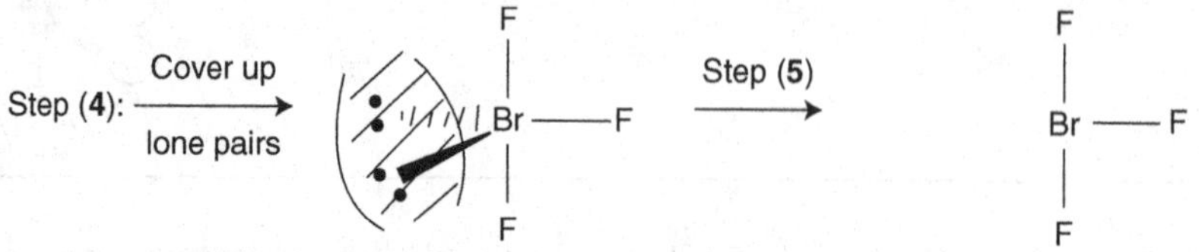

Step (4): Cover up lone pairs → Step (5) →

Step (**5**): The E.R. geometry contains 3 single bonds and two lone pairs. After covering up the lone pairs, the resulting atom geometry (shape of just the atoms) is termed **Tee-shape**.

Example: Complete steps (1) through (5); determine the electron region geometry and the atom geometry for bromine pentafluoride (BrF_5). Show lone pairs only for the central atom.

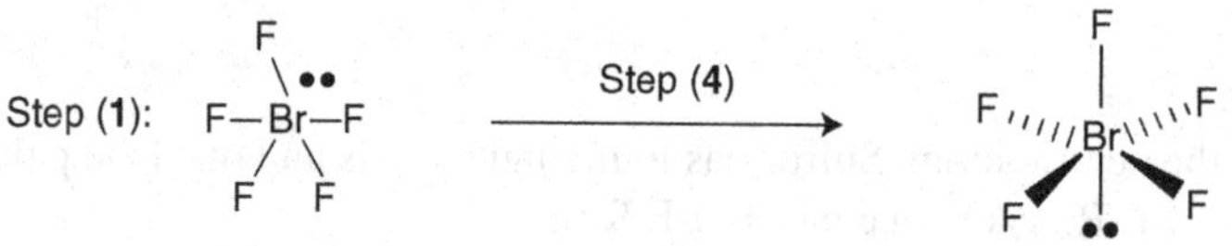

Step (**1**): → Step (4) →

Step (**2**): Bromine is the central atom. Bromine has five single bonds and one lone pair. Each single bond = 1 E.R.; each lone pair = 1E.R.:

5 single bonds plus **1** lone pair = **6 E.R.**

Step (**3**): The summary table states that 6 E.R. will take an **octahedral** shape.

Step (**4**): Adapt the general drawing of octahedral to this specific molecule. The bromine is drawn in the center of the figure. Five fluorines are placed at five corners of the figure. The lone pair may occupy any of the corners of an octahedron since all bond angles and all positions are equivalent; the choice of the specific corner is arbitrary and selected for easier visualization of the atom geometry.

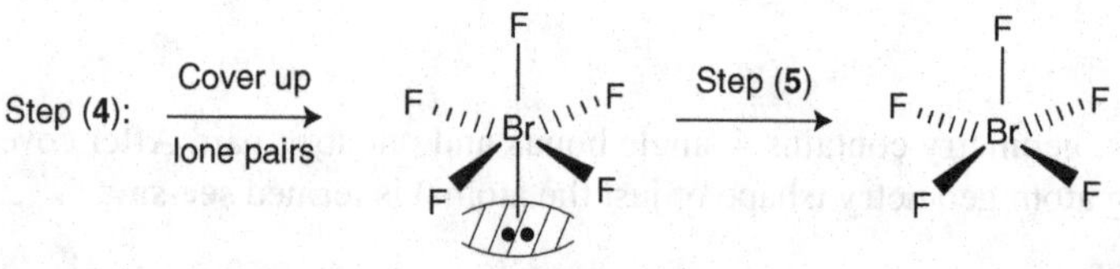

Step (4): Cover up lone pairs → Step (5) →

Step (**5**): The E.R. geometry contains 5 single bonds and one lone pair. After covering up the lone pair, the resulting atom geometry (shape of just the atoms) is termed **square pyramidal**.

Example: Complete steps (**1**) through (**5**); determine the electron region geometry and the atom geometry for the negative ion bromine tetrafluoride with a (–1) charge, $(BrF_4)^{-1}$. Show lone pairs only for the central atom.

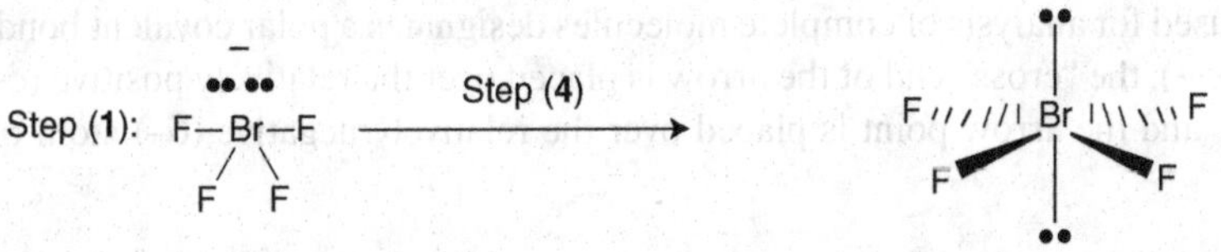

Step (2): Bromine is the central atom. Bromine has four single bonds and two lone pairs. Each single bond = 1 E.R.; each lone pair = 1 E.R.:

4 single bonds plus **2** lone pairs = **6 E.R.**

Step (3): The summary table states that 6 E.R. will take an **octahedral** shape.

Step (4): Adapt the general drawing of octahedral to this specific molecule. The bromine is drawn in the center of the figure. Four fluorines are placed at four corners of the figure. The first lone pair may occupy any of the corners of an octahedron. The second lone pair must be placed at a corner as far away as possible; the bond angle should be 180°; the choice of the specific corners is arbitrary and selected for easier visualization of the atom geometry.

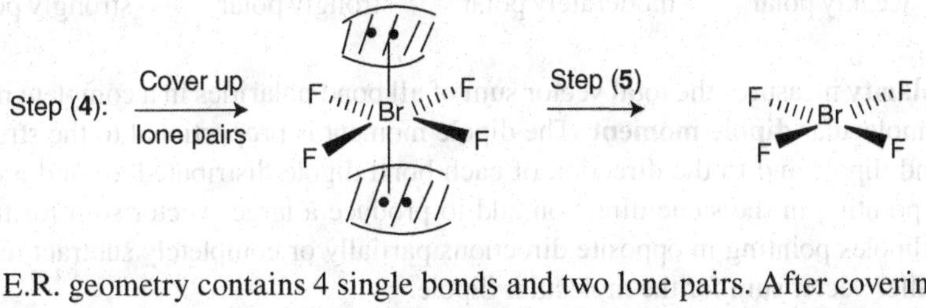

Step (5): The E.R. geometry contains 4 single bonds and two lone pairs. After covering up the lone pairs, the resulting atom geometry (shape of just the atoms) is termed **square planar**.

IV MOLECULAR POLARITY

Bond polarity (a polar covalent bond) is caused by unequal sharing of the electron pair in a covalent bond. The strength with which an atom holds a bonding electron pair in a covalent bond is measured by the atom's **electronegativity** (EN): the higher the value, the more tightly the atom holds the electron pair. The atom that holds the greater share of the electron pair is termed relatively negative (δ–); the atom that holds the lesser share of the electron pair is termed relatively positive (δ+).

The following is a list of typical values for electronegativity (EN).

F = 4.0	**O** = 3.5	**Cl** = 3.2	**N** = 3.0	**Br** = 3.0
I = 2.7	**S** = 2.6	**Se** = 2.6	**C** = 2.5	**P** = 2.2
H = 2.2	**As** = 2.2	**Te** = 2.1	**B** = 2.0	**Si** = 1.9

The degree of polarity of a specific bond (termed the strength of the **dipole**) is, in part, proportional to the difference in electronegativity (ΔEN) between the two bonded atoms. An approximate, arbitrary classification is:

$$(\Delta EN) < 0.5 = \text{weakly polar bond}$$
$$(\Delta EN) = 0.5 \text{ to } 1.0 = \text{moderately polar bond}$$
$$(\Delta EN) > 1.0 = \text{strongly polar bond}$$

A convention used for analysis of complete molecules designates a polar covalent bond with a dipole arrow = (+⟶); the "cross" end of the arrow is placed over the relatively positive (δ+) atom of the covalent bond, and the arrow point is placed over the relatively negative (δ–) atom of the covalent bond:

$$\overset{\delta+}{X}—\overset{\delta-}{Y} \quad = \quad \overset{+\!\!\longrightarrow}{X—Y}$$

Example: Determine the (ΔEN) for the following bonds and classify the bonds as weakly polar, moderately polar, or strongly polar: C—H; C—O; O—H; H—F.

	⟵+	+⟶	⟵+	+⟶
	C—H	C—O	O—H	H—F
EN:	2.5 2.2	2.5 3.5	3.5 2.2	2.2 4.0
	ΔEN = **0.3**	ΔEN = **1.0**	ΔEN = **1.3**	ΔEN = **1.8**
	weakly polar	moderately polar	strongly polar	strongly polar

Molecular polarity measures the total vector sum of all bond polarities in a complete molecule; this is termed the molecular **dipole moment**. The dipole moment is proportional to the strength of each individual bond dipole and to the direction of each bond dipole distributed around a central atom. Bond dipoles pointing in the same direction add to produce a larger vector sum for the molecular dipole. Bond dipoles pointing in opposite directions partially or completely subtract (**cancel out**) to produce a smaller vector sum for the molecular dipole.

The requirement for a molecule to be polar is that it must be electronically "lopsided," referring to the distribution of electron density. Relating to mass density, a ball will wobble in flight only if the mass around the center is distributed asymmetrically; i.e., if the ball is lopsided. A ball, such as a baseball or golf ball, has several different material densities representing the core, the winding, and the cover. The balls are not lopsided because each of the materials is distributed symmetrically around the center. A central atom may show varying electron densities through polar bonds, but, if the polar bonds are distributed symmetrically around the central atom, the electron density will not be lopsided and the central atom will not contribute to polarity.

Determination of the complete molecular dipole requires an analysis based on the 3-D shape (atom geometry) of the molecule. Only bonded atoms can contribute to a molecular dipole; lone electron pairs are never directly included. Each central atom must be analyzed individually; this is step (**6**) in a complete analysis of the Lewis structure, geometry and polarity of a covalent molecule.

Step (**6**): Complete the polarity analysis; a molecule can be polar to varying degrees as measured by the **net dipole**.

(**6a**): Use a table of atom electronegativities (EN) to determine the polarity of each bond in the molecule. Designate the bond polarities using the arrow notation.

(**6b**): Analyze each central atom separately; determine whether all the bond dipole arrows around a central atom show a net additive direction or whether they all exactly cancel.

(**1**) For a molecule to be polar (have a net dipole), at least one bond in the molecule must be polar (have a dipole) as determined by a difference in electronegativity (EN) of the bonding atoms. Bonds may range between a weak dipole (Δ EN approximately ≤ 0.4) up to a strong dipole (Δ EN approximately > 1.0).

(2) For a molecule to be polar (have a net dipole), for at least one central atom, all polar bonds (dipoles) around this central atom must not cancel due to a symmetrical arrangement of the bonded atoms. (A further requirement for certain molecules is that the central atoms themselves must not be symmetrically arranged.)

A complete mathematical vector analysis is not required to determine the polarity of a central atom or molecule. The following general guidelines can often be used to reach a reasonable conclusion.

Polarity Rule #1: Linear, trigonal planar, tetrahedral, trigonal bipyramidal, and octahedral electron region geometries are symmetrical structures. If all corners of these figures are occupied by identical atoms, any set of polar arrows involving the central atom in these shapes must automatically cancel. In this case, the central atom cannot contribute to a polar molecule.

Polarity Rule #2: For linear, trigonal planar, tetrahedral, trigonal bipyramidal, and octahedral electron region geometries, if any **one** of the corners is non-identical (assuming the bonds are polar), the set of polar arrows involving the central atom in these shapes cannot cancel. In this case, the central atom must contribute to a polar molecule to some degree. The complete molecule must also be polar unless, in certain cases, two identical central atoms cancel each other out.

Polarity Rule #3: A direct corollary to Rule #2: If any **one** corner of the linear, trigonal planar, tetrahedral, trigonal bipyramidal, or octahedral electron region geometries contains a lone-pair, the arrangement around the central atom is not symmetrical and this central atom must contribute to polarity (assuming the bonds are polar).

Polarity Rule #4: The polarity of molecules with two or three non-identical atoms or lone pairs depends on the specific E.R. geometry and must be analyzed directly. For tetrahedral E.R. geometry, any number of lone pairs or non-identical atoms will produce a non-symmetrical central atom. A tee-shape atom geometry (two equatorial lone pairs in trigonal bipyramidal) is non-symmetrical. A square planar atom geometry (two opposite lone pairs in octahedral) is symmetrical.

Example: Use vector dipole arrows to determine the polarity of (a) CO_2 (linear), (b) BF_3 (trigonal planar), and (c) CF_4 (tetrahedral).

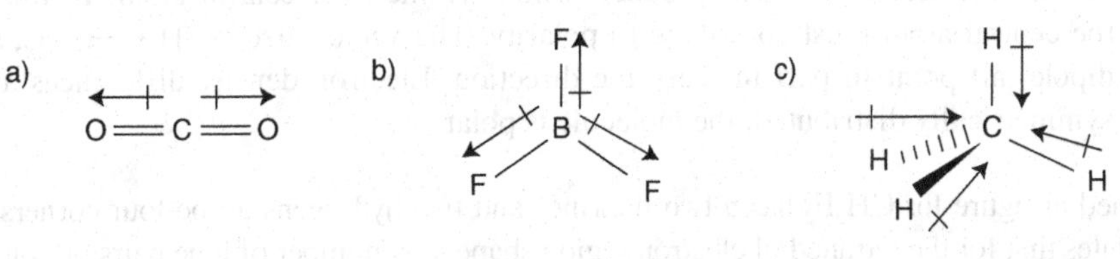

a) O═C═O b) BF_3 c) CH_4

(a) The linear figure for CO_2 shows identical atoms at the corners of the figure. Polarity Rule **#1** states that if all corners of any of the electron region geometrical figures are occupied by identical atoms, any set of polar arrows involving the central atom must automatically cancel. This observation is confirmed by noting that the (C+⟶O) vector arrow pointing to the right is exactly cancelled by the (O⟵+C) vector arrow pointing to the left. The C—O bond has a moderate dipole, but electron density differences are symmetrically distributed; the molecule is not polar.

(b) A similar conclusion results for BF_3. The trigonal planar figure shows identical fluorine atoms at all three corners; any set of polar arrows involving the central boron atom must automatically cancel. Although less obvious, a vector analysis of the equilateral triangle would show that the (B+⟶F) vector arrows exactly cancel. The B—F bond is strongly polar, but electron density differences are symmetrically distributed; the molecule is not polar.

(c) The vector sums for a tetrahedron are more difficult to read directly. However, the tetrahedral electron region shape for CF_4 shows identical fluorine atoms at all four corners. Polarity Rule **#1** applies; any set of polar arrows for bonds to carbon must cancel. The C—F bond is strongly polar, but electron density differences are symmetrically distributed; the molecule is not polar.

Example: Use vector dipole arrows to determine the polarity of (a) CH_2O, (b) NH_3, (c) H_2O, and (d) CH_2F_2.

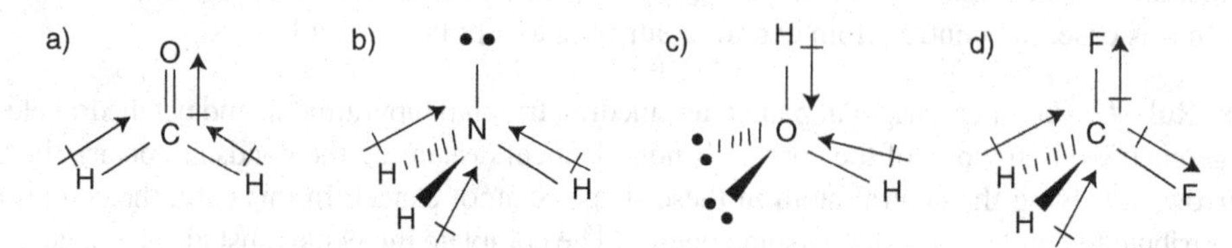

(a) The trigonal planar figure for CH_2O does not have identical atoms at all three corners. Polarity Rule **#2** states that if any **one** of the corners is non-identical (assuming the bonds are polar), the set of polar arrows involving the central atom cannot cancel. The vector arrows (H+⟶C; weak dipole) and (C+⟶O; moderate dipole) all point in part in the same direction. Electron density differences are not symmetrically distributed; the molecule is polar.

(b) The tetrahedral figure for NH_3 has a lone pair at one of the four corners. Polarity Rule **#3** states that if any **one** corner contains a lone-pair, the arrangement around the central atom is not symmetrical and this central atom must contribute to polarity (assuming the bonds are polar). The vector arrows (H+⟶N; moderate dipole) all point in part in the same direction. Electron density differences are not symmetrically distributed; the molecule is polar.

(c) The tetrahedral figure for H_2O has two lone pairs at two of the four corners. Polarity Rule **#4** states that for the tetrahedral electron region shape, any number of long pairs or non-identical atoms will produce a non-symmetrical central atom. In this case, the central atom must contribute to polarity. The vector arrows (H+⟶C; strong dipole) all point in part in the same direction. Electron density differences are not symmetrically distributed; the molecule is polar.

The tetrahedral figure for CH_2F_2 has a two fluorines and two hydrogens at the four corners. Polarity Rule **#4** states that for the tetrahedral electron region shape, any number of lone pairs or non-identical atoms will produce a non-symmetrical central atom. In this case, the central atom must contribute to polarity. The vector arrows (H+⟶C; weak dipole) and (C+⟶F; strong dipole) all point in the same general direction. Electron density differences are not symmetrically distributed; the molecule is polar.

Polarity Rule #5: The strength of a molecular dipole is, in general, a complicated function of all central atoms in a molecule. For smaller molecules, often one very polar central atom may dominate the strength of the measured dipole. In some cases, a molecule may have two central atoms, each of which contributes to polarity but is in a symmetrical arrangement in the complete molecule. In this case, the symmetrical polar central atoms may cancel each other out and produce a non-polar result.

Polarity Rule #6: The strength of a molecular dipole is an important consideration for the evaluation of many physical properties of a compound such as melting point, boiling point, or solution formation (see Chapter 21). Molecular dipoles can be classified in a very wide range of net polarity:

non-polar ←→ very weakly polar ←→ weak ←→ moderate ←→ strongly polar

Example: Complete steps (**1**) through (**6**) for C_2F_4; analyze both central atoms in the molecule. Show lone pairs for central atoms only.

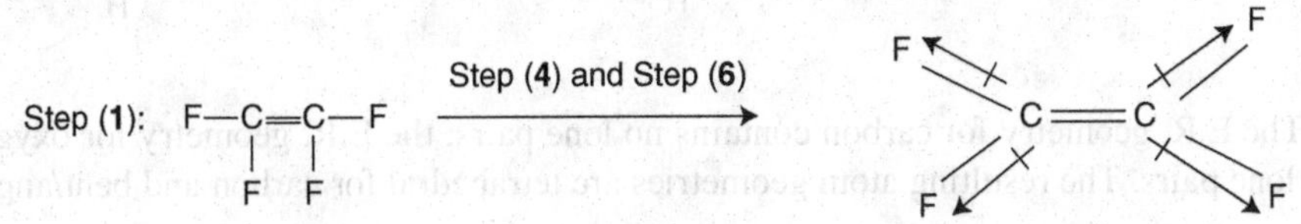

Step (**1**): F—C═C—F (each C also bonded to F below) → Step (**4**) and Step (**6**)

Step (**2**): Each carbon is a central atom. Both carbons have an identical bonding structure of two single bonds and one double bond.

Each single bond = 1 E.R.; each double = 1 E.R.:

2 single bonds plus **1** double bond = **3 E.R.**

Step (**3**): The summary table states that 3 E.R. will take a **trigonal planar** shape.

Step (**4**): Adapt the general drawing of trigonal planar for both carbons in this specific molecule. Each carbon is drawn in the center of each trigonal planar figure, with the two carbons double bonded. Two fluorines each are placed at the two remaining corners of each figure.

Step (**5**): The E.R. geometry contains no lone pairs; the atom geometry of the complete molecule is represented by the electron region drawing.

Step (**6**): Dipole arrows are shown to indicate that each central atom individually can contribute to polarity: the vector arrows (C+⟶F; strong dipole) both point in the same general direction for each central atom considered separately. However, the vector arrows also show that two central atoms cancel each other out. This is an example of a non-polar molecule containing symmetrical (mirror image) polar central atoms.

Example: Complete steps (**1**) through (**6**) for CH_4O; analyze both central atoms in the molecule.

Step (**1**): H—C—O—H (C also bonded to two H; O with two lone pairs) → Step (**4**)

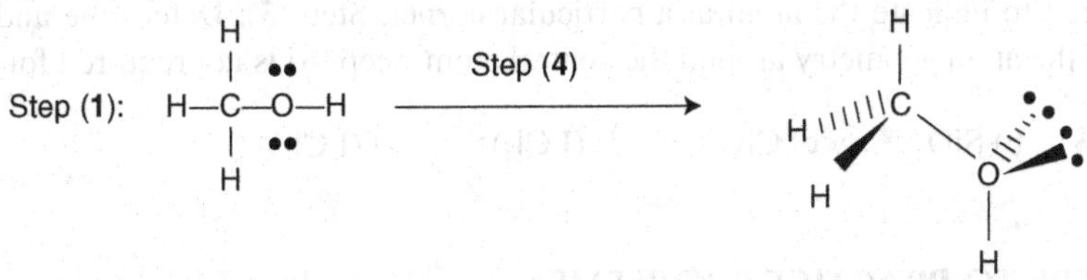

Step (**2**): One carbon and one oxygen are central atoms.

The carbon has 4 single bonds and no lone pairs: **4 E.R.**

The oxygen has 2 single bonds and two lone pairs: **4 E.R.**

Step (**3**): The summary table states that 4 E.R. will take a **tetrahedral** shape.

Step (**4**): Adapt the general drawing of tetrahedral for both the carbon and the oxygen; connect the two central atoms.

Step (4): Cover up lone pairs → Step (5) Step (6) →

Step (**5**): The E.R. geometry for carbon contains no lone pairs; the E.R. geometry for oxygen contains two lone pairs. The resulting atom geometries are tetrahedral for carbon and bent/angular for oxygen.

Step (**6**): The vector arrows (H+⟶C; weak dipole) point in part in the same direction. The vector arrow (H+⟶O; strong dipole) point in part in the same direction; electron density differences are not symmetrically distributed. The molecule is polar with the O—H bond portion contributing much more to the total strength of the dipole.

V PRACTICE PROBLEMS

1. Complete analysis steps (**1**) through (**6**) for the following molecules.
   Step (**1**): Determine the Lewis structure; show lone pairs only for the central atom; chlorine is octet bonded. Steps (**2**) through (**4**): Identify and draw the electron region geometry around the central atom; if the geometrical figure has double-bonded atoms, only one line is required to indicate the atom at a particular corner. Step (**5**): Determine and state the name of the atom geometry around the central atom. Step (**6**): Use the polarity rules and/or dipole arrows as necessary to determine the polarity result for the molecule.

   a) SO_2 b) SO_3 c) $SeCl_4$ d) AsF_3 e) AsF_5 f) $I\,Cl_3$ g) $I\,Cl_5$

2. Complete analysis steps (**1**) through (**5**) for the following ions.
   Step (**1**): Determine the Lewis structure; show lone pairs only for the central atom; chlorine is octet bonded. Steps (**2**) through (**4**): Identify and draw the electron region geometry around the central atom; if the geometrical figure has double-bonded atoms, only one line is required to indicate the atom at a particular corner. Step (**5**): Determine and state the name of the atom geometry around the central atom. Step (**6**) is not required for ions.

   a) SO_3^{-2} b) SiO_3^{-2} c) ClO_2^- d) $(I\,Cl_4)^-$ e) $(I\,Cl_4)^+$

VI ANSWERS TO PRACTICE PROBLEMS

1. a) Step (**4**) shows only one line for the double bonded oxygens.

(1): O=S=O —(4)→

(**2**): Sulfur has two double bonds and one lone pair = **3 E.R.**
(**3**): The electron region geometry is **trigonal planar**.
(**5**): The atom geometry is **bent/angular** with bond angles ≅ 120°.
(**6**): The (S+⟶O) bond has a moderate dipole. Polarity Rule **#3** indicates the molecule is polar.

1. b) Step (**4**) shows only one line for the double bonded oxygens.

(**1**):

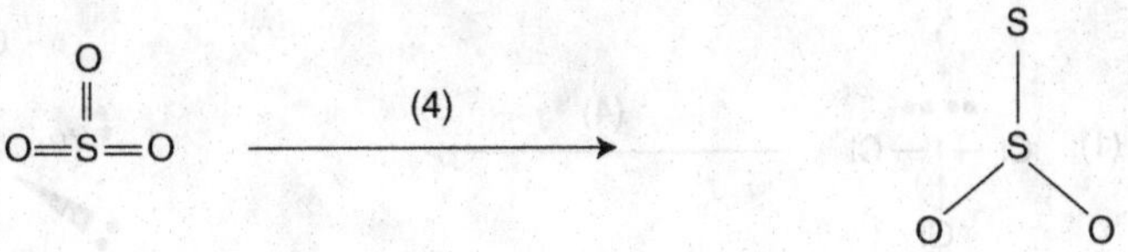

(**2**): Sulfur has three double bonds and no lone pairs = **3 E.R.**
(**3**): The electron region geometry is **trigonal planar**.
(**5**): The atom geometry is **trigonal planar**.
(**6**): Although the (S+⟶O) bond has a moderate dipole, Polarity Rule **#1** indicates the molecule is non-polar due to symmetry.

1. c)

(**1**):

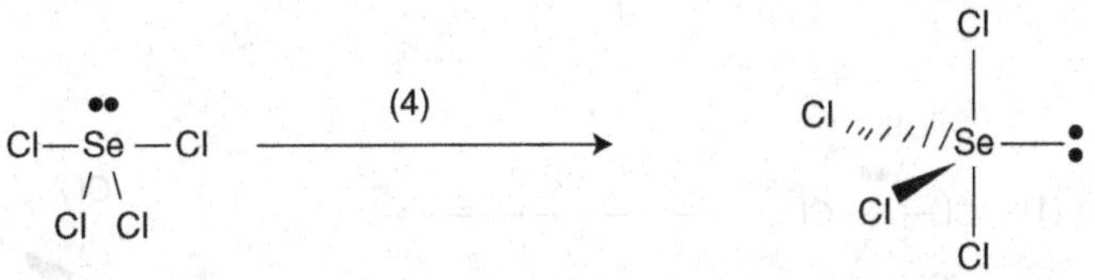

(**2**): Selenium has four single bonds and one lone pair = **5 E.R.**
(**3**): The electron region geometry is **trigonal bipyramidal**.
(**5**): The atom geometry is **see-saw**.
(**6**): The (Se+⟶Cl) bond has a moderate dipole. Polarity Rule **#3** indicates the molecule is polar.

1. d)

(**1**):

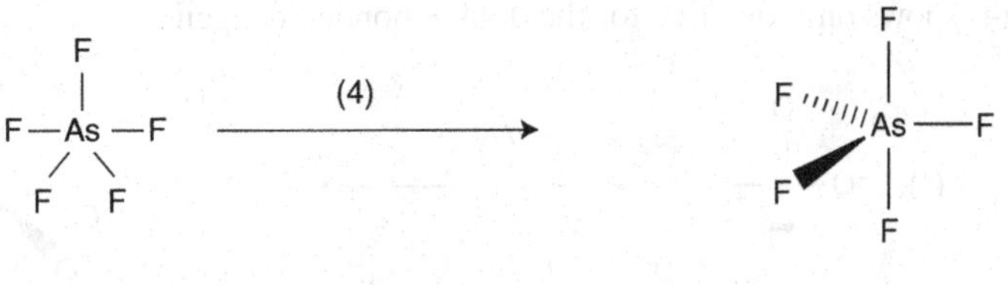

(**2**): Arsenic has five single bonds and no lone pairs = **5 E.R.**
(**3**): The electron region geometry is **trigonal bipyramidal**.
(**5**): The atom geometry is **trigonal bipyramidal**.
(**6**): Although the (As+⟶F) bond has a strong dipole, Polarity Rule **#1** indicates the molecule is non-polar due to symmetry.

1. e)

(**1**):

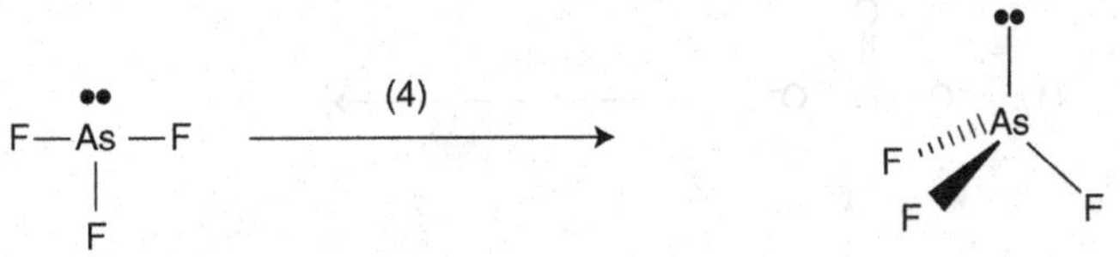

(2): Arsenic has three single bonds and one lone pairs = **4 E.R.**
(3): The electron region geometry is **tetrahedral**.
(5): The atom geometry is **trigonal pyramidal.**
(6): The (As +⟶F) bond has a strong dipole. Polarity Rule **#3** indicates the molecule is polar.

1. f)

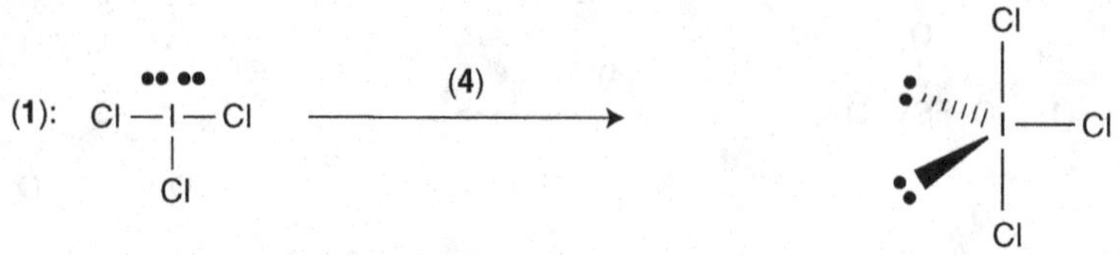

(1): Cl—I—Cl (with Cl below I and two lone pairs on I) —(4)→ 3D structure of ICl_3

(2): Iodine has three single bonds and two lone pairs = **5 E.R.**
(3): The electron region geometry is **trigonal bipyramidal.**
(5): The atom geometry is **tee-shape**.
(6): The (I+⟶Cl) bond has a weak-moderate dipole. Polarity Rule **#4** indicates the molecule is polar.

1. g)

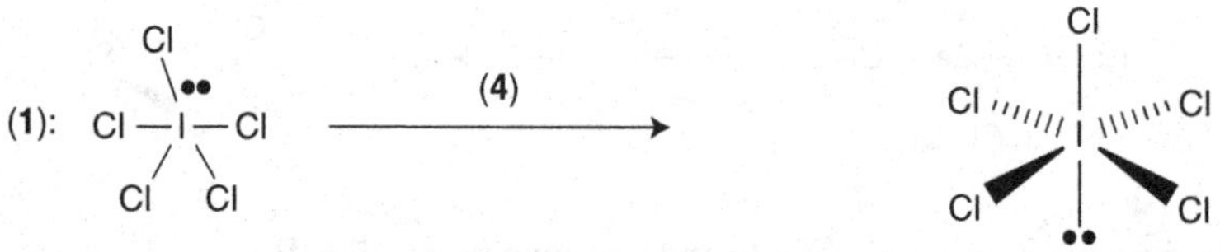

(1): Cl—I—Cl with three more Cl and one lone pair on I —(4)→ 3D structure of ICl_5

(2): Iodine has five single bonds and one lone pair = **6 E.R.**
(3): The electron region geometry is **octahedral.**
(5): The atom geometry is **square pyramidal.**
(6): The (I+⟶Cl) bond has a weak-moderate dipole. Polarity Rule **#3** indicates the molecule is polar.

2. a) Step (**4**) shows only one line for the double bonded oxygen.

(1): $^{-}$O—S(=O)—O^{-} (lone pair on S) —(4)→ 3D structure of S with three O

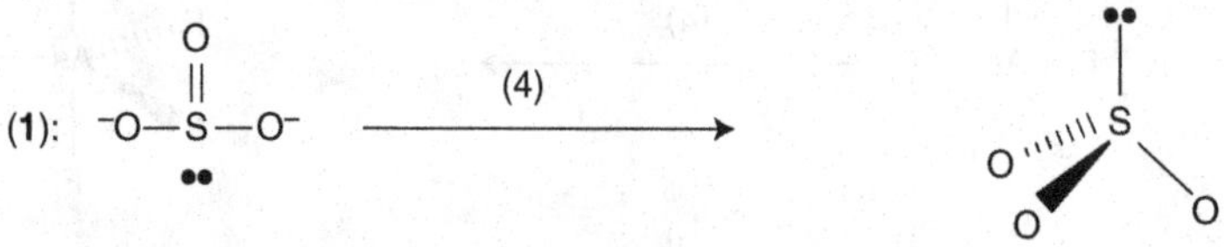

(2): Sulfur has one double bond, two single bonds, and one lone pair = **4 E.R.**
(3): The electron region geometry is **tetrahedral.**
(5): The atom geometry is **trigonal pyramidal.**

2. b) Step (**4**) shows only one line for the double bonded oxygen.

(1): $^{-}$O—Si(=O)—O^{-} —(4)→ Si with three O

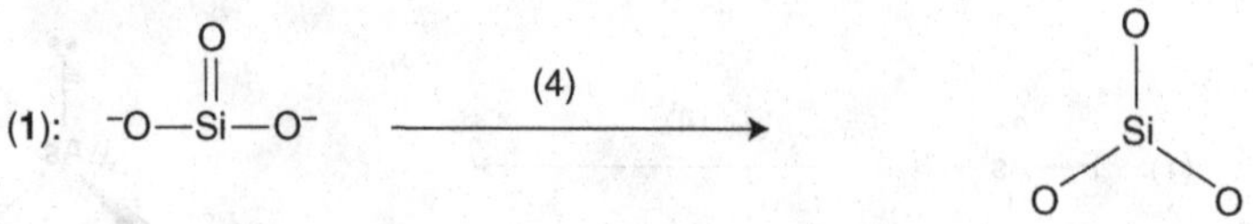

(**2**): Silicon has one double bond, two single bonds, and no lone pairs = **3 E.R.**
(**3**): The electron region geometry is **trigonal planar**.
(**5**): The atom geometry is **trigonal planar**.

2. c) Step (4) shows only one line for the double bonded oxygen.

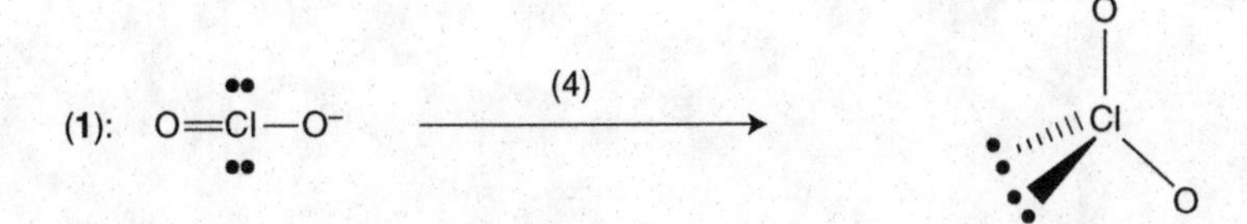

(**2**): Chlorine has one double bond, one single bond, and two lone pairs = **4 E.R.**
(**3**): The electron region geometry is **tetrahedral**.
(**5**): The atom geometry is **bent/angular** with bond angles ≅ 109°.

2. d)

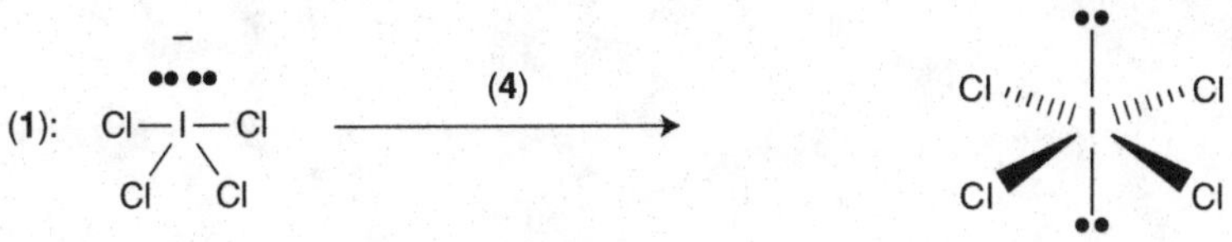

(**1**): Cl—I—Cl (with two more Cl on I; overall charge −) —(**4**)→ Cl, Cl, Cl, Cl around I, with lone pairs above and below

(**2**): Iodine has four single bonds and two lone pairs = **6 E.R.**
(**3**): The electron region geometry is **octahedral**.
(**5**): The atom geometry is **square planar**.

2. e)

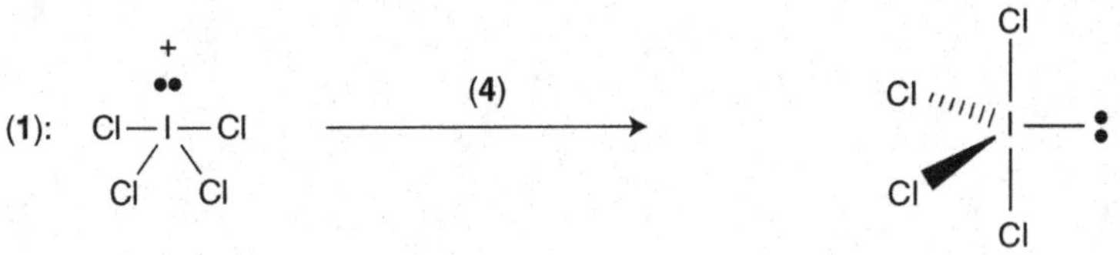

(**1**): Cl—I—Cl (with two more Cl on I; overall charge +) —(**4**)→ Cl, Cl, Cl, Cl around I, with one lone pair (I—⋮)

(**2**): Iodine has four single bonds and one lone pair = **5 E.R.**
(**3**): The electron region geometry is **trigonal bipyramidal**.
(**5**): The atom geometry is **see-saw**.

17 Summary Analysis of Central Atom Bonding, Hybridization, and Geometry

Each central atom in a molecule can be described by its bond types, hybridization, geometries, and resulting polarity. Chapter 16 presented six steps for analyzing each central atom in a molecule. This chapter adds a step (**7**) and a step (**8**). From Chapter 16, step (**1**) determined the Lewis structure; steps (**2**) through (**4**) identified and produced a drawing of the electron region geometry around each central atom. Step (**5**) determined the atom geometry around each central atom, which is derived from the electron region geometry. Step (**6**) determined the contribution of each central atom to a molecular dipole; the complete molecule was then classified as polar or non-polar. A new step (**7**) classifies the bond types, which are derived directly from the Lewis bonding structure. A new step (**8**) identifies the hybridization of each central atom, which is required by the number of electron regions.

I CONCEPT OF SIGMA BONDS AND PI BONDS

A **sigma bond** (σ-bond) is an **end-to-end overlap** of orbitals, which provides a spherically symmetrical electron-density distribution around the bond axis; it allows bonded atoms to rotate around bond axis. All orbital types can engage in end-to-end overlap: s, p, d, sp-hybridized, sp^2-hybridized, sp^3-hybridized, As well as other combinations. The first bond to form between atoms is always a sigma bond; end-to-end overlap of orbitals provides the most energetically favorable bonding molecular orbital and results in the relatively strongest bond (largest bond dissociation energy; see Chapter 18). All single bonds are sigma bonds.

A **pi bond** (π-bond) is a **side-to-side overlap** of orbitals, which produces electron density regions distributed above and below the bond axis. The overlap is **not** spherically symmetrical and does **not** allow rotation of bonded atoms around the bond axis. Pi bonds are usually formed through overlap of p-orbitals and sometimes d-orbitals. The second bond of a double bond and the second and third bonds of a triple bond are pi bonds. Only one sigma bond (end-to-end overlap) can occur between bonding atoms; additional bonds must be pi bonds and must be placed within the geometrical structure dictated by the sigma bonds and lone pairs (the sigma bonding framework).

The sigma bonding framework determines the electron region geometry. Step (**2**) for the analysis of a central atom counts the number of electron regions; the assignment of electron regions can be more specifically stated. The result, however, matches the count produced from the assignments defined for single, double, and triple bonds.

Step (**2**): Determine the number of electron regions based on the presence of lone pairs and sigma bonds.

(**1**) Assign one electron region (E.R.) to each lone electron pair.
(**2**) Assign one electron region (E.R.) to each sigma-bonding electron pair.
(3) Summary of electron region count based on assignments (1) and (2).

Each lone electron pair = **one** E.R.
Each single bond = **one** E.R. (represents **one** sigma bond)

Each double bond = **one** E.R. (represents **only one** sigma bond, the other is a pi bond)
Each triple bond = **one** E.R. (represents **only one** sigma bond, the other two are pi bonds)

A new step (**7**) is added to the analysis of a complete molecule; this process may be applied after determining the Lewis structure.

Step (**7**): Classify the bond type for each bond in the molecule according to the following identifications:

Single Bond = **one** sigma-type bond
Double Bond = **one** sigma-type bond **plus one** pi-type bond
Triple Bond = **one** sigma-type bond **plus two** pi-type bonds

Example: Complete step (**7**) for the following molecules; count the number of sigma bonds and pi bonds in each molecules.

a) F–C(F)=C(H)–C(=O)–O–H b) S=C=C=S c) N≡C–C≡C–C(H)(Br)–S–C(Cl)=O

a) There are four single bonds to the four outside atoms (two hydrogens and two fluorines). There is one single bond connecting one carbon to one oxygen and one single bond connecting two carbons. There are two double bonds in the molecule: 6 single bonds = 6 sigma bonds; 2 double bonds = 2 sigma bonds plus 2 pi bonds.

Total = 8 sigma bonds and 2 pi bonds.

b) The molecule contains three double bonds; 3 double bonds = 3 sigma bonds plus 3 pi bonds.
c) There are a total of seven single bonds, two triple bonds, and one double bond in the molecule: 7 single bonds = 7 sigma bonds; 2 triple bonds = 2 sigma bonds plus 4 pi bonds; 1 double bond = 1 sigma bond plus 1 pi bond.

Total = 10 sigma bonds and 5 pi bonds.

II CONCEPT OF HYBRIDIZATION

The necessity of the fundamental concept of atomic orbital hybridization can be illustrated with a consideration of the element carbon. As a neutral bonding element, carbon can form four bonds; for example, four single bonds in the molecule methane, CH_4. This molecule is known to have a tetrahedral geometry in which all four corners of the figure are symmetrical and all four C—H bonds have an equal bond length and have equivalent energy.

The difficulty arises when bond formation is viewed through the ground state electron configuration of the element carbon: $1s^22s^22p^2$; the potential energy diagram for the element electron configuration is shown on the left of the figure, which follows. Formation of four normal neutral single bonds in CH_4 requires carbon to have four equal energy bonding orbitals, each with one electron. The ground state shows only two p-orbitals with one electron; the 2s orbital is filled and the third 2p orbital is empty. In addition, these orbitals do not point to the corners of a tetrahedron with 109.5° bond angles. The three p-orbitals are orthogonal (perpendicular, 90° to each other); the s-orbital is spherical.

A solution to this dilemma is central atom hybridization. **Hybridization** refers to the mixing of the energy and direction properties of atomic orbitals (s, p, and d-orbitals) to achieve optimum bonding conditions in molecules. The result of hybridization for carbon in CH_4 is shown on the right of the following figure.

Electron Configuration and Hybridization of Carbon in CH_4

PE

2p ↑. ↑.

2s ↑↓

↑. ↑. ↑. ↑. sp^3-hybrids

Ground State of C

Four Hybridized Orbitals

Hybridization of one s-orbital with three p-orbitals results in four equivalent energy orbitals, which point to the corners of a tetrahedral shape. Since all four hybridized orbitals have equal energy, the four valence electrons are distributed singly in each orbital and are available for single bond formation. The notation used for hybridized orbitals shows the identity of each of the orbital types (subshell) used in the mix and the number of orbitals of each type used indicated by a superscript. The hybrid orbitals for carbon in this hybridization example are designated as **sp^3**, representing a combination of one s-orbital and three p-orbitals; the orbitals are stated as being 75% p-character and 25% s-character. An sp^2-hybridized orbital would be 67% p-character and 33% s-character; sp-hybridized orbitals are 50% p-character and 50% s-character.

Some general conclusions for determining the correct hybridization for any central atom can be identified.

(**1**) Hybridization is usually applied only to central atoms.

(**2**) One atomic orbital is required for each hybridized orbital; atomic orbitals and the resulting hybridized orbitals are in a one-to-one ratio.

(**3**) Hybridized orbitals are used for lone pairs in central atoms.

(**4**) Hybridized orbitals are used to form sigma bonds; hybridized orbitals are never used to form pi bonds.

(**5**) A central atom specifically requires **one** hybridized orbital for **each** sigma bond and **each** lone pair; this requirement is independent of the number of pi bonds needed in the molecule.

(**6**) The central atom sigma bonds and lone pairs are determinants of electron region count. A central atom specifically requires **one** hybridized orbital for **each** electron region; pi bonds always use unhybridized orbitals (p or d).

(**7**) Conclusion (**6**) states that there must be a one-to-one ratio between the number of hybridized orbitals and the number of electron regions. Hybridized orbitals are constructed by starting with the s-orbital and adding (in order) the proper number of p-orbitals, then (if needed) d-orbitals, to reach the required electron region count.

A new step (**8**) is added to the analysis of each central atom.

Step (**8**): Identify the hybridization of each central atom, determined from the electron region (E.R.) count.

Example: Complete step (**8**) for each of the following central atoms; determine the hybridization of each central atom.

(a) the nitrogen in NH_3; (b) the oxygen in H_2O; (c) the boron in BF_3; (d) the carbon in CO_2.

a) H—N—H (with H below N; lone pair on N)   b) H—O—H (two lone pairs on O)   c) F—B—F (with F below B)   d) O=C=O

a) Step (**8**): Nitrogen has three single bonds and one lone pair for **4 E.R.**
Four hybridized orbitals are required; therefore, four atomic orbitals are required for the mix. Begin with one s-orbital then add three p-orbitals to reach the necessary four atomic orbitals required to produce four hybridized orbitals; the result is **sp^3** hybridization.

b) Step (**8**): Oxygen has two single bonds and two lone pairs for **4 E.R.**
Four hybridized orbitals and, thus, four atomic orbitals are required for the mix. Begin with one s-orbital then add three p-orbitals to reach the necessary four atomic orbitals required to produce four hybridized orbitals; the result is **sp^3** hybridization.

c) Step (**8**): Boron has three single bonds and no lone pairs for **3 E.R.**
Three hybridized orbitals and, thus, three atomic orbitals are required for the mix. Begin with one s-orbital then add two p-orbitals to reach the necessary three atomic orbitals required to produce three hybridized orbitals; the result is **sp^2** hybridization.

d) Step (**8**): Carbon has two double bonds and no lone pairs for **2 E.R.**

Two hybridized orbitals and, thus, two atomic orbitals are required for the mix. Begin with one s-orbital then add one p-orbital to reach the necessary two atomic orbitals required to produce two hybridized orbitals; the result is **sp** hybridization.

Example: Complete step (**8**) for each of the following central atoms; determine the hybridization of each central atom.

a) The phosphorous in PF_5; the sulfur in SF_6; the sulfur in SF_4; the bromine in BF_5.

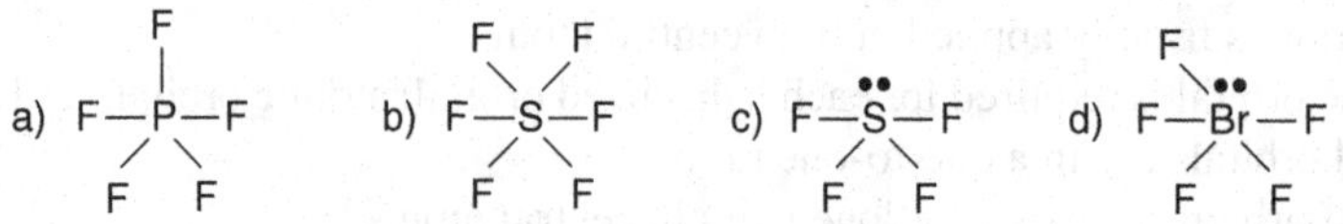

a) PF_5 b) SF_6 c) SF_4 d) BrF_5

a) Step (**8**): Phosphorous has five single bonds and no lone pairs for **5 E.R.**
Five hybridized orbitals and, thus, five atomic orbitals are required for the mix. Begin with one s-orbital, add three p-orbitals, then add one d-orbital to reach the necessary five atomic orbitals required to produce five hybridized orbitals; the result is **sp^3d** hybridization.

b) Step (**8**): Sulfur has six single bonds and no lone pairs for **6 E.R.**
Six hybridized orbitals and, thus, six atomic orbitals are required for the mix. Begin with one s-orbital, add three p-orbitals, then add two d-orbital to reach the necessary six atomic orbitals required to produce six hybridized orbitals; the result is **sp^3d^2** hybridization.

c) Step (**8**): Sulfur has four single bonds and one lone pair for **5 E.R.**
Five hybridized orbitals and, thus, five atomic orbitals are required for the mix. Begin with one s-orbital, add three p-orbitals, then add one d-orbital to reach the necessary five atomic orbitals required to produce five hybridized orbitals; the result is **sp^3d** hybridization.

d) Step (**8**): Bromine has five single bonds and one lone pair for **6 E.R.**
Six hybridized orbitals and, thus, six atomic orbitals are required for the mix. Begin with one s-orbital, add three p-orbitals, then add two d-orbital to reach the necessary six atomic orbitals required to produce six hybridized orbitals; the result is **sp^3d^2** hybridization.

Example: Use the two molecules shown to answer the following questions concerning the number labeled atoms; there are no formal charges on any of the atoms. Lone pairs are not shown in the structural formulas.

H–O–C(H)=C=C(H)–N=O (atoms ① O, ② C, ③ C, ④ N) H–O–C≡C–C(=O)–N(H)–H (atoms ⑤ C, ⑥ C, ⑦ C, ⑧ N)

1. What is the hybridization of the oxygen labeled #1?
2. What is the electron region geometry of the carbon labeled #2?
3. What is the hybridization of the carbon labeled #3?
4. What is the atom geometry of the nitrogen labeled #4?
5. What is the electron region geometry of the carbon labeled #5?
6. What is the hybridization of the carbon labeled #6?
7. What is the hybridization of the carbon labeled #7?
8. What is the atom geometry of the nitrogen labeled #8?

1. Neutral oxygen (①) with two bonds must have two lone pairs; two single bonds plus two lone pairs produces four electron regions. Four electron regions require four hybridized orbitals; the hybridization must be **sp³**.
2. Neutral carbon (②) has two single bonds plus one double bond; this produces three electron regions. Three electron regions require a **trigonal planar** E.R. geometry.
3. Neutral carbon (③) has two double bonds; this produces two electron regions. Two electron regions require two hybridized orbitals; the hybridization must be **sp**.
4. Neutral nitrogen (④) with three bonds must have one lone pair; one single bond plus one double bond plus one lone pair produces three electron regions. Three electron regions, one of which is a lone pair, result in an atom geometry of **bent/angular** with bond angles ≅ 120°.
5. Neutral carbon (⑤) has one single bond plus one triple bond; this produces two electron regions. Two electron regions require a **linear** E.R. geometry.
6. Neutral carbon (⑥) has one single bond plus one triple bond; this produces two electron regions. Two electron regions require two hybridized orbitals; the hybridization must be **sp**.
7. Neutral carbon (⑦) has two single bonds plus one double bond; this produces three electron regions. Three electron regions require three hybridized orbitals; the hybridization must be **sp²**.
8. Neutral nitrogen (⑧) with three bonds must have one lone pair; three single bond plus one lone pair produces four electron regions. Four electron regions, one of which is a lone pair, result in an atom geometry of **trigonal pyramidal**.

Example: Use the two molecules shown to answer the following questions concerning the number labeled atoms; there are no formal charges on any of the atoms. Phosphorous and sulfur are octet bonded; lone pairs are not shown in the structural formulas.

```
      H  H  H     H                    H   ⑦S—Cl
      |  |  |     |                    |    |
Br—P—C═C—O—C═S          F—P═C—S—C—O—C—Br
   ①        ② ③              ④ |  ⑤ |     |⑥
                                 H     H     H
```

1. What is the hybridization of the phosphorous labeled #1?
2. What is the atom geometry of the oxygen labeled #2?
3. What is the hybridization of the carbon labeled #3?
4. What is the electron region geometry of the phosphorous labeled #4?
5. What is the electron region geometry of the sulfur labeled #5?
6. What is the hybridization of the carbon labeled #6?
7. What is the hybridization of the sulfur labeled #7?

1. Neutral phosphorous (①) with three bonds must have one lone pair; three single bonds plus one lone pair produce four electron regions. Four electron regions require four hybridized orbitals; the hybridization must be **sp^3**.
2. Neutral oxygen (②) has two single bonds and two lone pairs; this produces four electron regions. Four electron regions, two of which are a lone pairs, result in an atom geometry of **bent/angular** with bond angles ≅ 109°.
3. Neutral carbon (③) has two single bonds plus one double bond; this produces three electron regions. Three electron regions require three hybridized orbitals; the hybridization must be **sp^2**.
4. Neutral phosphorous (④) with three bonds must have one lone pair; one single bond plus one double bond plus one lone pair produces three electron regions. Three electron regions, one of which is a lone pair, result in an atom geometry of **bent/angular** with bond angles ≅ 120°.
5. Neutral sulfur (⑤) with two bonds must have two lone pairs; two single bonds plus two lone pairs produce four electron regions. Four electron regions require a **tetrahedral** E.R. geometry.
6. Neutral carbon (⑥) has four single bonds; this produces four electron regions. Four electron regions require four hybridized orbitals; the hybridization must be **sp^3**.
7. Neutral sulfur (⑦) has two single bonds plus two lone pairs; this produces four electron regions. Four electron regions require four hybridized orbitals; the hybridization must be **sp^3**.

III SUMMARY TABLES

The following tables summarize some of the important results from Chapters 16 and 17.

Summary Table for Hybridization

# E.R.	Orbitals Used for Hybridization	Orbital Notation	# of Equivalent Hybridized Orbitals
2	s + p	sp	2
3	s + p + p	sp^2	3
4	s + p + p + p	sp^3	4
5	s + p + p + p + d	sp^3d	5
6	s + p + p + p + d + d	sp^3d^2	6

Summary Table for Central Atom Analysis

# Electron Regions (E.R.)	Bonded Atoms/Lone Pairs	Hybridization	E.R. Geometry	Atom Geometry
2	2 Bonded atoms 0 Lone pairs	sp	Linear	Linear
3	3 Bonded atoms 0 Lone pairs	sp^2	Trigonal planar	Trigonal planar
3	2 Bonded atoms 1 Lone pair	sp^2	Trigonal planar	Bent/angular
4	4 Bonded atoms 0 Lone pairs	sp^3	Tetrahedral	Tetrahedral
4	3 Bonded atoms 1 Lone pair	sp^3	Tetrahedral	Trigonal pyramidal

(Continued)

# Electron Regions (E.R.)	Bonded Atoms/Lone Pairs	Hybridization	E.R. Geometry	Atom Geometry
4	2 Bonded atoms 2 Lone pairs	sp^3	Tetrahedral	Bent/angular
5	5 Bonded atoms 0 Lone pairs	sp^3d	Trigonal bipyramidal	Trigonal bipyramidal
5	4 Bonded atoms 1 Lone pair	sp^3d	Trigonal bipyramidal	See-saw
5	3 Bonded atoms 2 Lone pairs	sp^3d	Trigonal bipyramidal	Tee-shape
6	6 Bonded atoms 0 Lone pairs	sp^3d^2	Octahedral	Octahedral
6	5 Bonded atoms 1 Lone pairs	sp^3d^2	Octahedral	Square pyramidal
6	4 Bonded atoms 2 Lone pairs	sp^3d^2	Octahedral	Square planar

The geometry and hybridization of carbon, nitrogen, and oxygen are important for the understanding of organic and biological molecules. A summary of their behavior as **central** atoms is shown in the table. The table refers to normal neutral bonding, with a formal charge = 0. Geometry of atoms with formal charges must be found by further electron region analysis described by bonding exceptions in Chapter 14.

General Summary for the Geometry and Hybridization for C, N, and O

Atom/# E.R.	Bond Types/ Lone Pairs	Sigma Bonds/ Pi Bonds	Hybridization	E.R. Geometry	Atom Geometry
Carbon 4 E.R.	4 Single bonds 0 Lone pairs	4 Sigma bonds 0 Pi bonds (0 lone pairs)	sp^3	Tetrahedral	Tetrahedral
Carbon 3 E.R.	2 Single bonds 1 Double bond 0 Lone pairs	3 Sigma bonds 1 Pi bond (0 lone pairs)	sp^2	Trigonal planar	Trigonal planar
Carbon 2 E.R.	2 Double bonds 0 Lone pairs	2 Sigma bonds 2 Pi bonds (0 lone pairs)	sp	Linear	Linear
Carbon 2 E.R.	1 Single bond 1 Triple bond 0 Lone pairs	2 Sigma bonds 2 Pi bonds (0 lone pairs)	sp	Linear	Linear
Nitrogen 4 E.R.	3 Single bonds 1 Lone pairs	3 Sigma bonds 0 Pi bonds (1 lone pair)	sp^3	Tetrahedral	Trigonal pyramidal
Nitrogen 3 E.R.	1 Single bond 1 Double bond 1 Lone pair	2 Sigma bonds 1 Pi bond (1 lone pair)	sp^2	Trigonal planar	Bent/angular
Oxygen 4 E.R.	2 Single bonds 2 Lone pairs	2 Sigma bonds 0 Pi bonds (2 lone pairs)	sp^3	Tetrahedral	Bent/angular

IV PRACTICE PROBLEMS

1. Use the partially condensed molecule shown to answer the following question and complete the table concerning the number labeled atoms. There are no formal charges on any of the atoms; sulfur is octet bonded.

```
              CH2Cl                         O                  Br
               |                            ||                 |
ClCH2—CH—CH=CH—C≡C—C—O—CH2—S—N—N=CH2
  ①     ②    ③       ④      ⑤  ⑥         ⑦   ⑧   ⑨    ⑩
```

a) How many pi bonds are in the molecule?
b) Complete the table by filling in all information concerning geometry and hybridization.

Specific Atom	Hybridization	Electron Region Geometry	Atom Geometry
Carbon #1			
Carbon #2			
Carbon #3			
Carbon #4			
Carbon #5			
Oxygen #6			
Sulfur #7			
Nitrogen #8			
Nitrogen #9			
Carbon #10			

2. Use the two partially condensed rings shown to answer the following question and complete the table concerning the number labeled atoms. There are no formal charges on any of the atoms; sulfur is octet bonded.

```
           ④
      H   N=O                      Br  Br  O
      |   |                        |   |   ||
    ①C=C                    H—C—C—C—O—CH2CH3
     /     \                     /   \  ⑧      ⑩
H—N②     O③                ⑦S     N—OH
     \     /                      \   /  ⑨
  H—C—C—OH              CH3—O—C=N
     |   |                         ⑤   ⑥
     F   H
```

a) How many pi bonds are in the both molecules combined?
b) Complete the table by filling in all information concerning geometry and hybridization.

Specific Atom	Hybridization	Electron Region Geometry	Atom Geometry
Carbon #1			
Nitrogen #2			
Oxygen #3			
Nitrogen #4			
Oxygen #5			
Carbon #6			
Sulfur #7			
Carbon #8			
Oxygen #9			
Carbon #10			

V ANSWERS TO PRACTICE PROBLEMS

1.

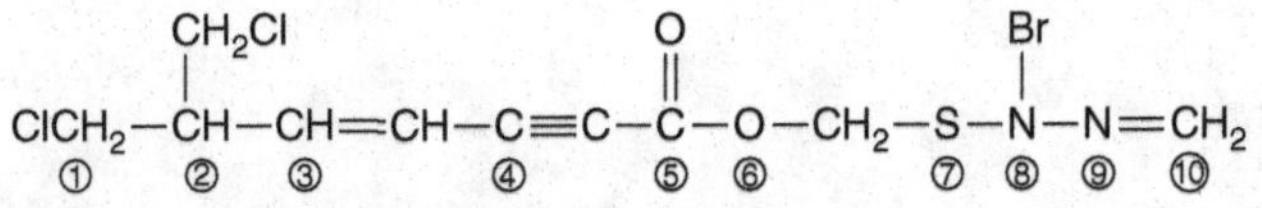

```
          CH2Cl                      O                  Br
           |                         ||                 |
ClCH2—CH—CH=CH—C≡C—C—O—CH2—S—N—N=CH2
  ①      ②   ③        ④      ⑤ ⑥       ⑦ ⑧ ⑨  ⑩
```

a) There are 5 pi bonds are in the molecule.

b)

Specific Atom	Hybridization	Electron Region Geometry	Atom Geometry
Carbon #1	sp^3	Tetrahedral	Tetrahedral
Carbon #2	sp^3	Tetrahedral	Tetrahedral
Carbon #3	sp^2	Trigonal planar	Trigonal planar
Carbon #4	sp	Linear	Linear
Carbon #5	sp^2	Trigonal planar	Trigonal planar
Oxygen #6	sp^3	Tetrahedral	Bent/angular
Sulfur #7	sp^3	Tetrahedral	Bent/angular
Nitrogen #8	sp^3	Tetrahedral	Trigonal pyramidal
Nitrogen #9	sp^2	Trigonal planar	Bent/angular
Carbon #10	sp^2	Trigonal planar	Trigonal planar

2.

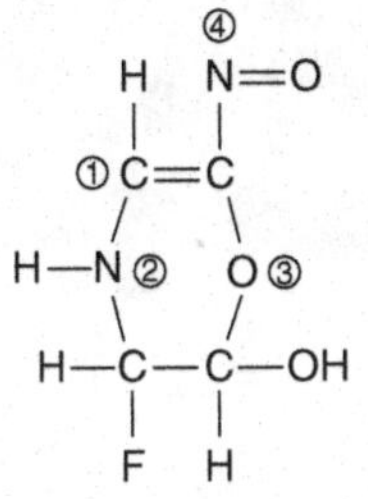

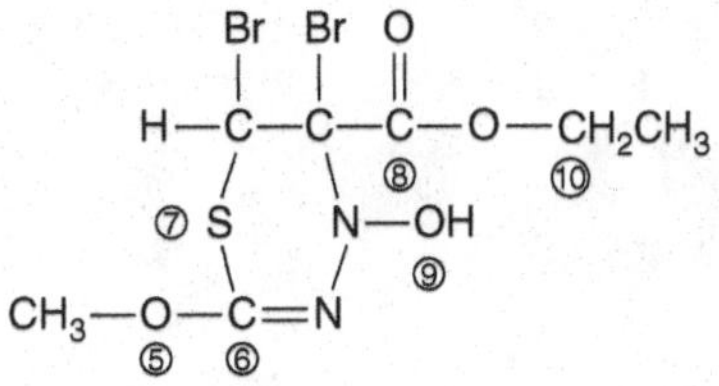

```
          ④
       H  N=O                         Br  Br  O
       |  |                           |   |   ||
     ①C=C                         H—C—C—C—O—CH2CH3
      /    \                         /    \ ⑧    ⑩
 H—N②     O③                    ⑦S      N—OH
      \    /                           \    /  ⑨
   H—C—C—OH               CH3—O—C=N
      |   |                          ⑤  ⑥
      F   H
```

a) There are 4 pi bonds in both molecules combined.

b)

Specific Atom	Hybridization	Electron Region Geometry	Atom Geometry
Carbon #1	sp^2	Trigonal planar	Trigonal planar
Nitrogen #2	sp^3	Tetrahedral	Trigonal pyramidal
Oxygen #3	sp^3	Tetrahedral	Bent/angular
Nitrogen #4	sp^2	Trigonal planar	Bent/angular
Oxygen #5	sp^3	Tetrahedral	Bent/angular
Carbon #6	sp^2	Trigonal planar	Trigonal planar
Sulfur #7	sp^3	Tetrahedral	Bent/angular
Carbon #8	sp^2	Trigonal planar	Trigonal planar
Oxygen #9	sp^3	Tetrahedral	Bent/angular
Carbon #10	sp^3	Tetrahedral	Tetrahedral

18 Concepts of Potential Energy, Enthalpy, and Bond Energy Calculations

I CONCEPT OF ENERGY

Energy is the capacity to do work and/or to produce heat. It can be conveniently divided into **potential energy** and **kinetic energy.**

Potential energy (**PE**) can be defined as energy stored based on the position or arrangement of matter. Potential energy can be released, for example as kinetic energy, in response to a force.

Kinetic energy (abbreviated as **kE**) is the energy of **motion** of matter. Kinetic energy includes both the motion of bulk matter and the production of heat. Heat is thermal energy and is a measure of the kinetic energy of motion of individual atoms, molecules, or ions in bulk matter.

Kinetic energy (**kE**) of matter is proportional to mass and the velocity: $\mathbf{kE = \frac{1}{2}mv^2}$
The metric units of the equation are (kg) × $(m/sec)^2$ = (kg-m^2/sec^2) = one **Joule**
Joule (**J**) is the metric unit of energy; one kiloJoule (**kJ**) = 1000 J; 4.184 Joules = 1 calorie.
Temperature (in °K) is a measure of the average kinetic energy of atoms and molecules.

The **law of conservation of energy** (a form of the **first law of thermodynamics**) states that energy can neither be created or destroyed during any process in the universe: total energy of the universe is constant. Forms of energy are interconvertible; one relationship is the interconversion of kinetic and potential energy.

Potential Energy $\rightleftarrows$ **Kinetic Energy**

An object at a height converts gravitational potential energy to kinetic energy of motion (bulk motion and heat) when it falls. Conversely kinetic energy can be converted back to potential energy if work is performed to lift the object back to its original height.

Chemical potential energy stored in the electromagnetic force of chemical bonds can be converted to kinetic energy during a chemical reaction. For example, an explosion converts chemical bond potential energy to work, heat, (and light). Conversely, kinetic energy can be converted to chemical potential energy when it is consumed to form certain types of chemical bonds.

The two diagrams below were introduced in Chapter 3 as visual examples to picture potential energy conversion based on response to a force. The falling boulder represents response to gravity; the analogous relationship depicts response to the electromagnetic force. For the electromagnetic force, opposite charges farther apart represents the higher potential energy position. As opposite charges approach, the decrease in potential energy can be observed as an increase in kinetic energy such as heat.

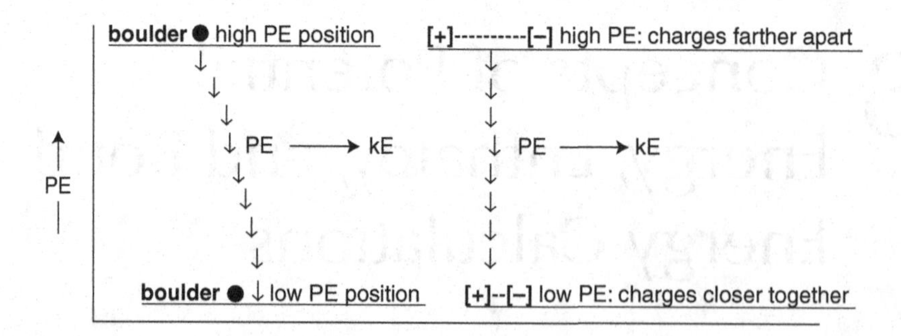

The electromagnetic force is responsible for the energy of chemical bonding. Molecules and compounds store **chemical/electrical potential energy** through the arrangements of electrons and nuclei relative to each other in chemical covalent or ionic bonds. Chemical potential energy released (or absorbed) through the response to the electromagnetic force during a chemical reaction is caused by rearrangement of atoms and electrons to form new combinations and new bonding/positional relationships.

Within the confines of potential and kinetic energy, the law of conservation of energy requires that any change in total potential energy (ΔPE) must be accompanied by an equal and opposite change in the kinetic energy.

II CHEMICAL BOND ENERGETICS

Recall, as discussed in Chapter 12, that the potential energy of an electron in an atom is based on the electromagnetic force of attraction to the positively charged nucleus; a greater strength of attractive force is equated to a lower potential energy. The force of attraction is proportional to the size of the positive and negative charges and inversely proportional to the square of the distance between the charges.

The strength of the attractive force and the stability of an electron is inversely proportional to the distance of the electron from the nucleus: the closer the electron to the nucleus, the greater the attractive force and thus the lower the potential energy.

Electron relationship to the nucleus:

greater attractive force ⟷ **less attractive force**
lower potential energy ⟷ **higher potential energy**
closer to the nucleus ⟷ **farther from nucleus**
more stable ⟷ **less stable**

The potential energy of a (covalent) chemical bond is based on the arrangements of the bonding electrons to the nuclei of the two bonded atoms. Bonding energy is derived through the electromagnetic force of attraction between both nuclei and the bonding pair of electrons.

The sum total of attractive forces for both electrons to both bonding nuclei is always greater for an electron pair in a covalent bond than for the same electrons in separated atoms. This increase in the attractive force is the basis for the strength of the covalent bond.

For a chemical bond to form, the total potential energy of atoms in a covalent bond must always be less than the total potential energy of the same atoms as separated atoms.

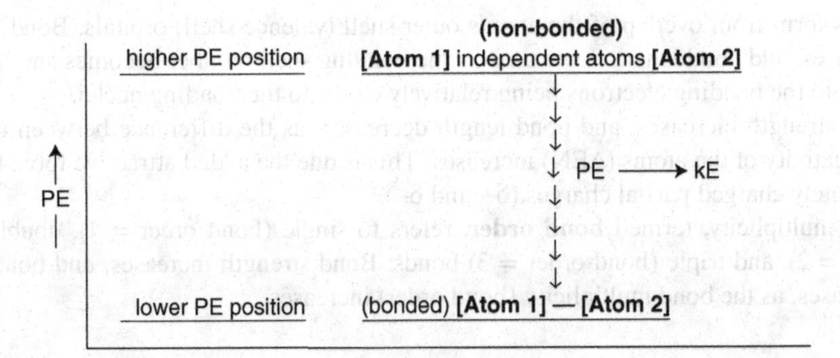

Bonded Atoms: **Separated Atoms:**
greater attractive force ←——less——→ **zero attractive force**
lower total potential energy ←——→ **higher total potential energy**
Atoms more stable ←——→ **Atoms less stable**

Total potential energy (PE) always decreases when a covalent bond is formed from the corresponding separated atoms. The energy released by the decrease in potential energy ($-\Delta$PE) is converted to kinetic energy (+ ΔkE in some form).

Total potential energy (PE) must always increase when a covalent bond is broken to reform the original separated atoms. The required increase in potential energy (+ ΔPE) must be supplied by kinetic energy (in some form). This means that energy must be supplied from an outside source to break a chemical bond (the outside source measures a $-\Delta$kE).

A specific potential energy associated with a specific bond, a **bond PE**, can be identified through the amount of potential energy decrease when the bond is formed, that is, the amount of energy released upon the formation of a specific covalent bond. This ΔPE must always be a negative value.

Conversely, the amount of energy required (i.e., must be supplied) to **break** a specific covalent bond must be equal in value, but opposite in sign, to the amount of energy released when the bond is formed, this is termed the **bond dissociation energy (B.D.E.** or **BDE)**. The value is always **positive** and is equal to the amount of **added** energy (as kE) required to break a specific covalent bond; a greater positive numerical value corresponds to a greater **strength** of the bond.

Bond Strengths and Bond Dissociation Energy (BDE)

Bond strength = **BDE** = energy required to break a bond, measured in Joules.
Bond length = average optimum distance between the bonding nuclei in a covalent bond (measured in meters, nanometers, or picometers).
Bond polarity = measure of separation of charge determined from differences in electronegativity (**ΔEN**).

Each specific covalent bond is characterized by a unique combination of specific atomic orbital overlap, size of the nuclear positive charges, and bonding electron distances. Thus, each covalent bond has a unique bond strength, a unique value of BDE.

The following general trends describe the influence of certain factors on bond strength; exceptions can always occur.

(1) Bonds form from overlap of the atom's outer shell (valence shell) orbitals. Bond strength increases, and bond length decreases, as the bonding shell number becomes smaller. This is due to the bonding electrons being relatively closer to the bonding nuclei.
(2) Bond strength increases, and bond length decreases, as the difference between the electronegativity of the atoms (ΔEN) increases. This is due the added attractive force between oppositely charged partial charges (δ+ and δ–).
(3) Bond multiplicity, termed **bond order**, refers to single (bond order = 1), double (bond order = 2), and triple (bond order = 3) bonds. Bond strength increases, and bond length decreases, as the bond multiplicity (bond order) increases.

forms stronger, shorter bond ⟷ **forms weaker, longer bond**

atom outer shell n = 1 ⟷ **atom outer shell n = 8**
higher atoms ΔEN ⟷ **lower atoms ΔEN**
Bond order = 3 ⟷ **Bond order = 1**
lower P.E. ⟷ **higher P.E.**

Example:		**Bond Length**	**BDE**
Rule (1):	H—H bond: H = row 1 (**n = 1**)	74 pm	436 J
	C—C bond: C = row 2 (**n = 2**)	154 pm	345 J
Rule (2):	H—C bond: **ΔEN = 0.4**	109 pm	415 J
	H—O bond: **ΔEN = 1.4**	96 pm	460 J
Rule (3):	C—C bond: **bond order = 1**	154 pm	345 J
	C=C bond: **bond order = 2**	133 pm	615 J
	C≡C bond: **bond order = 3**	120 pm	827 J

III ENERGY AND CHEMICAL REACTIONS

Chemical reactions occur through rearrangement of atoms in molecules and compounds and thus always involve bond changes. Certain bonds in the reactants are broken in order to from new bonds in the products.

Bonding changes in chemical reactions always involve a corresponding change in potential energy. The total **bond** potential energy of a complete molecule is the sum of bond potential energies of each of the individual bonds comprising the molecule:

bond PE (molecule) = sum of bond PE (each bond)

The total bond potential energy of all reactants or all products in a chemical reaction is equal to the sum of the bond potential energies of all reactant or product molecules:

bond PE (reactants) = sum of bond PE (each reactant molecule)

bond PE (products) = sum of bond PE (each product molecule)

The potential energy **change** in a chemical reaction, ΔPE (reaction), is measured by the **difference** between the potential energy sum of all the reactant molecules and the potential energy sum of all product molecules.

POTENTIAL ENERGY CHANGES AND ENTHALPY

Under a certain restricted set of conditions, the bond potential energy change in a reaction is defined by the term **enthalpy change**, (ΔH). The symbol **ΔH** can be thought of as representing heat, since most of the potential energy change can be measured as heat production. Although not exact (see Chapter 19), a useful working definition of enthalpy can be applied: change in enthalpy (ΔH) can be thought of as the change in the potential energy (ΔPE) of a chemical process as measured by heat transfer to the surroundings. In a limited sense, the "surroundings" essentially represent the matter that contains the chemical process: the solvent in a solution of reactants, the container holding the reactants, and potentially other factors.

Enthalpy, as a measure of potential energy change, has equivalent meaning for signs: ($-\Delta PE$ of reaction) equates to ($-\Delta H$ of reaction). In this case, the potential energy decreases, and energy is released to the surroundings; the reaction is termed **exothermic** ("heat out").

($+\Delta PE$ of reaction) equates to ($+\Delta H$ of reaction). In this case, the potential energy increases, and energy is required to be added from the surroundings; the reaction is termed **endothermic** ("heat in").

Stronger bonds equate to lower PE; molecules with stronger total bonds are more stable and are lower in potential energy than molecules with weaker total bonds. The BDE of a bond is always a positive value; it is equal in value but opposite in sign to the bonding potential energy PE Stronger bonds will have higher values for BDE; weaker bonds will have smaller values for BDE.

Molecules with:

total stronger bonds ⟷ **total weaker bonds**
lower PE ⟷ **higher PE**
larger BDE values ⟷ **smaller BDE values**
more stable ⟷ **less stable**

For calculation purposes, measurement of bond energies is usually expressed in terms of the bond dissociation energy, **BDE**. For calculations, the following equation, applying an enthalpy measurement for potential energy can be used.

ΔH (reaction) = [total bond BDE (reactants)] – [total bond BDE (products)]

A conventional chemical reaction describes the process of converting a specific set of reactants to a specific set of products. The corresponding balanced equation, read from left-to-right, describes what is termed the **forward** reaction.

Any specific forward chemical reaction can **theoretically** have an opposite direction reaction. The products of the original forward reaction act as the reactants in the opposite reaction; the reactants of the original forward reaction represent the products of the opposite reaction. Based on the original left-to-right form of the balanced equation, the original reaction is termed the forward reaction direction; the opposite reaction is termed the **reverse** reaction direction. An important concept (explored further in Chapter 19) is the enthalpy relationship between the forward and reverse reactions:

ΔH (forward reaction) = –ΔH (reverse reaction)

IV POTENTIAL ENERGY DIAGRAMS

Changes in reaction potential energy are very often displayed in the form of a reaction **potential energy diagram**. This is a pictorial description of a reaction based on plotting the **relative** potential energies of **all** reactants and products as a function of the general concept termed reaction progress.

Reaction progress is non-specific and represents some sequential description of how the reaction changes; reaction progress is shown along the horizontal axis.

Relative potential energies can be shown as PE or ΔH alone the vertical axis.

For the simplest form of the diagram, the **complete** set of initial **reactants** and the **complete** set of final **products** are each displayed as a separate energy "platform." Connecting the two in sequence is a smooth curve, which represents the energy changes of all the molecules for the reaction. The diagram follows the convention of balanced chemical equations: the reaction described by the equation, termed the forward reaction, is read from left-to-right; the reactants are on the left side and the products are on the right side.

The two potential energy diagrams below demonstrate **one** bond-breaking reaction and the corresponding bond formation reaction.

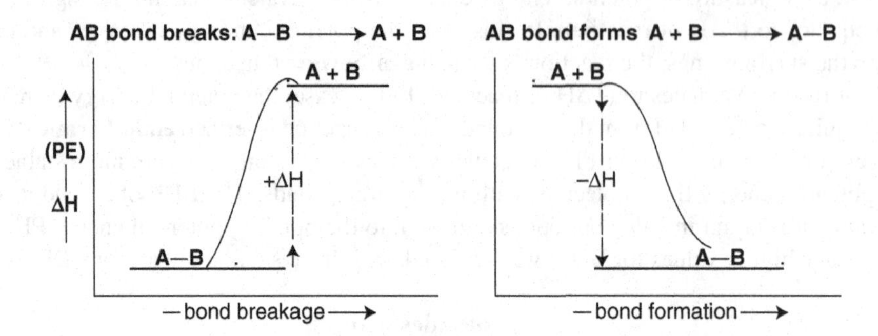

Bond breaking is always endothermic (+ΔH); energy must be added to break a bond. Bond formation is always exothermic (–ΔH); energy is released when a bond forms. The diagrams represent the same atoms A and B. The two potential energy positions for the bonded atoms must be equivalent (same "height"); the two potential energy platforms for the independent atoms must also be equivalent. For the specific A—B bond, the two enthalpy change (ΔH) values must be equal in value but opposite in sign.

The potential energy diagram shown below depicts a **reaction** in which **both** bond breaking (A—B) and bond making (A—C) occur.

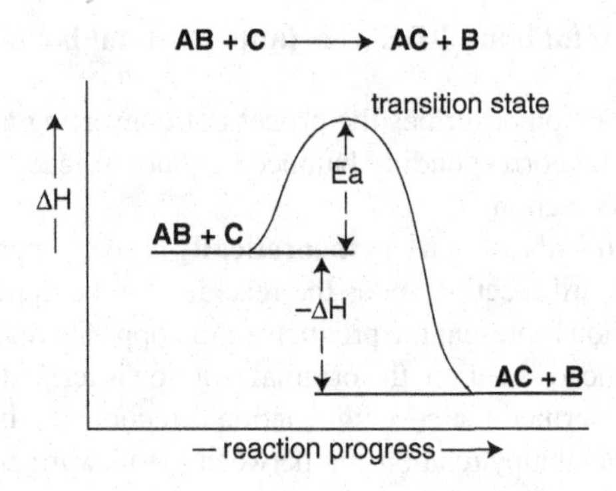

The relative heights of the energy platforms for products and reactants are determined by the relative bond energies of A—B (with C as a free atom) vs. A—C (with B as a free atom).

Activation Energy

The curve showing the change in potential energy as the reaction proceeds from reactants to products does not connect the two energy platforms with a straight line. All complete reactions show an **initial increase** in energy in the reaction progress; this is true regardless of the relative PE positions of reactants or products.

This initial increase in energy for all reactions is termed the **activation barrier** or **activation energy (Ea)**; it represents the energy that must be initially "invested" to allow a chemical reaction to proceed.

Reactions always involve some combination of bond making and/or bond breaking. Other than simple bond formation, all chemical reactions must begin by the initial breaking or partial breaking of one or more chemical bonds in the reactant molecules.

The initial breaking/partial breaking of chemical bonds requires added energy to increase the potential energy of the reactant molecules. The energy required to partially break bonds in the reactants is the basis for the activation barrier (activation energy) required to start any chemical reaction. The activation barrier exists regardless of whether the overall reaction is exothermic or endothermic: the initial investment of energy is required even if a net amount of energy is released in the overall process.

The presence of the activation barrier allows molecules of higher potential energy (less stable) to exist against the downward energy pull, which may exist to convert these reactants to lower potential energy (more stable) products. As an example, the combustion of wood in air has a highly negative value for ΔH; the reaction is highly exothermic. The presence of the activation energy barrier, however, allows wood to be stable unless the activation energy is supplied by heat or a spark, (or termites).

The activation barrier (Ea) for a reaction step where bond making and bond breaking occur simultaneously will depend on the relative degree of **partial** bond breaking and bond making. This bonding description at the highest potential energy is termed the **transition state**. The potential energy at the transition state is determined by the participating bond potential energies: how strong each bond is and to what degree is each broken or formed.

Example: The conversion of hydrogen and oxygen to form water according to the molar balanced equation is found to have an enthalpy change of negative 474 kiloJoules (ΔH = −474 kJ); this can be labeled as the forward reaction. Conversely, it is true that 474 kJ of energy must be added to the reaction to allow water to be converted back to hydrogen gas and oxygen gas from water; the reaction is labeled as the reverse reaction.

$$2H_{2\,(g)} + O_{2\,(g)} \longrightarrow 2\,H_2O_{(l)} \qquad \Delta H\ (\text{forward reaction}) = -474\ \text{kJ/mole-reaction}$$

$$2\,H_2O_{(l)} \longrightarrow 2H_{2\,(g)} + O_{2\,(g)} \qquad \Delta H\ (\text{reverse reaction}) = +474\ \text{kJ/mole-reaction}$$

The statement ΔH = −474 kJ per "mole-reaction" refers to a specific calculated value for enthalpy change based on the molar balanced equation as written: 2 moles of diatomic hydrogen gas reacts with one mole of diatomic oxygen gas to form 2 moles of liquid water.

Potential energy diagrams displaying the forward and reverse reactions interconverting hydrogen, oxygen, and water are shown below:

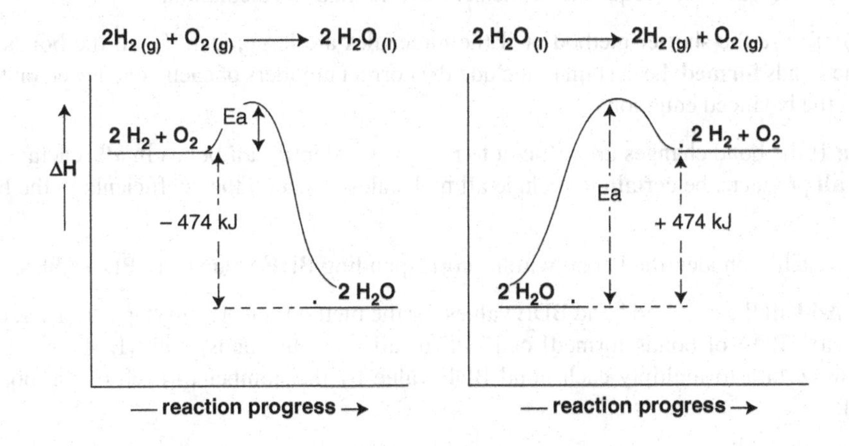

The positions of reactant and product platforms require that the ΔH separation in the forward direction must be exactly equal to the ΔH separation in the reverse direction. This is an additional way to visualize the requirement that:

$$\Delta H \text{ (forward reaction)} = -\Delta H \text{ (reverse reaction)}$$

$$\{-474 \text{ kJ (forward)}\} = -\{-474 \text{ kJ (reverse)}\}$$

The forward reaction releases kinetic energy, such as heat, light, or mechanical work; hydrogen can be used as a fuel with oxygen and can also explode in the presence of oxygen. The increase in potential energy required for the reverse reaction must be supplied by some form of added kinetic energy; the reaction can be accomplished through added electrical energy (electrolysis of water).

A **reversible reaction** is one in which both the forward reaction and the reverse reaction can occur at the same time. Reversible reactions can be read on a single potential energy diagram. In these cases the forward reaction is read from left-to-right and the reverse reaction is read from right-to-left.

V CALCULATING REACTION ENTHALPY VALUES

Reaction enthalpy calculations are performed using bond dissociation energies (**BDE**):

ΔH (reaction) = [total bond BDE (reactants)] – [total bond BDE (products)]

In many cases, the summation of all bonds in all reactants and products for large molecules is impractical. The enthalpy change in a chemical reaction is based only on the actual bonds that change; bonds that are not involved in the reaction will not affect the net enthalpy **change**. Therefore, an alternate form of the equation can be used:

ΔH (reaction) = [Sum of BDE for bonds broken] – [Sum BDE for bonds formed]

Both equational formats for the subtraction sequences ("reactants minus products" and "bonds broken minus bonds formed") are required to produce the correct sign for enthalpy based on the positive values used for BDEs.

Process for Calculation of ΔH (Reaction) for a Balanced Equation

Step (1): Draw Lewis structures of each reactant and product molecule to identify all necessary bonds; use the techniques from Chapter 14.

Step (2): Identify all bonds required for the selected method of calculation.

Step (2a): To use the shorter method or if the molecules are large, identify **all** the bonds **broken** and **all** the bonds **formed**; be certain to include the correct numbers of each bond based on the coefficients in the balanced equation.

Step (2b): If the bond changes are difficult to recognize, identify **all** bonds in **all** reactants and **all** bonds in **all** products; be certain to include all molecules based on the coefficients in the balanced equation.

Step (3): Match each identified bond with its corresponding **BDE** value from BDE tables.

Step (4): Add all the required bond BDE values for the method selected in step (2): [BDE of bonds broken] with [BDE of bonds formed] or [BDE of all reactant bonds] with [BDE of all product bonds]. Be certain to multiply each bond BDE value by the number of each of the bond types identified.

Step (5): Complete the calculation of ΔH (reaction) using the appropriate equation; be careful to keep the subtraction sequence in the correct order.

(2a): ΔH (reaction) = [Sum of BDE for bonds broken) − [Sum BDE for bonds formed]
(2b): ΔH (reaction) = [total bond BDE (reactants)] − [total bond BDE (products)]

The following is a **partial** list of **approximate** bond dissociation energies (BDE); other factors in different molecule types will affect the exact BDE.

Bond	BDE
C—H	**415 kJ**
C—C	**345 kJ**
C=C (double bond)	**615 kJ**
C≡C (triple bond)	**827 kJ**
C—Cl	**325 kJ**
C—I	**215 kJ**
C—O	**360 kJ**
C=O (double bond in most molecules)	**750 kJ**
C=O (for each double bond specifically contained in carbon dioxide)	**805 kJ**
C≡O (triple bond)	**1071 kJ**
O—H	**460 kJ**
O=O (double bond)	**494 kJ**
H—H	**436 kJ**
H—Cl	**428 kJ**
H—I	**295 kJ**
Cl—Cl	**240 kJ**

Example: Calculate the value for **ΔH** (reaction) for the following balanced equation; state whether the reaction is exothermic or endothermic:

$$CH_4 + O_2 \longrightarrow CH_2O + H_2O$$

(1) (Lone pairs are not involved in the calculation and are not shown in Lewis structures.)

$$H{-}\overset{H}{\underset{H}{C}}{-}H + O{=}O \longrightarrow H{-}\overset{O}{\overset{\|}{C}}{-}H + H{-}O{-}H$$

(2) Step (2a) is selected.

(2a) **Bonds broken**: Compare CH_4 with CH_2O: **two** of the reactant **C—H** bonds from CH_4 must be broken; the other two are retained in CH_2O and thus do not change. The reactant O=O double bond must be broken.
Bonds formed: The product CH_2O has a newly formed C=O bond; the product H_2O has **two** newly formed **O—H** bonds.

(3) Use the table provided.
Bonds broken: C—H = **415 kJ**; O=O = **494 kJ**
Bonds formed: C=O = **750 kJ**; O—H = **460 kJ**

(4)

Bonds Broken	Bonds Formed
2 × C—H = 2 × 415 = 830 kJ	1 × C=O = 1 × 750 = 750 kJ
1 × O=O = 1 × 494 = 494 kJ	2 × O—H = 2 × 460 = 920 kJ
1324 kJ	**1670 kJ**

(5) The equation used for calculations based on bonds broken and bonds formed is:

ΔH (reaction) = [Sum of BDE for bonds broken) – [Sum BDE for bonds formed]

ΔH (reaction) = [1324 kJ] – [1670 kJ] = **–346 kJ/mole-reaction = exothermic**

Example: Calculate the value for **ΔH** (reaction) for the same balanced equation using step (2b); demonstrate that the same result is found.

$$\mathbf{CH_4 + O_2 \longrightarrow CH_2O + H_2O}$$

(1) H—C(H)(H)—H + O=O ⟶ H—C(=O)—H + H—O—H

(2b) **Total reactant bonds**: The reactant CH_4 has **four C—H** bonds; the reactant has **one** double bond.

Total product bonds: The product CH_2O has **one** bond and **two C—H** bonds; the product H_2O has **two O—H** bonds.

(3) Bonds broken: C—H = **415 kJ**; O=O = **494 kJ**

Bonds formed: C=O = **750 kJ**; O—H = **460 kJ**

(4)

Total Reactant Bonds	Total Product Bonds
	2 × C—H = 2 × 415 = 830 kJ
4 × C—H = 4 × 415 = 1660 kJ	1 × C=O = 1 × 750 = 750 kJ
1 × O=O = 1 × 494 = 494 kJ	2 × O—H = 2 × 460 = 920 kJ
2154 kJ	**2500 kJ**

(5) The equation used for calculations based on total reactant and product bonds is:

ΔH (reaction) = [total bond BDE (reactants)] – [total bond BDE (products)]

The answers based on either method must be the same:

ΔH (reaction) = [2154 kJ] – [2500 kJ] = **–346 kJ/mole-reaction = exothermic**

Example: Draw a **very general** potential energy diagram based on the enthalpy change (potential energy change) information, which was calculated in the previous examples.

The symbol ΔH can be used for the vertical axis. The reactant platform is placed on the left side of the diagram; the formulas for the reactant compounds are written on the platform: **{CH_4+ O_2}**.; the product platform is labeled **[CH_2O + H_2O]**. The product platform is placed on the right; the formulas for the product compounds are written on the platform: **[CH_2O + H_2O]**. The calculated value of ΔH (reaction) is **negative**, and ΔH decreases and potential energy decreases; thus, the reactant platform must be higher than product platform. An initial increase in potential energy is included to indicate the activation energy, **Ea**. Note that the size of the activation barrier **cannot** be determined by calculated values of enthalpy.

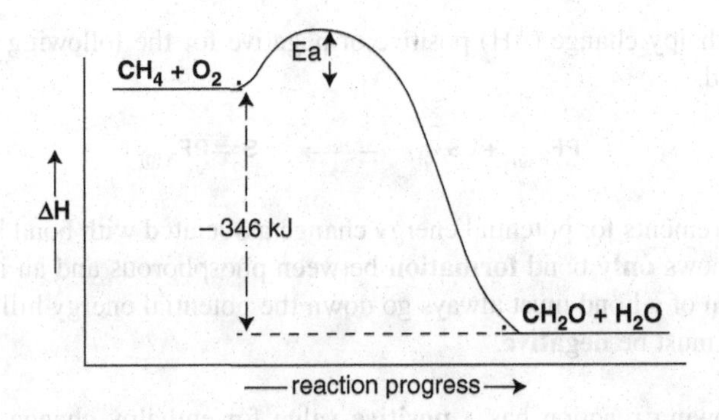

Example: Calculate the value for **ΔH** (reaction) for the following balanced equation; state whether the reaction is exothermic or endothermic:

$$C_2H_2 + H_2 + Cl_2 \longrightarrow CH_3CHCl_2$$

(1) The product molecule is written as a condensed structure (see Chapter 15) and, therefore, has only one possible Lewis structure. However, any valid Lewis structure produced by the techniques in Chapter 14 would result in the same calculation for ΔH.

H—C≡C—H + H—H + Cl—Cl ⟶ H—C(H)(H)—C(Cl)(Cl)—H

(2) Step (2a) is selected for this reaction, however step (2b) is no more difficult.

(2a) Bonds broken: Compare C_2H_2 with CH_3CHCl_2: the carbon-carbon triple bond (**C≡C**) must be broken; the two C—H bonds from C_2H_2 remain in the product molecule. The reactant **H—H** and **Cl—Cl** single bonds must be broken.

Bonds formed: The product molecule CH_3CHCl_2 has a newly formed **C—C** single bond two new **C—H** bonds and two new **C—Cl** bonds.

(3) Use the table provided.

Bonds broken: **C≡C = 827 kJ**; H—H = **436 kJ**; Cl—Cl = **240 kJ**
Bonds formed: C—C = **345 kJ**; C—H = **415 kJ**; C—Cl = **324 kJ**

(4)

Bonds Broken	Bonds Formed
1 × C≡O = 1 × 827 = 830 kJ	1 × C—C = 1 × 345 = 345 kJ
1 × H—H = 1 × 436 = 436 kJ	2 × C—H = 1 × 415 = 830 kJ
1 × Cl—Cl = 1 × 240 = 240 kJ	2 × C—Cl = 1 × 324 = 648 kJ
1503 kJ	**1823 kJ**

(5) The equation used for calculations based on bonds broken and bonds formed is:

ΔH (reaction) = [Sum of BDE for bonds broken) − [Sum BDE for bonds formed]

ΔH (reaction) = [1503 kJ] − [1823 kJ] = **−320 kJ = exothermic**

Example: Is the enthalpy change (ΔH) positive or negative for the following reaction? No values for BDE are provided.

$$PF_{3\,(g)} + S_{(g)} \longrightarrow S{=}PF_{3\,(g)}$$

Recognize the requirements for potential energy change associated with bond breaking and formation. The reaction shows **only** bond **formation** between phosphorous and an independent atom of sulfur. The formation of a bond must always go down the potential energy hill: high PE → low PE This means that ΔH must be **negative**.

Example: The following reaction has a **positive** value for enthalpy change (ΔH). What can be determined about the bonding in the molecules? No values for BDE are provided. Choose from the multiple-choice possibilities.

$$SeF_{3\,(g)} + H_2O_{\,(g)} \longrightarrow H{-}SeF_{3\,(g)} + OH_{\,(g)}$$

a) The Se—H bond is stronger than the O—H bond.
b) The O—H bond is stronger than the Se—H bond.
c) Energy is required to form the Se—H bond.
d) Energy is released when the O—H bond in water breaks.

Considering the requirements for the potential energy change associated with bond breaking and formation, determine which bonds are formed and which bonds are broken in this reaction. The **only** bond **broken** is an O—H bond in water. The **only** bond **formed** is the Se—H bond.

$$\mathbf{\Delta H\ (reaction)} = [\text{BDE O—H}] - [\text{BDE Se—H}] = \mathbf{+value}$$

For this equation to have a positive value, BDE of O—H must be greater than BDE of Se—H. Therefore: (b) The O—H bond is stronger than the Se—H bond. For (c): as described in the previous example, energy cannot be required to form any bond; and for (d): energy cannot be released in the process of bond breaking.

VI READING AN ENERGY DIAGRAM

A potential energy diagram displays information about a reaction, including its reactants, products, a relative size for the activation barrier, and a value for enthalpy change. The value for ΔH thus provides the energy direction: exothermic (down the potential energy hill) or endothermic (up the potential energy hill).

If specific energy values are shown in the diagram, specific energy measurements can be read or calculated. If the reaction is reversible, both the forward and reverse reaction can be read from the same diagram: the forward reaction is read from left-to-right; the reverse reaction is read from right-to-left.

Example: Answer the questions by reading the following energy diagram.

a) What is the **forward** reaction; i.e., the reaction formula equation for the **forward** reaction as shown in the diagram?
b) What is the value for ΔH, the potential energy change for this **forward** reaction?
c) What is the value for activation energy (activation barrier) for this **forward** reaction?

d) What is the **reverse** reaction; i.e., the reaction formula equation for the **reverse** reaction as shown in the diagram?
e) What is the value for ΔH, the potential energy change for this **reverse** reaction?
f) What is the value for activation energy (activation barrier) for this **reverse** reaction?
g) Which reaction is endothermic?
h) Which set of molecules, reactants, or products represent the greater **total** (i.e., all bonds in the molecules) bond strengths?

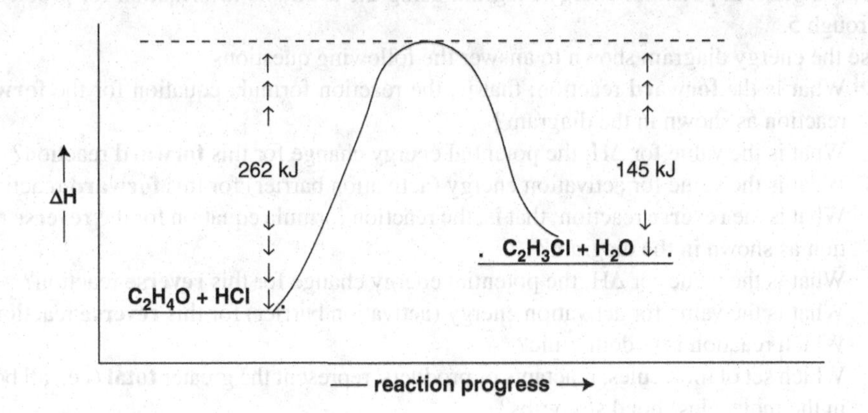

a) **Forward** reaction: read left-to-right = $\mathbf{C_2H_4O + HCl \longrightarrow C_2H_3Cl + H_2O}$
b) ΔH for **forward** reaction: the difference between the two energy platforms can be calculated as 262 kJ − 145 kJ = 117; the potential energy increases: ΔH = **+117 kJ**
c) Activation energy (activation barrier) for **forward** reaction: complete height of the barrier is read from left-to-right: Ea = **262 kJ**
d) **Reverse** reaction: read right-to-left = $\mathbf{C_2H_3Cl + H_2O \longrightarrow C_2H_4O + HCl}$
e) ΔH for **reverse** reaction: energy **difference** must be the same value as for the forward reaction; the potential energy decreases: **ΔH = −117 kJ**

$$\Delta H \text{ (forward reaction)} = -\Delta H \text{ (reverse reaction)}$$

f) Activation energy (activation barrier) for **reverse** reaction: complete height of the barrier is read from right-to-left: Ea = **145 kJ**
Note that Ea does **not** show the same relationship as ΔH.
g) The endothermic reaction is the one that has a **positive** ΔH; this is the **forward reaction**.
h) The **lower** PE represents the stronger bonds. The **reactants** are at the lower PE position; therefore the reactants (C_2H_4O + HCl) have the greater total bond strengths.

VII PRACTICE PROBLEMS

For problems 1–5: For **each** reaction 1–5, complete the following parts:

a) Calculate the value for **ΔH (reaction)** for each of the following balanced equations; use the table of BDEs provided.
b) Is the reaction exothermic or endothermic?
c) Which set of molecules, reactants, or products represents the greater **total** (i.e., all bonds in the molecules) bond strengths.

1. $CH_4 + Cl_2 \longrightarrow CH_3Cl + HCl$
2. $CH_4 + HCl \longrightarrow CH_3Cl + H_2$
3. $C_3H_8 + 5\,O_2 \longrightarrow 3CO_2 + 4\,H_2O$
4. $C_2H_6 \longrightarrow C_2H_4 + H_2$
5. $CH_3I + H_2O \longrightarrow CH_4O + HI$

6. Draw a **general** potential energy diagram using the available information for problems 1 through 5.
7. Use the energy diagram shown to answer the following questions:
   a) What is the **forward** reaction; that is, the reaction formula equation for the **forward** reaction as shown in the diagram?
   b) What is the value for ΔH, the potential energy change for this **forward** reaction?
   c) What is the value for activation energy (activation barrier) for this **forward** reaction?
   d) What is the **reverse** reaction; that is, the reaction formula equation for the **reverse** reaction as shown in the diagram?
   e) What is the value for ΔH, the potential energy change for this **reverse** reaction?
   f) What is the value for activation energy (activation barrier) for this **reverse** reaction?
   g) Which reaction is endothermic?
   h) Which set of molecules, reactants or products, represent the greater **total** (i.e., all bonds in the molecules) bond strengths?

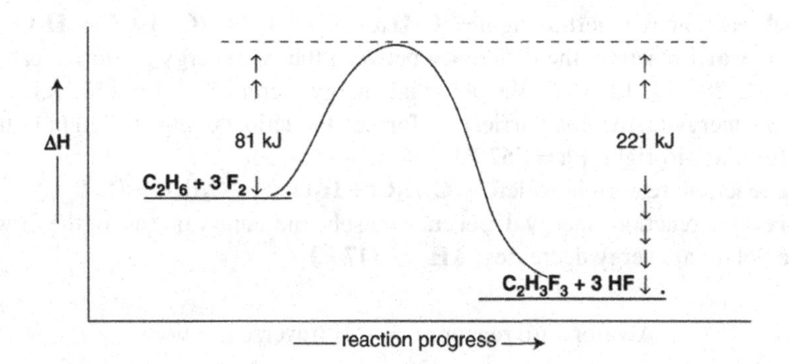

VIII ANSWERS TO PRACTICE PROBLEMS

1–5. Lone pairs are not shown in Lewis structures; step (2a) was used for all problems.

1. a)

```
      H                         H
      |                         |
  H—C—H  +  Cl—Cl  ———>   H—C—H     H—Cl
      |                         |
      H                         Cl
```

Bonds Broken	**Bonds Formed**
1 × C—H = 1 × 415 = 415 kJ	1 × C—Cl = 1 × 325 = 325 kJ
1 × Cl—Cl = 1 × 240 = 240 kJ	1 × H—Cl = 1 × 428 = 428 kJ
655 kJ	**753 kJ**

ΔH (reaction) = [655 kJ] – [753 kJ] = **–98 kJ/mole-reaction**

b) Reaction is **exothermic.** c) **Products** have greater total bond strengths.

2. a)

```
      H                              H
      |                              |
  H—C—H  +  H—Cl  ——————→    H—C—H  +  H—H
      |                              |
      H                              Cl
```

Bonds Broken	Bonds Formed
1 × C—H = 1 × 415 = 415 kJ	1 × C—Cl = 1 × 325 = 325 kJ
1 × H—Cl = 1 × 428 = 428 kJ	1 × H—H = 1 × 436 = 436 kJ
843 kJ	**761 kJ**

ΔH (reaction) = [843 kJ] − [761 kJ] = **+82 kJ/mole-reaction**

b) Reaction is **endothermic**. c) **Reactants** have greater total bond strengths.

3. a)

```
     H   H   H
     |   |   |
  H—C—C—C—H  +  5 O═O  ——————→  3 O═C═O  +  4 H—O—H
     |   |   |
     H   H   H
```

Bonds Broken	Bonds Formed
8 × C—H = 8 × 415 = 3320 kJ	6 × C═O = 6 × 805 = 4830 kJ
2 × C—C = 2 × 345 = 690 kJ	8 × O—H = 8 × 460 = 3680 kJ
5 × O═O = 5 × 494 = 2470 kJ	**8510 kJ**
6480 kJ	

ΔH (reaction) = [6480 kJ] − [8510 kJ] = **−2030 kJ/mole-reaction**

b) Reaction is **exothermic**. c) **Products** have greater total bond strengths.

4. a)

```
     H   H                     H   H
     |   |                     |   |
  H—C—C—H  ——————→   H—C═C—H  +  H—H
     |   |
     H   H
```

Bonds Broken	Bonds Formed
2 × C—H = 2 × 415 = 830 kJ	1 × C═C = 1 × 615 = 615 kJ
1 × C—C = 1 × 345 = 345 kJ	1 × H—H = 1 × 436 = 436 kJ
1175 kJ	**1051 kJ**

ΔH (reaction) = [1175 kJ] − [1051 kJ] = **+124 kJ/mole-reaction**

b) Reaction is **endothermic**. c) **Reactants** have greater total bond strengths.

5. a)

```
      H                          H
      |                          |
  H—C—I  +  H—O—H  ——————→  H—C—O—H  +  H—I
      |                          |
      H                          H
```

Bonds Broken	Bonds Formed
1 × C—I = 1 × 215 = 215 kJ	1 × C—O = 1 × 360 = 360 kJ
1 × O—H = 1 × 460 = 460 kJ	1 × H—I = 1 × 295 = 295 kJ
675 kJ	**655 kJ**

ΔH (reaction) = [675 kJ] − [655 kJ] = **+20 kJ/mole-reaction**

b) Reaction is **endothermic**. c) **Reactants** have greater total bond strengths.

6. Energy diagrams are general and not to scale.

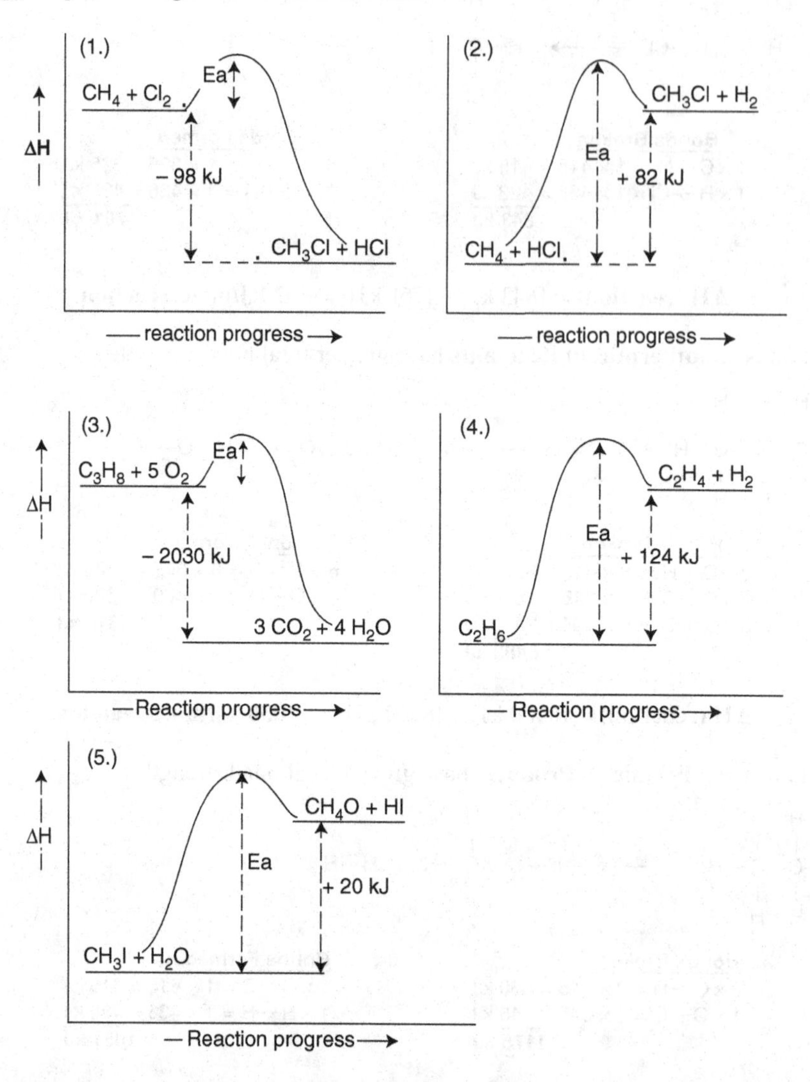

7. a) Forward reaction: read left-to-right = $C_2H_6 + 3\,F_2 \longrightarrow C_2H_3F_3 + 3\,HF$
   b) ΔH for forward reaction: the difference between the two energy platforms can be calculated as 81 kJ – 221 kJ = –140; the potential energy decreases: ΔH = –140 kJ
   c) Activation energy (activation barrier) for forward reaction: the complete height of the barrier is read from left-to-right: Ea = 81 kJ
   d) Reverse reaction: read right-to-left = $C_2H_3F_3 + 3\,HF \longrightarrow C_2H_6 + 3\,F_2$
   e) ΔH for reverse reaction: the energy difference must be the same value as the forward reaction; the potential energy increases: ΔH = +140 kJ
   f) Activation energy (activation barrier) for reverse reaction: the complete height of the barrier is read from right-to-left: Ea = 221 kJ
   g) Endothermic reaction is the one which has a positive ΔH = the reverse reaction.
   h) The lower potential energy represents the stronger bonds. The products are at the lower PE position; therefore, the products ($C_2H_3F_3$ + 3 HF) have the greater total bond strengths.

19 Thermochemistry Calculations
Heat Capacity and Enthalpy

I GENERAL CONCEPTS OF THERMODYNAMICS

Thermochemistry is a subset of the science of thermodynamics. **Thermodynamics** ("thermo" = heat; "dynamics" = motion or work) is the study of the **complete** energy changes **(ΔE)** for a chemical or physical system.

A **system** is any physical or chemical process or partial process that can be studied in **isolation.** A system could be a chemical reaction, a phase change, or a solution formation. There is no restriction to the size of a system as long as **all** energy changes can be isolated and completely identified. For example, a system could be a complete car; a system could be just the engine; a system could be just one cylinder; a system could be just one drop of oil.

The **surroundings** represent everything that can interact with the system. Once a system has been selected, everything else in the **universe** is considered the **surroundings** for measurement of ΔE:

system + surroundings = complete universe

The term **state** refers to a measured parameter describing a system. A **change of state** refers to any change in one or more parameters describing the system:

State (initial) ⟶ State (final)

Example:

For a change in the total energy of a system: E(initial) ⟶ E(final)
the change of state is measured as **(ΔE) = E(final) − E(initial)**.
For a change in the total pressure of a system: P(initial) ⟶ P(final)
the change of state is measured as **(ΔP) = P(final) − P(initial)**.
For a change in the total volume of a system: E (initial) ⟶ E (final)
the change of state is measured as **(ΔV) = V(final) − V(initial)**.

A chemical and physical change can be described by more than one change of state. The combustion of octane in an engine produces a ΔPE, a ΔkE, and a combination of ΔP and ΔV.

$$2\,C_8H_{18\,(l)} + 25\,O_{2\,(g)} \longrightarrow 16\,CO_{2\,(g)} + 18\,H_2O_{\,(g)}$$

A **state function** is one in which the measured values depend only on the state of the system and **not** on how the state occurred. A measured **change** in a **state function** depends only on the initial and final state and **not** on how the state change occurred. ΔE, ΔP, and ΔV are all state functions; changes of these states depend only on the initial and final state and are independent of how the change occurred.

In contrast, a **path function** depends on the **method** of the state change; the pathway, or how the change occurred. While ΔE for a reaction is a state function, the rate of a reaction (how fast or even if a reaction occurs) is a path function. The rate of a reaction directly depends on how the reaction occurs; that is, which bonds are broken and formed and in what sequence of events. (Reaction rates are the subject of kinetics and are found in Chapters 22 and 24.) An example of a state function and a path function is shown with a diagram of the now familiar boulder on a hill. The boulder can fall down the hill along the steep left path or the shallow right path.

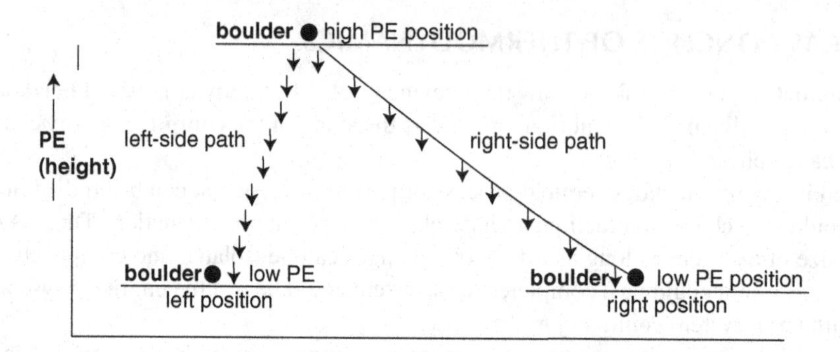

The change in **height** and the change in **potential energy** (based on the gravitational force) are **state** functions; they depend only on the boulder's starting height and the height of the positions where the boulder ends up. The bottom of the hill on the left path and the bottom of the hill on the right path are at the same height. Therefore, ΔPE and ΔHeight are independent of the path by which the boulder falls.

$$\Delta\text{Height and } \Delta\text{PE (left-side path)} = \Delta\text{Height and } \Delta\text{PE (right-side path)}$$

In contrast, the **distance** traveled is a **path** function; the boulder will travel a longer distance if it falls down the right-side shallow path than if it falls down the left-side steep path. ΔDistance (left-side path) is not equal to ΔDistance (right-side path).

The First Law of Thermodynamics

The **first law of thermodynamics** (the law of conservation of energy) states that energy can neither be created or destroyed during any process in the universe. The total energy of the universe is constant. This requires that the **change** in the energy of the universe must be zero:

$$\mathbf{\Delta E\ (universe) = 0}$$

The system plus the surroundings = the (complete) universe

$$\mathbf{thus{:}\ \Delta E\ (system) + \Delta E\ (surroundings) = \Delta E\ (universe)}$$

$$\mathbf{and\ therefore{:}\ \Delta E\ (system) + \Delta E\ (surroundings) = 0}$$

The equation specifying ΔE (system) + ΔE (surroundings) = 0 also implies that:

$$\mathbf{\Delta E\ (system) = -\Delta E\ (surroundings)}$$

The first law of thermodynamics requires that any change in the total energy of any system, ΔE (system), must be accompanied by an equal and opposite change in the total energy of the surroundings, ΔE (surroundings).

An additional component of the first law of thermodynamics states that the total energy change for a system (or surroundings) can be completely described by only two measured energy parameters: heat change (q) and work (w).

$$\Delta E = q + w$$

Energy is defined as the capacity to do work or to transfer heat. **Heat transfer (q)** represents a change in the kinetic energy of the atoms/molecules; this is measured by temperature. **Work (w)** is the conversion of kinetic energy to applied force, or to a change in potential energy. Heat transfer and work are path functions. For any specific change in the total energy, ΔE, the relative values for the components q and w depend on the circumstances of the change; that is, how the energy was transferred.

$$\Delta E \text{ (system)} = q \text{ (system)} + w \text{ (system)}$$

For any specific system, the signs are defined as:

+q = heat is **absorbed** by the system; system gains energy: ΔE (sys) is (+)
−q = heat is **removed** from the system; system loses energy: ΔE (sys) is (−)
+w = work is **done on** the system; system gains energy: ΔE (sys) is (+)
−w = work is **performed by** the system; system loses energy: ΔE (sys) is (−)

Work **done on** a system represents addition of energy to a system. As an example, pushing a boulder up a hill represents the addition of work (the pushing) to increase the potential energy of the rock. Work **performed by** a system represents removal of energy from a system. As an example, falling water in a hydroelectric plant represents a decrease in the potential energy of the water as it performs the work of turning the turbine.

II HEAT CAPACITY AND HEAT TRANSFER BETWEEN SUBSTANCES

The heat (thermal) energy of an object is proportional to the average kinetic energy of all atoms; this includes kE of translation (movement through space), kE of bond vibration (vibration of atoms in a molecule), and kE of molecular rotation. The average kE of matter is measured by temperature (**T**).

Heat transfer between matter (system and surroundings) at different temperatures follows the first law of thermodynamics; temperature change is designated as **ΔT**:

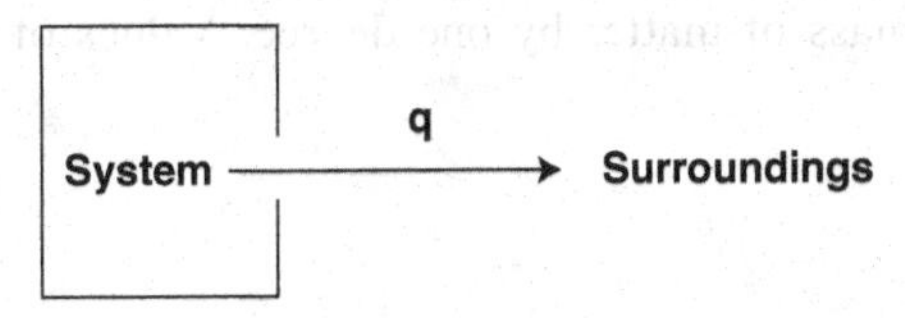

For heat transfer from **system to surroundings**, system $\xrightarrow{q}$ Surroundings:
Heat is removed from the system: $q_{system} = (-)$
Temperature of the system decreases: $\Delta T_{system} = (-)$

Heat is added to the surroundings: $q_{surroundings} = (+)$
Temperature of the surroundings increases: $\Delta T_{surroundings} = (+)$

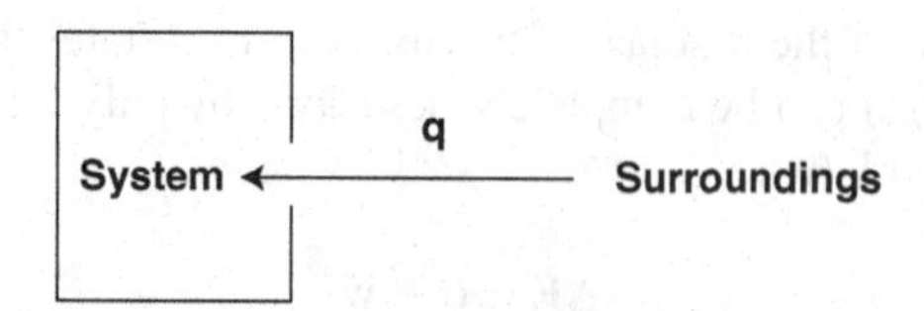

For heat transfer from surroundings to system, system $\overset{q}{\longleftarrow}$ surroundings:
Heat is removed from the surroundings: $q_{surroundings} = (-)$
Temperature of the surroundings decreases: $\Delta T_{surroundings} = (-)$
Heat is added to the system: $q_{system} = (+)$
Temperature of the system increases: $\Delta T_{system} = (+)$

The requirement is that $\mathbf{q_{system} = -q_{surroundings}}$, does not mean that q of the system is always positive, and q of the surroundings is always negative. The equation states that heat changes must be related as equal in value but opposite in sign; the actual sign of (q_{system}) or ($q_{Surroundings}$) depends on the specific direction of the heat transfer.

Simple heat transfer refers to related processes where no change in potential energy is measured. The mathematical relationship for simple heat transfer is:

$$\mathbf{q = (\Delta T)(mass\ in\ grams)(C)}$$

(q) = heat transferred, heat gained (+q) or lost (−q).

(ΔT) = change in temperature in °C or K; (ΔT) is proportional to the amount of heat gained (+q) with (ΔT = +) or lost (−q) with (ΔT = −). In this specific case, either temperature scale is acceptable. The size of the degree unit for °C and K are the same; they differ only in their zero point. Since (ΔT) measures a difference (subtraction) of temperature measurements, the differing zero point of the temperature scales is not carried in the calculation.

C (or C_p) is the heat capacity (or specific heat capacity) of the matter involved in heat transfer. The specific heat capacity (C) is defined as the amount of heat (q) required to raise the mass of exactly 1 gram of a substance by exactly 1 Kelvin (K) or 1 degree centigrade (°C). The mass of the material or object must be measured in grams to match the units of heat capacity.

The specific heat capacity measures the specific response of each compound or element to heat changes; units are Joules/gram-K or Joules/gram-°C. The degree Kelvin and the degree centigrade are the same size; for ΔT calculations, either temperature scale can be used. Each substance has a different response to temperature change as a function of heat changes. The higher the value of the heat capacity (C), the greater the amount of heat required to raise the temperature of a specific mass of matter by one degree. Values of specific heat capacity are found in tables.

Example:

C of Fe = 0.450 J/g-K
C of glass = 0.84 J/g-K
C of Al = 0.91 J/g-K
C of wood = 1.76 J/g-K
C of antifreeze = 2.42 J/g-K
C of H_2O = 4.184 J/g-K

Simple Heat Transfer for One Substance

The general equation **q = (ΔT)(m)(C)** can be solved for the other unknown variables; the resulting equations can be used for both the system and surroundings.

$$\Delta T = \frac{(q)}{(c)(m)} \qquad C = \frac{(q)}{(m)(T)} \qquad \text{mass (g)} = \frac{(q)}{(C)(T)}$$

Example: Calculate the value of q (heat transfer) required to heat a 2.50-kilogram iron pot from 26°C to 100°C (299 K to 373 K); C_{Fe} = 0.450 J/g-K or J/g-°C.

Step (**1**): Select the correct form of the heat transfer equation required to calculate the unknown variable. A general technique is to designate the chosen system or surroundings with a more specific label to help keep track of variables.

$$q_{system} = (T_{System})(m_{system})(C_{system}); \text{ designate as } q_{Fe} = (T_{Fe})(m_{Fe})(C_{Fe})$$

Step (**2**): Identify the values for the known variables.

$$m_{Fe} \text{ must be in grams} = 2.50 \text{ kg} \times \frac{1{,}000 \text{ g}}{1 \text{kg}} = 2{,}500 \text{ grams}$$

$$C_{Fe} = 0.450 \text{ J/g-K}$$

Temperature changes, ΔT, can be measured either in °C or K; the numerical value will always be the same.

$$\Delta T = \textbf{T(final)} - \textbf{T(initial)} = (100°C) - (26°C) = +74°C$$

$$\text{or } (373 \text{ K}) - (299 \text{ K}) = +74 \text{ K}$$

The correct sign for (q) will always be correct if ΔT is measured as T(final) − T(initial); always confirm the sign of (q) qualitatively by a conceptual analysis of the direction of heat transfer.

Step (**3**): Complete the calculation.

$$q_{Fe} = (\Delta T_{Fe})(m_{Fe})(C_{Fe})$$

$$q_{Fe} = (+74 \text{ K})(2500 \text{ g})(0.450 \text{ J/g-K}) = \mathbf{+83{,}250 \text{ J}} \text{ or } \mathbf{+83.3 \text{ kJ}}$$

A qualitative analysis shows that a positive sign for (q) matches the fact that heat must be added to iron to increase the temperature (positive ΔT).

Example: Calculate the value of q (heat transfer) required to heat a 2.50-kilogram glass pot from 26°C to 75°C; C_{glass} = 0.84 J/g-°C.

Step (**1**): $q_{system} = (\Delta T_{system})(m_{system})(C_{system})$; designate as $q_{glass} = (\Delta T_{glass})(m_{glass})(C_{glass})$

Step (**2**): m_{glass} = 2500 grams; C_{glass} = 0.84 J/g-°C

$$\Delta T = \text{T(final)} - \text{T(initial)} = (75°C) - (26°C) = +49°C$$

Step (**3**): $q_{glass} = (\Delta T_{glass})(m_{glass})(C_{glass})$

$$\mathbf{q_{glass}} = (+49°C)(2500\ g)(0.84\ J/g\text{-}°C) = \mathbf{+10{,}290\ J}\ \text{or}\ \mathbf{+103\ kJ}$$

(q) positive matches ΔT positive.

Example: Calculate the value of q (heat transfer) required to heat 2.50 kilograms of water from 26°C to 50°C; $C_{water} = 4.184$ J/g-°C.

Step (**1**): $q_{water} = (\Delta T_{water})(m_{water})(C_{water})$

Step (**2**): $m_{water} = 2.50\ kg \times \frac{1000\ g}{1 kg} = 2500\ grams$

$C_{water} = 4.184$ J/g-°C; ΔT = T(final) – T(initial) = (50°C) – (26°C) = +24°C

Step (**3**): $q_{water} = (\Delta T_{Water})(m_{water})(C_{water})$

$$\mathbf{q_{water}} = (+24°C)(2500\ g)(4.184\ J/g\text{-}°C) = \mathbf{+251{,}040\ J}\ \text{or}\ \mathbf{+251\ kJ}$$

(q) Positive matches ΔT positive.

Note that the input of 251 kJ of heat to the water raises its temperature by only 24°C. An input of one-third that amount of heat raises the equivalent mass of iron by 74°C. This is an example of the large difference in the two substances in response to heat transfer encoded in the value for heat capacity.

Example: Calculate the value of q (heat transfer) for an **engine** cooled by a radiator filled with 4.00 gallons of antifreeze; assume that only the engine and the antifreeze interact. The density of the antifreeze = 1.15 g/mL; C(antifreeze) = 2.42 J/g-°C; 1 gallon = 3.79 liters. T(initial) of the antifreeze = 12.0°C; T(final) of the antifreeze = 99.0°C.

The problem provides no information about the engine. The key is to use the concept of heat transfer between substances: $\mathbf{q_{system} = -q_{surroundings}}$. Information about the engine comes from data describing the radiator and antifreeze.

Step (**1**): $q_{antifreeze} = (\Delta T_{antifreeze})(m_{antifreeze})(C_{antifreeze})$

Step (**2**): $m_{antifreeze} = 4.00\ gal. \times \frac{3.79\ L}{1\ gal.} \times \frac{1000\ mL}{1\ L} \times \frac{1.15\ g}{1\ mL} = 1.74 \times 10^4\ g$

$C_{antifreeze} = 2.42$ J/g-°C; ΔT = T(final) – T(initial) = (99.0°C) – (12.0°C) = +87.0°C

Step (**3**): $q_{antifreeze} = (\Delta T_{antifreeze})(m_{antifreeze})(C_{antifreeze})$

$$\mathbf{q_{antifreeze}} = (+87°C)(1.74 \times 10^4\ g)(2.42\ J/g\text{-}°C) = \mathbf{+3.66 \times 10^6\ J\ or\ +3660\ kJ}$$

Step (**4**): $\mathbf{q_{engine} = -q_{antifreeze};\ q_{engine} = -3660\ kJ}$

Simple Heat Transfer between Two Substances

Heat transfer between two substances usually requires two sets of equations to determine all required known variables; one for the system and one for the surroundings. It generally does not matter which substance is the system and which is the surroundings; labels are used to avoid confusion.

Heat transfer always proceeds in only one direction (a form of the second law of thermodynamics, Chapter 25): heat must always flow from a hotter body to a colder body and never in reverse. Heat transfer calculations are always performed under conditions of thermal equilibrium: both substances reach the same temperature.

Example: A 15.5-gram bar of chromium has a temperature of 100.0°C (373 K). The chromium bar is then added to 55.5 grams of water at an initial temperature of 16.5°C (289.5 K). The final temperature of the two substances is 18.9°C (291.9 K). Calculate the value for the specific heat capacity of chromium without using a table of values. $\mathbf{C_{water}}$ = 4.184 J/g-°C.

Step (1): Select the correct form of the heat transfer equation required to calculate the unknown variable. $C_{Cr} = \dfrac{(q_{Cr})}{(m_{Cr})(\Delta T_{Cr})}$

Step (2): Identify the values for the known variables.

$$\Delta T_{Cr} = T(\text{final}) - T(\text{initial}) = (18.9°C) - (100.0°C) = \mathbf{-81.1}°C \text{ (or } \mathbf{-81.1} \text{ K)}$$

$$m_{Cr} = \mathbf{15.5\ grams}$$

$$\mathbf{q_{Cr}} = \text{not given in the problem.}$$

Analysis for step (**2**) requires an extrapolation based on the relationship for heat transfer between two substances: $\mathbf{q_{system}} = \mathbf{-q_{surroundings}}$ designated as $\mathbf{q_{cr}} = \mathbf{-q_{water}}$. (The choice of Cr or water as system or surroundings is not important; the exact same answer would be derived from: $\mathbf{q_{water}} = \mathbf{-q_{cr}}$).

The value for $\mathbf{q_{cr}}$ (as the system) must come from solving the heat transfer to the surroundings; all required known variables are provided for the surroundings.

Repeat steps (**1**), (**2**), and (**3**) for the surroundings.

step (**1surr**) $q_{water} = (\Delta T_{water})(m_{water})(\mathbf{C_{water}})$

step (**2surr**) C_{water} = 4.184 J/g-°C; m_{water} = 55.5 grams

$$\Delta T_{water} = (18.9°C) - (16.5°C) = \mathbf{+2.4°C} \text{ (temperature increases)}$$

step (**3surr**) $\mathbf{q_{water}}$ = (2.4°C)(55.5 g)(4.184 J/g-°C) = **+557.3 J**

Completion of step (**2**): $\mathbf{q_{Cr}} = -q_{water}$ = −(+557.3 J) = **−557.3 J**

Step (**3**): Complete the calculation for the original system.

$$C_{Cr} = \frac{(q_{Cr})}{(m_{Cr})(\Delta T_{Cr})} = \frac{(-557.3\ J)}{(15.5g)(-81.1°C)} = +\mathbf{0.443\ J/g\text{-}°C} \text{ (C must be a positive value.)}$$

Example: An aluminum bar at an initial temperature of 166.0°C is placed in 325 grams of liquid alcohol at an initial temperature of 22.0°C. The final temperature for both the aluminum bar and the alcohol (at thermal equilibrium) was 92.0°C. Calculate the **mass** of the aluminum bar in **grams.** C (Al) = 0.910 J/g-°C; C (alcohol) = 1.75 J/g-°C.

Step (**1**): $m_{Al} = \dfrac{(q_{Al})}{(C_{Al})(\Delta T_{Al})}$

Step (2): $C_{ai} = 0.910$ J/g-°C

$$\Delta T_{ai} = T(\text{final}) - T(\text{initial}) = (92.0°C) - (166.0°C) = -74.0°C$$

q_{AI} = not given in the problem.

$q_{system} = -q_{surroundigs}$ designated as $q_{AI} = -q_{alcohol}$.

step (1surr) $\mathbf{q_{alcohol} = (\Delta T_{alcohol})(m_{alcohol})(C_{alcohol})}$
step (**2surr**) $C_{alcohol} = 1.75$J/g-°C; $m_{alcohol} = 325$ grams

$$\Delta T_{alcohol} = (92.0°C) - (22.0°C) = \mathbf{+70.0°C}$$

step (**3surr**) $\mathbf{q_{alcohol}}$ = (70.0°C)(325 g)(1.75 J/g-°C) = **+39,813 J**
Completion of step (**2**): $\mathbf{q_{AI}} = -q_{alcohol}$ = –(+39813 J) = **–39,813 J**

Step (**3**): Complete the calculation for the original system.

$$\mathbf{m_{AI}} = \frac{(q_{Al})}{(C_{Al})(\Delta T_{Al})} = \frac{(-39813 \text{ J})}{(0.910 \text{ J/g}-°C)(-74.0°C)} = \mathbf{591\,g}$$

Example: A gold bar has a mass of 17.5 kg. It is placed in a bath of glycol, which contains 3.47 kg of glycol at an initial temperature of 28.7°C. The final temperature for both the gold bar and the glycol (at thermal equilibrium) was 112.0°C. Calculate the initial temperature **(T-initial)** of the gold.

C (Au) = 0.129 J/g-°C; C (glycol) = 2.24 J/g-°C.

To start this example, it must be recognized that if $\mathbf{\Delta T_{Au}}$ can be calculated, T(initial) for gold can be determined from the given final temperature. This points out the required equations for step (**1**).

Step (**1**): $T_{Au} = \dfrac{(q_{Au})}{(C_{Au})(m_{Au})}$; $\mathbf{T_{Au}}$ **(initial)** $=$ $\mathbf{T_{Au}}$ **(final)** – $\mathbf{(\Delta T_{Au})}$

Step (**2**): $C_{Au} = 0.129$ J/g-°C; mass of Au $= 17.5 \text{ kg} \times \dfrac{1000 \text{ g}}{1 \text{kg}} = 17500$ grams

q_{Au} = not given in the problem.

$\mathbf{q_{system} = -q_{surroundings}}$ designated as $\mathbf{q_{au} = -q_{glycol}}$.

Step (**1surr**) $q_{glycol} = (\Delta T_{glycol})(m_{glycol})(C_{glycol})$
step (**2surr**) $C_{glycol} = 2.24$ J/g-°C; $m_{glycol} = 3470$ grams

$$\Delta T_{glycol} = (112.0°C) - (28.7°C) = \mathbf{+83.3°C}$$

step (**3surr**) $\mathbf{q_{glycol}}$ = (83.3°C)(3470 g)(2.24 J/g-°C) = **+6.474 × 10⁵ J**
Completion of step (2): $\mathbf{q_{Au} = -q_{glycol}}$ = –(+6.474 × 10^5 J) = –6.474 × 10^5 J

Step (**3**): Complete the calculation for the original system.

$$\Delta T_{Au} = \frac{(q_{Au})}{(C_{Au})(m_{Au})} = \frac{(-6.474 \times 10^5 \text{J})}{(0.129 \text{ J/g}-°C)(17{,}500 \text{ g})} = \mathbf{-287\,°C}$$

Step (**4**): T_{Au} (initial) $= T_{Au}$ (final) – (ΔT_{Au}) = (112.0°C) – (–287°C) = **399°C**

Example: A 100-gram aluminum bar at 200°C is added to 100 grams of water at 25.0°C; the C of Al = 0.910 J/g-°C, C of water = 4.184 J/g-°C. Calculate the final temperature of the combination after both substances reach the same temperature. Check the answer by comparing q for both water and aluminum.

Calculation of final temperature after combination of two substances at different initial temperatures requires the same general techniques used in the previous examples; however, the exact format of the solution is different. The technique of solving simultaneous equations with two unknowns must be used directly.

Step (**1**): Equations for both $\mathbf{q_{system}}$ and $\mathbf{q_{surroundings}}$ must be set up simultaneously.

The value for ΔT is the unknown; it is convenient to use the original (q) equations rather than solving for the unknown variable in step (**1**). The result is two equations with two unknowns: $\mathbf{\Delta T_{Al}}$ and $\mathbf{\Delta T_{water}}$.

$$\mathbf{q_{Al} = (\Delta T_{Al})(m_{Al})(C_{Al})}$$

$$\mathbf{q_{water} = (\Delta T_{water})\ (m_{water})\ (C_{water})}$$

Step (**2**): Substitute the given values for the known variables:

$$\mathbf{q_{Al}} = (\mathbf{\Delta T_{Al}})(100\text{ g})(0.910\text{ J/g-°C})$$

$$\mathbf{q_{water}} = (\mathbf{\Delta T_{water}})(100\text{ g})(4.184\text{ J/g-°C})$$

Express ΔT_{ai} and ΔT_{water} as a function of the final temperature, **Tf** in °C, which must be the same for both system and surroundings at thermal equilibrium.

$$\Delta T_{ai}\text{ in °C} = (\mathbf{Tf - 200})\ \Delta T_{water}\text{ in °C} = (\mathbf{Tf - 25})$$

Use $\mathbf{q_{system}} = -\mathbf{q_{surroundings}}$ to produce **one** equation. Substitute the expressions for ΔT_{ai} and ΔT_{water} into the equation for $\mathbf{q_{Al}} = -\mathbf{q_{water}}$ to produces **one** unknown (Tf).

$$[(\mathbf{Tf - 200})(100\text{g})(0.910\text{ J/g-°C})] = -[(\mathbf{Tf - 25})(100\text{ g})(4.184\text{ J/g-°C})]$$

Step (**3**): Solve the equation for **Tf**:

$$[(\mathbf{Tf - 200})(91)] = -[(\mathbf{Tf - 25})(418.4)]$$

$$(91\text{ Tf}) - 18{,}200 = (-418.4\text{ Tf}) + 10{,}460$$

$$509.4\text{ Tf} = 28{,}660;\ \mathbf{Tf} = 28{,}660/509.4 = \mathbf{56.3°C}$$

Step (**4**): For this type of problem, check that $q_{system} = -q_{surroundigs}$.

$$\Delta T_{water} = 56.3 - 25.0 = +31.3°C$$

$$\mathbf{q_{water}} = (+31.\ 3°C)(100\text{ g})(4.184\text{ J/g-°C}) = \mathbf{+13{,}096\ J}$$

$$\Delta T_{ai} = 56.0 - 200 = -144°C$$

$$\mathbf{q_{ai}} = (-144°C)(100\text{ g})(0.910\text{ J/g-°C}) = \mathbf{-13{,}104\ J}$$

To three significant figures, $q_{system} = -q_{surroundings}$.

III ENERGY AND ENTHALPY IN CHEMICAL REACTIONS

The total energy change of a reaction system (ΔE) can be defined for specific conditions using the general equation $\Delta E = q + w$. The subscripts for each component indicate the variable that remains constant.

If the reaction occurs at constant volume: $\Delta E = q_v + w_v$
If the reaction occurs at constant pressure: $\Delta E = q_p + w_p$

Work (**w**) in chemical reactions is a function of **expansion,** usually of gases. Expansion is measured as a change in pressure (ΔP) or volume (ΔV) or both: $\Delta P \times \Delta V$ (=ΔPV).

If volume remains constant, ΔV = 0; pressure may increase but no expansion occurs.

Whenever volume is constant, work must be equal to zero:

$$w = \Delta P \times \Delta V = \Delta P \times 0 = 0;\ w_v = 0;$$

$\Delta E = q_v + w_v = q_v + 0$; $\Delta E = q_v$. The total energy change of a system under conditions of constant volume is measured by the heat transfer at constant volume.

Expansion in chemical reactions often occurs under conditions of constant pressure; in this case, the equation for work is: $w = -P\Delta V$. The negative sign indicates that as volume increases (+ΔV = expansion), work is performed by the system (−w); pressure is constant and must always be a positive value.

Whenever pressure is constant, work (w_p) = $-P\Delta V$;

$$\Delta E = q_p + w_p = q_p + (-P\Delta V);\ \Delta E = q_p - P\Delta V$$

Enthalpy, ΔH, was conceptually defined in Chapter 18 as the change in potential energy of a reaction (or process) measured as heat transfer to the surroundings. A specific definition of enthalpy is heat transfer at constant pressure, q_p: $\Delta H \equiv q_p$.

$$\Delta E = q_p - P\Delta V$$

$$\Delta E = \Delta H - P\Delta V;\ \text{or solving for } \Delta H\text{: } \Delta H = \Delta E + P\Delta V$$

Enthalpy is the change in potential energy (ΔPE) of a chemical process measured as heat transfer under conditions of constant pressure; work energy change (expansion or contraction of volume) is not included. Enthalpy, however, is a useful measure of energy change for a wide variety of chemical processes and is often a close approximation of total energy change. For many chemical reactions, such as solubility reactions, reactions in solution, or reactions involving only solids and liquids, volume expansion at constant pressure is very small (**$\Delta V \cong 0$**). In these cases, enthalpy and total energy change are approximately equal: **$\Delta E \cong \Delta H$**.

Volume expansion or contraction can be significant whenever gases are formed or consumed in a reaction; the number of moles of gas then changes from reactants to products. Even in many of these cases, however, the total energy contribution from the work term (−PΔV) can often be small as compared to the enthalpy term (ΔH).

Example:

$$C_2H_8N_{2\,(l)} + 2\,N_2O_{4(g)} \longrightarrow 3\,N_{2\,(g)} + 2\,CO_{2\,(g)} + 4\,H_2O_{(g)}$$

2 moles of gas ⟶ **9** moles of gas

At constant pressure, the work of gas expansion ($w = -P\Delta V$) equals −22 kJ/mole. (Properties of gases and energy are described in Chapter 20.) ΔH for this reaction is −1772 kJ/mole. The total energy change can be calculated as:

$$\mathbf{\Delta E = (\Delta H) + (-P\Delta V) = (-1{,}772\ kJ/mole) + (-22\ kJ/mole) = -1{,}794\ kJ/mole}$$

The difference between ΔE and ΔH is only about 1%.

IV CALORIMETRY

Calorimetry is the experimental method for measurement of heat transfer. **A calorimeter** is an insulated container that isolates both the system and surroundings (the calorimeter plus contents) from the rest of the universe; heat transfer to outside the calorimeter is designed to be approximately zero. **A constant pressure** calorimeter is an open insulated container that allows the pressure to remain constant at atmospheric pressure; in this case, volume can expand. Under these conditions, heat transfer is a measure of ΔH (q_p). **A constant volume** calorimeter is a sealed insulated container that does not allow volume to expand; in this case, pressure may change. Under these conditions, heat transfer is a measure of ΔE (q_v).

The calorimeter **container** is often involved in heat transfer; calculations require a heat capacity for the container. The heat capacity is measured through a calibration experiment; the result is expressed as Joules per degree °C per the **entire container** (J/°C or K); the mass of the container does **not** appear in the heat capacity.

Example: A calorimeter is calibrated by transferring 1.53 kJ of electrical energy to the container. The entire calorimeter increases in temperature from 20.50°C to 21.85°C. Calculate the heat capacity of the complete calorimeter in kJ/°C.

Step (1): $q_{cal} = (\Delta T_{cal})(C_{cal})$; $C_{cal} = \frac{(q_{cal})}{(\Delta T_{cal})}$ Mass of the calorimeter is not required

Step (2): $\Delta T_{cal} = T(\text{final}) - T(\text{initial}) = (21.85°C) - (20.50°C) = +1.35°C$

Step (3): $C_{cal} = \frac{(q_{cal})}{(\Delta T_{cal})} = \frac{1.53\ kJ}{1.35\ °C} = 1.13\ kJ/°C\ (\text{or} = 1130\ J/°C)$

The term calorimeter sometimes refers to both the actual insulated container plus certain contents; for example, a solvent in which the reaction is performed. The calorimeter plus certain contents are classified as the surroundings.

Example: A 505-gram copper tube at 99.9°C is added to a calorimeter containing 59.8 grams of water; the container and water are at 24.8°C. Calculate the **final** temperature of the calorimeter plus water plus copper tube. The water plus the calorimeter are the surroundings. $C_{cu} = 0.387$ J/g-°C, $C_{water} = 4.184$ J/g-°C.; $C_{calorimeter} = 10.0$ J/°C.

Steps **(1)** and **(2):** $\mathbf{q_{Cu}} = (\mathbf{\Delta T_{Cu}})(505\ g)(0.387\ J/g\text{-}°C)$

$\mathbf{q_{water}} = (\mathbf{\Delta T_{water}})(59.8\ g)(4.184\ J/g\text{-}°C)$

$\mathbf{q_{Calorimeter}} = (\mathbf{\Delta T_{Calorimeter}})(10.0\ J/°C)$

ΔT_{Cu} in °C = **(Tf − 99.9)**;

ΔT_{water} in °C = **(Tf − 24.8)**; $\Delta T_{calorimeter}$ in °C = **(Tf − 24.8)**

$q_{cu} = -q_{surroundings}; q_{cu} = -[q_{calorimeter} + q_{Water}]$

Sum the terms for the surroundings for clarity:

$$[(\mathbf{Tf} - \mathbf{24.8})(10.0\ \text{J/°C}) + (\mathbf{Tf} - \mathbf{24.8})(59.8\ \text{g})(4.184\ \text{J/g-°C})]$$
$$= [(\mathbf{Tf} - \mathbf{24.8})(10) + (\mathbf{Tf} - \mathbf{24.8})(250.2)] = [(\mathbf{Tf} - \mathbf{24.8})(260.2)]$$
$$q_{cu} = -[q_{Calorimeter} + q_{water}]$$

Step (**3**): $[(\mathbf{Tf} - \mathbf{99.9})(505\ \text{g})(0.387\ \text{J/g-°C})] = -[(\mathbf{Tf} - \mathbf{24.8})(260.2)]$

$(\mathbf{Tf} - \mathbf{99.9})(195.4) = -[(\mathbf{Tf} - \mathbf{24.8})(260.2)]$

$(195.4\ \text{Tf}) - 19520 = (-260.2\ \text{Tf}) + 6453$

$455.6\ \text{Tf} = 25{,}973;\ \mathbf{Tf} = \dfrac{25973}{455.6} = \mathbf{57.0°C}$

Step (**4**): A check shows that q_{Cu} (−8384 J) $= -q_{surroundigs}(+8378\ \text{J})$

Determining ΔH and ΔE from Calorimetry

Calorimetry experiments are designed to measure values for ΔH/mole-reaction or ΔE/mole-reaction. The term "mole-reaction" refers to an energy value based on the molar coefficients of all reactants and products indicated by the balanced equation.

$$\Delta\text{H/mole-reaction} = \Delta\text{H/per 1 balanced equation; written as: } \frac{(\Delta\mathbf{H})}{\mathbf{1\ bal\ eq}}$$

$$\Delta\text{E/mole-reaction} = \Delta\text{E/per 1 balanced equation; written as: } \frac{(\Delta\mathbf{E})}{\mathbf{1\ bal\ eq}}$$

Heat transfer measurements will always provide a numerical value of energy per mole of the limiting reagent in the reaction. To interconvert these values, an equation can be developed based on the coefficient of the limiting reagent in the balanced equation.

$$\Delta\text{H or }\Delta\text{E/mole-reaction} = \frac{(\Delta\text{H or }\Delta\text{E})}{1\text{ bal eq}} = \frac{(\Delta\mathbf{H}\textbf{ or }\Delta\mathbf{E})}{\textbf{(mole reactant)}} \times \frac{\textbf{moles reactant (coefficient)}}{\textbf{1 bal eq}}$$

General Procedures for Solving Calorimetry Problems

Step (**1**) Since a calorimetry experiment can only measure energy changes per mole of the limiting reagent, the desired result for step (**1**) is ΔH or ΔE/mole-limiting reagent.

$\Delta H = q_p$ (reaction); $\Delta E = q_v$ (reaction); therefore, ΔH or ΔE/mole-limiting reagent is solved as $\mathbf{q_p}$ **or** $\mathbf{q_v}$**/mole-limiting reagent.**

Step (**2**) ΔH or ΔE measures potential energy change of a reaction. A change in **potential** energy is measured as **heat** transfer to the surroundings. Therefore, solving for $\mathbf{q_p}$ or $\mathbf{q_v}$ requires an analysis of the surroundings. Step (**2**) is based on solving:

$$q_{system} = -\ q_{surroundings} \text{ or } q_{reaction} = -\ q_{surroundings}$$

Step (**3**) Complete the calculation of q/mole-limiting reagent and designate as ΔH or ΔE /mole-limiting reagent.

Step (4) ΔH or ΔE/mole-limiting reagent is converted to ΔH or ΔE/mole-reaction by applying the equation relating moles of reactant to moles in the balanced equation.

Example: An open (constant pressure) insulated container calorimeter was used to determine the ΔH of the reaction shown. **All** heat from the reaction was transferred to the solution; no heat was lost to the container.

$$HCl_{(aq)} + NaOH_{(S)} \longrightarrow H_2O_{(l)} + NaCl_{(aq)}$$

5.00 grams of NaOH as the limiting reagent was added to 200 grams of HCl (aqueous solution). T (initial) of the complete solution = 29.5°C; the T (final) of the complete solution after reaction = 38.6°C. Calculate ΔH°/mole-rxn based on the limiting reagent. Assume no heat is lost to the container or outside; all heat of reaction is transferred to the aqueous solution. The total mass of solution equals the original HCl solution plus the NaOH solute; C (complete solution) = 4.20 J/g-°C; MM (NaOH) = 40.01 g/mole. Constant pressure conditions measure q_p and enthalpy.

Step (1): ΔH = q_p (reaction); therefore, ΔH/mole NaOH = **q_p/mole NaOH.**

Step (2): Use $q_{system} = -q_{surroundings}$ to calculate q_p (reaction)

The system = reaction, the surroundings = aqueous solution: $q_{(reaction)} = -q_{(solution)}$.
Complete the calculation of: $\mathbf{q_{solution} = (\Delta T_{solution})(m_{solution})(C_{solution})}$

Step (2): $C_{solution}$ = 4.20 J/g-°C

$$m_{solution} = 200.\text{ g HCl(aq)} + 5.00\text{ grams NaOH} = 205\text{ g solution}$$

$$\Delta T_{(solution)} = T(\text{final}) - T(\text{initial}) = (38.6°C) - (29.5°C) = \mathbf{9.1°C}$$

$$q_{solution} = (\Delta T_{solution})(m_{solution})(C_{solution})$$

$$q_{solution} = (9.1°C)(205\text{ g})(4.20\text{ J/g-°C}) = 7835\text{ J}$$

$$q_{(reaction)} = -q_{(solution)};$$

$$q_{(reaction)} = -(7835\text{ J}) = \mathbf{-7835\text{ J}}$$

$$\text{moles NaOH} = \frac{5.00\text{ g}}{40.01\text{ g/mole}} = 0.125\text{ mole}$$

Step (3): $q_p\text{/mole NaOH} = \dfrac{-7835\text{ J}}{0.124\text{ mole}} = \mathbf{-62680\text{ J/mole or }-62.7\text{ kJ/mole}}$

Step (4): Apply the equation to interconvert ΔH/per mole to ΔH/mole-reaction:

$$\Delta H\text{/mole-reaction} = \frac{(-62.7\text{ kJ})}{(\text{mole NaOH})} \times \frac{1\text{ NaOH}}{1\text{ bal eq}} = \mathbf{-62.7\text{ kJ/mole-rxn}}$$

Example: An open insulated calorimeter was used to determine the ΔH of the reaction shown. All heat from the reaction was transferred to the solution; no heat was lost to the calorimeter (container).

$$MM = 28.06\ g/mole$$
$$3\,C_2H_{4\,(g)} \xrightarrow{\text{solvent}} C_6H_{12\ (solution)}$$

62.0 grams of C_2H_4 was reacted in the calorimeter, which contained 300 grams of solvent. T (initial) of reactants plus solvent = 47.0°C; the T (final) of product plus solvent = 81.0°C; C (solution) = 1.30 J/g-°C. Calculate ΔH°/mole-rxn. Total mass of the solution equals solute plus solvent; note the coefficient of C_2H_4 is 3 in the balanced equation. An open container represents constant pressure conditions.

Step (1): $\Delta H = q_p$ (reaction); therefore, ΔH/mole C_2H_4 = **q_p/mole C_2H_4**

Step (2): Use $q_{system} = -q_{surroundings}$ to calculate q_p (reaction)

The system = reaction, the surroundings = aqueous solution: $q_{(reaction)} = -q_{(solution)}$.
Complete the calculation of: $\mathbf{q_{solution} = (\Delta T_{solution})(m_{solution})(C_{solution})}$

Step (2): $C_{solution} = 1.30$ J/g-°C

$$m_{solution} = 300.\ g\ solvent + 62.0\ grams\ C_2H_4 = 362\ g\ solution$$

$$\Delta T_{(solution)} = T(final) - T(initial) = (81.0°C) - (47.0°C) = 34.0°C$$

$$q_{solution} = (\Delta T_{solution})(m_{solution})(C_{solution})$$

$$q_{solution} = (34.0°C)(362\ g)(1.30\ J/g\text{-}°C) = 16{,}000\ J$$

$$q_{(reaction)} = -q_{(solution)}$$

$$\mathbf{q_{(reaction)}} = -(16{,}000\ J) = \mathbf{-16{,}000\ J};\ moles\ C_2H_4 = \frac{62.0\ g}{28.06\ g/mole} = 2.21\,mole$$

Step (3): $q_p/mole\ C_2H_4 = \dfrac{-16000\ J}{2.21\ mole} = \mathbf{-7241\ J/mole}$ or $-$ **7.24 kJ/mole**

Step (4): Apply the equation to interconvert ΔH/per mole to ΔH/mole-reaction: (Note that the unit "kj/mole = rxn" means kiloJoules per mole per complete reaction)

$$\Delta H/mole\text{-}reaction = \frac{(-7.24\,kJ)}{(mole\ C_2H_4)} \times \frac{3\ C_2H_4}{1\ bal\ eq} = \mathbf{-21.7\,kJ/mole\text{-}rxn}$$

Example: A 3.56-gram sample of sulfur as the limiting reagent is burned in a constant volume calorimeter according to the equation shown; this type of calorimeter will measure q_v and ΔE. The temperature readings were: T(initial) = 25.93°C; T(final) = 33.56°C. All the heat was transferred to the calorimeter; $C_{calorimeter}$ = 4.32 kJ/°C Calculate ΔE/mole-reaction for this reaction. MM of S = 32.1 g/mole

$$S_{(S)} + O_{2\,(g)} \longrightarrow SO_{2\,(g)}$$

Step (1): $\Delta E = q_v$ (reaction); therefore, ΔE/mole S = **q_v/mole S.**

Step (2): Use $q_{system} = -q_{surroundings}$ to calculate q_v (reaction).

The system = reaction, the surroundings = calorimeter: $q_{(reaction)} = -q_{(calorimeter)}$.

$$C_{calorimeter} = 4.32\ kJ/°C$$

$$\Delta T_{(calorimeter)} = T(final) - T(initial) = (33.56°C) - (25.93°C) = 7.63°C$$

$$q_{(calorimeter)} = (\Delta T_{(calorimeter)})(C_{calorimeter}) = (7.36°C)(4.32\ kJ/°C) = \mathbf{+33.0\ kJ}$$

Note that the mass of the calorimeter is already accounted for in $C_{calorimeter}$.

$$q_{(reaction)} = -q_{(calorimeter)};$$

$$\mathbf{q_{(reaction)}} = -(\mathbf{+33.0\ kJ}) = \mathbf{-33.0\ kJ};\ mole\ S = \frac{3.56\ g}{32.1 g/mole} = 0.111\ mole$$

Step (**3**): $q_v / mole\ S = \frac{-33.0\,kJ}{0.111\,mole} = \mathbf{-297\,kJ/mole}$

Step (**4**): $\Delta E/mole\text{-}reaction = \frac{(-297\,kJ)}{(mole\ S)} \times \frac{1\ S}{1\ bal\ eq} = \mathbf{-297\,kJ/mole\text{-}rxn}$

Example: A 1.500-gram sample of benzoic acid as the limiting reagent is burned in a constant volume calorimeter containing 775 grams of water as a solvent. The equation is shown below; this reaction will measure q_v and ΔE. The temperature readings were: T(initial) = 22.50°C; T(final) = 31.69°C. The heat in this reaction was transferred to both the calorimeter (container) plus the water solvent. $C_{calorimeter} = 893$ J/°C;

$C_{water} = 4.184$ J/g-°C. Calculate ΔE/mole-reaction for this reaction.

$MM = 122.13\,g/mole$

$$C_7H_6O_{2(S)} + 7\frac{1}{2}O_{2(g)} \longrightarrow 7\,CO_{2(g)} + 3\,H_2O_{(l)}$$

Step (**1**): $\Delta E = q_v$ (reaction); therefore, ΔE/mole $C_7H_6O_2$ = $\mathbf{q_v/mole\ C_7H_6O_2}$.

Step (**2**): Use $q_{system} = -q_{surroundings}$ to calculate q_v (reaction).

The system = reaction; the surroundings = calorimeter plus the water solvent:

$$q_{(reaction)} = -[q_{(calorimeter)} + q_{(water)}].$$

$$C_{calorimeter} = 893\ J/°C,\ C_{water} = 4.184\ J/g\text{-}°C;\ m_{water} = 775\,g$$

The water and the calorimeter must undergo the same temperature change:

$$\Delta T_{(calorimeter)} = T(final) - T(initial) = (31.69°C) - (22.50°C) = +9.19°C$$

$$\Delta T_{(water)} = T(final) - T(initial) = (31.69°C) - (22.50°C) = +9.19°C$$

$$q_{(calorimeter)} = (\Delta T_{(calorimeter)})(C_{calorimeter}) = (9.19°C)(893\ J/°C) = +8207\ J$$

$$q_{(water)} = (\Delta T_{(water)})(C_{water})(m_{water}) = (9.19°C)(4.184\ J/g\text{-}°C)(775\ g) = +29{,}800\ J$$

$$[\mathbf{q_{(calorimeter)}} + \mathbf{q_{(water)}}] = [(8207\ J) + (29{,}800\ J)] = \mathbf{+38{,}010\ J}$$

$$\mathbf{q_{(reaction)}} = -[\mathbf{q_{(calorimeter)}} + \mathbf{q_{(water)}}]$$

$$\mathbf{q_{(reaction)}} = \mathbf{-[+38{,}010\ J] = -38{,}010\ J\ or\ -38.01\ kJ}$$

$$\text{mole } C_7H_6O_2 = \frac{1.500 \text{ g}}{122.13 \text{ g/mole}} = 0.01228 \text{ mole}$$

Step (3): $q_v/\text{mole } C_7H_6O_2 = \dfrac{-38.01\,\text{kJ}}{0.01228\,\text{mole}} = \mathbf{-3095\,kJ/mole}$

Step (4): $\Delta E/\text{mole-reaction} = \dfrac{(-3095\,\text{kJ})}{(\text{mole } C_7H_6O_2)} \times \dfrac{1C_7H_6O_2}{1 \text{ bal eq}} = \mathbf{-3095\,kJ/\,mole\text{-}rxn}$

V ENTHALPY AND HESS'S LAW

Hess's law states that for any sequence of reactions that add to a specific final reaction, the addition of the enthalpies for this sequence of reactions must equal the enthalpy of the final reaction.

Example: For the general reaction sequence shown, add the reactions and corresponding values for ΔH:

$$A \xrightarrow{\Delta H_{AB}} B \xrightarrow{\Delta H_{BC}} C \xrightarrow{\Delta H_{CD}} D$$

$A \longrightarrow B$ ΔH (reaction) symbolized as ΔH_{ab}
$B \longrightarrow C$ ΔH (reaction) symbolized as ΔH_{bc}
$C \longrightarrow D$ ΔH (reaction) symbolized as ΔH_{cd}

reactions add to: $A \longrightarrow D$ ΔH (reaction) symbolized as ΔH_{ad}

The sequence produces the reaction $A \longrightarrow D$; Hess's law requires that the enthalpy of the net reaction, ΔHAD must be equal to the sum of the enthalpies of the three component reactions: $\mathbf{\Delta H_{AD} = \Delta H_{AB} + \Delta H_{BC} + \Delta H_{CD}}$.

The validity of addition of enthalpy changes (Hess's law) is required based on the nature of enthalpy as a state function. The enthalpy change for the reaction $A \longrightarrow D$ must depend **only** on the initial state (A) and the final state (D), **not** on how the change occurred. Thus, ΔH for a direct change of $A \longrightarrow D$ (as ΔH_{ad}) must be equal to the ΔH produced by the **identical** change through any different path:

$$A \longrightarrow B \longrightarrow C \longrightarrow D \text{ (as } \Delta H_{ab} + \Delta H_{bc} + \Delta H_{cd}).$$

Analogy: A person takes the elevator from floor A to floor D in the diagram below. The height this person traveled is $\mathbf{\Delta Ht}_{AD}$. The height change need not be measured directly down the elevator shaft. Another person could walk up the (inefficient) stairs to determine $\mathbf{\Delta Ht_{AD}}$ (measured as upward steps). Following the steps sequentially from A to B records the $\mathbf{\Delta Ht_{AB}}$. The next sequence B to C records $\mathbf{\Delta Ht_{BC}}$; since the steps lead downward, the sign for height change must be negative. Finally, the series from C to D records $\mathbf{\Delta Ht_{CD}}$.

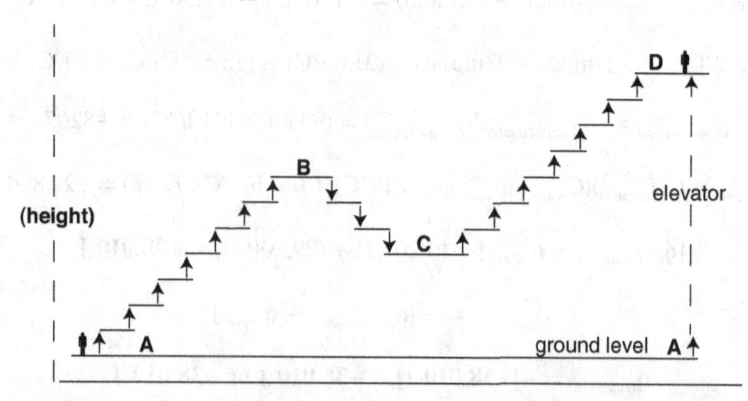

The net height change ΔHt_{Ad} is found by adding all the steps, being certain to use a negative sign for all downward steps. This calculation applies for the same reason that Hess's law is valid: height change is a state function and depends only on the initial and final heights and not on the path taken.

$$Ht_{AD} = Ht_{AB} + Ht_{BC} + Ht_{CD}$$

$$H_{AD} = (+7\,\text{steps}) + (-3\,\text{steps}) + (7\,\text{steps}) = +11\,\text{steps}$$

The process for adding individual reactions to produce a net reaction requires summation of all reactants and all products for all equations to yield the total of all elements and compounds in the final net equation; stoichiometric coefficients are included for addition of identical species. Exactly equivalent species that appear on both sides of the final equation (i.e., they are both reactants and products) are then cancelled on a one-to-one basis based on the stoichiometric coefficients. A more direct process is to perform the cancellation in the set up of component reactions before final summation.

Process for Applying Hess's Law Calculations

The requirement is to adapt the given component equations to produce a combination that will add up to the desired equation:

Step (**1**): Write the direction of each component equation to correctly place reactants and products to match the desired equation. The selection of reaction direction also depends on the reactants or products to be cancelled in the final equation. Reverse the component equation as necessary.

Step (**2**): The total equation may have a different stoichiometric coefficient for a specific reactant or product. If required, multiply a component equation through by a correct multiplier to produce the correct coefficient.

Step (**3**): When required, adjust the ΔH for each given equation to match the altered equation from step (**1**) and step (**2**). If the forward equation is used without change, no change in the corresponding ΔH is necessary. If the reverse equation is used, change the sign of the ΔH as required by: ΔH (forward) = −ΔH (reverse).

If applicable, multiply any specific value of ΔH by the corresponding multiplier used to adapt the equation from step (**2**).

Step (**4**): Add the equation sequence and the corresponding values for ΔH:

Directly add the sequence of component equations to produce the desired net equation. Then directly add all corresponding adjusted values of ΔH based on the adjusted component equations in the sequence, being certain to include the correct sign value. If the equation sequence sums to the correct final total equation, the addition of all equation ΔH must add to the correct ΔH of the final equation.

Example: Use the component equations #1 and #2 to determine ΔH for the following reaction: $\mathbf{Ca_{(S)} + \frac{1}{2}O_{2(g)} + CO_{2(g)} \longrightarrow CaCO_{3(s)}}$

The two component equations with provided ΔH are:

Equation #1: $CaO_{(s)} \longrightarrow Ca_{(s)} + \frac{1}{2}O_{2(g)}$ $\quad \Delta H = +\ 635.1$ kJ/mole-rxn

Equation #2: $CaCO_{3(s)} \longrightarrow CaO_{(s)} + CO_{2(g)}$ $\quad \Delta H = +\ 178.3$ kJ/mole-rxn

Step (1): For the component equation #1, $CaO_{(s)}$ must be eventually cancelled out and $Ca_{(s)} + \frac{1}{2}O_{2(g)}$ are reactants in the final equation. Since they are products in the given component equation, reverse the equation: $\mathbf{Ca_{(s)} + \frac{1}{2}O_{2(g)} \longrightarrow CaO_{(s)}}$ For the component equation #2, $CaCO_{3(s)}$ is a product in the final equation but a reactant in the given component equation. Reverse the equation: $\mathbf{CaO_{(s)} + CO_{2(g)} \longrightarrow CaCO_{3(s)}}$

Step (2): All reactants and products have correct numbers based on equation coefficients; no multiplications are required.

Step (3): Use ΔH (forward) = –ΔH (reverse) for the reverse of component equations #1 and #2; these values are shown below. $CaO_{(s)}$ is cancelled out.

step (3)

$$\mathbf{Ca}_{(s)} + 1/2O_{2(g)} \longrightarrow CaO_{(s)} \quad H = -635.1\,\text{kJ/mole-rxn}$$

$$CaO_{(s)} + \mathbf{CO_{2(g)}} \longrightarrow \mathbf{CaCO_{3(s)}} \quad \Delta H = -178.3\ \text{kJ/mole-rxn}$$

Step (4) Add: $\mathbf{Ca_{(s)} + 1/2O_{2(g)} + CO_{2(g)} \longrightarrow CaCO_{3(s)} \quad \Delta H = -813.4\ kJ/mole\text{-}rxn}$

Example: Use the component equations #1 and #2 to determine ΔH for the following reaction: $\mathbf{2\,NOCl_{(g)} \longrightarrow N_{2(g)} + O_{2(g)} + Cl_{2(g)}}$

The two component equations with provided ΔH are:

Equation #1: $NO_{(g)} \longrightarrow 1/2N_{2(g)} + 1/2O_{2(g)} \quad \Delta H = -90.3\ \text{kJ/mole-rxn}$

Equation #2: $NOCl_{(g)} \longrightarrow NO_{(g)} + \frac{1}{2}Cl_{2(g)} \quad \Delta H = +38.6\ \text{kJ/mole-rxn}$

Step (1): For component equation #1, both N_2 and O_2 are products in the component reaction and products in the final equation; no reversal is required. For component equation #2, Cl_2 is a product in the component reaction and a product in the final equation. In addition, NO must be on the product side to cancel out the NO on the reactant side of equation #1; no reversal is required.

Step (2): The final equation has coefficients exactly twice that of the component equations for all final products and reactants. Multiply each component equation by 2.

Step (3): Multiply the ΔH for each component reaction by 2; the multiplied equations and the corresponding ΔH are shown below.

step (3)

$$2\,NO_{(g)} \longrightarrow N_{2(g)} + O_{2(g)} \quad \Delta H = -180.6\ \text{kJ/mole-rxn}$$

$$\mathbf{2\,NOCl_{(g)}} \longrightarrow 2\,NO_{(g)} + Cl_{2(g)} \quad \Delta H = +77.2\ \text{kJ/mole-rxn}$$

Step (4): Add: $2\,NOCl_{(g)} \longrightarrow N_{2(g)} + O_{2(g)} + Cl_{2(g)} \quad \Delta H = -103.4\ \text{kJ/mole-rxn}$

VI CALCULATION OF ΔH (REACTION) FROM ENTHALPIES OF FORMATION

The ΔH for a specific reaction can be calculated by direct experimental measurement of heat transfer (calorimetry), calculation from bond energies, or calculation from a correct sequence of additive equations (using Hess's law). General applicability of Hess's law to AH calculation involves the use of specific equations termed **formation equations;** the corresponding enthalpy values for these equations are termed **enthalpies of formation** or **heats of formation**.

The **standard** enthalpy (heat) of formation for any species (compound or element) is equal to the enthalpy of the reaction that describes the formation of this species from the requisite elements, as measured in their most stable states under standard conditions. The symbol is **ΔH°f;** the letter **(f)** indicates formation and the (°) superscript indicates standard conditions. Standard conditions of temperature and pressure (STP) are defined as 25°C (298 K) and one atmosphere of pressure (approximately equal to the average air pressure at sea level, this is defined in Chapter 20).

Elements in their standard states refers to the elemental bonding arrangement and correct phase (gas, liquid, or solid) for each element at 298 K and 1 atmosphere. Standard formation equations are always written for the formation of one mole of a compound or element: termed **molar** enthalpies (heats) of formation. Many compounds have a AH°f for more than one phase; these different values are reflected in the ΔH°f data tables. Half-coefficients are used for diatomic elements.

Example: The formation equation for $CH_{4(g)}$ is $C_{(s)} + 2\,H_{2(g)} \longrightarrow CH_{4(g)}$ Carbon and hydrogen are the elements that form CH_4; carbon is a solid under standard conditions; hydrogen is a gas and a diatomic molecule under standard conditions. The ΔH for this specific reaction = –74.8 kJ/mole. Therefore, the ΔH°f for $CH_{4\,(g)}$ is specifically –74.8 kJ/mole.

Example: The formation equation for $CO_{(g)}$ is $C_{(s)} + 1/2O_{2(g)} \longrightarrow CO_{(g)}$ Carbon and oxygen are the elements that form CO; carbon is a solid under standard conditions; oxygen is a gas and a diatomic molecule (½ coefficient is used) under standard conditions. The ΔH for this specific reaction = –110.5 kJ/mole. Therefore, the ΔH°f for $CO_{(g)}$ is specifically –110.5 kJ/mole.

Example: The formation equation for $CO_{2(g)}$ is $C_{(s)} + O_{2(g)} \longrightarrow CO_{2(g)}$.

The ΔH for this specific reaction = –393.5 kJ/mole. Therefore, the ΔH°f for $CO_{2(g)}$ is specifically –393.5 kJ/mole.

Example: The formation equation for $H_2O_{(g)}$ is $H_{2(g)} + \frac{1}{2}O_{2(g)} \longrightarrow H_2O_{(g)}$.

Hydrogen and oxygen are diatomic gases under standard conditions; oxygen has a (½) coefficient. The ΔH for this specific reaction = –241.8 kJ/mole. Therefore, the ΔH°f for $H_2O_{(g)}$ is specifically –241.8 kJ/mole. Note that the ΔH°f for $H_2O_{(l)}$ (liquid) is –285.8 kJ/mole reflecting the different potential energy of another phase.

Example: The "formation" equation for $O_{2(g)}$ is $O_{2(g)} \longrightarrow O_{2(g)}$

ΔH of this reaction is zero, since it represents no change. ΔH°f of $O_{2(g)}$ = O

By definition, elements already in their standard states will not have a formation equation that depicts any reaction change: ΔH°f for all elements in their standard states is defined as zero; ΔH°f = 0. Note that ΔH°f for an element **not** in its standard state must have a value.

Example: The formation equation for $O_{(g)}$ is $\frac{1}{2}O_{2(g)} \longrightarrow O_{(g)}$

Oxygen exists as a diatomic molecule in its standard state; the value for oxygen as a single atom must reflect the enthalpy required to break the O==O double bond. The ΔH of this reaction is +249 kJ/mole. Therefore, ΔH°f of $O_{(g)}$ = 249 kJ/mole.

Example: Determine which of the following substances has a ΔH°f = 0.

a) $O_{3(g)}$ b) $Hg_{(s)}$ c) $N_{2(g)}$ d) $F_{(g)}$

a) $O_{3(g)}$ is not the standard form of the element oxygen; ΔH°f of $O_{3(g)}$ = 142.7 kJ/mole.
b) $Hg_{(s)}$ is not the standard phase of Hg; Hg is a liquid under standard conditions; the ΔH°f of Hg (solid) is not zero.
c) $N_{2(g)}$ is the standard state for the element nitrogen; ΔH°f of $N_{2(g)}$ = 0.
d) $F_{(g)}$ is not the standard state for element fluorine. Fluorine is a diatomic molecule; the ΔH°f would reflect the bond energy of the F—F bond.

Hess's law can be used to calculate the **standard** enthalpy (ΔH°) for any reaction for which the ΔH°f for all reactants and products are known. All complete reactions can be conceptually broken down into two sets of related formation equations:

(1) Conceptually break all reactants down into their elements under standard conditions; this is the **reverse** of the formation reaction. The ΔH for these reactions are the reverse of the ΔH°f; each reaction is (–ΔH°f).
(2) Conceptually form all products from the resulting reactant elements in their standard states; this is the standard formation equation. The ΔH for these reactions is the ΔH°f for each reaction. Since the complete reaction equation must be balanced, all atom rearrangements to form products from reactants must be accounted for with correct coefficients.

Example: Calculate the value for ΔH° for the reaction by displaying steps **(1)** and **(2)**.

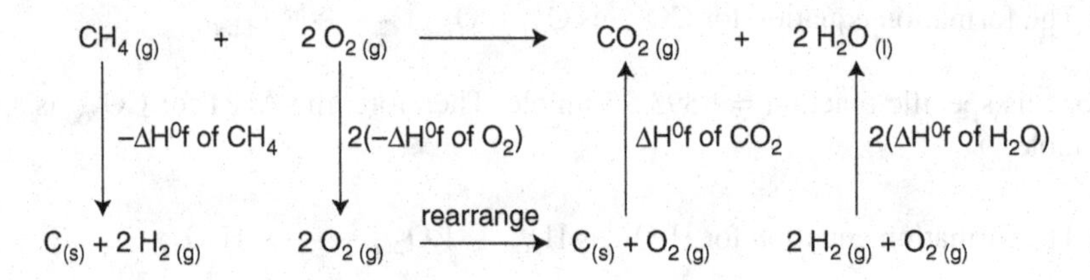

The ΔH° of this reaction is determined through Hess's law.

$$\Delta H°(\text{reaction}) = \Delta H°f \text{ of } CO_{2(g)} + 2(\Delta H°f \text{ of } H_2O_{(l)}) + (-\Delta H°f \text{ of } CH_{4(g)}) + 2(-\Delta H°f \text{ of } O_{2(g)})$$

The values for ΔH°f can be found in thermodynamic data tables:

$$\Delta H°f \text{ of } CO_{2(g)} = -393.5 \text{ kJ/mole } \Delta H°f \text{ of } H_2O_{(l)} = -285.8 \text{ kJ/mole}$$

$$\Delta H°f \text{ of } CH_{4\,(g)} = -74.8 \text{ kJ/mole } \Delta H°f \text{ of } O_{2(g)} = 0 \text{ kJ/mole}$$

$$\Delta H° \text{ (reaction)} = (-393.5 \text{ kJ}) + 2(-285.8 \text{ kJ}) + \{-(-74.8 \text{ kJ})\} + 2(0 \text{ kJ})$$

$$\Delta H° \text{ (reaction)} = -890.3 \text{ kJ/mole-reaction}$$

Example: Calculate the value for $\Delta H°$ for the reaction by displaying steps (1) and (2).

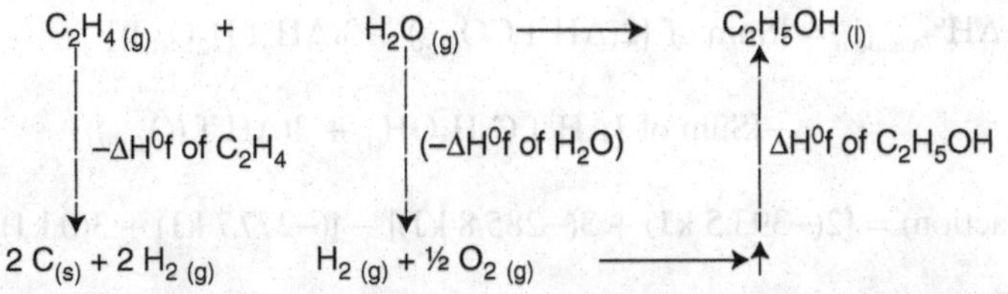

$$C_2H_{4\,(g)} + H_2O_{(g)} \longrightarrow C_2H_5OH_{(l)}$$

$-\Delta H^0f$ of C_2H_4 $(-\Delta H^0f$ of $H_2O)$ ΔH^0f of C_2H_5OH

$$2\,C_{(s)} + 2\,H_{2\,(g)} \qquad H_{2\,(g)} + \tfrac{1}{2}\,O_{2\,(g)} \longrightarrow$$

$\Delta H^0(\text{reaction}) = (\Delta H^0f \text{ of } C_2H_5OH_{(l)}) + (-\Delta H^0f \text{ of } H_2O_{(g)}) + (-\Delta H^0f \text{ of } C_2H_{4\,(g)})$

ΔH^0f of $C_2H_5OH_{(l)} = -277.7$ kJ/mole ΔH^0f of $H_2O_{(g)} = -241.8$ kJ/mole
ΔH^0f of $C_2H_{4\,(g)} = +52.3$ kJ/mole

$\Delta H^0(\text{reaction}) = (-277.7\text{ kJ}) + \{-(-241.8\text{ kJ})\} + \{-(+52.3\text{ kJ})\}$
$\Delta H^0_{(\text{reaction})} = -88.2$ kJ/mole-reaction

A general equation can be written that summarizes the process displayed in the previous examples. Calculation of $\Delta H°$ for the complete reaction involves summing the enthalpies for formation of all products using $\Delta H°f$ directly. Summation of the enthalpies for breakdown of the reactants requires the reverse of $\Delta H°f$. The reactant summation is placed in the final equation with a negative sign and, thus, is a subtraction.

$\Delta H°$ (reaction) = [Sum of $\Delta H°f$ for each mole of **products**]

− [Sum of $\Delta H°f$ for each mole of **reactants**]

The summation of each mole indicates that the value of each $\Delta H°f$ for each product or reactant must be multiplied by the coefficient in the balanced equation. This is symbolized by the following version of the same equation: i = each possible reactant or product in the summation; $\mathbf{n_i}$ = the stoichiometric coefficient of each reactant or product.

$\Delta H°$ (reaction) = Sum [n_i ($\Delta H°f_i$) **products**] − Sum [n_i ($\Delta H°f_i$) **reactants**]

	ΔH°f kJ/mole		ΔH°f kJ/mole
$(C_2H_{2(g)})$	+266.7 kJ	$(H_2O_{(g)})$	−241.8 kJ
$(C_2H_{6(g)})$	−84.7 kJ	$(CO_{(g)})$	−110.5 kJ
$(H_{2(g)})$	0 kJ	$(C_6H_{6(l)})$	+49.0 kJ
$(C_2H_5OH_{(l)})$	−277.7	$(Fe_{(S)})$	0 kJ
$(O_{2(g)})$	0 kJ	$(Fe_3O_{4(S)})$	−1118.4 kJ
$(CO_{2(g)})$	−393.5 kJ	$(H_2O_{(l)})$	−285.8 kJ

Example: Use the table shown to calculate $\Delta H°$ for the reaction using the $\Delta H°f$ equation format.

$$C_2H_5OH_{(l)} + 3\,O_{2(g)} \longrightarrow 2\,CO_{2(g)} + 3\,H_2O_{(l)}$$

Step (**1**): From the tables, identify all required values for $\Delta H°f$; be certain to read the correct phase of the compound or element.

Reactants: $\Delta H°f\,(C_2H_5OH_{(l)}) = -277.7$ **kJ**; $\Delta H°f\,(O_{2(g)}) = 0$ **kJ**

Products: $\Delta H°f(CO_{2(g)}) = -393.5$ **kJ**; $\Delta H°f(H_2O_{(l)}) = -285.8$ **kJ**

Step (**2**): Complete the calculation using the equation given:

$$\Delta H°_{(reaction)} = [\text{Sum of } \{2(\Delta H°f\ CO_{2(g)}) + 3(\Delta H°f\ H_2O_{(l)})\}]$$

$$-[\text{Sum of } \{\Delta H°f\ C_2H_5OH_{(l)} + 3(\Delta H°f\ (O_{2(g)})\}]$$

$$\Delta H°(\text{reaction}) = [2(-393.5\ kJ) + 3(-285.8\ kJ)] - [(-277.7\ kJ) + 3(0\ kJ)]$$

$$\mathbf{\Delta H°(reaction) = -1366.7\ kJ}$$

Example: Use the table shown to calculate ΔH° for the reaction using the ΔH°f equation format.

$$C_2H_{2\,(g)} + 2\ H_{2\,(g)} \longrightarrow C_2H_{6\,(g)}$$

Step (**1**): From the tables, identify all required values for ΔH°f

Reactants: $\Delta H°f\ (C_2H_{2(g)}) = +227\ kJ$; $\Delta H°f\ (H_{2(g)}) = 0\ kJ$

Products: $\Delta H°f\ (C_2H_{6(g)}) = -85\ kJ$

Step (**2**): Complete the calculation using the equation given:

$$\Delta H°(\text{reaction}) = [\text{Sum of } \Delta H°f\ (C_2H_{6(g)})] - [\text{Sum of } \{(\Delta H°f\ (C_2H_{2(g)})) + 2(\Delta H°f\ (H_{2(g)})\}]$$

$$\Delta H°(\text{reaction}) = [(-85\ kJ)] - [(227\ kJ) + 2(0\ kJ)]$$

$$\mathbf{\Delta H°(reaction) = -312\ kJ/mole\text{-}reaction}$$

Example: Use the table shown to calculate ΔH° for the reaction using the ΔH°f equation format.

$$2\ C_6H_{6\,(l)} + 9\ O_{2(g)} \longrightarrow 12\ CO_{(g)} + 6\ H_2O_{(g)}$$

Step (**1**): From the tables, identify all required values for ΔH°f

Reactants: $\Delta H°f\ (C_6H_{6(l)}) = +49.0\ kJ$; $\Delta H°f\ (O_{2(g)}) = 0\ kJ$

Products: $\Delta H°f\ (CO_{(g)}) = -110.5\ kJ$; $\Delta H°f(H_2O_{(g)}) = -241.8\ kJ$

Step (**2**): Complete the calculation using the equation given:

$$\Delta H°_{(reaction)} = [\text{Sum of } \{12(\Delta H°f\ CO_{(g)}) + 6(\Delta H°f\ H_2O_{(g)})\}]$$

$$-[\text{Sum of } \{2(\Delta H°f\ C_6H_{6(l)}) + 9(\Delta H°f\ O_{2(g)})\}]$$

$$\Delta H°_{(reaction)} = [12(-110.5\ kJ) + 6(-241.8\ kJ] - [2(+49.0\ kJ) + 9(0\ kJ)]$$

$$\Delta H°_{(reaction)} = [(-2776.8\ kJ)] - [(98.0\ kJ)]$$

$$\mathbf{\Delta H°_{(reaction)} = -2874.8\ kJ/mole\text{-}reaction}$$

Example: Use the table shown to calculate ΔH° for the reaction using the ΔH°f equation format.

$$3Fe_{(s)} + 4H_2O_{(l)} \longrightarrow Fe_3O_{4(s)} + 4H_{2(g)}$$

Step (1): From the tables, identify all required values for $\Delta H°f$

Reactants: $\Delta H°f\ (Fe_{(s)}) = 0$ kJ; $\Delta H°f\ (H_2O_{(l)}) = -285.8$ kJ

Products: $\Delta H°f\ (Fe_3O_{4(s)}) = -1118.4$ kJ; $\Delta H°f(H_{2(g)}) = 0$ kJ

Step (2): Complete the calculation using the equation given:

$$\Delta H°_{(reaction)} = [\text{Sum of } \{(\Delta H°f\ Fe_3O_{4(s)}) + 4(\Delta H°f\ (H_{2(g)})\}]$$

$$-[\text{Sum of } \{3(\Delta H°f\ Fe_{(s)}) + 4(\Delta H°f\ H_2O_{(l)})]$$

$$\Delta H°_{(reaction)} = [(-1118.4\text{ kJ}) + 4(0\text{ kJ})] - [3(0\text{ kJ}) + 4(-285.8\text{ kJ})]$$

$$\mathbf{\Delta H°_{(reaction)} = [(-1118.4\text{ kJ})] - [(-1143.2\text{ kJ})] = +24.8\text{ kJ/mole-reaction}}$$

VII ENTHALPY AND STOICHIOMETRIC CALCULATIONS

Energy and enthalpy can be considered as a product or a reactant in a chemical reaction. If $\Delta H\ (q_p)$ is **positive**, heat is required for the reaction to occur; heat is a **reactant**. If $\Delta H\ (q_p)$ is **negative**, heat is released in the reaction; heat is a **product**.

Example: For the following reversible reaction, write the forward and reverse reactions separately with energy as a product or reactant.

$$2H_{2(g)} + O_{2(g)} \rightleftarrows 2\,H_2O_{(g)} \quad \Delta H\ (\text{forward reaction}) \quad = \quad -474\text{ kJ/mole-reaction}$$

$$2\,H_{2(g)} + O_{2(g)} \longrightarrow 2\,H_2O_{(g)} + 474\text{ kJ} \quad \Delta H \quad (\text{forward}) \quad = \quad (-);\ \text{heat is a product}$$

$$2\,H_2O_{(g)} + 474\text{ kJ} \longrightarrow 2\,H_{2(g)} + O_{2(g)} \quad \Delta H \quad (\text{reverse}) \quad = \quad (+);\ \text{heat is a reactant}$$

Enthalpy can be included in reaction stoichiometric calculations; the methods are essentially the same as for other stoichiometry. The relationship developed for calorimetry can be used. Recall that ΔH/mole-reaction = ΔH/1 balanced equation.

$$\Delta H/\text{mole-reaction} = \frac{(\Delta H)}{(\text{mole compound})} \times \frac{\text{moles compound (coefficient)}}{1\text{bal eq}}$$

Solving this for ΔH/mole of a compound:

$$\Delta H/\text{mole compound} = \frac{(\Delta H)}{1\text{bal eq}} \times \frac{1\text{ bal eq}}{\text{moles compound (coefficient)}}$$

Energy stoichiometry requires relating the total heat transfer (q) for a specific reaction to the number of moles of one or more compounds in the balanced equation.

To calculate a total (q) based on moles of a specific compound:

$$\mathbf{q\ (total) = (\Delta H/\text{mole compound}) \times (\text{moles of compound})}$$

To calculate moles of a specific compound related to a total (q):

$$\textbf{moles compound} = \frac{\textbf{q (total)}}{(\boldsymbol{\Delta}\textbf{H / mole compound})}$$

Example: Use the reaction shown to calculate the heat released {q (total)} when 25.0 grams of $H_{2(g)}$ reacts with excess oxygen.

$$2\ H_{2(g)} + O_{2(g)} \longrightarrow 2\ H_2O_{(g)} \qquad \Delta H = -474\ \text{kJ/mole-reaction}$$

Step (**1**): q (total) = (ΔH/mole H_2) × (moles of H_2)

Step (**2**): Use the balanced equation ratio to calculate ΔH/mole H_2.

$$\Delta H/\text{mole } H_2 = \frac{-474\ \text{kJ}}{1\text{bal eq}} \times \frac{1\text{bal eq}}{2\text{mole } H_2} = \mathbf{-237\ kJ/mole\ H_2}$$

$$\text{moles } H_2 = \frac{25.0\ \text{g}}{2.2\ \text{g/mole}} = \mathbf{12.38\ mole\ H_2}$$

Step (**3**): Complete the calculation:

$$\text{q (total)} = (\Delta H/\text{mole } H_2) \times (\text{moles of } H_2)$$

$$\mathbf{q\ total = (12.38\ mole\ H_2) \times (-237\ kJ/mole\ H_2) = -2933\ kJ\ (-2930\ kJ)}$$

Example: Use the same reaction to calculate the mass of water formed during the production of 800. kJ of energy (q total = –800 kJ).

$$\text{Step (1): Moles } H_2O = \frac{\text{q (total)}}{(\Delta H/\text{mole } H_2O)}$$

Step (**2**): Use the balanced equation ratio to calculate ΔH/mole H_2O.

$$\Delta H/\text{mole } H_2O = \frac{-474\ \text{kJ}}{1\text{ bal eq}} \times \frac{1\text{ bal eg}}{2\text{ mole } H_2O} = \mathbf{-237\ kJ/mole}\ H_2O$$

$$\text{q total} = -800\ \text{kJ}$$

$$\text{Step (3): Moles } H_2O = \frac{\text{q (total)}}{(\Delta H/\text{mole } H_2O)} = \frac{-800\ \text{kJ}}{(-237\ \text{kJ/mole } H_2O)} = \mathbf{3.376\ mole}$$

Step (**4**): Mass H_2O = (3.376 mole H_2O) × (18.02 g/mole) = **60.8 g**

Example: Use the reaction shown to complete the following parts.

$$C_2H_{4(g)} + 3\ O_{2\ (g)} \longrightarrow 2\ CO_{2(g)} + 2\ H_2O_{(g)} \qquad \Delta H = -1411\ \text{kJ/mole-reaction}$$

a) Write the reaction with heat as a product or reactant.
b) Calculate q (total) for reaction of 1.00 kg of C_2H_4 (excess oxygen).
c) Calculate q (total) for production of 500. g of CO_2.
d) Calculate the mass of O_2 consumed during the production of 600. kJ of energy; (q total = –600. kJ).

a) $C_2H_{4(g)} + 3\,O_{2(g)} \longrightarrow 2\,CO_{2(g)} + 2\,H_2O_{(g)} + 1411\text{ kJ}$

b) **(1)**: q (total) = (ΔH/mole C_2H_4) × (moles of C_2H_4)

$$\text{(2): } \Delta H/\text{mole } C_2H_4 = \frac{-1411\text{kJ}}{1\text{ bal eq}} \times \frac{1\text{ bal eg}}{1\text{ mole } C_2H_4} = -1411\text{kJ/mole } C_2H_4$$

$$\text{moles } C_2H_4 = \frac{1000\text{ g}}{28.4\text{ g/mole}} = 35.66\text{ mole}$$

(3) q total = (35.66 mole C_2H_4) × (–1411 kJ/mole C_2H_4) = **-5.03×10^4 kJ**

c) **(1)** q (total) = (ΔH/mole CO_2) × (moles of CO_2)

$$\text{(2) } \Delta H/\text{mole } CO_2 = \frac{-1411\text{kJ}}{1\text{ bal eq}} \times \frac{1\text{ bal eg}}{2\text{ mole } CO_2} = -705.5\text{kJ/mole } CO_2$$

$$\text{moles } CO_2 = \frac{500\text{ g}}{44.01\text{ g/mole}} = 11.36\text{mole}$$

(3): **q total** = (11.36 mole CO_2) × (–705.5 kJ/mole CO_2) = **-8.02×10^3 kJ**

$$\text{(1): moles } O_2 = \frac{q\text{ (total)}}{(\Delta H/\text{mole } O_2)}$$

$$\text{(2): } \Delta H/\text{mole } O_2 = \frac{-1411\text{kJ}}{1\text{ bal eq}} \times \frac{1\text{ bal eg}}{3\text{ mole } O_2} = -470.3\text{kJ/mole } O_2$$

q total = –600. kJ

$$\text{(3): moles } O_2 = \frac{q\text{ (total)}}{(\Delta H/\text{mole } O_2)} = \frac{-600\text{kJ}}{(-470.3\text{kJ/mole } O_2)} = 1.276\text{mole}$$

(4): mass O_2 = (1.276 mole O_2) × (32.00 g/mole) = **40.8 g**

Example: Various fossil fuels have different levels of CO_2 emission per energy delivered. The following balanced equations represent models of certain fossil fuels:

$CH_{4(g)}$ (natural gas); $C_8H_{18(l)}$ (oil); $C_{14}H_{10(s)}$ (coal: best case); $C_{(s)}$ (coal: worst case).

a) $CH_{4(g)} + 2\,O_{2(g)} \longrightarrow CO_{2(g)} + 2\,H_2O_{(l)}$ $\Delta H = -890.3$ kJ/mole-rxn

b) $2\,C_8H_{18(l)} + 25\,O_{2(g)} \longrightarrow 16CO_{2(g)} + 18H_2O_{(l)}$ $\Delta H = -11{,}390$ kJ/mole-rxn

c) $2C_{14}H_{10(s)} + 33\,O_{2(g)} \longrightarrow 28\,CO_{2(g)} + 10H_2O_{(l)}$ $\Delta H = -14040$ kJ/mole-rxn

d) $C_{(s)} + O_{2(g)} \longrightarrow CO_{2(g)}$ $\Delta H = -393.5$ kJ/mole-rxn

For each of the four fuels, calculate the mass of carbon dioxide released during the production of one million kilojoules of energy ($q = -1.00 \times 10^6$ kJ) measured as ΔH.

$$\text{(1): moles } CO_2 = \frac{q\text{ (total)}}{(\Delta H/\text{mole } CO_2)}\text{ for each fuel.}$$

(2): $q = -1.00 \times 10^6$kJ

$$\text{a) } \Delta H/\text{mole } CO_2 = \frac{-890.3\text{ kJ}}{1\text{bal eg}} \times \frac{1\text{bal eg}}{1\text{ mole } CO_2} = -890.3\text{ kJ/mole } CO_2(\text{for } CH_4)$$

b) $\Delta H/\text{mole } CO_2 = \frac{-11390\,kJ}{1\,\text{bal eg}} \times \frac{1\,\text{bal eg}}{16\text{ mole } CO_2} = -711.9\,kJ/\text{mole } CO_2(\text{for } C_8H_{18})$

c) $\Delta H/\text{mole } CO_2 = \frac{-14040\,kJ}{1\,\text{bal eg}} \times \frac{1\,\text{bal eg}}{28\text{ mole } CO_2} = -501.5\,kJ/\text{mole } CO_2(\text{for } C_{14}H_{10})$

d) $\Delta H/\text{mole } CO_2 = \frac{-393.5\,kJ}{1\,\text{bal eg}} \times \frac{1\,\text{bal eg}}{1\text{ mole } CO_2} = -393.5\,kJ/\text{mole } CO_2(\text{for } C)$

(3) and (4):

a) $\text{moles } CO_2 = \frac{-1.00\times10^6\,kJ}{(-890.3\,kJ/\text{mole } CO_2)} = 1123\,\text{mole}\times 44.01\text{ g/mole}$

$= 4.94\times10^4\,g\ (49.4\,kg \text{ for } CH_4)$

b) $\text{moles } CO_2 = \frac{-1.00\times10^6\,kJ}{(-711.9\,kJ/\text{mole } CO_2)} = 1405\,\text{mole}\times 44.01\text{ g/mole}$

$= 6.18\times10^4\,g\ (61.8\,kg \text{ for } C_8H_{18})$

c) $\text{moles } CO_2 = \frac{-1.00\times10^6\,kJ}{(-501.4\,kJ/\text{mole } CO_2)} = 1994\,\text{mole}\times 44.01\text{ g/mole}$

$= 8.78\times10^4\,g\ (87.8\,kg \text{ for } C_{14}H_{10})$

d) $\text{moles } CO_2 = \frac{-1.00\times10^6\,kJ}{(-393.5\,kJ/\text{mole } CO_2)} = 2541\,\text{mole}\times 44.01\,g/\text{mole}$

$= 1.12\times10^5\,g\ (112\,kg \text{ for } C)$

VIII PRACTICE PROBLEMS

1. A mass of 25.64 grams of an unknown solid is heated to 100.0°C. The solid is added to 50.00 grams of water, which has an initial temperature of 25.10°C; the combined objects eventually reach a temperature of 28.49°C. Determine the heat capacity (C) of the unknown metal. C_{water} = 4.184 J/g-°C)
2. A molten glass sculpture at an initial temperature of 455.0°C is cooled in a vat of 1650 grams of water at an initial temperature of 27.0°C. The final temperature for both the glass and the water (at thermal equilibrium) was 44.0°C. Calculate the mass of the glass sculpture. C (glass) = 0.840 J/g-°C; C (water) = 4.184 J/g-°C.
3. A nickel bar at an initial temperature of 423.0°C is placed in 890 grams of a certain liquid ether at an initial temperature of 38.5°C. The final temperature for both the nickel bar and the ether (at thermal equilibrium) was 77.0°C. Calculate the mass of the nickel bar.

   C (Ni) = 0.440 J/g-°C; C (ether) = 3.42 J/g-°C.
4. A 450-gram tungsten ball at an unknown initial temperature is placed in 720 grams of water at an initial temperature of 22.0°C. The final temperature for both the tungsten ball and the water (at thermal equilibrium) was 30.5°C. Calculate the initial temperature of the tungsten ball. C (W) = 0.130 J/g-°C; C (water) = 4.184 J/g-°C.
5. A 2.05-gram (10.25 carats) diamond (pure carbon solid) is heated to 74.21°C. This object is then added to 26.05 grams of water, which has an initial temperature of 27.20°C. Calculate the final temperature of the combined two objects.

   C_{carbon} = 0.519 J/g-°C; C_{water} = 4.184 J/g-°C)

6. A 0.8650-gram sample of carbon graphite is burned in a constant volume calorimeter; $C_{(calorimeter)} = 10.85$ kJ/°C. The calorimeter has a temperature change (ΔT) of + 2.613°C. Calculate ΔE/mole-rxn. The calorimeter has no other contents.

$$C_{(s)} + O_{2(g)} \longrightarrow CO_{2(g)}$$

7. An open container calorimeter was used to determine the ΔH of the reaction shown. All heat from the reaction was transferred to the solution; no heat was lost to the container.

$$2\ C_3H_{4(aq)} + 2\ HBr_{(aq)} \longrightarrow C_6H_{10}Br_{2(aq)}$$

66.0 grams of C_3H_4 (limiting reagent) was reacted in the calorimeter, which contained 305 grams of aqueous HBr solution (excess). T (initial) of reactants plus solution = 22.4°C; the T (final) of product plus solution = 51.6°C. Total mass of solution equals solute plus solvent. Calculate

$\Delta H°$/mole-rxn. C (solution) = 4.40 J/g-°C;

MM of C_3H_4 = 40.07 g/mole

8. An open container calorimeter was used to determine the ΔH of the synthesis of ether. All heat from the reaction was transferred to the solution plus solutes; no heat was lost to the container.

$$2\ C_2H_6O \xrightarrow{\text{solvent}} C_4H_{10}O$$

5.50 grams of C_2H_6O (MM = 46.08 g/mole) was reacted in the 125 grams of solvent. T (initial) of reactants plus solution = 28.2°C; the T (final) of product plus solution = 31.6°C. Calculate $\Delta H°$/mole-rxn. Total mass of solution equals solute plus solvent. C (solution) = 1.76 J/g-°C.

9. The following reaction is evaluated in a calorimetry experiment to determine the ΔH of the reaction. An open insulated container calorimeter was used to measure the heat change at constant pressure. The reaction was performed in solution such that all heat from the reaction was transferred to the solution.

$$2\ CHO_2Cl \xrightarrow{\text{solvent}} C_2HO_4Cl + H_2O$$

A mass of 32.8 grams of CHO_2Cl (MM = 80.5 g/mol) was added to 665 grams of solvent to make the solution. The initial temperature of the solution was 31.9°C; the final temperature of the solution after the reaction was completed was 47.4°C. The heat capacity of the solution (C) = 2.24 J/g-°C. Calculate ΔH /mole-reaction for this reaction. The total mass of the solution must equal the mass of the solvent plus the added mass of CHO_2Cl.

10. The following is evaluated in a calorimetry experiment to determine the ΔH of the reaction. An open insulated container calorimeter was used to measure the heat change at constant pressure. The reaction was performed in a solvent solution such that the total heat from the reaction was transferred to the solution plus the calorimeter.

$$C_{10}H_{8(s)} + 2\ C_2H_{4(g)} + \xrightarrow{\text{solvent}} C_{14}H_{16}$$

A mass of 52.5 grams of $C_{10}H_8$ (MM = 128.2g/mol) was added to the calorimeter with 960 grams of solution containing the C_2H_4. The initial temperature of the solution plus calorimeter was 31.5°C; the final temperature of the solution plus calorimeter after the reaction was completed was 44.3°C. Calculate ΔH/mole-reaction for this reaction. The total mass of the solution equals the mass of the C_2H_4 solution plus the added mass of $C_{10}H_8$. The total q for the surroundings must include both the solution and the calorimeter. The heat capacity of the solution (C) = 1.55 J/g-°C. The heat capacity of the calorimeter C (cal) = 875 J/°C.

11. Exactly one mole of $CUSO_4$ (aq) and one mole of Zn metal are reacted in a constant pressure (open) calorimeter:

$$CuSO_{4\,(aq)} + Zn_{(s)} \longrightarrow Cu_{(s)} + ZnSO_{4\,(aq)}$$

Initial temperature of the solution = 25.00°C; final temperature of the solution after reaction = 59.54°C. The mass of reaction solution is 1300 grams; this includes the mass of one mole of $CuSO_4$. Heat is transferred to both the calorimeter and the aqueous solution; the small amount absorbed by the solid metal is ignored. Calculate ΔH/mole-rxn. $C_{(calorimeter)}$ = 756 J/°C, $C_{(solution)}$ = 4.184 J/g-°C.

12. Calculate ΔH°/mole-rxn for the following reactions using (ΔH°f).

	ΔH°f kJ/mole		ΔH°f kJ/mole
($NH_{3(g)}$)	−46.1 kJ	($H_2O_{(g)}$)	−241.8 kJ
($C_2H_{6(g)}$)	−84.7 kJ	($C_{(s)}$)	0 kJ
($HCN_{(g)}$)	+135.1 kJ	($C_6H_{6(l)}$)	+49.0 kJ
($C_3H_3N_{(l)}$)	+172.9 kJ	($Fe_{(s)}$)	0 kJ
($O_{2\,(g)}$)	0 kJ	($Fe_2O_{3(S)}$)	−824.2 kJ
($CO_{2(g)}$)	−393.5 kJ	($H_2O_{(l)}$)	−285.8 kJ
($HNO_{3(g)}$)	−135.1 kJ	($NO_{(g)}$)	+90.3 kJ
($C_2H_{2(g)}$)	+226.7 kJ	($C_2H_{4(g)}$)	+52.3 kJ
($N_2O_{(g)}$)	+82.1 kJ	($NO_{(g)}$)	+33.2 kJ

a) $2\,Fe_2O_{3(s)} + 3\,C_{(S)} \longrightarrow 4\,Fe_{(s)} + 3\,CO_{2\,(g)}$
b) $12\,NH_{3(g)} + 21\,O_{2(g)} \longrightarrow 8\,HNO_{3(g)} + 4\,NO_{(g)} + 12\,H_2O_{(l)}$
c) $C_3H_3N_{(l)} + 2\,H_2O_{(l)} \longrightarrow C_2H_{6\,(g)} + O_{2\,(g)} + HCN_{(g)}$
d) $2\,C_2H_{2(g)} + 2\,NO_{(g)} + C_2H_{4\,(g)} \longrightarrow C_6H_{6\,(l)} + N_2O_{(g)} + H_2O_{(l)}$
e) $C_2H_{6\,(g)} + 2\,NO_{2(g)} + O_{2\,(g)} \longrightarrow 2\,C_3H_3N_{(l)} + 6\,H_2O_{(l)}$

13. Calculate total q for reaction of 10.1 g H_2 with 8.00 g O_2. This is a limiting reagent stoichiometry problem. Solve a standard limiting reagent problem to determine the correct limiting reagent; then complete the enthalpy stoichiometry.

$$2\,H_2 + O_2 \longrightarrow 2H_2O \qquad \Delta H^{\circ} = -474\text{kJ/mole-rxn}$$

14. Use the following equation with the given value for ΔH° to solve the energy stoichiometry problems.

$$3\,C_2H_{6(g)} \longrightarrow C_6H_{8(l)} + 2\,H_{2(g)} \qquad \Delta H^{\circ} = +95.6\text{ kJ/mole-reaction}$$

a) What mass (grams) of H_2 can be formed for a reaction in which 1800 kJ was added to the reaction system (i.e., the total q = +1800 kJ)
b) A chemical plant reacts 500 grams of C_2H_6. Calculate the amount of energy in kJ (total q), which must be added to react the 500 grams of C_2H_6.

15. Use the reaction shown to solve the following energy stoichiometry problems.

$$2\ C_6H_{6(l)} + 5\ Br_{2(g)} \longrightarrow C_{12}H_6Br_{4(s)} + 6\ HBr_{(g)}$$

MM = 78.12 g/mol MM = 141.8 g/mol

$\Delta H = -145.8$ kJ/mole-reaction

a) A chemical plant reacts 350 grams of C_6H_6 in a reactor. Calculate the amount of energy in kJ that would be released during this reaction.
b) Calculate the mass of Br_2 required to produce 885 kilojoules of energy (q total = −885 kilojoules).

16. Use the following equation to solve the energy stoichiometry problems.

$$4\ C_2H_{4(g)} + O_{4(g)} \longrightarrow 2\ C_4H_{6\,(g)} + 2\ H_2O_{(l)} \qquad \Delta H° = -155 \text{ kJ/mole-reaction}$$

a) What mass (grams) of C_2H_4 is consumed if a total of 675 kJ was released by the reaction (total q = −675 kJ).
b) A chemical plant produces 800 grams of C_4H_6. Calculate the amount of energy in kJ (total q) that was released during the formation of the 800 grams of C_4H_6.

IX ANSWERS TO PRACTICE PROBLEMS

1. Mass of unknown solid = 25.64 grams; T(initial) = 100.0°C. Added to 50.00 grams of water at T(initial) = 25.10°C; T(final) = 28.49°C. Determine the heat capacity (C) of the unknown metal. C_{water} = 4.184 J/g-°C)

Step (1): $C_{metal} = \dfrac{q_{metal}}{(m_{metal})(\Delta T_{metal})}$

(2): ΔT_{metal} = (28.94°C) − (100.0°C) = −71.51°C; m_{metal} = 25.64 grams

$$\mathbf{q_{metal} = -q_{water}}$$

(1surr): $q_{water} = (\Delta T_{water})(m_{water})(C_{water})$

(2surr): C_{water} = 4.184 J/g-K; m_{water} = 50.0 grams

$$\Delta T_{water} = (28.49°C) - (25.1°C) = +3.39°C$$

(3surr): q_{water} = (3.39°C)(50.0 g)(4.184 J/g-°C) = +709.2 J

$$\mathbf{q_{metal} = -q_{water} = -(+709.2\ J) = -709.2\ J}$$

(3): $\mathbf{C_{metal}} = \dfrac{(-709.2\ J)}{(25.64\ g)(-71.51\ °C)} = \mathbf{+0.387\ J/g\text{-}K}$

2. Glass sculpture T(initial) = 455.0°C; cooled in 1650 grams of water at T(initial) = 27.0°C. T(final) = 44.0°C. Calculate the mass of the glass sculpture.

C (glass) = 0.840 J/g-°C; C (water) = 4.184 J/g-°C.

Step (1): $m_{glassl} = \dfrac{q_{glass}}{(C_{glass})(\Delta T_{glass})}$

(2): $\Delta T_{glass} = (44.0°C) - (455°C) = -411°C$; $C_{glass} = 0.840$ J/g-°C

$$\mathbf{q_{glass} = -q_{water}}$$

(1surr): $q_{water} = (\Delta T_{water})(m_{water})(C_{water})$

(2surr): $C_{water} = 4.184$ J/g-K; $m_{water} = 1{,}650$ grams

$$\mathbf{\Delta T_{water} = (44.0°C) - (27.0°C) = +17.0°C}$$

(3surr): $q_{water} = (17.0°C)(1650\ g)(4.184\ J/g\text{-}°C) = +1.174 \times 10^5\ J$

$$q_{glass} = -q_{water} = -(+1.174 \times 10^5\ J) = -1.174 \times 10^5$$

(3): $m_{glass} = \dfrac{(-1.174 \times 10^5 J)}{(0.840\ J/g\text{-}°C)(-411°C)} = \mathbf{340.g}$

3. Nickel bar at T(initial) = 423.0°C is placed in 890 grams of ether T(initial) = 38.5°C. T(final) = 77.0°C. Calculate the mass of the nickel bar.

C (Ni) = 0.440 J/g-°C; C (ether) = 3.42 J/g-°C.

Step (1): $m_{Ni} = \dfrac{q_{Ni}}{(C_{Ni})(\Delta T_{Ni})}$

(2): $\Delta T_{Ni} = (77.0°C) - (423°C) = -346°C$; $C_{Ni} = 0.440$ J/g-°C

$$\mathbf{q_{Ni} = -q_{ether}}$$

(1surr). $\mathbf{q_{ether}} = (\Delta T_{ether})(m_{ether})(C_{ether})$

(2surr): $C_{ether} = 3.42$ J/g-K; $m_{ether} = 890$ grams

$$\Delta T_{ether} = (77.0°C) - (39.5°C) = +37.5°C$$

(3surr): $q_{ether} = (37.5°C)(890\ g)(3.42\ J/g\text{-}°C) = +1.172 \times 10^5 J$

$$q_{Ni} = -q_{ether} = -(+1.172 \times 10^5\ J) = -1.172 \times 10^5$$

(2): $m_{Ni} = \dfrac{(-1.172 \times 105 J)}{(0.440 J/g\text{-}°C)(-346\ °C)} = \mathbf{770\ g}$

4. A 450-gram tungsten ball at an unknown initial temperature is placed in 720 g of water T(initial) = 22.0°C. T(final) = 30.5°C. Calculate the initial temperature of the tungsten ball. C (W) = 0.130 J/g-°C; C (water) = 4.184 J/g-°C.

(1): $\Delta Tw = \dfrac{(qw)}{(C_w)(m_w)}$

(2): Cw = 0.130 J/g-°C; $m_w = 450$ g; $\mathbf{q_w = -q_{water}}$

(1surr) $q_{water} = (\Delta T_{water})(m_{water})(C_{water})$

(2surr) $C_{water} = 4.184$ J/g-°C, $m_{water} = 720$ grams

$$\Delta T_{water} = (30.5°C) - (22.0°C) = +8.5°C$$

(3surr) $q_{water} = (8.5°C)(720\ g)(4.184\ J/g\text{-}°C) = +2.56 \times 10^4 J$

$$qw = -q_{water} = -(+2.56 \times 10^4\ J) = -2.56 \times 10^4\ J$$

(3): $\Delta T_W = \dfrac{(q_w)}{(C_w)(m_w)} = \dfrac{(-2.56 \times 10^4\ J)}{(0.130\ J/g-°C)(450\ g)} = \mathbf{-437°C}$

(4): T(initial) for W = T(final) − (ΔT_W) = (30.5°C) − (−437°C) = **468°C**

5. A 2.05-gram diamond at T(initial) = 74.21°C is added to 26.05 grams of water at T(initial) = 27.20°C. Calculate the final temperature of the combined two objects.

$$C_{carbon} = 0.519\ J/g\text{-}°C;\ C_{water} = 4.184\ J/g\text{-}°C)$$

(1): $q_{diamond} = (\Delta T_{diamond})(m_{diamond})(C_{diamond})$

$$q_{water} = (\Delta T_{water})\ (m_{Water})(C_{water})$$

(2): $q_{diamond} = (\Delta T_{diamond})(2.05\ g)(0.519\ J/g\text{-}K)$

$$q_{water} = (\Delta T_{water})(26.05\ g)(4.184\ J/g\text{-}K)$$

$$\Delta T_{diamond} \text{ in °C} = (Tf - 74.21)\ \Delta T_{water} \text{ in °C} = (Tf - 27.20)$$

$$[(Tf - 74.21)(2.05\ g)(0.519\ J/g\text{-}°C)] = -[(Tf - 27.20)(26.05\ g)(4.184\ J/g\text{-}°C)]$$

(3): $[(Tf - 74.21)(1.064)] = -[(Tf - 27.20)(109.0)]$

$$(1.064\ Tf) - 78.96 = (-109Tf) + 2964.8$$

$$110.064\ Tf = 3043.76;\ Tf = 3043.76/110.064 = 27.7°C$$

(4): $\Delta T_{water} = 27.654 - 27.20 = +0.454°C$

$$q_{water} = (+0.454\ K)(26.05\ g)(4.184\ J/g\text{-}°C) = +49.48\ J$$

$$\Delta T_{diamond} = 27.654 - 74.21 = -46.556°C$$

$$q_{diamond} = (-46.556\ K)(2.05\ g)(0.519\ J/g\text{-}°C) = -49.53\ J$$

$$\mathbf{q_{diamond} = -q_{water}}$$

6. A 0.8650-gram sample of carbon graphite is burned in a constant volume calorimeter; $C_{(calorimeter)}$ = 10.85 kJ/°C. The calorimeter has a temperature change (ΔT) of +2.613°C. Calculate ΔE/mole-rxn. The calorimeter has no other contents.

$$C_{(S)} + O_{2(g)} \longrightarrow CO_{2(g)}$$

(1): ΔE/mole C = q_v/ mole C.

(2): $q_{(reaction)} = q_{(calorimeter)}$; $C_{calorimeter} = 10.85\ kJ/°C$

$$\Delta T_{(calorimeter)} = \text{is given as } +2.613°C$$

$$q_{(calorimeter)} = (\Delta T_{(calorimeter)})(C_{calorimeter}) = (2.613\ C)(10.85\ kJ/°C) = +28.35\ kJ$$

$$q_{(reaction)} = -q_{(calorimeter)};$$

$$q_{(reaction)} = -\ (+28.35\ kJ) = -28.35\ kJ;\ mole\ C = \frac{0.8650\,g}{12.01\,g/mole} = 0.0720\,mole$$

(3): $q_v/mole\ C = \frac{-28.35\,kJ}{0.0720\,mole} = \mathbf{-393.8\,kJ\,/\,mole}$

(4): $\Delta E/mole\text{-}reaction = \frac{(-393.8\ kJ)}{(mole\ C)} \times \frac{1C}{1\,bal\,eq} = \mathbf{-393.8\,kJ/mole\text{-}rxn}$

7. $2C_3H_{4(aq)} + 2\ HBr_{(aq)} \longrightarrow C_6H_{10}Br_{2(aq)}$
   66.0 grams of C_3H_4 was reacted in 305 grams of aqueous HBr solution. T (initial) of reactants plus solution = 22.4°C; the T (final) of product plus solution = 51.6°C.
   Calculate ΔH°/mole-rxn. C (solution) = 4.40 J/g-°C; MM of C_3H_4 = 40.07 g/mole

(1): $\Delta H = q_p$ (reaction); therefore, $\Delta H/mole\ C_3H_4 = q_p/mole\ C_3H_4$.
(2): $q_{(reaction)} = -q_{(solution)}$; $q_{solution} = (\Delta T_{solution})(m_{solution})(C_{solution})$

$$C_{solution} = 4.40\ J/g\text{-}°C;\ m_{solution} = 305\ g\ solvent + 66.0\ grams\ C_3H_4 = 371\ g\ solution$$

$$\Delta T_{(solution)} = T(final) - T(initial) = (51.6°C) - (22.4°C) = 29.2°C$$

$$q_{solution} = (\Delta T_{solution})(m_{solution})(C_{solution}) = (29.2°C)(371\ g)(4.40\ J/g\text{-}°C) = 47666\ J$$

$$q_{(reaction)} = -q_{(solution)};$$

$$q_{(reaction)} = -(47666\ J) = -47666\ J;\ moles\ C_3H_4 = \frac{66.0\,g}{40.07\,g/mole} = 1.647\,mole$$

(3): $q_p/mole\ C_3H_4 = \frac{-47666\ J}{1.647\,mole} = \mathbf{-28951\,J/mole}$ or $\mathbf{-28.95\,kJ/mole}$

(4): $\Delta H/mole\text{-}reaction = \frac{(-28.95\ kJ)}{(mole\,C_3H_4)} \times \frac{2C_3H_4}{1\,bal\,eq} = \mathbf{-57.9\,kJ/mole\text{-}rxn}$

8. $2\ C_2H_6O \longrightarrow C_4H_{10}O$
   50 grams of C_2H_6O was reacted in the 125 grams of solvent. T(initial) of reactants plus solution = 28.2°C; the T (final) of product plus solution = 31.6°C.
   Calculate ΔH°/mole-rxn. C (solution) = 1.76 J/g-°C

(1): $\Delta H = q_p$ (reaction); therefore, $\Delta H/mole\ C_2H_6O = q_p/mole\ C_2H_6O$.
(2): $q_{(reaction)} = -q_{(solution)}$, $q_{solution} = (\Delta T_{solution})(m_{solution})(C_{solution})$

$$C_{solution} = 1.76\ J/g\text{-}°C;\ m_{solution} = 125\ g\ solvent + 5.50\ grams\ C_2H_6O = 130.5\ g\ solution$$

$$\Delta T_{(solution)} = T(final) - T(initial) = (31.6°C) - (28.2°C) = 3.4°C$$

$$q_{solution} = (\Delta T_{solution})(m_{solution})\ (C_{solution}) = (3.4°C)(130.5\ g)(1.76\ J/g\text{-}°C) = 780.9\ J$$

$$q_{(reaction)} = -q_{(solution)};$$

$$q_{(reaction)} = -(780.9\ J) = -780.9\ J;\ \text{moles } C_2H_6O = \frac{5.50\,g}{46.08\,g/mole} = 0.1194\,mole$$

(3): $q_p/\text{mole } C_2H_6O = \dfrac{-780.9\,J}{0.1194\,mole} = \mathbf{-6540\,J/mole}$ or $\mathbf{-6.54\,kJ/mole}$

(4): $\Delta H/\text{mole-reaction} = \dfrac{(-6.54\ kJ)}{(\text{mole } C_2H_6O)} \times \dfrac{2\ C_2H_6O}{1\,bal\,eq} = \mathbf{-13.1\,kJ/mole\text{-}rxn}$

9. $2\ CHO_2Cl \longrightarrow C_2HO_4Cl + H_2O$
   A mass of 32.8 grams of CHO_2Cl was added to 665 grams of solvent. T(initial) of the solution was 31.9°C; the T (final) of the solution was 47.4°C.
   Calculate ΔH/mole-reaction. C (solution) = 2.24 J/g-°C.

(1): $\Delta H = q_p$ (reaction); therefore, $\Delta H/\text{mole } CHO_2Cl = q_p/\text{mole } CHO_2Cl$.
(2): $q_{(reaction)} - q_{(solution)};\ q_{solution} = (\Delta T_{solution})(m_{solution})(C_{solution})$

$$C_{solution} = 2.24\ J/g\text{-}°C;\ m_{solution} = 665\ g\ \text{solvent} + 32.8\ \text{grams } CHO_2Cl = 698.8\ g\ \text{solution}$$

$$\Delta T_{(solution)} = T(final) - T(initial) = (47.4°C) - (31.9°C) = 15.5°C$$

$$q_{solution} = (\Delta T_{solution})(m_{solution})(C_{solution}) = (15.5°C)(698.8\ g)(2.24\ J/g\text{-}°C) = 24262\ J$$

$$q_{(reaction)} = -q_{(solution)};$$

$$q_{(reaction)} = -(24262\ J) = -24262\ J;\ \text{moles } CHO_2Cl = \frac{32.8\,g}{80.5\,g/mole} = 0.4075\,mole$$

(3): $q_p/\text{mole } CHO_2Cl = \dfrac{-24262\,J}{0.4075\ mole} = \mathbf{-59539\,J/mole}$ or $\mathbf{-59.54\,kJ/mole}$

(4): $\Delta H/\text{mole-reaction} = \dfrac{(-59.54\ kJ)}{(\text{mole } CHO_2Cl)} \times \dfrac{2\ CHO_2Cl}{1\,bal\,eq} = \mathbf{-119\,kJ/mole\text{-}rxn}$

10. $C_{10}H_8(s) + 2\ C_2H_4(g) + \longrightarrow C_{14}H_{16}$
    A mass of 52.5 grams of $C_{10}H_8$ was added to the calorimeter with 960 grams of solution containing the C_2H_4. T(initial) of the solution plus calorimeter was 31.5°C; T(final) was 44.3°C. Calculate ΔH/mole-reaction for this reaction. Total q for the surroundings must include both the solution and the calorimeter. C (solution) = 1.55 J/g-°C; $C_{(cal)}$ = 875 J/°C.

(1): $\Delta H = q_{p(reaction)}$; therefore, $\Delta H/\text{mole } C_{10}H_8 = q_p/\text{mole } C_{10}H_8$.
(2): $q_{(reaction)} = -[q_{(calorimeter)} + q_{(solution)}];\ q_{(calorimeter)} = (\Delta T_{(calorimeter)})(C_{calorimeter})$

$$q_{solution} = (\Delta T_{solution})(m_{solution})(C_{solution})$$

$$C_{solution} = 1.55\ J/g\text{-}°C;\ m_{solution} = 960\ g\ \text{solution} + 52.5\ \text{grams } C_{10}H_8 = 1012.5\ g\ \text{solution}$$

$$\Delta T_{(solution)} \text{ and } \Delta T_{(calorimeter)} = T(final) - T(initial) = (44.3°C) - (31.5°C) = 12.8°C$$

$$q_{(calorimeter)} = (\Delta T_{(calorimeter)})(C_{calorimeter}) = (12.8°C)(875\ J/°C) = 11200\ J$$

$$q_{solution} = (\Delta T_{solution})(m_{solution})(C_{solution}) = (12.8°C)(1012.5\ g)(1.55\ J/g\text{-}°C) = 20{,}088\ J$$

$$q_{(reaction)} = -[q_{(calorimeter)} + q_{(water)}]$$

$$q_{(reaction)} = -[(11200\,J) + (20088\,J)] = 31288\ J$$

$$q_{(reaction)} = -[31288\ J] = \mathbf{-31288\ J} \text{ or } \mathbf{-31.3\ kJ}$$

$$\text{moles } C_{10}H_8 = \frac{52.5\ g}{128.2\ g/mole} = 0.4095\ mole$$

(3): $q_p/\text{mole } C_{10}H_8 = \dfrac{-31.3\,J}{0.4095\,mole} = \mathbf{-76.4\,kJ/mole}$

(4): $\Delta H/\text{mole-reaction} = \dfrac{(-76.4\,kJ)}{(mole\,C_{10}H_8)} \times \dfrac{1\ C_{10}H_8}{1\ bal\ eq} = \mathbf{-76.4\ kJ/mole\text{-}rxn}$

11. $CuSO_{4(aq)} + Zn\ (s) \longrightarrow Cu\ (s) + ZnSO_4\ (aq)$
Initial temperature of the solution = 25.00°C; final temperature of the solution after reaction = 59.54°C. The mass of reaction solution with the $CuSO_4$ is 1300 grams. Heat is transferred to both the calorimeter and the aqueous solution; the small amount absorbed by the solid metal is ignored. Calculate ΔH/mole-rxn. $C_{(calorimeter)} = 756\ J/°C$;

$$C_{(solution)} = 4.184\ J/g\text{-}K$$

(1): Each reactant is exactly 1 mole; either can be used for the general equation: ΔH/mole $CuSO_4 = q_p/\text{mole } CuSO_4$.

(2): $q_{(reaction)} = -[q_{(calorimeter)} + q_{(solution)}]$; $q_{(calorimeter)} = (\Delta T_{(calorimeter)})(C_{calorimeter})$

$$q_{solution} = (\Delta T_{solution})(m_{solution})(C_{solution})$$

$$C_{solution} = 4.184\ J/g\text{-}°C,\ m_{solution} = 1300.\ g\ solution$$

$$\Delta T_{(solution)} \text{ and } \Delta T_{(calorimeter)} = T(final) - T(initial) = (59.54°C) - (25.00°C) = 34.54°C$$

$$q_{(calorimeter)} = (\Delta T_{(calorimeter)})(C_{calorimeter}) = (34.54°C)(756\ J/°C) = 26112\ J$$

$$q_{solution} = (\Delta T_{solution})(m_{solution})(C_{solution}) = (34.54°C)(1300\ g)(4.184\ J/g\text{-}°C) = 187{,}870\ J$$

$$q_{(reaction)} = -[q_{(calorimeter)} + \mathbf{q_{(water)}}]$$

$$q_{(reaction)} = -[(26{,}112\ J) + (187{,}870\ J)] = -213982\ J$$

$$\mathbf{q_{(reaction)} = -[213{,}982\ J] = -213{,}982\ J \text{ or } -214\ kJ}$$

$$\text{moles } CuSO_4 = 1 \text{ mole; coefficient of each reactant} = 1$$

(3) and **(4)**: $q_p/\text{mole } CuS0_4 = \Delta H/\text{mole-reaction} = \mathbf{-214\ kJ/mole\text{-}rxn}$

12. a) $2\ Fe_2O_{3(s)} + 3\ C_{(s)} \longrightarrow 4\ Fe_{(s)} + 3\ CO_{2\,(g)}$

Reactants: $\Delta H°f\ (Fe_2O_{3(s)}) = -824.2\ kJ$; $\Delta H°f\ (C_{(s)}) = 0\ kJ$

Products: $\Delta H°f\ (Fe_{(s)}) = 0\ kJ$; $\Delta H°f\ (CO_{2(g)}) = -393.5\ kJ$

$$\Delta H° \text{ (reaction)} = [\text{Sum of } \{4(\Delta H°f\ (Fe_{(s)}) + 3(\Delta H°f\ CO_{2(g)})\}]$$

$$-[\text{Sum of } \{2(\Delta H°f\ Fe_2O_{3(s)}) + 3(\Delta H°f\ C_{(s)})\}]$$

$$\Delta H^0 \text{ (reaction)} = [4\ (0\ kJ) + 3\ (-393.5\ kJ)] - [2\ (-824.2\ kJ) + 3\ (0\ kJ)]$$

$$\mathbf{\Delta H^0 \text{ (reaction)} = +467.9\ kJ}$$

b) $12NH_{3\,(g)} + 21\,O_{2(g)} \longrightarrow 8\,HNO_{3(g)} + 4\,NO_{(g)} + 12\,H_2O_{\,(l)}$

Reactants: $\Delta H°f\ (NH_{3(g)}) = -46.1$ kJ; $\Delta H°f\ (O_{2(g)}) = 0$ kJ

Products: $\Delta H°f\ (HNO_{3(g)}) = -135.1$ kJ; $\Delta H°f\ (NO_{(g)}) = +90.3$ kJ

$\Delta H°f\ (H_2O_{(l)}) = -285.8$ kJ

$$\Delta H^0 \text{ (reaction)} = [\text{Sum of } \{\mathbf{8}(\Delta H°f\ HNO_{3(g)}) + 4(\Delta H°f\ NO_{(g)}) + 12(\Delta H°f\ H_2O_{(l)})\}]$$

$$-[\text{Sum of } \{12(\Delta H°f\ NH_{3(g)}) + 21(\Delta H°f\ O_{2(g)})\}]$$

$$\Delta H°_{rxn} = [8(-135.1\ kJ) + 4(90.3\ kJ) + 12(-285.8\ kJ)] - [12(-46.1\ kJ) + 21(0)]$$

$$\mathbf{\Delta H_{(reaction)} = -3597\ kJ/mole\text{-}reaction}$$

c) $C_3H_3N_{(l)} + 2H_2O_{\,(l)} \longrightarrow C_2H_{6\,(g)} + O_2 + HCN_{(g)}$

Reactants: $\Delta H°f\ (C_3H_3N_{(l)} = 172.9$ kJ; $\Delta H°f\ (H_2O_{(l)}) = -285.8$ kJ

Products: $\Delta H°f\ (C_2H_{6(g)}) = -84.7$ kJ; $\Delta H°f\ (O_{2(g)}) = 0$ kJ

$\Delta H°f\ (HCN_{(g)}) = 135.1$ kJ

$$\Delta H°\text{(reaction)} = [\text{Sum of } \{(\Delta H°f\ C_2H_{6(g)}) + (\Delta H°f\ O_{2(g)}) + (\Delta H°f\ HCN_{(g)})\}]$$

$$-[\text{Sum of } \{(\Delta H°f\ C_3H_3N_{(l)}) + 2(\Delta H°f\ H_2O_{(l)})\}]$$

$$\Delta H°_{rxn} = [(-84.7\ kJ) + (0) + (135.1\ kJ)] - [(172.9\ kJ) + 2(-285.8\ kJ)]$$

$$\mathbf{\Delta H°_{(reaction)} = +449\ kJ/mole\text{-}reaction}$$

d) $2\,C_2H_{2(g)} + 2\,NO_{(g)} + C_2H_{4(g)} \longrightarrow C_6H_{6(l)} + N_2O_{(g)} + H_2O_{(l)}$

Reactants: $\Delta H°f\ (C_2H_{2(g)}) = 226.7$ kJ; $\Delta H°f\ (NO_{(g)}) = 90.3$ kJ

$\Delta H°f\ (C_2H_{4(g)}) = 52.3$ **kJ**

Products: $\Delta H°f\ (C_6H_{6\,(l)}) = 49.0$ **kJ;** Δ**H**$°f\ (N_2O_{(g)}) = 82.1$ **kJ**

$\Delta H°f\ (H_2O_{(l)}) = -285.8$ **kJ**

$$\Delta H° \text{ (reaction)} = [\text{Sum of } \{(\Delta H°f\ C_6H_{6(l)}) + (\Delta H°f\ N_2O_{(g)}) + (\Delta H°f\ H_2O_{(l)})\}]$$

$-[\text{Sum of } \{2(\Delta H°f\ C_2H_{2(g)}) + 2(\Delta H°f\ NO_{(g)}) + (\Delta H°f\ C_2H_{2\ (g)})\}]$

$\Delta H°_{rxn} = [(49.0\ kJ) + (82.1kJ) + (-285.8\ kJ)] - [2(226.7\ kJ) + 2(90.3\ kJ) + (52.3)]$

$\mathbf{\Delta H°_{(reaction)} = -841\ kJ/mole\text{-}reaction}$

e) $3\ C_2H_{6(g)} + 2NO_{2(g)} + O_{2(g)} \longrightarrow 2\ C_3H_3N_{(l)} + 6H_2O_{(l)}$

Reactants: $\Delta H°f\ (C_2H_{6(g)}) = -84.7\ kJ$; $\Delta H°f\ (NO_{2(g)}) = 33.2\ kJ$

$\Delta H°f\ (O_{2(g)}) = 0\ kJ$

Products: $\Delta H°f\ (C_3H_3N_{(l)}) = 172.9\ kJ$; $\Delta H°f\ (H_2O_{(l)}) = -285.8\ kJ$

$\Delta H°\ (\text{reaction}) = [\text{Sum of } \{2(\Delta H°f\ C_3H_3N_{(l)} + 6(\Delta H°f\ H_2O_{(l)})\}]$

$-[\text{Sum of } \{3(\Delta H°f\ C_2H_{6(g)}) + 2(\Delta H°f\ NO_{2(g)}) + (\Delta H°f\ O_{2(g)})\}]$

$\Delta H°_{rxn} = [2(172.9\ kJ) + 6(-285.8\ kJ)] - [3(-84.7\ kJ) + 2(33.2\ kJ) + (0)]$

$\mathbf{\Delta H°_{(reaction)} = -1{,}182\ kJ/mole\text{-}reaction}$

13. $2\ H_2 + O_2 \longrightarrow 2\ H_2O \Delta H° = -474\ kJ/mole\text{-}rxn$

$\text{mole } H_2 = 10.1\ g/2.02\ g/mole = 5.00\ mole$

$$\text{mole of } H_2O \text{ from } H_2 = 5.00\ \text{mole } H_2 \times \frac{2\ H_2O}{2\ H_2} = 5.00\ \text{mole } H_2O$$

$\text{mole } O_2 = 8.00\ g/32.0\ g/mole = 0.250\ \text{mole } O_2$

$$\text{mole of } H_2O \text{ from } O_2 = 0.250\ \text{mole } O_2 \times \frac{2\ H_2O}{1\ O_2} = 0.500\ \text{mole } H_2O$$

O_2 is the limiting reagent; this mole number must be used in the energy calculation.

(1): q (total) = (ΔH/mole O_2) × (moles of O_2)

(2): $\Delta H/\text{mole } O_2 = \dfrac{-474\ kJ}{1\ \text{bal eq}} \times \dfrac{1\ \text{bal eq}}{1\ \text{mole } O_2} = -474\ kJ/\text{mole } O_2$

$\text{mole } O_2 = \dfrac{8.00\ g}{32.0\ g/mole} = 0.250\ \text{mole } O_2$

(3): q total = (0.250 mole O_2) × (−474 kJ/mole O_2) = **−118.5 kJ**

14. $3\ C_2H_{6(g)} \longrightarrow C_6H_{8(l)} + 2\ H_{2(g)} \quad \Delta H^0 = +95.6\ kJ/mole\text{-}reaction$

a) **(1):** $\text{moles } H_2 = \dfrac{q\ (\text{total})}{(\Delta H/\text{mole } H_2)}$

(2): $\Delta H/\text{mole } H_2 = \dfrac{95.6\ kJ}{1\ \text{bal eq}} \times \dfrac{1\ \text{bal eq}}{2\ \text{mole } H_2} = 47.8\ kJ/\text{mole } H_2$

q total = 1800. kJ

(**3**): moles $H_2 = \frac{q\ (total)}{\Delta H/mole\ H_2} = \frac{1800\ kJ}{(47.8\ kJ/mole\ H_2)} = 37.66\ mole$

(**4**): mass H_2 = (37.66 mole H_2) × (2.02 g/mole) = **76.1 g**

b) (**1**): q (total) = (ΔH/mole C_2H_6) × (moles of C_2H_6)

(**2**): $\Delta H/mole\ C_2H_6 = \frac{95.6\ kJ}{1\ bal\ eq} \times \frac{1\ bal\ eq}{3\ mole\ C_2H_6} = 31.87\ kJ/mole\ C_2H_6$

moles $C_2H_6 = \frac{500\ g}{30.08\ g/mole} = 16.62\ mole$

(**3**): q total = (16.62 mole C_2H_6) × (31.87 kJ/mole C_2H_6) = **530 kJ**

15. $2\ C_6H_{6\ (l)} + 5\ Br_{2\ (g)} \longrightarrow C_{12}H_6Br_{4(s)} + 6\ HBr_{(g)}$

$\Delta H = -145.8$ kJ/mole-reaction

a) (**1**): q (total) = (ΔH/mole C_6H_6) × (moles of C_6H_6)

(**2**): $\Delta H/mole\ C_6H_6 = \frac{-145.8\ kJ}{1\ bal\ eq} \times \frac{1\ bal\ eq}{2\ mole\ C_6H_6} = -72.9\ kJ/mole\ C_6H_6$

moles $C_6H_6 = \frac{350\ g}{78.12\ g/mole} = 4.480\ mole$

(**3**): q total = (4.480 mole C_6H_6) × (–72.9 kJ/mole C_6H_6) = **–327 kJ**

b) (**1**): moles $Br_2 = \frac{q\ (total)}{(\Delta H/mole\ Br_2)}$

(**2**): $\Delta H/mole\ Br_2 = \frac{-145.8\ kJ}{1\ bal\ eq} \times \frac{1\ bal\ eq}{5\ mole\ Br_2} = -29.16\ kJ/mole\ Br_2$

q total = –885 kJ

(**3**): moles $Br_2 = \frac{q\ (total)}{(\Delta H/mole\ Br_2)} = \frac{-885\ kJ}{(-29.16\ kJ/mole\ Br_2)} = 30.35\ mole$

(**4**): mass Br_2 = (30.35 mole Br_2) × (141.8 g/mole) = **4.30 × 10³ g**

16. $4\ C_2H_{4(g)} + O_{2\ (g)} \longrightarrow 2\ C_4H_{6\ (g)} + 2\ H_2O_{\ (l)}$ $\Delta H° = -155$ kJ/mole-rxn

a) (**1**): moles $C_2H_4 = \frac{q\ (total)}{\Delta H/mole\ C_2H_4}$

(**2**): $\Delta H/mole\ C_2H_4 = \frac{-155\ kJ}{1\ bal\ eq} \times \frac{1\ bal\ eq}{4\ mole\ C_2H_4} = -38.75\ kJ/mole\ C_2H_4$

q total = –675 kJ

(**3**): moles $C_2H_4 = \frac{q\ (total)}{\Delta H/mole\ C_2H_4} = \frac{-675\ kJ}{(-38.75\ kJ/mole\ C_2H_4)} = 17.42\ mole$

(**4**): mass C_2H_4 = (17.42 mole C_2H_4) × (28.06 g/mole) = **489 g**

b) (**1**): q (total) = (ΔH /mole C_4H_6) × (moles of C_4H_6)

(**2**): $\Delta H/\text{mole } C_4H_6 = \frac{-155 \text{ kJ}}{1 \text{ bal eq}} \times \frac{1 \text{ bal eq}}{2 \text{ mole } C_4H_6} = -77.5 \text{ kJ/mole } C_6H_6$

$$\text{moles } C_4H_6 = \frac{800 \text{ g}}{54.1 \text{ g / mole}} = 14.79 \text{ mole}$$

(**3**): q total = (14.79 mole C_4H_6) × (−77.5 kJ/mole C_4H_6) = **−1146 kJ (−1150 kJ)**

20 Working with Gas Laws

I KINETIC THEORY OF GASES

Phases of a compound or element (solid, liquid, gas) are determined by the balance between the strength of intermolecular forces and the average kinetic energy of the molecules. The gas phase is characterized by very weak attractive forces during which the average kinetic energy of motion dominates the physical properties of gases. For a restricted set of conditions, analysis of a gas is characterized by assuming that attractive forces between individual gas molecules are zero and that the volume occupied by the physical size of the molecules is essentially zero as compared to the volume of empty space between the molecules. Under these conditions, the gas, termed an **ideal gas**, is analyzed through the average statistical behavior of rapidly moving independent particles.

Gases have very low density as there is a relatively large amount of empty space between individual molecules. Gases can have **variable volumes**; they can expand (molecules become farther apart) or be compressed (molecules are squeezed closer together). The **pressure** of a gas is produced by the kinetic energy of molecular collisions on the walls of the container. Gases can **diffuse** into each other; the rapidly moving molecules of two distinct gases can occupy the empty spaces between each other and form a gas solution (mixing at the molecular level).

The kinetic energy of a gas molecule (or atom) in units of Joules is found from $\mathbf{kE = \frac{1}{2} mv^2}$ where **m** = the molecular (or atomic) mass in **kilograms** and **v** = velocity of the molecule in **meters per second**. The **average** kinetic energy of any sample of a specific molecular gas is determined by the molecular mass and the average or mean velocity: $\mathbf{kE_{(average)} = \frac{1}{2} m(v_{(average)})^2}$; $v_{(average)} \cong v_{(mean)}$

The average kinetic energy **per mole** in units of Joules of any sample of a specific molecular gas is determined by the molar mass expressed in kilograms (MM in kg) and the average velocity: $\mathbf{kE_{(average)}/mole = \frac{1}{2}\,(MM\ in\ kg)(v_{(average)})^2}$

Example: Calculate the average kinetic energy per mole in Joules of hydrogen gas (H_2) with a mean (average) velocity of 15.7×10^2 m/sec at 300 K.

$$\text{MM in kg} = (2.02\ \text{g/mole}) \times (1\ \text{kg}/1000\ \text{g}) = 2.02 \times 10^{-3}\ \text{kg/mole}$$

$$\text{kE}\ (H_2) = \tfrac{1}{2}\,(2.02 \times 10^{-3}\ \text{kg/mole})(15.7 \times 10^2\ \text{m/sec})^2 = \mathbf{2.49 \times 10^3}\ \text{kg-m}^2/\text{sec}^2\ \text{(J)}$$

Example: Calculate the average kinetic energy per mole in Joules of methane gas (CH_4) with a mean (average) velocity of 5.57×10^2 m/sec at 300 K.

$$\text{MM in kg} = (16.04\ \text{g/mole}) \times (1\ \text{kg}/1000\ \text{g}) = 16.04 \times 10^{-3}\ \text{kg/mole}$$

$$\text{kE}\ (CH_4) = \tfrac{1}{2}\,(16.04 \times 10^{-3}\ \text{kg/mole})(5.57 \times 10^2\ \text{m/sec})^2 = \mathbf{2.49 \times 10^3\ kg\text{-}m^2/sec^2\ (J)}$$

Kinetic Energy, Temperature, and Graham's Law for Diffusion

Temperature (**T**) is a measure of the average kinetic energy of a sample of matter. For **any** gas sample, $\mathbf{kE_{(average)}} \propto \mathbf{T}$. Specifically for the temperature scale Kelvin (**K**), the corresponding equation is: $\mathbf{kE_{(average)}}$ = **(constant)** × **(T in K)**. As the previous examples indicate, this equation states that all gases, regardless of identity, have the same average value for kinetic energy at any specific fixed temperature.

For one mole of gas, the constant in the equation is **(3/2)R** where **R** is the molar gas constant with a value of **8.314 J/mole-K**. For any sample of gas:

$$\mathbf{kE_{(average)}/mole = (3/2)\ RT};\ T\ (K) = °C + 273$$

Example: Calculate the $kE_{(average)}$/mole for steam (water) and the $kE_{(average)}$/mole for argon at a temperature of 120°C.

(1) $kE_{(average)}$/mole = (3/2) RT
(2) Temperature in K: 120°C + 273 = 393 K; R = 8.314 J/mole-K
(3) $kE_{(average)}$/mole = (3/2) RT = (3/2) × (8.314 J/mole-K) × (393 K) = **4901 J**
(4) This value applies to both steam and argon; the average kinetic energy does **not** depend on the identity of the gas.

Example: A gas is measured at an average kinetic energy per mole of 10.0 kJ/mole. Calculate the temperature of the gas.

(1) $kE_{(average)}/\text{mole} = (3/2)\ RT;\quad \text{solve for T}: T = \dfrac{2\,kE_{(average)}}{3R}$

(2) $kE_{(average)} = 10.0\ kJ \times \dfrac{1000\ J}{1\ kJ} = 1.00 \times 10^4\ J;\ R = 8.314\ J/mole\text{-}K$

(3) $T = \dfrac{2\,kE_{(average)}}{3R} = \dfrac{2(1.00 \times 10^4\ J)}{3(8.314\ J/mole\text{-}K)} = \mathbf{802\ K}$

Diffusion is the movement of molecules of one gas through another gas; **effusion** is the movement of gas molecules through empty space (vacuum). The rate of diffusion or effusion of a gas at a specific temperature is directly proportional to the velocity of the gas molecules. The average velocity of any gas can be calculated from the average kinetic energy through $kE_{(average)}$/mole = ½ (MM in kg)$(v_{(average)})^2$. The average kinetic energy can be determined from the temperature using $kE_{(average)}$/mole = (3/2) RT.

Example: Calculate the average velocity of the molecules of steam at 393 K. The average kinetic energy/mole at 393 K was previously calculated as 4901 J.

(1) ($kE_{(average)}$/mole) = ½ (MM in kg)$(v_{(average)})^2$. Solve the equation for $v_{(average)}$:

$$\left(v_{(average)}\right) = \left(\frac{kE_{(average)}}{(\tfrac{1}{2})\ (MM\ in\ kg)}\right)^{1/2}$$

The equation is written as: ($v_{(average)}$) = [($kE_{(average)}$/(½) (MM in kg)]^{½}; the power (½) means square root; the units of v = m/sec.

(2) MM in kg = (18.02 g/mole) × (1 kg/1000g) = 0.0182 kg/mole; $kE_{(average)}$ = 4901 J
(3) ($v_{(average)}$) = [(4901 J)/(½) (0.01802 kg/mole)]^{½} = **738 m/sec**

Example: Calculate the average velocity of the atoms of argon at 393 K. The average kinetic energy/mole at 393 K was previously calculated as 4901 J.

(1) ($v_{(average)}$) = [($kE_{(average)}$/(½) (MM in kg)]^{½}
(2) MM in kg = (4.00 g/mole) × (1 kg/1000 g) = 0.00400 kg/mole; $kE_{(average)}$ = 4901 J
(3) ($v_{(average)}$) = [(4901 J)/(½) (0.00400 kg/mole)]^{½} = **1565 m/sec**

The previous examples demonstrate that the average velocity of particles of any gas sample is inversely proportional to the size of the molecules, specifically: $\mathbf{v_{(average)} \propto 1/(MM)^{1/2}}$. For any specific value of $kE_{(average)}$ (any specific temperature), the larger the molecule or atom the slower the average velocity. The rate of diffusion is proportional to the average velocity. **Graham's law** states that the rate of diffusion (average velocity) of a specific gas at a specific temperature is inversely proportional to the square root of the molar mass:

$$\textbf{(rate of diffusion)} \propto (\mathbf{v_{(average)}} \textbf{ at a specific T}) \propto \mathbf{(1/MM)^{1/2}}$$

The ratio of the diffusion rate for two different gases can be derived from this relationship. Comparing any two different gases at a specific temperature:

For gas **A: rate of diffusion of A** $\propto$ **(1/MM of A)**$^{1/2}$
For gas **B: rate of diffusion of B** $\propto$ **(1/MM of B)**$^{1/2}$

Solve for the ratio of the rate of diffusion for gas A of diffusion for gas B. Constants are not required for a ratio and units of MM can be in g/mole since units cancel in the ratio.

$$\frac{\textbf{rate of diffusion of A}}{\textbf{rate of diffusion of B}} \propto \frac{(\textbf{1/MM of A})^{1/2}}{(\textbf{1/MM of B})^{1/2}}$$

Rewrite the ratio of inverses: (MM of A) is placed on the bottom of the new fraction; (MM of B) is an inverse of an inverse and, thus, comes back up to the top:

$$\frac{(1/\text{MM of A})^{1/2}}{(1/\text{MM of B})^{1/2}} = \frac{\{\text{MM of B}\}^{1/2}}{\{\text{MM of A}\}^{1/2}} = \mathbf{\{MM\ (B) / MM\ (A)\}^{1/2}}$$

$$\textbf{rate diffusion of A/rate diffusion of B} = \mathbf{\{MM\ (B)/MM\ (A)\}^{1/2}}$$

Example: Deuterium (D) is the isotope hydrogen-2 (1 p^+ and 1 n^o); MM = 2.008 g/mole. Deuterium was extracted and concentrated for nuclear experiments by taking advantage of the diffusion rate difference between "normal" hydrogen-1 water (H_2O) and D_2O ("heavy" water). Calculate the ratio of diffusion of H_2O to D_2O.

Step (**1**): Rate diffusion H_2O/rate diffusion D_2O = {MM (D_2O)/MM (H_2O)} $^{1/2}$

Step (**2**): MM D_2O = 20.02 g/mole; MM H_2O = 18.02 g/mole

Step (**3**): Rate diffusion H_2O/rate diffusion D_2O = [(20.02 g/more)/(18.02 g/more)]$^{1/2}$
= (1.111) $^{1/2}$ = **1.054**

Example: Calculate the ratio of diffusion of argon to xeon.

Step (**1**): Rate diffusion Ar/rate diffusion Xe = {MM (Xe)/MM (Ar)} $^{1/2}$

Step (**2**): MM Ar = 39.95 g/mole; MM Xe = 131.3 g/mole

Step (**3**): Rate diffusion Ar/rate diffusion Xe = [(131.3 g/more)/(39.95 g/more)]$^{1/2}$
= (3.287)$^{1/2}$ = **1.81**

Example: Calculate the rate of diffusion of N_2 gas at a specific temperature if the diffusion rate for O_2 gas at this same temperature is measured to be 600. m/sec.

Step (1): Solve the diffusion equation for the unknown variable.

$$\text{rate diffusion } N_2/\text{rate diffusion } O_2 = \{MM\ (O_2)/MM\ (N_2)\}^{1/2}$$

$$\text{rate diffusion } N_2 = \{MM\ (O_2)/MM\ (N_2)\}^{1/2} \times (\text{rate diffusion } O_2)$$

Step (2): MM O_2 = 32.00 g/mole; MM N_2 = 28.02 g/mole; rate diffusion O_2 = 600 m/sec

Step (3): **Rate diffusion N_2** = $[(32.00 \text{ g/mole})/(28.02 \text{ g/mole})]^{1/2} \times (600 \text{ m/sec})$

$= (1.142)^{1/2} \times (600 \text{ m/sec})$

$= (1.069) \times (600 \text{ m/sec}) =$ **641 m/sec**

II THE IDEAL GAS LAW

The analysis of an **ideal gas** requires that gas molecules behave as independent particles: attractive forces between gas molecules are zero, and the volume occupied by the molecules is essentially zero as compared to the volume of the container that contains the gas (mostly empty space).

A gas is described by the variables of temperature (**T**), volume (**V**), number of moles (**n**), and pressure (**P**). Gas **pressure** is produced by molecular collisions on the walls of the container; pressure increases with increasing energy of collisions and increasing number of collisions. Pressure is affected by the following variables.

(1) Pressure is proportional to the energy of molecular collisions ($kE_{(average)}$): $\mathbf{P \propto kE_{(average)}}$. Since $kE_{(average)} \propto RT$, and this produces the relationship: $\mathbf{P \propto RT}$. Charles's law states that the pressure of a gas is directly proportional to temperature.

(2) Pressure is proportional to the number of collisions per time. As the volume of a container decreases, the number of collisions with the container walls increases; pressure is inversely proportional to the volume (Boyle's law): $\mathbf{P \propto 1/V}$

(3) The number of collisions increases as the number of molecules in a container increases; pressure is proportional to the number of moles (symbol = n) of gas in a container: $\mathbf{P \propto n}$

Combining all variables: $\mathbf{P \propto (n) \times (RT) \times (1/V)}$ or $\mathbf{P \propto nRT/V}$

The constant (R) when expressed in the correct units is the only constant required; the equation produced is

$$\mathbf{P = nRT/V} \text{ or } \mathbf{PV = nRT}$$

The units of volume are **liters (L)**; the pressure is measured in **atmospheres (atm)**; the temperature must be in Kelvin (**K**). The molar gas constant (**R**) has a different value for the different units of this equation. The units of R must be derived from the other variables:

$$\mathbf{R} = \frac{PV}{nT} = \frac{\text{atmospheres} \times \text{liters}}{\text{moles} \times \text{Kelvin}}; \mathbf{R = 0.0821\ Liter\text{-}atm/mole\text{-}K}$$

MEASUREMENTS OF GAS PRESSURE

Pressure is measured as a force per unit area. In the metric system, force is measured in units of kg-m/sec^2 and area is in square meters (m^2). Force per (i.e., divided by) area is: $\frac{\text{kg-m/sec}^2}{\text{m}^2}$. The resulting unit is kg/m-sec^2 and is defined as the **Pascal (Pa)**.

In this English system, force per unit area is stated as **pounds per square inch (psi)**. Pressure can be measured by a manometer or barometer. Gas pressure is indicated by the height of a column of a liquid that can be supported by the pressure of the gas. The average pressure of the atmosphere at sea level will support a column of **mercury** liquid to a height of 760 millimeters (**mm**). This specific pressure, termed 760 **mmHg** (or 760 **torr**), is standardized as equal to **one atmosphere (atm)** of pressure. An English unit barometer reports this height in inches; 760 mmHg = 29.9 inches Hg. Other units of pressure are converted as:

1 atmosphere (atm) = 760 mmHg = 760 torr = 101,300 Pa = 101.3 kPa = 14.7 psi

Example: One atmosphere of pressure supports a column of mercury to 760 mm. Calculate the corresponding height of a column of water that one atmosphere would support; density

H_2O = 1.00 g/ml; density of Hg = 13.6 g/ml

The height of a column of liquid supported by a fixed pressure is inversely proportional to its density. Thus $\frac{\text{height } H_2O \text{ column}}{\text{height Hg column}} = \frac{\text{density of Hg}}{\text{density of } H_2O}$

$$\text{Height } H_2O \text{ column} = \frac{13.6}{1.0} \times (\text{height Hg column}) = 13.6(760 \text{ mm}) = 10336 \text{ mm}$$

10336 mm = 10.3 meters ≅ 33 feet. Each 33 feet of fresh water (≅ 30 feet salt water) adds 1 atmosphere, or 14.7 psi, to a diver's body or a submarine's hull.

Solving Ideal Gas Law Problems

Step (**1**): Algebraically solve the PV = nRT equation for the unknown variable.

Step (**2**): Identify or calculate all required known variables in correct units. V in liters; P in atmospheres; n = moles; T in Kelvin; R = L-atm/mole-K.

Step (**3**): Complete the calculation using the correct solved equation from step (**1**).

Step (**4**): Use the information from step (**3**) to solve an additional related problem if necessary.

Example: Calculate the pressure (P) in units of atm, mmHg, psi, and kPa for 12.0 grams of helium contained in a 65.0 mL bottle at −24°C.

(1) Solve PV = nRT for P: $\mathbf{P} = \frac{\mathbf{nRT}}{\mathbf{V}}$

(2) Calculate n: $n = \frac{12.0 \text{ g He}}{4.00 \text{ g/mole}} = \mathbf{3.00 \text{ mole}}$

R is a given constant: R = **0.0821 L-atm/mole-K**
Convert T to Kelvin: T = (−24°C) + 273 = **249 K**
Convert V to liters: $V = (65.0 \text{ mL}) \times \frac{1 \text{ L}}{1000 \text{ mL}} = \mathbf{0.0650 \text{ L}}$

(3) $\mathbf{P} = \frac{\mathbf{nRT}}{\mathbf{V}} = \frac{(3.00 \text{ mole}) \times (0.0821 \text{ L-atm/mole-K}) \times (249 \text{ K})}{0.0650 \text{ L}} = \mathbf{944 \text{ atm}}$

(4) Convert 944 atm to mmHg: $(944 \text{ atm}) \times \frac{760 \text{ mmHg}}{1 \text{ atm}} = \mathbf{717000 \text{ mmHg}}$

$$\text{Convert 944 atm to psi}: (944\text{ atm}) \times \frac{14.7\text{ psi}}{1\text{ atm}} = \mathbf{1390\ psi}$$

$$\text{Convert 944 atm to kPa}: (944\text{ atm}) \times \frac{101.3\text{ kPa}}{1\text{ atm}} = \mathbf{95600\ kPa}$$

Example: Calculate the temperature (T) if 88.0 g of CO_2 in a 20.0 L container, which was found to have a pressure of 2800 torr (mmHg).

(1) Solve PV = nRT for T: $\mathbf{T} = \frac{\mathbf{PV}}{\mathbf{nR}}$

(2) Calculate n: $n = \frac{88.0\text{ g}}{44.01\text{ g/mole}} CO_2 = \mathbf{2.00\ mole}$

R is a given constant: R = **0.0821 L-atm/mole-K**
V in liters: V = 20.0 L

$$\text{Calculate P: } P = 2800\text{ mmHg} \times \frac{1\text{ atm}}{760\text{ mmHg}} = 3.684\text{ atm}$$

T is the unknown: T = ?

$$(3)\ \mathbf{T} = \frac{\mathbf{PV}}{\mathbf{nR}} = \frac{(3.684\text{ atm}) \times (20.0\text{ L})}{(2.00\text{ mole}) \times (0.0821\text{ L-atm/mole-K})} = \mathbf{449\ °K}$$

Example: Calculate the mass of C_2H_4 gas in a 20.5 L tank at 2,000 psi and 50°C.

(1) Solve PV = nRT for n: $\mathbf{n} = \frac{\mathbf{PV}}{\mathbf{RT}}$

(2) Convert P to atm: $P = 2{,}000\text{ psi} \times \frac{1\text{ atm}}{14.7\text{ psi}} = \mathbf{136\ atm}$

V is given in liters: V = **20.5 L**
R is a given constant: R = **0.0821 L-atm/mole-K**
Convert T to Kelvin: T = (50°C) + 273 = **323 K**
n is unknown: n = ?

$$(3)\ \mathbf{n} = \frac{\mathbf{PV}}{\mathbf{RT}} = \frac{(136\text{ atm}) \times (20.5\text{ L})}{(0.0821\text{ L-atm/mole-K}) \times (323\text{ K})} = \mathbf{105\ moles}$$

(4) m(g) = (105 moles)(28.06 g/mole) = 2950 grams

Example: Calculate the volume of 5.00 moles of argon at 450 K and 12.0 atm.

(1) Solve PV = nRT for V: $\mathbf{V} = \frac{\mathbf{nRT}}{\mathbf{P}}$

(2) n is **5.00** mole; R = **0.0821 L-atm/mole-K**; T = **450 K**; P = **12.0 atm**

$$(3)\ \mathbf{V} = \frac{\mathbf{nRT}}{\mathbf{P}} = \frac{(5.00\text{ mole}) \times (0.0821\text{ L-atm/mole-K}) \times (450\text{ K})}{12.0\text{ atm}} = \mathbf{15.4\ L}$$

Example: Repeat the previous example for the gas krypton.

The concept of a **molar volume of a gas** states that a specific number of moles of **any** ideal gas will occupy the same volume under the same conditions of temperature and pressure, regardless of the identity of the gas. The volume occupied by 5.00 moles of krypton at 450 K and 12.0 atm is **15.4 L**.

The density of a gas is measured as $\frac{m(g)}{V(L)}$; m(g) = moles × MM;

The mass density of a gas can be calculated from the molar density: $\frac{\textbf{moles}}{\textbf{V(L)}}$

Example: Calculate the density of C_2H_6 gas in a tank at 2,700 psi and 45°C.

(1) Solve PV = nRT for molar density: $\frac{n}{V} = \frac{P}{RT}$

(2) Convert P to atm: $P = 2{,}700\ \text{psi} \times \frac{1\ \text{atm}}{14.7\ \text{psi}} = \mathbf{183.7\ atm}$

R is a given constant: R = **0.0821 L-atm/mole-K**
Convert T to Kelvin: T = (45°C) + 273 = **318 K**
V is liters and is an unknown; **n** is unknown; the **ratio** as one variable is the unknown.

$$(3)\ \frac{n}{V} = \frac{P}{RT} = \frac{(183.7\ \text{atm})}{(0.0821\ \text{L-atm/mole-K}) \times (318\ \text{K})} = \mathbf{7.036 \frac{moles}{L}}$$

$$(4)\ \text{density in (g/L)} = \frac{(7.036\ \text{moles})}{\text{L}} \times \frac{(30.08\ \text{g})}{\text{mole}} = \mathbf{212\ g/L}$$

III CONDITION CHANGES FOR A SPECIFIC GAS SAMPLE

One specific gas sample can undergo a change in pressure, temperature, and/or volume. In this case, the number of moles of a specific sample of gas does not change; the number of moles in the sample under the initial conditions must equal the number of moles in the sample under the final conditions:

$$\mathbf{n_{(initial)} = n_{(final)}}$$

Example: A sample of methane, CH_4, occupies 260 mL at 32°C and 0.500 atm. These are the initial set of conditions. What volume will the **same sample** occupy at 1.58 atm and 150°C? These are the final set of conditions.

The final (**f**) set of conditions lists only the pressure and the temperature; the final volume and the # of moles are unknowns. PV = nRT cannot be used directly to solve for the final conditions with two missing variables. The technique is to solve the initial (**i**) set of conditions for the # of moles in the sample, then use this value to solve for the final volume.

(1) Solve PV = nRT for the initial n: $\mathbf{n_i} = \frac{\mathbf{P_iV_i}}{\mathbf{RT_i}}$

(2) $P_i = 0.500$ atm; $V_i = 0.260$ L; $T_i = 305$ K; R = 0.0821 L-atm/mole-K

$$(3)\ \mathbf{n_i} = \frac{\mathbf{P_iV_i}}{\mathbf{RT_i}} = \frac{(0.500\ \text{atm}) \times (0.260\ \text{L})}{(0.0821\ \text{L-atm/mole-K}) \times (305\ \text{K})} = \mathbf{0.00519\ moles}$$

$$\mathbf{n_{(final)} = n_{(initial)} = 0.00519\ moles}$$

(1) Solve PV = nRT for V_f: $\mathbf{V_f} = \frac{\mathbf{n_fRT_f}}{\mathbf{P_f}}$

(2) $n_f = 0.00519$ mole; R = 0.0821 L-atm/mole-K; $T_f = 423$ K; $P_f = 1.58$ atm

$$(3)\ \mathbf{V_f} = \frac{\mathbf{n_fRT_f}}{\mathbf{P_f}} = \frac{(0.00519\ \text{mole}) \times (0.0821\ \text{L-atm/mole-K}) \times (423\ \text{K})}{1.58\ \text{atm}} = \mathbf{0.114\ L}$$

A more direct method is to solve two PV = nRT equations for the initial and final sets of conditions: $\mathbf{P_iV_i = n_{(initial)}RT_i}$ and $\mathbf{P_fV_f = n_{(final)}RT_f}$. Solve each for n:

$$\mathbf{n_{(initial)}} = \frac{\mathbf{P_iV_i}}{\mathbf{RT_i}} \quad \text{and} \quad \mathbf{n_{(final)}} = \frac{\mathbf{P_fV_f}}{\mathbf{RT_f}} \quad \text{Set the expressions equal to each other:}$$

$$\mathbf{n_{(initial)}} = \mathbf{n_{(final)}} : \frac{\mathbf{P_iV_i}}{\mathbf{RT_i}} = \frac{\mathbf{P_fV_f}}{\mathbf{RT_f}} \quad \text{The constant R cancels out:} \frac{\mathbf{P_iV_i}}{\mathbf{T_i}} = \frac{\mathbf{P_fV_f}}{\mathbf{T_f}}$$

Example: Solve the previous example using the equation for initial and final conditions.

(1) Solve $P_iV_i/T_i = P_fV_f/T_f$ for V_f: $\mathbf{V_f} = \frac{\mathbf{P_iV_iT_f}}{\mathbf{P_fT_i}}$

(2) The required known variables must be in consistent units. T must always be in Kelvin; however, since the equation uses ratios, the requirement for V in liters and P in atmospheres need not be applied as long as consistent units are used for initial and final conditions. In general, converting all units to the same ones used for PV = nRT problems ensures a correct setup.

P_i = **0.500 atm** P_f = **1.58 atm**
V_i = **0.260 L** V_f = unknown
T_i = 32°C + 273 = **305 K** T_f = 150°C + 273 = **423 K**

(3) $\mathbf{V_f} = \frac{\mathbf{P_iV_iT_f}}{\mathbf{P_fT_i}} = \frac{(0.500\text{ atm})\times(0.260\text{ L})\times(423\text{ K})}{(1.58\text{ atm})\times(305\text{ K})} = \mathbf{0.14\ L}$

Example: A sample of nitrogen gas, N_2 is contained in a 122.0 L tank at 30.0°C and 5,000 psi. The gas is transferred to a new 22,090 L tank, which results in a pressure of 2000 torr (mmHg). What must be the final temperature of the gas in the new tank?

(1) Solve $P_iV_i/T_i = P_fV_f/T_f$ for T_f: $\mathbf{T_f} = \frac{\mathbf{P_fV_fT_i}}{\mathbf{P_iV_i}}$

(2) $P_i = 5{,}000\text{ psi}\times\frac{1\text{ atm}}{14.7\text{ psi}} = \mathbf{340\ atm}$ $\quad P_f = 2000\text{ mmHg}\times\frac{1\text{ atm}}{760\text{ mmHg}} = \mathbf{2.63\ atm}$

V_i = **122.0 L** V_f = **22090 L**
T_i = 30°C + 273 = **303 K** T_f = unknown

(3) $\mathbf{T_f} = \frac{\mathbf{P_fV_fT_i}}{\mathbf{P_iT_i}} = \frac{(2.63\text{ atm})\times(22090\text{ L})\times(303\text{ K})}{(340\text{ atm})\times(122.0\text{ L})} = \mathbf{424\ K}$

A specific gas sample may undergo a change in only one known variable. In this case, the other known variable is constant and can be cancelled out of the general equation. If T remains constant, use $\mathbf{P_iV_i = P_fV_f}$; if P remains constant, use $\mathbf{V_i/T_i = V_f/T_f}$; if V remains constant, use $\mathbf{P_i/T_i = P_f/T_f}$.

Example: A sample of helium has a volume of 400 mL at a pressure of 760 mmHg. What would be the volume of the sample at 2.00 atmospheres?

(1) No temperature change is stated; therefore, T remains constant so use $P_iV_i = P_fV_f$. Solve for V_f: $\mathbf{V_f} = \frac{\mathbf{P_iV_i}}{\mathbf{P_f}}$

(2) $P_i = (760\text{ mmHg})\times\frac{1\text{ atm}}{760\text{ mmHg}} = \mathbf{1.00\ atm}$ $\quad P_f = \mathbf{2.00\ atm}$

$V_i = \mathbf{0.400\ L}$ $\quad V_f$ = unknown

(3) $V_f = \dfrac{P_iV_i}{P_f} = \dfrac{(1.00\ \text{atm}) \times (0.400\ \text{L})}{(2.00\ \text{atm})} = \mathbf{0.200\ L}$

Example: A sample of carbon dioxide has a volume of 100 mL at a temperature of 25°C. What would be the volume of the sample at 50°C?

(1) No pressure change is stated; therefore, P remains constant.

Use $\dfrac{V_i}{T_i} = \dfrac{V_f}{T_f}$ and solve for V_f : $V_f = \dfrac{T_fV_i}{T_i}$

(2) $T_i = 25°C + 273 = \mathbf{298\ K}$ $T_f = 50°C + 273 = \mathbf{323\ K}$

$V_i = 0.100\ \text{L}$ $V_f = \text{unknown}$

(3) $V_f = \dfrac{T_fV_i}{T_i} = \dfrac{(323\ \text{K}) \times (100\ \text{ml})}{(298\ \text{K})} = \mathbf{0.108\ L}$

IV SOLVING ADDITIONAL PROBLEMS USING GAS LAWS

A common use of the ideal gas law is to determine the molar mass (MM) of any compound that is a gas or can be converted to a gas. The MM of a compound can be defined as equal to the mass of any sample of the compound divided by the number of moles of the same sample:

$$\text{MM} = \frac{\text{mass of any sample}}{\text{moles of same sample}}$$

If the mass of a sample is directly measured, moles of the same sample can be found by measurement of gas volume under a specific set of conditions.

Example: A sample of an unknown gas was shown to have a volume of 546 mL at 48.5°C and 26.4 psi pressure; the mass of the sample was 1.464 grams. Determine the **MM** of the unknown gas.

(1) $\text{MM} = \dfrac{\text{mass of sample}}{\text{moles of sample}}$; moles of sample must be calculated: Solve PV = nRT for n: $\mathbf{n} = \dfrac{\mathbf{PV}}{\mathbf{RT}}$

(2) Convert P to atm: $P = 29.4\ \text{psi} \times \dfrac{1\ \text{atm}}{14.7\ \text{psi}} = \mathbf{1.796\ atm}$

Convert V to liters: $V = (546\ \text{ml}) \times \dfrac{1\ \text{L}}{1000\ \text{mL}} = \mathbf{0.546\ L}$

R is a given constant: R = **0.0821 L-atm/mole-K**
Convert T to Kelvin: T = 48.5°C + 273 = **321.5 K**
n is unknown: n = ?

(3) $\mathbf{n = PV/RT} = \dfrac{(1.796\ \text{atm}) \times (0.546\ \text{L})}{(0.0821\ \text{L-atm/mole-K}) \times (321.5\ \text{K})} = \mathbf{0.03715\ moles}$

(4) $\text{MM} = \dfrac{\text{mass of sample}}{\text{moles of sample}} = \dfrac{1.464\ \text{grams}}{0.03715\ \text{moles}} = \mathbf{39.4\ g/mole}$

Stoichiometric Calculations Using Gas Laws

Many chemical reactions have gases as either products or reactants. Stoichiometric problems involving gases may require gas laws to perform mole calculations. Stoichiometry techniques are the same as described in Chapter 8.

Example: Calculate the **volume** in liters of CO_2 measured at 1.45 atmospheres and 55°C (328 K), which can be prepared from 1.50 kg of Fe_2O_3 based on the balanced equation.

$$\mathbf{Fe_2O_3} + \mathbf{3CO} \longrightarrow \mathbf{2Fe} + \mathbf{3CO_2}$$

Fe_2O_3: mass = 1500 g ↓ CO_2: volume = **524 L** ↑

(1)

$$\text{moles } Fe_2O_3 = \frac{1500\text{ g}}{159.6\text{ g/mole}} = 9.40\text{ mole}$$

(4)

$$\text{V of } CO_2 = \frac{nRT}{P} = \frac{(28.2\text{ mol})\times(0.0821)\times(328\text{ K})}{1.45\text{ atm}} = \mathbf{524\ L}$$

$$\downarrow 9.40\text{ mole} \xrightarrow{(2)} \times\frac{3\ CO_2}{1\ Fe_2O_3} \xrightarrow{(3)} = 28.2\text{ mole } CO_2 \uparrow$$

Example: Calculate the **volume** in liters of H_2 gas measured at 2.00 atmospheres and 400°C (673 K), which is required to exactly react with 2.00 kg of C_6H_6 based on the balanced equation.

$$\mathbf{C_6H_6} + \mathbf{3H_2} \longrightarrow \mathbf{C_6H_{12}}$$

C_6H_6: mass = 2000 g ↓ H_2: volume = **2120 L** ↑

(1)

$$\text{moles } C_6H_6 = \frac{2000\text{ g}}{78.12\text{ g/mole}} = 25.60\text{ mole}$$

(4)

$$\text{V of } H_2 = \frac{nRT}{P} = \frac{(76.80\text{ mol})\times(0.0821)\times(673\text{ K})}{(2.00\text{ atm})} = \mathbf{2122\ L\ (2120\ L)}$$

$$\downarrow 25.60\text{ mole} \xrightarrow{(2)} \times\frac{3\ H_2}{1\ C_6H_6} \xrightarrow{(3)} = 76.80\text{ mole } H_2 \uparrow$$

V GAS MIXTURES AND LAW OF PARTIAL PRESSURES

Gas mixtures are solutions (mixing is at the molecular level). All the different gas molecules in an **ideal** gas mixture behave as independent particles. As independent particles, the volume of each gas type are additive in that all individual volumes can be added directly to produce a total volume; each independent volume must be measured under identical conditions:

$$\mathbf{V_{(total\ in\ mixture)} = V_{(gas\ A)} + V_{(gas\ B)} + V_{(gas\ C)} \cdots\cdots}$$

Pressures of each gas type are additive to produce a total pressure if all gases are measured in the same container volume:

$$\mathbf{P_{(total\ in\ mixture)} = P_{(gas\ A)} + P_{(gas\ B)} + P_{(gas\ C)} \cdots\cdots}$$

The individual pressure of each gas type in a mixture is termed the **partial pressure** of the gas; the partial pressure of gas A is symbolized as $P_{(gas\ A)}$. The partial pressure of a gas component of a mixture is the pressure that gas would have if it were the only gas present in the volume of a specific container.

Example: 1.00 mole of H_2 and 0.500 mole of O_2 are placed in a 2.50 L container at 273 K. What are $P_{(H2)}$, $P_{(O2)}$, and $P_{(total)}$?

(1) $P_{(H2)}$ = partial pressure of H_2; $P_{(O2)}$ = partial pressure of O_2

$$P_{(H2)} = \frac{n_{(H2)}RT}{V_{(container)}} \qquad P_{(O2)}\frac{n_{(O2)}RT}{V_{(container)}}$$

(2) $V_{(container)} = 2.50$ L; $T = 273$ K; $n_{(H2)} = 1.00$ mole; $n_{(O2)} = 0.500$ mole;

$$R = 0.0821 \text{ L-atm/mole-K}$$

(3) $$\mathbf{P_{(H2)}} = \frac{(1.00 \text{ mole}) \times (0.0821) \times (273 \text{ K})}{(2.50 \text{ L})} = \mathbf{8.96\ atm}$$

$$\mathbf{P_{(O2)}} = \frac{(0.500 \text{ mole}) \times (0.0821) \times (273 \text{ K})}{(2.50 \text{ L})} = \mathbf{4.48\ atm}$$

$$P_{(total)} = P_{(O2)} + P_{(H2)} = 4.48 \text{ atm} + 8.96 \text{ atm} = \mathbf{13.44\ atm}$$

For a gas mixture in a specific container at a specific temperature, V, T, and R are all constants; therefore, the partial pressure of a specific gas in a mixture is proportional to its number of moles: $\mathbf{P_{(gas\ A)} \propto n_{(gas\ A)}}$ and therefore $\mathbf{P_{(total)} \propto n_{(total)}}$.

Pressures, volumes, and number of moles are all additive. An alternative method for determining the total pressure in the previous example is to use:

$$\mathbf{P_{(total)}\ V_{(container)} = n_{(total)}RT}$$

$$n_{(total)} = n_{(H2)} + n_{(O2)} = 1.00 \text{ moles} + 0.500 \text{ moles} = 1.50 \text{ moles}$$

$$P_{(total)} = \frac{n_{(total)}RT}{V_{(total)}} = \frac{(1.50 \text{ mole}) \times (0.0821) \times (273 \text{ K})}{(2.50 \text{ L})} = \mathbf{13.44\ atm}$$

Since $\mathbf{P_{(gas\ A)} \propto n_{(gas\ A)}}$ and $\mathbf{P_{(total)} \propto n_{(total)}}$, it can be derived that the partial pressure of a gas in a mixture is equal to the total pressure multiplied by the ratio of the number of moles of A divided by the total number of moles in the gas solution.

$$\mathbf{P_{(gas\ A)}} = (\mathbf{P_{(total)}}) \times \frac{(\mathbf{n_{(gas\ A)}})}{(\mathbf{n_{(total)}})}$$

This equation can be solved for $\frac{(\mathbf{n_{(gas\ A)}})}{(\mathbf{n_{(total)}})}$: $\frac{(\mathbf{n_{(gas\ A)}})}{(\mathbf{n_{(total)}})} = \frac{\mathbf{n_{(gas\ A)}}}{\mathbf{n_{(total)}}}$

The ratio of the # of moles of A divided by the total # of moles in the solution (mixture) is defined as the **mole fraction of component A** and abbreviated with the symbol ($\chi_{(gas\ A)}$). This is a solution concentration unit. $\chi_{(gas\ A)} \equiv \frac{(\mathbf{n_{(gas\ A)}})}{(\mathbf{n_{(total)}})}$

$$P_{(gas\ A)} = (P_{(total)}) \times \frac{(n_{(gas\ A)})}{(n_{(gas\ A)})} \quad \text{is phrased as:} \quad P_{(gas\ A)} = P_{(total)} \times \chi_{(gas\ A)}$$

$$\frac{(n_{(gas\ A)})}{(n_{(total)})} = \frac{P_{(gas\ A)}}{P_{(total)}} \quad \text{is phrased as:} \quad \chi_{(gas\ A)} = \frac{(n_{(gas\ A)})}{(n_{(total)})}$$

Example: Air is 79% by mole N_2 and 21% by mole O_2. What is the $P_{(N2)}$ and $P_{(O2)}$ in a sample of air at 785 mmHg?

Since 79% by mole and 21% by mole add up to 100% mole total, the mole fraction of N_2 ($\chi_{(N2)}$) would be 0.79 and the mole fraction of O_2 ($\chi_{(O2)}$) would be 0.21.

$$P_{(N2)} = P_{(total)} \times \chi_{(N2)} = (785\text{ mmHg}) \times (0.79) = \mathbf{620\ mmHg}$$
$$P_{(O2)} = P_{(total)} \times \chi_{(O2)} = (785\text{ mmHg}) \times (0.21) = \mathbf{165\ mmHg}$$

Example: A diver's air tank is enriched with oxygen; it is 40% mole percent O_2. The tank has a volume of 22.0 L at a pressure of 26.5 atm and a temperature of 10.5°C (283.5 K). Calculate the number of moles of oxygen in the tank.

(1) Solve $P_{(O2)}V = n_{(O2)}RT$ for $n_{(O2)}$: $\mathbf{n_{(O2)}} = \dfrac{\mathbf{P_{(O2)}V}}{\mathbf{RT}}$

(2) V = 22.0 L;R = 0.0821 L-atm/mole-K; T = 283.5 K

$P_{(O2)}$ is not given but $P_{(total)} = 26.5$ atm; use $P_{(O2)} = P_{(total)} \times \chi_{(O2)}$
40% O_2 by mole produces $\chi_{(O2)} = 0.40$
$P_{(O2)} = P_{(total)} \times \chi_{(O2)} = (26.5\text{ atm}) \times (0.40) = 10.6$ atm

$$\text{(3) } \mathbf{n_{(O2)}} = \frac{\mathbf{P_{(O2)}V}}{\mathbf{RT}} = \frac{(10.6\text{ atm}) \times (22.0\text{ L})}{(0.0821\text{ L-atm/mole-K}) \times (283.5\text{ K})} = \mathbf{10.0\ moles\ O_2}$$

An alternative method is to solve for moles of O_2 from the definition of mole fraction: $(n_{(O2)}) = (n_{(total)})\chi_{(O2)}$; Solve $P_{(total)}V = n_{(total)}RT$ for $n_{(total)}$:

$$\text{(1) } \mathbf{n_{(total)}} = \frac{\mathbf{P_{(total)}V}}{\mathbf{RT}}$$

(2) V = 22.0 L; T = 283.5 K; $P_{(total)} = 26.5$ atm

$$\text{(3) } \mathbf{n_{(total)}} = \frac{(26.5\text{ atm}) \times (22.0\text{ L})}{(0.0821\text{ L-atm/mole-K}) \times (283.5\ K)} = \mathbf{25.0\ moles}$$

(4) $\mathbf{n_{(O2)}} = n_{(total)} \times \chi_{(O2)} = (25.0\text{ moles}) \times (0.40) = \mathbf{10.0\ moles\ O_2}$

Example: A contaminated sample of air was taken from a house with a defective furnace. A 225 mL sample bottle was filled with contaminated air at 746 mmHg at T = 19°C. The sample was analyzed for mass of carbon monoxide (CO). The following parts lead to a calculation for the mole fraction of CO in the air.

a) Calculate the total number of moles of gas in the sample. Note that the **identity** of all the gases is not required.

(1) $\mathbf{n} = \dfrac{\mathbf{PV}}{\mathbf{RT}}$

(2) $P = (746\text{ mmHg}) \times \dfrac{1\text{ atm}}{760\text{ mmHg}} = 0.982\text{ atm};\ T = 292\text{ K};\ V = 0.225\text{ L}$

(3) $\mathbf{n} = \dfrac{(0.982\text{ atm}) \times (0.225\text{ L})}{(0.0821\text{ L-atm/mole-K}) \times (292\text{ K})} = \mathbf{9.22 \times 10^{-3}}$ total mole of gas

b) The sample was found to contain 7.23 **mg** of carbon monoxide (CO). Calculate moles of CO from the mass and MM.

$$\text{mass in grams of CO} = (7.23\ \text{mg}) \times \frac{1\ \text{g}}{1000\ \text{mg}} = 0.00723\ \text{g}$$

$$\textbf{mole CO} = \frac{0.00723\ \text{g}}{28.01\ \text{g/mole}} = \mathbf{2.58 \times 10^{-4}\ moles\ of\ CO}$$

c) Complete the calculation of mole fraction of CO.

$$\chi_{(CO)} = \frac{\mathbf{n_{(CO)}}}{\mathbf{n_{(total)}}} = \frac{(2.58 \times 10^{-4}\ \text{moles CO})}{(9.22 \times 10^{-3}\ \text{moles total})} = \mathbf{0.0280}$$

d) Separately calculate $P_{(CO)}$ and calculate mole fraction of CO from $\chi_{(CO)} = \frac{P_{(CO)}}{(P_{(total)})}$

$$P_{(CO)} = \frac{\mathbf{n_{(CO)}RT}}{\mathbf{V}} = \frac{(2.58 \times 10^{-4}\ \text{mole CO}) \times (0.0821\ \text{L-atm/mole-K}) \times (292\ \text{K})}{0.225\ \text{L}}$$

$$= \mathbf{0.0275\ atm}$$

$$\chi_{(CO)} = \frac{\mathbf{P_{(CO)}}}{\mathbf{P_{(total)}}} = \frac{0.0275\ \text{atm}}{0.982\ \text{atm}} = \mathbf{0.0280}$$

VI PRACTICE PROBLEMS

1. Calculate the total kinetic energy at 0°C of 1.00 g of N_2 gas.
2. Calculate the ratio of gas diffusion rates for CH_4 vs. H_2S.
3. A 5.00 L tank contains 132.3 g of propane (C_3H_8) at a temperature of 37.0°C. Calculate the **pressure** in the tank.
4. Calculate the **mass** of helium in a balloon of 55.0 L at 10°C and 720 mmHg.
5. A tank contains 0.500 **kilograms** of oxygen gas at 20°C. The pressure in the tank is measured as 742.4 psi. Calculate the **volume** of the tank.
6. A sample of unknown gas has a volume of 273 mL at a pressure of 710 torr and a temperature of 79°C. The mass of the unknown sample is measured to be 0.406 grams. Determine the **molar mass** of the unknown gas.
7. A sample of hydrogen gas at 25°C and 745 mmHg fills a 100 mL container. A piston compresses the volume of the container to 50.0 mL, and the gas is heated to 100°C. What is the new **pressure?**
8. A 35.0 L tank stores acetylene at 2250 psi pressure at −10°C. What **volume** must a new tank be if the gas is transferred for storage at a desired pressure of 700 torr at a temperature of 30°C.
9. A sample of argon has a pressure of 1.00 atm. At 25°C. What is the **pressure** if the same sample volume is heated to 50°C?
10. How many **liters** of oxygen gas, measured at 300°C and 740 mmHg are required to completely burn 100 g of carbon according to the equation shown?

$$C_{(s)} + O_{2\,(g)} \longrightarrow CO_{2\,(g)}$$

11. How many **grams** of CO_2 are produced by the burning of a full tank of propane? The propane tank holds 6.20 L at 200 psi at 25°C.

$$C_3H_{8\,(g)} + 5\,O_{2\,(g)} \longrightarrow 3\,CO_{2\,(g)} + 4\,H_2O_{(g)}$$

12. A certain reaction produces 0.5648 grams of an unknown gas which is collected in a 500 mL glass bottle filled with air at 122°C. The partial pressure of the unknown gas in the bottle was $P_{(unknown\ gas)} = 4.62$ psi. Calculate the molar mass of the unknown gas.

13. 5.00 g of NH_4NO_3 is placed in helium filled 2.00 L container at 25.0°C and a pressure of 1.00 atm. The ammonium nitrate explodes according to the equation shown, and the temperature increases to 200°C. Complete the following parts to calculate the total pressure in the container after explosion.

$$2\,NH_4NO_{3\,(s)} \longrightarrow O_{2\,(g)} + 2\,N_{2\,(g)} + 4\,H_2O_{(g)}$$

a) Calculate the number of moles of He in the container before explosion.
b) Calculate the number of moles of each of the gas products of the reaction using stoichiometry based on complete consumption of the NH_4NO_3 limiting reagent.
c) Calculate the total pressure based on the final conditions; determination of n(total) is the shortest method.

VII ANSWERS TO PRACTICE PROBLEMS

1. Calculate the kinetic energy at 0°C of 1.00 gram of N_2 gas:

$$\text{Mole } N_2 = \frac{1.00\text{ g}}{28.02\text{ g/mole}} = 0.0357\text{ mole; T} = 273\text{ K; R} = 8.314\text{ J/mole-K}$$

$$kE_{(average)}/\text{mole} = (3/2)\,RT = (3/2) \times (8.314\text{ J/mole-K}) \times (273\text{ K}) = 3405\text{ J/mole}$$

$$\textbf{total kE}_{\textbf{(average)}} = (3405\text{ J/mole}) \times (0.0357\text{ mole } \mathbf{N_2}) = \mathbf{122\ J}$$

2. Calculate the ratio of gas diffusion rates for CH_4 vs. H_2S:

$$\frac{\text{rate diffusion } CH_4}{\text{rate diffusion } H_2S} = \{MM(H_2S)/MM(CH_4)\}^{1/2}$$

$$= \{(34.12\text{ g/mole})/(16.05\text{ g/mole})\}^{1/2} = (2.126)^{1/2} = \mathbf{1.458}$$

3. 5.00 L tank; 132.3 grams of propane (C_3H_8); T = 37.0°C.

(1) $P = \dfrac{nRT}{V}$ (2) R = 0.0821 L-atm/mole-K; V = 5.00 L

$$n = \frac{132.3\text{ g}}{44.11\text{ g/mole}} = 3.00\text{ mole; T} = (37°\text{C}) + 273 = 310\text{ K}$$

(3) $P = \dfrac{nRT}{V} = \dfrac{(3.00\ \text{mole}) \times (0.0821\ \text{L-atm/mole-K}) \times (310\ \text{K})}{5.00\ \text{L}} = \mathbf{15.3\ atm}$

4. Mass of helium in a balloon of 55.0 L at 10°C and 720 mmHg.

(1) $n = \dfrac{PV}{RT}$ (2) $V = 55.0\ \text{L};\ T = (10\ °\text{C}) + 273 = 283\ \text{K}$

$$P = 720\ \text{mmHg} \times \frac{1\ \text{atm}}{760\ \text{mmHg}} = 0.947\ \text{atm};\ R = 0.0821\ \text{L-atm/mole-K}$$

(3) $n = \dfrac{PV}{RT} = \dfrac{(0.947\ \text{atm}) \times (55.0\ \text{L})}{(0.0821\ \text{L-atm/mole-K}) \times (283\ \text{K})} = \mathbf{2.24\ moles}$

(4) **mass** He = (2.24 moles) × (4.00 g/mole) = **8.96 g**

5. Tank with 0.500 kg of oxygen gas at 20°C and 742.4 psi

(1) $V = \dfrac{nRT}{P}$ (2) $T = (20\ °\text{C}) + 273 = 293\ \text{K};\ R = 0.0821\ \text{L-atm/mole-K}$

$$n = \frac{500\ \text{g}}{32.00\ \text{g/mole}} = 15.6\ \text{mole};\ P = (742.4\ \text{psi}) \times \frac{1\ \text{atm}}{14.7\ \text{psi}} = 50.5\ \text{atm}$$

(3) $V = \dfrac{nRT}{P} = \dfrac{(15.6\ \text{mole}) \times (0.0821\ \text{L-atm/mole-K}) \times (293\ \text{K})}{50.5\ \text{atm}} = \mathbf{7.43\ L}$

6. Unknown gas: V = 273 mL; P = 710 torr; T = 79°C; sample mass = 0.406 grams.

(1) $n = \dfrac{PV}{RT}$

(2) $T = (79\ °\text{C}) + 273 = 352\ \text{K};\ R = 0.0821\ \text{L-atm/mole-K}$

(3) $P = (710\ \text{torr} = 710\ \text{mmHg}) \times \dfrac{1\ \text{atm}}{760\ \text{mmHg}} = 0.934\ \text{atm};\ V = 0.273\ \text{L}$

(4) $n = \dfrac{PV}{RT} = \dfrac{(0.934\ \text{atm}) \times (0.273\ \text{L})}{(0.0821\ \text{L-atm/mole-K}) \times (352\ \text{K})} = \mathbf{0.00882\ moles}$

(5) $MM = \dfrac{\text{mass of sample}}{\text{moles of sample}} = \dfrac{0.406\ \text{grams}}{0.00882\ \text{moles}} = 46.0\ \text{g/mole}$

7. A sample of hydrogen gas at initial conditions of 25°C, 745 mmHg, V = 100 mL. Final conditions: V = 50.0 mL; T = 100°C

(1) $P_f = \dfrac{P_i V_j T_f}{V_f T_i}$

(2) $P_i = (745\ \text{mmHg}) \times \dfrac{1\ \text{atm}}{760\ \text{mmHg}} = 0.980\ \text{atm}$ P_f = unknown

$V_i = 0.100\ \text{L}$ $V_f = 0.0500\ \text{L}$

$T_i = 25°\text{C} + 273 = 298\ \text{K}$ $T_f = 100°\text{C} + 273 = 373\ \text{K}$

(3) $P_f = \frac{P_i V_j T_f}{V_f T_i} = \frac{(0.980\ atm)\times(0.100\ L)\times(373\ K)}{(0.0500\ L)\times(298\ K)} = \mathbf{2.45\ atm}$

8. Initial conditions: 35.0 L tank of acetylene at 2250 psi and −10°C. Final conditions: 700 torr and 30°C; (700 torr = 700 mmHg)

(1) $V_f = \frac{P_i V_i T_f}{P_f T_i}$

(2) $P_i = (2250\ psi) \times \frac{1\ atm}{14.7\ psi} = 153\ atm \quad P_f = (700\ mmHg) \times \frac{1\ atm}{760\ mmHg} = 0.921\ atm$

$V_i = 35.0\ L\ V_f = unknown$

$T_i = (-10°C) + 273 = 263\ K\ T_f = 30°C + 273 = 303\ K$

(3) $V_f = \frac{P_i V_i T_f}{P_f T_i} = \frac{(153\ atm)\times(35.0\ L)\times(303\ K)}{(0.921\ atm)\times(263\ K)} = \mathbf{6627\ L}$

9. A sample of argon; initial conditions: pressure of 1.00 atm at 25°C. Final conditions: 50°C.

(1) No volume is state; use $\frac{P_i}{T_i} = \frac{P_f}{T_f}$; Solve for P_f : $P_f = \frac{P_i T_f}{T_i}$

(2) $P_i = 1.00\ atm\ P_f = unknown$

$T_i = 25°C + 273 = 298\ K \qquad T_f = 50°C + 273 = 323\ K$

(3) $P_f = \frac{P_i T_f}{T_i} = \frac{(1.00\ atm)\times(323\ K)}{(298\ K)} = \mathbf{1.08\ atm}$

10. a) How many **liters** of oxygen, measured at 300°C and 740 mmHg are required to completely burn 100 g of carbon according to the equation shown.
T = 300°C + 273 = 573 K P = (740 mmHg) × (1 atm/760 mmHg) = 0.974 atm

$C_{(s)}$	+	$O_{2\ (g)}$	$\longrightarrow$	$CO_{2\ (g)}$
mass = 100 g		volume = **402 L**		

↓ (1)

moles C = $\frac{100\ g}{12.01\ g/mole}$ = 8.33 mole

(4) ↑ $V = \frac{nRT}{P} = \frac{(8.33\ mol)\times(0.0821)\times(573\ K)}{(0.974\ atm)} = \mathbf{402\ L}$

↓ 8.33 mole $\xrightarrow{(2)} \times \frac{1\ O_2}{1\ C} \xrightarrow{(3)}$ = 8.33 mole O_2 ↑

11. How many **grams** of CO_2 are produced by the burning of a full tank of propane? The propane tank holds 6.20 L at 200 psi at 25°C.
T = 25°C + 273 = 298 K; P = (200 psi) × (1 atm/14.7 psi) = 13.6 atm

$$\mathbf{C_3H_{8\,(g)} \quad + \quad 5\,O_{2\,(g)} \quad \longrightarrow \quad 3\,CO_{2\,(g)} \quad + \quad 4\,H_2O_{(g)}}$$

V = 6.20 L — mass = **456 g**

↓ (1) — (4) ↑

$$n = PV/RT = \frac{(13.6\text{ atm})\times(6.20\text{ L})}{(0.0821)\times(298\text{ K})} = \mathbf{3.45\ moles}$$

$$\text{mass } CO_2 = (\text{moles})\times(\text{MM}) = (10.4\text{ mole})\times(44.01\text{ g/mole}) = \mathbf{456\ grams}$$

$$\downarrow\ 3.45\text{ mole} \xrightarrow{(2)} \times \frac{3\,CO_2}{1\,C_3H_8} \xrightarrow{(3)} = 10.4\text{ mole } CO_2 \uparrow$$

12. 0.5648 grams of an unknown gas; T = 122°C; V = 500 mL; $P_{(unknown\ gas)}$ = 4.62 psi. Calculate the molar mass of the unknown gas.

(1) $\text{MM} = \dfrac{\text{mass of sample}}{\text{moles of sample}}$; $n_{(unknown\ gas)} = \dfrac{P_{(unknown\ gas)}V}{RT}$

(2) $P_{(unknown\ gas)} = 4.62\text{ psi} \times \dfrac{1\text{ atm}}{14.7\text{ psi}} = 0.314\text{ atm}$; V = 0.500 L;

T = 122°C + 273 = 395 K; R = 0.0821 L-atm/mole-K

(3) $n_{(unknown\ gas)} = \dfrac{(0.314\text{ atm})\times(0.500\text{ L})}{(0.0821\text{ L-atm/mole-K})\times(395\text{ K})} = \mathbf{0.00484\ moles}$

(4) $\text{MM} = \dfrac{\text{mass of sample}}{\text{moles of sample}} = \dfrac{0.5648\text{ grams}}{0.00484\text{ moles}} = \mathbf{117\ g/moles}$

13. 5.00 grams of NH_4NO_3 in a helium filled 2.00 L container at 25.0°C and a pressure of 1.00 atm. The final temperature after explosion is 200°C.

$$2\,NH_4NO_{3\,(s)} \longrightarrow O_{2\,(g)} + 2\,N_{2\,(g)} + 4\,H_2O_{(g)}$$

a) Calculate the number of moles of He in the container before explosion.

(1) $n = \dfrac{PV}{RT}$ (2) P = 1.00 atm; T = 298 K; V = 2.00 L; R = 0.0821 L-atm/mole-K

(3) $\mathbf{n} = \dfrac{(1.00\text{ atm})\times(2.00\text{ L})}{(0.0821\text{ L-atm/mole-K})\times(298\text{ K})} = \mathbf{0.0817\ moles}$

b) Calculate the number of moles of each of the gas products.

$$\text{mole } NH_4NO_3 = \frac{5.00\text{ g}}{80.0\text{ g/mole}} = \mathbf{0.0652\ moles}$$

$$\text{mole } O_2 = 0.0625\text{ mole} \times \frac{1\,O_2}{2\,NH_4NO_3} = 0.0313\text{ mole } O_2$$

$$\text{mole } N_2 = 0.0625\text{ mole} \times \frac{2\,N_2}{2\,NH_4NO_3} = 0.0625\text{ mole } N_2$$

$$\text{mole } H_2O = 0.0625\text{ mole} \times \frac{4\,H_2O}{2\,NH_4NO_3} = 0.125\text{ mole } H_2O$$

c) Calculate the total pressure based on the final conditions; determination of n(total) is the shortest method.

(1) $P_{(total)} = \dfrac{n_{(total)}RT}{V}$ **(2)** $T = 473\text{ K};\ V = 2.00\text{ L};\ R = 0.0821\text{ L-atm/mole-K}$

$$\mathbf{n_{(total)}} = (0.0817\text{ mole He}) + (0.0313\text{ mole } O_2) + (0.0625\text{ mole } N_2) + (0.125\text{ mole } H_2O)$$

$$= 0.301\text{ mole total}$$

$$\textbf{(3) } \mathbf{P} = \frac{\mathbf{n_{(total)}RT}}{\mathbf{V}} = \frac{(0.301\text{ mole}) \times (0.0821\text{ L-atm/mole-K}) \times (473\text{ K})}{2.00\text{ L}} = \mathbf{5.84\ atm}$$

21 Guideline for Analyzing Intermolecular Forces and Calculating Phase Change Enthalpies

I OVERVIEW OF INTERPARTICLE AND INTERMOLECULAR FORCES

Interparticle forces are the forces of attraction between atoms, ions, or molecules that hold bulk matter together.

Ionic compounds do not exist in individual molecules but as extended three-dimensional arrays of ions held together by **ionic forces**; these are the same electromagnetic forces of attraction that hold any individual positive and negative ions together. Ionic forces are extremely strong; ionic compounds are all solids at 25°C and have very high melting points.

Metals in the elemental form are extended three-dimensional arrays of metal atoms held together by **metallic bonds**. The metallic bond is formed by the overlap of many individual metal atomic orbitals producing very extended molecular orbitals, "delocalized" orbitals.

Network covalent compounds are extended three-dimensional arrays of atoms held together by **covalent bonds**. The structure can be completely covalent; examples are diamond (carbon solid) or quartz/glass (SiO_2). The structure may also be a "hybrid" of large covalent sections partially held together by other intermolecular forces; an example is graphite, another form of solid carbon.

Molecular covalent compounds exist in the form of discrete molecules. Individual molecules are formed through covalent bonding between the atoms in one molecule. However, discrete individual molecules cannot be covalently bonded to each other to assemble into larger amounts of matter. In this case, large amounts of molecular matter must be held together through non-covalent attractive forces termed intermolecular forces. Intermolecular forces are non-bonding attractive forces between separate individual molecules; they are usually classified as (permanent) dipole force, hydrogen bonding, and dispersion forces (induced dipole).

Phase Changes and Temperature

All particles of solid, liquid, or gas have kinetic energy of motion: kE of translation (movement through space), kE of vibration (vibration of atoms in bonds), or kE of rotation. The average kE of motion is proportional to temperature; **kE α T**.

The physical state of matter (solid, liquid, or gas) depends on the balance between the total strength of the interparticle force holding the particles together and the average kinetic energy of motion. Therefore, the physical state of matter changes as a function of temperature.

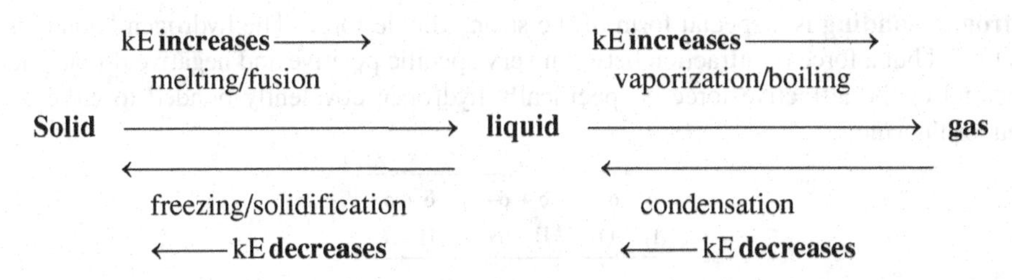

The normal boiling point of a liquid is defined as the temperature at which the vapor pressure of the liquid reaches one atmosphere (see Chapter 20). A visual interpretation of this might be that the boiling point is the temperature at which a significant amount of liquid is bubbling and evaporating. The boiling point represents the temperature at which the kinetic energy of motion overcomes the strength of (breaks) the total interparticle forces. The stronger interparticle force is associated with the higher boiling point.

The normal melting point of a solid is defined as the temperature at which the solid and liquid are at equilibrium at one atmosphere of pressure. A visual interpretation of this might be that the melting point is the temperature at which a significant amount of sold is melting to form liquid. The melting point represents the temperature at which the kinetic energy of motion partially overcomes the strength of (partially breaks) the interparticle forces holding the solid. The stronger interparticle force is associated with the higher melting point.

Descriptions of Interparticle Forces

The interparticle force for ionic compounds is the ionic bond; the corresponding interparticle force for network covalent elements or compounds is the covalent bond. Molecular covalent compounds must be held together in bulk matter by intermolecular forces.

The **dipole force** (or permanent dipole force) is the electromagnetic force of attraction between oppositely charged ends of permanent molecular dipoles. The strength of the dipole force is proportional to the degree of polarity of the entire molecule reflected in the size of the permanent molecular dipole.

Example: $^{\delta+}$H—Br$^{\delta-}$-------$^{\delta+}$H—Br$^{\delta-}$ (------- = dipole force of attraction)

The size of the permanent molecular dipole (described in Chapter 16) is proportional to the electronegativity difference (**ΔEN**) between bonded atoms and the degree of asymmetry of the central atoms involved in polar bonds. The vector result of Individual bond dipoles determines whether symmetric central atoms will completely or mostly have cancelling bond dipoles.

Very generally, the greater the electronegativity difference between atoms in a covalent bond, the larger the potential dipole (related to degree of charge separation). For practical experience, ranges of ΔEN can be thought of as leading to certain approximate dipole strengths:

ΔEN $\leq$ 0.4… usually **weak** dipoles
ΔEN of 0.5 to 1.0… usually leads to **moderate** dipoles
ΔEN > 1.0… usually leads to **strong** dipoles

The strength of the molecular dipoles can range in a continuum.

very weak ⟷ weak ⟷ moderate ⟷ strong ⟷ very strong

Very weak molecular dipoles will provide only very weak total dipole intermolecular attractive forces; the strength of the total dipole attractive forces will increase with the strength of the molecular dipole.

Hydrogen bonding is a special form of the strong dipole force. The hydrogen "bond" is **not** a covalent bond but a force of attraction between very specific positive and negative dipole ends. The **positive** end of the attractive force is specifically hydrogen covalently bonded to either oxygen, nitrogen, or fluorine:

δ + δ−	δ + δ−	δ + δ−
H—O	H—N	H—F

The **negative** end of the hydrogen bond can be the lone pairs of oxygen, nitrogen, or fluorine, regardless of whether these atoms are bonded to a hydrogen:

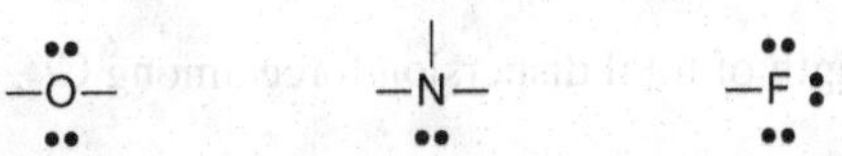

$$-\ddot{\underset{\cdot\cdot}{O}}- \qquad -\underset{\cdot\cdot}{\overset{|}{N}}- \qquad -\ddot{\underset{\cdot\cdot}{F}}:$$

Hydrogen bonding is stronger than other dipole forces due to the small sizes and high relative charge concentrations of hydrogen and the row-2 elements (O, N, F). Hydrogen bonding can occur in a variety of ways between molecules of the **same** compound. The positive end is very specific but the negative end can be more variable; the oxygen or nitrogen lone pair negative end need not come from the same type of bonding structure producing the hydrogen positive end.

Example: Arrowed/dashed lines (← – →) indicate hydrogen bonding. Examples (a) and (b) show hydrogen bonding between symmetrical H—O bonding sections of the molecules; that is, the negative end oxygen is the same atom bonded to the positive end hydrogen. Example (c) shows the same molecule as (b) participating in a different hydrogen bonding combination that involves the double-bonded oxygen. Example (d) shows two versions of hydrogen bonding in one molecule.

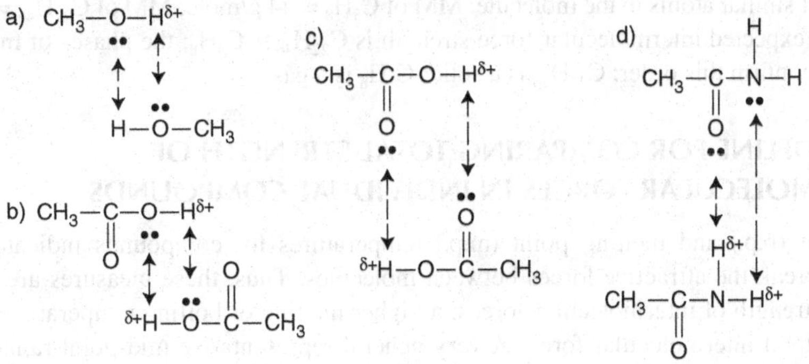

a) $CH_3-\ddot{O}-H^{\delta+}$ ↕ $H-\ddot{O}-CH_3$

b) $CH_3-C(=O)-\ddot{O}-H^{\delta+}$ ↕ $^{\delta+}H-\ddot{O}-C(=O)-CH_3$

c) $CH_3-C(=\ddot{O})-O-H^{\delta+}$ ↕ $^{\delta+}H-O-C(=\ddot{O})-CH_3$

d) $CH_3-C(=\ddot{O})-\ddot{N}(H)-H$ ↕ $CH_3-C(=O)-N(H^{\delta+})-H^{\delta+}$

Dispersion forces (induced dipole) do not involve a permanent dipole, but instead are based on a temporary or induced dipole. An induced dipole occurs because electrons in orbitals (electron density) are not in fixed positions, but are described as probability distributions. In response to a temporary neighboring charge separation, a symmetrical electron density across a non-polar molecule can distort, providing a temporary positive and negative molecular end; this is termed an induced dipole.

The induced dipole can then cause a distortion of electron density in another neighboring molecule. The process can then continue and the induced dipoles are temporarily held together by the attractive forces. Although the temporary dipoles rapidly decompose back to a symmetrical structure, new induced dipoles are in the process of constantly forming and breaking; the **net** result is an attractive force.

Dispersion forces are the weakest of the intermolecular forces **per individual attraction**, but the **total** force can be large due to the **quantity** of the induced dipole attractions. They are present in all molecules, and are the major attractive force in non-polar or weakly polar molecules. The total dispersion force between molecules of the same compound is proportional to the size and number of the atoms in the molecule, which itself is proportional to molar mass. This relationship is composed of two considerations.

Very generally, as the **size** of the **atoms** in a molecule increase, the strength of the induced dipole increases; this is related to the concept of polarizability. Electron densities are more easily distorted (larger induced dipole) as the number of electrons and the distance to the outer shell increases, especially true when comparing atoms with similar chemical properties. These criteria relate generally to the atom size (atomic number).

Total size of the **molecule** (for molecules containing similar atoms) affects dispersion forces because the larger the molecule, with a greater number of total atoms and total bonds, generally increases the number of possible locations for forming induced dipole attractive forces. Based on total number of atoms and/or size of the atoms, as a general rule, for most molecular comparisons, the **total strength** of the molecular **dispersion** force is proportional to the **molar mass** of the compound.

Example: Compare the strength of total dispersion force among Cl_2, Br_2, and I_2.

Each of these diatomic molecules are non-polar; dispersion force is the only intermolecular force operating. In addition they are all from the same periodic family. The total strength of intermolecular force is predicted to be proportional to the molar mass of each molecule based on atomic size: MM of Cl_2 = 71 g/mole; MM of Br_2 = 160 g/mole; MM of I_2 = 254 g/mole. The order of expected Intermolecular force strength is $I_2 > Br_2 > Cl_2$; the phases of matter at room temperature confirm this order: I_2 is a solid, Br_2 is a liquid, Cl_2 is a gas.

Example: Compare the strength of total dispersion force between C_3H_8 and $C_{22}H_{46}$.

Each of these molecules are very weakly polar and comparison is between hydrocarbons of similar chemical structure. Dispersion force is the dominant intermolecular force operating. The total strength of intermolecular force is predicted to be proportional to the molar mass of each based on the number of similar atoms in the molecule: MM of C_3H_8 = 44 g/mole; MM of $C_{22}H_{46}$ = 310 g/mole. The order of expected intermolecular force strength is $C_{22}H_{46} > C_3H_8$; the phases of matter at room temperature confirm this order: $C_{22}H_{46}$ is a solid; C_3H_8 is a gas.

II A GUIDELINE FOR COMPARING TOTAL STRENGTH OF INTERMOLECULAR FORCES IN INDIVIDUAL COMPOUNDS

Boiling point (b.p.) and melting point (m.p.) temperatures for compounds indicate the energy required to break the attractive forces between molecules. Thus, these measures are proportional to the total strength of intermolecular forces: a higher melting or boiling temperature corresponds to a greater total intermolecular force. A very general representative mid-point range for relative strengths of intermolecular force is shown below.

Strengths of Interparticle Forces

Ionic Bonding	|←	Intermolecular Forces		→|
Covalent Bonding	>>	Hydrogen Bonding >	Dipole Forces >	Dispersion Forces
200–400 kJ/mole		10–30 kJ	5–15 kJ	1–3 kJ

Other factors can influence total intermolecular attractive force; for example, the three-dimensional molecular shape, especially in the solids.

Process for Comparing Total Intermolecular Forces for Different Compounds

Boiling point and melting point are used as direct measures.

b.p. and **m.p.** α to the **T** required to break attractive forces.
b.p. and **m.p.** α to the **total** strength of interparticle and intermolecular forces.

A list of typical values for electronegativity (**EN**):

F = 4.0;	**O** = 3.5;	**Cl** = 3.2;	**N** = 3.0;	**Br** = 3.0;
I = 2.7;	**S** = 2.6;	**Se** = 2.6;	**C** = 2.5;	**P** = 2.2;
H = 2.2;	**As** = 2.2;	**Te** = 2.1;	**B** = 2.0;	**Si** = 1.9.

Step (**1**): Determine the bonding structure of the compounds to the level required to identify bond dipoles and molecular geometry. In some cases, the molecular formula alone may be sufficient. In other cases, a full Lewis structure must be generated from a condensed structure or from techniques described in Chapters 14 and 15.

Step (**2**): Calculate the molar masses of the compounds to be compared. Determine as a general result whether the compounds have a MM within 10% of each other.

Step (**3**): Select the most polar (largest dipole) 1–3 (as appropriate) bonds in the molecule that will contribute to a molecular dipole; as a guideline, use the value of ΔEN. Determine if symmetry cancels the bond dipoles.

Step (**4**) Based on steps (**1**) and (**3**), determine the **force types** for each compound and identify the **dominant force type**. The dominant force type is the **strongest** force type based on the strengths of interparticle forces provided. For the intermolecular forces, these are hydrogen bonding (H-bonding) > dipole force > dispersion force. In most cases, for molecules with weak or very weak dipoles, the dominant force type is the dispersion force.

(**4a**) if the dominant force type is the permanent dipole force, distinguish the strength of the force in the range from very weak to very strong using the ΔEN:
$\Delta EN \leq 0.4$ = weak dipoles; ΔEN 0.5 – 1.0 = moderate dipoles; $\Delta EN > 1.0$ = strong dipoles.

(**4b**) Identify the **number of sites** in the molecule for hydrogen bonding or dipole forces.

Step (**5**): For molecules of approximately the same size (measured by molar masses within approximately 10% of each other), the compound with the greater total intermolecular forces will be the one with the stronger dominant force type or the greater number of sites of attraction.

Step (**6**): All molecules have operating dispersion forces. For molecules with similar force types and the same number of sites, the compound with the greater total intermolecular forces will be the molecule with the larger molar mass. This conclusion is most accurate when comparing molecules of very similar general molecular structure.

Step (**7**): Comparing molecules with different force types and greatly different molar masses leads to two results. If the trend in force type and molar mass are in agreement, use the previous rules. If the two trends lead to opposite conclusions, experience is necessary to estimate the net effect; this case is not covered by these rules.

Example: Compare the total strength of total intermolecular force among C_4H_{10}, C_5H_{12}, and $C_{10}H_{22}$.

Step (**1**): The molecules to be compared are all hydrocarbons of varying size. It is not necessary to draw out detailed structures of each; the only bonds in each molecule are C—C and C—H. The table below describes the analysis for step (**2**) through (**4**):

	Step (2)	**Step (3)**		**Step (4)**
	MM	**Bond Example**	**Bond ΔEN**	**Force Types**
C_4H_{10}	58	C—H	0.3	dispersion; very weak dipole
C_5H_{12}	72	C—H	0.3	dispersion; very weak dipole
$C_{10}H_{22}$	142	C—H	0.3	dispersion; very weak dipole
				Dispersion is dominant for cases of weak or very weak dipole.

Step (**5**) does not apply in this example.

Step (**6**): For molecules with similar force types the compound with the greater total intermolecular forces will be the molecule with the larger molar mass. The prediction in order of increasing intermolecular force and, therefore, increasing boiling point is:

$$C_4H_{10} < C_5H_{12} < C_{10}H_{22}$$

The analysis is confirmed by the boiling points:

$$C_4H_{10}\ (0°C) < C_5H_{12}\ (36°C) < C_{10}H_{22}\ (174°C)$$

Example: Compare the total strength of total intermolecular force among CH_3OH, C_2H_5OH, and C_3H_7OH.

Step (**1**): Expand the condensed structures to show the C—OH portion. The molecules to be compared have a C—O—H bonding arrangement connected to a hydrocarbon portion of varying size. The partially condensed structures are sufficient to complete the analysis.

	Step (2)	Step (3)		Step (4)
	MM	Bond Example	Bond ΔEN	Force Types
CH_3—OH	32	C—O	1.0	strong dipole; H-bonding dispersion
		O—H	1.3	
C_2H_5—OH	46	C—O	1.0	strong dipole; H-bonding dispersion
		O—H	1.3	
C_3H_7—OH	60	C—O	1.0	strong dipole; H-bonding dispersion
		O—H	1.3	

Hydrogen bonding is the dominant force type followed by strong dipole.

Step (**5**): To apply this step, the molecules must have approximately the same molar mass. In addition, to distinguish the compound with the greater total intermolecular force requires some difference in the type or number of dominant force type interactions. This is an example where even though dispersion is not the dominant force type, it must still be used as the "tie-breaker."

Step (**6**): For molecules with similar force types the compound with the greater total intermolecular forces will be the molecule with the larger molar mass. The prediction in order of increasing intermolecular force and, therefore, increasing boiling point is:

$$CH_3\text{—}OH\ (78°C) < C_2H_5\text{—}OH\ (89°C) < C_3H_7\text{—}OH\ (97°C).$$

Example: Compare the total strength of intermolecular force among C_4H_{10}, $CH_3OCH_2CH_3$, $CH_3C(O)CH_3$, C_3H_7OH, $HOCH_2CH_2OH$.

Step (**1**): The condensed structures for the molecules to be compared are expanded to indicate the major bonding arrangements according the guidelines described in Chapter 15.

Step (1)	Step (2)	Step (3)		Step (4)
	MM	Bond Example	Bond ΔEN	Force Types
C_4H_{10}	58	C—H	0.3	very weak dipole; dispersion
CH_3—O—CH_2CH_3	60	C—O	1.0	moderate dipole; dispersion
CH_3—C(=O)—CH_3	58	C=O	1.0	(moderate) to strong dipole

(Continued)

Step (1)	Step (2)	Step (3)		Step (4)
	MM	Bond Example	Bond ΔEN	Force Types
$CH_3CH_2CH_2$—OH	60	C—O	1.0	strong dipole; dispersion H-bonding
		O—H	1.3	
HO—CH_2CH_2—OH	62	C—O	1.0	strong dipole; dispersion H-bonding: 2 sites
		O—H	1.3	

Step (**5**): These molecules are all approximately the same size (the molar masses are within approximately 3% of each other); dispersion cannot be used to provide any difference in total intermolecular force. The molecules differ greatly in the dominant force type; apply step (**5**): the compound with the greater total intermolecular forces will be the one with the stronger dominant force type or the greater number of sites of attraction. The order first follows the dipole strength then includes hydrogen bonding. Note that the dipole created by the double bond ($^{\delta+}$C=O$^{\delta-}$) is stronger than the single bond (C—O) dipole; this is not predicted solely from the ΔEN value.

C_4H_{10} (very weak dipole, 0°C) < $CH_3OCH_2CH_3$ (moderate dipole, 8°C) < $CH_3C(O)CH_3$ (strong dipole, 56°C), < C_3H_7OH (strong dipole; hydrogen bonding, 97°C) < $HOCH_2CH_2OH$ (strong dipole; two sites for hydrogen bonding, 198°C).

The tables below summarize the results from the examples.

Comparing Boiling Point to Different MM in Molecules with Same Force Types

Compound	MM	Bond Example	Bond ΔEN	Force Types	Boiling Pt.
C_4H_{10}	58	C—H	0.3	**dispersion** very weak dipole	0°C
C_5H_{12}	72	C—H	0.3	**dispersion** very weak dipole	36°C
$C_{10}H_{22}$	142	C—H	0.3	**dispersion** very weak dipole	174°C
CH_3—OH	32	C—O	1.0	strong dipole	64°C
		O—H	1.3	H-bonding **dispersion**	
C_2H_5—OH	46	C—O	1.0	strong dipole	78°C
		O—H	1.3	H-bonding **dispersion**	
C_3H_7—OH	60	C—O	1.0	strong dipole	97°C
		O—H	1.3	H-bonding **dispersion**	

Comparing Boiling Point to Different Force Types in Molecules with Approximately the Same Molar Mass (MM)

Compound	MM	Bond Example	Bond ΔEN	Force Types	Boiling Pt.
C_4H_{10}	58	C—H	0.3	**very weak dipole** dispersion	0°C
CH_3—O—CH_2CH_3	60	C—O	1.0	**moderate dipole** dispersion	8°C
CH_3—C(=O)—CH_3	58	C=O	1.0	**strong dipole**	56°C
$CH_3CH_2CH_2$—OH	60	C—O	1.0	**strong dipole H-bonding**	97°C
		O—H	1.3	dispersion	
HO—CH_2CH_2—OH	62	C—O	1.0	**strong dipole H-bonding: 2 sites**	198°C
		O—H	1.3	dispersion	

Example: Compare the total strength of intermolecular force between C_5H_{12} and $CH_3C(O)CH_3$. Repeat the analysis for $C_{10}H_{22}$ and $CH_3C(O)CH_3$.

These molecules have been analyzed in previous examples. The comparisons illustrate cases where an unambiguous conclusion cannot be reached using the stated rules. $CH_3C(O)CH_3$ (MM = 58; strong dipole; b.p. 56°C) has a stronger force type but a lower molar mass than C_5H_{12} (MM = 72; very weak dipole; b.p. 36°C). The boiling point data show that the stronger force type overcomes the 20% difference in molar mass. The reverse occurs when comparing $CH_3C(O)CH_3$ with $C_{10}H_{22}$ (MM = 144; very weak dipole; b.p. 174°C); the 250% additional molar mass for the hydrocarbon overcomes the difference in force type strength.

Example: Compare the total strength of intermolecular force between CaF_2 and SF_2.

Step (**1**): CaF_2 is an ionic compound. SF_2 has tetrahedral electron region geometry with a bent (angular) atom geometry; the bond dipoles do not cancel.

	Step (2)	Step (3)		Step (4)
	MM	Bond Example	Bond ΔEN	Force Types
CaF_2	78	ionic		ionic
SF_2	70	S—F	1.4	dispersion; strong dipole

Step (**5**) applies. SF_2 << CaF_2. The ionic bond in CaF_2 is much stronger than any intermolecular force involving SF_2. CaF_2 is a solid at room temperature with a melting point of 1423°C; SF_2 is a gas at room temperature.

Example: Compare the total strength of intermolecular force between CH_3Cl and CH_2Cl_2.

Step (**1**): Both compounds have tetrahedral electron region geometry and atom geometry; the bond dipoles do not cancel.

	Step (2)	Step (3)		Step (4)
	MM	Bond Example	Bond ΔEN	Force Types
CH_3Cl	50	C—Cl	0.7	dispersion; moderate dipole
CH_2Cl_2	84	C—Cl	0.7	dispersion; moderate dipole

Step (**6**) applies. $CHCl_3$ (b.p. −24°C) < CH_2Cl_2 (b.p. 65°C). Although not identical, the force types are similar; dispersion force through the larger molar mass favors CH_2Cl_2.

Example: Compare the total strength of total intermolecular force between the molecules.

$$CH_3-\overset{\overset{\large O}{\|}}{C}-CH_3 \quad \text{and} \quad CH_3-\overset{\overset{\large O}{\|}}{C}-NH_2$$

Step (**1**): The structures shown for the molecules to be compared are sufficient to indicate the major bonding arrangements.

	Step (2)	Step (3)		Step (4)
	MM	Bond Example	Bond ΔEN	Force Types
$CH_3-\overset{\overset{O}{\|}}{C}-CH_3$	58	C=O	1.0	strong dipole; dispersion
$CH_3-\overset{\overset{O}{\|}}{C}-NH_2$	59	C=O C—N N—H	1.0 0.5 0.8	moderate and strong dipole H-bonding

Step (**5**): applies. $CH_3C(O)CH_3 < CH_3C(O)NH_2$. The greater number of moderate dipoles plus hydrogen bonding results in much stronger intermolecular force for $CH_3C(O)NH_2$. $CH_3C(O)NH_2$ is a solid at room temperature with a melting point of 82°C; $CH_3C(O)CH_3$ is a liquid with a boiling point of 56°C.

III ENTHALPY OF PHASE CHANGES

During the process of a phase change, intermolecular forces must be broken or formed. Although not full covalent bonds, breaking or forming intermolecular forces must therefore be accompanied by a change in potential energy of the system, usually measured as a change in enthalpy at constant temperature.

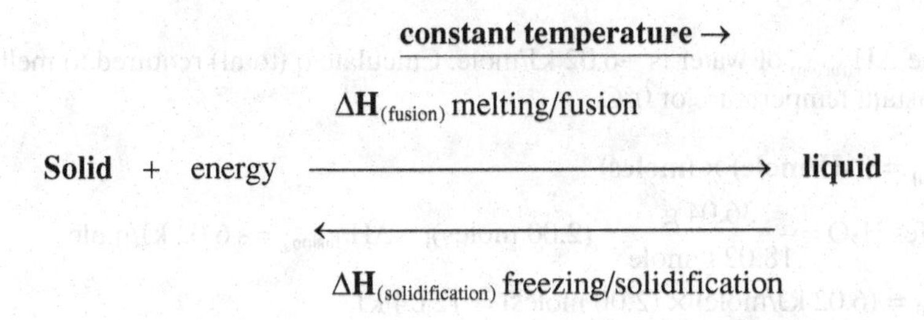

The energy that must be added to partially break intermolecular forces in the solid is termed the enthalpy of fusion: $\Delta \mathbf{H}_{(fusion)}$. $\Delta \mathbf{H}_{(fusion)}$ must always be a positive value since it represents energy added to the system. (The term "fusion" for this enthalpy direction seems counterintuitive to the typical English language usage; in this case, the definitional sense is derived from the Latin *fundere/fusio* meaning "to melt" or "to pour out.") The symmetry of enthalpy change requires that $\Delta \mathbf{H}_{(fusion)} = -\Delta \mathbf{H}_{(solidification)}$

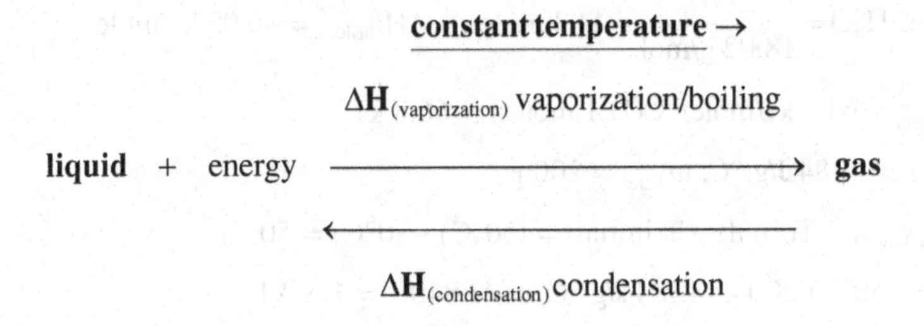

The corresponding energy, which must be added to break the liquid intermolecular forces to form the gas, is termed the enthalpy of vaporization: $\Delta \mathbf{H}_{(vaporization)}$. $\Delta \mathbf{H}_{(vaporization)}$ must also be always positive since it represents energy added to the system. The symmetry of enthalpy change requires that $\Delta \mathbf{H}_{(vaporization)} = -\Delta \mathbf{H}_{(condensation)}$. Note also that since q(system) = – q(surroundings), if q(vaporization) is positive, q(surroundings) must be negative. This demonstrates the concept of evaporative cooling.

The values for $\Delta H_{(fusion)}$ and $\Delta H_{(vaporization)}$ are measured at constant temperature. The enthalpies represent the potential energy associated with changes in intermolecular forces; the energy is not applied to raising the temperature of the substance. The following picture, though not exact and to scale, represents the distinction between the two concepts for the phase changes in water.

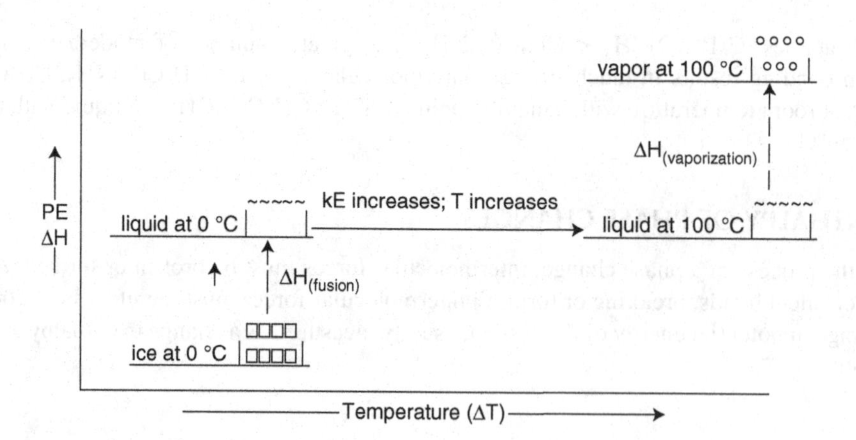

Example: The $\Delta H_{(fusion)}$ of water is +6.02 kJ/mole. Calculate q (total) required to melt 36.04 grams of ice at a constant temperature of 0°C.

(1) $\mathbf{q_{total} = (\Delta H/mole) \times (moles)}$

(2) moles $H_2O = \dfrac{36.04\ g}{18.02\ g/mole} = (2.00\ moles)$; $\Delta H_{(fusion)} = +6.02\ kJ/mole$

(3) $q_{total} = (6.02\ kJ/mole) \times (2.00\ moles) = 12.04\ kJ$

Example: The $\Delta H_{(fusion)}$ of water is +6.02 kJ/mole. Calculate q (total) required to convert 200 grams of ice at 0°C to liquid water at 50°C; $C_{water} = 4.184$ J/g-°C.
The problem requires two separate calculations:

$$H_2O_{(solid)}\ at\ 0°C \xrightarrow{(a)\Delta H_{(fusion)}} H_2O_{(liquid)}\ at\ 0°C \xrightarrow{(b)q_{(\Delta T)}} H_2O_{(liquid)}\ at\ 50°C$$

(1a) $q_{fusion} = (\Delta H_{(fusion)}/mole) \times (moles)$; **(1b)** $q_{\Delta T} = (\Delta T_{water})(m_{water})(C_{water})$

(2a) moles $H_2O = \dfrac{200\ g}{18.02\ g/mol} = 11.1\ moles$; $\Delta H_{(fusion)} = +6.02\ kJ/mole$

(3a) $q_{fusion} = (6.02\ kJ/mole) \times (11.1\ moles) = 66.8\ kJ$

(2b) $C_{water} = 4.184\ J/g\text{-}°C$; $m_{water} = 200\ g$

$\Delta T_{(solution)} = T(final) - T(initial) = (50°C) - (0°C) = 50°C$

(3b) $q_{\Delta T} = (50°C)(200\ g)(4.184\ J/g\text{-}°C) = 41{,}840\ J = 41.9\ kJ$

(4) $q_{total} = [q_{fusion} + q_{\Delta T}] = 66.8\ kJ + 41.9\ kJ = 109\ kJ$

Example: The $\Delta H_{(vaporization)}$ of water is + 40.8 kJ/mole. Calculate q (total) required to convert 200 grams of liquid water at 25°C to vapor (gas) at 100°C; $C_{water} = 4.184$ J/g-°C.

$$H_2O_{(liquid)}\ at\ 25\ °C \xrightarrow{(b)q_{(\Delta T)}} H_2O_{(liquid)}\ at\ 100\ °C \xrightarrow{(a)\Delta H_{(vaporization)}} H_2O_{(vapor)}\ at\ 100\ °C$$

(1a) $q_{vaporization} = (\Delta H_{(vaporization)}/mole) \times (moles)$; **(1b)** $q_{\Delta T} = (\Delta T_{water})\ (m_{water})\ (C_{water})$

(2a) moles $H_2O = \dfrac{200\ g}{18.02\ g/mol} = 11.1\ moles$; $\Delta H_{(vaporization)} = +40.8\ kJ/mole$

(3a) $q_{vaporization} = (40.8\ kJ/mole) \times (11.1\ moles) = 453\ kJ$

(2b) $(C_{water}) = 4.184$ J/g-°C; $m_{water} = 200$ g

$\Delta T_{(solution)} = T(final) - T(initial) = (100°C) - (25°C) = 75°C$

(3b) $q_{\Delta T} = (75°C)(200\ g)(4.184\ J/g\text{-}°C) = 62{,}760\ J = 62.7\ kJ$

(4) $q_{total} = [q_{vaporization} + q_{\Delta T}] = 453\ kJ + 62.7\ kJ = 516\ kJ$

Example: Calculate the energy required to convert a 500 gram gold coin from an initial temperature of 25.0°C to melted gold at its melting temperature of 1063°C.

$$C_{Au} = 0.130\ \text{Joules/g-°C};\ \Delta H_{fusion} = 12.7\ \text{kJ/mole}$$

$$Au_{(s)}\ \text{at}\ 25.0°C \xrightarrow{(b)\,q_{(\Delta T)}} Au_{(s)}\ \text{at}\ 1063°C \xrightarrow{(a)\,\Delta H_{(fusion)}} Au_{(l)}\ \text{at}\ 1063°C$$

(1): $q_{total} = q_{fusion} + q_{\Delta T}$

(1a) $q_{fusion} = (\Delta H_{fusion}/\text{mole})(\text{moles})$; **(1b)** $q_{\Delta T\ total} = (\Delta T_{(Au)})\ (m_{Au})\ (C_{Au})$

(2a) $\text{moles Au} = \dfrac{500\ g}{196.97\ g/mol} = 2.54\ \text{mole};\quad \Delta H_{fusion} = 12.7\ \text{kJ/mole}$

(3a) $q_{fusion} = (12.7\ \text{kJ/mole})(2.54\ \text{moles}) = 32.3\ kJ$

(2b) $m_{Au} = 500$ g; $C_{Au} = 0.130$ Joules/g-°C

$\Delta T_{(Au)} = T(final) - T(initial) = (1063°C) - (25.0°C) = 1038°C$

(3b) $q_{\Delta T} = (1038°C)\ (500\ g)\ (0.130\ \text{Joules/g-°C}) = 67{,}470\ J = 67.5\ kJ$

(4) $q_{total} = 32.3\ kJ + 67.5\ kJ =$ **99.8 kJ**

Example: A radiator is filled with 4.00 gallons of antifreeze ($C_2H_6O_2$) and left unsealed. The antifreeze absorbs 25,000 kJ from the engine at an initial temperature of 20.0°C. Will all the antifreeze evaporate at its boiling point? $C_{Anti} = 2.42$ Joules/g-°C;

$\Delta H_{vap} = 49.6$ kJ/mole; $d_{Anti} = 1.113$ g/mL; b.p. $_{Anti} = 198.0°C$; 1 gallon = 3.79 liters

(1) Calculate q_{total} and compare to 25,000 kJ: $q_{total} = q_{vaporization} + q_{\Delta T}$

(1a) $q_{vaporization} = (\Delta H_{vaporization}/\text{mole})(\text{moles})$; **(1b)** $q_{\Delta T} = (\Delta T_{(Anti)})\ (m_{Anti})\ (C_{Anti})$

(2a) $m_{Anti} = (4.00\ gal)(3.79\ L/gal)(1000\ mL/L)(1.113\ g/mL) = 16{,}873$ grams

$\text{moles} = \dfrac{16873\ g}{62.08\ g/mole} = 271.8\ \text{mole};\quad \Delta H_{vap} = 49.6\ \text{kJ/mole}$

(3a) $q_{vaporization} = (49.6\ \text{kJ/mole})(271.8\ \text{mole}) = 1.35 \times 10^4\ kJ$

(2b) $m_{Anti} = 16{,}873$ g; $C_{Anti} = 2.42$ Joules/g-°C

$\Delta T_{(Anti)} = T(final) - T(initial) = (198.0°C) - (20.0°C) = +\ 178°C$

(3b) $q_{\Delta T\ total} = (178°C)\ (16{,}873\ g)\ (2.42\ \text{Joules/g-°C})$

$= 7.27 \times 10^6\ J = 7.27 \times 10^3\ kJ$

(4) $q_{total} = 13.5 \times 10^3\ kJ + 7.27 \times 10^3\ kJ = 20.8 \times 10^3\ kJ = 20{,}800\ kJ$

20,800 kJ < 25,000 kJ: All the antifreeze evaporates.

IV INTERPARTICLE AND INTERMOLECULAR FORCES FOR SOLUTIONS

Solutions are homogeneous mixtures of compounds; mixing must occur at the molecular or ionic level. General terminology describes the **solute** as the compound being dissolved and the **solvent** as the compound acting as the dissolver. The term **solubility** refers to a measure of a specific amount of solute, which is dissolved in a particular amount of solvent or solution. This measure is expressed as a concentration unit; for example, molarity (moles solute/liter solution) or grams solute/mL solution (Chapter 11). A **saturated** solution describes the **maximum** solubility of a solute in a solvent.

SOLUBILITY REQUIREMENTS

Molecules (or particles) of a pure solvent and a pure solute are each held together by existing compound interparticle or intermolecular attractive forces. Formation of a solution requires molecular intermixing: some of these attractive forces must be broken to "expand" the solute and solvent compounds to allow interpenetration.

The solution formation process involves:

(1) **Breaking** of some solvent—solvent intermolecular force interactions; this requires an input of energy into the system (increase in PE).
(2) **Breaking** of solute—solute interparticle (intermolecular) force interactions; this requires an input of energy into the system (increase in PE).
(3) **Formation** of solute—solvent interparticle (intermolecular) attractive forces. This represents an energy release (decrease in PE) and must compensate the system to a large degree for the energy input necessary to break existing solute—solute and solvent—solvent interparticle forces.

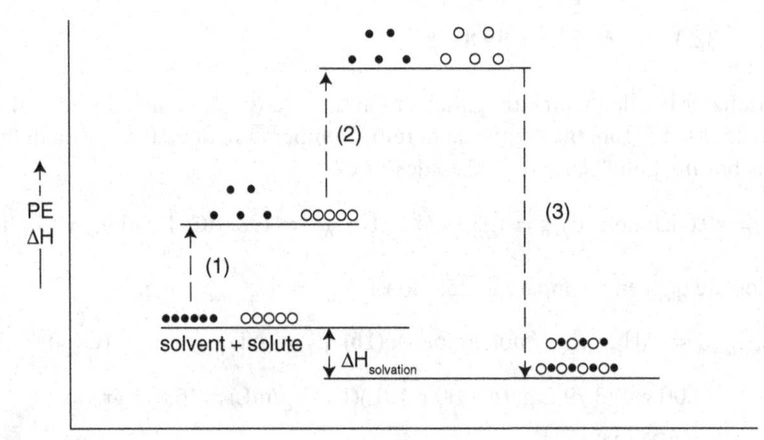

In the optimum circumstance, the sum of energies of solute—solvent attractive forces formed will be greater than the sum of energies of solute—solute and solvent—solvent attractive forces broken. This would lead to a negative value for ΔH (solvation).

A negative value for ΔH (solvation) is the most favorable energy balance for solution formation. Good solubility is possible for systems showing a positive ΔH (solution) due to the effects of **entropy** discussed in Chapter 25.

The more negative ΔH (solvation) and better (relative) solubility are most observed for circumstances where step **(3)**, forming solute to solvent attractive interactions, represents the greatest potential energy decrease. The key to achieving optimum solubility is the maximization of the solute—solvent interparticle and intermolecular attractive forces; the energy balance for solvation then becomes most favorable. This requirement forms the basis of solubility analysis: solute—solvent attractive forces and, therefore, solubility will be maximized whenever the force types in both solvent and solute are equivalent or as similar as possible, often phrased as "like dissolves like."

Matching of force types does **not** guarantee solubility. This process only ensures that the solute will have the best possible chance of dissolving in a properly selected solvent. The wide variety of interparticle or intermolecular forces, and the effect of entropy, produce variable maximum solubilities of solutes in solvents, even under optimum force type matching.

Interparticle and Intermolecular Forces between Solutes and Solvents

The force types for solute—solvent interactions are the same as for single compounds, but with the possibilities for additional combinations: dipole—dipole; hydrogen bonding; induced dipole—induced dipole (dispersion); dipole—induced dipole; ion—dipole. Particles in solutions may not have equivalent force types; closely similar force types may represent the best "match."

Example: Which of the two solvents listed would be expected to be the best solvent for NaCl? Solvents: hexane (C_6H_{12}) or water.

The dominant force type in a hydrocarbon (**Section II**) is dispersion (based on a very weak dipole). The dominant force types in water are strong dipole and hydrogen bonding. Solvation of ions in water is possible because ion—strong dipole is the closest match possible. The Ion—induced dipole for C_6H_{12} with predominantly dispersion forces would be a poor match and would not be expected to provide good solubility. Ionic compounds are not generally soluble in non-polar, dispersion force based hydrocarbons.

Example: Which of the two solvents listed would be expected to be the best solvent for benzene, C_6H_6? Solvents: "gasoline" (C_8H_{18}) or water.

The dominant force type in C_8H_{18} is dispersion (very weak dipole); water has strong dipole and hydrogen bonding. The dominant force type in benzene, C_6H_6, is dispersion and has a much better force match with gasoline than with water and would be much more soluble in C_8H_{18} than in H_2O.

Process for Comparing Total Intermolecular Forces for Solutions

The general process will apply the rule that solubility and total attractive force will be maximized by the closest match of solute—solvent interparticle and intermolecular forces.

The stated rule is used for deciding which solvent would be best for dissolving a particular solute or deciding which solute would have the greater solubility in a particular solvent.

Step (**1**): Use the techniques in **Section II** to determine the major force types present in the pure solute; that is, identify the forces between particles or molecules of the specific solute compound or possible solute compounds.

Step (**2**): Use the techniques in **Section II** to determine the major force types present in the pure solvent; that is, identify the forces between molecules of the specific solvent compound or possible solvent compounds.

Step (**3**): Compare the force types of all possible combinations of solute/solvent. The closest force type match will represent the best solvent for a specific solute or the solute with higher solubility in a specific solvent.

Example: Which of the two solvents listed would be expected to be the best solvent for the sugar glucose, $HOCH_2(CHOH)_4CHO$? Solvents: C_8H_{18} or water.

Step (**1**): Expand the structure of the sugar glucose:

```
HO—CH2—CH—CH—CH—CH—C—H
        |   |   |   |  ||
       HO  HO  HO  HO  O
```

A straightforward application of the analysis techniques indicates that glucose has many sites of strong dipole and hydrogen bonding.

Step (**2**): The dominant force type in C_8H_{18} is dispersion (very weak dipole). Water has strong dipole and hydrogen bonding.

Step (**3**): Glucose has a much better force match with water than with C_8H_{18}; glucose is much more soluble in H_2O.

Example: Which of the two solvents listed would be expected to be the best solvent for diatomic bromine, Br_2? Solvents: C_8H_{18} or water.

Step (**1**): Br_2 (Br—Br) is a non-polar molecule; the force type is dispersion only.

Step (**2**): The dominant force type in C_8H_{18} is dispersion (very weak dipole). Water has strong dipole and hydrogen bonding.

Step (**3**): Br_2 has a much better force match with C_8H_{18} than with water and is much more soluble in C_8H_{18}.

Example: Which of the two solutes shown would be expected to be most soluble in the solvent ethylene glycol (antifreeze): HO—CH_2CH_2—OH? The solutes to choose from are carbon tetrachloride (CCl_4) and acetic acid CH_3COOH.

Step (**1**): CCl_4 is a symmetric tetrahedral structure with all four corners occupied by the same atoms; it is non-polar and the force type is dispersion only. Acetic acid can be expanded to show strong dipole and hydrogen bonding.

$$CH_3-\underset{\underset{O}{\|}}{C}-OH$$

Step (**2**): In a previous example, the dominant force type in ethylene glycol was determined to be strong dipole and hydrogen bonding.

Step (**3**): Of the two solutes, acetic acid would be more soluble in ethylene glycol; carbon tetrachloride would be less soluble in the solvent.

Example: List the order of solubility from most soluble to least soluble for the following solutes in water as a solvent: C_3H_7—OH; C_4H_9—OH; C_5H_{11}—OH; C_6H_{13}—OH

Step (**1**): The **—OH** portion of each molecule contributes to strong dipole and hydrogen bonding. The $\mathbf{C_xH_y}$ (hydrocarbon) portion of each molecule contributes to dispersion forces.

Step (**2**): Water has strong dipole/hydrogen bonding with a very small contribution from dispersion forces.

Step (**3**): The strong dipole/hydrogen bonding (—OH) portion of the solutes is a very good match to the solvent water but the hydrocarbon/dispersion portion is a very poor match with water. As the dispersion (hydrocarbon) portion of the solute molecule increases in proportion to the strong dipole/hydrogen bonding portion, the solubility in water decreases. The hydrocarbon contribution is indicated by the number of carbons and hydrogens in the molecule while the

[—OH] portion remains constant at one per molecule. The following order is the result; the solubility values are shown as verification:

Solute Molecule	Maximum Solubility in Water
C_3H_7—OH	soluble in all concentrations
C_4H_9—OH	1.1 Molar
C_5H_{11}—OH	0.30 Molar
C_6H_{13}—OH	0.056 Molar

V ENTHALPIES OF SOLUTIONS

Solution formation (solvation) requires changes in the nature of the attractive forces between solute and solvent. Changes in the balance of attractive forces demonstrate the same potential energy and heat transfer processes as chemical reactions; solvation has an associated value for enthalpy change, ΔH(solvation). $\Delta H_{(solvation)}$ can be calculated from values of ΔH°f or from calorimetry experiments.

Example: Calculate $\Delta H_{(solvation)}$/mole for RbCl dissolving in water from the ΔH°f values provided. ΔH°f of $RbCl_{(solid)}$ = −435.4 kJ/mole; ΔH°f of $RbCl_{(aq)}$ = −418.3 kJ/mole

The solvation (solution forming) equation is : $RbCl_{(solid)} \xrightarrow{H_2O\ solvent} RbCl_{(aqueous)}$

(1): **Reactant:** ΔH°f ($RbCl_{(solid)}$) = −435.4 kJ

Product: ΔH°f ($RbCl_{(aqueous)}$) = −418.3 kJ

(2) $\Delta H°_{(reaction)} = [\Delta H°f\ RbCl_{(aqueous)}] - [\Delta H°f\ RbCl_{(solid)}]$

$\Delta H°_{(reaction)} = [(-418.3\ kJ)] - [(-435.4\ kJ)] = +$ **17.1 kJ/mole**

Example: $CaCl_2$ is sometimes used as a road salt to help melt ice. An open, constant pressure calorimeter is used to measure $\Delta H_{(solvation)}$. 6.56 grams of $CaCl_2$ is dissolved in 55.0 grams of water at an initial temperature of 26.5°C. The final temperature after the solution is formed is 45.5°C. Calculate $\Delta H_{(solvation)}$ per mole for $CaCl_2$ in water. To avoid confusion, $\Delta H_{(solvation)}$ is designated as $\Delta H_{(reaction)}$; C(solution) = 4.20 J/g-°C.

(1): $\Delta H = q_p$ (reaction); ΔH/mole $CaCl_2$ = q_p/mole $CaCl_2$.

(2): $q_{(reaction)} = -q_{(solution)}$; $q_{solution} = (\Delta T_{solution})(m_{solution})(C_{solution})$

$C_{solution}$ = 4.20 J/g-°C; $m_{solution}$ = 55.0 g solvent + 6.56 grams $CaCl_2$ = 61.56 g solution

$\Delta T_{(solution)}$ = T(final) − T(initial) = (45.5°C) − (26.5°C) = 19.0°C

$q_{solution} = (\Delta T_{solution})(m_{solution})(C_{solution})$ = (19.0°C) (61.56 g) (4.20 J/g-°C) = 4912 J

$q_{(reaction)} = -q_{(solution)}$;

$$q_{(reaction)} = -(4912\ J) = -4912\ J; \quad \text{moles } CaCl_2 = \frac{6.56\ g}{111.0\ g/mole} = 0.0591\ mole$$

$$(3): \Delta H_{(solvation)} = q_p/\text{mole } CaCl_2 = \frac{-4912\ J}{0.0591\ mole} = \mathbf{-83100\ J/mole} \quad \text{or} - \mathbf{83.1\ kJ/mole}$$

Example: The solvation of NH_4NO_3 in water can be used in cold packs to "ice" injuries. Calculate the final temperature of a solution if 25.0 grams of ammonium nitrate is dissolved in 250.0 g water, which has an initial temperature of 25.0°C.

$$C_{solution} = 4.184 \text{ J/g-°C. } \Delta H_{solvation} \text{ of } NH_4NO_3 = +21.1 \text{ kJ/mole}$$

(1): $T_{(final\ solution)} = T_{(initial\ solution)} + \Delta T_{(solution)}$; calculate $\Delta T_{(solution)}$.

$$\Delta T_{(solution)} = \frac{(q_{(solution)})}{(C_{solution})(m_{solution})}$$

(2): $C_{solution} = 4.184$ J/g-°C; $m_{solution} = 250.0$ g solvent + 25.0 grams NH_4NO_3 = 275 g

$$q_{(solution)} = -q_{(reaction)};\ q_{(reaction)} = (\Delta H_{(solvation)}/\text{mole}) \times (\text{moles})$$

$$\text{moles } NH_4NO_{3\,(s)} = \frac{25.0 \text{ g}}{80.06 \text{ g/mole}} = 0.312 \text{ mole}$$

$$q_{(reaction)} = (21.1 \text{ kJ/mole}) \times (0.312 \text{ mole}) = +6.59 \text{ kJ} = \mathbf{6590\ J}\ (\mathbf{J} \text{ for step 3})$$

$$q_{(solution)} = -q_{(reaction)} = -\ (6590 \text{ J}) = -6590 \text{ J}$$

(3) $$\Delta T_{(solution)} = \frac{(q_{(solution)})}{(C_{solution})(m_{solution})} = \frac{-6590\text{J}}{(4.184\text{J/g-°C}))(275\text{g})} = -5.73\text{°C}$$

(4) T(final) = T(initial) + ΔT = 25.0°C + (–5.73°C) = 19.3°C

VI PRACTICE PROBLEMS

1. Select the one compound from the following pairs that has the stronger total interparticle or intermolecular force, and, thus, the higher boiling point or melting point.
   a) PCl_3 or $MgCl_2$
   b) CH_3NH_2 or CH_3F
   c) CH_3Br or CH_3Cl
   d) CH_3CH_2—OH or CH_3—O—CH_3
   e) BH_3 or HF
   f) H_2NNH_2 or PH_3
   g) $CH_3CH_2CH_3$ or CH_3Cl
   h) C_6H_6 or $C_{10}H_8$
   i) $CH_3CH_2C(=O)-OCH_3$ or $CH_3CH_2CH_2C(=O)-OH$
   j) $HO-CH_2CH(OH)CH_2-OH$ or $Cl-CH_2CH_2-Cl$

2. Which of the two compounds listed below would be expected to be more soluble in water? Which of the two compounds would be expected to be more soluble in hexane, (C_6H_{14})?
   a) PCl_3 or $MgCl_2$
   b) CH_3NH_2 or CH_3Br

c) CH_3CH_2—OH or CH_3—O—CH_3
d) H_2NNH_2 or PH_3
e) $CH_3CH_2CH_3$ or CH_3Cl

f) $CH_3CH_2C(=O)-OCH_3$ or $CH_3CH_2CH_2C(=O)-OH$

g) $HO-CH_2CH(OH)CH_2-OH$ or $Cl-CH_2CH_2-Cl$

3. 1.75 liters of liquid Hg is heated from 25.0°C to its boiling point of 357°C; It is then completely evaporated at 357°C. Calculate the total amount of heat, q(total), required to completely evaporate the mercury; density of Hg = 13.6 g/mL; C_{Hg} = 0.140 J/g-°C; $\Delta H_{(vaporization)}$/mole for Hg = +59.6 kJ/mole
4. 908.0 kilograms of $Ti_{(s)}$ is heated from 28.0°C to its melting point of 1668°C; It is then completely melted at 1668°C. Calculate the total amount of heat, q(total), required to completely melt the titanium; C_{Ti} = 0.520 J/g-°C; $\Delta H_{(fusion)}$/mole for Ti = +20.9 kJ/mole
5. 558 mL of liquid alcohol is heated from 29.6°C to its boiling point of 98.4°C; It is then completely evaporated at 98.4°C. Calculated the total amount of heat, q(total), required to completely evaporate the alcohol; density of alcohol = 1.18 g/mL; $C_{alcohol}$ = 2.06 J/g-°C; MM of the alcohol = 144 grams/mole; $\Delta H_{(vaporization)}$/mole for the alcohol = +34.5 kJ/mole
6. 32.6 grams of $Ba(OH_2)$ (MM = 171.3 g/mole) is dissolved in a calorimeter containing 625.0 grams of water at an initial temperature of 27.3°C; the final temperature of the solution after solvation was 44.6°C. Calculate the $\Delta H_{(solvation)}$/mole for $Ba(OH)_2$. C(solution) = 4.15 J/g-°C
7. Carbon tetrachloride (CCl_4) is a common solvent for organic compounds. Calculate the $\Delta H_{(solvation)}$/mole for $C_{10}H_8$ if 28.2 grams of the $C_{10}H_8$ (MM = 128.2 g/mole) solute is dissolved in 643.6 grams of the CCl_4 solvent at an initial temperature of 27.5°C. The final temperature of the solution was 31.2°C; $C_{(solution)}$ = 0.857 J/g-°C.

VII ANSWERS TO PRACTICE PROBLEMS

1. The explanations in these answers summarizes the results of steps **(1)** through **(6)**
   a) PCl_3 or $MgCl_2$: **$MgCl_2$** has the higher melting point.
   PCl_3 has a polar P—Cl bond. The molecule has pyramidal atom geometry and, thus, is polar with a polarity estimated to be moderate. $MgCl_2$ is ionic; interparticle forces are ionic bonds. Ionic bonds are much stronger than moderate dipole force.
   b) CH_3NH_2 or CH_3F: **CH_3NH_2** has the higher boiling point.
   CH_3NH_2 (MM 31) has N—H bonds and can therefore form hydrogen bonds. CH_3F (MM 36) has all atoms bonded to carbon and, thus, has no hydrogen bonding. The MM of the two molecules are reasonably close, but force types are significantly different due to hydrogen bonding; apply step **(5)**: select the molecule with the strongest intermolecular force type.
   c) CH_3Br (MM = 97) or CH_3Cl (MM = 52): **CH_3Br** has the higher boiling point.
   Both molecules have weak to moderate dipoles and, thus, have similar force types. The deciding factor is the total size of the dispersion force, which is proportional to MM; apply step **(6)**. Select the larger molecule with the larger molar mass: CH_3Br (MM = 97).

d) CH_3CH_2—OH or CH_3—O—CH_3: **CH_3CH_2—OH** has the higher boiling point.
The two molecules have identical MM. CH_3CH_2—OH has strong dipole/hydrogen bonding. CH_3—O—CH_3 has a moderate dipole and no hydrogen bonding. Apply step (**5**): select the compound with the stronger force type.

e) BH_3 or HF: **HF** has the higher boiling point.
HF (MM = 20) can hydrogen bond; BH_3 (MM = 13) cannot. Apply step (**5**): select the molecule with the stronger force type. (HF also has a higher MM.)

f) H_2NNH_2 (MM = 32) or PH_3(MM = 34): **H_2NNH_2** has the higher boiling point.
The two molecules have approximately the same MM. **H_2NNH_2** can hydrogen bond; PH_3 cannot. Apply step (**5**): select the molecule with the stronger force type.

g) $CH_3CH_2CH_3$ (MM = 44) or CH_3Cl (MM = 50): **CH_3Cl** has the higher boiling point.
The two molecules have approximately the same MM. The C—Cl bond provides a larger bond dipole than C—H bonds. CH_3Cl would have a stronger dipole force than $CH_3CH_2CH_3$; apply step (**5**): select the molecule with the stronger force type.

h) C_6H_6 (MM = 78) or $C_{10}H_8$ (MM = 128): **$C_{10}H_8$** has the higher boiling point.
Both molecules have essentially the same force types: very weak dipole with predominant dispersion force. Apply step (**6**): select the molecule with the highest MM (greater total dispersion force).

i) $CH_3CH_2COOCH_3$ or $CH_3CH_2CH_2COOH$ (MM of both = 88): **$CH_3CH_2CH_2COOH$** has the higher boiling point.
The molecules are isomers of each other and, thus, have the same MM. Both molecules have strong dipole; however, $CH_3CH_2CH_2COOH$ can participate in hydrogen bonding. Apply step (**5**): select the molecule with the stronger force type.

j) $HOCH_2CH(OH)CH_2OH$ (MM = 92) or $ClCH_2CH_2Cl$ (MM = 99)
$HOCH_2CH(OH)CH_2OH$ has the higher boiling point.
The molecules have approximately the same MM. Both molecules have a dipole; however, $HOCH_2CH(OH)CH_2OH$ has a much stronger dipole and can participate in hydrogen bonding at potentially three sites.

2. Step (**1**), solute intermolecular force analysis, was completed in the previous problem or in previous examples. Step (**2**), solvent intermolecular force analysis, was completed previously with the results:

water = strong dipole/hydrogen bonding

hexane = very weak dipole/dispersion predominant

a) Step (**3**): PCl_3 would be the more soluble of the two compounds in hexane; weak dipole matches hexane. $MgCl_2$ would be the more soluble of the two compounds in water; ion-dipole force matches water.

b) Step (**3**): CH_3NH_2 would be the more soluble of the two compounds in water; hydrogen bonding matches water. CH_3Br would be the more soluble of the two compounds in hexane; despite the C—Br bond, dispersion forces are predominant and dispersion matches hexane.

c) Step (**3**): CH_3CH_2—OH would be the more soluble of the two in water; hydrogen bonding matches water. CH_3—O—CH_3 would be the more of the two in hexane; moderate dipole/dispersion matches hexane the best.

d) Step (**3**): H_2NNH_2 would be the more soluble of the two in water; hydrogen bonding matches water. PH_3 would be the more soluble of the two in hexane; weak dipole/dispersion dominant matches hexane.

e) Step (**3**): $CH_3CH_2CH_3$ would be the more soluble of the two in hexane; very weak dipole/dispersion matches hexane. CH_3Cl would be the more soluble of the two in water.

Although this compound is not very soluble in water, the dipole provided by the C—Cl bond would be a better match for water than the very weak dipole of $CH_3CH_2CH_3$.

f) Step (**3**): $CH_3CH_2CH_2COOH$ would be the more soluble of the two in water; hydrogen bonding provides the difference compared to the other molecule. $CH_3CH_2COOCH_3$ would be the more soluble of the two in hexane. Although this compound would also be soluble in water due to its strong dipole, it would be more soluble in hexane than the previous molecule. This is because it cannot participate in hydrogen bonding; hydrogen bonding is a bad match for hexane.

g) Step (**3**): $HOCH_2CH(OH)CH_2OH$ would be the more soluble of the two in water due to the extensive sites for hydrogen bonding. $ClCH_2CH_2Cl$ would be the more soluble of the two in hexane; moderate dipole/dispersion is a better match for hexane than the extensive hydrogen bonding of the other compound.

3. $Hg_{(liquid)}$ at 25 °C $\xrightarrow{q(\Delta T)}$ $Hg_{(liquid)}$ at 357 °C $\xrightarrow{\Delta H(vaporization)}$ $Hg_{(vapor)}$ at 357 °C

(**1a**) $q_{(vaporization)} = (\Delta H_{(vaporization)}/\text{mole}) \times (\text{moles})$; (**1b**) $q_{\Delta T} = (\Delta T_{Hg})(m_{Hg})(C_{Hg})$

(**2a**) mass of Hg = 1750 mL × 13.6 g/mL = 23,800 g

$$\text{moles Hg} = \frac{23{,}800\,\text{g}}{200.59\,\text{g/mole}} = 118.6\,\text{moles}; \quad \Delta H_{(vaporization)} = +59.6\ \text{kJ/mole}$$

(**3a**) $q_{(vaporization)} = (59.6\ \text{kJ/mole}) \times (118.6\ \text{moles}) = 7072\ \text{kJ}$

(**2b**) $C_{Hg} = 0.140$ J/g-°C; $m_{Hg} = 23{,}800$ g
$\Delta T_{(solution)} = T(\text{final}) - T(\text{initial}) = (357°C) - (25°C) = 332°C$

(**3b**) $q_{\Delta T} = (332°C)(23{,}800\ \text{g})(0.140\ \text{J/g-°C}) = 1.106 \times 10^6\ \text{J} = 1106\ \text{kJ}$

(**4**) $q_{total} = [q_{(vaporization)} + q_{\Delta T}] = 7{,}072\ \text{kJ} + 1{,}106\ \text{kJ} =$ **8178 kJ (8180 kJ)**

4. $Ti_{(s)}$ at 28.0 °C $\xrightarrow{q(\Delta T)}$ $Ti_{(s)}$ at 1668 °C $\xrightarrow{\Delta H(vaporization)}$ $Ti_{(l)}$ at 1668 °C

(**1**) $q_{total} = q_{fusion} + q_{\Delta T}$

(**1a**) $q_{fusion} = (\Delta H_{fusion}/\text{mole})(\text{moles})$; (**1b**) $q_{\Delta T\ total} = (\Delta T_{(Ti)})\ (m_{Ti})\ (C_{Ti})$

$$(\textbf{2a})\ \text{moles Ti} = \frac{908000\ \text{g}}{47.88\ \text{g/mole}} = 18{,}960\ \text{mole}; \quad \Delta H_{fusion} = 20.9\ \text{kJ/mole}$$

(**3a**) $q_{fusion} = (20.9\ \text{kJ/mole})(18{,}960\ \text{moles}) = 3.963 \times 10^5\ \text{kJ}$

(**2b**) $m_{Ti} = 908{,}000$ g; $C_{Ti} = 0.520$ Joules/g-°C
$\Delta T_{(Ti)} = T(\text{final}) - T(\text{initial}) = (1668°C) - (28.0°C) = 1{,}640°C$

(**3b**) $q_{\Delta T} = (1640°C)\ (908{,}000\ \text{g})\ (0.520\ \text{Joules/g-°C}) = 7.743 \times 10^8\ \text{J} = 7.743 \times 10^5\ \text{kJ}$

(**4**): $q_{total} = 3.963 \times 10^5\ \text{kJ} + 7.743 \times 10^5\ \text{kJ} = \mathbf{1.17 \times 10^6\ kJ}$

5. alcohol $_{(liquid)}$ at 29.6 °C $\xrightarrow{q(\Delta T)}$ alcohol $_{(liquid)}$ at 98.4 °C $\xrightarrow{\Delta H(vap)}$ alcohol$_{(vapor)}$ at 98.4 °C

(**1a**) $q_{(vaporization)} = (\Delta H_{(vaporization)}/\text{mole}) \times (\text{moles})$; (**1b**) $q_{\Delta T} = (\Delta T_{alc})(m_{alc})(C_{alc})$

(**2a**) mass of alc = 558 mL × 1.18 g/mL = 658.4 g

$$\text{moles alc} = \frac{658.4\,\text{g}}{144\,\text{g/mole}} = 4.573\ \text{moles}; \quad \Delta H_{(vapourization)} = +34.5\ \text{kJ/mole}$$

(**3a**) $q_{vaporization} = (34.6\ \text{kJ/mole}) \times (4.573\ \text{moles}) = 158.2\ \text{kJ}$

(2b) $C_{alc} = 2.06$ J/g–°C; $m_{alc} = 658.4$ g

$\Delta T_{(solution)} = T(final) - T(initial) = (98.4°C) - (29.6°C) = 68.8°C$

(3b) $q_{\Delta T} = (68.8°C)\ (658.4\ g)\ (2.06\ J/g\text{-}°C) = 9.33 \times 10^4\ J = 93.3\ kJ$

(4) $q_{total} = [q_{vaporization} + q_{\Delta T}] = 158.2\ kJ + 93.3\ kJ =$ **252 kJ**

6. 32.6 grams of $Ba(OH)_2$ (MM = 171.3 g/mole) is dissolved in a calorimeter containing 625.0 grams of water at an initial temperature of 27.3°C; the final temperature of the solution after solvation was 44.6°C. Calculate the $\Delta H_{(solvation)}$/mole for $Ba(OH)_2$.

$$C_{(solution)} = 4.15\ J/g\text{-}°C$$

(1) $\Delta H = q_p$ (reaction); ΔH/mole $Ba(OH)_2 = q_p$/mole $Ba(OH)_2$.

(2) $q_{(reaction)} = -q_{(solution)}$; $q_{(solution)} = (\Delta T_{solution})\ (m_{solution})\ (C_{solution})$

$C_{solution} = 4.15$ J/g-°C; $m_{solution} = 625.0$ g water + 32.6 grams $Ba(OH)_2 = 657.6$ g

$\Delta T_{(solution)} = T(final) - T(initial) = (44.6°C) - (27.3°C) = 17.3°C$

$q_{(solution)} = (\Delta T_{solution})\ (m_{solution})\ (C_{solution}) = (17.3°C)\ (657.6\ g)\ (4.15\ J/g\text{-}°C) = 47{,}212\ J$

$q_{(reaction)} = -q_{(solution)}$;

$$q_{(reaction)} = -(47212\ J) = -47212\ J;\quad \text{moles } Ba(OH)_2 = \frac{32.6\ g}{171.3\ g/mole} = 0.1903\ mole$$

$$(3){:}\ \Delta H_{(solvation)} = q_p/\text{mole } Ba(OH)_2 = \frac{-47212\ J}{0.1903\ mole} = \mathbf{-2.48 \times 10^5\ J/mole\ (-248\ kJ/mole)}$$

7. Carbon tetrachloride (CCl_4) is a common solvent for organic compounds. Calculate the $\Delta H_{(solvation)}$/mole for $C_{10}H_8$ if 28.2 grams of the $C_{10}H_8$ (MM = 128.2 g/mole) solute is dissolved in 643.6 grams of the CCl_4 solvent at an initial temperature of 27.5°C. The final temperature of the solution was 31.2°C; $C_{(solution)} = 0.857$ J/g-°C.

(1) $\Delta H = q_p$ (reaction); ΔH/mole $C_{10}H_8 = q_p$/mole $C_{10}H_8$.

(2) $q_{(reaction)} = -q_{(solution)}$; $q_{(solution)} = (\Delta T_{solution})\ (m_{solution})\ (C_{solution})$

$C_{solution} = 0.857$ J/g-°C; $m_{solution} = 643.6$ g CCl_4 + 28.2 grams $C_{10}H_8 = 671.8$ g

$\Delta T_{(solution)} = T(final) - T(initial) = (31.2°C) - (27.5°C) = +3.7°C$

$q_{(solution)} = (\Delta T_{solution})\ (m_{solution})\ (C_{solution}) = (3.7°C)\ (671.8\ g)\ (0.857\ J/g\text{-}°C) = 2130\ J$

$q_{(reaction)} = -q_{(solution)}$;

$$q_{(reaction)} = -(2130\ J) = -2130\ J;\quad \text{moles } C_{10}H_8 = \frac{28.2\ g}{128.2\ g/mole} = 0.220\ mole$$

$$(3)\ \Delta H_{(solvation)} = q_p/\text{mole } C_{10}H_8 = \frac{-2130\ J}{0.220\ mole} = \mathbf{-9683\ J/mole\ (-9.68\ kJ/mole)}$$

22 Kinetics Part 1 *Rate Laws, Rate Equations, and an Introduction to Reaction Mechanisms*

I GENERAL CONCEPT OF KINETICS

Kinetics is the study of rates of complete reactions by mathematical and experimental analysis to determine reaction mechanisms. A reaction mechanism is a description of the bonding events that occur along the pathway reactants follow to form products.

Rates of reaction are measured by quantifying **concentration changes**, either the disappearance of a reactant or the appearance of a product as a function of time **(t)**:

$$\text{rate}(r) = -\Delta[\text{reactant}]/\Delta t \quad \text{or} \quad \text{rate}(r) = +\Delta[\text{product}]/\Delta t$$

Rates of reactions are generally proportional to the concentration of one or more of the reactants raised to a specific power (exponent); This relationship is termed the **rate expression** or **rate law**:

$$\text{rate}(r) = k\,[\text{reactant 1}]^X\,[\text{reactant 2}]^Y \ldots.$$

The constant in the expression is called the **rate constant, k**; the numerical values of the exponents x, y, and so on, are called the **reactant orders**. For example, if x = 2 in the above general expression, the reaction is "second order in reactant 1."

The **sum** of all the **exponents** x + y... is called the **reaction order**. For example, if x = 2 and y = 1 in the above general expression, the reaction would be described as a "third-order reaction." The values of k, x, y... depend on the reaction mechanism.

ADDITIONAL VARIABLES AFFECTING REACTION RATES

Rates are **inversely** proportional to the energy barriers that prevent a specific reaction or reaction step from occurring; this barrier is called the **activation energy**, or **activation barrier**, for the reaction or step (general symbol: **Ea**).

The activation energy, **Ea**, represents the energy that must be added to reactants to allow bonding changes to occur.

A reaction rate will increase as the activation energy (barrier) decreases. Ea is a path function and does not depend directly on the value of the energy change for the complete reaction. The specific Ea, which is most important for a complete reaction containing many steps, depends on the reaction mechanism.

Rates of reaction are directly proportional to **temperature**; reaction **rates increase** as **temperature increases**.

Reaction rates increase with the presence of a **catalyst**.

INFORMATION RELATIONSHIPS FOR KINETICS AND MECHANISMS

One type of kinetic experiments analyze specific functions of reactant concentration (or product) vs. time: **f{[A]} vs. t.; t** = time; **[A]** = concentration of **any** reactant (or product) molecule; **f{}** can be certain functions.

The experimental information (if complete) yields the **rate expression (rate law)**:

$$\text{rate} = k\,[\text{reactant 1}]^x\,[\text{reactant 2}]^y.....$$

The complete rate expression with reactant exponents, value of k, (and other information) can result in the determination of an acceptable description of the reaction mechanism.

A mechanism cannot be **proven**, but can be shown to provide a correct description of the experimental facts.

II INTRODUCTION TO REACTION MECHANISMS

A **reaction mechanism** is an accepted sequence of elementary reaction steps, which describe all (based on available information) bond-making and bond-breaking events characterizing the change of reactant molecules to product molecules.

A **reaction step** (or **elementary step**) is the smallest observable change in molecular bonding, an individual **bond-making, bond-breaking**, or **combination** event (simultaneous bond making and bond breaking), which can be distinguished experimentally from other such events.

The complete reaction mechanism may be composed of only **one** step or **many** steps depending on the overall (complete) reaction and the conditions.

A reaction mechanism, along with the parameters that describe it such as rate, activation energies, intermediates (Chapter 24) is a **path function.** A **path** function is dependent on the "pathway" or method by which a change occurs. Regardless of the numerical value or sign of the energy change, a reaction can only occur if there exists an available pathway by which reactant molecules can be converted to product molecules.

Path functions are distinguished from **state functions** such as ΔH or ΔE. These depend **only** on the initial and final states of the system: the total energies of the reactants vs. the products. State functions do **not** depend on how the reaction changes occur.

Each reactant and product in a complete reaction or in a single reaction step **must** exist as an independent species for some measurable amount of time. This existence is due to the presence of energy barriers blocking "instant" decomposition. The compound is considered to be in a "potential energy well," (i.e. a stable energy "valley" similar to a rock sitting in a hole) termed a **local energy minimum**. A "deep" hole represents a very stable molecule (slow to react) because the energy barriers on each side are high. A "shallow" hole represents a **relatively** unstable molecule, (faster reacting) because the energy barriers to reaction are low.

A **reaction intermediate** is the product of an individual reaction step, which is later consumed (used as a reactant) in a subsequent step. Since it is an actual product of one reaction step and an actual reactant in a following step, an intermediate is a detectable independent species at a **local energy minimum**.

A **transition state** is a description of atom arrangements showing partial bonds formed or broken for the required molecular changes involved in a reaction step. The partial bonds indicate **how** the reactant atoms are rearranging to form the correct product molecules.

The transition state specifically shows bonding changes at the highest energy point, **potential energy maximum**, of the reaction step. Atom bonding arrangements at the energy maximum transition state **cannot** represent an intermediate and cannot be isolated as an individual molecule.

III DESCRIPTION OF REACTIONS BY MECHANISMS

Complete reactions, indicated by a balanced equation, can be composed of one or more elementary reaction steps. When **all** bonds are broken and/or made simultaneously (at least to the experimental ability to distinguish) then the complete reaction mechanism is **one-step**. A one-step reaction can have one bond-making, one bond-breaking, or any multiple combination if the events cannot be distinguished.

Examples:

$N_2O_4 \longrightarrow 2NO_2$ **(N—N bond broken)**

$O_2 \longrightarrow 2O$ **(O = O bond broken)**

$Br + Br \longrightarrow Br_2$ **(Br—Br bond formed)**

$CH_3Br + Cl \longrightarrow CH_3Cl + Br$ **(C—Cl formed; C—Br broken)**

In the last example, all of these bond changes occur simultaneously or the events cannot be distinguished; thus, they are considered one-step.

Multi-step reactions must involve **more than one** bonding change during the complete reaction and must have bond-making and bond-breaking events, which are **not simultaneous** and **can be distinguished** from each other.

Example: More than one step is detected for the following reaction:

$\mathbf{CH_4 + Cl_2 \longrightarrow CH_3Cl + HCl}$ (balanced equation of complete reaction)

Distinguishable events:

Step 1 $Cl_2 \longrightarrow 2Cl$ **Cl—Cl bond is broken**

Step 2 $Cl + CH_4 \longrightarrow CH_3 + HCl$ **C—H broken, H—Cl formed**

Step 3 $CH_3 + Cl \longrightarrow CH_3Cl$ **C—Cl bond formed**

Cl and CH_3 are **intermediates** in this mechanism: they are generated in a previous step and consumed in a subsequent step. They can be detected by instruments or analytical reactions.

Variability of Mechanisms

Except for the simplest cases, **a mechanism cannot be determined directly from the balanced equation**. A specific reaction given by a balanced equation can potentially proceed by a number of different paths (mechanisms).

General Example:

A—B + C ⟶ A—C + B (balanced equation of complete reaction)

Possible (suggested) mechanism #1 to accomplish this complete reaction could be **one-step**:

The only Step A—B + C ⟶ A—C + B

Since the complete mechanism is only **one-step**, it must be depicted as equivalent to the balanced equation. The **transition state** of this **one** step would have to show both the A—B bond breaking and the A—C bond forming simultaneously. A suggested **pictorial model**:

C-------------A--------------B
(forming) (breaking)

Possible (suggested) mechanism #2 to accomplish this complete reaction could be **two-steps**:

Step 1	A—B ⟶ A + B	**(A—B bond breaks)**
Step 2	A + C ⟶ A—C	**(A—C bond forms)**
Add Steps:	**A—B + C ⟶ A—C + B**	Steps 1 + 2 **must** add to = balanced equation

"A" is an **intermediate**: it is produced in step 1 and consumed in step 2. Recall that addition of reactions requires cancellation of equivalent species appearing on each side of the equation. Addition of all molecules in steps 1 plus 2, after cancellation of the intermediate **"A," must** equal the balanced equation.

Possible (suggested) mechanism #3 to accomplish this complete reaction could be an alternative **two-step**:

Step 1	A—B + C ⟶ C—A—B	**(A—C bond forms)**
Step 2	C—A—B ⟶ C—A + B	**(A—B bond breaks)**
Add Steps:	**A—B + C ⟶ C—A + B**	Steps 1 + 2 must **add** = balanced equation

"C—A—B" is an **intermediate** and is cancelled from both sides. In this case, the order of bond making and bond breaking has been reversed.

IV EXPERIMENTAL KINETICS: DETERMINING REACTANT ORDERS AND RATE CONSTANTS FROM INTEGRATED RATE EQUATIONS

The **rate expression (rate law)** mathematically describes the rate of the complete reaction as a function of **all possible** reactant concentrations.

$$\textbf{rate (r)} = \textbf{k [reactant 1]}^{x} \textbf{ [reactant 2]}^{Y}\ldots\ldots$$

This relationship depends on the type and sequence of reaction steps: the **mechanism**. Since the balanced equation does not reveal the mechanism, the rate expression cannot be determined directly from the balanced equation. (However, the rate expression can be predicted for any reaction if the mechanism is known or suggested; this technique is described in Chapter 24.)

The dependence of the reaction rate on the concentration of each reactant can be found experimentally. The data is generated by measuring changes in reactant concentration as a function of time.

Reaction rates are rates of concentration changes (concentration changes per unit time). Changes in a concentration of any specific reactant may or may not affect how fast the reaction occurs.

Mathematical analysis of **f{concentrations}** vs. **time** can provide the exponent (order) for each reactant (the order may be zero, equivalent to no relationship) and a calculation for the rate constant, k.

Zero-Order Reactants

The simplest case, possible for multistep reactions, is a **rate** that is **independent** of a **specific reactant** concentration. This does not mean that the reactant is unused, just that the specific molecule being analyzed has no role in the reaction steps that affect the rate of the overall (i.e. complete) reaction.

The rate of the reaction is independent of **[A]**; the reaction rate expression **for this specific reactant** for any reaction would be:

$$\textbf{rate (r)} = \textbf{k [A]}^0$$

This is equivalent to rate (r) = rate constant × 1 or rate (r) = k.

(Any number to the zero power = 1.) The reaction is said to be **zero order** in this component and the rate remains constant with changing concentration of [A].

Using the general letter **A** for any specific reactant, the plot of **[A]** vs. **t** (time) would be a straight line with a single (constant) slope. (**[A] means "concentration of A"**)

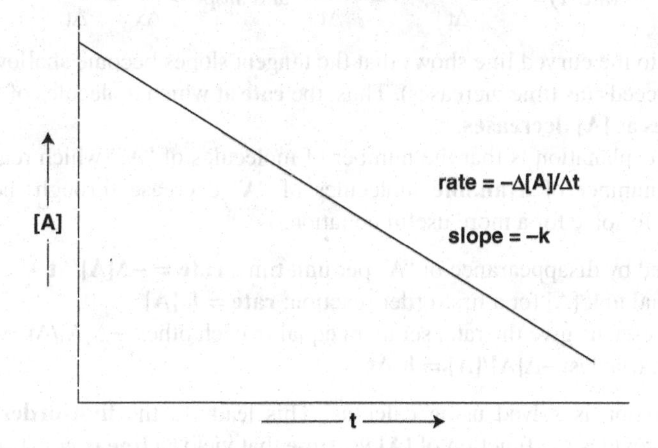

Do not confuse **rate** with **concentration**. The **concentration** of "A" changes continuously with time, but the **rate of change** of the concentration remains the same (constant slope). The number of molecules of "A" consumed by reaction per unit time does not change as the concentration of "A" decreases.

A kinetic analysis that yields a **straight-line** plot of **reactant concentration** vs. **time** ([reactant] vs. t), specifically characterizes that reactant as being **zero order** in the overall rate expression.

First-Order Reactants

A more common situation occurs when the rate of the reaction changes as a direct function of the concentration of a specific reactant. In this case, the reaction is termed **first order in component** "A."

$$\textbf{rate (r)} = \textbf{k [A]}^1 \text{ (equivalent to } \textbf{rate (r)} = \textbf{k [A]})$$

Exponential decrease (exponential decay) can occur for a chemical reaction where "A" is a **reactant** being **consumed**. To derive useful equations, consider a simple one-step reaction with "A" as the only reactant: An example would be a rearrangement of bonding pattern using all the same atoms; termed an isomerization reaction.

$$\textbf{A} \longrightarrow \textbf{B}$$

The change in **[A]** vs. **time** shows an exponential "decay" (decrease); the slope of the curve becomes continuously less negative as concentration of reactant "A" decreases (as the reaction proceeds).

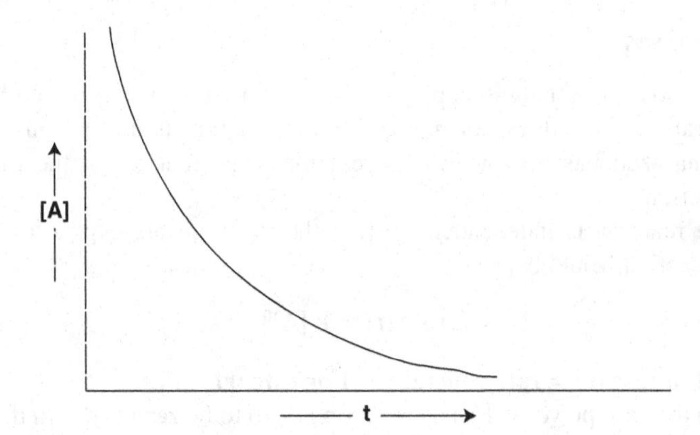

The slope of the line at any point measures the rate at that particular time (t) since

$$\text{rate (r)} = \frac{-\Delta[\text{reactant}]}{\Delta t} = \frac{-\Delta[A]}{\Delta t} \text{ and slope} = \frac{\Delta y}{\Delta x} = \frac{\Delta[A]}{\Delta t}$$

Drawing tangents to the curved line shows that the tangent slopes become shallower and shallower as the reaction proceeds (as time increases). Thus, the **rate** at which molecules of "A" disappear (by reaction) **decreases** as **[A] decreases.**

The molecular explanation is that the number of molecules of "A," which react **per unit time**, decreases as the number of **available** molecules of "A" decrease through their conversion to molecules of "B." To solve for a more useful equation:

Rate is measured by disappearance of "A" per unit time: **rate = –Δ[A]/Δt**
Rate is also equal to k [A] for a first-order reaction: **rate = k [A]**
Since both expressions give the rate, set them equal to each other: **–Δ[A]/Δt = k [A]**
Rearrange the variables: **–Δ[A]/[A] = k Δt**

The above equation is solved using calculus. This leads to the **first-order integrated rate equation**, which provides the function of **[A]** vs. **time** that yields a **linear** equation:
ln [A] = –kt + ln [A_0]
Compare to general linear equation: **y = mx + b**
ln [A] is the natural log (log to the base e) of the concentration of **A** at any time (t).
[A_0] in the equation is the initial concentration of "A" (i.e., at time = 0)

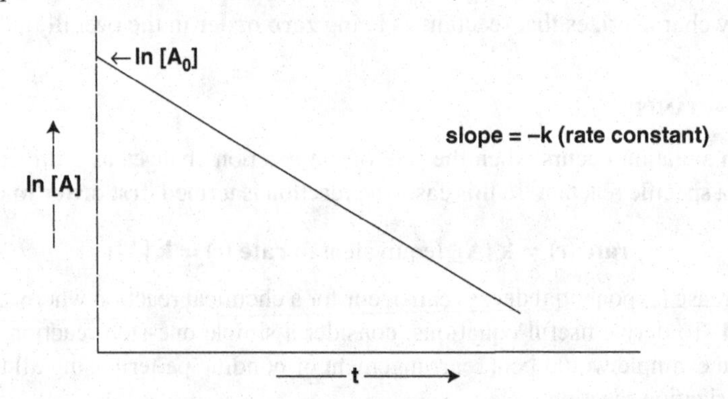

The integrated rate equation, **ln [A] = –kt + ln [A_0]** provides a straight-line graph. Comparison to the general linear equation shows that the **slope** of the line is = **–k**, and the **y-intercept** is **ln [A_0]**.

A kinetic analysis that yields a **straight-line** plot of the **natural log** of **reactant concentration** vs. **time** (ln [reactant] vs. t) specifically characterizes that reactant as being **first order** in the overall rate expression.

Do not confuse **change** in **[A]** (Δ[A]) with **percent change** in **[A]**. The rate equation measures the actual number of molecules (i.e., moles) of "A," which react as a function of time. In fact, it can be shown that the **percentage** of available "**A**," which reacts per unit time, **remains constant**. However, the actual number of reacting "**A**" per time decreases, because the same percentage is multiplying successively smaller concentration numbers.

Example: The "**half-life**" of a reactant is the time it takes for 1/2 (50%) of the reactant to be used up in a reaction. For any **starting concentration** of "**A**" (**[A_0]**), use the equation to determine the **half-life** of "A" if the rate constant, k, is found to be 0.0025 sec^{-1}.

The "units" of rate constants are whatever is necessary to produce an answer with the correct units for the particular situation. In the case of a first-order equation, the units are inverse time, 1/second or sec^{-1}.

(1) ln [A] = −kt + ln [A_0], rearrange variables to isolate **t:** ln [A] − ln [A_0] = **−kt.**
For clarity, isolation of the variable **t** by dividing through by (−k) will be performed last. The subtraction of logs can be expressed as the log of a division; the term ln [A] − ln [A_0] can be written as ln {[A]/[A_0]} producing the equation:

$$\mathbf{\ln\frac{[A]}{[A_0]} = -kt}$$

(2) Use the fact that the half-life is defined as the time required for the concentration [A] to be reduced to ½ of the original amount ([A_0]); thus, **[A] = (0.5) [A_0]**; substitution shows that the term ln {[A]/[A_0]} becomes:

$$\ln\frac{\{(0.5)\,[A_0]\}}{[A_0]} = \ln\,(\mathbf{0.5})$$

(3) Substitute the known rate constant and the value for the ratio {[A]/[A_0]} into the integrated rate equation; **be careful of signs** (units are left out for clarity).

$$\mathbf{\ln\,(0.5) = -0.0025\;t}$$

Complete the calculation: The value of ln (0.5) = − 0.693; substitute (**−0.693**) into the equation for ln (0.5):

$$-0.693 = -0.0025\;\mathbf{t;\ t =} -0.693/-0.0025/\text{sec};\ \mathbf{t = 277\ seconds}$$

The half-life of a first-order process is constant: Decreasing the amount by 50% will always take the same amount of time (277 seconds in the example) regardless of the starting amount. A general expression for half-life is: **t for 1/2 life = 0.693/k.**

Example: The reaction shown is **first order** in the reactant molecule termed enol.

$$\underset{\text{enol}}{\mathbf{CH_2CHOH}} \longrightarrow \underset{\text{keto}}{\mathbf{CH_3CHO}} \qquad \mathbf{rate\ (r) = k[enol]^1}$$

The starting concentration of enol ([enol]) is 6.0 M; the starting concentration of keto ([keto]) is 0 M. The rate constant is 0.25 min^{-1}.

a) Calculate [enol] after the reaction has proceeded for 15 minutes.
   1. ln [A] = −kt + ln [A_0]; ln [enol] = −kt + ln [$enol_0$]; ln [enol] is isolated:
      ln [enol] = −k t + ln [$enol_0$]
   2. k = 0.25 min^{-1}; [$enol_0$] = 6.0 M; t = 15 minutes

3. **$\ln [enol] = -(0.25\ min^{-1})(15\ min) + \ln [6.0]$**
   $= -3.75 + 1.79 = -1.96$
4. $[enol] = e^{-1.96}$ = anti(natural log) $(-1.96) = 0.14$ M

b) Calculate the time required for [keto] to equal 4.5 M.
First calculate the concentration of enol [enol] required for the problem. A more complete discussion of this concept for more complicated balanced equations is in section (V).

The simple balanced equation, CH_2CHOH**(enol)** $\longrightarrow$ CH_3CHO**(keto)** shows that one keto is formed for every one enol reacted.

$$+\Delta[keto] = -\Delta[enol];\ \Delta[keto] \text{ required} = 4.5\ M;\ \text{thus, } (-\Delta[enol]) = 4.5\ M$$

$$[enol] \text{ at the required time} = [enol_0] - \Delta[enol] = 6.0\ M - 4.5\ M = 1.5\ M$$

1. $\ln [enol] = -kt + \ln [enol_0]$; isolate kt:
   $\ln [enol] - \ln [enol_0] = -k\,t$
2. $[enol] = 1.5$ M; $k = 0.25\ min^{-1}$; $[enol_0] = 6.0$ M
3. $\ln [\mathbf{1.5}] - \ln [\mathbf{6.0}] = -(\mathbf{0.25}\ min^{-1})(\mathbf{t}\ min)$
   $0.405 - 3.75 = -0.25\ \mathbf{t}$

$$t = \frac{-1.385}{-0.25} = \mathbf{5.5\ min}$$

Second-Order Reactants

The rate of reaction may be proportional to the concentration of a reactant to the second power. The reaction is termed **second order in reactant "A"**:

$$\mathbf{rate = k\,[A]^2}$$

To derive more useful equations, consider the simplest example for analysis; a complete reaction that has only one-step with only one reactant.

$$\mathbf{A} + \mathbf{A} \longrightarrow \mathbf{A_2}$$
$$NO_2 + NO_2 \longrightarrow N_2O_4$$
$$Br + Br \longrightarrow Br_2$$

A plot of **[A]** vs. **time** is shown for a second-order reaction. The number of **"A"** molecules decreases with reaction progress (time) and the **rate of change** in the concentration of "A" is not constant. As reactant "A" is used up, the **rate** of reaction, as measured by the slope of the line, continuously **decreases;** the curve is more acute than for a first-order reaction.

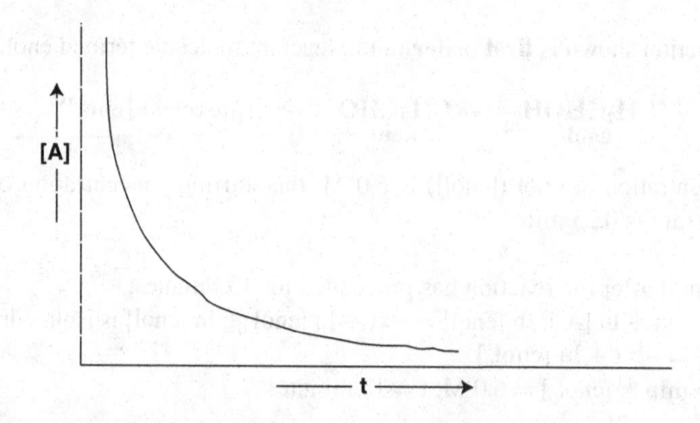

The molecular reason for a dependence on the **square** of the concentration is based on the simple collision requirements of the reaction. A molecule of "A" must collide with another molecule of itself in order to achieve a bonding overlap; the rate of the reaction will be proportional to the number of collisions per unit time (frequency of collision). Although the two molecules of **A** in the general equation are identical for this example, consider them individually. Doubling the concentration of the "**first**" **A** will double the frequency of collisions. Independently doubling the concentration of the "**second**" **A** will also double the frequency. Since **A** is actually the same molecule (or atom), generally doubling the concentration of "A" ([A]) will increase the frequency of collisions, hence the rate by **four** times (statistically, **2 squared**).

To solve for the integrated rate equation, set up the same relationships as for the first-order example.
Rate is measured by disappearance of "A" per unit time: **rate** $= -\Delta[A]/\Delta t$
Rate is also equal to **k** $[A]^2$ for a second-order reaction: **rate** $= k[A]^2$
Since both expressions give the rate, set them equal to each other: $-\Delta[A]/\Delta t = k[A]^2$
Rearrange the variables: $-\Delta[A]/[A]^2 = k\,\Delta t$

The equation is solved using calculus. This leads to the **second-order integrated rate equation**, which provides the function of **[A]** vs. **time** that yields a **linear** equation:

$$1/[A] = kt + 1/[A_0]$$

Compare to the general linear equation: $y = mx + b$

This equation provides a straight-line graph. The **y-axis** is "**1/[A]**," an inverse function of the concentration of **A** at any time during the reaction; the **x-axis** is **time** (general units).

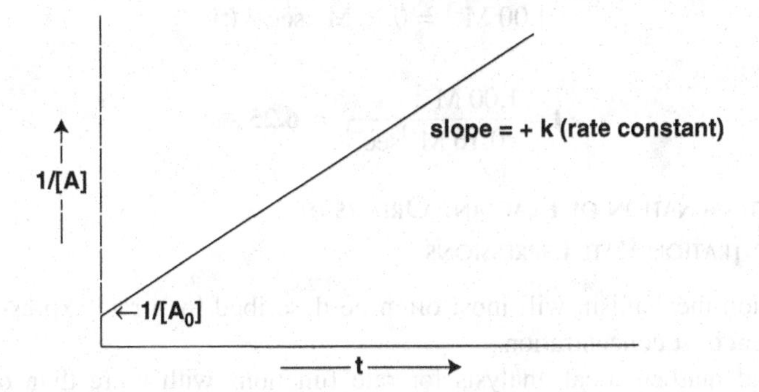

Comparison to the general linear equation shows that the **slope** of the line is = **+k**, and the **y-intercept** is $1/[A_0]$, 1/(initial concentration of "A").

A kinetic analysis that yields a **straight-line plot** of the **inverse** of **reactant concentration** vs. **time** (1/[reactant] vs. t) specifically characterizes that reactant as being **second order** in the overall rate expression.

Example: The reaction shown is **second order** in the reactant ozone (O_3).

$$2\,O_3 \longrightarrow 3\,O_2 \qquad \text{rate (r)} = k[O_3]^2$$

The rate constant is 0.16 $M^{-1}sec^{-1}$; the units are stated as "per molar per second."

a. The starting concentration of ozone is 2.00 M; calculate $[O_3]$ after the reaction has proceeded for 5 seconds.

1. $\frac{(1)}{[A]} = kt + \frac{(1)}{[A_0]}$; $\frac{(1)}{[O_3]} = kt + \frac{(1)}{[O_3]_0}$
2. $k = 0.16\ M^{-1}sec^{-1}$; $[O_3]_0 = 2.0$ M; $t = 5$ seconds

3. $\frac{(1)}{[O_3]} = \mathbf{(0.16\,M^{-1}sec^{-1})\,(5\,sec)} + \frac{(1)}{\mathbf{2.00}}$
$= 0.80 + 0.5 = 1.3$

4. $\frac{(1)}{[O_3]} = 1.3\,M^{-1}$; $[O_3] = \frac{1}{1.3\,M^{-1}} = \mathbf{0.77\,M}$

b. The starting concentration of ozone is 3.00 M; calculate the time required for $[O_3]$ to reach 0.75 M.

1. $\frac{(1)}{[O_3]} = kt + \frac{(1)}{[O_3]_0}$; isolate kt:

$$\frac{\mathbf{(1)}}{\mathbf{[O_3]}} - \frac{\mathbf{(1)}}{\mathbf{[O_3]_0}} = \mathbf{kt}$$

2. $k = 0.16\,M^{-1}sec^{-1}$; $[O_3]_0 = 3.0\,M$; $[O_3] = 0.75M$

3. $\frac{(1)}{\mathbf{0.75}} - \frac{(1)}{\mathbf{3.0}} = \mathbf{0.16\,M^{-1}sec^{-1}(t)}$

$$1.33\,M^{-1} - 0.33\,M^{-1} = 0.16\,M^{-1}sec^{-1}\,(t)$$

$$1.00\,M^{-1} = 0.16\,M^{-1}sec^{-1}\,(t)$$

$$t = \frac{1.00\,M^{-1}}{0.16\,M^{-1}sec^{-1}} = \mathbf{6.25\,sec}$$

Experimental Determination of Reactant Orders in Multiple-Concentration Rate Expressions

A complete reaction mechanism will most often be described by a rate expression that contains **more than one reactant** concentration.

The kinetic and mathematical analysis for rate functions with more than one concentration variable is complex. Solving can be performed by isolation of each of the reactants in experiments, so that (usually) one of the three previous equations will apply to each reactant concentration (**[reactant]**) in turn. (The term for this is "**pseudo-order**" analysis.)

The experiment is carried out simply by keeping the concentrations of all reactants nearly constant, **except** the one being studied. The design usually requires a reaction in which the concentration of the studied reactant is **much smaller** than all the others.

The studied concentration will undergo a relatively **large percentage change** during the reaction while the others will change by only a very small percentage; that is, the others will stay **nearly constant**.

The zero-order, first-order, and second-order integrated rate equations, plus the technique of isolating variables, can be used to determine the rate expression for a complete reaction mechanism.

Example: Consider a simple one-step, two reactant reaction:

$$\mathbf{A + B \longrightarrow AB}$$

The **rate expression** for this one-step reaction may include a dependence term for **both A** and **B**; none of the integrated rate equations will provide a correct mathematical relationship if the concentrations of A and B are **both measurably changing**.

By design, let **[A]** = 0.1 M and **[B]** = 10 M. At the completion of the reaction, **[A]** will approach **zero** but **[B]** will still be 9.9 M, a total change of only about **1%**. A similar experiment could be performed where **[B]** = 0.1 M and **[A]** = 10.0 M.

Example: Assume experiments show that **ln [A]** vs. **time** is a straight line whenever **[B]** is held nearly constant. A similar experiment, exchanging the roles of "A" and "B," indicates that **ln [B]** vs. **time** is a straight line whenever **[A]** is held nearly constant. Determine the complete rate expression.

Analyze the relationship between a particular function that provides a straight-line plot and the corresponding reactant order. If **ln [A]** vs. **t** is a straight line, the reactant order for "A" is **1**: **first-order**. The straight-line plot of **ln [B]** vs. **time** shows that the reactant order for "B" is also **1**. Combine the information to give:

$$\text{rate} = k\,[A]^1\,[B]^1;\ (\text{or rate} = k\,[A]\,[B]) = \text{second-order } \textbf{overall}$$

Example: Determine the complete rate expression for the following reaction based on the data from integrated rate equation experiments. Experiments show that $1/[CH_4]$ vs. time is a straight line whenever $[O_2]$ is held nearly constant, and In $[O_2]$ vs. time is a straight line whenever $[CH_4]$ is held nearly constant.

$$2CH_4 + O_2 \longrightarrow CH_3CH_2OH + H_2O$$

If $1/[CH_4]$ vs. time is a straight line whenever $[O_2]$ is held nearly constant, the reactant order for CH_4 is **2**. A straight-line plot of In $[O_2]$ vs. time whenever $[CH_4]$ is held nearly constant indicates a reactant order of **1** for O_2.

$$\textbf{rate} = \mathbf{k\,[CH_4]^2[O_2]^1}$$

Example: Determine the complete rate expression for the following reaction based on the data from integrated rate equation experiments. Assume experiments show that $1/[CH_2CHCH_2Br]$ vs. time is a straight line whenever $[H_2O]$ is held nearly constant and a **direct** plot of $[H_2O]$ vs. time is a straight line whenever $[CH_2CHCH_2Br]$ is held nearly constant.

$$CH_2CHCH_2Br + H_2O \longrightarrow CH_2CHCH_2OH + HBr$$

Based on these experiments, a $1/[CH_2CHCH_2Br]$ vs. time straight-line plot whenever $[H_2O]$ is held nearly constant would indicate the reactant order for CH_2CHCH_2Br is **2**. A straight-line plot of $[H_2O]$ directly vs. time whenever $[CH_2CHCH_2Br]$ is held nearly constant would indicate a reactant order of **0** for H_2O.

$$\textbf{rate} = \mathbf{k\,[CH_2CHCH_2Br]^2\,[H_2O]^0} \quad \text{or} \quad \textbf{rate} = \mathbf{k\,[CH_2CHCH_2Br]^2}$$

Example: Determine the complete rate expression for the following reaction based on the data from integrated rate equation experiments. Experiments show that In $[C_6H_6]$ vs. time is a straight line whenever $[HNO_3]$ is held nearly constant, and In $[HNO_3]$ vs. time is a straight line whenever $[C_6H_6]$ is held nearly constant.

$$C_6H_6 + HNO_3 \longrightarrow C_6H_5NO_2 + H_2O$$

If ln $[C_6H_6]$ vs. time is a straight line whenever $[HNO_3]$ is held nearly constant, the reactant order for C_6H_6 is **1**. A straight-line plot of ln $[HNO_3]$ vs. time whenever $[C_6H_6]$ is held nearly constant indicates a reactant order of **1** for HNO_3.

$$\textbf{rate} = \mathbf{k\,[C_6H_6]^1\,[HNO_3]^1}$$

V DETERMINING RATE EXPRESSIONS FROM INITIAL RATE DATA CALCULATIONS

An **initial rate** measures the rate of a reaction **only** at the very start of the reaction, **before** the **starting concentrations** of **reactants can change by any significant amount**. This is the equivalent of determining the initial slope (initial tangent) of the curve plotting the change in concentration of a reactant vs. time.

The previous section showed that reactant orders for a rate expression can be determined by following concentration changes as a function of time. Using integrated rate equations, a single reaction experiment can theoretically provide a specific reactant order. Since initial rates measure only the initial slope of the concentration vs. time plot, determination of the correct rate expression by an initial rate method requires a **series** of experiments, which systematically change the **initial concentrations** of reactants individually.

The advantage of using initial rate determinations is that only the easily measured **starting** concentrations of reactants are used in the calculations of reactant orders. Once the rate expression is determined, only **starting** concentrations are required to calculate rate constants and rates at any other set of conditions.

Relative Rate Measurements in the Balanced Equation

Rates can be measured as an appearance of a product as a function of time **(t)**:

$$\textbf{rate(r)} = +\Delta\,\textbf{[product]}/\Delta \textbf{t}$$

Rates can be measured as a disappearance of a reactant as a function of time **(t)**:

$$\textbf{rate(r)} = -\Delta\,\textbf{[reactant]}/\Delta \textbf{t}$$

Rates must always be **positive** values: signs refer to the changes in concentration as the reaction proceeds. A reaction rate can be expressed as a change in concentration of any reactant or product; each can be related through coefficients in the balanced equation.

Example: Use the balanced equation shown to determine the relative rates for the components. The rate of the reaction as measured by disappearance of the reactant Br_2 is: **rate(r) = –1.0 M [Br$_2$]/minute.**

$$\mathbf{2\,NO + Br_2 \longrightarrow 2\,NOBr}$$

a) Use equation stoichiometery to calculate the reaction rate as measured by **[NO]**.
b) Use equation stoichiometery to calculate the reaction rate as measured by **[NOBr]**.

a) Calculation of the reaction rate as measured by **[NO]**.
NO is a reactant; [NO] **decreases** as the reaction proceeds in the forward direction. Use the coefficients in the balanced equation to form the correct ratio:

$$\frac{\Delta\,[NO]}{\Delta t} = \frac{\Delta\,[Br_2]}{\Delta t} \times \frac{2\,NO}{1\,Br_2}$$

$$\text{rate (r)} = -\frac{\Delta\,[Br_2]}{\Delta t} \times \frac{2\,NO}{1\,Br_2} \qquad \text{rate (r)} = -\frac{1.0\,M[Br_2]}{min} \times \frac{2\,NO}{1\,Br_2}$$

$$\textbf{rate (r) = –2.0 M [NO]/min}$$

b) Calculation of the reaction rate as measured by **[NOBr]**.
NOBr is a product; [NOBr] **increases** as the reaction proceeds in the forward direction.
Use the coefficients in the balanced equation to form the correct ratio:

$$+\frac{\Delta\,[NOBr]}{\Delta t} = -\frac{\Delta\,[Br_2]}{\Delta t} \times \frac{2\,NOBr}{1\,Br_2}$$

$$\text{rate (r)} = \frac{\Delta\,(-[Br_2])}{\Delta t} \times \frac{2(+NOBr)}{1(-Br_2)} \qquad \text{rate (r)} = \frac{1.0\,M(-[Br_2])}{\text{min}} \times \frac{2(+NOBr)}{1(-Br_2)}$$

$$\textbf{rate (r) = + 2.0M [NOBr]/min}$$

Rates must always be **positive** values: signs **need not** be explicitly used in stoichiometric calculation. Regardless of specific rate values, reactant concentrations will always decrease and product concentrations will always increase for reactions in the forward direction.

Working with Initial Rate Data Tables

Example: The data table shown applies to the general reaction:

$$2A + B \longrightarrow A_2B$$

Initial rates are measured for a varied set of A and B reactant concentrations. The rate of the reaction is measured by the appearance of the reactant A_2B:

$$\textbf{rate (r)} = +\Delta\,[A_2B\ M]/\textbf{minute}$$

Data Set #	Initial [A]	Initial [B]	(+Δ [A_2B M]/minute) Initial Rate (M/min)
1	2.6 M	3.6 M	1.75
2	6.7 M	3.6 M	11.6
3	10.4 M	3.6 M	28.0
4	5.2 M	3.6 M	7.0
5	5.2 M	7.2 M	14.0
6	5.2 M	1.8 M	3.5
7	2.0 M	2.0 M	?

Based on the general format for the rate expression, the data shown can be used to find the following information. General format for the rate expression: **rate (r) = k** $[A]^x\,[B]^y$

a) Determine the value for **x**: the reactant order for **[A]**.
b) Determine the value for **y**: the reactant order for **[B]**.
c) Write the complete rate expression for this reaction.
d) Calculate the rate constant: **k**.
e) Calculate the specific initial rate for data set **#7**.

Techniques for Solving for Reactant Orders by Comparing Data Sets

The general technique for solving for reactant orders requires selecting two data sets for which the initial concentration of only **one** reactant **changes** while the **other** initial concentration(s) remain **constant**. Thus, the comparison of rate ratios vs. concentration ratios for the selected two sets shows rate variation as a function of only one reactant.

Solve for the value for **x**: the reactant order for **[A]**.

(1) To solve for rate as a function of any specific concentration, [A], use the part of the table for which [A] varies while concentrations of the other reactants stay constant.
In this demonstration example, the letters A and B signify general reactants. To solve for the reactant order for [A], use the part of the table for which [A] varies while [B] stays constant; for example, select data sets **#3** and **#4**:

Data Set #	Initial [A]	Initial [B]	Initial Rate (M/min.)
3	10.4 M	3.6 M	28.0
4	5.2 M	3.6 M	7.0

rate = k $[A]^x$ $[B]^y$; [B] is constant at 3.6 M for comparing the two data sets:
rate = k $[A]^x \times (3.6 = \text{constant})^y$; $(3.6)^y$ remains constant

(2) The initial rate ratios are a function of [A] ratios only. Compare the two data sets by forming two ratios: one comparing the rate and one comparing the corresponding concentrations of A, [A].

$$\frac{\text{data set \#3}}{\text{data set \#4}} \quad \frac{\text{rate } 28.0}{\text{rate } 7.0} = \frac{4}{1} \qquad \frac{[A]10.4}{[A]5.2} = \frac{2}{1}$$

(3) Substitute the numerical values for the ratios determined in step (2) into the rate proportionality determined from the data sets selected in step (1):

With [B] constant: **rate** $\alpha[A]^x$;
thus: **(rate ratio) = ([A] ratio)**^{x}

(4) Substitute the numerical values for these ratios: $(4) = (2)^x$

(5) Solve the equation for **x**, the reactant order. In this case, simple inspection (see next subsections) shows that the value of **x** must be **2** since $4 = 2^2$. (In other words, solve the problem "2 to what power equals 4.")

(6) Confirm the result by a second comparison if the data is available.

For example, select data sets **#1** and **#4;** note that it is convenient to place the data set with the faster rate as the numerator to create the simpler version of the ratios.

Data Set #	Initial [A]	Initial [B]	Initial Rate (MI min)
1	2.6 M	3.6 M	1.75
4	5.2 M	3.6 M	7.0

$$\frac{\text{data set \#4}}{\text{data set \#1}} \quad \frac{\text{rate } 7.0}{\text{rate } 1.75} = \frac{4}{1} \qquad \frac{[A]5.2}{[A]2.6} = \frac{2}{1}$$

(rate ratio) = ([A] ratio)^x

$$(4) = (2)^x$$

the value of **x** must be **2** since: $4 = 2^2$.

Reactant Order Determination by Simple Inspection

Sequential initial reactant concentrations in controlled experiments can be designed to be even multiples of each other; in these cases, simple "inspection" can yield reactant orders.

The general expression for an isolated concentration is rate α $[A]^x$ (or rate = $k[A]^x$).

If **x = 0, rate = k[A]°**; since [A]°=1, **rate = k** and rate is unchanged.
For data set comparisons, if the **rate** remains **unchanged as [A] changes**, then the order of the reactant A must be **zero**: reactant is **zero order**.
For the general expression for an isolated concentration, rate α[A]^x (or rate = k[A]^x):
If x = 1, rate α [A]^1 = rate α [A]
For data set comparisons, if the **rate changes** by the **same ratio as the [A] ratio changes**, then the order of the reactant A must be **one**: reactant is **first order**.
Example: If [A] doubles, rate must double.
If [A] triples, rate must triple.
If [A] is halved, rate must be halved.
For the general expression for an isolated concentration rate α [A]^x (or rate = k[A]^x):
If x = 2, rate α [A]^2
For data set comparisons, if the rate changes by the square of the [A] ratio change, then the order of the reactant A must be **two**: reactant is **second order**.
Example: If [A] doubles (2x), rate must quadruple (2^2x = 4x).
If [A] triples (3x), rate must increase nine times (3^2x = 9x).
If [A] is halved (1/2x), rate must decrease by one-quarter ($1/2^2$ = 1/4x).

Additional mathematical analysis is usually not necessary for reactant orders of 0, 1, or 2 with simple multiples of [A].

Reactant Order Determination by Logarithmic Analysis

For data set comparisons that do not have simple concentration multiples, the general equation relationship can be solved by using logarithms. The general expression for an isolated concentration, rate α [A]^x, is expressed as (rate ratio) = ([A] ratio)^x for solving initial rate problems.

$$\textbf{(rate ratio)} = \textbf{([A] ratio)}^{x}$$

Take the log of both sides to simplify: **log (rate ratio) = (x) log ([A] ratio)**

(1) The example data table indicates that data sets **#2** and **#4** show [B] constant but do not show the change in [A] in even multiples:

Data Set #	Initial [A]	Initial [B]	Initial Rate (M/min)
2	6.7 M	3.6 M	11.6
4	5.2 M	3.6 M	7.0

(2) $\dfrac{\text{data set \#2}}{\text{data set \#4}} \quad \dfrac{\text{rate } 11.6}{\text{rate } 7.0} = \dfrac{1.66}{1} \qquad \dfrac{[A]\,6.7}{[A]5.2} = \dfrac{1.29}{1}$

(3) **(rate ratio) = ([A] ratio)^x**

$$1.66 = (1.29)^x$$

(4) Simple inspection does not yield an obvious value for x in the equation. Trial and error could be used: x = 0 and x = 1 do not fit the equation; x = 2 could be tried, alternatively, the logarithm equation can be solved:

$$\textbf{log (rate ratio) = (x) log ([A] ratio)}$$

$$\log (1.66) = x \log (1.29)$$

$$0.2201 = x\ (0.1106)$$

$$x = 0.2201/0.1106 \cong \mathbf{2}$$

The value of $x \cong \mathbf{2}$ agrees with the other data set comparisons. The log equation can be used for all comparisons. For example, the original comparison for data sets **#2** and **#4** must yield the same results:

$$\frac{\text{data set \#3}}{\text{data set \#4}} \quad \frac{\text{rate } 28.0}{\text{rate } 7.0} = \frac{4}{1} \qquad \frac{[A]\,10.4}{[A]\,5.2} = \frac{2}{1}$$

log (rate ratio) = (x) log ([A] ratio)

$$\log(4) = x \log(2)$$

$$0.60206 = x\ (0.30103)$$

$$x = 0.60206/0.30103 = 2$$

To solve for the value for **y**, the reactant order for **[B]** follow steps (**1**) through (**5**).

(**1**) Select the part of the table for which [B] varies while [A] stays constant; for example select data sets **#4** and **#5**:

Data Set #	Initial [A]	Initial [B]	Initial Rate (M/min.)
4	5.2 M	3.6 M	7.0
5	5.2 M	7.2 M	14.0

rate = k $\mathbf{[B]^y \times [A]^x}$ The value of "x" was determined to be 2; however, this value is not required to analyze the selected data sets.
rate = k $\mathbf{[B]^y \times (5.2 = \text{constant})^x}$; initial rate ratio is a function of [B] ratio only.

(**2**) Compare the two data sets by forming a rate ratio and a ratio of [B]:

$$\frac{\text{data set \#5}}{\text{data set \#4}} \quad \frac{\text{rate } 14.0}{\text{rate } 7.0} = \frac{2}{1} \qquad \frac{[B]\,7.2}{[B]\,3.6} = \frac{2}{1}$$

with [A] constant: **rate** $\boldsymbol{\alpha[B]^y}$;
thus: (rate ratio) = ([B] ratio)$^{\mathbf{y}}$

Step (**3**) $(2) = (2)^y$

Step (**4**) value of **y** must be **1** since: $2 = 2^1$

5. Confirm the result by a second comparison; for example, data sets #4 and #6:

Data Set #	Initial [A]	Initial [B]	Initial Rate (M/min)
4	5.2 M	3.6 M	7.0
6	5.2 M	1.8 M	3.5

$$\frac{\text{data set \#4}}{\text{data set \#6}} \quad \frac{\text{rate } 7.0}{\text{rate } 3.5} = \frac{2}{1} \qquad \frac{[B]\,3.6}{[B]\,1.8} = \frac{2}{1}$$

$$\textbf{(rate ratio)} = \textbf{([B] ratio)}^{y}$$

$$\textbf{(2)} = \textbf{(2)}^{y}$$

the value of **y** must be **1** since: $2 = 2^1$.

Determining the Complete Rate Expressions

The complete rate expression for any reaction, if the data is complete, is found by combining the reactant order information determined for each isolated concentration. In the demonstration example, the complete rate expression results by combining reactant orders for [A] and [B].

The reactant order for [A] is 2; second order: **rate α $[A]^2$**
The reactant order for [B] is 1; first order: **rate α $[B]^1$**

$$\textbf{rate} \ \alpha \ [A]^2 [B]^1$$

$$\textbf{rate} = \textbf{k} \ [A]^2 [B]$$

Determining the Value for the Rate Constant

The initial rate data can be combined with the completed rate expression to calculate the value of the rate constant: select **any** data set and substitute the known values for concentrations and initial rate directly in the rate expression. For example, for data set **#4**:

Data Set #	Initial [A]	Initial [B]	Initial Rate (M/min.)
4	5.2 M	3.6 M	7.0

Substitute the values for initial concentrations of all reactants and the value for the initial rate into the determined correct rate expression:

$$\textbf{rate} = \textbf{k} \ [A]^2 [B]$$

$$7.0 \text{ M/min} = k \ (5.2 \text{ M})^2 \ (3.6 \text{ M})$$

$$7.0 \text{ M/min} = k \ (97.3 \text{ M}^3)$$

$$k = \frac{7.0 \text{ M/min}}{97.3 \text{ M}^3} = \mathbf{7.19 \times 10^{-2} \ M^{-2} \ min^{-1}}$$

Units of "per molar squared per minute" indicate a overall third-order reaction.

Determining the Value for Any Unknown Rate

The completed rate expression combined with the value of the rate constant can be used to calculate the rate for any concentration combination by direct substitution. For the demonstration example, calculate the initial rate for data set #7.

Data Set #	Initial [A]	Initial [B]	Initial Rate (M/min.)
7	2.0 M	2.0 M	?

$$\textbf{rate} = \textbf{k} [\textbf{A}]^2 [\textbf{B}]$$

$$\text{rate} = (7.19 \times 10^{-2}\ \text{M}^{-2}\ \text{min}^{-1})(2.0\ \text{M})^2\ (2.0\ \text{M})$$

$$\textbf{rate} = \textbf{0.58 M min}^{-1}$$

Example: Write the general format of the rate expression, then use the rate data shown for the following reaction to determine the complete rate expression, the rate constant, and the initial rate for data set #4.

$$CO_{(g)} + NO_{2(g)} \longrightarrow CO_{2(g)} + NO_{(g)}$$

Data Set #	Initial [CO]	Initial [NO_2]	Initial Rate (M/hour)
1	0.50 M	0.36 M	3.4 10^{-4}
2	0.50 M	0.18 M	1.7 10^{-4}
3	1.0 M	0.36 M	6.8 10^{-4}
4	1.5 M	0.72 M	?

(1) rate = k [CO]^x [NO_2]^y
To solve for the reactant order for CO, select data sets **#1** and **#3**:
rate = k[CO]^x × (0.36 = constant)^y; initial rate ratio is a function of [CO] ratio only.

(2) Compare the two data sets by forming a rate ratio and a ratio of [CO]:

$$\frac{\text{data set \#3}}{\text{data set \#1}} \quad \frac{\text{rate } 6.8\times10^{-4}}{\text{rate } 3.4\times10^{-4}} = \frac{2}{1} \qquad \frac{[\text{CO}]\,1.0}{[\text{CO}]\,0.5} = \frac{2}{1}$$

With [NO_2] constant: **rate α [CO]^x**;
thus: (rate ratio) = ([CO] ratio)^x

Step (**3**) (2) = (2)^x

Step (**4**) value of **x** must be **1** since: 2 = 2^1

Alternatively, simple inspection shows that the rate of the reaction doubles as the concentration of CO doubles; the CO reactant must be **first order**.

Solving for the reactant order of NO_2, select data sets **#1** and **#2**:

rate = k [0.5 = constant]^x × (NO_2)^y; initial rate ratio is a function of [NO_2] ratio only.
(**2**) Compare the two data sets by forming a rate ratio and a ratio of [NO_2].

$$\frac{\text{data set \#1}}{\text{data set \#2}} \quad \frac{\text{rate } 3.4\times10^{-4}}{\text{rate } 1.7\times10^{-4}} = \frac{2}{1} \qquad \frac{[\text{NO}_2]\,0.36}{[\text{NO}_2]\,0.18} = \frac{2}{1}$$

With [CO] constant: **rate α** [NO_2]^y;
thus: **(rate ratio)** = ([NO_2] **ratio**)^y

Step (**3**): (2) = (2)^x

Step (**4**): Value of **y** must be **1** since: 2 = 2^1

Alternatively, simple inspection shows that the rate of the reaction doubles as the concentration of NO_2 doubles; the NO_2 reactant must be **first order**.

To determine the rate constant, select any data set, for example data set **#1**:

Data Set #	Initial [A]	Initial [B]	Initial Rate (M/hour)
1	0.50 M	0.36 M	3.4×10^{-4}

$$\textbf{rate} = \textbf{k [CO] [NO}_2\textbf{]}$$

$$3.4 \times 10^{-4}\,\text{M/hour} = \text{k}\,(0.50\text{ M})\,(0.36\text{ M})$$

$$3.4 \times 10^{-4}\,\text{M/hour} = \text{k}\,(0.18\text{ M}^2)$$

$$\text{k} = \frac{3.4 \times 10^{-4}\text{ M/hour}}{0.18\text{M}^2} = \mathbf{1.89 \times 10^{-3}\ M^{-1}\ hour^{-1}}$$

The rate for data set **#4** is found by direct substitution.

Data Set #	Initial [CO]	Initial [NO_2]	Initial Rate (M/hour)
4	1.5 M	0.72 M	?

$$\textbf{rate} = \textbf{k [CO] [NO}_2\textbf{]}$$

$$\text{rate} = (1.89 \times 10^{-3}\,\text{M}^{-1}\,\text{hour}^{-1})(1.5\text{ M})\,(0.72\text{ M})$$

$$\mathbf{rate = 2.0 \times 10^{-3}\,M\,hour^{-1}}$$

Determining Reactant Orders from Limited Data Sets

Example: The following data table shows limited data sets; no two data sets are found that keep the [CH_3Br] constant. However, the data table does allow a complete solution.

$$\mathbf{CH_3Br + Cl^- \longrightarrow CH_3Cl + Br^-\ \ rate\ (r) = +\Delta[CH_3Cl\ M]/\,minute}$$

Data Set #	Initial [CH_3Br]	Initial [Cl]	Initial Rate
1	0.25 M	1.0 M	0.15
2	0.50 M	1.0 M	0.30
3	1.0 M	2.0 M	1.2

(1) No two data sets show a constant value for [CH_3Br]; solve for **[CH_3Br]^x** first since two data sets have constant [Cl^-]: select data sets **#1** and **#2**

rate = k [CH_3Br]^x[Cl^-]^y; [Cl^-] is constant at 1.0 M for comparing the two data sets:
rate = k [CH_3Br]^x × (1.0 = constant)^y; initial rate ratio is a function of [CH_3Br] ratio only.

$$\text{step } \mathbf{(2)}\ \frac{\text{data set \#2}}{\text{data set \#1}}\quad \frac{\text{rate } 0.30}{\text{rate } 0.15} = \frac{2}{1}\qquad \frac{[\text{CH}_3\text{Br}]\,0.50}{[\text{CH}_3\text{Br}]\,0.25} = \frac{2}{1}$$

With [Cl^-] constant: **rate α[CH_3Br]^x;**
thus: (rate ratio) = ([CH_3Br] ratio)^x

step **(3)** (2) = $(2)^x$

step **(4)** value of x must be **1** since: 2 = 2^1

Now solve for **$[Cl^-]^y$**. Select data sets for a changing $[Cl^-]$: data sets **#2** and **#3** show $[CH_3Br]$ is **not** constant for these data sets but **x** was determined to = **1**; this is sufficient information to complete the analysis.

Modified step **(1)**:

Data Set #	Initial $[CH_3Br]$	Initial $[Cl^-]$	Initial Rate
2	0.50 M	1.0 M	0.30
3	1.0 M	2.0 M	1.2

$$\textbf{rate} = \textbf{k} [CH_3Br]^x[Cl^-]^y;$$

Modified step **(2): rate = k $[CH_3Br]^1 [Cl^-]^y$**
Modified step **(3)**:

$$\frac{\text{rate}\#3}{\text{rate}\#2} = \frac{([CH_3Br]^1{}_{\#3}) \times ([Cl^-]^y{}_{\#3})}{([CH_3Br]^1{}_{\#2}) \times ([Cl^-]^y{}_{\#2})}$$

$$\frac{\text{data set}\#3}{\text{data set}\#2} \quad \frac{\text{rate}1.2}{\text{rate}0.3} = \frac{[1.0]^1 \times (2.0)^y}{[0.5]^1 \times (1.0)^y}$$

$$\textbf{(rate ratio)} = \textbf{([CH}_3\textbf{Br] ratio)([Cl}^-\textbf{]ratio)}^y$$

$$(4) = (2) \times (2)^y$$

Step **(4)** **y = 1** since: 4 = (2) × 2^1
or solve the above equation for $[Cl^-]^y$:

$$\frac{\textbf{(rate ratio)}}{\textbf{([CH}_3\textbf{Br] ratio)}} = \textbf{([Cl}^-\textbf{]ratio)}^y : \quad \frac{4}{2} = (2)^y \quad ; \quad 2 = (2)^1$$

VI PRACTICE PROBLEMS

1. A first-order reaction is shown:

$$\underset{\textbf{(CP)}}{\textbf{Cyclopropane}} \longrightarrow \textbf{propene} \quad \textbf{rate} = \textbf{k[CP]}$$

Assume an experiment starts with $[CP_0]$ (initial concentration) equal to 1.0 M. The value for k = 0.010/sec.

a) What will be **[CP]** after 300 seconds?
b) How many seconds must the reaction proceed to reduce the **[CP]** to 0.020 M?

2. The reaction shown is **second order** in the reactant nitrogen dioxide (NO_2).

$$\textbf{2 NO}_2 \longrightarrow \textbf{2 NO} + \textbf{O}_2 \quad \textbf{rate (r)} = \textbf{k[NO}_2\textbf{]}^2$$

The rate constant is 3.40 $M^{-1}min^{-1}$.

a) The starting concentration of NO_2 is 6.00 M; calculate $[NO_2]$ after the reaction has proceeded for 30 seconds.
b) The starting concentration of NO_2 is 2.00 M; calculate the time required for $[NO_2]$ to reach 1.50 M.

3. Consider the following general reaction: **AB + CD ⟶ AC + BD**
Use the **kinetic experimental** information to determine the rate expressions for the separate mechanisms #1, #2, and #3 based on the general equation shown above. No mechanistic steps are given for any mechanism.

Mechanism #1: A plot of [AB] vs. time is a straight line whenever [CD] is held very high and nearly constant. A plot of 1/[CD] vs. t is a straight line.
Mechanism #2: A plot of In [AB] vs. t is a straight line when [CD] is held nearly constant. A plot of In [CD] vs. t is a straight line when [AB] is held nearly constant.
Mechanism #3: A plot of In [AB] vs. t is a straight line with a slope of –3.5 and a y intercept of 0.2. A plot of [CD] vs. t is a straight line when [AB] is held nearly constant. Include the actual numerical value for rate constant k in the expression.

4. The rate of the reaction shown is measured by the rate of ammonia formation:

$$\text{rate (r)} = +\,0.15\ M\ [NH_3]/\text{second}$$

a) Use the balanced equation to determine the relative rate of $[N_2]$ disappearance compared to the measured rate of ammonia formation.
b) Repeat the calculation for the disappearance of $[H_2]$.

$$N_2 + 3\,H_2 \longrightarrow 2\,NH_3$$

5. Use the data table shown below to complete the following parts:

$$2\,H_2 + 2\,NO \longrightarrow N_2 + 2H_2O$$

$$\textbf{rate (r)} = +\Delta[N_2\,M]/\textbf{second}$$

Data Set #	Initial $[H_2]$	Initial [NO]	Initial Rate (M/sec)
1	0.122 M	0.420 M	0.136
2	0.122 M	0.210 M	0.0339
3	0.244 M	0.210 M	0.0678
4	0.488 M	0.105 M	0.0339
5	0.205 M	0.350 M	?

a) Write the general format for the rate expression.
b) Determine the value for **x**: the reactant order for $[H_2]$.
c) Determine the value for **y**: the reactant order for **[NO]**.
d) Write the complete rate expression for this reaction.
e) Calculate the rate constant: **k**.
f) Use the rate constant to prove that data sets **#2** and **#4** should have the same rate.
g) Calculate the specific initial rate for data set **#5**.

VII ANSWERS TO PRACTICE PROBLEMS

1. A first-order reaction is shown:

Cyclopropane (CP) ⟶ propene rate = k[CP]

Assume an experiment starts with $[CP_0]$ (initial concentration) equal to 1.0 M.
The value for k = 0.010/sec

a) What will be **[CP]** after 300 seconds?
$\ln [CP] = -kt + \ln [CP_0]$
$\ln [CP] = -(0.010/sec)(300\ sec) + \ln (1.0)$
$\ln [CP] = -3 + 0 = -3$
[CP] = antiln $(-3) = e^{-3}$ **= 0.0498 M**

b) How many seconds must the reaction proceed to reduce the **[CP]** to 0.020 M?
$\ln (0.02) = -(0.010)(\mathbf{t}) + \ln (1.0)$
$-3.91 = -(0.010)(\mathbf{t}) + 0$

$$t = \frac{-3.91}{-0.010\,sec^{-1}} = \mathbf{391\ sec}$$

2. The reaction shown is **second order** in the reactant nitrogen dioxide (NO_2).

$$\mathbf{2\,NO_2 \longrightarrow 2\,NO + O_2 \qquad rate\ (r) = k\,[NO_2]^2}$$

The rate constant is 3.40 $M^{-1}min^{-1}$.

a) The starting concentration of NO_2 is 6.00 M; calculate $[NO_2]$ after the reaction has proceeded for 30 seconds. 30 seconds = 0.50 minutes

$$\frac{(1)}{[NO_2]} = kt + \frac{(1)}{[NO_2]_0}$$

$$\frac{(1)}{[NO_2]} = (3.40\,M^{-1}\,min^{-1})(0.50\,min) + \frac{(1)}{6.00} = 1.7 + 0.167 = 1.87$$

$$[NO_2] = \frac{(1)}{1.87\,M^{-1}} = \mathbf{0.53M}$$

b) The starting concentration of NO_2 is 2.00 M; calculate the time required for $[NO_2]$ to reach 1.50 M.

$$\frac{(1)}{[NO_2]} - \frac{(1)}{[NO_2]_0} = kt$$

$$\frac{(1)}{1.50} - \frac{(1)}{2.00} = 3.40\,M^{-1}\,min^{-1}\,(\mathbf{t})$$

$0.667M^{-1} - 0.500\,M^{-1} = 3.40\,M^{-1}min^{-1}\,(\mathbf{t})$

$$t = \frac{0.167\,M^{-1}}{3.40\,M^{-1}\,min^{-1}} = \mathbf{0.049\ min}$$

3. Consider the following general reaction: $\mathbf{AB + CD \longrightarrow AC + BD}$
Use the **kinetic experimental** information to determine the rate expressions for the separate mechanisms #1, #2, and #3 based on the general equation shown above.

Mechanism #1: A plot of [AB] vs. time is a straight line whenever [CD] is held very high and nearly constant. A plot of 1/[CD] vs. t is a straight line.

Analysis: Whenever [CD] is constant, if [AB] vs. t = straight line, reaction is zero order in [AB]. Since [AB] is zero order, the reaction rate depends **only** on [CD]. The straight-line plot of 1/[CD] vs. time, regardless of [AB], indicates that the reaction is second-order with respect to [CD]. Combine the information:

$$\text{rate} = k[AB]^0\,[CD]^2 \text{ or } \mathbf{rate = k[CD]^2}$$

Mechanism #2: A plot of In [AB] vs. t is a straight line when [CD] is held nearly constant. A plot of In [CD] vs. t is a straight line when [AB] is held nearly constant.
Analysis: A straight-line plot of In [AB] vs. time (with [CD] constant) indicates that the reaction is first order in [AB]. The equivalent result for [CD] indicates the reaction is first order in [CD]. Combine the information:

$$\text{rate} = k[AB]^1\,[CD]^1 \quad \text{or} \quad \mathbf{rate = k}\,[AB]\,[CD]$$

Mechanism #3: A plot of ln [AB] vs. t is a straight line with a slope of −3.5 and a y-intercept of 0.2. A plot of [CD] vs. t is a straight line when [AB] is held nearly constant. Include the actual numerical value for rate constant k in the expression.

Analysis: The reaction is zero order in [CD] based on the straight line for [CD] vs. time. The reaction is first order in [AB] based on the straight line for ln [AB] vs. time. The integrated rate equation for first order shows that the slope = −k. Thus, k = − (slope) for the data given, or k = −(−3.5) = 3.5/sec (or 3.5 sec^{-1}) if time is in seconds.

$$\mathbf{rate = 3.5\,sec^{-1}[AB]}$$

4. The rate of the reaction shown is measured by the rate of ammonia formation:

$$\text{rate (r)} = +\ 0.15\ M\ [NH_3]/\text{second}$$

$$\mathbf{N_2 + 3\ H_2 \longrightarrow 2\,NH_3}$$

a) Use the balanced equation to determine the relative rate of $[N_2]$ disappearance compared to the measured rate of ammonia formation.
$\Delta[N_2]/\Delta t = -\Delta[NH_3]/\Delta t \times 1N_2/2\ NH_3$
rate (r) = $-\Delta[NH_3]/\Delta t \times 1N_2/2\ NH_3$
rate (r) = $-0.15\ M\ [NH_3]/sec \times 1\ N_2/2\ NH_3$ =
rate (r) = $-0.075\ M\ [N_2]/sec$

b) Repeat the calculation for the disappearance of $[H_2]$.
H_2 is a reactant; $[H_2]$ decreases as the reaction proceeds in the forward direction. Use the coefficients in the balanced equation to form the correct ratio:
$\Delta[H_2]/\Delta t = -\Delta[NH_3]/\Delta t \times 3\ H_2/2\ NH_3$
rate (r) = $-\Delta[NH_3]/\Delta t \times 3\ H_2/2\ NH_3$
rate (r) = $-0.15\ M\ [NH_3]/sec \times 3\ H_2/2\ NH_3$
rate (r) = $-0.23\ M\ [H_2]/sec$

5. Use the data table shown below to complete the following parts:

$$\mathbf{2\,H_2 + 2\,NO \longrightarrow N_2 + 2\,H_2O}$$

$$\mathbf{rate\ (r) = +\Delta[N_2\,M]/second}$$

Data Set #	Initial $[H_2]$	Initial[NO]	Initial Rate (M/sec)
1	0.122 M	0.420 M	0.136
2	0.122 M	0.210 M	0.0339
3	0.244 M	0.210 M	0.0678
4	0.488 M	0.105 M	0.0339
5	0.205 M	0.350 M	?

a) The general format for the rate expression is **rate (r) = k** $[H_2]^x$ $[NO]^y$

b) To solve for the value for **x**, the reactant order for $[H_2]$, select data sets with constant [NO]; data sets **#2** and **#3**

(1)

Data Set #	Initial $[H_2]$	Initial [NO]	Initial Rate (M/sec)
2	0.122 M	0.210 M	0.0339
3	0.244 M	0.210 M	0.0678

rate = k $[H_2]^x$ $[NO]^y$; [NO] is constant at 0.210 M
rate = k $[H_2]^x \times$ **(0.210 = constant)**^{y}

(2) $\frac{\text{data set \#3}}{\text{data set \#2}} \quad \frac{\text{rate } 0.0678}{\text{rate } 0.0339} = \frac{2}{1} \quad \frac{[H_2]0.244}{[H_2]0.122} = \frac{2}{1}$

With [NO] constant: **rate α** $[H_2]^x$;
thus: **(rate ratio) = (**$[H_2]$**ratio)**^{x}

(3) --(2) = $(2)^x$

(4) The value of **x** must be **1** since: $2 = 2^1$

Or by inspection: Rate doubles when $[H_2]$ doubles; x must =1, first order.

c) To solve for the value for **y**, the reactant order for **[NO]**, select data with constant $[H_2]$; data sets **#1** and **#2**:

(1)

Data Set #	Initial $[H_2]$	Initial [NO]	Initial Rate (M/sec)
1	0.122 M	0.420 M	0.136
2	0.122 M	0.210 M	0.0339

rate = k $[NO]^y[H_2]^x$; $[H_2]$ is constant at 0.122 M
rate = k $[NO]^y \times$ **(0.122 = constant)**^{x}

The value of "x" was determined to be 1; however, this value is not required to analyze the selected data sets.

(2) $\frac{\text{data set \#1}}{\text{data set \#2}} \quad \frac{\text{rate } 0.136}{\text{rate } 0.0339} = \frac{4}{1} \quad \frac{[NO]0.420}{[NO]0.210} = \frac{2}{1}$

With $[H_2]$ constant: **rate α** $[NO]^y$;
Thus: **(rate ratio) = ([NO] ratio)**^{y}

(3) (4) = $(2)^y$

(4) The value of **y** must be **2** since: $4 = 2^2$
Or by inspection: Rate quadruples when [NO] doubles; y must = 2, second order.

d) The complete rate expression is **rate (r) = k** $[H_2]^1$ $[NO]^2$

e) If data set #1 is used:

$$\textbf{rate (r)} = \mathbf{k\,[H_2]^1\,[NO]^2}$$

$$0.136\ \mathbf{M/sec} = \mathbf{k}\ (0.122\ \text{M})(0.420\ \text{M})^2$$

$$0.136\ \mathbf{M/sec} = \mathbf{k}\ (0.02152\ \mathbf{M^3})$$

$$k = 0.136\ \text{M/sec}/0.02152\ \text{M}^3$$

$$\mathbf{k = 6.32\ M^{-2}\,sec^{-1}}$$

f) Rate for data set **#2**: rate = $(6.32\ M^{-2}\,sec^{-1})(0.122\ M)(0.210\ M)^2 = 0.0340$ M/sec
g) Rate for data set **#4**: rate = $(6.32\ M^{-2}\,sec^{-1})(0.488\ M)(0.105\ M)^2 = 0.0340$ M/sec
h) To solve for the unknown rate, use data set **#5**:

Data Set #	Initial $[H_2]$	Initial [NO]	Initial Rate (M/sec)
5	0.205 M	0.350 M	?

$$\mathbf{rate = k\,[H_2]^1\,[NO]^2}$$

$$\text{rate} = (6.32\ \text{M}^{-2}\text{sec}^{-1})\ (0.205\ \text{M})\ (0.350\ \text{M})^2$$

$$\mathbf{rate = 0.159\ M\ sec^{-1}}$$

23 General Techniques for Solving Equilibrium Problems

I CONCENTRATION QUOTIENT AND EQUILIBRIUM EXPRESSION

Each reversible reaction has two sets of molecules: the set of molecules on the left represents reactants in the forward reaction and products in the reverse reaction; the set of molecules on the right represents products in the forward reaction and reactants in the reverse reaction.

$$\mathbf{a\,A + b\,B \rightleftharpoons c\,C + d\,D}$$

A, B, C, D = general reactants and products; **a, b, c, d** = stoichiometric coefficients of A, B, C, D respectively in the balanced equation.

Viewed from the perspective of the forward reaction, the **reactants** are **A** and **B**; the **products** are **C** and **D**. A specific ratio of products to reactants, termed the concentration quotient (symbol = **Q**), is defined for **any** actual values for concentrations of components at any point in a reversible reaction. The form of the ratio expression uses the specific concentration of each species, usually as molarity (**M**) or gas partial pressure ($\mathbf{P_{gas}}$), raised to the power of its coefficient in the balanced equation:

$$Q_{(\text{forward reaction})} = \frac{[C]^c\,[D]^d}{[A]^a\,[B]^b} \quad \alpha \quad \frac{[\text{products}]}{[\text{reactants}]}$$

The rate of a reaction is proportional to the concentrations of the molecules reacting (Chapter 22). The rate of the **forward** reaction is proportional to the concentrations of the reactants **A** and **B**; the rate of the **reverse** reaction is proportional to the concentrations of the products **C** and **D**.

$$\text{rate}_{(\text{forward})} = k_{(\text{forward})}\,[A]^x\,[B]^y \quad \text{rate}_{(\text{reverse})} = k_{(\text{reverse})}\,[C]^x\,[D]^y$$

Viewed from the forward direction, as the reaction proceeds, the concentrations of the reactants [A] and [B] decrease; the rate of the forward reaction correspondingly decreases. Concurrently, the concentrations of the products [C] and [D] increase, and the rate of the reverse reaction correspondingly increases. Regardless of the numerical values of the forward and reverse rate constants, there must come a point in the progress of the reaction when the net forward and net reverse rates are equal. (At the extreme, as concentration of reactants approach zero, the forward rate approaches zero.)

$$\text{rate}_{(\text{forward})} = \text{rate}_{(\text{reverse})};\ \text{thus, } k_{(\text{forward})}[A]^x\,[B]^y = k_{(\text{reverse})}\,[C]^x[D]^y$$

Under this condition, the reversible reaction reaches a state termed **equilibrium**. This means that reactant molecules are forming product molecules at the same rate that product molecules are being reconverted to reactant molecules; a system at equilibrium shows no **net** change in concentrations of products or reactants.

Since the concentrations of all products and reactants remain in balance (no change) at equilibrium, the ratio determined by the concentration quotient, Q, remains a constant value:

$$Q_{(\text{forward reaction})} = \frac{[C]^c\,[D]^d}{[A]^a\,[B]^b} = \text{constant value at equilibrium.}$$

The concentration ratio defined by Q is the equilibrium concentration ratio when applied to reactions at equilibrium; the numerical value for Q at equilibrium is the equilibrium constant, **K** (or $\mathbf{K_{(subscript)}}$ to indicate a specific type of reaction). One conceptual view of equilibrium is described through the general rate expressions for the forward and reverse reactions (a more exact derivation can be developed using the techniques described in Chapter 24):

$\mathbf{k_{(forward)}\,[A]^x\,[B]^y = k_{(reverse)}\,[C]^x\,[D]^y}$; this equation can be rewritten as:

$$\frac{k_{(forward)}}{k_{(reverse)}} = \frac{[C]^x[D]^y}{[A]^x[B]^y} \text{ where } \frac{k_{(forward)}}{k_{(reverse)}} \text{ is related to the equilibrium constant } \mathbf{K}.$$

An equation that expresses the correct equilibrium concentration ratio equal to the equilibrium constant (either the symbol or its actual numerical value) is termed the **equilibrium expression:** $\mathbf{K_{(forward\ reaction)} = [C]^c\,[D]^d/[A]^a\,[B]^b}$

The equilibrium constant, K, has a specific numerical value determined by the concentrations of each molecule in the balanced equation under specific conditions.

Example: Use the general definitions to write the complete equilibrium expression for the following reversible reaction; the symbol, **K**, is used since no numerical value is given.

$$CH_4 + 2\,Cl_2 \rightleftarrows CH_2Cl_2 + 2\,HCl$$

$$\mathbf{K = \frac{[CH_2Cl_2]\,[HCl]^2}{[CH_4]\,[Cl_2]^2}}$$

Example: Assume that under certain conditions, the concentrations **at equilibrium** of each species in the previous equation are: $[CH_2Cl_2] = 0.850$ M; $[HCl] = 0.145$ M

$[CH_4] = 0.0522$ M; $[Cl_2] = 0.166$ M. Calculate the numerical value of K.

$$K = \frac{[CH_2Cl_2]\,[HCl]^2}{[CH_4]\,[Cl_2]^2} = \frac{(0.850\text{ M})(0.145\text{ M})^2}{(0.0522\text{ M})(0.166\text{ M})^2} \qquad \mathbf{K = 12.4}$$

The expression for the reverse reaction could be stated as:

$$\mathbf{K_{(reverse\ reaction)} = \frac{[A]^a\,[B]^b}{[C]^c\,[D]^d}}$$

The symmetry of the expressions for the forward and reverse reactions shows that $\mathbf{K_{(forward\ reaction)} = 1/K_{(reverse\ reaction)}}$; this applies to both the format of the concentration ratio and the corresponding numerical value for K.

Equilibrium Types and Expressions

Equilibrium can be established for a variety of chemical reactions, physical processes, and phase changes for all states of matter. The equilibrium expression uses a number of different concentration units; for concentrations other than molarity, the proper symbol for the specific concentration unit is used.

Gas concentration can be measured as partial pressure of the gas, $\mathbf{P_{(gas)}}$.

The concentration unit mole fraction was introduced in Chapter 20. The ratio of the number of moles of solute divided by the total number of moles in the solution (mixture) is defined as the **mole fraction of solute** and abbreviated with the symbol ($\chi_{(solute)}$).

$$\chi_{(solute)} \equiv \frac{(n_{(solute)})}{(n_{(total)})}$$

(1) Solutes in most solutions are measured in **molarity (M)**; specific problems will designate this unit in concentration brackets: **[Solute]**
(2) Gases in solutions may be shown in moles/liter (**molarity**) or as **partial pressure** in atmospheres, indicated with the pressure symbol, $P_{(gas)}$.
(3) For a pure liquid or solid, **mole fraction** is used ($\chi_{(compound)}$). The mole fraction of a pure solid or liquid is defined as being equal to exactly one: $\chi_{(pure\ compound)} = 1$.
(4) The solvent in most solutions is measured as a **mole fraction,** ($\chi_{(solvent)}$). The mole fraction of a solvent in most solutions is usually very close to one: $\chi_{(solvent)} \cong 1$.

Example: Write the equilibrium expression for the following solution phase reaction; note that the (aq) notation is often not written in the concentration brackets if it is understood that the solution is in water.

$$C_2H_3Br_{3\,(aq)} + 3\,NaCl_{(aq)} \rightleftharpoons C_2H_3Cl_{3\,(aq)} + 3\,NaBr_{(aq)}$$

$$K = \frac{[C_2H_3Cl_3]\ [NaBr]^3}{[C_2H_3Br_3]\ [NaCl]^3}$$

Example: Write the equilibrium expression for the following gas-phase reaction in partial pressures; the symbol K_p is used to specify this form. Then write the expression using molarity; The symbol K_c is used to aid in distinguishing the two forms.

$$4\,NH_{3(g)} + 5\,O_{2(g)} \rightleftharpoons 4\,NO_{(g)} + 6\,H_2O_{(g)}$$

$$K_p = \frac{(P_{(NO)})^4\ (P_{(H_2O)})^6}{(P_{(NH3)})^4\ (P_{(O2)})^5} \quad \text{and} \quad K_c = \frac{[NO]^4\ [H_2O]^6}{[NH_3]^4\ [O_2]^5}$$

The numerical values for K_p and K_c are not the same. The molarity of a gas = n/V and, thus, n(gas)/V = P(gas)/RT.

The numerical value of K_p will be some function of 1/RT depending on the number of gases and coefficients in the expression.

Example: Write an equilibrium expression for the following mixed concentration unit reaction; use molarity for Cl_2.

$$PCl_{5\,(s)} \rightleftharpoons PCl_{3\,(l)} + Cl_{2\,(g)}$$

$$K = \frac{[Cl_2](\chi_{(PCl3)})}{(\chi_{(PCl5)})} = \frac{[Cl_2](\chi_{(PCl3)} = 1)}{(\chi_{(PCl5)} = 1)}; \quad K = [Cl_2]$$

The mole fraction of pure solids and pure liquids, and most solvents, is equal to exactly one or approximately one; in these cases, the numerical value of one is not included in the expression. Only those compounds that can change concentration remain in the equilibrium expression.

Example: Write the equilibrium expression for the following solution formation equilibrium (solubility equilibrium):

$$CaCl_{2(s)} \underset{}{\overset{H_2O\ solvent}{\rightleftarrows}} Ca^{+2}_{(aq)} + 2\,Cl^-_{(aq)}$$

$$\mathbf{K_{sp}} = \frac{[\mathbf{Ca^{+2}}]\ [\mathbf{Cl^-}]^2}{(\chi_{(\mathbf{Cacl_2})})} = \frac{[\mathbf{Ca^{+2}}]\ [\mathbf{Cl^-}]^2}{(\chi_{(\mathbf{CaCl_2})} = \mathbf{1})} = [\mathbf{Ca^{+2}}]\ [\mathbf{Cl^-}]^2$$

Solution solubility uses the symbol (**K_{sp}**) for solubility product constant. The determination of equilibrium position involves using an excess amount of solid in the water solvent and measuring ion concentration at maximum solubility. The solid $CaCl_{2(s)}$ is a pure compound and has a mole fraction equal to one; $\chi_{(CaCl_2)} = \mathbf{1}$. Since the mole fraction does not appear in the equilibrium expression, it does not matter how much excess is used. The water solvent **could** have been included in the expression; however, as a solvent ($\chi_{(solvent)} \cong \mathbf{1}$), it would not have appeared.

Example: Write the equilibrium expression for the following acid/base dissociation equilibrium shown in two different formats (see Chapter 9).

$$HCN_{(aq)} \rightleftarrows H^+_{(aq)} + CN^-_{(aq)}$$

$$\mathbf{K_a} = \frac{[\mathbf{H^+_{(aq)}}][\mathbf{CN^-_{(aq)}}]}{[\mathbf{HCN_{(aq)}}]}$$

$$HCN_{(aq)} + H_2O_{(l)} \rightleftarrows H_3O^+_{(aq)} + CN^-_{(aq)}$$

$$\mathbf{K_a} = \frac{[\mathbf{H_3O^+_{(aq)}}][\mathbf{CN^-_{(aq)}}]}{[\mathbf{HCN_{(aq)}}]\ (\chi_{(\mathbf{H_2O})} \cong \mathbf{1})} = \frac{[\mathbf{H_3O^+_{(aq)}}][\mathbf{CN^-_{(aq)}}]}{[\mathbf{HCN_{(aq)}}]}$$

The symbol **K_a** is used for acid dissociation. The solvent water has $\chi_{(solvent)} \cong \mathbf{1}$ and $H^+_{(aq)} = H_3O^+_{(aq)}$; the same equilibrium expression is obtained from either reaction format.

II TECHNIQUES FOR PERFORMING EQUILIBRIUM CALCULATIONS: CALCULATING K

The techniques for calculations using the equilibrium expression involve determining all variables (concentrations or numerical value for K) in terms of **one** unknown variable to be solved. In all cases, stoichiometric analysis of the balanced equation is required.

One type of problem involves calculating the numerical value of the equilibrium constant (K) from stoichiometrically determined concentrations of all reactants and products. The other major problem type involves calculating the concentrations of all reactants and products from the numerical value of K based on stoichiometric analysis of the balanced equation.

Process for Calculating the Numerical Value of the Equilibrium Constant (K) from Concentrations

Step (**1**): Identify or write the balanced equation.

Step (**2**): Write the correct form of the equilibrium expression based on the balanced equation; use the appropriate concentration units. This is the format of the equation to be solved.

Step (**3**): The numerical value of the equilibrium is the **unknown** variable. Therefore, the form of the equilibrium expression equation identifies all reactant and product equilibrium concentrations as the required **known** variables; the numerical values for all equilibrium concentrations must be determined.

Step (**3a**): Identify all reactant and product concentrations specified in the problem. Specifically distinguish **equilibrium** concentrations from **starting** concentrations. Place this information in a data organizing table similar to that used for stoichiometric calculations. This table is a valuable tool for avoiding confusion.

Step (**3b**): Use stoichiometric analysis of the balanced equation to calculate all reactant and product **equilibrium** concentrations from the corresponding known **starting** concentrations. This requires determining a **change** in concentration for one of the components and then relating the change in one concentration to changes in all other compounds in the reaction. Complete the stoichiometric reaction table to organize the equilibrium data.

Step (**4**): Substitute all reactant and product equilibrium concentrations from step (**3**) into the equilibrium expression equation and complete the calculation. The equilibrium constant can be expressed with the units that result from the contribution of each concentration unit to the calculation. To avoid confusion, equilibrium constants are usually written only as the numerical value; the correct units for each case are implied.

Example: At 986°C, the following reaction at equilibrium shows 0.11 mole of CO, 0.11 mole of H_2O, 0.087 mole of H_2, and 0.087 CO_2 in a 1.00-liter container. Calculate the numerical value of the equilibrium constant at this temperature.

$$\mathbf{H_{2\,(g)} + CO_2 \rightleftarrows H_2O_{\,(g)} + CO_{\,(g)}}$$

Step (**1**): The balanced equation is provided.

Step (**2**): Moles/liter (molarity) is the required unit: $\mathbf{K_c} = [H_2O]\,[CO]/[H_2]\,[CO_2]$

Step (**3a**): All concentrations are equilibrium concentrations; step (**3b**) is not required and a data table need not be used.

$$[CO] = \frac{0.11\ \text{mole}}{1.00\ \text{L}} = 0.11\ \text{M}; \qquad [H_2O] = \frac{0.11\ \text{mole}}{1.00\ \text{L}} = 0.11\ \text{M};$$

$$[H_2] = \frac{0.087\ \text{mole}}{1.00\ \text{L}} = 0.087\ \text{M}; \qquad [CO_2] = \frac{0.087\ \text{mole}}{1.00\ \text{L}} = 0.087\ \text{M}$$

Step (4): $K_c = \dfrac{[H_2O]\,[CO]}{[H_2]\,[CO_2]} = \dfrac{(0.11\text{ M})(0.11\text{ M})}{(0.087\text{ M})(0.087\text{ M})} = \mathbf{1.6}$

Example: The following reaction is allowed to reach equilibrium in a 2.0 L flask at 700°C. At equilibrium, the flask contains 0.10 mole of $CO_{(g)}$ and 0.20 mole of $CO_{2(g)}$; the flask also contains 0.40 mole of the solid carbon. Calculate the numerical value of the equilibrium constant at this temperature.

$$C_{(s)} + CO_{2(g)} \rightleftarrows 2\,CO_{(g)}$$

Step (**1**): The balanced equation is provided.

Step (**2**): Moles/liter (molarity) is the required unit: $\mathbf{K_c = [CO]^2/[CO_2](\chi_{(c(s))} = 1)}$

Carbon solid is a pure compound; the solid phase does not mix with the gas phase to form a solution, ($\chi_{(c(s))} = 1$). The mole fraction of one does not appear in the final expression.

Step (**3a**): All concentrations are equilibrium concentrations; step (**3b**) is not required and a data table need not be used.

$$[CO] = \frac{0.10\text{ mole}}{2.0\text{ L}} = 0.050\text{ M}; \qquad [CO_2] = \frac{0.20\text{ mole}}{2.0\text{ L}} = 0.10\text{ M}$$

The solid carbon does not appear in the final equilibrium expression. The number of moles of carbon is not used in the problem; it does not matter whether the container has 2 grams or 20 grams of solid carbon on the bottom. The carbon is not part of the gas solution and has no concentration component for the calculation.

Step (**4**): $K = \dfrac{[CO]^2}{[CO_2]} = \dfrac{(0.050\,M)^2}{(0.10\,M)} = \mathbf{0.025}$

Note that, in this case, the units of the equilibrium constant are $M^2/M = M$.

Example: The reaction shown is allowed to reach equilibrium at a certain temperature. The **starting** concentration of [HI] is 2.0 M. After the reaction reaches equilibrium, it is found that 20.0% of the HI has reacted to form the products. Calculate the numerical value of the equilibrium constant at this temperature.

$$2\,HI_{(g)} \rightleftarrows H_{2\,(g)} + I_{2\,(g)}$$

Step (**1**): The balanced equation is provided.

Step (**2**): The concentration unit of **molarity** is stated: $\mathbf{K_c = [H_2]\,[I_2]/[HI]^2}$ All gases form a solution.

Step (**3a**): A starting concentration of 2.0 M is given for the reactant HI; any starting concentrations not specifically given in the problem are always assumed to be zero. No equilibrium concentrations are directly given; the **change** in the concentration of [HI] to achieve equilibrium is 20.0%. The data table is started; the question marks used for missing information help direct the problem-solving process to the next steps.

		$2HI_{(g)}$	$\longrightarrow$	$H_{2(g)}$	+	$I_{2(g)}$
Starting conc.	:	2.0 M		0 M		0 M
Change, Δ	:	−?M		+?M		+?M
Equilibrium conc.:		?M		?M		?M

Step (**3b**): The only additional information provided is the fact that 20.0% of the HI reacts to form products; this is used to calculate the molar change in concentration for this reactant. The delta symbol (Δ) designates "change in;" 20.0% = 0.200.

$$\Delta\,[\mathbf{HI}] = (2.0\text{ M}) \times (0.200) = |\mathbf{0.40\ M}|$$

The **sign** of a change must always be determined by analyzing the progress of the reaction through the balanced equation. The reactant HI reacts to form products and, thus, **decreases** in concentration. This is confirmed by the fact that the concentration of each product is zero; the reverse reaction cannot form more HI. The change in concentration of HI must be negative:

$$\Delta\,[\mathbf{HI}] = (2.0\text{ M}) \times (-0.200) = \mathbf{-0.40\ M}$$

Solve for the changes in product concentrations through a stoichiometric analysis of the balanced equation. HI reacts to form H_2 and I_2; the concentrations of the products **increase** and the change must be **positive**.

$$\Delta[\mathbf{H_2}] = (0.40\text{ M }\Delta\text{ HI}) \times \frac{1\text{ H}_2}{2\text{ HI}} = \mathbf{+0.20\ M};\ \ \Delta[\mathbf{I_2}] = (0.40\text{ M }\Delta\text{ HI}) \times \frac{1\text{ H}_2}{2\text{ HI}} = \mathbf{+0.20\ M}$$

Complete the data table by inserting the change information; add or subtract the changes to determine all equilibrium concentrations.

		$2HI_{(g)}$	$\longrightarrow$	$H_{2(g)}$	+	$I_{2(g)}$
Starting conc.	:	2.0 M		0 M		0 M
Change, Δ	:	−0.40 M		+0.20 M		+0.20 M
Equilibrium conc. :		1.60 M		0.20 M		0.20 M

Step (4): $\mathbf{K} = \dfrac{[H_2]\,[I_2]}{[HI]^2} = \dfrac{(0.20\text{ M})(0.20\text{ M})}{(1.60\text{ M})^2} = \mathbf{0.0156}$

Example: The following reaction is allowed to reach equilibrium. The **starting** concentration of [CO] is 0.0102 M; **starting** concentration of [Cl_2] is 0.00609 M. After the reaction reaches equilibrium, it is found that the **equilibrium** concentration of [Cl_2] is 0.00301 M. Calculate the numerical value of the equilibrium constant.

$$\mathbf{CO_{(g)} + Cl_{2(g)} \rightleftarrows COCl_{2(g)}}$$

Step (**1**): The balanced equation is provided.

Step (**2**): The concentration unit of **molarity** is stated: $\mathbf{K_c = [COCl_2]/[CO][Cl_2]}$
All gases form a solution.

Step (**3a**): The starting concentrations for the two reactants are provided; the starting concentration of $COCl_2$ is not specifically given in the problem and is assumed to be zero. One equilibrium concentration is directly given; the information is inserted into a data table to direct the next steps.

		$CO_{(g)}$	+	$Cl_{2(g)}$	$\longrightarrow$	$COCl_{2(g)}$
Starting conc.	:	0.0102 M		0.00609 M		0 M
Change, Δ	:	−? M		−? M		+? M
Equilibrium conc.	:	? M		0.00301 M		? M

Step (**3b**): The **change** in the concentration of the Cl_2 can be found by difference from the starting and equilibrium concentration information.

$$\Delta[Cl_2] = [\textbf{final M} \text{ (equilibrium)}] - [\textbf{initial M} \text{ (starting)}]$$

$$= (0.00301\text{ M}) - (0.00609\text{ M}) = \mathbf{-0.00308\text{ M}}$$

Relate the change in $[Cl_2]$ to changes for the other compounds.

CO is the other reactant and must be consumed with the Cl_2 in the reaction; the change must be negative:

$$\Delta[CO] = (0.00308\ \Delta\text{ M } Cl_2) \times \frac{1\text{ CO}}{1\text{ Cl}_2} = -0.00308\text{ M}$$

$COCl_2$ is the product and must be formed from the Cl_2; the change is positive:

$$\Delta[COCl_2] = (0.00308\text{ M }\Delta\text{ Cl}_2) \times \frac{1\text{ CO}}{1\text{ Cl}_2} = +0.00308\text{ M}$$

Complete the data table by inserting the change information; add or subtract the changes to determine all equilibrium concentrations.

		$CO_{(g)}$	+	$Cl_{2(g)}$	$\longrightarrow$	$COCl_{2(g)}$
Starting conc.	:	0.0102 M		0.00609 M		0 M
Change, Δ	:	−0.0.00308 M		−0.00308 M		+ 0.00308 M
Equilibrium conc.	:	0.00712 M		0.00301 M		0.00308 M

Step (4): $\mathbf{K} = \dfrac{[COCl_2]}{[CO][Cl_2]} = \dfrac{(0.00308\text{ M})}{(0.00712\text{ M})(0.00301\text{ M})} = \mathbf{144}$ (units are M^{-1})

Example: A solubility equilibrium reaction is shown below; the excess of the solid dissolves until a saturated (equilibrium) solution is formed. Calculate the K_{sp} for $Ca_3(AsO_4)_2$ if, from the excess solid, exactly 0.036 grams of the $Ca_3(AsO_4)_2$ actually dissolves in 1.00 L to form the equilibrium saturated solution.

$$\mathbf{Ca_3(ASO_4)_{2(s)} \xrightarrow{H_2O} 3\,Ca^{+2}_{(aq)} + 2\,ASO_4{}^{-3}_{(aq)}}$$

(1): The balanced equation is provided.
(2): Molarity is used for the aqueous ions in aqueous solution; (aq) is implied:

$$\mathbf{K_{sp} = \frac{[Ca^{+2}]^3[AsO_4{}^{-3}]^2}{(\chi_{Ca_3(AsO_4)^2(s)} = 1)} \quad or \quad K_{sp} = [Ca^{+2}]^3[AsO_4{}^{-3}]^2}$$

The $Ca_3(AsO_4)_2$ is a pure solid with a mole fraction of exactly one; the excess solid never appears in the equilibrium expression or in any calculation.

Step (**3a**): The equilibrium concentrations of the ions are not directly given; the information provided states the mass of $Ca_3(AsO_4)_2$ actually dissolved; convert this to the molarity: moles of $Ca_3(AsO_4)_2 = \dfrac{0.0360\text{ g}}{398\text{ g/mole}} = 9.05 \times 10^{-5}\text{mole}$

$$\text{Molarity (M) of } Ca_3(AsO_4)_2 \text{ aqueous} = \frac{9.05 \quad 10^{-5}\text{ mole}}{1.00\text{ L}} = \mathbf{9.05 \times 10^{-5}\,M}$$

This molarity represents the amount of $Ca_3(AsO_4)_2$ dissolved; this is a **negative** change in $Ca_3(AsO_4)_2$ solid; the starting concentrations of the aqueous ions are zero.

		$\mathbf{Ca_3(AsO_4)_{2\,(s)}}$	$\longrightarrow$	$\mathbf{3\,Ca^{+2}_{(aq)}}$	+	$\mathbf{2\,ASO_4{}^{-3}_{(aq)}}$
Starting conc.	:	excess		0 M		0 M
Change, Δ	:	-9.05×10^{-5} M		+ ? M		+? M
Equlibrium conc.:		excess		? M		? M

Step (**3b**): The changes in ion concentration must be **positive** from dissolved solid:

$$\Delta[Ca^{+2}{}_{(aq)}] = (9.05 \times 10^{-5}M\ \Delta\ Ca_3(AsO_4)_2) \times \frac{3\ Ca^{+2}}{1\ Ca_3(AsO_4)_2} = \mathbf{+2.72 \times 10^{-4}M}$$

$$\Delta[AsO_4{}^{-3}{}_{(aq)}] = (9.05 \times 10^{-5}M\ \Delta\ Ca_3(AsO_4)_2) \times \frac{2\ AsO_4{}^{-3}}{1\ Ca_3(AsO_4)_2} = \mathbf{+1.81 \times 10^{-4}M}$$

Complete the table:

		$\mathbf{Ca_3(AsO_4)_{2\,(s)}}$	$\longrightarrow$	$\mathbf{3\,Ca^{+2}_{(aq)}}$	+	$\mathbf{2\,AsO_4{}^{-3}_{(aq)}}$
Starting conc.	:	excess		0 M		0 M
Change, Δ	:	-9.05×10^{-5} M		$+2.72 \times 10^{-4}$ M		$+1.81 \times 10^{-4}$M
Equilibrium conc.:		excess		$+2.72 \times 10^{-4}$M		$+1.81 \times 10^{-4}$M

(4) $\mathbf{K_{sp}} = [Ca^{+2}]^3\,[AsO_4{}^{-3}]^2 = (2.72 \times 10^{-4}\ M)^3(1.81 \times 10^{-4}M)^2 = \mathbf{6.74 \times 10^{-19}}$

III TECHNIQUES FOR PERFORMING EQUILIBRIUM CALCULATIONS

Process for Calculating Equilibrium Concentrations of All Reactants and Products from the Numerical Value of the Equilibrium Constant (K)

Step (**1**): Identify or determine the balanced equation.

Step (**2**): Write the correct form of the equilibrium expression based on the balanced equation; use the appropriate concentration units. This is the format of the equation to be solved.

Step (**3**): The numerical value of the equilibrium constant is given in these problems. The equilibrium expression equation must be solved for unknown equilibrium concentrations. However, a solution requires that **only one** concentration variable be unknown.

Step (**3a**): Identify all reactant and product **starting** concentrations specified in the problem and set up a stoichiometric data table. The problem solution requires expressing all equilibrium concentrations as a function of **one** unknown variable labeled (**x**); set one of the concentration changes equal to the unknown variable (**x**) and insert it in the table. The selection of which concentration change to be the unknown is not critical but is usually done to simplify the mathematics; this is usually done by working with the reactant or product with the smallest balanced equation coefficient.

Step (**3b**): Use stoichiometric analysis to relate the change in the one concentration expressed in terms of the variable, (**x**), to the changes in all other compounds in the reaction.

Step (**3c**): Add or subtract all changes expressed as a function of (**x**) to/from the starting concentrations to produce all reactant and product equilibrium concentrations in terms of the one unknown variable.

Step (**4**): Substitute all reactant and product concentrations expressed in terms of the one variable, (**x**), into the equilibrium expression equation; use the numerical value of the equilibrium constant, **K**, to complete the equation. Solve for (**x**) and use this value to calculate all equilibrium concentrations.

Step (**5**): As a check on the mathematical solution found in step (**4**), substitute all of the equilibrium concentrations back into the equilibrium expression equation. Check to see that the given numerical value of the equilibrium constant can be reproduced.

Example: Calculate the molar solubility of $BaSO_4$, the number of moles of $BaSO_4$ that will dissolve in 1 liter of water at equilibrium, and all molar concentrations of aqueous ions. The K_{sp} for the solubility equilibrium of $BaSO_4$ is 1.10×10^{-10}.

$$\mathbf{BaSO_{4\,(s)}} \underset{}{\overset{H_2O}{\rightleftarrows}} \mathbf{Ba^{+2}_{(aq)} + SO_4{}^{-2}{}_{(aq)}}$$

Step (**1**): The balanced equation is provided.

Step (**2**): Molarity is used for the aqueous ions in aqueous solution; (aq) is implied.

$$\mathbf{K_{sp} = \frac{[Ba^{+2}][SO_4{}^{-2}]}{(\chi_{BaSO_{4(s)}} = 1)} \quad or \quad K_{sp} = [Ba^{+2}][SO_4{}^{-2}]}$$

$BaSO_4$ is a pure solid with a mole fraction of exactly one; it will not appear in the final equilibrium expression. The equilibrium formation requires an excess of $BaSO_4$ solid.

Step (**3a**): Set (**x**) equal to the molarity change of $BaSO_4$ required to reach solubility equilibrium; this is the molarity of the solid $BaSO_4$, which will dissolve, and, therefore, the change must be **negative**. The starting concentrations of the aqueous ions are zero for this experiment; the starting $BaSO_4$ solid is listed as excess.

		$BaSO_{4(s)}$ $\longrightarrow$	$Ba^{+2}_{(aq)}$ +	$SO_4^{-2}_{(aq)}$
Starting conc.	:	excess	0 M	0 M
Change, Δ	:	−x M	+ ? M	+ ? M
Equilibrium conc.	:	excess	? M	? M

Step (**3b**): Express the concentration changes of the aqueous ions in terms of the variable (**x**); both changes must be **positive**.

$$\Delta[Ba^{+2}_{(aq)}] = (\mathbf{x\ M\ \Delta\ BaSO_4}) \times \frac{1\ Ba^{+2}}{1\ BaSO_4} = \mathbf{+x\ M}$$

$$\Delta[SO_4^{-2}{}_{(aq)}] = (\mathbf{x\ M\ \Delta\ BaSO_4}) \times \frac{1 SO_4^{-2}}{1 BaSO_4} = \mathbf{+x\ M}$$

Step (**3c**):

		$BaSO_{4(s)}$ $\longrightarrow$	$Ba^{+2}_{(aq)}$ +	$SO_4^{-2}_{(aq)}$
Starting conc.	:	excess	0 M	0 M
Change, Δ	:	−x M	+ x M	+ x M
Equilibrium conc.	:	excess	x M	x M

Step (**4**): $K_{sp} = [Ba^{+2}]\ [SO_4^{-2}]$; $1.10 \times 10^{-10} = (x\ M)(x\ M) = x^2$

$$x^2 = 1.10 \times 10^{-10}\ x = (1.10 \times 10^{-10})^{1/2} = 1.05 \times 10^{-5}\ M$$

Molar solubility (molarity of $BaSO_4$ dissolved) = $\mathbf{x = 1.05 \times 10^{-5}\ M}$

$$\mathbf{[Ba^{+2}] = x = 1.05 \times 10^{-5}\ M}$$

$$\mathbf{[SO_4^{-2}] = x = 1.05 \times 10^{-5}\ M}$$

		$BaSO_{4(s)}$ $\longrightarrow$	$Ba^{+2}_{(aq)}$ +	$SO_4^{-2}_{(aq)}$
Starting conc.	:	excess	0 M	0 M
Change, Δ	:	− x M	+ x M	+ x M
Equilibrium conc.	:	excess	x M	x M
[result]	:		1.05×10^{-5} M	1.05×10^{-5} M

Step (**5**): Ksp = $[Ba^{+2}]\ [SO_4^{-2}]$

$$1.10 \times 10^{-10}\ ?=?\ (1.05 \times 10^{-5}\ M)(1.05 \times 10^{-5}\ M)$$

$$(1.05 \times 10^{-5}\ M)(1.05 \times 10^{-5}\ M) = 1.10 \times 10^{-10};\ \text{values} \checkmark$$

Example: Calculate the molar solubility of $Mg_3(AsO_4)_2$ and all molar concentrations of aqueous ions. The K_{sp} for the solubility equilibrium of $Mg_3(AsO_4)_2$ is 4.00×10^{-16}.

$$\mathbf{Mg_3(AsO_4)_{2(s)}} \underset{\longleftarrow}{\xrightarrow{H_2O}} \mathbf{3\ Mg^{+2}_{(aq)} + 2\ AsO_4^{-3}{}_{(aq)}}$$

Step (**1**): The balanced equation is provided.

Step (**2**): Molarity is used for the aqueous ions in aqueous solution; (aq) is implied.

$$\mathbf{K_{sp}} = \frac{\mathbf{[Mg^{+2}]^3[AsO_4^{-3}]^2}}{(\chi_{Mg_3(AsO_4)_{2(s)}} = 1)} \quad \textbf{or} \quad \mathbf{K_{sp} = [Mg^{+2}]^3[AsO_4^{-3}]^2}$$

Step (**3a**): Set (**x**) equal to the molarity change (dissolved) of $Mg_3(AsO_4)_2$ required to reach solubility equilibrium; the change must be **negative**. The starting concentrations of the aqueous ions are zero; the starting $Mg_3(AsO_4)_2$ solid is listed as excess.

		$\mathbf{Mg_3(AsO_4)_{2(s)}}$	$\longrightarrow$	$\mathbf{3\ Mg^{+2}_{(aq)}}$	+	$\mathbf{2\ AsO_4^{-3}{}_{(aq)}}$
Starting conc.	:	excess		0 M		0 M
Change, Δ	:	– x M		+? M		+ ? M
Equilibrium conc.:		excess		? M		?M

Step (**3b**): Express the concentration changes of the aqueous ions in terms of the variable (**x**); both changes must be **positive**.

$$\Delta[\mathbf{Mg^{+2}{}_{(aq)}}] = (x\ M\ \Delta\ Mg_3(AsO_4)_2) \times \frac{3\ Mg^{+2}}{1\ Mg_3(AsO_4)_2} = \mathbf{+3x\ M}$$

$$\Delta[\mathbf{AsO_4^{-3}{}_{(aq)}}] = (x\ M\ \Delta\ Mg_3(AsO_4)_2) \times \frac{2\ AsO_4^{-3}}{1\ Mg_3(AsO_4)_2} = \mathbf{+2x\ M}$$

Step (**3c**):

		$\mathbf{Mg_3(AsO_4)_{2(s)}}$	$\longrightarrow$	$\mathbf{3\ Mg^{+2}_{(aq)}}$	+	$\mathbf{2\ AsO_4^{-3}{}_{(aq)}}$
Starting conc.	:	excess		0 M		0 M
Change, Δ	:	– x M		+3x M		+2x M
Equilibrium conc.:		excess		3x M		2x M

Step (4): $K_{sp} = [Mg^{+2}]^3 [AsO_4^{-3}]^2$; $4.00 \times 10^{-16} = (3x\ M)^3(2x\ M)^2$

$$4.00 \times 10^{-16} = (27x^3)(4x^2) = 108x^5;\ x^5 = \frac{4.00 \times 10^{-16}}{108};\ x^5 = 3.70 \times 10^{-18}$$

$$x = (3.70 \times 10^{-18})^{1/5} = \mathbf{3.26 \times 10^{-4}}$$

Note that taking the fifth root is made easier by converting the exponent to a number divisible by 5 and changing the 3.70 number accordingly:

$$x = (3.70 \times 10^{-18})^{1/5} = (370 \times 10^{-20})^{1/5} = (370)^{1/5} \times 10^{-4}$$

Molar solubility (molarity of $Mg_3(AsO_4)_2$ dissolved) = x = $\mathbf{3.26 \times 10^{-4}\ M}$

$$[Mg^{+2}] = 3x = 3(3.26 \times 10^{-4}\ M) = \mathbf{9.78 \times 10^{-4}\ M}$$

$$[AsO_4^{-3}] = 2x = 2(3.26 \times 10^{-4}\ M) = \mathbf{6.52 \times 10^{-4}\ M}$$

		$Mg_3(AsO_4)_{2(s)}$	$\longrightarrow$	$3\ Mg^{+2}_{(aq)}$	+	$2\ AsO_4^{-2}{}_{(aq)}$
Starting conc.	:	excess		0 M		0 M
Change, Δ	:	− x M		+ 3x M		+2x M
Equilibrium conc.	:	excess		3x M		2x M
[result]	:			9.78×10^{-4} M		6.52×10^{-4}M

Step (5): $K_{sp} = [Mg^{+2}]^3 [AsO_4^{-3}]^2$

$$4.00 \times 10^{-16}\ ? = ?\ (9.78 \times 10^{-4}\ M)^3(6.52 \times 10^{-4}\ M)^2$$

$$(9.78 \times 10^{-4}\ M)^3(6.52 \times 10^{-4}\ M)^2 = 3.98 \times 10^{-16};\ \approx\checkmark$$

In the previous examples, the equilibrium expression had a value of one for the denominator; this simplified the mathematical solution in step (**4**). The following examples describe progressive techniques for solving more complicated equilibrium expressions.

Example: Calculate the equilibrium molar concentrations for the reactant and product. The starting amount of C_6H_{12} (cyclohexane) is 0.045 mole in a 2.8 L container

$$K_c = 0.12$$

$$C_6H_{12(g)} \rightleftarrows C_5H_9CH_{3\,(g)}$$

Step (**1**): The balanced equation is provided.

Step (**2**): Molarity is the unit required in the problem: $\mathbf{K_c = [C_5H_9CH_3]/[C_6H_{12}]}$

Step (**3a**): Starting $[C_6H_{12}] = 0.045$ mole/2.8 L; $\mathbf{M = 0.016\ M}$

Set (**x**) equal to the molarity change of C_6H_{12}; the change must be **negative** since the product $C_5H_9CH_3$ has a **starting** concentration of zero.

		$\mathbf{C_6H_{12(g)}}$	$\longrightarrow$	$\mathbf{C_5H_9CH_{3(g)}}$
Starting conc.	:	0.016 M		0 M
Change, Δ	:	−x M		+? M
Equilibrium conc.:		? M		?M

Step (**3b**): The change for the product $C_5H_9CH_3$ must be **positive**.

$$\mathbf{\Delta[C_5H_9CH_3] = (x\ M\ \Delta}\ C_6H_{12}) \times 1\ C_5H_9CH_3/1\ C_6H_{12} = \mathbf{+x\ M}$$

Step (**3c**):

		$\mathbf{C_6H_{12(g)}}$	$\longrightarrow$	$\mathbf{C_5H_9CH_{3\,(g)}}$
Starting conc.	:	0.016 M		0 M
Change, Δ	:	−x M		+x M
Equilibrium conc.:		(0.016 − x) M		x M

Step (**4**): $K = \dfrac{[C_5H_9CH_3]}{[C_6H_{12}]}$; $0.12 = \dfrac{(X)}{(0.016 - X)}$; $(0.12)(0.016 - X) = X$

multiply factors: $\{(0.12)(0.016) - (0.12)(x)\} = x$; $0.00192 - 0.12x = x$

$$0.00192 = 1.12x;\ \mathbf{x} = \frac{0.00192}{1.12} = \mathbf{0.0017}$$

$$\mathbf{[C_5H_9CH_3]} = x = \mathbf{0.0017\ M}; \mathbf{[C_6H_{12}]} = (0.016 - x) = (0.016) - (0.0017) = \mathbf{0.0143\ M}$$

Step (**5**): $K = \dfrac{[C_5H_9CH_3]}{[C_6H_{12}]}$; $0.12\ ? = ?\ \dfrac{0.0017}{0.0143} = 0.12$; value ✓

IV TECHNIQUES FOR PERFORMING EQUILIBRIUM CALCULATIONS

Using the Quadratic Formula

Example: Calculate the molar concentrations of all compounds in the following equilibrium. The starting amounts are: 1.50 gram of PCl_5 and 1.50 gram of PCl_3 reacted in a 0.0363 L container; $K_c = 33.3$

$$\mathbf{PCl_{5(g)}} \underset{\longleftarrow}{\longrightarrow} \mathbf{PCl_{3\,(g)} + Cl_{2(g)}}$$

Step (**1**): The balanced equation is provided.

Step (**2**): Molarity is the required unit: $\mathbf{K_c = [PCl_3][Cl_2]/[PCl_5]}$

Step (**3a**): Calculate the starting molarities of the two compounds given:

$$[\mathbf{PCl_3}]: \text{mole} = \frac{1.50\ \text{g}}{137.32\ \text{g/mole}} = 0.0109\ \text{moles};\quad M = \frac{0.0109\ \text{moles}}{0.0363\ \text{L}} = \mathbf{0.300\ M}$$

$$[\mathbf{PCl_5}]: \text{mole} = \frac{1.50\ \text{g}}{208.22\ \text{g/mole}} = 0.00720\ \text{moles};\quad M = \frac{0.00720\ \text{moles}}{0.0363\ \text{L}} = \mathbf{0.200\ M}$$

Set (**x**) equal to the molarity change of PCl_5; the change must be **negative** since some PCl_5 must react to form Cl_2, which has a starting concentration of zero.

		$\mathbf{PCl_{5(g)}}$	$\longrightarrow$	$\mathbf{PCl_{3(g)}}$	+	$\mathbf{Cl_{2(g)}}$
Starting conc.	:	0.200 M		0.300 M		0 M
Change, Δ	:	−x M		+? M		+? M
Equilibrium conc.	:	?M		? M		?M

Step (**3b**): The concentration changes of PCl_3 and Cl_2 must be **positive** since the change in PCl_5 was determined to be negative.

$$[\mathbf{PCl_3}] = (\text{x M}\ \Delta\ PCl_5) \times \frac{1\ PCl_3}{1\ PCl_5} = \mathbf{+x\ M}$$

$$[\mathbf{Cl_2}] = (\mathbf{x\ M}\ \Delta\ PCl_5) \times \frac{1\ Cl_2}{1\ PCl_5}\ \mathbf{+x\ M}$$

Step (**3c**)		$\mathbf{PCl_{5(g)}}$	$\longrightarrow$	$\mathbf{PCl_{3(g)}}$	+	$\mathbf{Cl_{2(g)}}$
Starting conc.	:	0.200 M		0.300 M		0 M
Change, Δ	:	−x M		+x M		+x M
Equilibrium conc.	:	(0.2 − x) M		(0.3 + x) M		x M

Step (**4**):

$$K = \frac{[PCl_3][Cl_2]}{[PCl_5]};\ 33.3 = \frac{(0.3 + x)(x)}{(0.2 - x)};\ 33.3\ (0.2 \quad x) = (0.3 + x)(x)$$

Multiply the left and right side factors: $6.66 - 33.3\,x = 0.3\,x + x^2$
Group terms to produce the standard quadratic equation: $x^2 + 33.6\,x - 6.66 = 0$

This equation must be solved using the quadratic formula: $\mathbf{ax^2 + bx + c = 0}$; **a** = coefficient term for x^2; **b** = coefficient term for x; **c** = the equation constant

The quadratic formula produces the solution for (**x**): $\mathbf{x} = -b \pm \{(b^2 - 4ac)\}^{1/2}/2a$
(The symbol (±) means **plus** or **minus** = **+ or −**)

Substitute the values for **a**, **b**, and **c** from the specific equation into the general quadratic formula: a = (1); b = (33.6); c = (−6.66)

$$x = \frac{-(33.6) - \{(33.6)^2 - 4(1)(-6.66)\}^{1/2}}{2(1)};\quad x = \frac{-(33.6) - \{(1128.96) - (26.64)\}^{1/2}}{2}$$

The two solutions for (**x**) are $x = -33.6 - \{33.994\}/2$ and $\mathbf{x = -33.6 + \{33.994\}/2}$

x cannot be a negative number; this would lead to a **negative** concentration for Cl_2. The one correct answer is: $x = \dfrac{-33.6 + \{33.994\}}{2} = \dfrac{+0.394}{2} = 0.197$

$$[PCl_3] = 0.300\ M + x = 0.300 + 0.197 = \mathbf{0.497\ M};\ [Cl_2] = x = \mathbf{0.197\ M}$$

$$[PCl_5] = 0.200\ M - x = 0.200 - 0.197 = \mathbf{0.003\ M}$$

		$PCl_{5(s)}$	$\longrightarrow$	$PCl_{3\,(g)}$	+	$Cl_{2\,(g)}$
Starting conc.	:	0.200 M		0.300 M		0 M
Change, Δ	:	−x M		+x M		+x M
Equilibrium conc.	:	0.2 − x M		0.3 + x M		x M
[result]	:	0.003 M		0.497 M		0.197 M

Step (**5**):

$$K = \frac{[PCl_3][Cl_2]}{[PCl_5]};\ 33.3\ ?=?\ \frac{(0.497)(0.197)}{(0.003)} = \mathbf{32.6}$$

The two values agree to two significant figures; this result is acceptable considering the precision of the final calculated value for $[PCl_5]$.

Example: Calculate the molar concentrations of all compounds in the following equilibrium. The starting amounts are: $H_2 = 3.00$ atm; $F_2 = 3.00$ atm; HF = 0 atm.

$$\mathbf{H_{2\,(g)} + F_{2(g)} \rightleftarrows 2\,HF_{(g)} \qquad K_p = 115}$$

Step (**1**): The balanced equation is provided.

Step (**2**): Partial pressure is the required unit: $\mathbf{K_p = \{P_{(HF)}\}^2/\{P_{(H2)}\}\{P_{(F2)}\}}$

Step (**3a**): The starting partial pressures are given for the reactants.

Set (**x**) equal to the molarity change of H_2 (or F_2); the change must be **negative** since the product has a starting concentration of zero.

		$H_{2(g)}$	+	$F_{2(g)}$	$\longrightarrow$	$2\,HF_{(g)}$
Starting conc.	:	3.00 atm		3.00 atm		0 atm
Change, Δ	:	− x atm		− ? atm		+? atm
Equilibrium conc.	:	? atm		? atm		? atm

Step (**3b**): The concentration change of the other reactant F_2 must be **negative**; the change in the product HF must be **positive**.

$$\mathbf{\Delta P_{(F2)} = (x\ M\ \Delta\ P_{(H2)}) \times \frac{1\,P_{(F2)}}{1\,P_{(H2)}} = -x;\quad \Delta\ P_{(HF)} = (x\ M\ \Delta\ P_{(H2)}) \times \frac{2\,P_{(HF)}}{1\,P_{(H2)}} = +2x}$$

Step (**3c**):		$H_{2(g)}$	+	$F_{2\,(g)}$	$\longrightarrow$	$2HF_{(g)}$
Starting conc.	:	3.00 atm		3.00 atm		0 atm
Change, Δ	:	−x atm		− x atm		+ 2x atm
Equilibrium conc.:		(3 − x) atm		(3 − x atm)		2x atm

Step (**4**): $K = \dfrac{\{P_{(HF)}\}^2}{\{P_{(H_2)}\}\{P_{(F_2)}\}}$; $115 = \dfrac{(2x)^2}{(3-x)(3-x)}$; $115(3-x)(3-x) = 4x^2$

Multiply the left side (3 − x) factors: $115(9 - 6x + x^2) = 4x^2$

Multiply the left side by 115 and group terms: $111x^2 - 690x + 1035 = 0$

Substitute the values for **a**, **b**, and **c** from the specific equation into the general quadratic formula: a = (111); b = (−690); c = (1085)

$$x = \frac{-(-690) \pm \{(-690)^2 - 4(111)(1035)\}^{1/2}}{2\,(111)};\ x = \frac{690 \pm \{(4.761\times 10^5) - (4.595\times 10^5)\}^{1/2}}{222}$$

$$x = \frac{690 \pm \{(1.66\times 10^4)\}^{1/2}}{222}$$ The two solutions for (**x**) are:

$$\mathbf{x} = \frac{(690 - 128.84)}{222} = \mathbf{2.53} \text{ and } \mathbf{x} = \frac{(690 + 128.84)}{222} = 3.69$$

x cannot be a 3.69 since this would produce a negative pressure for the reactants: (3.00 − 3.69) = −0.69. The one correct value is **x = 2.53 atm.**

$$\mathbf{P_{(H2)}} = 3.00 \text{ atm} - x = 3.00 - 2.53 = \mathbf{0.47\ atm}$$

$$\mathbf{P_{(F2)}} = 3.00 \text{ atm} - x = 3.00 - 2.53 = \mathbf{0.47\ atm}$$

$$\mathbf{P_{(HF)}} = 2x = 2(2.53) = \mathbf{5.06\ atm}$$

		$H_{2\,(g)}$	+	$F_{2\,(g)}$	$\longrightarrow$	$2\,HF_{(g)}$
Starting conc.	:	3.00 atm		3.00 atm		0 atm
Change, Δ	:	−x atm		−x atm		+2x atm
Equilibrium conc.:		3 − x atm		3 − x atm		**2x** atm
[result]	:	0.47 atm		0.47 atm		5.06 atm

Step (**5**): $K_p = \dfrac{\{P_{(HF)}\}^2}{\{P_{(H2)}\}\{P_{(F2)}\}}$; 115 ? = ? $\dfrac{(5.06)^2}{(0.47)(0.47)} = \mathbf{116}$; values reasonably ✓

Alternate step (**4**): Although use of the quadratic formula for this example provides valuable practice, the mathematical solution can be simplified.

$115 = (2x)^2/(3-x)(3-x)$ can be written as $115 = (2x)^2/(3-x)^2$. Since both the numerator and denominator represent quantities squared, the square root of both sides of the equation can be taken:

$(115)^{1/2} = \dfrac{2x}{(3-x)}$; $10.72 = \dfrac{2x}{(3-x)}$; $10.72\,(3\quad x) = 2x$

$$32.17 \quad 10.72x = 2x;\ 32.17 = 12.72x;\ x = \frac{32.17}{12.72} = \mathbf{2.53}$$

V TECHNIQUES FOR PERFORMING EQUILIBRIUM CALCULATIONS

Using the Simplification Technique

The following example is solved first using the complete quadratic equation and the quadratic formula. The same example is then solved employing a common simplification technique; the results show very close correlation between the final values for final concentrations.

Example: Calculate the molar concentrations of all compounds in the following acetic acid dissociation equilibrium. The starting [CH_3COOH] is 2.50 M. The K_a for CH_3COOH is 1.80×10^{-5}.

$$\mathbf{CH_3COOH_{(aq)} \underset{}{\overset{H_2O}{\rightleftarrows}} H^+_{(aq)} + CH_3COO^-_{(aq)}}$$

Step (**1**): The balanced equation is provided.

Step (**2**): Molarity is used: $\mathbf{K_a = [H^+_{(aq)}][CH_3COO^-_{(aq)}]/[CH_3COOH_{(aq)}]}$

Step (**3a**): Set (**x**) equal to the molarity change of CH_3COOH. The change must be **negative** since CH_3COOH reacts to form the aqueous ions; the starting concentrations of the aqueous ions are assumed to be zero.

		$\mathbf{CH_3COOH_{(aq)}}$	$\longrightarrow$	$\mathbf{H^+_{(aq)}}$	+	$\mathbf{CH_3COO^-_{(aq)}}$
Starting conc.	:	2.50 M		0 M		0 M
Change, Δ	:	− x M		+? M		+?M
Equilibrium conc.	:	? M		? M		? M

Step (**3b**): The concentration changes of the aqueous ions must be **positive** since the starting concentrations are zero.

$$\mathbf{\Delta[H^+_{(aq)}] = (x\ M\ \Delta}\ CH_3COOH_{(aq)}) \times \frac{1\ H^+}{1\ CH_3COOH}\ \mathbf{+x\ M}$$

$$\mathbf{\Delta[CH_3COO^-_{(aq)}] = (x\ M\ \Delta}\ CH_3COOH) \times \frac{1\ CH_3COO^-}{1\ CH_3COOH} = \mathbf{+\,x\ M}$$

Step (3c)		$\mathbf{CH_3COOH_{(aq)}}$	$\longrightarrow$	$\mathbf{H^+_{(aq)}}$	+	$\mathbf{CH_3COO^-_{(aq)}}$
Starting conc.	:	2.50 M		0 M		0 M
Change, Δ	:	− x M		+x M		+x M
Equilibrium conc.	:	(2.50 − x) M		x M		x M

Step (**4**): $K_a = \dfrac{[H^+][CH_3COO^-]}{[CH_3COOH]}$; $1.80 \times 10^{-5} = \dfrac{(x)(x)}{(2.50 - x)}$; $(1.8 \times 10^{-5})\,(2.5\ \ x) = x^2$

$$4.5 \times 10^{-5} - (1.8 \times 10^{-5})x = x^2;\ x^2 + (1.8 \times 10^{-5})x - 4.5 \times 10^{-5} = 0$$

For demonstration purposes, the solution using the quadratic formula will be carried to (an incorrect) five significant figures, a = (1); b = (1.8×10^{-5}); c = (-4.5×10^{-5})

$$x = \frac{-(1.8 \times 10^{-5}) - \{(1.8 \times 10^{-5})^2 - 4(1)(-4.5 \times 10^{-5})\}^{1/2}}{2(1)}$$

Only the positive value can lead to a correct solution.

$$x = \frac{-(1.8 \times 10^{-5}) + (1.34164 \times 10^{-2})}{2}; \; \mathbf{x} = \frac{1.33984 \times 10^{-2}}{2} = \mathbf{6.6992 \times 10^{-3} M}$$

$$[H^+] = x = \mathbf{6.6992 \times 10^{-3} M};$$

$$[CH_3COO^-] = x = \mathbf{6.6992 \times 10^{-3}\ M}$$

$$[CH_3COOH] = 2.50 - x = (2.50) - (6.6992 \times 10^{-3}\ M) = \mathbf{2.4933\ M}$$

		$CH_3COOH_{(aq)}$ $\longrightarrow$	$H^+_{(aq)}$ +	$CH_3COO^-_{(aq)}$
Starting conc.	:	2.50 M	0 M	0 M
Change, Δ	:	−x M	+x M	+x M
Equilibrium conc.	:	(2.50 − x) M	x M	x M
[result]	:	2.4933 M	6.6992×10^{-3}M	6.6992×10^{-3}M

Step (**5**): $Ka = \frac{[H^+][CH_3COO\ ]}{[CH_3COOH]}; 1.80 \times 10^{\ 5}\ ? = ?\ \frac{(6.6992 \times 10^{-3})(6.6992 \times 10^{\ 3})}{2.4933}$

$$= 1.7999 \times 10^{\ 3}$$

Many of the equations resulting from equilibrium expressions can be simplified by using an approximation. It is very often true that the numerical value of (x), as the **change** in a concentration, is **much smaller** than the **starting** concentrations of the reactants or products. Whenever this is the case, equation concentration terms that show (x) being added to or subtracted from a starting concentration can be simplified.

(1) If (**x**) (or 2**x** or 3**x**, etc.) is a very small value as compared to the numerical value, which it is being added to or subtracted from, ignore this **x** (cancel it out) in that **specific** addition or subtraction term:
{[very large number] + or − [very small number]} ? [very large number].

(2) This technique applies only to each individual concentration term that meets the required criteria. The (**x**) cannot be ignored if it is a term by itself; that is, if it is not being added to or subtracted from a much larger number.

(3) The simplification usually results in a final answer that is accurate to at least two significant figures, depending on the true difference between (**x**) and the starting concentration to which it is related.

(4) As a general guideline, the simplification technique can be used if the numerical value of **K** is **less** than approximately **1 × 10⁻⁴**, and if the appropriate **starting concentration** is greater than approximately **0.1 M**; other combinations that show the same relative differential are also applicable.

(5) The simplification technique is especially valuable for solubility equilibria and acid dissociation equilibria; K_{sp} and K_a values are typically less than 1×10^{-4}.

The acetic acid example can be solved using the simplification technique:

	$CH_3COO_{(aq)}$	$\longrightarrow$	$H^+_{(aq)}$	+	$CH_3COO^-_{(aq)}$
Starting conc. :	2.50 M		0 M		0 M
Change, Δ :	−x M		+x M		+x M
Equilibrium conc.:	(2.50 − x) M		x M		x M

$$\text{Alternate Step (4):}\quad K_a = \frac{[H^+][CH_3COO^-]}{[CH_3COOH]} : 1.80 \times 10^{-5} = \frac{(x)(x)}{(2.50-x)}$$

(x) is subtracted from the value 2.50 in the concentration term for $[CH_3COOH]$. If it can be assumed that the value of **(x)** is much smaller than 2.50, then subtracting the very small value of x from 2.50 results in a value that should be very close to the original 2.50; $(2.50 - x) \cong (2.50)$. This is the only individual concentration term that can be simplified; all other terms for x in the final equation must be kept.

$$1.80 \times 10^{-5} \cong (x)^2/2.50;\ (1.8 \times 10^{-5})\,(2.50) = x^2;\ 4.5 \times 10^{-5} = x^2$$

$$\mathbf{x} = (4.5 \times 10^{-5})^{1/2} = \mathbf{6.708 \times 10^{-3}\ M}$$

$$\mathbf{[CH_3COOH]} = 2.50 - x = (2.50) - (6.7 \times 10^{-3}\ M) = \mathbf{2.4933\ M}$$

These values agree very closely to the ones calculated from the quadratic equation. The result also confirms that 6.7×10^{-3} M is very small when compared to 2.50 M.

The Common Ion Effect

The common ion effect describes conditions for aqueous equilibria in which a solution already contains a measurable starting concentration of one of the aqueous ion products. This circumstance most often occurs for solubility or acid/base equilibria.

Example: Calculate the molar solubility of $Al(OH)_3$ in an aqueous solution that contains 0.200 M of soluble NaOH. Then determine the molar concentration of each aqueous ion; first write the equation. The K_{sp} for $Al(OH)_3$ is 1.9×10^{-33}.

Step (1): $\mathbf{Al(OH)_{3\,(s)}} \overset{H_2O}{\rightleftharpoons} \mathbf{Al^{+3}_{(aq)} + 3\,OH^-_{(aq)}}$

Step (2): $\mathbf{K_{sp} = [Al^{+3}][OH^-]^3}$

Step (**3a**): Set (**x**) equal to the molarity change of $Al(OH)_3$ (molarity of the solid $Al(OH)_3$ which will dissolve); this must be negative. The starting concentration of $Al^{+3}_{(aq)}$ is zero. Starting

$$[OH^-_{(aq)}] = (0.200\text{ M NaOH}) \times \frac{1\text{ OH}}{1\text{ NaOH}} = \mathbf{0.200\text{ M}}$$

		$Al(OH)_{3(s)}$ $\longrightarrow$	$Al^{+3}_{(aq)}$ +	$3\ OH^-_{(aq)}$
Starting conc.	:	excess	0 M	0.200 M
Change, Δ	:	– x M	+ ? M	+ ? M
Equilibrium conc.:		excess	? M	? M

Step (**3b**): The changes in the dissolved ions as products must be **positive**; although the $[OH^-]$ is not zero, the starting concentration of the other ion is zero.

$$\mathbf{\Delta[Al^{+3}_{(aq)}] = (x\ M\ \Delta\ Al(OH)_{3(s)}) \times 1\,Al^{+3}/1\,Al(OH)_3 = +x\ M}$$

$$\mathbf{\Delta[SO_4^{-2}{}_{(aq)}] = (x\ M\ \Delta\ Al(OH)_{3(s)}) \times 3\,OH\ /1\,Al(OH)_3 = +3x\ M}$$

Step (**3c**): $\mathbf{Al(OH)_{3\,(s)} \longrightarrow Al^{+3}{}_{(aq)} + 3\ OH\ _{(aq)}}$

Starting conc.	:	excess	0 M	0.200 M
Change, Δ	:	– x M	+ x M	+3x M
Equilibrium conc.:		excess	x M	(0.200 + 3x M

Step (**4**): $K_{sp} = [Al^{+3}][OH^-]^3$; $1.9 \times 10^{-33} = (x)\ (0.2 + 3x)^3$. Multiplication of all factors produces the equation: $27x^4 + 3.6x^3 + 0.24x^2 + 0.008x - 1.9 \times 10^{-33} = 0$ (!)

Fortunately, the very small value of the K_{sp} clearly indicates that the simplification technique can be used: (x) <<< 0.2; {0.2 + 3x} ≅ {0.2}

$$K_{sp} = [Al^{+3}][OH^-]^3;\ 1.9 \times 10^{-33} \cong (x)(0.2)^3;\ 1.9 \times 10^{-33} \cong 0.008x$$

$$x = 1.9 \times 10^{-33}/0.008 = 2.375 \times 10^{-31} = \mathbf{2.4 \times 10^{-31}}$$

Molar solubility = molarity of $Al(OH)_3$ dissolved = x = $\mathbf{2.4 \times 10^{-31}\ M}$

$$\mathbf{[Al^{+3}]} = x = \mathbf{2.4 \times 10^{-31}\ M}$$

$$\mathbf{[OH^-]} = (0.200\text{ M} + 3x) = \{0.200\text{ M} + 3(2.4 \times 10^{-31}\text{ M})\} \cong \mathbf{0.200\text{ M}}$$

Step (**5**): $K_{sp} = [Al^{+3}][OH^-]^3$; 1.9×10^{-33} ? = ? $(2.4 \times 10^{-31})\ (0.2)^3 = 1.92 \times 10^{-33}$✓

VI SHIFTS IN EQUILIBRIUM

A reaction at equilibrium means that reactant molecules are forming product molecules at the same rate that product molecules are being reconverted to reactant molecules; concentrations of products or reactants cannot change if the system is isolated. However, an **outside** change can be applied to a reaction after it has reached equilibrium. In this case, the concentrations of all reactants and products must change in order to re-establish the required equilibrium ratios.

The outside changes may be an addition or removal of some of the moles of a reactant or product; this will then change the concentration of the compound added or removed and, thus, **temporarily** will change the [product] to [reactant] ratio. The outside change may be an addition (temperature increase) or removal (temperature decrease) of some heat. Heat can be considered a product or reactant based on the sign of the enthalpy (ΔH). A change in temperature results in the change of the heat product or reactant and will also produce a **temporary** change in the actual concentrations described by the [product] to [reactant] ratio.

Le Chatelier's principle states that if an additional outside change is induced on an established equilibrium reaction, the reaction will respond by attempting to counteract the outside change.

(1) If reactants are added to or products are removed from an equilibrium, the **forward** reaction **increases** its rate to convert some of the reactant molecules to product molecules; the result is to remove some of the extra reactant molecules or replace some of the depleted product molecules. This response **decreases** the concentration of the reactants, **increases** the concentration of product molecules, and, thus, reestablishes the original required equilibrium ratio. If the rate of the forward reaction must increase, the reaction is said to **shift to the right** or to **shift toward products**.

(2) If reactants are removed from or products are added to an equilibrium, the **reverse** reaction **increases** its rate to convert some of the product molecules to reactant molecules; the result is to remove some of the extra product molecules or replace some of the depleted reactant molecules. This response **increases** the concentration of the reactants, **decreases** the concentration of product molecules, and, thus, re-establishes the original required equilibrium ratio. If the rate of the reverse reaction must increase, the reaction is said to **shift to the left** or to **shift toward reactants**.

(3) If ΔH is positive, heat is a reactant; if ΔH is negative, heat is a product. A change in temperature results in the change of the heat product or reactant. The response of an equilibrium to a temperature change follows the same principle: a reaction will counteract a temperature **increase** by increasing the rate of the direction that consumes heat (direction of ΔH positive); a reaction will counteract a temperature **decrease** by increasing the rate of the direction that produces heat (direction of ΔH negative).

The response of an equilibrium to an outside change can be determined by analyzing the balanced equation and predicting the reaction direction, which must increase its rate to offset the outside change.

Step (**1**): Identify or write the balanced equation.

Step (**2**): Identify heat as part of the balanced equation; determine whether heat is a reactant (ΔH positive) or product (ΔH negative). For clarity, add the label "heat" to the balanced equation (see Chapter 19).

Step (**3**): Use Le Chatelier's principle to determine which reaction direction must increase its rate to counteract or undo the outside change.

Example: For the following reversible reaction forming ammonia, show the equation to include heat as a reactant or product; then state which direction the equilibrium will shift for each of the following induced changes:

(a) Hydrogen gas is removed; (b) Nitrogen gas is added; (c) Ammonia is removed;
(d) Ammonia is added; (e) Temperature is increased; (f) Temperature is decreased.

$$\mathbf{N_{2\,(g)} + 3\,H_{2\,(g)} \rightleftarrows 2\,NH_{3\,(g)} \quad \Delta H = -92\ kJ/mole\text{-}reaction}$$

ΔH is negative; heat is a product of the forward reaction:

Step (1): $\mathbf{N_{2(g)} + 3\,H_{2(g)} \rightleftharpoons 2\,NH_{3(g)} + heat}$ Step (2)

(a) Hydrogen gas is removed. Step (3): Hydrogen gas is a reactant; the reverse reaction must increase its rate to replace some of the hydrogen molecules that were removed: the reaction **shifts to the left** (or shifts toward reactants).
(b) Nitrogen gas is added. Step (3): Nitrogen gas is a reactant; the forward reaction must increase its rate to remove some of the extra nitrogen molecules that were added: the reaction **shifts to the right** (or shifts toward products).
(c) Ammonia is removed. Step (3): Ammonia is a product; the forward reaction must increase its rate to replace some of the ammonia molecules that were removed: the reaction **shifts to the right** (or shifts toward products).
(d) Ammonia is added. Step (3): Ammonia is a product; the reverse reaction must increase its rate to remove some of the extra ammonia molecules that were added: the reaction **shifts to the left** (or shifts toward reactants).
(e) Temperature is increased. Step (3): Heat is a product; the reverse reaction must increase its rate to remove some of the extra heat that was added: the reaction **shifts to the left** (or shifts toward reactants); i.e., the reaction shifts in the direction of the ΔH positive.
(f) Temperature is decreased. Step (3): The forward reaction must increase its rate to replace some of the heat that was removed: the reaction **shifts to the right** (or shifts toward products); i.e., the reaction shifts in the direction of the ΔH negative.

VII PRACTICE PROBLEMS

1. 3.00 moles of SO_3 gas is placed in an 8.00-liter container with no other initial gases. At equilibrium, 0.58 mole of O_2 gas was formed according to the equation shown. Calculate the value of K_c.

$$2\,SO_{3(g)} \rightleftharpoons 2\,SO_{2(g)} + O_{2(g)}$$

2. The starting pressure of HCl is 2.3 atm; the starting pressure of O_2 is 1.0 atm; all other gases have a starting pressure of zero. When the reaction reaches equilibrium, the pressure of the Cl_2 is 0.93 atm. Calculate the value of $\mathbf{K_p}$.

$$4\,HCl_{(g)} + O_{2(g)} \rightleftharpoons 2\,Cl_{2(g)} + 2\,H_2O_{(g)}$$

3. For the acid dissociation, the starting concentration of HCNO is 0.200 M; the concentration of the aqueous ions is assumed to be zero. After equilibrium is reached, the concentration of $H^+_{(aq)}$ is 6.5×10^{-3} M; calculate the value for K_a.

$$HCNO_{(aq)} \rightleftharpoons H^+_{(aq)} + CNO^-_{(aq)}$$

4. Determine the K_{sp} for the solubility equilibrium of $AuCl_3$ if it is found that 1.00 **milligram** of $AuCl_3$ will dissolve in 10.0 L of water at equilibrium; first write the correct solubility equation.

5. Calculate the molar equilibrium concentrations of NH_3 gas and H_2S gas if the pure compound NH_4HS solid as excess decomposes (reacts) in an otherwise empty flask. $Kc = 1.8 \times 10^{-4}$.

$$NH_4HS_{(S)} \rightleftarrows NH_{3\,(g)} + H_2S_{(g)}$$

6. Calculate the molar solubility of $Sr_3(PO_4)_2$ (number of moles of $Sr_3(PO_4)_2$ that will dissolve in one liter of water at equilibrium); then determine the molar concentration of each of the aqueous ions. First write the correct solubility equation.

The K_{sp} for $Sr_3(PO_4)_2$ is 1.00×10^{-31}.

7. Calculate the molar solubility of $Sr_3(PO_4)_2$ in a solution already containing 0.220 M K_3PO_4 (moles that will dissolve in the 0.220 M solution at equilibrium); then determine the molar concentration of each of the aqueous ions. For this problem, use the simplification technique: x will be much smaller than the starting concentration of $[PO_4^{-3}]$. The K_{sp} for $Sr_3(PO_4)_2$ is 1.00×10^{-31}.

8. Calculate the molar concentrations of all compounds if 6.70 grams of SO_2Cl_2 is initially placed in a 1.00-liter flask and then allowed to reach equilibrium; initial concentrations of both products are zero. The $K_c = 0.045$.

$$SO_2Cl_{2\,(g)} \rightleftarrows SO_{2\,(g)} + Cl_{2\,(g)}$$

9. Calculate the molar concentrations of all compounds in the following equilibrium. The starting $[H_2]$ is 1.00 M; the starting [HCl] is 1.00 M. The starting concentration of Cl_2 is zero. The $K_c = 25.0$; the simplification technique cannot be used.

$$H_{2\,(g)} + Cl_{2\,(g)} \rightleftarrows 2HCl_{(g)}$$

10. Calculate all equilibrium molar concentrations for the following reaction at 450°C. The starting amount of NH_3 is 3.60 mole in a 2.00 L container; the concentrations of N_2 and H_2 are zero. The simplification technique cannot be used for this problem; however, look for a way to mathematically simplify the initial equilibrium equation. $K_c = 6.30$

$$2\,NH_{3\,(g)} \rightleftarrows N_{2\,(g)} + 3H_{2\,(g)}$$

11. Calculate the molar concentrations of all compounds in the following acid dissociation equilibrium of hydrogen peroxide (H_2O_2). The starting $[H_2O_2]$ is 0.650 M; the starting concentrations of the aqueous ions are assumed to be zero. For this problem, use the simplification technique: x will be much smaller than the starting concentration of $[H_2O_2]$. The K_a for H_2O_2 is 2.40×10^{-12}

$$H_2O_{2\,(aq)} \xrightleftharpoons{H_2O} H^+_{(aq)} + HO_2^-{}_{(aq)}$$

12. Calculate the concentrations of all compounds if 1.00 M initial concentration of benzoic acid (C_6H_5OH) is allowed to reach equilibrium; starting concentrations of the aqueous ions are zero. Use the simplification technique for this problem.

The K_a for C_6H_5OH is 1.3×10^{-10}

$$C_6H_5OH_{(aq)} \underset{}{\overset{H_2O}{\rightleftharpoons}} H^+_{(aq)} + C_6H_5O^-_{(aq)}$$

13. For the following three equilibria, show the equilibrium reaction equation to include heat as a reactant or product; then state which direction the equilibrium will shift for each of the following induced changes: (1) Hydrogen gas is removed; (2) Carbon monoxide is added; (3) Temperature is increased; (4) Temperature is decreased.

a) $CO_{(g)} + H_2O_{(l)} \rightleftharpoons CO_{2(g)} + H_{2(g)}$ $\Delta H = -55$ kJ/mol-rxn

b) $CO_{(g)} + H_{2(g)} \rightleftharpoons CH_2O_{(g)}$ (formaldehyde) $\Delta H = -73$ kJ/mol-rxn

c) $CO_{(g)} + 2H_{2(g)} \rightleftharpoons CH_3OH_{(l)}$ (methanol) $\Delta H = -122$ kJ/mol-rxn

VIII ANSWERS TO PRACTICE PROBLEMS

1. **(1)** $2\,SO_{3(g)} \rightleftharpoons 2\,SO_{2(g)} + O_{2(g)}$ **(2)** $K_c = [SO^2]^2[O_2]/[SO_3]^2$

(3a) Initial $[SO_3] = 3.00$ mole/8.00 L = 0.375 M; equilibrium $[O_2] = 0.58$ mole/8.00 L = 0.0725 M

		$2\,SO_{3(g)}$	$\longrightarrow$ $2\,SO_{2(g)}$	+ $O_{2(g)}$
Starting conc.	:	0.375 M	0 M	0 M
Change, Δ	:	−? M	+? M	+? M
Equilibrium conc.:		? M	? M	0.0725 M

(3b) $\Delta[O_2] = (0.0725\text{ M}) - (0\text{ M}) = +0.0725$ M
SO_3 is a reactant consumed in the reaction; change is negative:

$$\Delta [SO_3] = (0.0725\text{ M }\Delta O_2) \times 2SO_3/1O_2 = -0.145\text{ M}$$

SO_2 is a product formed with the O_2; change is positive:

$$\Delta [SO_2] = (0.0725\text{ M }\Delta O_2) \times 2SO_2/1O_2 = +0.145\text{ M}$$

		$2\,SO_{3(g)}$	$\longrightarrow$ $2\,SO_{2(g)}$	+ $O_{2(g)}$
Starting conc.	:	0.375 M	0 M	0 M
Change, Δ	:	−0.145 M	+0.145 M	+0.0725 M
Equilibrium conc.:		0.23 M	0.145 M	0.0725 M

(4) $K_c = \frac{[SO_2]^2[O_2]}{[SO_3]^2} = \frac{(0.145)^2(0.0725)}{(0.23)^2} = \mathbf{0.0288}$ (M)

2. (1) $4\ HCl_{(g)} + O_{2(g)} \longrightarrow 2\ Cl_{2\,(g)} + 2\ H_2O_{(g)}$ (2) $K_p = \frac{(P_{Cl2})^2(P_{H_2O})^2}{(P_{HCl})^4(P_{O_2})}$

(3a)

		$4\ HCl_{(g)}$	+	$O_{2(g)}$	$\longrightarrow$	$2\ Cl_{2(g)}$	+	$2\ H_2O_{(g)}$
Starting conc.	:	2.3 atm		1.0 atm		0 atm		0 atm
Change, Δ	:	− ? atm		− ? atm		+ ? atm		+ ? atm
Equilibrium conc.	:	? atm		? atm		0.93 atm		? atm

(3b) $\Delta P_{Cl2} = (0.93\ \text{atm}) - (0\ \text{atm}) = 0.93\ \text{atm}$
HCl and O_2 are reactants and consumed in the reaction; change is negative.

$$\Delta P_{HCl} = (0.93\ \text{atm}\ \Delta\ Cl_2) \times \frac{4\ HCl}{2 Cl_2} = -1.86\ \text{atm}$$

$$\Delta P_{O_2} = (0.93\ \Delta\ Cl_2) \times \frac{1\ O_2}{2\ Cl_2} = -0.465\ \text{atm}$$

H_2O is a product formed with the Cl_2; change is positive:

$$\Delta P_{H_2O} = (0.93\ \Delta\ Cl_2) \times \frac{2\ H_2O}{2\ Cl_2} = +0.93\ \text{atm}$$

		$4\ HCl_{(g)}$	+	$O_{2(g)}$	$\longrightarrow$	$2\ Cl_{2(g)}$	+	$2\ H_2O_{(g)}$
Starting conc.	:	2.3 atm		1.0 atm		0 atm		0 atm
Change, Δ	:	−1.86 atm		−0.465 atm		+0.93 atm		+0.93 atm
Equilibrium conc.	:	0.44 atm		0.535 atm		0.93 atm		0.93 atm

(4) $K_p = \frac{(P_{Cl2})^2(P_{H2O})^2}{(P_{HCl})^4(P_{O2})} = \frac{(0.93\,\text{atm})^2(0.93\,\text{atm})^2}{(0.44\,\text{atm})^4(0.535\,\text{atm})} = \mathbf{37.3}\ (\text{atm}^{-1})$

3. (1) $HCNO_{(aq)} \rightleftharpoons H^+_{(aq)} + CNO^-_{(aq)}$ (2) $K_a = \frac{[H^+_{(aq)}][CNO^-_{(aq)}]}{[HCNO_{(aq)}]}$

(3a) Starting concentrations of H^+ and CNO^- are assumed to be zero.

		$HCNO_{(aq)}$	$\longrightarrow$	$H^+_{(aq)}$	+	$CNO^-_{(aq)}$
Starting conc.	:	0.200 M		0 M		0 M
Change, Δ	:	− ? M		+ ? M		+ ? M
Equilibrium conc.	:	? M		6.5×10^{-3} M		? M

(3b) $\Delta[H^+] = (6.5 \times 10^{-3}\ \text{M}) - (0\ \text{M}) = +6.5 \times 10^{-3}\ \text{M}$

HCNO is a reactant consumed in the reaction; change is negative:

$$\Delta [HCNO] = (6.5 \times 10^{-3} \text{ M } \Delta H^+) \times 1\, HCNO/1\, H^+ = -6.5 \times 10^{-3} \text{ M}$$

$CNO^-_{(aq)}$ is a product formed with the H^+; change is positive:

$$\Delta [CNO^-] = (6.5 \times 10^{-3} \text{M } \Delta H^+) \times 1\, CNO^-/1\, H^+ = +6.5 \times 10^{-3} \text{ M}$$

		$\mathbf{HCNO_{(aq)}}$ $\longrightarrow$	$\mathbf{H^+_{(aq)}}$ +	$\mathbf{CNO^-_{(aq)}}$
Starting conc.	:	0.200 M	0 M	0 M
Change, Δ	:	-6.5×10^{-3} M	$+6.5 \times 10^{-3}$ M	$+6.5 \times 10^{-3}$M
Equilibrium conc.:		0.1935 M	6.5×10^{-3}M	6.5×10^{-3}M

$$\textbf{(4) } \mathbf{K_a} = \frac{[H^+][CNO^-]}{[HCNO]} = \frac{(6.5 \times 10^{-3} \text{ M})(6.5 \times 10^{-3} \text{ M})}{(0.1935 \text{ M})} = \mathbf{2.18 \times 10^{-4}}$$

4. **(1)** $AuCl_{3(s)} \underset{\longleftarrow}{\xrightarrow{H_2O}} Au^{+3}_{(aq)} + 3\, Cl^-_{(aq)}$ **(2)** $K_{sp} = [Au^{+3}_{(aq)}]\,[Cl^-_{(aq)}]^3$

(3a) mass of $AuCl_3 = 1.00 \text{ mg} \times 1\text{g}/1000\text{mg} = 1.00 \times 10^{-3}$ gram

moles of $AuCl_3 = 1.00 \times 10^{-3}$ g/303.32 g/mole $= 3.30 \times 10^{-6}$ mole

M $AuCl_3 = 3.30 \times 10^{-6}$ mole/10.0 L $= 3.30 \times 10^{-7}$M = $AuCl_3$ dissolved ($\Delta AuCl_3$)

		$\mathbf{AuCl_{3(s)}}$ $\longrightarrow$	$\mathbf{Au^{+3}_{(aq)}}$ +	$\mathbf{3\,Cl^-_{(aq)}}$
Starting conc.	:	excess	0 M	0 M
Change, Δ	:	-3.30×10^{-7} M	+?M	+? M
Equilibrium conc.:		excess	? M	?M

(3b) Changes in the aqueous ions must be positive:

$$\Delta[Au^{+3}_{(aq)}] = (3.30 \times 10^{-7} \text{ M } \Delta\, AuCl_3) \times 1\, Au^{+3}/1\, AuCl_3 = +3.30 \times 10^{-7} \text{ M}$$

$$\Delta[Cl^-_{(aq)}] = (3.30 \times 10^{-7} \text{ M } \Delta\, AuCl_3) \times 3\, Cl^-/1\, AuCl_3 = +\,9.90 \times 10^{-7} \text{ M}$$

		$\mathbf{AuCl_{3(s)}}$ $\longrightarrow$	$\mathbf{Au^{+3}_{(aq)}}$ +	$\mathbf{3\,Cl^-_{(aq)}}$
Starting conc.	:	excess	0 M	0 M
Change, Δ	:	-3.30×10^{-7} M	$+3.30 \times 10^{-7}$ M	$+9.90 \times 10^{-7}$ M
Equilibrium conc.:		excess	$+3.30 \times 10^{-7}$ M	$+9.90 \times 10^{-7}$ M

$$\textbf{(4) } \mathbf{K_{sp}} = [Au^{+3}][Cl^-]^3 = (3.30 \times 10^{-7} \text{ M})\,(9.90 \times 10^{-7} \text{ M})^3 = \mathbf{3.20 \times 10^{-25}}$$

5. (**3a**) The starting concentrations of the products are zero. Set (**x**) equal to the molarity change of NH_3 (or H_2S); as a product the change must be positive

		$NH_4HS_{(s)}$	$\longrightarrow$	$NH_{3(g)}$	+	$H_2S_{(g)}$
Starting conc.	:	excess		0 M		0 M
Change, Δ	:	−? M		+ x M		+ ? M
Equilibrium conc.	:	? M		? M		? M

(**3b**) The change for H_2S as a product must be positive; the change for NH_4HS as a reactant must be negative (but will not be used in the calculation).

$$\Delta[H_2S] = (\mathbf{x}\ \mathbf{M}\ \Delta\ NH_3) \times 1H_2S/1NH_3 = +x\ M$$

$$\Delta\ (NH_4HS) = (\mathbf{x}\ \mathbf{M}\ \Delta\ NH_3) \times 1NH_4HS/1NH_3 = -x$$

(**3c**)		$NH_4HS_{(s)}$	$\longrightarrow$	$NH_{3(g)}$	+	$H_2S_{(g)}$
Starting conc.	:	excess		0 M		0 M
Change, Δ	:	− x		+x M		+x M
Equilibrium conc.	:	excess		x M		x M

(**4**) $K_c = \dfrac{[NH_3][H_2S]}{(\chi_{NH_4HS(solid)} = 1)}$; $1.8 \times 10^{-4} = \dfrac{(x)\ (x)}{1}$; $x^2 = 1.8 \times 10^{-4}$; $x = (1.8 \times 10^{-4})^{1/2}$

$$\times = \mathbf{0.0134;\ [NH_3] = \times = 0.0134;\ [H_2S] = \times = 0.0134}$$

(**5**) $K_c = [NH_3][H_2S]$; 1.8×10^{-4} ?=? $(0.0134)(0.0134) = 1.795 \times 10^{-4}$✓

6. (**1**) $Sr_3(PO_4)_{2(s)} \underset{\longleftarrow}{\xrightarrow{H_2O}} 3\,Sr^{+2}{}_{(aq)} + 2PO_4{}^{-3}{}_{(aq)}$ (**2**) $K_{sp} = [Sr^{+2}]^3\,[PO_4{}^{-3}]^2$

(**3a**) Set (**x**) equal to the molarity change of $Sr_3(PO_4)_2$ (molarity of the solid $Sr_3(PO_4)_2$, which will dissolve). The starting concentrations of aqueous ions are zero.

		$Sr_3(PO_4)_{2(s)}$	$\longrightarrow$	$3\,Sr^{+3}{}_{(aq)}$	+	$2\,PO_4{}^{-3}{}_{(aq)}$
Starting conc.	:	excess		0 M		0 M
Change, Δ	:	−x M		+ ? M		+ ? M
Equilibrium conc.	:	excess		? M		? M

(**3b**) The concentration changes of the aqueous ions must be positive:

$$\Delta[Sr^{+2}{}_{(aq)}] = (\mathbf{x}\ \mathbf{M}\ \Delta\ Sr_3(PO_4)_2) \times \frac{3Sr^{+2}}{1Sr_3(PO_4)_2} = +3x\ M$$

$$\Delta[PC_4{}^{-3}{}_{(aq)}] = (\mathbf{x}\ \mathbf{M}\ \Delta\ Sr_3(PO_4)_2) \times \frac{2PO_4{}^{-3}}{1Sr_3(PO_4)_2} = +2x\ M$$

(3c)		**$Sr_3(PO_4)_{2(s)}$**	$\longrightarrow$	**$3\,Sr^{+3}{}_{(aq)}$**	+	**$2\,PO_4{}^{-3}{}_{(aq)}$**
Starting conc.	:	excess		0 M		0 M
Change, Δ	:	−x M		+3 x M		+2 x M
Equilibrium conc.	:	excess		3x M		2x M

(4) $K_{sp} = [Sr^{+2}]^3\,[PO_4{}^{-3}]^2$; $1.00 \times 10^{-31} = (3x\ M)^3(2x\ M)^2 = 108\ x^5$

$$108\ x^5 = 1.00 \times 10^{-31}\quad x^5 = \frac{(1.00\times10^{-31})}{108} = 9.26 \times 10^{-34}$$

$$\mathbf{x} = (9.26 \times 10^{-34})^{1/5} = \mathbf{2.47 \times 10^{-7}\ M}$$

Molar solubility = molarity of $Sr_3(PO_4)_2$ dissolved = x = **2.47×10^{-7} M**

$$\mathbf{[Sr^{+2}]} = 3x = 3(2.47 \times 10^{-7}\ M) = \mathbf{7.41 \times 10^{-7}\ M}$$

$$\mathbf{[PO_4{}^{-3}]} = 2x = 2(2.47 \times 10^{-7}\ M) = \mathbf{4.94 \times 10^{-7} M}$$

(5) $K_{sp} = [Sr^{+2}]^3[PO_4{}^{-3}]^2$; $1.00 \times 10^{-31}\ ? = ?\ (7.41 \times 10^{-7})^3\ (4.94 \times 10^{-7})^2$

$$= 0.993 \times 10^{-31}\checkmark$$

7. **(1)** $Sr_3(PO_4)_{2(s)} \overset{H_2O}{\rightleftarrows} 3\,Sr^{+2}{}_{(aq)} + 2\,PO_4{}^{-3}{}_{(aq)}$ **(2)** $K_{sp} = [Sr^{+2}]^3[PO_4{}^{-3}]^2$

(3a) Set (**x**) equal to the molarity change of $Sr_3(PO_4)_2$ (molarity of the solid $Sr_3(PO_4)_2$, which will dissolve). The starting concentration of $Sr^{+2}{}_{(aq)}$ is zero; the starting concentration of $PO_4{}^{-3}{}_{(aq)}$ is: $[PO_4{}^{-3}{}_{(aq)}] = (0.220\ K_3PO_4) \times 1\ PO_4{}^{-3}/1\ K_3PO_4 = 0.220$ M

		$Sr_3(PO_4)_{2(s)}$	$\longrightarrow$	**$3\,Sr^{+3}{}_{(aq)}$**	+	**$2\,PO_4{}^{-3}{}_{(aq)}$**
Starting conc.	:	excess		0 M		0.220 M
Change, Δ	:	−x M		+ ? M		+ ? M
Equilibrium conc.	:	excess		?M		?M

(3b) The concentration changes of the aqueous ions must be positive:

$$\Delta[Sr^{+2}{}_{(aq)}] = (\mathbf{x}\,M\,\Delta\,Sr_3(PO_4)_2) \times \frac{3Sr^{+2}}{1Sr_3(PO_4)_2} = +3x\ M$$

$$\Delta[PO_4{}^{-3}{}_{(aq)}] = (\mathbf{x}\,M\,\Delta\,Sr_3(PO_4)_2) \times \frac{2PO_4{}^{-3}}{1Sr_3(PO_4)_2} = +2x\ M$$

(3c)		**$Sr_3(PO_4)_{2(s)}$**	$\longrightarrow$	**$3\,Sr^{+3}{}_{(aq)}$**	+	**$2\,PO_4{}^{-3}{}_{(aq)}$**
Starting conc.	:	excess		0 M		0.220 M
Change, Δ	:	−x M		+ 3x M		+ 2x M
Equilibrium conc.	:	excess		3x M		(0.220 + 2x)

(**4**) $K_{sp} = [Sr^{+2}]^3\ [PO_4^{-3}]^2$; $1.00 \times 10^{-31} = (3x\ M)^3\ (0.220 + 2x\ M)^2$

$1.00 \times 10^{-31} = (3x)^3\ (0.220 + 2x)^2$; assume that x is much smaller than 0.220:

$(0.220 + 2x) \cong (0.220)$; $1.00 \times 10^{-31} = (3x)^3(0.220)^2$; $1.00 \times 10^{-31} = 27x^3\ (0.0484)$

$1.00 \times 10^{-31} = 1.307x^3$; $1.307x^3 = 1.00 \times 10^{-31}$ $x^3 = (1.00\times10^{-31})/1.307 = 7.65 \times 10^{-32}$

$x = (7.65 \times 10^{-32})^{1/3} = \mathbf{4.25 \times 10^{-11}\ M}$

Molar solubility = molarity of $Sr_3(PO_4)_2$ dissolved = x = $\mathbf{4.25 \times 10^{-11}\ M}$

$[\mathbf{Sr^{+2}}] = 3x = 3(4.25 \times 10^{-11}\ M) = \mathbf{1.28 \times 10^{-10}\ M}$

$[\mathbf{PO_4^{-3}}] = 0.220\ M + 2x = 0.220\ M + 8.50 \times 10^{-11}\ M) = \mathbf{0.220\ M}$

(**5**) $K_{sp} = [Sr^{+2}]^3\ [PO_4^{-3}]^2$; $1.02 \times 10^{-31}\ ? = ?\ (1.28 \times 10^{-10})^3\ (0.220)^2 = 1.02 \times 10^{-31}$✓

8. (**1**) $SO_2Cl_{2(g)} \rightleftharpoons SO_{2(g)} + Cl_{2(g)}$ (**2**) $K_c = \dfrac{[SO_2][Cl_2]}{[SO_2Cl_2]}$

(**3a**) mole SO_2Cl_2 = 6.70 g/135.0 g/mole = 0.0496 mole; initial $[SO_2Cl_2]$ = 0.0496 mole/1.00 L = 0.0496 M

The starting concentrations of the products are zero. Set (**x**) equal to the molarity change of SO_2Cl_2; as a reactant the change must be negative.

		$\mathbf{SO_2Cl_{2(g)}}$	$\longrightarrow$	$\mathbf{SO_{2(g)}}$	+	$\mathbf{Cl_{2(g)}}$
Starting conc.	:	0.0496 M		0 M		0 M
Change, Δ	:	− x M		+ ? M		+ ? M
Equilibrium conc.	:	?M		?M		? M

(**3b**) The change for SO_2 and Cl_2 as products must be positive.

$$\Delta[SO_2] = (\mathbf{x\ M\ \Delta}\ SO_2Cl_2) \times \frac{1SO_2}{1SO_2Cl_2} = +x\ M$$

$$\Delta[Cl_2] = (\mathbf{x\ M\ \Delta}\ SO_2Cl_2) \times \frac{1Cl_2}{1SO_2Cl_2} = +x\ M$$

(**3c**)		$\mathbf{SO_2Cl_{2(g)}}$	$\longrightarrow$	$\mathbf{SO_{2(g)}}$	+	$\mathbf{Cl_{2(g)}}$
Starting conc.	:	0.0496 M		0 M		0 M
Change, Δ	:	− x M		+ x M		+x M
Equilibrium conc.	:	(0.0496 − x M)		x M		x M

(**4**) $K_c = \dfrac{[SO_2][Cl_2]}{[SO_2Cl_2]}$; $0.045 = \dfrac{(x)(x)}{(0.0496 - x)}$; $0.045(0.0496 - x) = x^2$; $0.00223 - 0.045x = x^2$

$$x^2 + 0.045x - 0.00223 = 0;$$

$$x = \frac{-(0.045) \pm \{(0.045)^2 - 4(1)(-0.00223)\}^{1/2}}{2(1)}$$

Only the positive value for x is valid: $\mathbf{x} = -(0.045) + (0.1046)/2 = \mathbf{0.0298}$

$$\mathbf{[SO_{2\,(g)}] = x = 0.0298\ M;\ [Cl_{2(g)}] = x = 0.0298\ M;}$$

$$\mathbf{[SO_2Cl_{2(g)}]} = (0.0496 - x) = (0.0496 - 0.0298) = \mathbf{0.0198\ M}$$

$$\textbf{(5)}\ \mathrm{K} = \frac{[SO_2][Cl_2]}{[SO_2Cl_2]};\ 0.045\ ? = ?\ \frac{(0.0298)(0.0298)}{(0.0198)} = 0.0449\checkmark$$

9. **(1)** $H_{2(g)} + Cl_2(g) \rightleftarrows 2\,HCl_{(g)}$ **(2)** $K_c = \dfrac{[HCl]^2}{[H_2][Cl_2]}$

(3a) Set **(x)** equal to the molarity change of H_2 (or Cl_2); since the starting concentration of Cl_2 is zero, the reverse reaction must occur. Cl_2 and therefore also H_2 must be products and the change must be positive. Note that correct assignment of the signs for changes allows the technique for solving the equilibrium to be the same regardless of the direction of reaction.

		$\mathbf{H_{2(g)}}$	+	$\mathbf{Cl_{2(g)}}$	$\longrightarrow$	$\mathbf{2\,HCl_{(g)}}$
Starting conc.	:	1.00 M		0 M		1.00 M
Change, Δ	:	−x M		+ ? M		+ ? M
Equilibrium conc.	:	?M		?M		? M

(3b) The change for HCl as a reactant must be negative.

$$\Delta[Cl_2] = (\mathbf{x\ M\ \Delta\ H_2}) \times \frac{1Cl_2}{1H_2} = +x\ M;\ \Delta[HCl] = (\mathbf{x\ M\ \Delta\ H_2}) \times \frac{2HCl}{1H_2} = -2x\ M$$

(3c)		$\mathbf{H_{2(g)}}$	+	$\mathbf{Cl_{2(g)}}$	$\longrightarrow$	$\mathbf{2\,HCl_{(g)}}$
Starting conc.	:	1.0 M		0 M		1.0 M
Change, Δ	:	+ x M		+ x M		− 2x M
Equilibrium conc.	:	(1.0 + x) M		x M		(1.0 − 2x) M

$$\textbf{(4)}\ K_c = \frac{[HCl]^2}{[H_2][Cl_2]};\ 25.0 = \frac{(1.0-2x)^2}{(1.0+x)(x)};\ 25 = \frac{(1-4x+4x^2)}{x+x^2}$$

$$25x + 25x^2 = 1 - 4x + 4x^2;\ 21x^2 + 29x - 1 = 0;\ x = \frac{-(29) \pm \{(29)^2 - 4(21)(-1)\}^{1/2}}{2(21)}$$

Only the positive value for x is valid: $\mathbf{x} = \dfrac{-(29) + (30.414)}{42} = \mathbf{0.0337}$

$\mathbf{[Cl_{2(g)}]} = x = \mathbf{0.0337\ M}$ (0.034 M);

$\mathbf{[H_{2\,(g)}]} = 1.0 + x = (1.0 + 0.0337) = \mathbf{1.0337\ M}$ (1.03 M);

$\mathbf{[HCl_{(g)}]} = (1.0 - 2x) = 1.0 - 2(0.0337) = \mathbf{0.9326\ M}$ (0.933 M)

(5) $K_c = [HCl]^2/[H_2][Cl_2];\ 25.0\ ? = ?\ (0.9326)^2/(1.0337)(0.0337) = 24.97\ \checkmark$

10. **(1)** $2\ NH_{3(g)} \rightleftarrows N_{2(g)} + 3\ H_{2(g)}$ **(2)** $K_c = \frac{[N_2][H_2]^3}{[NH_3]^2}$

(**3a**) $[NH_3] = 3.60\ mole/2.00\ L = \mathbf{1.80\ M}$

Set (**x**) equal to the molarity change of N_2; applying (**x**) to the smallest coefficient molecule simplifies the mathematics. Since the starting concentration is zero, the change in $[N_2]$ must be positive.

	$\mathbf{2\ NH_{3(g)}}$	$\longrightarrow$	$\mathbf{N_{2(g)}}$	+	$\mathbf{3\ H_{2(g)}}$
Starting conc. :	1.80 M		0 M		0 M
Change, Δ :	+ ? M		+ x M		+ ? M
Equilibrium conc.:	?M		?M		? M

(**3b**) The change in NH_3 must be negative since some NH_3 must react to form N_2 and H_2, which have a starting concentration of zero; the concentration changes of H_2 as a product must be positive.

$$\Delta[NH_3] = (\mathbf{x\ M\ \Delta\ N_2}) \times \frac{2\,NH_3}{1\,N_2} = -2x\ M;\ \Delta[H_2] = (\mathbf{x\ M\ \Delta\ N_2}) \times \frac{3\,H_2}{1\,N_2} = +3x\ M$$

(**3c**)	$\mathbf{2\ NH_{3(g)}}$	$\longrightarrow$	$\mathbf{N_{2(g)}}$	+	$\mathbf{3\ H_{2(g)}}$
Starting conc. :	1.80 M		0 M		0 M
Change, Δ :	− 2x M		+ x M		+ 3x M
Equilibrium conc.:	(1.80 − 2x) M		x M		3x M

(**4**) $K = \frac{[N_2][H_2]^3}{[NH_3]^2};\ 6.30 = \frac{(x)(3x)^3}{(1.80-2x)^2};\ 6.30 = \frac{27x^4}{(1.80-2x)^2}$

This equation can be reduced to a simpler form by taking the square root of both sides:

$$(6.30)^{1/2} = \frac{\{27x^4\}^{1/2}}{\{(1.80-2x)^2\}^{1/2}};\ 2.51 = \frac{5.20x^2}{(1.80-2x)};\ 2.51(1.80-2x) = 5.20x^2$$

$4.52 - 5.02x = 5.20x^2;\ 5.20x^2 + 5.02x - 4.52 = 0$; select the positive value.

$$x = \frac{-5.02 \pm \{(5.02)^2 - 4(5.20)(-4.52)\}^{1/2}}{2(5.20)};\ \text{select}: \mathbf{x} = \frac{-(5.02)+(10.92)}{10.4} = \mathbf{0.567}$$

$$[N_2] = x = \mathbf{0.567\ M};\ [H_2] = 3x = 3(0.567) = \mathbf{1.701\ M}$$

$$[NH_3] = (1.80\ M - 2x) = 1.80 - 2(0.567) = \mathbf{0.666\ M}$$

(**5**) $K = \frac{[N_2][H_2]^3}{[NH_3]^2};\ 6.30\ ?\ =\ ?\frac{(0.567)(1.701)^3}{(0.666)^2} = 6.28$ ✓

11. **(1)** $H_2O_{2(aq)} \rightleftarrows H^+_{(aq)} + HO_2^-{}_{(aq)}$ **(2)** $K_a = \frac{[H^+_{(aq)}][HO_2^-{}_{(aq)}]}{[H_2O_{2(aq)}]}$

(3a) Set **(x)** equal to the molarity change of H_2O_2. The change must be negative since H_2O_2 reacts to form the aqueous ions; the starting concentrations of the aqueous ions are assumed to be zero.

		$H_2O_{2\,(aq)}$	$\longrightarrow$	$H^+_{(ag)}$	+	$HO_2^-{}_{(ag)}$
Starting conc.	:	0650 M		0 M		0 M
Change, Δ	:	− x M		+ ? M		+ ? M
Equilibrium conc.:		? M		? M		? M

(3b) The concentration changes of the aqueous ions must be positive.

$$\Delta[H^+_{(aq)}] = (\mathbf{x}\ M\ \Delta\ H_2O_{2(aq)}) \times 1H^+/1H_2O_2 = +x\ M$$

$$\Delta[HO_2^-{}_{(aq)}] = (\mathbf{x}\ M\ \Delta\ H_2O_{2(aq)}) \times 1HO_2^-/1H_2O_2 = +x\ M$$

(3c)

		$H_2O_{2\,(aq)}$	$\longrightarrow$	$H^+_{(ag)}$	+	$HO_2^-{}_{(ag)}$
Starting conc.	:	0.650 M		0 M		0 M
Change, Δ	:	− x M		+ x M		+ x M
Equilibrium conc.:		(0.650 − x) M		x M		x M

(4) $K_a = \dfrac{[H^+][HO_2^-]}{[H_2O_2]}$; $2.40 \times 10^{\ 12} = \dfrac{(x)(x)}{(0.650 - x)}$; $2.40 \times 10^{-12} = \dfrac{(x)^2}{(0.650 - x)}$

Use the simplification technique: Assume that x is much smaller than 0.650.

In this case: $(0.650 - x) \cong (0.650)$; $2.40 \times 10^{-12} = x^2/(0.650)$

$$(2.40 \times 10^{-12})(0.650) = x^2$$

$$1.56 \times 10^{-12} = x^2;\ \mathbf{x} = (1.56 \times 10^{-12})^{1/2} = \mathbf{1.25 \times 10^{-6}}$$

$$\mathbf{[H^+] = x = 1.25 \times 10^{-6}M;\ [HO_2^-] = x = 1.25 \times 10^{-6}M}$$

$$\mathbf{[H_2O_2]} = (0.650 - x) = (0.650) - (1.25 \times 10^{-6}\ M) = \mathbf{0.650\ M}$$

(5) $K_a = \dfrac{[H^+][HO_2^-]}{[H_2O_2]}$; $2.40 \times 10^{\ 12}\ ? = ?\ \dfrac{(1.25 \times 10^{-6})(1.25 \times 10^{-6})}{(0.650)} = 2.40 \times 10^{\ 12}$ ✓

12. **(1)** $C_6H_5OH_{(aq)} \longrightarrow H^+_{(aq)} + C_6H_5O^-_{(aq)}$ **(2)** $K_a = \dfrac{[H^+_{(aq)}][C_6H_5O^-_{(aq)}]}{[C_6H_5OH_{(aq)}]}$

(3a) Set **(x)** equal to the molarity change of $C_6H_5OH_{(aq)}$. The change must be negative; the starting concentrations of the aqueous ions are zero.

		$C_6H_5OH_{(aq)}$	$\longrightarrow$	$H^+_{(ag)}$	+	$C_6H_5O^-_{(ag)}$
Starting conc.	:	1.00 M		0 M		0 M
Change, Δ	:	− x M		+ ? M		+ ? M
Equilibrium conc.:		? M		? M		? M

(**3b**) The concentration changes of the aqueous ions must be positive.

$$\Delta[H^+_{(aq)}] = (\mathbf{x}\ M\ \Delta\ C_6H_5OH) \times 1\ H^+/1\ C_6H_5OH = +x\ M$$

$$\Delta[HO_2^-{}_{(aq)}] = (\mathbf{x}\ M\ \Delta\ C_6H_5OH) \times 1\ C_6H_5O\ /1\ C_6H_5OH = +x\ M$$

(3c)		$C_6H_5OH_{(aq)}$	$\longrightarrow$	$H^+_{(ag)}$	+	$C_6H_5O^-_{(ag)}$
Starting conc.	:	.00 M		0 M		0 M
Change, Δ	:	− x M		+ x M		+ x M
Equilibrium conc.:		(1.00 − x) M		x M		x M

(**4**) $K_a = \dfrac{[H^+][C_6H_5O^-]}{[C_6H_5OH]}$; $1.3 \times 10^{-10} = \dfrac{(x)(x)}{(1.00-x)}$; $1.3 \times 10^{-10} = \dfrac{(x)^2}{(1-x)}$

Use the simplification technique: assume that x is much smaller than 1.00.
In this case: $(1 - x) \cong (1)$; $1.3 \times 10^{-10} = x^2/(1)$

$$\mathbf{x} = (1.3 \times 10^{-10})^{1/2} = \mathbf{1.14 \times 10^{-5}}$$

$$\mathbf{[H^+]} = x = \mathbf{1.14 \times 10^{-5}\ M;\ [C_6H_5O^-]} = x = \mathbf{1.14 \times 10^{-5}\ M}$$

$$\mathbf{[C_6H_5OH]} = 1.00 - x = (1.00) - (1.14 \times 10^{-5}M) \cong \mathbf{1.00\ M}$$

(**5**) $K_a = \dfrac{[H^+][C_6H_5O^-]}{[C_6H_5OH]}$; $1.3 \times 10^{\ 10} ? = ? \dfrac{(1.14\times10^{-5})(1.14\times10^{-5})}{(1.00)} = 1.3\times10^{\ 10}$ ✓

13. a) $CO_{(g)} + H_2O_{(l)} \rightleftarrows CO_{2(g)} + H_{2(g)} + \text{heat}$

(**1**) Hydrogen gas is removed. Hydrogen is a product; the **forward** reaction must increase its rate: the reaction shifts to the right (or shifts toward products).
(**2**) Carbon monoxide is added. Carbon monoxide is a reactant; the **forward** reaction must increase its rate: the reaction shifts to the right (or shifts toward products).
(**3**) Temperature is increased. Heat is a product; the **reverse** reaction must increase its rate: the reaction shifts to the left (shifts in the direction of the ΔH positive).
(**4**) Temperature is decreased. Heat is a product; the **forward** reaction must increase its rate: the reaction shifts to the right (shifts in the direction of the ΔH negative).

b) $CO_{(g)} + H_{2\,(g)} \rightleftarrows CH_2O_{(g)} + \text{heat}$

(**1**) Hydrogen gas is removed. Hydrogen is a reactant; the **reverse** reaction must increase its rate: the reaction shifts to the left (or shifts toward reactants).
(**2**) Carbon monoxide is added. Carbon monoxide is a reactant; the **forward** reaction must increase its rate: the reaction shifts to the right (or shifts toward products).
(**3**) Temperature is increased. Heat is a product; the **reverse** reaction must increase its rate: the reaction shifts to the left (shifts in the direction of the ΔH positive).
(**4**) Temperature is decreased. Heat is a product; the **forward** reaction must increase its rate: the reaction shifts to the right (shifts in the direction of the ΔH negative).

c) $CO_{(g)} + 2\,H_{2\,(g)} \rightleftarrows CH_2OH_{(l)} + \text{heat}$

(1) Hydrogen gas is removed. Hydrogen is a reactant; the **reverse** reaction must increase its rate: the reaction shifts to the left (or shifts toward reactants).
(2) Carbon monoxide is added. Carbon monoxide is a reactant; the **forward** reaction must increase its rate: the reaction shifts to the right (or shifts toward products).
(3) Temperature is increased. Heat is a product; the **reverse** reaction must increase its rate: the reaction shifts to the left (shifts in the direction of the ΔH positive).
(4) Temperature is decreased. Heat is a product; the **forward** reaction must increase its rate: the reaction shifts to the right (shifts in the direction of the ΔH negative).

24 Kinetics Part 2

Application of Rate Laws and Rate Variables to Reaction Mechanisms

I PREDICTING RATE EXPRESSIONS FROM MECHANISMS

An important application of kinetic information is the investigation and, as a goal, ultimate determination of valid reaction mechanisms. A knowledge of how a reaction proceeds is critical to understanding and/or predicting the role of a chemical process in, for example, synthesis, biological systems, or the environment.

A common method is to test the validity of a hypothetical mechanism against kinetic experimental results. To accomplish this, one technique follows the sequence:

(1) For a reaction proceeding through an unknown pathway, propose a suggested mechanism based on known factors, such as properties of reactants, known reactivities of reactant combinations, and known or suggested mechanisms of related reactions.
(2) Predict what the rate expression should be based on the proposed (suggested) mechanistic steps for the complete reaction.
(3) Compare the predicted rate expression of the proposed mechanism to the rate expression derived from the kinetic analysis (reactant orders, rate constants, activation energies).

The experimental kinetic results must match the predicted rate expression for the mechanism to be valid.

II GENERAL CONCEPTS FOR RATE EXPRESSION COMPARISON: IDENTIFICATION OF THE RATE DETERMINING STEP

Reaction mechanisms may be composed of only a single step. However, in a complete multi-step reaction, one of the steps is very often much slower than the rest of the steps. (This is the molecular equivalent of a "bottleneck" in a manufacturing plant.) The slowest chemical step in a complete multi-step mechanism is termed the **rate determining step** (abbreviated **r.d.s.**).

The rate determining step is identified through kinetic experiments. When identified it can be specifically marked in a mechanism description as "**slow**" or "**r.d.s.**"

The slowest step is usually the one with the highest energy barrier (**Ea**) and can, therefore, be identified on a **potential energy diagram**.

The rate of the complete reaction depends on the rate of the **rate determining step** plus all steps occurring **before** the rate determining step (**r.d.s.**). **Previous** steps affect the overall rate because these are the steps that produce the intermediates required for reaction in the slow (r.d.s.) step. By implication, the overall rate of the reaction is **independent** of steps occurring **after** the rate determining step.

In less complicated reactions, it is very often true that a fast step, which precedes a slow step, reaches **equilibrium**. This is because the product of the fast step (an intermediate) is used as a reactant in the slow step. Because it is used slowly, the concentration of this intermediate builds-up and increases the rate of the **reverse** reaction in the preceding fast step. (Recall that equilibrium is achieved when the rate of the forward and reverse reactions are equivalent.)

The overall rate depends on all reactants and intermediates, which participate at or before the rate determining step; thus, the overall rate expression depends on, and provides information about, the mechanism.

III RULES AND PROCEDURES FOR PREDICTING RATE EXPRESSIONS

The rate expression for a complete reaction can be predicted based on information about specific steps in the suggested reaction mechanism. A complete **(overall) rate expression** is derived from individual rate expressions for appropriate elementary steps. The key connecting concept is (arbitrary terminology) the **reaction step rate rule**:

Rule 1: The **rate** of any **elementary** reaction step is proportional to the concentrations of each reactant species involved in that step, raised to the power of the number of molecules of the specific reactant involved in that reaction step. This power will equal the reactant coefficient in the specific reaction step balanced chemical equation.

The rule applies **only** to reaction steps, **not** to general balanced equations. Rate expressions for multi-step reactions **cannot** be analyzed by applying **Rule 1** to the balanced equation. Multi-step reactions **must** be derived from elementary step rate expressions and the additional rules given in the next section.

Example: Apply **Rule 1** to the following proposed elementary reaction steps:

$$A + B + C \longrightarrow ABC \quad \textbf{rate of the step} = \mathbf{k[A][B][C]}$$

$$NO_2 + NO_2 \longrightarrow N_2O_4 \quad \textbf{rate of the step} = \mathbf{k[NO_2]^2}$$

$$2\,C + O_2 \longrightarrow 2\,CO \quad \textbf{rate of the step} = \mathbf{k[C]^2\,[O_2]}$$

$$CH_4 + Cl \longrightarrow CH_3 + HCl \quad \textbf{rate of the step} = \mathbf{k\,[CH_4][Cl]}$$

Rule 2: If the complete reaction mechanism consists of **only one (elementary) step**, then the rate expression for the overall (complete) reaction is identical to the rate expression for the single step that describes the mechanism. In this case, the reaction step rate rule is used directly; nothing else is required.

Rate expressions for multiple-step mechanisms **cannot** be predicted **directly** from the reaction step rate rule: the rule applies only to reaction steps, **not** to general balanced equations.

Example: The balanced equation for the synthesis of ammonia is shown.

$$N_2 + 3H_2 \longrightarrow 2NH_3$$

Experiments have determined that this is a **multi-step** reaction; this could **not** be inferred simply from the equation. A multi-step reaction implies that the rule **cannot** be used directly: rate (overall) is **not equal** to k $[N_2]\,[H_2]^3$

Rule 3: Multi-step reactions require identification of the rate determining step (r.d.s.). The rate expression for the overall reaction mechanism is derived by first writing the rate expression for the r.d.s. based on **Rule 1.**

Rule 4: Intermediates are found only in multi-step reactions. The overall rate expression for a multi-step reaction must contain as few intermediate concentration terms as possible. It is not always possible to eliminate all intermediate concentrations; some pre-equilibrium concentrations may sometimes be included. Elimination of intermediate concentrations provides the basis for determining the overall rate expression for a multi-step reaction.

Rule 4 Procedure for Predicting Rate Expressions for Multi-Step Reactions

Step (1): Identify the critical step in the reaction mechanism, the **rate determining step** (r.d.s.). This can be marked as **slow,** or **r.d.s.,** in the mechanism step outline or can be identified in a potential energy diagram as the step with the highest activation (Ea).

Step (2): Use **Rule 1** (reaction step rate rule) to write the rate expression specifically and only for the rate determining step. This expression will be valid because the rate expression generated in this case is specific only for the single step, not for the complete reaction.

Step (3): Determine if the rate expression for the **r.d.s.** contains any intermediate concentrations ([intermediate]).

Step (4): If the rate expression for the **r.d.s.** contains **no** intermediate concentrations, then the process is finished. In this case, the **overall** rate expression is equivalent to the **r.d.s.** rate expression and no further analysis is required.

rate expression (r.d.s.) = rate expression (overall/complete reaction)

Step (5): If the rate expression for the r.d.s. contains one or more intermediate concentrations [intermediate], they must be solved and substituted for into the rate expression for the **r.d.s.** An **overall** rate expression is derived from the expression for the **r.d.s.** by mathematically solving for **[intermediate]** in terms of reactant concentrations (**[reactant]**).

1. Select the fast/equilibrium step in the reaction mechanism that involves **both** the **intermediate** to be eliminated and the **reactants** from which it is produced; this step must precede the **r.d.s.**
2. Set up an equilibrium equation (or rate equivalence equation) using this **[intermediate]** and all appropriate **[reactant]**.
3. Solve (isolate) the required **[intermediate]** term as a function of the corresponding **[reactant]** terms.
4. Substitute the solved [reactant] terms for [intermediate] into the r.d.s. rate expression determined in **step (2)**. The **overall** rate expression will be the rate expression produced when the [reactant] terms have replaced [intermediate] in rate (r.d.s.).

IV DETERMINING RATE EXPRESSIONS FROM MECHANISM DESCRIPTIONS

Notation for Mechanisms Written in Equation Form

(1) Each step in the reaction shows the step reactants and products. For mechanisms written with step equations, only the specific step reactants and products are indicated. Elements or compounds that may appear in the complete reaction but are not used or formed in the specific step are not shown.

(2) Each step (usually) has an indicated associated rate constant or equilibrium constant, generally written above and/or below the arrow for each step.

(3) Rate constants are designated by a lower case "**k.**" A subscript number identifies the step number to which this constant applies: k_1, k_2, k_3. A negative sign for the number indicates that the constant applies to the reverse reaction: k_{-1}, k_{-2}, k_{-3}.
(4) A capital "**K**," with a number subscript, indicates an equilibrium constant for a fast/equilibrium step in the mechanism.

General Example: Consider a reaction in which the overall balanced equation is shown. The reaction may proceed by a number of different mechanisms. For each mechanism, use **Rules 1** through **4** and any necessary steps to determine the overall rate expression.

$$\mathbf{2A + B \longrightarrow A_2B}$$

Possible Mechanism #1: The reaction proceeds in only one step:

Step 1 = only step: $2A + B \xrightarrow{k_1} A_2B$

If the complete reaction occurs in only one step, all bonds must be formed simultaneously (at least to the best ability to distinguish). In this case, apply Rules 1 and 2:

$$\textbf{rate (overall)} = \mathbf{k_1 [A]^2[B]} \text{ or } \textbf{rate (overall)} = \mathbf{k [A]^2[B]}$$

Possible Mechanism #2

Step 1	$A + A \xrightarrow{k_1} A_2$	**slow (r.d.s.)**
Step 2	$A_2 + B \xrightarrow{k_2} A_2B$	**fast**
Add Steps:	$2A + B \longrightarrow A_2B$	**Steps add to balanced equation.**

The steps in the proposed mechanism must add to the balanced equation for the mechanism to be valid; the molecule "$\mathbf{A_2}$" is an **intermediate.**

The reaction mechanism is multi-step; follow the steps from **Rule 4:**

(1) Identify the rate determining step.
This is shown as **reaction step (1):** $A + A \longrightarrow A_2$ **Slow (r.d.s.)**
(2) Write the rate expression for **r.d.s.** using **Rule 1** (reaction step rate rule); use the specific symbol for the rate constant for that reaction step.

$$\textbf{rate (r.d.s.)} = \mathbf{k_1 [A]^2}$$

(3) Determine whether any of the concentrations in the rate expression for **r.d.s.** are intermediates ([intermediate]). There are **no** [intermediate]; in this reaction, **"A"** is a reactant.
(4) Whenever the rate expression for the **r.d.s.** contains **no** [intermediate], then rate (r.d.s.) equals rate (overall); no further analysis is necessary:

$$\textbf{rate (overall)} = \mathbf{k_1 [A]^2} \text{ or } \textbf{rate (overall)} = \mathbf{k[A]^2}$$

Possible Mechanism #3

Step 1	$A + A \underset{k_{-1}}{\overset{k_1}{\rightleftarrows}} A_2$	**fast and equilibrium;** $\mathbf{k_1 / k_{-1} \equiv K_1}$
Step 2	$A_2 + B \xrightarrow{k_2} A_2B$	**slow (r.d.s.)**
Add Steps:	$2A + B \longrightarrow A_2B$	**Steps add to balanced equation.**

The steps in the proposed mechanism must add to the balanced equation for the mechanism to be valid; the molecule "A_2" is an **intermediate.**

The reaction mechanism is multi-step; follow the steps from **Rule 4:**

(1) Identify the rate determining step.
This is shown as **reaction step (2)**: $A_2 + B \longrightarrow A_2B$ **slow (r.d.s.)**

(2) Write rate expression for **r.d.s.** using **Rule 1** (reaction step rate rule); use the specific symbol for the rate constant for that **reaction step**.

$$\textbf{rate (r.d.s.)} = \mathbf{k_2[A_2][B]}$$

(3) Determine whether any of the concentrations in the rate expression for **r.d.s.** are intermediates. "A_2" is an intermediate. In this mechanism, $[A_2]$ is included in the rate expression for the r.d.s.; the rate (r.d.s.) is not equivalent to rate (overall).

(4) Rule 4, step (4) does not apply in this case; go to **Rule 4, step (5)**:

(5) Solve for $[A_2]$ in terms of [reactant]

1. The only possible equation that will provide the proper relationship is **reaction step (1)**, which contains **both $[A_2]$** (intermediate) and **[A]** (reactant from which it is produced). This follows mechanistic concepts: reaction steps **before** the r.d.s. are expected to influence the rate of the overall reaction; thus, affecting the rate (overall) expression.

   In this mechanism, **reaction step (1)** is a fast, reversible reaction immediately preceding the slow step; it is, in consequence for this example, an equilibrium. This is indicated by showing both the forward (k_1) and reverse (k_{-1}) rate constants. The equilibrium constant is defined as the forward rate divided by the reverse rate; for one **elementary** step: $\mathbf{K_1 = (k_1)/(k_{-1})}$

$$\textbf{Step 1} \quad A + A \underset{k_{-1}}{\overset{k_1}{\rightleftarrows}} A_2 \quad \textbf{fast and equilibrium; } \mathbf{k_1 / k_{-1} \equiv K_1}$$

2. Set up the corresponding equilibrium expression: $\mathbf{K_1} = \dfrac{[A_2]}{[A]^2}$
3. Isolate the [intermediate]; isolate $[A_2]$: $\mathbf{[A_2] = K_1[A]^2}$
4. **rate (r.d.s.)** $= \mathbf{k_2[A_2][B]}$; complete the process by substituting the term $\mathbf{(K_1[A]^2)}$ for the term $\mathbf{([A_2])}$ in the rate (r.d.s.):

$$\text{Substitute } \mathbf{(K_1[A]^2)} \text{ for } \mathbf{([A_2])}\text{: } \textbf{rate} = \mathbf{k_2\{K_1[A]^2\}[B]}$$

Substitution produces the rate (overall) expression: **rate (overall)** $= \mathbf{k_2K_1[A]^2[B]}$.

The term $\mathbf{k_2K_1}$ represents a multiplication of constants; this combination can be represented by any general rate constant **k** in the expression: **rate (overall)** $= \mathbf{k[A]^2[B]}$.

Comparison of Rate Expressions for Mechanisms #1, #2, #3

The rate expression for **Mechanism #2** does **not** contain a term for **[B]**. The rate of the reaction does not depend on the concentration of "**B**" because this reactant is not involved in the reaction at or before the **r.d.s.**

Mechanism #1 contains **[B]** in the rate expression; the reactant "**B**" is required in the one-step reaction. **Mechanism #3** contains **[B]** in the rate (overall) expression; the reactant "**B**" is present in the slow step.

Kinetic experiments that produce rate expressions could distinguish between mechanism #2 compared to the other two but could not distinguish between mechanism #1 and mechanism #3.

Example: The following three **hypothetical** reaction mechanisms refer to the same **net** balanced equation. Complete the eight questions for these mechanisms; see Chapter 22 for some of the concepts required.

Kinetic experiments were performed to determine the experimental rate expression. Kinetic analysis shows that a plot of ln $[CH_2O]$ vs. time is straight line whenever $[NH_3]$ is held approximately constant.

Kinetic analysis also shows that a plot of ln$[NH_3]$ vs. time is a straight line whenever $[CH_2O]$ is held approximately constant.

Reaction Mechanism #1:

Step 1 $CH_2O + NH_3 \xrightarrow{k_1} CH_2OH^+ + NH_2^-$ **slow / r.d.s.**

Step 2 $CH_2OH^+ + NH_2^- \xrightarrow{k_2} CH_2NH_2^+ + OH^-$ **fast**

Step 3 $CH_2NH_2^+ + OH^- \xrightarrow{k_3} CH_2NH + H_2O$ **fast**

Complete the eight questions for these mechanisms.

1. What is the balanced equation for reaction mechanism #1?
2. What are the intermediates for reaction mechanism #1?
3. What is the predicted rate expression for the overall rate?
4. Is mechanism #1 a possible valid mechanism for the reaction, based on the kinetic experimental results?

Reaction Mechanism #2: The reaction determined from the **balanced equation** for mechanism #1 occurs in **one step**.

5. What is the predicted rate expression for the overall rate for mechanism #2?
6. Is mechanism #2 a possible valid mechanism for the reaction, based on the kinetic experimental results?

Reaction Mechanism #3:

The balanced equation and intermediates are the same as for mechanism #1; mechanistic steps differ as to fast vs. slow.

Step 1 $CH_2O + NH_3 \underset{k_{-1}}{\overset{k_1}{\rightleftarrows}} CH_2OH^+ + NH_2^-$ **fast / equilibrium** $k_1/k_{-1} \equiv K_1$

Step 2 $CH_2OH^+ + NH_2^- \xrightarrow{k_2} CH_2NH_2^+ + OH^-$ **slow / r.d.s.**

Step 3 $CH_2NH_2^+ + OH^- \xrightarrow{k_2} CH_2NH + H_2O$ **fast**

7. What is the predicted rate expression for the overall rate for mechanism #3?
8. Is mechanism #3 a possible valid mechanism for the reaction, based on the kinetic experimental results?

Reaction Mechanism #1:

1. Add the mechanism steps to determine the balanced equation; cancel out intermediates appearing on each side of the net equation.

Step 1	$CH_2O+NH_3 \xrightarrow{k_1} CH_2OH^+ + NH_2^-$	**slow / r.d.s.**
Step 2	$CH_2OH^+ + NH_2^- \xrightarrow{k_2} CH_2NH_2^+ + OH^-$	**fast**
Step 3	$CH_2NH_2^+ + OH^- \xrightarrow{k_3} CH_2NH + H_2O$	**fast**
	$CH_2O+NH_3 \longrightarrow CH_2NH+H_2O$	**balanced equation**

2. Each of these species is formed in a prior reaction step and consumed in a subsequent reaction step; they are intermediates: CH_2OH^+; NH_2^-; $CH_2NH_2^+$; OH^-
3. The reaction is multi-step; **Rule 4** is required.

Rule 4/(1). The rate determining step is step (1), labeled as slow/r.d.s.

$$CH_2O + NH_3 \xrightarrow{k_1} CH_2OH^+ + NH_2^- \quad \textbf{slow/r.d.s.}$$

Rule 4/(2) states that the predicted rate expression for the r.d.s. is found by **Rule 1** (reaction step rate rule): rate (r.d.s.) = $k_1[CH_2O][NH_3]$

Rule 4/(3) CH_2O is a reactant, not an intermediate; NH_3 is a reactant, not an intermediate. Rate (r.d.s.) expression contains no intermediate concentrations;

Rule 4/(4) rate (r.d.s.) = rate (overall) = $k_1[CH_2O][NH_3]$

4. A kinetic analysis that shows that a straight line plot of $\ln[CH_2O]$ vs. time whenever $[NH_3]$ is held approximately constant indicates a reactant order of 1 (first order) for CH_2O. A kinetic analysis of $[NH_3]$ whenever $[CH_2O]$ is held approximately constant also shows a straight-line plot of $\ln [NH_3]$ vs. time; reactant order for NH_3 is also 1 (first order). The rate expression based on kinetic analysis is:

$$\textbf{rate} = \mathbf{k[CH_2O][NH_3]}$$

The **kinetic** rate expression agrees with the predicted rate expression for mechanism #1; mechanism #1 is a possible valid mechanism based on kinetics.

Reaction Mechanism #2: The reaction determined from the balanced equation for mechanism #1 occurs in one step.

5. **Rule 2:** The predicted rate expression for mechanism #2 based on the balanced equation as one step is found by applying **Rule 1** (reaction step rate rule) directly:

$$\text{rate } (\textbf{overall}) = \text{rate (one-step)} = k[CH_2O][NH_3]$$

6. The kinetic data was analyzed in question #4; The predicted rate expression for mechanism #2 agrees with the kinetic data; the rate expression based on **kinetic** analysis was rate = $k[CH_2O][NH_3]$. Mechanism #2 agrees with the kinetic result; it is a possible valid mechanism.

Reaction Mechanism #3: The balanced equation and intermediates are the same as for mechanism #1; mechanistic steps differ as to fast vs. slow.

$$\textbf{Step 1}\quad CH_2O + NH_3 \underset{k_{-1}}{\overset{k_1}{\rightleftarrows}} CH_2OH^+ + NH_2^- \qquad \textbf{fast / equilibrium } k_1/k_{-1} \equiv K_1$$

$$\textbf{Step 2}\quad CH_2OH^+ + NH_2^- \xrightarrow{k_2} CH_2NH_2^+ + OH^- \qquad \textbf{slow / r.d.s.}$$

$$\textbf{Step 3}\quad CH_2NH_2^+ + OH^- \xrightarrow{k_3} CH_2NH + H_2O \qquad \textbf{fast}$$

7. The predicted rate expression for the overall rate for mechanism #3 requires the application of **Rule 4**:

   Rule 4/(1). The rate determining step is step (2), labeled as slow/r.d.s.

$$CH_2OH^+ + NH_2^- \xrightarrow{k_2} CH_2NH_2^+ + OH^- \quad \textbf{slow / r.d.s.}$$

   Rule 4/(2). Rate (r.d.s.) = $k_2[CH_2OH^+][NH_2^-]$

   Rule 4/(3). The concentration $[CH_2OH^+]$ is an intermediate; the concentration $[NH_2^-]$ is an intermediate. The rate (r.d.s.) expression contains intermediate concentrations.

   Rule 4/(5). Solve for the intermediate concentrations as a function of starting reactants.

1. Mechanism step (1) is an equilibrium preceding the r.d.s.
2. The equilibrium expression is: $k_1/k_{-1} = K_1 = \dfrac{[CH_2OH^+][NH_2^-]}{[CH_2O][NH_3]}$
3. Isolate $[CH_2OH^+][NH_2^-]$: $[CH_2OH^+][NH_2^-] = K_1[CH_2O][NH_3]$
4. Rate (r.d.s.) = $k_2[CH_2OH^+]\,[NH_2^-]$

   Substitute $\{[K_1[CH_2O][NH_3]\}$ for $[CH_2OH^+][NH_2^-]$ into rate (r.d.s.)

$$\text{rate} = k_2\{K_1[CH_2O][NH_3]\}$$

   Substitution produces the rate (overall) expression:

$$\text{rate (overall)} = k_2\,K_1\,[CH_2O][NH_3];\ k_2\,K_1 = \text{(general) } k$$

$$\text{rate (overall)} = k[CH_2O][NH_3]$$

8. The kinetic data was analyzed in question #4; the predicted rate expression for mechanism #3 agrees with the kinetic data; the rate expression based on **kinetic** analysis was rate = $k[CH_2O][NH_3]$. Mechanism #3 agrees with the kinetic result; it is a possible valid mechanism.

This example shows that the investigation into a reaction mechanism may (and usually does) require multiple experimental techniques to discover a reasonable pathway for a reaction.

V GENERAL CONSTRUCTION OF POTENTIAL ENERGY DIAGRAMS

Overall reaction thermodynamics, plus the mechanistic information developed through kinetics, are very often displayed in the form of a **potential energy diagram**. This is a pictorial description of a reaction based on plotting the relative potential energies of all reactants, intermediates, products, and activation barriers (estimated or measured) as a function of the general concept termed reaction progress.

Reaction progress is non-specific and represents some sequential description of events, such as time, bonding changes or atom movements, charge distribution changes or electron movements or a number of other measurements.

Relative potential energies are shown as ΔG or ΔH values along the vertical axis; reaction progress is shown along the horizontal axis. The energy measurement described as Gibb's free energy, ΔG, mathematically relates enthalpy (ΔH) and entropy (ΔS) in one potential energy term (see Chapter 25). Either ΔG or ΔH can be used for potential energy diagrams, but ΔG is more common.

The **complete** set of Initial reactants, the **complete** set of final products, and every **complete** set of intermediates are each displayed as a separate energy "platform." Connecting each platform in sequence is a smooth curve, which represents the energy changes of all the molecules for the appropriate mechanistic step.

Each energy platform represents a **local energy minimum** (the requirement for a measurably **stable** reactant, intermediate, or product). The concept of **"local"** means that (in a 2-dimensional diagram) the connecting smooth curves are higher on both sides of the platform, indicating that energy must be added to "move" the set of molecules toward the forward or reverse reaction. The actual potential energy of the platform itself does **not** depend on the energy **barrier** on either side. (The height of a boulder somewhere on a mountain is not related to the depth of the hole it sits in; even a boulder in a shallow depression on the top of the mountain will resist falling for an indeterminate time.)

The **transition state** of a reaction step is **not** a stable species; it represents a **local energy maximum**, identified at the top of the smooth curve connecting any two platforms. The concept of "transition" comes from the fact that at this point, the partially-bonded molecular composite can "fall down" in either direction to either of the connecting platforms. The transition state can be drawn showing the partial bonding changes.

The potential energy diagram for a complete **multi-step reaction** is put together by describing each of the steps in turn. The diagram follows the convention of chemical reaction equations: the forward reaction is read from left-to-right, the reverse reaction (**if** it exists using the same pathway) is read from right-to-left.

The general concepts of potential energy diagrams and activation energy were introduced in Chapter 18. The following potential energy diagrams illustrate how the information is displayed.

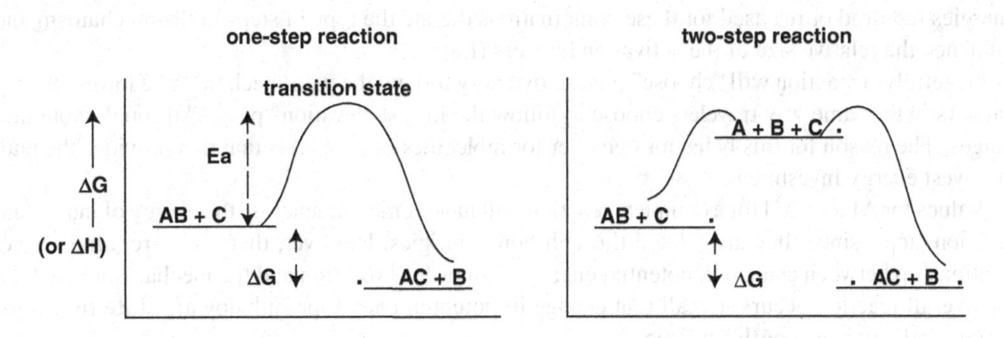

Both diagrams display the net reaction: $\mathbf{AB + C \longrightarrow AC + B}$

The curves that show the change in potential energy as the reaction proceeds from reactants to products do not connect consecutive energy platforms with a straight line. All complete reactions or reaction steps show an initial **increase** in energy in the reaction progress; this is true regardless of the relative PE positions of reactants or products.

This initial increase in energy for all reactions or steps is termed the **activation barrier** or **activation energy (Ea)**; it represents the energy that must be initially "invested" to allow a chemical reaction to proceed (see Chapter 18).

The diagram labeled a **one-step** reaction shows **both** bond-breaking (A—B) and bond-making (A—C) events occurring simultaneously. Therefore, the relative heights of the energy platforms for products and reactants are determined by the relative bond energies of A—B (with C as a free atom) vs. A—C (with B as a free atom). The activation barrier (Ea) for a reaction step where bond making and bond breaking occur simultaneously will depend on the relative **partial** bond energies represented by the **transition state**: how strong each bond is and to what degree each is broken or formed.

The diagram labeled a **two-step** reaction shows the **same net** reaction but with a **different** mechanism; this mechanism depicts the two reaction steps:

Step 1	$AB + C \longrightarrow A + B + C$	(A B bond breaks)
Step 2	$A + C \longrightarrow AC$	(A C bond forms)
Net	$\mathbf{AB + C \longrightarrow AC + B}$	(adds to balanced equation)

"A" is an intermediate in the reaction. Recall that for a complete reaction, ΔG or ΔH are state functions and are determined by the initial and final states only. **Regardless of the path**, ΔG (or ΔH) must be the same for both diagrams.

Note that the species "B" is carried through all of the platforms in the diagram, even though it has no role in step (2) (remains unchanged). A potential energy diagram must show the complete energies of all species at each energy platform; all atoms and all compounds must be displayed through each step. In contrast, displaying of a mechanism with step equations excludes any species that does not participate in that step. This is an important difference when writing rate expressions for a particular step from a potential energy diagram. Since "B" does not participate in the step (2) reaction, application of **Rule 1** indicates that the rate expression for step (2) is rate = k[A][C], **not** rate = k[A][B][C].

VI ENERGY, TEMPERATURE, AND CHEMICAL REACTIONS

Bonding Changes and Activation Energy

Reaction steps always involve some combination of bond making and/or bond breaking. The net energies required or released for these combinations dictate the type of steps in the mechanism and influence the relative size of the activation barriers (Ea).

Generally, a reaction will "choose" (i.e., statistically follow) the path/mechanism of lowest energy barriers in the same way travelers choose to follow the lowest elevation "passes" through mountain ranges. The reason for this behavior (whether for molecules or people) is that this provides the path of lowest energy investment.

Values for ΔPE or ΔH for a complete reaction will have some influence on the energy of individual reaction steps, since they are related through bond energies. However, there is no required direct relationship between change in potential energy of an overall reaction and the mechanism by which the overall reaction occurs. Recall that change in potential energy or enthalpy are **state** functions, while mechanism is a **path** function.

Bonding Changes and Temperature

The energy barrier (Ea) for a bond-breaking step is essentially the bond dissociation energy (**B.D.E.**). Bond breaking can occur through **unimolecular** decomposition: the bond in a molecule is broken with no outside molecule participating:

$$A\ \ B \longrightarrow A + B$$

The energy to overcome the **B.D.E.** barrier is supplied by heat, visible light, UV radiation etc. The energy is added to the internal **bond-vibration** energy of the molecule. Sufficient energy will cause the required bond to break in some statistical fraction of the reactant molecules. Since bond (heat) vibration energy is proportional to temperature, unimolecular bond breaking increases with increasing temperature.

Bond breaking can also occur through **bimolecular** decomposition: the bond in a molecule is broken through the (kinetic) energy of **collision** with another molecule:

$$\text{A B} + \text{C} \longrightarrow \text{A} + \text{B} + \text{C}$$

$$\text{A B} + \text{C} \longrightarrow \text{A C} + \text{B}$$

The energy to overcome the **B.D.E.** barrier comes from the kinetic energy of both molecules. Since molecular kinetic energy is proportional to temperature, bimolecular bond breaking increases with increasing temperature.

Bond-making steps do not require additional (bond) energy, since bond formation always decreases the total potential energy of the species involved (energy is released).

$$\text{A} + \text{B} \longrightarrow \text{A B}$$

In this case, the "barrier" (termed the "diffusion barrier") simply represents the statistical probability that the appropriate molecules will collide with each other in the correct geometry to allow orbital overlap. Since rates of diffusion are proportional to temperature, bond formation through collision increases with increasing temperature.

The resulting general conclusion is that the **rates** of all chemical reactions **increase with increasing temperature**.

Relationship between Rate Constant, Temperature, and Ea

The general rate expression is in the form **rate = k [reactant 1]^X [reactant 2]^Y**. Excluding the complication of gases, concentrations are essentially constant with temperature. Therefore, the relationship between rate, temperature, and activation energy must exist through the **rate constant, k.**

1. Rates of all chemical reactions increase with increasing temperature.
2. Rates of one-step chemical reactions are inversely proportional to the activation barrier, Ea. In multi-step reactions, rates are inversely proportional to the energy barrier of the slowest step (r.d.s.).

Rates increase as **Ea** (activation energy) **decreases.**

Rates decrease as **Ea increases.**

The two relationships are summed up in the **Arrhenius equation:**

$$\mathbf{k = A\,e^{-Ea/RT} \text{ or (alternate form): } \ln(k) = -Ea/RT + \ln A}$$

T = Temperature in Kelvin; **R** = 0.008314 kJ/Mole-K; **e** and **ln** represent the natural log function; **A** is a constant (called the pre-exponential factor).

The mathematical directional trends for the equation follow the concepts stated: Ea is in the numerator of the term (–Ea/RT); as Ea increases the term (–Ea/RT) becomes a larger negative value, the value of k decreases. As Ea decreases, the term (–Ea/RT) becomes a smaller negative value, the value of k increases.

Temperature is in the denominator of the term (–Ea/RT). As T decreases, the term (–Ea/RT) becomes a larger negative value, the value of k decreases. As T increases, the term (–Ea/RT) becomes a smaller negative value, the value of k increases.

The constant "A" represents the maximum rate of the reaction, when **all** molecules possess sufficient energy to overcome the energy barrier Ea. (Mathematically, as Ea approaches zero or T approaches infinity, {–Ea/RT} approaches zero: the equation reduces to $k = A$ or $\ln k = \ln A$.) The value of "A" is different for every reaction and depends on diffusion rates, geometries of molecular collisions, and so on.

The Arrhenius equation can be used to determine the **activation barrier** (Ea) for a reaction or a rate determining step. Rate constants, k, are found from kinetic experiments described in Chapter 22. The experiments are performed at a sequence of different temperatures; values of "k" as a function of "T" are generated. A plot of ln (**k**) vs. **1/T** yields a straight line with **slope = –Ea/R** and a **y-intercept of ln A**.

VII READING POTENTIAL ENERGY DIAGRAMS

Example: Use the potential energy diagram shown to determine the following information. The diagram is labeled with suggested values for potential energy as measured by ΔG. The hypothetical reaction is reversible.

a) What is the overall **forward** reaction?
b) Specify each step and classify as **fast** or **slow** (rate-determining); try to identify a probable equilibrium step.
c) Calculate **Ea** for the rate-determining step (r.d.s.).
d) Determine the sign and value for ΔG for the complete **forward** reaction.
e) What is the specific ΔG for reaction **step (1)**?
f) What is the overall **reverse** reaction?
g) Calculate **Ea** for the rate-determining step (r.d.s.) for the **reverse** reaction.
h) Determine the sign and value for ΔG for the complete **reverse** reaction.

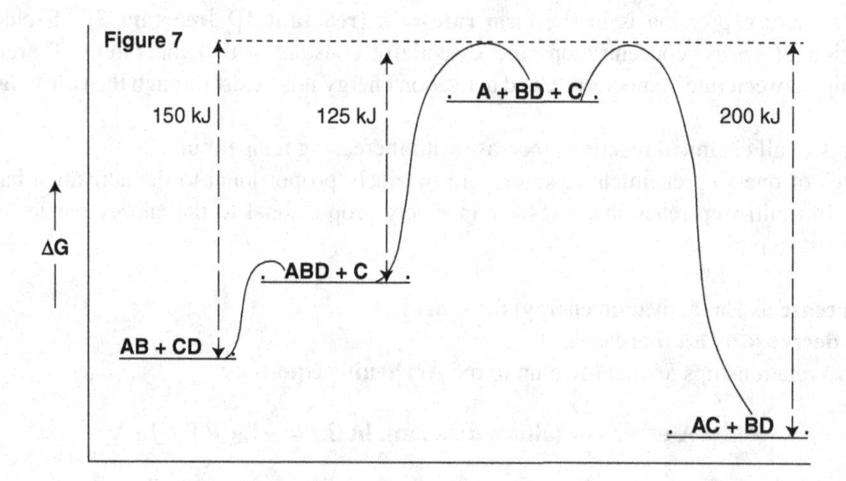

a) Analysis: Overall Forward Reaction

The overall forward reaction is read from left-to-right and identifies the compounds on the first platform as initial reactants and the last platform as final products; intermediates appear on platforms 2 and 3.

Overall forward reaction: $AB + CD \longrightarrow AC + BD$

b) Analysis: Mechanism Steps

Step (1) is **relatively** fast because it has a much smaller **Ea** than **step (2)** for the **forward** reaction. The **reverse of step (1)** ($ABD + C \longrightarrow AB + CD$) has an **Ea** that is even smaller than the **forward Ea**; thus, it is **probable** that this step is at equilibrium. (This conclusion cannot be proved from the diagram alone.)

Step (2) is the slow step because it has the highest activation energy (**Ea**). Note that the compound "**C**" was not included on either side of the equation for **step (2)**, even though it must appear in the diagram. The energy diagram depicts total potential energy and must include all species that appear from reactants to products. By convention, identical species are cancelled from each side of a balanced step reaction equation; in this case, "**C**" undergoes no change.

Step (3) is relatively fast because it has a much smaller Ea than step (2) for the **forward** reaction.

Mechanism: steps for the **forward** reaction.

Step 1	$AB + CD$	$\rightleftharpoons$	$ABD + C$	**fast/equilibrium**
Step 2	ABD	$\longrightarrow$	$A + BD$	**slow**
Step 3	$A + C$	$\longrightarrow$	AC	**fast**
Net Reaction	$\mathbf{AB + CD}$	$\longrightarrow$	$\mathbf{AC + BD}$	

c), d), e), **Analysis of Energy Using Values Provided in Diagram**

c. The rate determining step was identified as step (2). The complete energy difference between the step (2) reactant platform and the top energy of the step (2) transition state is labeled as 125 kJ; therefore, **Ea** for the **r.d.s. (forward)** is **125** kJ/mole.

d. The ΔG for the complete forward reaction is the energy difference between the initial reactant potential energy and the final product potential energy. Each energy platform has an energy value difference labeled to the same top potential energy "line."

$$\mathbf{\Delta G\ (forward)} = 150\ \text{kJ (reactant platform)} - 200\ \text{kJ (product platform)} = \mathbf{-50\ kJ}$$

e. The ΔG for step (1) is the energy difference between the initial reactant potential energy and the potential energy represented by energy platform 2. Each energy platform also has an energy value difference labeled to the same top potential energy "line."

$$\mathbf{\Delta G\ (step\ 1)} = 150\ \text{kJ (reactant platform)} - 125\ \text{kJ (step 2 platform)} = \mathbf{+25\ kJ}$$

Analysis: Overall Reverse Reaction

f. The overall reverse reaction is read from right-to-left. Initial reactants are AC and BD; final products are AB and CD.

Overall reverse reaction: $AC + BD \longrightarrow AB + CD$

g. The rate determining step for the reverse reaction is found by following the energy curves in the direction of right-to-left. Reading in this direction shows that the **reverse step**

(1) has the highest activation energy. The complete energy difference between the AC + BD platform and the top energy of the next transition state is labeled as 200 kJ; therefore, **Ea** for the **r.d.s. (reverse)** is **200** kJ/mole.

h. ΔG (forward) was calculated to –50 kJ. The requirement of a state function is that

$$\Delta G\ (\text{reverse}) = -\ \Delta G\ (\text{forward})$$

$$\Delta G\ (\text{reverse}) = -\ (-50\ \text{kJ}) = \mathbf{+50\ kJ}$$

Example: The potential energy diagram shown describes a hypothetical suggested mechanism for a chemical reaction. Use the diagram and the suggested labeled energy values (only very approximately to scale) to answer the following questions.

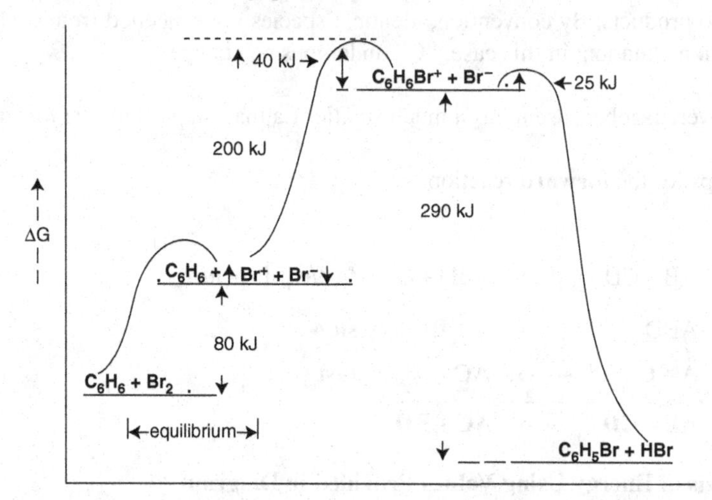

Forward Reaction:

1. What is the overall **forward** reaction (balanced equation)?
2. List the forward mechanism: Specify each step and classify as fast or slow; Identify the r.d.s.
3. Identify all the intermediates.
4. Calculate Ea for the forward rate-determining step (r.d.s.).
5. Calculate ΔG for the complete forward reaction.
6. Calculate ΔG for the forward r.d.s. step.
7. Determine the rate expression for the forward reaction.

Reverse Reaction:

8. What is the overall reverse reaction?
9. Write the reverse mechanism steps; label r.d.s. reverse.
10. Calculate ΔG for the reverse r.d.s. step.
11. Calculate Ea for the rate-determining step (r.d.s.).
12. Determine ΔG for the complete reverse reaction.
13. Determine the rate expression for the reverse reaction overall.

1. Overall forward reaction (balanced equation) is read from left-to-right. The first platform shows the initial reactants, the last platform shows the final products.

$$C_6H_6 + Br_2 \longrightarrow C_6H_5Br + HBr$$

2. Mechanism:
Reaction steps: Each reaction step in the forward direction is read from left-to-right. All atoms or molecules are carried through the diagram; atoms or molecules that do not change for a particular step are eliminated from each side of the step equation: Exclude (C_6H_6) from step #1, since it is not involved (remains unchanged) in this step. Exclude (Br^-) from step #2, since it is not involved (remains unchanged) in this step. The first step in the forward reaction is labeled as an equilibrium.

Step 1 $Br_2 \underset{k_{-1}}{\overset{k_1}{\rightleftharpoons}} Br^+ + Br^-$ **fast/equilibrium** $k_1 / k_{-1} \equiv K_1$

Step 2 $C_6H_6 + Br^+ \xrightarrow{k_2} (C_6H_6Br)^+$ **slow/r.d.s.** { }

Step 3 $(C_6H_6Br)^+ + Br^- \xrightarrow{k_3} C_6H_5Br + HBr$ **fast**

Activation energies for the forward reaction are read from left-to-right. Assume the PE diagram is sufficiently to scale for activation energy determination. The Ea for the first step, although not specifically marked, can be determined to be 80 kJ plus ≈ 20 kJ extra energy to reach the transition state peak. The activation energy for the third step is marked as 25 kJ. The rate determining step (highest activation energy) is the second step with the activation energy directly marked as 200 kJ. Step #2 is labeled as slow/r.d.s. and the other two steps are labeled as fast; step #1 is labeled as an equilibrium.

3. The intermediates are identified as: $(C_6H_6Br)^+$; Br^+; Br^-
4. The r.d.s. was determined to be the second step. The Ea (activation energy) was read directly from the PE diagram = 200 kJ.
5. The ΔG for the complete forward reaction is the energy difference for the balanced equation; the difference between the reactant (first) platform and the product (last) platform. The numerical value must be calculated from the given marked energy values, using the constant energy position at the top of the diagram:

$$\Delta G^0 = [80 + 200] - [40 + 290] = -50 \text{ kJ/mole-reaction}$$

6. ΔG for the forward r.d.s. step: $\Delta G = [200] - [40] = +160$ kJ/mole-reaction
7. The predicted rate expression for the overall rate is found by using the techniques from

Rule 4:

$$\text{rate (r.d.s.)} = k_2[C_6H_6][Br^+]$$

The rate (r.d.s.) expression contains an intermediate concentration; solve for the intermediate concentration as a function of starting reactants.

$$K_1 = \frac{[Br^+][Br^-]}{[Br_2]} \qquad \text{Solve for } [Br^+]: \ [Br^+] = \frac{K_1[Br_2]}{[Br^-]}$$

Substitute $\frac{\{K_1[Br_2]\}}{[Br^-]}$ into rate (r.d.s.): rate $= k_2[C_6H_6]\frac{\{K_1[Br_2]\}}{[Br^-]}$

$$\text{rate (overall)} = k_2K_1\frac{[C_6H_6][Br_2]}{[Br^-]} = k\frac{[C_6H_6][Br_2]}{[Br^-]}$$

$[Br^-]$ could not be eliminated from the expression

8. Overall reverse reaction (balanced equation) is read from right-to-left:

$$C_6H_5Br + HBr \longrightarrow C_6H_6 + Br_2$$

9. Reverse Mechanism.

Step 1 $C_6H_5Br + HBr \xrightarrow{k_1} (C_6H_6Br)^+ + Br^-$ **slow/r.d.s.**

Step 2 $(C_6H_6Br)^+ \xrightarrow{k_2} C_6H_6 + Br^+$ **fast**

Step 3 $Br^+ + Br^- \xrightarrow{k_3} Br_2$ **fast**

Activation energies for the reverse reaction are read from right-to-left. Step (1) in the reverse direction clearly has the highest Ea and is labeled as slow/r.d.s.; the other two steps are labeled as fast:

10. The r.d.s. was determined to be the first step. ΔG for the reverse r.d.s. step is marked as +290 kJ/mole-reaction.
11. The Ea (activation energy) can be calculated as [290 + 25] = 315 kJ
12. The ΔG (reverse) = $-\Delta G$ (forward) = $-(-50)$ = +50 kJ/mole-reaction.
13. The predicted rate expression for the overall rate is found by using the techniques from **Rule 4** step (1) (reverse) is the rate determining step:

$$\text{rate (r.d.s. for reverse reaction)} = k[C_6H_5Br][HBr]$$

The predicted rate expression for the reverse r.d.s. contains no intermediate concentrations. Therefore, rate (r.d.s. reverse) = rate (reverse overall). rate (reverse overall) = $k[C_6H_5Br]$ [HBr].

VIII RATES AND CATALYSIS

A **catalyst** is an agent that increases the rate of a chemical reaction without itself being **net** consumed in the reaction. The catalyst may act without undergoing observable change in chemical form. This is typical of **heterogeneous** catalysis; the catalyst and reactant(s) are in **different phases**. This is the dominant technology used in the chemical industry. Examples are solid metal surfaces catalyzing reactions of liquid, solution, or gaseous reactants.

The catalyst may be directly involved **as a reactant** in a chemical reaction but is **regenerated** to its initial form at the end of the reaction; there is no **net** change in the catalyst structure or concentration. This is typical of **homogeneous** catalysis in which catalyst and reactants are in the **same phase** (most often in solution). Examples of homogeneous catalysts are acids, bases, and enzymes.

The concept of a **homogeneous catalyst** used as a reactant but **regenerated** at the end of the reaction can be illustrated:

Step 1 $CH_2{=}CH_2$ + $\mathbf{H^+_{(aq)}}$ **(catalyst)** $\longrightarrow$ $H{-}CH_2{-}CH_2^+$

Step 2 $H{-}CH_2{-}CH_2^+$ + $H{-}O{-}H$ $\longrightarrow$ $H{-}CH_2{-}CH_2{-}\overset{+}{O}(H){-}H$

Step 3 $H{-}CH_2{-}CH_2{-}\overset{+}{O}(H){-}H$ $\longrightarrow$ $H{-}CH_2{-}CH_2{-}O{-}H$ + $H^+_{(aq)}$

Net Rxn: $CH_2{=}CH_2 + H_2O \longrightarrow CH_3{-}CH_2{-}OH$

Note that the H^+ formed as a product in step (3) is not the same H^+ as consumed in step (1); however, all H^+ are chemical identical. Thus, the catalyst is consumed in step (1); its role is to increase the rate of step (2). It is then regenerated in step (3) as another H^+ and can be used over again; there is no **net** consumption.

Do not confuse the role of the catalyst in this type of reaction with that of an intermediate. Both can be cancelled from the net equation, but a **catalyst** can be an **initial** reactant and a **final** product, while an intermediate is always formed as a product in a prior reaction step and must be consumed by reaction end. A catalyst can be (and usually is) included in an overall rate expression; rate usually depends on catalyst concentration. Rates of most biochemical reactions in living systems are ultimately controlled by catalytic enzyme concentrations.

The mechanistic role of the catalyst is to lower the r.d.s. activation barrier (Ea) for a reaction by providing an **alternative lower energy pathway** not otherwise available; the catalyst changes the mechanism (path function). A catalyst **cannot** change the value of ΔPE or ΔH (state functions) for the overall reaction.

The role of a catalyst can be visualized through a potential energy diagram for a hypothetical reaction, which can occur through an uncatalyzed one-step pathway or a catalyzed three-step mechanism. The uncatalyzed reaction has a high activation barrier and, therefore, a relatively slow rate. The catalyst converts the same reaction to a multi-step alternative pathway, which has a much lower activation energy for the rate determining step, thus increasing the rate. The catalyst, in effect, uses a different mechanism to tunnel through the high activation energy of the one-step reaction.

The initial potential energy of the reactants and the final potential energy of the products cannot be changed by a catalyst; catalysts cannot change a state function.

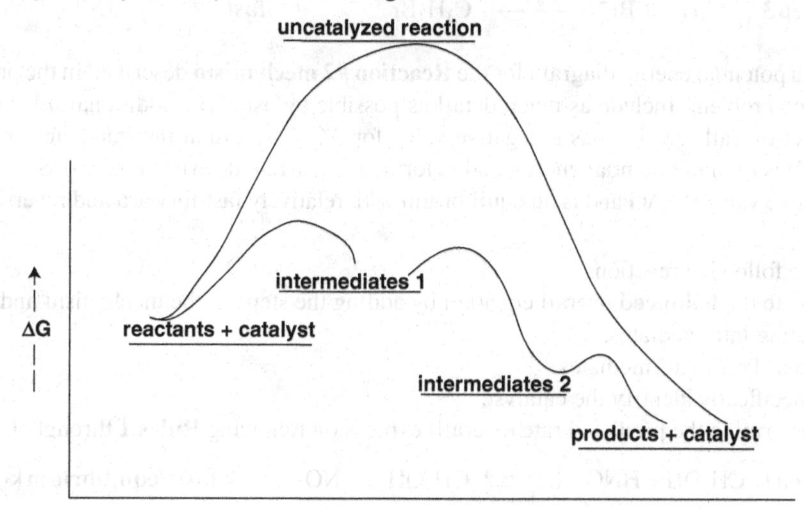

IX PRACTICE PROBLEMS

1. Use the reaction step rate rule to predict the rate expression for each of the following hypothetical one-step reactions; that is, use Rule 1 to write the rate expression if each of these processes occurred in one step.

a) $O_3 \longrightarrow O_2 + O$

b) $CH_3{-}CH_3 \longrightarrow CH_2{=}CH_2 + H_2$

c) $CO + O \longrightarrow CO_2$

d) $CH_3I + F \longrightarrow CH_3F + I$

e) $CH_2 + Cl + Cl \longrightarrow CH_2Cl_2$

f) $CO + Cl_2 \longrightarrow COCl_2$

g) $C_4H_6 + C_4H_6 \longrightarrow C_8H_{12}$

h) $CH_2 + H + Cl \longrightarrow CH_3Cl$

2. For the following two reactions:
   a) Write the **balanced** overall equation by adding the steps in the mechanism and eliminating intermediates.
   b) Determine the predicted rate (overall) expression following **Rules 1** through **4**.

Reaction #1 Mechanism:

Step 1 $C_4H_9Br \xrightarrow{k_1} C_4H_9^+ + Br^-$ **slow**

Step 2 $C_4H_9^+ + Cl^- \xrightarrow{k_2} C_4H_9Cl$ **fast**

Reaction #2 Mechanism:

Step 1 $C_3H_8O + HBr \underset{k_{-1}}{\overset{k_1}{\rightleftharpoons}} C_3H_9O^+ + Br^-$ **fast/equilibrium** $k_1/k_{-1} \equiv K_1$

Step 2 $C_3H_9O^+ \xrightarrow{k_2} C_3H_7^+ + H_2O$ **slow**

Step 3 $C_3H_7^+ + Br^- \xrightarrow{k_3} C_3H_7Br$ **fast**

3. Draw a potential energy diagram for the **Reaction #2 mechanism** described in the previous practice problem. Include as much detail as possible by using the additional information: The net overall reaction has a negative value for ΔG. The cation intermediate formed in step (2) is high in potential energy and is formed in the rate determining step. Step (1) has a positive value for ΔG and is an equilibrium with relatively fast-forward and reverse rates.

4. For the following reaction:
   a) Write the **balanced** overall equation by adding the steps in the mechanism and eliminating intermediates.
   b) Identify all intermediates.
   c) Specifically identify the catalyst.
   d) Determine the predicted rate (overall) expression following **Rules 1** through **4**.

Step 1 $CH_3OH + HNO_3 \underset{k_{-1}}{\overset{k_1}{\rightleftharpoons}} CH_3OH_2^+ + NO_3^-$ **fast / equilibrium** $k_1/k_{-1} \equiv K_1$

Step 2 $CH_3OH_2^+ + CH_3OH \xrightarrow{k_2} CH_3O^+HCH_3 + H_2O$ **slow**

Step 3 $CH_3O^+HCH_3 + NO_3^- \xrightarrow{k_3} CH_3OCH_3 + HNO_3$ **fast**

5. The three **hypothetical** reaction mechanisms in this problem refer to the same **net** balanced equation. Kinetic analysis shows that a plot of 1/[Br] vs. time is straight line whenever $[F_2]$ is held approximately constant. Kinetic analysis shows that a plot of $\ln[F_2]$ vs. time is straight line whenever [Br] is held approximately constant. Complete the ten question for these mechanisms.
   a) What is the balanced equation for possible mechanism #1?
   b) What are the intermediates for possible mechanism #1?

c) What is the predicted rate expression for the overall rate for possible mechanism #1?
d) What is the rate expression for the reaction based on the kinetic experiments?
e) Is possible mechanism #1 a possible valid mechanism for the reaction, based on the kinetic experimental results?
f) Is possible mechanism #2 a possible valid mechanism for the reaction, based on the kinetic experimental results?
g) Confirm that possible mechanism #3 has the same balanced equation.
h) What are the intermediates for possible mechanism #3?
i) What is the predicted rate expression for the overall rate for possible mechanism #3?
j) Is mechanism #3 a possible valid mechanism for the reaction, based on the kinetic experimental results?

Possible Mechanism #1:

Step 1 $2Br + F_2 \xrightarrow{k_1} Br_2F_2$ **slow / r.d.s.**

Step 2 $Br + Br_2F_2 \xrightarrow{k_2} BrF_2 + Br_2$ **fast**

Possible Mechanism #2: The reaction occurs in one-step based on the balanced equation determined for possible mechanism #1.

Possible Mechanism #3:

Step 1 $3Br + F_2 \underset{k_{-1}}{\overset{k_1}{\rightleftharpoons}} Br_3F_2$ **fast / equilibrium** $k_1/k_{-1} \equiv k_1$

Step 2 $Br_3F_2 \xrightarrow{k_2} BrF_2 + Br_2$ **slow / r.d.s.**

6. The potential energy diagrams shown describe two different mechanisms for the **same** reaction. Answer the following questions for **each** mechanism.
a) What is the **overall** reaction?
b) Specify each step and classify as **fast** or **slow** (rate-determining). Do not include molecules that are unchanged in the step. Hint: Acid dissociation in water is always an equilibrium.
c) Calculate **Ea** for the rate-determining step (r.d.s.).
d) Determine the sign and value for ΔG for the **overall** reaction.
e) What is the rate expression specifically for the **r.d.s.**?
f) What is the rate expression for the **overall** reaction?
g) What might the transition state of the **r.d.s.** look like?

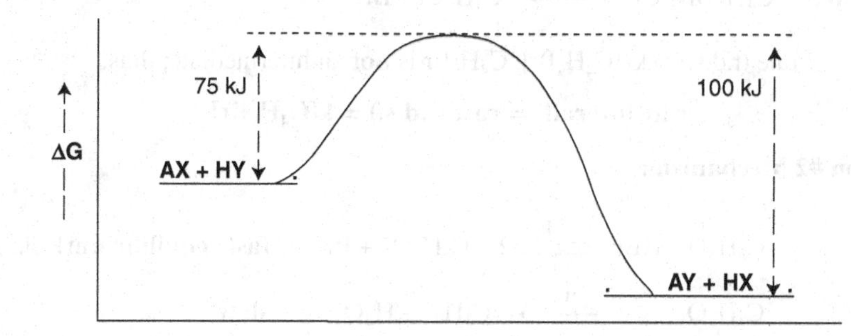

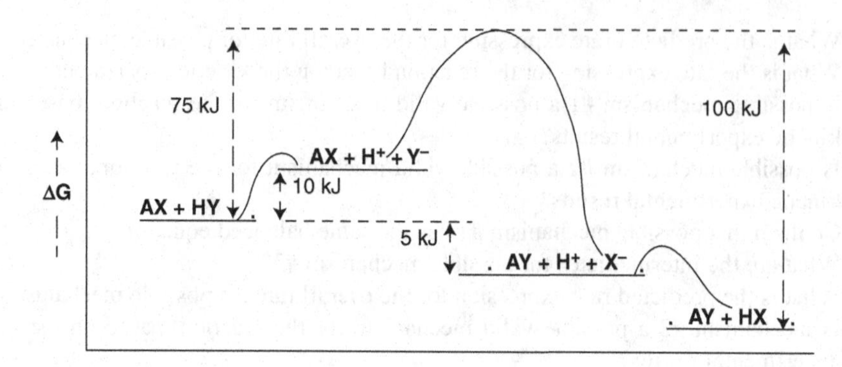

X ANSWERS TO PRACTICE PROBLEMS

1. Use the reaction step rate rule to predict the rate expression for each of the following hypothetical one-step reactions; that is, use Rule 1 to write the rate expression if each of these processes occurred in one step.

   a) $O_3 \longrightarrow O_2 + O$ rate = k$[O_3]$

   b) $CH_3{-}CH_3 \longrightarrow CH_2{=}CH_2 + H_2$ rate = k $[CH_3CH_3]$

   c) $CO + O \longrightarrow CO_2$ rate = k[CO][O]

   d) $CH_3I + F \longrightarrow CH_3F + I$ rate = k$[CH_3I][F]$

   e) $CH_2 + Cl + Cl \longrightarrow CH_2Cl_2$ rate = k$[CH_2][Cl]^2$

   f) $CO + Cl_2 \longrightarrow COCl_2$ rate = k$[CO][Cl_2]$

   g) $C_4H_6 + C_4H_6 \longrightarrow C_8H_{12}$ rate = k$[C_4H_6]^2$

   h) $CH_2 + H + Cl \longrightarrow CH_3Cl$ rate = k$[CH_2][H][Cl]$

2. For the following two reactions:

   a) Write the **balanced** overall equation by adding the steps in the mechanism and eliminating intermediates.

   b) Determine the predicted rate (overall) expression following **Rules 1** through **4**.

   Reaction #1 Mechanism:

Step 1	$C_4H_9Br \xrightarrow{k_1} C_4H_9^+ + Br^-$	**slow**
Step 2	$C_4H_9^+ + Cl^- \xrightarrow{k_2} C_4H_9Cl$	**fast**
Equation:	$C_4H_9Br + Cl^- \longrightarrow C_4H_9Cl + Br^-$	

rate (r.d.s.) = k_1 $[C_4H_9Br]$; C_4H_9Br is **not** an intermediate; thus,

rate (overall) = rate (r.d.s.) = k$[C_4H_9Br]$

   Reaction #2 Mechanism:

Step 1	$C_3H_8O + HBr \underset{k_{-1}}{\overset{k_1}{\rightleftarrows}} C_3H_9O^+ + Br^-$	**fast / equilibrium** $k_1/k_{-1} \equiv K_1$
Step 2	$C_3H_9O^+ \xrightarrow{k_2} C_3H_7^+ + H_2O$	**slow**
Step 4	$C_3H_7^+ + Br^- \xrightarrow{k_3} C_3H_7Br$	**fast**
Equation:	$C_3H_8O + HBr \longrightarrow C_3H_7Br + H_2O$	

rate (r.d.s.) = $k_2[C_3H_9O^+]$; $C_3H_9O^+$ **is an intermediate.**

$$\text{Define } K_1 = \frac{[C_3H_9O^+][Br^-]}{[C_3H_8O][HBr]}$$

$$[C_3H_9O^+] = \frac{K_1[C_3H_8O][HBr]}{[Br^-]}$$ **($[Br^-]$ cannot be eliminated from expression)**

Substitute expression for $[C_3H_9O^+]$ into rate (r.d.s.)

$$\textbf{rate (overall)} = \frac{k_2\{K_1[C_3H_8O][HBr]\}}{[Br^-]} \quad \text{or} \quad \textbf{rate (overall)} = \frac{k[C_3H_8O][HBr]}{[Br^-]}$$

3. Draw a potential energy diagram for the **Reaction #2 mechanism** described in the previous practice problem. Include as much detail as is possible by using the additional information.

Step 1 $C_3H_8O + HBr \underset{k_{-1}}{\overset{k_1}{\rightleftharpoons}} C_3H_9O^+ + Br^-$ **fast/equilibrium**

Step 2 $C_3H_9O^+ \xrightarrow{k_2} C_3H_7^+ + H_2O$ **slow**

Step 3 $C_3H_7^+ + Br^- \xrightarrow{k_3} C_3H_7Br$ **fast**

Additional information conclusions:

a) The net overall reaction has a negative value for ΔG: products must be shown lower than reactants on the diagram.
b) The cation intermediate formed in step (2) is high in potential energy and is formed in the rate determining step: The energy platform for the products of step (2) should be relatively high on the diagram.
c) Step (1) has a positive value for ΔG and is an equilibrium with relatively fast-forward and reverse rates.

The energy platform for the products of step (1) should be higher than the reactants. An equilibrium for this step indicates that the activation energies for the forward and reverse reactions for step (1) should be smaller than for step (2), which is the r.d.s.

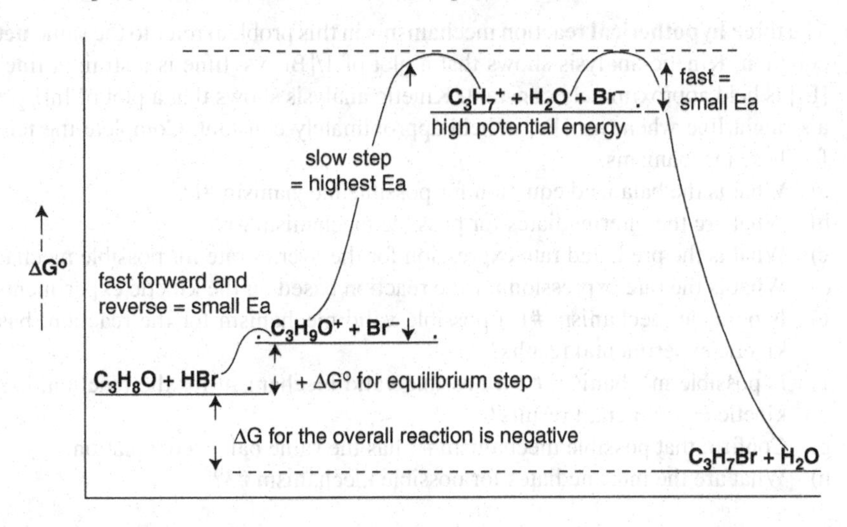

4. For the following reaction:
   a) Write the **balanced** overall equation by adding the steps in the mechanism and eliminating intermediates.
   b) Identify all intermediates.
   c) Specifically identify the catalyst.
   d) Determine the predicted rate (overall) expression following **Rules 1** through **4**.

$$\textbf{Step 1}\ CH_3OH + HNO_3 \underset{k_{-1}}{\overset{k_1}{\rightleftharpoons}} CH_3OH_2^+ + NO_3^- \qquad \textbf{fast/equilibrium}\ k_1/k_{-1} \equiv K_1$$

$$\textbf{Step 2}\ CH_3OH_2^+ + CH_3OH \xrightarrow{k_2} CH_3O^+HCH_3 + H_2O\ \textbf{slow}$$

$$\textbf{Step 3}\ CH_3O^+HCH_3 + NO_3^- \xrightarrow{k_3} CH_3OCH_3 + HNO_3\ \textbf{fast}$$

a) **Equation:** $2CH_3OH \xrightarrow{HNO_3} CH_3OCH_3 + H_2C$
b) Intermediates: $CH_3OH_2^+$; $CH_3O^+HCH_3$; NO_3^-
c) HNO_3 is not an intermediate. HNO_3 is a catalyst; it is present as an initial reactant and regenerated as a final product. It can be cancelled from the balanced equation; a common practice is to place the identity of the catalyst above the reaction arrow.
d) Rate (r.d.s.) = $k_2[CH_3OH_2^+][CH_3OH]$; $CH_3OH_2^+$ is an intermediate.

$$\text{Define } K_1 = \frac{[CH_3OH_2^+][NO_3^-]}{[CH_3OH][HNO_3]}$$

$$[CH_3OH_2^+] = \frac{K_1[CH_3OH][HNO_3]}{[NO_3^-]}$$ ($[NO_3^-]$ cannot be eliminated from expression)

Substitute expression for $[CH_3OH_2^+]$ into rate (r.d.s.):

$$\textbf{rate (overall)} = \frac{k_2\{K_1[CH_3OH][HNO_3]\}[CH_3OH]}{[NO_3^-]}$$

or

$$\textbf{rate (overall)} = \frac{k[CH_3OH]^2[HNO_3]}{[NO_3^-]}$$

5. The three **hypothetical** reaction mechanisms in this problem refer to the same **net** balanced equation. Kinetic analysis shows that a plot of 1/[Br] vs. time is a straight line whenever $[F_2]$ is held approximately constant. Kinetic analysis shows that a plot of $\ln[F_2]$ vs. time is a straight line whenever [Br] is held approximately constant. Complete the ten questions for these mechanisms.
   a) What is the balanced equation for possible mechanism #1?
   b) What are the intermediates for possible mechanism #1?
   c) What is the predicted rate expression for the overall rate for possible mechanism #1?
   d) What is the rate expression for the reaction based on the kinetic experiments?
   e) Is possible mechanism #1 a possible valid mechanism for the reaction, based on the kinetic experimental results?
   f) Is possible mechanism #2 a possible valid mechanism for the reaction, based on the kinetic experimental results?
   g) Confirm that possible mechanism #3 has the same balanced equation.
   h) What are the intermediates for possible mechanism #3?

i) What is the predicted rate expression for the overall rate for possible mechanism #3?
j) Is for possible mechanism #3 a possible valid mechanism for the reaction, based on the kinetic experimental results?

Possible Mechanism #1:

Step 1	$2\,Br + F_2 \xrightarrow{k_1} Br_2F_2$	**slow/r.d.s.**
Step 2	$Br + Br_2F_2 \xrightarrow{k_2} BrF_2 + Br_2$	**fast**
Equation:	$3\,Br + F_2 \longrightarrow BrF_2 + Br_2$	

b) Intermediate: Br_2F_2
c) Rate (r.d.s.) = $k_1\,[Br]^2[F_2]$; both Br and F_2 are **not** intermediates; thus,

$$\textbf{rate (overall) = rate (r.d.s.) = k[Br]}^2\textbf{[F}_2\textbf{]}$$

d) A straight-line plot of 1/[Br] vs. time whenever $[F_2]$ is held approximately constant indicates that [Br] is second order. A straight-line plot of $\ln[F_2]$ vs. time whenever [Br] is held approximately constant indicates $[F_2]$ is first order.

The kinetic rate expression: **rate = k[Br]^2[F$_2$]**

e) Possible mechanism #1 agrees with the kinetic results; it is possible valid mechanism for the reaction.

Possible Mechanism #2: The reaction occurs in one-step based on the balanced equation determined for possible mechanism #1.

f) The balanced equation: $3\,Br + F_2 \xrightarrow{\text{one-step}} BrF_2 + Br_2$

$$\textbf{Rate (one-step reaction) = k [Br]}^3\textbf{[F}_2\textbf{]}$$

Possible mechanism #2 does not agree with the kinetic results; it is not a possible valid mechanism for the reaction.

Possible Mechanism #3:

Step 1	$3\,Br + F_2 \underset{k_{-1}}{\overset{k_1}{\rightleftarrows}} Br_3F_2$	**fast/equilibrium**
Step 2	$Br_3F_2 \xrightarrow{k_2} BrF_2 + Br_2$	**slow/r.d.s.**
Equation:	$3\,Br + F_2 \longrightarrow BrF_2 + Br_2$	

h) Intermediate: Br_3F_2
i) Rate (r.d.s.) = $k_2[Br_3F_2]$; Br_3F_2 is an intermediate.

$$\text{Define } K_1 = \frac{[Br_3F_2]}{[Br]^3[F_2]}$$

$[Br_3F_2] = K_1[Br]^3[F_2]$
Substitute expression for $[Br_3F_2]$ into rate (r.d.s.):

rate (overall) = $k_2K_1[Br]^3[F_2]$ or rate (overall) = $k[Br]^3[F_2]$

j) Possible mechanism #3 does not agree with the kinetic results; it is not a possible valid mechanism for the reaction.

6. The energy level diagrams shown describe two different mechanisms for the same reaction.

Mechanism #1

a) What is the overall reaction: **$AX + HY \longrightarrow AY + HX$**?
b) Specify each step and classify as fast or slow: **One-step only**
c) Calculate **Ea** for the rate-determining step (only step): **Ea = 75 kJ/mole** as shown.
d) Determine the sign and value for ΔG for the overall reaction:

$$\Delta G° = 75 - 100 = \mathbf{-25\ kJ/mole\text{-}reaction}$$

e) What is the rate expression specifically for the **r.d.s.: rate (r.d.s.) = k [AX] [HY]?**
f) What is the rate expression for the overall reaction.

rate (overall) = rate (r.d.s.) = k[AX][HY]?

g) What might the transition state of the **r.d.s.** look like:

A—X and H—Y are being broken; A—Y and H—X are being formed? Transition state must show all changing bonds as partial bonds. A possible configuration (dashed lines = partial bonding):

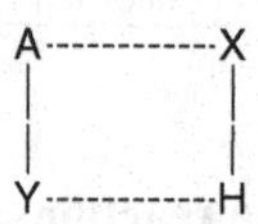

Mechanism #2

a) What is the overall reaction? b) Specify each step and classify as fast or slow:

Step 1	$HY \underset{k_{-1}}{\overset{k_1}{\rightleftharpoons}} H^+ + Y^-$	**fast / equilibrium $K_1 = k_1/k_{-1}$**
Step 2	$AX + Y^- \xrightarrow{k_2} AY + X^-$	**slow / r.d.s.**
Step 3	$H^+ + X^- \xrightarrow{k_3} HX$	**fast**
Balanced equation	$HY + AX \longrightarrow AY + HX$	

c) Calculate **Ea** for the rate-determining step (r.d.s.): r.d.s. is step (2)

$$\text{Ea (Step 2)} = 75 - 10 = \mathbf{65\ kJ/mole}$$

d) Determine the sign and value for ΔG for the overall reaction:

$$\Delta G = 75 - 100 = \mathbf{-25\ kJ/mole\text{-}reaction}$$

f) What is the rate expression specifically for the **r.d.s.**:
rate (r.d.s.) = $k_2[AX][Y^-]$?

What is the rate expression for the overall reaction;

$[Y^-]$ is an intermediate and must be substituted for:

$$K_1 = \frac{[H^+][Y^-]}{[HY]} \qquad [Y^{-1}] = \frac{K_1[HY]}{[H^+]}$$

substitute into rate (r.d.s.):

$$\textbf{rate (overall)} = \frac{k_2[AX]\{K_1[HY]\}}{[H^+]}$$

$$\textbf{rate (overall)} = \frac{k_2K_1[AX][HY]\}}{[H^+]}$$

$$\textbf{rate (overall)} = \frac{k[AX][HY]\}}{[H^+]}$$

g) What might the transition state of the **r.d.s.** look like:
Step (2) is the r.d.s.? In step (2), the A—X bond is breaking; the A—Y bond is forming.
A representation of the transition state might be (dashed lines show partial bonds):

Y--------A--------X

25 Thermodynamics
Entropy and Free Energy

I GENERAL CONCEPTS OF ENTROPY

The entropy of a system (symbol = $S_{(system)}$) is a measure of the disorder, or randomness, of that system. The larger the entropy value the greater the randomness, or disorder, of the system; the smaller the entropy value, the lesser the randomness or disorder (or, equivalently, the smaller the entropy value, the more ordered the system).

Entropy measures the statistical probabilities of specific system configurations vs. all possible configurations (termed the **degrees of freedom**).

Randomness, or disorder, always represents the **more probable** arrangement of matter. The result is that **entropy** always **increases** in the direction of **increasing probability** of the arrangement of matter within a system.

Entropy is a state function, the change in entropy of any system depends only on the initial and final states and not on how the change occurs.

Randomness, or disorder, in a system of matter is related to the number of arrangements in which the matter can exist. General trends for a specific amount of matter are:

(1) A specific amount of matter has more probable arrangements and is more disordered (higher entropy) if it exists in a greater number of small, simpler molecules rather than a fewer number of more complex, larger molecules. Formation of large, complex molecules produces a more ordered, less random (lower entropy) system.

(2) For more than one substance, a specific amount of matter has more probable arrangements and is more disordered (higher entropy) if it exists as a mixture or solution rather than if the components are separated as pure substances.

(3) A specific amount of matter has more probable arrangements and is more disordered (higher entropy) in proportion to the degree of independence of the molecules based on phase; gas is more disordered (higher entropy) than liquid, which is more disordered (higher entropy) than solid.

(4) A specific amount of matter has more probable arrangements and is more disordered (higher entropy) in proportion to the amount of molecular motion, as measured by average kinetic energy; disorder (entropy) increases as temperature increases.

PREDICTING ENTROPY CHANGES FOR A CHEMICAL PROCESS

A chemical process is a chemical reaction or physical change. The **change** in entropy for any process relates to the degree of disorder of the reactants compared to the products. The change in the entropy of a system (**ΔS**) is a **positive** value whenever entropy **increases**; the system becomes more disordered and tends toward a more probable arrangement of matter. **ΔS** is a **negative** value whenever entropy **decreases**; the system becomes less disordered (i.e., more ordered) and tends toward a less probable arrangement of matter.

The general probability trends for a specific amount of matter can be used to analyze a process based on a **balanced** equation, phase change, or solution change. General concepts can be used to predict the direction of entropy change (sign of **ΔS**) in the absence of any calculations.

(1) **Entropy increases** (disorder increases) for reactions in which a lesser fewer number of larger or more complex molecules are broken up into a greater number of smaller, simpler molecules. This can be measured as a greater number of moles of product molecules forming from a fewer number of moles of reactant molecules. Relative numbers of reactant and product molecules can be measured directly by examination of coefficients in the balanced equation. A larger effect occurs whenever the number of moles of **gas** increases or decreases during the reaction.

Example: Determine the entropy change for the reaction shown.

$$C_6H_{12}O_{6\,(s)} + 6\,O_{2\,(g)} \longrightarrow 6\,CO_{2(g)} + 6\,H_2O_{(g)}$$

The reaction shows 7 moles of reactants, one mole representing a complex reactant molecule decomposing to 12 moles of smaller products; in addition, the number of moles of gas molecules increase. The entropy change of the reaction is positive: entropy increases.

$$C_6H_{12}O_{6\,(s)} + 6\,O_{2\,(g)} \xrightarrow[\text{7 moles form12 moles}]{\text{Entropy increases } (\Delta S = +)} 6\,CO_{2(g)} + 6\,H_2O_{(g)}$$

Conversely, **entropy decreases** (disorder decreases, i.e., system becomes **more ordered**) if a fewer number of larger, more complex molecules are produced by combination of smaller, simpler molecules. This is measured as fewer moles of product molecules forming from a greater number of moles of reactant molecules and can be measured through the coefficients in the balanced equation.

Example: The reverse of the previous example: 7 moles of products with one complex molecule are formed by combination of 12 moles of simpler reactant molecules. The entropy change of the reaction is negative: entropy decreases.

$$6\,CO_{2\,(g)} + 6H_2O_{(g)} \xrightarrow[\text{12 moles from 7 moles}]{\text{Entropy decreases } (\Delta S = -)} C_6H_{12}O_{6\,(s)} + 6\,O_{2(g)}$$

(2) **Entropy increases** (disorder increases) during the process of **solution** formation from pure substances, or during the **dilution** of a solution to a lower concentration.

General Example:

$$\text{pure substances} \xrightarrow[\text{solution or mixture formation}]{\text{Entropy increases } (\Delta S = +)} \text{solution or mixture}$$

General Example:

$$\textbf{high}\text{ concentration solution} \xrightarrow[\text{dilution of solution}]{\text{Entropy increases } (\Delta S = +)} \textbf{low}\text{ concentration solution}$$

Entropy decreases (disorder decreases, i.e., system becomes **more ordered**) in processes that purify or separate mixtures into pure substances or during the concentration of a solution to higher concentration.

General Example:

$$\text{Solution or mixture} \xrightarrow[\text{separaton/purification}]{\text{Entropy decreases } (\Delta S = -)} \text{pure substances}$$

General Example:

$$\textbf{low}\text{ concentration solution} \xrightarrow[\text{solution concentration}]{\text{Entropy decreases } (\Delta S = -)} \text{high concentration solution}$$

(3) **Entropy increases** (disorder increases) during phase changes in the sequence solid melting to liquid and liquid vaporization to a gas; **entropy decreases** (disorder decreases, i.e., system becomes **more ordered**) for condensation of a gas to a liquid and freezing of a liquid to a solid.

General Example:

$$\text{solid} \xrightarrow{\text{Entropy increases } (\Delta S = +)} \text{liquid} \xrightarrow{\text{Entropy increases } (\Delta S = +)} \text{gas}$$

$$\text{gas} \xrightarrow{\text{Entropy decreases } (\Delta S = -)} \text{liquid} \xrightarrow{\text{Entropy decreases } (\Delta S = -)} \text{solid}$$

(4) For any specific type of matter, **entropy increases** (disorder increases) as the temperature (average kinetic energy) **increases; entropy decreases** (disorder decreases, i.e., system becomes **more ordered**) as the temperature (average kinetic energy) **decreases.** ΔS is directly proportional to **temperature.**

General Example:

$$\text{matter at \textbf{lower} temperature} \xrightarrow{\text{Entropy increases } (\Delta S = +)} \text{matter at \textbf{higher} temperature}$$

$$\text{matter at \textbf{higher} temperature} \xrightarrow{\text{Entropy decreases } (\Delta S = -)} \text{matter at \textbf{lower} temperature}$$

Example: Predict the sign of the entropy.

$$\text{ice at } 0\ ^\circ\text{C} \longrightarrow \text{liquid water at } 0\ ^\circ\text{C}$$

$$\text{solid} \longrightarrow \text{liquid at constant temperature : concept } \mathbf{(3)}\ \Delta S = (+)$$

Example: Predict the sign of the entropy.

$$\text{steam (water vapor) at } 100\ ^\circ\text{C} \longrightarrow \text{liquid water at } 100\ ^\circ\text{C}$$

$$\text{gas} \longrightarrow \text{liquid at constant temperature : concept } \mathbf{(3)}\ \Delta S = (-)$$

Example: Predict the sign of the entropy.

$$H_2SO_{4(l)} + H_2O_{(l)} \longrightarrow H_2SO_{4(aq)}$$

$$\text{pure substances} \longrightarrow \text{solution: concept } \mathbf{(2)}\ \Delta S = (+)$$

Example: Predict the sign of the entropy.

$$CaCl_{2(aq)}\ (1.0\ M) \longrightarrow CaCl_{2\ (aq)}\ (0.5\ M)$$

$$\text{Higher concentration} \longrightarrow \text{lower concentration : concept } \mathbf{(2)}\ \Delta S = (+)$$

Example: Predict the sign of the entropy.

$H_2O_{(l)}$ at 25 °C ⟶ $H_2O_{(l)}$ at 50 °C

lower temperature ⟶ higher temperature: concept **(4)** $\mathbf{\Delta S = (+)}$

Example: Predict the sign of the entropy.

$CH_3OH_{(l)} \longrightarrow CO_{(g)} + 2\,H_{2(g)}$

1 mole (liquid) ⟶ 2 moles (gas): concept **(1)** $\mathbf{\Delta S = (+)}$

Example: Predict the sign of the entropy.

$2\,H_{2(g)} + O_{2(g)} \longrightarrow 2\,H_2O_{(l)}$

3 moles (gas) ⟶ 2 moles (liquid): concept **(1)** $\mathbf{\Delta S = (-)}$

Although changes in number of moles of products and reactants can be read from the coefficients in the balanced equation, various combinations of solids, liquids, and gases can sometimes obscure a clear conclusion.

Example: Predict the sign of the entropy.

$3Fe_{(s)} + 4H_2O_{(l)} \longrightarrow Fe_3O_{4(s)} + 4\,H_{2(g)}$

7 mole (liquid and solid) ⟶ 5 moles (4 moles are gas)

Concept **(1)** does not provide an unambiguous answer. In this case, a calculation (Section III) indicates the formation of gaseous products contributes more toward a positive entropy than the small decrease in the number of moles of product. In this case, the result is a positive entropy change.

II ENTROPY AND THE SECOND LAW OF THERMODYNAMICS

The second law of thermodynamics states that the total entropy of the **universe** is always increasing. All processes must show: $\mathbf{\Delta S_{(universe)} > 0}$ or $\mathbf{\Delta S_{(universe)}} = (+)$.

The universe is composed of the system plus surroundings:

$$\mathbf{\Delta S_{(universe)} = [\Delta S_{(system)} + \Delta S_{(surroundings)}] > 0}$$

Since $\Delta S_{(universe)}$ must have a positive value, the second law requires that for any specific **decrease** in entropy of a system or the surroundings, the other component must show an opposite **increase** in entropy with a greater numerical value.

Example:

If $\mathbf{\Delta S_{(system)}} = (-)$, $\mathbf{\Delta S_{(surroundings)}}$ must be (+) with a greater numerical value.

If $\mathbf{\Delta S_{(system)}} = (+)$, $\mathbf{\Delta S_{(surroundings)}}$ must be (–) with a lesser numerical value.

Entropy change is often measured under conditions where temperature is constant (**$\Delta T = 0$**). Under this condition, entropy change can be measured as heat transfer at constant temperature (**q_T**) divided by temperature in Kelvins (**T in K**):

$$\mathbf{\Delta S = \frac{(q_T)}{T}}$$

The units of entropy (in all cases) are Joules/K; this emphasizes the relationship between entropy and temperature.

Example: Calculate $\Delta S_{(system)}$, $\Delta S_{(surroundings)}$, and $\Delta S_{(universe)}$ for a system of 2.00 moles of ice melting at a constant T = 273 K in a room (surroundings) held constant at 298 K. q_T for this problem is measured as: $\Delta H_{(fusion)} = +6010$ J/mole

$$q_{(system)} = q_{(water)} = (+\,6010 \text{ J/mole}) \times (2.00 \text{ mole}) = +12{,}020 \text{ J}$$

$$\Delta S_{(system)} = \Delta S_{(water)} = \frac{q_{(water)}}{T_{(water)}} = \frac{+12{,}020\,J}{273\,K} = \mathbf{+44.0\,J/K}$$

$$3\{q_{(surroundings)} = -q_{(system)}\}:$$

$$q_{(surroundings)} = q_{(room)} = -q_{(water)} = -(+12{,}020\,J) = \mathbf{-12{,}020\ J}$$

$$\Delta S_{(surroundings)} = \Delta S_{(room)} = \frac{q_{(room)}}{T_{(room)}} = \frac{-12{,}020\,J}{298\,K} = \mathbf{-40.3\,J/K}$$

$$\Delta S_{(universe)} = [\Delta S_{(system)} + \Delta S_{(surroundings)}] = (+44.0 \text{ J/K}) + (-40.3 \text{ J/K}) = \mathbf{+3.7\ J/K}$$

Example: Calculate $\Delta S_{(System)}$, $\Delta S_{(surroundings)}$, and $\Delta S_{(universe)}$ for a system of 2.00 moles of water freezing at a constant T = 273 **K** in a freezer (surroundings) held constant at 258 **K**. q_T for this problem is measured as: $\Delta H_{(solidification)}$ for water = $-\,\Delta H_{(fusion)} = -\,6010$ J/mole

$$q_{(system)} = q_{(water)} = (-6010 \text{ J/mole}) \times (2.00 \text{ mole}) = -12{,}020 \text{ J}$$

$$\Delta S_{(system)} = \Delta S_{(water)} = \frac{q_{(water)}}{T_{(water)}} = \frac{-12{,}020\,J}{273\,K} = \mathbf{-44.0\,J/K}$$

$$\{q_{(surroundings)} = -q_{(system)}\}:$$

$$q_{(surroundings)} = q_{(freezer)} = -q_{(water)} = -(-12{,}020 \text{ J}) = +12{,}020 \text{ J}$$

$$\Delta S_{(surroundings)} = \Delta S_{(freezer)} = \frac{q_{(freezer)}}{T_{(freezer)}} = \frac{+12{,}020\,J}{258\,K} = \mathbf{+\,46.6\,J/K}$$

$$\Delta S_{(universe)} = [\Delta S_{(system)} + \Delta S_{(surroundings)}] = (-44.0 \text{ J/K}) + (+46.6 \text{ J/K}) = \mathbf{+2.6\ J/K}$$

It might appear that ΔS of the universe is these examples is always positive due to a favorable selection of temperature conditions. What would happen if a hotter body could remove heat from a colder body?

Example: Calculate $\Delta S_{(system)}$, $\Delta S_{(surroundings)}$, and $\Delta S_{(universe)}$ for a system of 2.00 moles of water freezing at a constant **T** = 273 **K** in a room (surroundings) held constant at 298 K. q_T for this problem is measured as: $\Delta H_{(solidification)}$ for water = – $\Delta H_{(fusion)}$ = –6010 J/mole

$$q_{(system)} = q_{(water)} = (-6010 \text{ J/mole}) \times (2.00 \text{ mole}) = -12{,}020 \text{ J}$$

$$\Delta S_{(system)} = \Delta S_{(water)} = \frac{q_{(water)}}{T_{(water)}} = \frac{-12{,}020\,\text{J}}{273\,\text{K}} = \mathbf{-44.0\,J/K}$$

$$\{q_{(surroundings)} = -q_{(system)}\}:$$

$$q_{(surroundings)} = q_{(room)} = -q_{(water)} = -(-12{,}020 \text{ J}) = +12{,}020 \text{ J}$$

$$\Delta S_{(surroundings)} = \Delta S_{(room)} = \frac{q_{(room)}}{T_{(room)}} = \frac{+12{,}020\,\text{J}}{298\,\text{K}} = \mathbf{+\,40.3\,J/K}$$

$$\Delta S_{(universe)} = [\Delta S_{(system)} + \Delta S_{(surroundings)}] = (-44.0 \text{ J/K}) + (+40.3 \text{ J/K}) = \mathbf{-3.7\ J/K}$$

The requirement for $\Delta S_{(universe)}$ = (+) emphasizes a corollary to the second law of thermodynamics. Heat transfer always proceeds in only one direction: heat must always flow from a hotter body to a colder body and never in reverse.

III CALCULATION OF ENTROPY AND THE THIRD LAW OF THERMODYNAMICS

The third law of thermodynamics states that the entropy of a pure, perfectly ordered crystal is zero at a temperature of absolute zero Kelvin: $\mathbf{S_{(substance)}}$ **at 0 K = 0.**

Unlike enthalpy (H), entropy (S) has a true zero point. This allows the entropy of any substance to be expressed in the form of an **absolute molar entropy (S°)** in units of Joules/Kelvin (J/K) rather than a change as used for enthalpy **(ΔH°f).**

Absolute standard molar entropies are provided under **standard conditions**: one atmosphere of pressure and 298 K. Since S = 0 only at 0 K, all values for absolute molar entropies at 298 K must be **positive** values (J/K); they are provided in standard tables similar to listings for **ΔH°f.**

The standard entropy **change** (ΔS°) for a reaction (or other process) measures the general **increase** (positive value for entropy change) or **decrease** (negative value for entropy change) in the total disorder present in all products and reactants under standard conditions.

The standard equation for calculating entropy change has a format similar to that used for enthalpy change.

$$\Delta S^\circ_{(reaction)} = \textbf{[Sum of the absolute molar entropies of all products]}$$
$$- \textbf{[Sum of the absolute molar entropies of all reactants]}$$

or

$$\Delta S^\circ_{(reaction)} = \textbf{[Sum of the absolute molar entropies of each mole of products]}$$
$$- \textbf{[Sum of the absolute molar entropies of each mole of reactants]}$$

or

$$\Delta S^\circ_{(reaction)} = \textbf{[Sum } \mathbf{n_i\, S^\circ_i} \textbf{ (products)]} - \textbf{[Sum } \mathbf{n_i\, S^\circ_i} \textbf{ (reactants)]}$$

The terms $S°_i$ are the **absolute** molar entropies of each species, while $\mathbf{n_i}$ designates the **individual** coefficients in the balanced equation. The table below contains values for S° used in the examples that follow.

	S°		S°
$(C_2H_{2(g)})$	201 J/k	$(H_2O_{(g)})$	188.8 J/K
$(C_2H_{6(g)})$	230 J/K	$(CO_{(g)})$	197.7 J/K
$(H_{2(g)})$	131 J/K	$(C_6H_{6(l)})$	173.3 J/K
$(C_2H_5OH_{(l)})$	160.7 J/K	$(Fe_{(s)})$	27.3 J/K
$(O_{2(g)})$	205.1 J/K	$(Fe_3O_{4(s)})$	146.4 J/K
$(CO_{2(g)})$	213.7 J/K	$(H_{2(g)})$	130.7 J/K
$(H_2O_{(l)})$	69.9 J/K		

Example: Predict the direction of entropy change for the following reaction, then use tables of **S°** values to calculate $\Delta S°_{(reaction)}$ for the reaction.

$$C_2H_{6(g)} + 2\,H_{2(g)} \longrightarrow C_2H_{6(g)}$$

Three moles of gas reactants combine to form 1 mole of gas product. This represents the direction of a greater number of smaller, simpler molecules combining to form fewer more complex molecules (concept **1**): entropy should decrease; $\Delta S°_{(reaction)} = (-)$

To perform a calculation, follow steps **(1)** and **(2)**:

Step **(1)**: From the table, identify all required values for S°

Reactants: $S°(C_2H_{2(g)}) = 201$ J/k; $S°(H_{2(g)}) = 131$ J/K

Products: $S°(C_2H_{6(g)}) = 230$ J/K;

Note that all substances, including elements, have a positive value for S°.
Step **(2)**: Complete the calculation using the equation given:

$$\Delta S°_{(reaction)} = [\text{Sum of } (S°\,(C_2H_{6\,(g)})] - [\text{Sum of } \{(S°(C_2H_{2\,(g)})) + 2(S°\,(H_{2\,(g)})\}]$$

$$\Delta S°_{(reaction)} = [230\text{ J/K})] - [(201\text{ J/K}) + 2(131\text{ J/K})]$$

$$\Delta S°_{(reaction)} = [230\text{ J/K})] - [(463\text{ J/K})] = \mathbf{-233\ J/K} \text{ or } \mathbf{-0.233\ kiloJoules/K\ (kJ/K)}$$

Example: Predict the direction of entropy change for the reverse reaction of the previous example, then use tables of S° values to calculate $\Delta S°_{(reaction)}$ for the reaction.

$$C_2H_{6(g)} \longrightarrow C_2H_{2(g)} + 2H_{2(g)}$$

Entropy change is a **state** function; the prediction for the reverse reaction must be opposite to the forward direction: $\Delta S°(\text{reverse}) = (+)$. The symmetry of the components in the reverse direction indicates that all values for S° used in a calculation for the reverse direction would be the same; the direction of subtraction is reversed.

$$\Delta S°_{(reverse)} = \mathbf{+\ 233\ J/K\ or\ +\ 0.233\ kiloJoules/K\ (kJ/K)}$$

The general result is: $\Delta S^0_{(forward\ reaction)} = -\Delta S^0_{(reverse\ reaction)}$

Example: Predict the direction of entropy change for the following reaction, then use tables of **S°** values to calculate $\Delta S°_{(reaction)}$ for the reaction.

$$C_2H_5OH_{(l)} + 3\ O_{2(g)} \longrightarrow 2\ CO_{2(g)} + 3\ H_2O_{(l)}$$

Four moles of reactants (3 moles of gas) react to form 5 moles products (2 moles of gas). This reaction presents a difficult prediction. As in a previous example the difference in change in moles of gas usually contributes more than the small difference in total number of moles; entropy should decrease; $\Delta S°_{(reaction)} = (-)$

Step (**1**): From the table, identify all required values for **S°**

Reactants: $S°\ (C_2H_5OH_{(l)}) = 160.7$ J/K; $S°\ (O_{2\,(g)}) = 205.1$ J/K

Products: $S°(CO_{2(g)}) = 213.7$ J/K; $S°(H_2O_{(l)}) = 69.9$ J/K

Step (**2**): Complete the calculation using the equation given:

$$\Delta S°_{(reaction)} = [\text{Sum of } \{2(S°\ CO_{2\,(g)}) + 3(S°\ (H_2O_{(l)})\}]$$

$$- [\text{Sum of } \{(S°\ C_2H_5OH_{(l)}) + 3(S°\ O_{2\,(g)})]$$

$$\Delta S°_{(reaction)} = [2(213.7\ J/K) + 3(69.9\ J/K)] - [(160.7\ J/K) + 3(205.1\ J/K)]$$

$$\Delta S°_{(reaction)} = [637.1\ J/K)] - [(776\ J/K)] = \mathbf{-138.9\ J/K} \text{ or } \mathbf{-0.1389\ kJ/K}$$

Example: Predict the direction of entropy change for the following reaction, then use tables of S° values to calculate $\Delta S°_{(reaction)}$ for the reaction.

$$2\ C_6H_{6\,(l)} + 9\ O_{2(g)} \longrightarrow 12\ CO_{(g)} + 6\ H_2O_{(g)}$$

Eleven moles of reactants (9 moles of gas) react to form 18 moles gaseous products. This represents the direction of fewer more complex molecules forming a greater number of smaller, simpler molecules with an increase in gas moles (concept **1**): entropy should increase; $\Delta S°_{(reaction)} = (+)$

Step (**1**): From the table, identify all required values for S°

Reactants: $S°\ (C_6H_{6\,(l)}) = 213.7$ J/K; $S°\ (O_{2\,(g)}) = 205.1$ J/K

Products: $S°\ (CO_{(g)}) = 197.7$ J/K; $S°(H_2O_{(g)}) = 188.8$ J/K

Step (**2**): Complete the calculation using the equation given:

$$\Delta S°_{(reaction)} = [\text{Sum of } \{12(S°\ CO_{(g)}) + 6(S°\ (H_2O_{(g)})\}]$$

$$- [\text{Sum of } \{2(S°\ C_2H_5OH_{(l)}) + 9(S°\ O_{2\,(g)})]$$

$$\Delta S°_{(reaction)} = [12(197.7\ J/K) + 6(188.8\ J/K)] - [2(173.3\ J/K) + 9(205.1\ J/K)]$$

$$\Delta S°_{(reaction)} = [3505.2\ J/K)] - [(2192.5\ J/K)] = \mathbf{+1312.7\ J/K} \text{ or } \mathbf{+1.313\ kJ/K}$$

Example: Predict the direction of entropy change for the following reaction and then use tables of S° values to calculate $\Delta S^\circ_{(reaction)}$ for the reaction.

$$3\ Fe_{(S)} + 4\ H_2O_{(l)} \longrightarrow Fe_3O_{4(S)} + 4\ H_{2\ (g)}$$

It was suggested previously that the formation of gaseous products contributes more toward a positive entropy than the small decrease in the number of moles of product. Entropy change was predicted to be positive.

Step **(1)**: From the table, identify all required values for **S°**

Reactants: $S^\circ(Fe_{(S)}) = 27.3$ J/K; $S^\circ(H_2O_{(l)}) = 69.9$ J/K

Products: $S^\circ(Fe_3\ O_{4(s)}) = 146.4$ J/K; $S^\circ(H2_{(g)}) = 130.7$ J/K

Step **(2)**: Complete the calculation using the equation given:

$$\Delta S^\circ_{(reaction)} = [\text{Sum of } \{(S^\circ\ Fe_3\ O_{4\ (S)}) + 4(S^\circ\ (H_{2\ (g)})\}]$$

$$- [\text{Sum of } \{3(S^\circ\ Fe_{(s)}) + 4(S^\circ H_2O_{(l)})\}]$$

$$\Delta S^\circ_{(reaction)} = [(146.4\ J/K) + 4(130.7\ J/K)] - [3(27.3\ J/K) + 4(69.9\ J/K)]$$

$$\Delta S^\circ_{(reaction)} = [(669.2\ J/K)] - [(361.5\ J/K)] = \mathbf{+307.7\ J/K} \text{ or } + \mathbf{0.308\ kJ/K}$$

IV ENTHALPY, ENTROPY, AND CHEMICAL SPONTANEITY

A conventional chemical reaction describes the process of converting a specific set of reactants to a specific set of products. The corresponding balanced equation, read from left-to-right, describes the **forward** reaction.

Any specific forward chemical reaction can **theoretically** have an opposite direction reaction. The products of the original forward reaction act as the reactants in the opposite direction; the reactants of the original forward reaction represent the products of the opposite reaction. Based on the original left-to-right form of the balanced equation, the original reaction is the forward reaction direction; the opposite reaction is the **reverse** reaction direction.

Many chemical processes are theoretically **reversible**; the forward and reverse reactions or processes can occur at the same time. Reactants and products can exchange roles; solids, liquids and gases can be interconvertible. It is valuable to determine the **expected direction** of a reversible chemical process. The **expected direction** for a reversible chemical reaction refers to determining which molecules will be favored as being products and which will be favored as being reactants under a defined set of conditions. **Chemical Spontaneity** refers to the direction (forward or reverse) a chemical reaction proceeds as a function of certain criteria.

A **spontaneous** chemical process is one which, once started, will continue on its own (that is, by itself) without any additional stimulus of any type, such as the addition of required energy; this is also termed **product favored.**

A **non-spontaneous** chemical process is one which, even once started, will **not** continue its own; this process requires continuous input of required energy in order to proceed; this is also termed **reactant favored.**

Spontaneity does **not** indicate **if** the reaction will occur or **how fast** a reaction will occur (the **rate** of reaction). It answers the question: "**If** a reaction starts, what is the probable result and what is the resultant energy change?"

Enthalpy and Chemical Spontaneity

Potential energy change is one main criterion for determining the spontaneous direction of a chemical process. Potential energy change is most conveniently measured as the enthalpy change (**ΔH**) for a reaction. One general measure of enthalpy is the difference in bond strengths between products and reactants. The value can be calculated from bond dissociation energies (BDE) (Chapter 18):

$$\Delta H^\circ_{(reaction)} = [\text{Sum } n_i \text{ BDE}_i, \text{(bonds broken)}] - [\text{Sum } n_i \text{ BDE}_i \text{ (bonds formed)}]$$

The value for ΔH° can also be calculated from ΔH°f (Chapter 19):

$$\Delta H^\circ_{(reaction)} = [\text{Sum } n_i\ \Delta H^\circ f_i \text{ (products)}] - [\text{Sum } n_i\ \Delta H^\circ f_i \text{ (reactants)}]$$

Based on the enthalpy (potential energy) criterion (i.e., ignoring other factors), all chemical processes are **spontaneous** in the direction of **potential energy decrease**: the direction of higher PE to lower PE as measured by enthalpy change. The spontaneous direction is indicated by a **negative** value for enthalpy change: **ΔH** = (–); this can be thought of as "down the energy hill." A related viewpoint is that a negative value for ΔH always **contributes** to the spontaneity of a process.

The corollary is that based on the enthalpy (potential energy) criterion, all chemical processes are **non-spontaneous** in the direction of **potential energy increase**: the direction of lower PE to higher PE; this is indicated by a **positive** value for enthalpy change: **ΔH** = (+). This can then be thought of as "up the energy hill." A related viewpoint is that a positive value for **ΔH** always **contributes** to the to the non-spontaneity of a process.

Entropy and Chemical Spontaneity

For any specific chemical process (system), the entropy is the second main criterion for determining the spontaneous direction. Entropy change can often be predicted and can usually be calculated through element or compound absolute molar entropies:

$$\Delta S^\circ_{(reaction)} = [\text{Sum } n_i\ S^\circ_i \text{ (products)}] - [\text{Sum } n_i\ S^\circ_i \text{ (reactants)}]$$

Based on the entropy criterion (i.e., ignoring other factors), all chemical processes are **spontaneous** in the direction of **entropy increase**; that is, in the direction of greater randomness or disorder. This direction is indicated by a **positive** value for entropy change (**ΔS** = +). A positive value for **ΔS** always **contributes** to the spontaneity of a process.

The direction of spontaneity as a function of entropy is driven by probability. Disorder or randomness (as compared to order) is the most **probable** arrangement for a system. Therefore, all chemical systems, based on the entropy criterion, always tend toward the direction of the most probable molecular configuration.

The corollary is that based on the entropy criterion, all chemical processes are **non-spontaneous** in the direction of entropy **decrease**; that is, in the direction of lower randomness or disorder (or equivalently in the direction of greater order), indicated by a **negative** value for entropy (**ΔS** = –). Thus, a negative value for **ΔS** always **contributes** to the non-spontaneity of a process. The relationship between a non-spontaneous process and a decrease in entropy can be recognized by noting that energy is required to produce order from disorder.

Combining Enthalpy and Entropy as Spontaneity Measures

ΔH = (–) contributes to a spontaneous process.
ΔH = (+) contributes to a non-spontaneous process.

ΔS = (+) contributes to a spontaneous process.
ΔS = (–) contributes to a non-spontaneous process.

Combining enthalpy and entropy to determine spontaneity produces four results:

1. **ΔH** = (–) contributes to spontaneity; **ΔS** = (+) contributes to spontaneity. **Result: ΔH** = (–) and **ΔS** = (+) indicates a process is **always** spontaneous **Always spontaneous** indicates that all possible values of enthalpy and entropy produce spontaneous criteria; both enthalpy and entropy show the same trend toward spontaneous.
2. **ΔH** = (+) contributes to **non**-spontaneity; **ΔS** = (–) contributes to **non**-spontaneity. **Result: ΔH** = (+) and **ΔS** = (–) indicates a process is **never** spontaneous. **Never spontaneous** indicates that there is **no** possible combination of enthalpy and entropy values, which will produce spontaneous criteria; both enthalpy and entropy show the same trend toward non-spontaneous.
3. **ΔH** = (–) contributes to spontaneity; **ΔS** = (–) contributes to **non**-spontaneity. **Result: ΔH** = (–) and **ΔS** = (–) indicates that the spontaneity depends on the balance between enthalpy and entropy.
4. **ΔH** = (+) contributes to **non**-spontaneity; **ΔS** = (+) contributes to spontaneity. **Result:** ΔH = (+) and ΔS = (+) indicates that the spontaneity depends on the balance between enthalpy and entropy.

Spontaneity and Temperature

The balance between the enthalpy contribution and the entropy contribution to a reaction or process is in part determined by the initial numerical values. The balance can also be greatly affected by **temperature.** The enthalpy change (ΔH) is nearly independent of temperature, but the enthalpy change (**ΔS**) always increases as temperature increases. Therefore, the contribution of the **entropy** change in a process continues to **increase** as the **temperature increases.**

For reactions (or processes) where the spontaneous direction depends on the balance between enthalpy and entropy, there will be (theoretically) a specific temperature above which the entropy will dictate the spontaneity result and below which the enthalpy will dictate the spontaneity result. In practice, this requires that a forward or reverse reaction is available at the required temperature.

High temperature favors the spontaneity direction indicated by the **entropy** contribution of a process. At a sufficiently high temperature, the direction (forward or reverse) indicated by the entropy (ΔS) will be spontaneous.

Low temperature favors the spontaneity direction indicated by the **enthalpy** contribution of a process. At a sufficiently low temperature, the direction (forward or reverse) indicated by the enthalpy (ΔH) will be spontaneous.

Example: Predict the sign of ΔS for the reaction then use the provided information to determine the spontaneity result. Choose from:

a) The process is always spontaneous.
b) The process is never spontaneous.
c) The process is spontaneous at a sufficiently high temperature.
d) The process is spontaneous at a sufficiently low temperature.

$$2C_8H_{18(l)} + 25O_{2(g)} \longrightarrow 16CO_{2(g)} + 18\,H_2O_{(g)} \quad \Delta H = -10{,}900 \text{ kJ/mol-rxn}$$

Twenty-seven moles of reactants (25 moles of gas) react to form 34 moles gaseous products. This represents the direction of fewer more complex molecules forming a greater number of smaller, simpler molecules with an increase in gas moles: **ΔS** = (+)

ΔH is shown as (–); ΔS = (+).

Result: ΔH = (–) and ΔS = (+) indicates a process is always spontaneous.

Example: Predict the sign of **ΔS** for the reaction then use the provided information to determine the spontaneity result. Choose from:

a) The process is always spontaneous.
b) The process is never spontaneous.
c) The process is spontaneous at a sufficiently high temperature.
d) The process is spontaneous at a sufficiently low temperature.

$$6\,CO_{2(g)} + 6\,H_2O_{(l)} \longrightarrow C_6H_{12}O_{6\,(aq)} + 6O_{2(g)} \quad \Delta H = +800 \text{ kJ/mol-rxn}$$

The reaction describes photosynthesis: 12 moles of gas reactants react to form 7 moles of products (6 moles of gas). This represents the direction of a greater number of smaller, simpler molecules forming fewer more complex molecules with a decrease in gas moles: **ΔS** = (–)

ΔH is shown as (+); ΔS = (–).

Result: ΔH = (+) and ΔS = (–) indicates a process is never spontaneous.

Example: Predict the sign of **ΔS** for the reaction then use the provided information to determine the spontaneity result. Choose from:

a) The process is always spontaneous.
b) The process is never spontaneous.
c) The process is spontaneous at a sufficiently high temperature.
d) The process is spontaneous at a sufficiently low temperature.

$$2\,Fe_{(S)} + 3O_{2\,(g)} \longrightarrow 2\,Fe_2O_{3(S)} \quad \Delta H = -400 \text{ kJ/mol-rxn}$$

Five moles of reactants (3 moles of gas) react to form 2 moles of solid product. This represents the direction of a greater number of smaller, simpler elements forming fewer more complex compounds with a decrease in gas moles: **ΔS** = (–)

ΔH is shown as (–); ΔS = (–).

Result: ΔH = (–) and ΔS = (–) indicates that the spontaneity depends on the balance between enthalpy and entropy. When the spontaneity depends on the balance, temperature is the deciding factor. ΔH = (–) contributes to a spontaneous reaction but ΔS = (–) contributes to non-spontaneous reaction. The contribution of ΔH is maximized at low temperature: the process is spontaneous at a sufficiently low temperature.

Example: Predict the sign of **ΔS** for the process then use the provided information to determine the spontaneity result. Choose from:

a) The process is always spontaneous.
b) The process is never spontaneous.

c) The process is spontaneous at a sufficiently high temperature.
d) The process is spontaneous at a sufficiently low temperature.

$$NH_4NO_{3(s)} + H_2O_{(l)} \longrightarrow NH_4NO_{3(aq)} \quad \Delta H = +\,28\ \text{kJ/mol}$$

The process indicates the formation of a solution from pure components: **ΔS** = (+)

ΔH is shown as (+); ΔS = (+).

Result: ΔH = (+) and ΔS = (+) indicates that the spontaneity depends on the balance between enthalpy and entropy. ΔH = (+) contributes to a non-spontaneous process but ΔS = (+) contributes to spontaneous process. The contribution of ΔS is maximized at high temperature: the process is spontaneous at a sufficiently high temperature.

V REACTION SPONTANEITY AND FREE ENERGY

Reversible reactions are those that can have both the forward and reverse reactions occurring simultaneously. The general reversible reaction (Chapter 24) is shown as the equation:

$$\mathbf{a\,A + b\,B \rightleftharpoons c\,C + d\,D}$$

The uppercase letters (**A, B, C, D**) represent the general reactants and products. The lower-case letters (**a, b, c, d**) represent the stoichiometric coefficients of A, B, C, D respectively in the balanced equation.

Determining a **quantitative** measurement of reaction spontaneity requires combining the functions of enthalpy, entropy, temperature, plus reactant and product concentrations. The calculation terms are the standard enthalpy (**ΔH°**) of the reaction, the standard entropy (**ΔS°**) for the reaction, the temperature (**T**), and the starting concentrations of all compounds (**[A], [B], [C], [D]**, etc.).

DEFINITION OF FREE ENERGY (ΔG)

The second law of thermodynamics provides a **general** definition for spontaneity under all possible circumstances. The entropy of the **universe** must increase for all processes. Since non-spontaneous processes based on the **universe** do not occur, all spontaneous processes must show an increase in the entropy of the universe:

$$\Delta S_{(universe)} > 0;\ \Delta S_{(universe)} = (+).$$

The derivation for free energy starts with conditions of constant temperature and pressure. Temperature is measured in Kelvins.

$$\Delta S_{(universe)} = \Delta S_{(system)} + \Delta S_{(surroundings)}$$

For a system at constant temperature, $\Delta S_{(surroundings)} = \dfrac{q_{(surroundings)}}{T}$

$$q_{(surroundings)} = -q_{(system)};\quad \Delta S_{(surroundings)} = \frac{-q_{(system)}}{T}$$

For a system at constant pressure, $q_{(system)} = \Delta H_{(system)}$

$$\Delta S_{(surroundings)} = \frac{-\Delta H_{(system)}}{T}$$

$$\Delta S_{(universe)} = \Delta S_{(surroundings)} + \Delta S_{(system)}$$

$$\Delta S_{universe} = \frac{-\Delta H_{system}}{T} + \Delta S_{system}; \quad \text{multiply both sides by} (-T)$$

$$-T\Delta S_{(universe)} = \Delta H_{(system)} - T\Delta S_{(system)}$$

$-T\Delta S_{(universe)}$ is defined as Gibb's free energy: $\mathbf{\Delta G_{(system)}}$

Free energy incorporates the values for enthalpy (ΔH), entropy (ΔS), and temperature in Kelvins (T in K). The dependence on product/reactant concentration is included in the total entropy. The general form of the equation is:

$$\mathbf{\Delta G = \Delta H - T\Delta S}$$

The requirement for a spontaneous process is $\Delta S_{(universe)} = (+)$. Since temperature in Kelvin is always positive, a positive value for $\Delta S_{(universe)}$ requires that the term $(-T\Delta S_{(universe)}) = \Delta G$ must be negative.

A forward reaction is **spontaneous** if the free energy change (ΔG) for the direction as written is **negative**: ΔG = (–). This means that the free energy of the system **decreases**: the reaction (based on the definition of free energy) is energetically "downhill." A spontaneous reaction need not proceed at a measurable rate; rate is a function of the path through which the reaction occurs and can be influenced but does not directly depend on the free energy change (a state function).

A forward reaction is **non-spontaneous** if the free energy change (ΔG) is **positive**: **ΔG** = (+). This means that the free energy of the system **increases**: the reaction (based on the definition of free energy) is energetically "uphill." A non-spontaneous reaction can still occur if energy is supplied to the reaction system from the surroundings. However, for this to occur, a path (sequence of bond-making and bond-breaking steps) must be available for energy input even when sufficient energy is provided.

Free energy, based on the state functions of enthalpy and entropy, is also a **state function**; measurements for the forward and reverse reactions must be symmetrical. If the forward reaction is spontaneous (ΔG = (–)), the reverse reaction must be non-spontaneous (ΔG = (+)). If the forward reaction is non-spontaneous (ΔG = (+)), the reverse reaction must be spontaneous

(ΔG = (–)).

Based on the symmetry of calculations for ΔH and ΔS, it is always true that:

$$\mathbf{\Delta G_{(forward\ reaction)} = -\Delta G_{(reverse\ reaction)}}$$

Enthalpy, Entropy, and Temperature Contributions to Free Energy

The conceptual relationships between spontaneity and the signs for enthalpy and entropy can be confirmed through an analysis of the free energy equation.

Based on the equation (ΔG = ΔH – T ΔS), a **negative** value for reaction enthalpy change contributes to a **negative** value for **ΔG: ΔH** = (–) contributes to **ΔG** = (–), which contributes to reaction spontaneity.

The equation for free energy (**ΔG** = **ΔH** – **T ΔS**) shows a negative sign in front of the complete term for entropy (including the temperature): **[–T ΔS]**. Since temperature (T) in K is always a positive value, the **[–T ΔS]** term must be **negative** for all **positive** values of [ΔS]. A **positive** value for reaction entropy change contributes to a **negative** value for ΔG: **ΔS** = (+) contributes to **ΔG** = (–), which contributes to reaction spontaneity.

The entropy term is the only variable in the general equation for ΔG, which is multiplied by the temperature; the greater the value of the temperature, the greater the contribution of the entropy term [$-T\Delta S$] to the numerical value of ΔG.

The general trends for temperature and spontaneity described previously can be confirmed. High temperature favors the spontaneity direction indicated by the entropy contribution because a higher numerical value for temperature results in a relatively higher numerical value for the [$-T\Delta S$] term compared to the Δ**H** term.

Low temperature favors the spontaneity direction indicated by the enthalpy contribution because a lower numerical value for temperature results in a relatively lower numerical value for the [$-T\Delta S$] term compared to the Δ**H** term.

Standard Free Energy ($\Delta G°$)

Calculations for free energy are based on all possible variables of enthalpy, entropy, temperature, and concentrations. The value and sign of AG is usually calculated under varying specific combinations of conditions.

Standard conditions for free energy calculations are defined as a temperature of 298 **K**, pressure equal to one atmosphere and concentrations of all components equal to exactly **1 Molar** (gas concentrations may be expressed as partial pressures). The thermodynamic function that determines spontaneity under standard conditions is termed the **standard free energy**; symbol = $\Delta G°$.

The equation for standard free energy includes the **standard** enthalpy (Δ**H°**) and **standard** entropy (Δ**S°**); in this case, the temperature (**T**) is **298 K**:

$$\Delta G^0 = \Delta H° - T\Delta S^0$$

Values for **standard free energies of formation (ΔG°f)** are defined in a manner similar to standard enthalpies of formation (Δ**H°f**, Chapter 19); values are listed in the same tables. The standard free energy of a reaction at 298 K can be calculated from values for the standard free energies of formation ($\Delta G°f$) for products and reactants following a similar equation:

$$\Delta G°_{(reaction)} = [\text{Sum } n_i\, \Delta G°f_i \text{ (products)}] - [\text{Sum } n_i\, \Delta G°f_i \text{ (reactants)}]$$

	$\Delta G°f$		$\Delta G°f$
($C_2H_{2\,(g)}$)	+ 209 kJ	($C_2H_{6\,(g)}$)	−33 kJ
($H_{2\,(g)}$)	0 kJ	($C_2H_5OH_{\,(l)}$)	−174.8 kJ
($O_{2\,(g)}$)	0 kJ	($CO_{2\,(g)}$)	−394.4 kJ
($H_2O_{\,(l)}$)	−237.1 kJ		

Example: Calculate $\Delta G°_{(reaction)}$ for the following reaction using table values for $\Delta G°f$.

$$C_2H_{2(g)} + 2H_{2(g)} \longrightarrow C_2H_{6(g)}$$

Step (**1**): From the tables, identify all required values for $\Delta G°f$.

Reactants: $\Delta G°f\ (C_2H_2(g)) = +209$ kJ; $\Delta G°f\ (H_{2(g)}) = 0$ kJ

Products: $\Delta G°f\ (C_2H_{6(g)}) = -33$kJ

Step (**2**): Complete the calculation using the equation given:

$$\Delta G°(\text{reaction}) = [\text{Sum of } [\Delta G°f\ (C_2H_{6\,(g)})]$$

$$- [\text{Sum of } \{[(\Delta G°f(C_2H_{2(g)})) + 2(\Delta G°f\ (H_{2\ (g)})\}]$$

$$\Delta G°\ (\text{reaction}) = [(-33\ kJ)] - [(209\ kJ) + 2(0\ kJ)]$$

$$\Delta G°\ (\text{reaction}) = [(-33\ kJ)] - [(209\ kJ)] = \mathbf{-242\ kJ/mole\text{-}reaction}$$

Example: Calculate ΔG°(reaction) for the following reaction using table values for ΔG°f.

$$C_2H_5OH_{(l)} + 3O_{2(g)} \longrightarrow 2CO_{2(g)} + 3H_2O_{(l)}$$

Step (**1**): From the tables, identify all required values for ΔG°f.

Reactants: $\Delta G°f\ (C_2H_5OH_{\ (l)}) = -174.8$ kJ; $\Delta G°f(O_{2(g)}) = 0$ kJ

Products: $\Delta G°f\ (CO_{2\ (g)}) = -394.4$ kJ; $\Delta G°f\ (H_2O_{\ (l)}) = -237.1$ kJ

Step (**2**): Complete the calculation using the equation given:

$$\Delta G°(\text{reaction}) = [\text{Sum of } \{2(\Delta G°f\ (CO_{2\ (g)}) + 3(\Delta G°f\ H_2O_{\ (l)})\}]$$

$$- [\text{Sum of } \{\Delta G°f(C_2H_5OH_{(l)} + 3(\Delta G°f\ O_{2\ (g)})\}]$$

$$\Delta G°\ (\text{reaction}) = [2(-394.4\ kJ) + 3(-237.1\ kJ] - [(-174.8\ kJ) + 3(0\ kJ)]$$

$$\Delta G°\ (\text{reaction}) = [(-1500.1\ kJ)] - [(-174.8\ kJ)] = \mathbf{-1325.3\ kJ/mole\text{-}reaction}$$

An alternative method of calculation involves the separate calculation of enthalpy and entropy using T = 298 K in the equation for standard free energy.

Example: Calculate $\Delta G°_{(reaction)}$ for the reaction at 298 K from ΔH° and ΔS° values.

$$C_2H_{2\ (g)} + 2H_{2\ (g)} \longrightarrow C_2H_{6\ (g)}$$

Calculation of $\Delta H°_{(reaction)}$ from tables of ΔH°f was described in Chapter 19.

Step (**1**): Calculate the value of $\Delta H°_{(reaction)}$.

Reactants: $\Delta H°f\ (C_2H_{2\ (g)}) = +227$ kJ; $\Delta H°f\ (H_{2\ (g)}) = 0$ kJ

Products: $\Delta H°f\ (C_2H_{6\ (g)}) = -85$ kJ

$$\Delta H°_{(reaction)} = [\text{Sum of } \Delta H°f\ (C_2H_{6\ (g)})] - [\text{Sum of } \{(\Delta H°f\ (C_2H_{2\ (g)})) + 2(\Delta H°f(H_{2\ (g)})\}]$$

$$\Delta H°_{(reaction)} = [(-85\ kJ)] - [(227\ kJ) + 2(0\ kJ)]$$

$$\Delta H°_{(reaction)} = \mathbf{-312\ kJ/mole\text{-}reaction}$$

Step (**2**): Calculate the value for $\Delta S°_{(reaction)}$ from tables of S°.

This calculation was shown previously; the result was:

$$\Delta S°_{(reaction)} = [\text{Sum of } S°\ (C_2H_{6\ (g)})] - [\text{Sum of}\{(S°(C_2H_{2\ (g)})) + 2(S°\ (H_{2\ (g)})\}]$$

$$\Delta S^\circ_{(reaction)} = [230\ J/K)] - [(201\ J/K) + 2(131\ J/K)]$$

$$\Delta S^\circ_{(reaction)} = \mathbf{-233\ J/K} \text{ or } \mathbf{-0.233\ kJ/K}$$

Enthalpy and entropy must be expressed in the same energy units. The conversion of all entropy values to units of kJ/K from J/K was performed immediately to forestall unit inconsistency.

Step (3): Complete the calculation for ΔG°_T using the given equation and the stated temperature. $\mathbf{\Delta G^\circ = \Delta H^\circ - T\Delta S^\circ}$

$$\Delta G^\circ_{298} = [(-312\ kJ)] - [(298\ K)(-0.233\ kJ/K)] = [(-312 kJ)] - [-69.4 kJ]$$

$$= \mathbf{-242.6\ kJ/mole\text{-}rxn}$$

This value agrees with the one from the $\Delta G^\circ f$ method.

Example: Calculate $\Delta G^\circ_{(reaction)}$ for the reaction at 298 K from ΔH° and ΔS° values.

$$C_2H_5OH_{(l)} + 3O_{2(g)} \longrightarrow 2CO_{2(g)} + 3H_2O_{(l)}$$

Calculation of $\Delta H^\circ_{(reaction)}$ from tables of $\Delta H^\circ f$ was described in Chapter 19.

Step (1): Calculate the value of $\Delta H^\circ_{(reaction)}$.

Reactants: $\Delta H^\circ f\ (C_2H_5OH_{(l)}) = -277.7$ kJ;$\Delta H^\circ f(O_{2\,(g)}) = 0$ kJ

Products: $\Delta H^\circ f\ (CO_{2\,(g)}) = -393.5$ kJ;$\Delta H^\circ f(H_2O_{(l)}) = -285.8$ kJ

$$\Delta H^\circ(\text{reaction}) = [\text{Sum of } \{2(\Delta H^\circ f\ (CO_{2\,(g)}) + 3(\Delta H^\circ f\ (H_2O_{(l)})\}]$$
$$- [\text{Sum of } \{\Delta H^\circ f\ (C_2H_5OH_{(l)}) + 3(\Delta H^\circ f\ O_{2\,(g)})\}]$$

$$\Delta H^\circ\ (\text{reaction}) = [2(-393.5\ kJ) + 3(-285.8\ kJ] - [(-277.7\ kJ) + 3(0\ kJ)]$$

$$\Delta H^\circ\ (\text{reaction}) = [(-1644.4\ kJ)] - [(-277.7\ kJ)]$$

$$\Delta H^\circ_{(reaction)} = \mathbf{-1366.7\ kJ/mole\text{-}reaction}$$

Step (2): Calculate the value for $\Delta S^\circ_{(reaction)}$ from tables of S°.

This calculation was shown previously; the result was:

$$\Delta S^\circ_{(reaction)} = [\text{Sum of } \{2(S^\circ\ CO_{2\,(g)}) + 3(S^\circ(H_2O_{(l)})\}]$$

$$- [\text{Sum of } \{(S^\circ\ C_2H_5OH_{(l)}) + 3(S^\circ\ O_{2\,(g)})\}]$$

$$\Delta S^\circ_{(reaction)} = [2(213.7\ J/K) + 3(69.9\ J/K)] - [(160.7\ J/K) + 3(205.1\ J/K)]$$

$$\Delta S^\circ_{(reaction)} = [637.1\ J/K)] - [(776\ J/K)] = \mathbf{-138.9\ J/K} \text{ or } \mathbf{-0.1389\ kJ/K}$$

Step (**3**): Complete the calculation for $\Delta G°_T$ using the given equation and the stated temperature.

$$\Delta G° = \Delta H° - T\,\Delta S°$$

$$\Delta G°_{298} = [(-1366.7\text{ kJ})] - [(298\text{ K})(-0.1389\text{ kJ/K})] = [(-1366.7\text{ kJ})] - [-41.4\text{kJ}]$$

$$= \mathbf{-1325.3\ kJ/mole\text{-}rxn}$$

This value agrees with the one from the $\Delta G°f$ method.

CALCULATION OF $\Delta G°$ AT VARIABLE TEMPERATURES

Use of standard free energies of formation ($\Delta G°f$) limits the temperature to 298 K. The equation for $\Delta G°$ can be adapted to use **temperature** as a variable while keeping the standard conditions of one atmosphere of pressure and all initial concentrations at exactly 1 Molar (or partial pressures of 1 atmosphere).

$$\Delta G°_T = \Delta H° - T\,\Delta S°$$

In the adapted equation, $\mathbf{\Delta G°_T}$ is the standard free energy under standard conditions of 1 molar concentration and 1 atmosphere pressure, but at a **variable** temperature specified by the subscript "**T**." The values for standard enthalpy and standard entropy at 298 K are used in the equation: $\mathbf{\Delta H°}$ means $\mathbf{\Delta H°_{298}}$ and $\mathbf{\Delta S°}$ means $\mathbf{\Delta S°_{298}}$.

The key value of this equation is that the standard free energy at **any** temperature (within a certain range) can be found from enthalpy and entropy changes measured at 298 K. Enthalpy is only a very weak function of temperature; $\Delta H°$ at any temperature is approximately equal to $\Delta H°_{298}$. Although entropy (S) is a strong function of temperature, entropy **changes** (**$\Delta S°$**) tend to remain approximately constant for both reactants and products as reaction temperature changes (within a certain range); $\Delta S°_{(reaction)}$ at any temperature is approximately equal to $\Delta S°_{298}$.

In contrast, the complete equation, $\Delta G°_T = \Delta H° - T\Delta S°$, shows that the standard free energy measured at different specified potential temperatures remains a strong function of temperature. This is based on the entropy term [$-T\Delta S°$] in which [$\Delta S°$] itself is multiplied by T in the equation; the temperature affects the value of $\Delta G°_T$ and thus the spontaneity of a reaction. These effects are especially important if the number of **gas**-phase molecules change. The effect of entropy is much smaller when the number of molecules of reactant and product remain the same and/or if gases are not involved. When the numerical value of $\Delta S°$ is large, the value of $\Delta G°_T$ can change dramatically with temperature variation.

Example: Calculate $\Delta G°_{(reaction)}$ for the following reaction at 400 K, 800 K, and 1400 K using the equation for $\Delta G°_T$.

$$C_2H_{2(g)} + 2H_{2(g)} \longrightarrow C_2H_{6(g)}$$

Step (**1**): The value for $\Delta H°_{(reaction)}$ was completed in a previous example.

$$\mathbf{\Delta H°_{(reaction)} = -312\ kJ/mole\text{-}reaction}$$

Step (**2**): The value for $\Delta S°_{(reaction)}$ was completed in a previous example.

$$\mathbf{\Delta S°_{(reaction)} = -233\ J/K}\text{ or }\mathbf{-0.233\ kJ/K}$$

Step (**3**): Complete the calculation for $\Delta G°_T$ using the given equation and the stated temperature. $\Delta G°_T = \Delta H° - T\,\Delta S°$; entropy is in units of kJ/K.

$$\Delta G°_{400} = (-312\text{ kJ}) - [(400\text{ K})(-0.233\text{ kJ/K})] = (-312\text{ kJ}) - [-93.2\text{ kJ}]$$

$$= \mathbf{-218.8\ kJ/mole\text{-}rxn}$$

$$\Delta G°_{800} = (-312\text{ kJ}) - [(800\text{ K})(-0.233\text{ kJ/K})] = (-312\text{ kJ}) - [-186.4\text{ kJ}]$$

$$= \mathbf{-125.6\ kJ/mole\text{-}rxn}$$

$$\Delta G°_{1400} = (-312\text{ kJ}) - [(1400\text{ K})(-0.233\text{ kJ/K})] = (-312\text{ kJ}) - [-326.2\text{ kJ}]$$

$$= \mathbf{+\,14.2\ kJ/mole\text{-}rxn}$$

Whenever $\Delta H°$ is (−) with $\Delta S°$ = (−) or $\Delta H°$ is (+) with $\Delta S°$ = (+), the temperature at which $\Delta G°_{(reaction)}$ changes sign can be calculated; this corresponds to the temperature at which the spontaneous direction changes. The value of $\Delta G°_{(reaction)}$ changes sign when its value goes through zero; the corresponding equation can be solved for the temperature at which $\Delta G°_{(reaction)} = 0$.

$$\Delta G°_T = \Delta H° - T\Delta S°;\ \text{for } \Delta G°_T = 0\text{: } \mathbf{0 = \Delta H° - T\,\Delta S°}$$

$$\text{Solve for T: } \mathbf{T_{\Delta G\to 0}} = \frac{\mathbf{\Delta H°}}{\mathbf{\Delta S°}}$$

Example: For the following reaction, calculate the temperature at which $\Delta G°_{(reaction)}$ changes sign; that is, the temperature at which $\Delta G°_T = 0$.

$$C_2H_{2(g)} + 2H_{2(g)} \longrightarrow C_2H_{6(g)}$$

The values for $\Delta H°$ and $\Delta S°$ for the reaction are shown in the previous example.

$$\Delta H°_{(reaction)} = -\,312\text{ kJ/mole-reaction}$$

$$\Delta S°_{(reaction)} = -\,0.233\text{ kJ/K}$$

$$\mathbf{T_{\Delta G\to 0}} = \frac{\mathbf{\Delta H^0}}{\mathbf{\Delta S^0}} = \frac{(-312\text{ kJ})}{(-0.233\text{ kJ/K})} = 1339\text{K};\quad \Delta G°_{1339} = 0$$

Example: Calculate $\Delta G°_{(reaction)}$ for the following reaction at 400 K (compare to the value calculated for 298 K) and then calculate the **hypothetical** temperature at which the reaction becomes non-spontaneous.

$$C_2H_5OH_{(l)} + 3O_{2(g)} \longrightarrow 2CO_{2(g)} + 3H_2O_{(l)}$$

Step (**1**): The value for $\Delta H°_{(reaction)}$ was completed in a previous example.

$$\mathbf{\Delta H°_{(reaction)} = -1366.7\ kJ/mole\text{-}reaction}$$

Step (**2**): The value for $\Delta S°_{(reaction)}$ was completed in a previous example.

$$\mathbf{\Delta S°_{(reaction)} = -138.9\ J/K} \text{ or } \mathbf{-0.1389\ kJ/K}$$

Step (**3**): Complete the calculation for $\Delta G°_T$ using the given equation and the stated temperature. $\mathbf{\Delta G°_T = \Delta H° - T\ \Delta S°}$; entropy is in units of kJ/K.

$$\mathbf{\Delta G°_{400}} = (-1366.7\ kJ) - [(400\ K)(-0.1389\ kJ/K)] = (-1366.7\ kJ) - [-55.6\ kJ]$$
$$= \mathbf{-1311.1\ kJ/mole\text{-}rxn}$$

The value of $\Delta S°_{(reaction)}$ is negative; therefore, as expected, the value for $\Delta G°_{(reaction)}$ becomes less negative with increasing temperature (less spontaneous).

With a highly negative value for standard free energy, this reaction is an example of one that is not readily reversible; the reaction would not exist at the large value for the temperature at which $\Delta G°$ goes to zero. A hypothetical result can be calculated:

$$T_{\Delta G \to 0} = \frac{\Delta H°}{\Delta S°} = \frac{(-1366.7\ kJ)}{(-0.1389\ kJ/K)} = \mathbf{9839\,K}; \quad \Delta G°_{9839} = 0$$

Example: Calculate $\Delta G°_{(reaction)}$ for the following reaction at 298 K and 400 K.

$$2C_6H_{6\,(l)} + 9O_{2(g)} \longrightarrow 12CO_{(g)} + 6H_2O_{(g)}$$

Calculation of $\Delta H°_{(reaction)}$ from tables of $\Delta H°f$ was described in Chapter 19.

Step (**1**): Calculate the value of $\Delta H°_{(reaction)}$

Reactants: $\Delta H°f\ (C_6H_{6\,(l)}) = +49.0$ kJ; $\Delta H°f\ (O_{2\,(g)}) = 0$ kJ

Products: $\Delta H°f\ (CO_{(g)}) = -110.5$ kJ; $\Delta H°f\ (H_2O_{(g)}) = -241.8$ kJ

$$\Delta H°_{(reaction)} = [\text{Sum of } \{12(\Delta H°f\ CO_{(g)}) + 6(\Delta H°f\ H_2O_{(g)})\}]$$
$$- [\text{Sum of } \{2(\Delta H°f\ C_6H_{6\,(l)}) + 9(\Delta H°f\ O_{2\,(g)})\}]$$

$$\Delta H°_{(reaction)} = [12(-110.5\ kJ) + 6(-241.8\ kJ] - [2(+49.0\ kJ) + 9(0\ kJ)]$$

$$\Delta H°_{(reaction)} = [(-2776.8\ kJ)] - [(98.0\ kJ)]$$

$$\mathbf{\Delta H°_{(reaction)} = -2874.8\ kJ/mole\text{-}reaction}$$

Step (**2**): Calculate the value for $\Delta S°_{(reaction)}$ from tables of S°.

This calculation was shown previously; the result was:

$$\Delta S°_{(reaction)} = [\text{Sum of } \{12(S°\ CO_{(g)}) + 6(S°\ (H_2O_{(g)})\}]$$
$$- [\text{Sum of } \{2(S°\ C_2H_5OH_{(l)}) + 9(S°\ O_{2\,(g)})]$$

$$\Delta S°_{(reaction)} = [12(197.7\ J/K) + 6(188.8\ J/K)] - [2(173.3\ J/K) + 9(205.1\ J/K)]$$

$$\Delta S°_{(reaction)} = [3505.2\ J/K)] - [(2192.5\ J/K)] = \mathbf{+1312.7\ J/K} \text{ or } + \mathbf{1.313\ kJ/K}$$

Step (**3**): Complete the calculation for $\Delta G°_T$ using the given equation and the stated temperature. $\mathbf{\Delta G°_T = \Delta H° - T\Delta S°}$; entropy is in units of kJ/K.

$$\mathbf{\Delta G°_{298}} = (-2874.8\ kJ) - [(298\ K)(+\ 1.313\ kJ/K)] = (-2874.8\ kJ) - [+391.3\ kJ]$$

$$= \mathbf{-3266\ kJ/mole\text{-}\ rxn}$$

$$\mathbf{\Delta G°_{400}} = (-2874.8\ kJ) - [(400\ K)(-1.313\ kJ/K)] = (-2874.8\ kJ) - [+525.2\ kJ]$$

$$= \mathbf{-3400\ kJ/mole\text{-}rxn}$$

Reaction spontaneity increases with temperature based on a positive value for $\Delta S°_{(reaction)}$. A negative value for $\Delta H°$ and a positive value for $\Delta S°$ indicates that the reaction is always spontaneous.

Example: Calculate $\Delta G°_{(reaction)}$ for the following reaction at 298 K and 73 K; then calculate the **hypothetical** temperature at which the reaction becomes non-spontaneous.

$$3Fe_{(s)} + 4H_2O_{(l)} \longrightarrow Fe_3O_{4(s)} + 4H_{2(g)}$$

Calculation of $\Delta H°_{(reaction)}$ from tables of $\Delta H°f$ was described in Chapter 19.

Step (**1**): Calculate the value of $\Delta H°_{(reaction)}$.

Reactants: $\Delta H°f\ (Fe_{(s)}) = 0$ kJ; $\Delta H°f\ (H_2O_{(l)}) = -\ 285.8$ KJ

Products: $\Delta H°f\ (Fe_3O_{4\ (s)}) = -\ 1118.4$ kJ; $\Delta H°f\ (H_{2\ (g)}) = 0$ kJ

$$\Delta H°_{(reaction)} = [\text{Sum of } \{(\Delta H°f\ Fe_3O_{4\ (s)}) + 4(\Delta H°f\ (H_{2\ (g)})\}]$$

$$- [\text{Sum of } \{3(\Delta H°f\ Fe_{(s)}) + 4(\Delta H°fH_2O_{\ (l)})]$$

$$\Delta H°_{(reaction)} = [(-\ 1118.4\ kJ) + 4(0\ kJ)] - [3(0\ kJ) + 4(-285.8\ kJ)]$$

$$\Delta H°_{(reaction)} = [(-1118.4\ kJ)] - [(-1143.2\ kJ)] = \mathbf{+24.8\ kJ/mole\text{-}reaction}$$

Step (**2**): Calculate the value for $\Delta S°_{(reaction)}$ from tables of S°.
This calculation was shown previously; the result was:

$$\Delta S°_{(reaction)} = [\text{Sum of } \{(S°\ Fe_3O_{4\ (s)}) + 4(S°\ (H_{2\ (g)})\}]$$

$$- [\text{Sum of } \{3(S°\ Fe_{\ (s)}) + 4(S°\ H_2O_{\ (l)})]$$

$$\Delta S°_{(reaction)} = [(146.4\ J/K) + 4(130.7\ J/K)] - [3(27.3\ J/K) + 4(69.9\ J/K)]$$

$$\Delta S°_{(reaction)} = [(669.2\ J/K)] - [(361.5\ J/K)] = + \mathbf{307.7\ J/K} \text{ or } + \mathbf{0.308\ kJ/K}$$

Step (**3**): Complete the calculation for $\Delta G°_T$ using the given equation and the stated temperature. $\mathbf{\Delta G°_T = \Delta H° - T\ \Delta S°}$; entropy is in units of kJ/K.

$$\Delta G^\circ_{298} = (+24.8\text{ kJ}) - [(298\text{ K})(-0.308\text{ kJ/K})] = (+24.8\text{ kJ}) - [+91.8\text{kJ}]$$

$$= \mathbf{-67.0\ kJ/mole\text{-}rxn}$$

The reaction is spontaneous at 298 K based on a positive value for $\Delta S^\circ_{(reaction)}$; the value for $\Delta G^\circ_{(reaction)}$ becomes more negative with increasing temperature (more spontaneous).

$$\Delta G^\circ_{73} = (+24.8\text{ kJ}) - [(73\text{ K})(-0.308\text{ kJ/K})] = (+24.8\text{ kJ}) - [+22.5\text{ kJ}]$$

$$= \mathbf{+2.3\ kJ/mole\text{-}rxn}$$

At a sufficiently low temperature, theoretically, the spontaneity direction indicated by the enthalpy will dictate the result; the positive value for ΔH° contributes to non-spontaneity. The reaction is not practical at these temperatures

$$T_{\Delta G \to 0} = \frac{\Delta H^\circ}{\Delta S^\circ} = \frac{(+24.8\text{ kJ})}{(+0.308\text{ kJ/K})} = \mathbf{80.5\ K;}\quad \Delta G^\circ_{80.5} = 0$$

VI NON-STANDARD FREE ENERGY (ΔG) AND CONCENTRATIONS

Standard free energy, even at variable temperatures (ΔG°_T), describes reactions, including reversible reactions, under specific concentration conditions. All reactant and product molecules are present in the mixture at the identical concentrations of exactly **1 Molar** (gas concentrations may be expressed as 1 atmosphere partial pressures).

A complete analysis of spontaneity requires the consideration of **variable concentrations** of product and reactant species; the calculation of free energy under variable concentrations is termed the **non-standard free energy** (Δ**G** without the 0 superscript). The concentration dependence for the non-standard free energy is produced through the **non-standard concentration entropy** of the reaction product/reactant mixture. Entropy changes for concentrations were stated as concept **(2)**:

$$\text{pure substances} \xrightarrow[\text{solution or mixture formation}]{\text{Entropy increase } (\Delta S = +)} \text{solution or mixture}$$

$$\textbf{high}\text{ concentration solution} \xrightarrow[\text{dilution of solution}]{\text{Entropy increase } (\Delta S = +)} \textbf{low}\text{ concentration solution}$$

Specific concentrations of products and reactants depend on all possible conditions of the reaction mixture for a reversible reaction. The general trend for concentration effects on spontaneity can be observed.

The contribution of the non-standard concentration entropy only (i.e., ignoring standard enthalpy and standard entropy), indicates that the spontaneous direction of a reversible reaction is to consume (use as reactants) the set of molecules in greater concentration to produce (form as products) the set of molecules in lower concentration.

The net spontaneity result of the non-standard concentration entropy is to "even out" all concentrations: to favor the direction that decreases the higher concentrations and increases the lower concentrations.

Calculating Non-Standard Free Energy (ΔG_T)

Non-standard free energy (ΔG_T) takes into account the variable concentration entropy; it applies to all non-standard conditions, including concentrations other than 1 Molar (or gases at partial pressures other than 1 atmosphere).

The general reversible reaction describing equilibrium (Chapter 23) was:

$$\mathbf{a\,A + b\,B \rightleftarrows c\,C + d\,D}$$

A, B, C, D are general reactants and products; **a, b, c, d** are the stoichiometric coefficients of A, B, C, D, respectively, in the balanced equation.

$$Q_{(\text{forward reaction})} = \frac{[\mathbf{C}]^c\,[\mathbf{D}]^d}{[\mathbf{A}]^a\,[\mathbf{B}]^b} \propto \frac{[\textbf{products}]}{[\textbf{reactants}]}$$

A specific ratio of products to reactants, termed the **concentration quotient (Q)**, is defined for any actual values for concentrations of components at any point in the reaction. The form of the ratio expression uses the specific concentration of each species, usually as **molarity (M)**, raised to the power of its coefficient in the balanced equation.

The numerical value of the **non-standard** $\mathbf{\Delta G_T}$ under all circumstances is given by the general equations:

$$\mathbf{\Delta G_T = \Delta G^\circ_T + RT\,In(Q)}$$

$$\mathbf{\Delta G_T = \Delta G^\circ_T + 2.303\,RT\,log_{10}(Q)}$$

$\mathbf{\Delta G^\circ_T}$ = standard free energy under standard conditions of pressure and concentration but at a variable temperature specified by "T." **R** = **gas** constant measured in units of kJ/mole-K = **0.008314 kJ/mole-K; T** is temperature in units of **K**. The term **"In"** is the natural log function. ΔG_T for a reaction system is dependent on the actual **starting** concentrations of all reactant and product species.

Note that if all concentrations are exactly 1 M, the calculation for the ratio **Q** becomes **exactly 1**. Since $In(1) = 0$ or $\log_{10}(1) = 0$, the expected equivalence results: $\Delta G_T = \Delta G^\circ_T + RT \times (0)$ or $\mathbf{\Delta G_T = \Delta G^\circ_T}$ when all concentrations = 1 M

$$Q \propto \frac{[\textbf{products}]}{[\textbf{reactants}]}$$

If the general concentrations of reactants [reactants] are greater than the general concentrations of products [products], the numerical value defined by the ratio **Q** is **less** than **1** and RTIn(Q) or RTlog(Q) is **negative**; ΔG_T for the forward reaction becomes more negative and more spontaneous. If concentration of [products] is greater than [reactants], the numerical value defined by the ratio **Q** is **greater** than **1** and RTIn(Q) or RTlog(Q) is **positive**; ΔG_T for the forward reaction becomes more positive and less spontaneous (reverse reaction becomes more spontaneous).

Example: Use the value for standard free energy (ΔG°_T) calculated in a previous example to calculate the value for non-standard free energy (ΔG_T) under the stated conditions for the following reaction at a temperature of 400 K.

$$C_2H_{2\,(g)} + 2\,H_{2(g)} \longrightarrow C_2H_{6(g)}$$

Partial pressure of C_2H_2 = 0.10 atm; partial pressure of H_2 = 0.010 atm; partial pressure of C_2H_6 = 22 atm; $\Delta G°_{400} = -218.8$ kJ/mole-rxn

Step (**1**): $\Delta G_{400} = \Delta G°_{400} + 2.303\ RT\ \log_{10}(Q)$

Step (**2**): $\Delta G°_{400} = -218.8$ kJ/mole-rxn; R = 0.008314 kJ/mole-K; T = 400 K

$$Q = \frac{(P_{C2H6})}{(P_{C2H2})(P_{H2})^2} = \frac{(22)}{(0.10)(0.01)^2} = 2.2 \times 10^6$$

Step (**3**)

$$\Delta G_{400} = (-218.8\ \text{kJ/mole-rxn}) + 2.303\ (0.008314\ \text{kJ/mole-K})(400\ \text{K})\log_{10}(2.2 \times 10^6)$$

$$\Delta G_{400} = (-218.8\ \text{kJ/mole-rxn}) + 2.303\ (0.008314\ \text{kJ/mole-K})(400\ \text{K})(6.342)$$

$$\Delta G_{400} = (-218.8\ \text{kJ/mole-rxn}) + (+48.6\ \text{kJ/mole}) = -170.2\ \text{kJ/mole-rxn}$$

The much greater partial pressure of the product compared to the partial pressures of the reactants has produced a decrease in the negative value of the non-standard free energy and has decreased the spontaneity of the forward reaction.

Example: Use the value for standard free energy ($\Delta G°_T$) calculated in a previous example to calculate the value for non-standard free energy (ΔG_T) under the stated conditions for the following reaction at a temperature of 1400 K.

$$C_2H_{2(g)} + 2H_{2(g)} \longrightarrow C_2H_{6(g)}$$

Partial pressure of C_2H_2 = 10 atm; partial pressure of H_2 = 10 atm; partial pressure of C_2H_6 = 1.0atm; $\Delta G°_{1400}$ = = +14.2 kJ/mole-rxn

Step (**1**): $\Delta G_{1400} = \Delta G°_{1400} + 2.303\ RT\ \log_{10}(Q)$

Step (**2**): $\Delta G°_{1400} = +14.2$ kJ/mole-rxn; R = 0.008314 kJ/mole-K; T = 1400 K

$$Q = \frac{(P_{C2H6})}{(P_{C2H2})(P_{H2})^2} = \frac{(1.0)}{(10)(10)^2} = 1.0 \times 10^{-3}$$

Step (**3**)

$$\Delta G_{400} = (+14.2\ \text{kJ/mole-rxn}) + 2.303\ (0.008314\ \text{kJ/mole-K})(1400\ \text{K})\log_{10}(1.0 \times 10^{-3})$$

$$\Delta G_{400} = (+14.2\ \text{kJ/mole-rxn}) + 2.303\ (0.008314\ \text{kJ/mole-K})(1400\ \text{K})(-3)$$

$$\Delta G_{400} = (+14.2\ \text{kJ/mole-rxn}) + (-80.4\ \text{kJ/mole}) = -66.2\ \text{kJ/mole-rxn}$$

The much greater partial pressures of the reactants compared to the partial pressure of the product has produced a change of sign in the value of the non-standard free energy and has converted the reaction with these **initial** concentrations to be spontaneous in the forward reaction.

Example: Use the value for standard free energy ($\Delta G°_T$) calculated in a previous example to calculate the value for non-standard free energy (ΔG_T) under the stated conditions for the following reaction at a temperature of 298 K.

$$C_2H_5OH_{(l)} + 3O_{2(g)} \longrightarrow 2CO_{2(g)} + 3H_2O_{(l)}$$

Mole fraction (χ) of C_2H_5OH (pure liquid) = 1.0; partial pressure of O_2 = 0.025 atm; partial pressure of CO_2 = 50 atm; mole fraction (χ) of H_2O (pure liquid) = 1.0

$$\Delta G°_{298} = -1325.3\text{kJ/mole-rxn}$$

Step (1): $\Delta G_{298} = \Delta G°_{298} + 2.303\ RT\ \log_{10}(Q)$

Step (2): $\Delta G°_{298} = -1325.3$ kJ/mole-rxn; R = 0.008314 kJ/mole-K; T = 298 K

$$Q = \frac{(P_{CO2})^2(\chi H_2O)^3}{(P_{O2})^3\ \chi C_2H_5OH} = \frac{(50)^2(1)^3}{(0.025)^3(1)} = 1.6\times10^8$$

Step (3)

$$\Delta G_{298} = (-1325.3\text{ kJ/mole-rxn}) + 2.303\ (0.008314\text{ kJ/mole-K})(298\text{ K})\log_{10}(1.6\times10^8)$$

$$\Delta G_{298} = (-1325.3\text{ kJ/mole-rxn}) + 2.303\ (0.008314\text{ kJ/mole-K})(298\text{ K})(8.204)$$

$$\Delta G_{298} = (-1325.3\text{ kJ/mole-rxn}) + (+46.8\text{ kJ/mole}) = -1279\text{ kJ/mole-rxn}$$

The much greater partial pressure of the product compared to the partial pressure of the reactant reduced the negative value of the non-standard free energy and has decreased the spontaneity of the forward reaction.

VII NON-STANDARD FREE ENERGY (ΔG) AND EQUILIBRIUM CONSTANTS

As developed in Chapter 24, a reversible reaction can reach a state of equilibrium. Equilibrium for a reversible reaction occurs whenever the forward and reverse reactions proceed at the same rate. This means that reactant molecules are forming product molecules at the same rate that product molecules are being reconverted to reactant molecules; a system at equilibrium shows no net change in concentrations of products or reactants.

Since the concentrations of all products and reactants remain in balance (no change) at equilibrium, the ratio determined by the concentration quotient, Q remains a constant value:

$$Q_{(\text{forward reaction})} = \frac{[C]^c[D]^d}{[A]^a[B]^b} = \text{constant value at equilibrium.}$$

The concentration ratio defined by Q is the equilibrium concentration ratio when applied to reactions at equilibrium; the numerical value for Q at equilibrium is the equilibrium constant (K).

Since a system at equilibrium shows no net change in concentrations of products or reactants, a reaction at equilibrium requires that neither the forward nor reverse reactions are spontaneous. The mathematical requirement for a system at equilibrium is: $\Delta G_T = 0$.

The numerical value of Q at equilibrium is defined as the equilibrium constant:

$$\mathbf{Q = K \text{ at equilibrium}}$$

The non-standard free energy at equilibrium must be zero:

$$\mathbf{\Delta G_T(\text{equilibrium}) = 0}$$

The complete equation for non-standard free energy is:

$$\mathbf{\Delta G_T = \Delta G^\circ_T + 2.303\ RT\ \log(Q)}$$

Specifically for equilibrium, substitute $\Delta G_T = 0$ and $Q = K$ in the equation:

$$\mathbf{0 = \Delta G^\circ_T + 2.303RT\ \log K}$$

Solve for ΔG°_T and log K:

$$\mathbf{\Delta G^\circ_T = -2.303RT \log K} \quad \text{and} \quad \mathbf{\log K = \frac{\Delta G^\circ_T}{-2.303RT}}$$

Any mixture of exactly 1 Molar reactants and products for a reversible reaction will proceed in the spontaneous direction given by the standard enthalpy and entropy, expressed as a value for ΔG°_T. The reaction will continue until the concentration of products in the spontaneous direction build up to the point where the concentration entropy exactly cancels the ΔG°_T: this is the point of equilibrium. The equilibrium constant defines the corresponding concentrations.

Example: Calculate the numerical value for the equilibrium constants, **K**, at the temperatures of 400 Kelvin, 800 Kelvin, and 1400 Kelvin for the following reaction.

$$C_2H_{2(g)} + 2H_{2(g)} \longrightarrow C_2H_{6(g)}$$

Step (**1**): $\log_{10} K$: $\log_{10} K = \dfrac{\Delta G^\circ_T}{-2.303\ RT}$

Step (**2**): R = 0.008314 kJ/mole-K; T = 400 K or 800 K or 1400 K
When required, calculate the value for ΔG°_T using the methods described.
The values for ΔG°_T were calculated in previous examples:

$$\Delta G^\circ_{400} = -218.8 \text{ kJ/mole-rxn}$$

$$\Delta G^\circ_{800} = -125.6 \text{ kJ/mole-rxn}$$

$$\Delta G^\circ_{1400} = +14.2 \text{ kJ/mole-rxn}$$

Step (**3**): Complete the calculation for $\log_{10}K$ and then take the antilog to find K.

$$\log_{10} K_{100} = \frac{\Delta G^\circ_T}{(-2.303\ RT)} = \frac{(-218.8\text{ kJ})}{(-2.303)(0.008314\text{ kJ/mole-K})(400\text{ K})}$$

$$\log_{10} K_{400} = 28.6;\ K = \text{antilog}(28.6) = 10^{28.6} = 3.98 \times 10^{28}$$

$$\log_{10} K_{800} = \frac{\Delta G^\circ_T}{(-2.303\ RT)} = \frac{(-125.6\text{ kJ})}{(-2.303)(0.008314\text{ kJ/mole-K})(800\text{ K})}$$

$$\log_{10} K_{800} = 8.20;\ K = \text{antilog}(8.20) = 10^{8.20} = 1.58 \times 10^{8}$$

$$\log_{10} K_{1400} = \frac{\Delta G^{\circ}_{T}}{(-2.303\ RT)} = \frac{(+14.2\ \text{kJ})}{(-2.303)(0.008314\ \text{kJ/mole-K})(1400\ \text{K})}$$

$$\log_{10} K_{1400} = -0.530;\ K = \text{antilog}(-0.530) = 10^{-0.530} = 2.95 \times 10^{-1}$$

Example: Calculate the **hypothetical** numerical value for the equilibrium constant, **K**, at the temperature of 298 Kelvin for the following reaction if it were reversible and could reach equilibrium.

$$C_2H_5OH_{(l)} + 3O_{2(g)} \longrightarrow 2CO_{2(g)} + 3H_2O_{(l)}$$

Step (**1**): $\log_{10} K$: $\log_{10} K = \frac{\Delta G^{\circ}_{T}}{-2.303\ RT}$

Step (**2**): R = 0.008314 kJ/mole-K; T = 298; the value for ΔG°_{T} was calculated in previous **example:** $\Delta G^{\circ}_{298} = -1325.3$ kJ/mole-rxn

Step (**3**): Complete the calculation for $\log_{10}K$ and then take the antilog to find K.

$$\log_{10} K_{298} = \frac{\Delta G^{\circ}_{T}}{(-2.303 RT)} = \frac{(-1325.3\text{kJ})}{(-2.303)(0.008314\,\text{kJ/mole-K})(298\,\text{K})}$$

$$\log_{10} K_{298} = 232.3;\ K = \text{antilog}(232.3) = 10^{232.3} = 2.0 \times 10^{232}$$

This huge value for an "equilibrium" constant provides a way to visualize the concept of essentially non-reversible reactions with large negative values of ΔG°.

Example: Calculate the numerical value for the equilibrium constants, K, at the temperatures of 298 Kelvin and 73 K (hypothetically) for the following reaction.

$$3\,Fe_{(s)} + 4\,H_2O_{(l)} \longrightarrow Fe_3O_{4\,(s)} + 4\,H_{2\,(g)}$$

Step (**1**): $\log_{10} K$: $\log_{10} K = \frac{\Delta G^{\circ}_{T}}{-2.303\ RT}$

Step (**2**): R = 0.008314 kJ/mole-K; T = 298 K or 73 K
The values for AG°T were calculated in previous examples:

$$\Delta G^{\circ}_{298} = -66.9\ \text{kJ/mole-rxn;}$$

$$\Delta G^{\circ}_{73} = +2.3\ \text{kJ/mole-rxn}$$

Step (**3**): Complete the calculation for $\log_{10}K$ and then take the antilog to find K.

$$\log_{10} K_{298} = \frac{\Delta G^{\circ}_{T}}{(-2.303\ RT)} = \frac{(-66.9\ \text{kJ})}{(-2.303)(0.008314\ \text{kJ/mole–K})(298\ \text{K})}$$

$$\log_{10} K_{298} = 11.7;\ K = \text{antilog}(11.7) = 10^{11.7} = 5.31 \times 10^{11}$$

$$\log_{10} K_{73} = \frac{\Delta G°_T}{(-2.303\ RT)} = \frac{(+2.3\ kJ)}{(-2.303)(0.008314\ kJ/mol\text{–}K)(73\ K)}$$

$$\log_{10} K_{73} = -1.65;\ K = \text{antilog}(-1.65) = 10^{-1.65} = 2.2 \times 10^{-2}$$

VIII COMPREHENSIVE EXAMPLES

	ΔH°f	ΔS°
($N_{2\,(g)}$)	0 kJ/mole	191.61 J/K
($H_{2\,(g)}$)	0 kJ/mole	131 J/K
($NH_{3\,(g)}$)	−46.11 kJ/mole	192.45 J/K
($C_3H_{6\,(aq)}$)	+20.4 kJ/mole	226.9 J/K
($H_2O_{(l)}$)	−285.8 kJ/mole	69.9 J/K
($C_3H_8O_{(aq)}$)	−316.7 kJ/mole	194.6 J/K

1. Complete the following parts for the common reaction forming ammonia shown.

$$\mathbf{N_{2\,(g)} + 3\,H_{2\,(g)} \rightleftarrows 2\,NH_{3\,(g)}}$$

a) Predict the sign of the entropy change for the forward reaction without calculation.
b) Calculate the numerical values for standard enthalpy and standard entropy from enthalpies of formation and absolute entropies.
c) Calculate the numerical values for standard free energy, $\Delta G°_T$ at temperatures of 298 K and 770 K; state which direction is spontaneous at each of these temperatures.
d) Calculate the temperature at which $\Delta G°_T$ changes sign; that is, T at which $\Delta G°_T = 0$
e) Calculate the value for ΔG_T at 770 K under the conditions of partial pressure of N_2 = 15 atm; partial pressure of H_2 = 10 atm; partial pressure of NH_3 = 0.20 atm;
f) Calculate the numerical value for K_{298} and K_{770} for this reaction.

a) The balanced equation shows 4 moles of gas reactants combining to form 2 moles of gas products. The standard entropy change is predicted to be negative through concept (1): greater number of smaller molecules combine to form a fewer number of product molecules; the system becomes more ordered (less disordered).
b) $\Delta H°_{(reaction)} = [\text{Sum of } 2(\Delta H°f\ NH_{3\,(g)})] - [\text{Sum of } \{(\Delta H°f\ N_{2\,(g)}) + 3(\Delta H°f\ H_{2\,(g)})\}]$

$$\Delta H°_{(reaction)} = [2(-46.11\ kJ)] - [(0\ kJ) + 3(0kJ)]$$

$$\Delta H°_{(reaction)} = \mathbf{-92.2\ kJ/mole\text{-}reaction}$$

$$\Delta S°_{(reaction)} = [\text{Sum of } 2(S°\ NH_{3\,(g)})] - [\text{Sum of } \{(S°\ N_{2\,(g)}) + 3(S°\ H_{2\,(g)})\}]$$

$$\Delta S°_{(reaction)} = [2(192.45)\ J/K)] - [(191.61\ J/K) + 3(131\ J/K)]$$

$$\Delta S°_{(reaction)} = \mathbf{-199.7\ J/K = -200\ J/K\ or\ -0.200\ kJ/K}$$

c) $\Delta G°_T = \Delta H° - T\Delta S°$

$$\Delta G°_{(298)} = -92.2\ kJ - [(298\ K)(-0.200\ kJ/K)]$$

$$= -92.2kJ + 59.6\ kJ = \mathbf{-32.6\ kJ/mole\text{-}rxn}$$

Forward reaction is spontaneous at 298 K.

$$\Delta G^\circ_{(770)} = -92.2\ \text{kJ} - [(770\ \text{K})(-0.200\ \text{kJ/K})]$$
$$= -92.2\text{kJ} + 154.0\ \text{kJ} = \mathbf{+61.8\ kJ/mole\text{-}rxn}$$

Forward reaction is non-spontaneous; reverse reaction is spontaneous at 770 K.

d) $T_{\Delta G = 0} = \Delta H^\circ/\Delta S^\circ = (-92.2\ \text{kJ})/(-0.200\ \text{kJ/K}) = \mathbf{461\ K}$; $\Delta G^\circ_{(461)} = 0$

e) $\Delta G_{770} = \Delta G^\circ_{770} + 2.303\ \text{RT}\ \log_{10}(Q)$

$\Delta G^\circ_{770} = +61.6$ kJ/mole-rxn; R = 0.008314 kJ/mole-K; T = 770K

$$Q = \frac{(P_{NH3})^2}{(P_{N2})(P_{H2})^3} = \frac{(0.2)^2}{(15)(10)^3} = 2.7\times10^{-6}$$

$$\Delta G_{770} = (+61.8\ \text{kJ/mole-rxn}) + 2.303\ (0.008314\ \text{kJ/mole-K})(400\ \text{K})\log_{10}(2.7\times10^{-6})$$

$$\Delta G_{770} = (+61.8\ \text{kJ/mole-rxn}) + 2.303\ (0.008314\ \text{kJ/mole-K})(400\ \text{K})(-5.57)$$

$$\Delta G_{770} = (+61.8\ \text{kJ/mole-rxn}) + (-82.1\ \text{kJ/mole}) = \mathbf{-20.3\ kJ/mole\text{-}rxn}$$

f)

$$\log_{10} K = \Delta G^\circ_T/-2.303\ \text{RT}$$

$$\log K_{(298)} = \Delta G^\circ_{(298)}/-2.303\ \text{R}\ (298)$$

$$= -32.6\ \text{kJ}/-2.303\ (0.008314\ \text{kJ/K})\ (298\ \text{K}) = +\ 5.71$$

$$K_{(298)} = \text{antilog}\ (5.71) = 10^{5.71} = \mathbf{5.1\times10^5}$$

$$\log K_{(770)} = \Delta G^\circ_{(770)}/-2.303\ \text{R}\ (770)$$

$$+61.8\ \text{kJ}/2.303\ (0.008314\ \text{kJ/K})\ (770\ \text{K}) = -4.19$$

$$K_{(770)} = \text{antilog}\ (-4.19) = 10^{-4.19} = \mathbf{6.4\times10^{-5}}$$

2. Complete the following parts for the reaction shown.

$$\mathbf{C_3H_{6(aq)} + H_2O_{(l)} \rightleftarrows C_3H_8O_{\ (aq)}}$$

a) Predict the sign of the entropy change for the forward reaction without calculation.
b) Calculate ΔH° from ΔH°f.
c) Use the predicted sign of entropy change and the calculated value for enthalpy to determine the spontaneity result for the reaction: always spontaneous, never spontaneous, spontaneous at a sufficiently low temperature or spontaneous at a sufficiently high temperature.
d) Calculate ΔS° from S°.
e) Calculate the numerical values for standard free energy, ΔG°_T at temperatures of 298 K and 750 K; state which direction is spontaneous at each of these temperatures.
f) Calculate the temperature at which $\Delta G^\circ_T = 0$.
g) Calculate the value for ΔG_T at 298 K under the conditions of $[C_3H_6] = 0.50$ M; $[C_3H_8O] = 2.0$ M.
h) Calculate the numerical value for K_{298} and K_{750} for this reaction.

i) The structure of C_3H_6 is CH_3—CH==CH_2 and the structure of C_3H_8O is $CH_3CH_2CH_2OH$. Calculate ΔH from BDE. Use the table of BDE contained in Chapter 18; review Chapter 15 on condensed structures if necessary.

a) The balanced equation shows 2 moles of reactants combining to form 1 mole of product. The standard entropy change is predicted to be negative through concept (**1**): greater number of smaller molecules combine to form fewer product molecules; the system becomes more ordered (less disordered).

b) $\Delta H°_{(reaction)}$ = [Sum of ($\Delta H°f\ C_3H_8O_{(aq)}$)] − [Sum of {($\Delta H°f\ C_3H_{6\ (aq)}$) + ($\Delta H°f\ H_2O_{(l)}$}]

$$\Delta H°_{(reaction)} = [(-316.7\ kJ)] - [(+20.4\ kJ) + (-285.8\ kJ)]$$

$$\Delta H°_{(reaction)} = \mathbf{-51.3\ kJ/mole\text{-}reaction}$$

c) $\Delta H°_{(reaction)}$ is negative and $\Delta S°_{(reaction)}$ is negative; the reaction is spontaneous at sufficiently low temperature.

d) $\Delta S°_{(reaction)}$ = [Sum of ($S°\ C_3H_8O_{(aq)}$)] − [Sum of {($S°\ C_3H_{6\ (aq)}$) + ($S°\ H_2O_{(l)}$)}]

$$\Delta S°_{(reaction)} = [(194.6)\ J/K)] - [(226.9\ J/K) + (69.9\ J/K)]$$

$$\Delta S°_{(reaction)} = \mathbf{-102.2\ J/K = -0.102\ kJ/K}$$

e) $\Delta G°_T = \Delta H° - T\,\Delta S°$

$$\Delta G°_{(298)} = -51.3\ kJ - [(298\ K)(-0.102\ kJ/K)]$$

$$= -51.3kJ + 30.4\ kJ = \mathbf{-20.9\ kJ/mole\text{-}rxn}$$

Forward reaction is spontaneous at 298 K.

$$\Delta G°_{(750)} = -51.3\ kJ - [(750\ K)(-0.102\ kJ/K)]$$

$$= -51.3kJ + 76.5\ kJ = \mathbf{+\ 25.2\ kJ/mole\text{-}rxn}$$

Forward reaction is non-spontaneous; reverse reaction is spontaneous at 750 K.

f) $T_{\Delta G\ =\ 0} = \Delta H°/\Delta S° = (-51.3\ kJ)/(-0.102\ kJ/K) = \mathbf{503\ K}$; $\Delta G°_{(503)} = 0$

g) $\Delta G_{298} = \Delta G°_{298} + 2.303\ RT\ \log_{10}(Q)$

$\Delta G°_{298} = -20.9$ kJ/mole-rxn; R = 0.008314 kJ/mole-K; T = 298 K

$$Q = \frac{[C_3H_8O]}{[C_3H_6]\chi_{H2O}} = \frac{(2.0)}{(0.50)(1)} = 4$$

$$\Delta G_{298} = (-20.9\ kJ/mole\text{-}rxn) + 2.303\ (0.008314\ kJ/mole\text{-}K)(298\ K)\log_{10}(4)$$

$$\Delta G_{298} = (-20.9\ kJ/mole\text{-}rxn) + 2.303\ (0.008314\ kJ/mole\text{-}K)(298\ K)(0.602)$$

$$\Delta G_{298} = (-20.9\ kJ/mole\text{-}rxn) + (3.43\ kJ/mole) = \mathbf{-17.5\ kJ/mole\text{-}rxn}$$

h) $\log_{10} K = \Delta G°_T/-2.303\ RT$

$$\log K_{(298)} = \Delta G°_{(298)}/-2.303\ R\ (298)$$

$$= -20.9\ kJ/-2.303\ (0.008314\ kJ/K)\ (298\ K) = +3.66$$

$$K_{(298)} = \text{antilog}\ (3.66) = 10^{3.66} = \mathbf{4.60 \times 10^3}$$

$$\log K_{(750)} = \Delta G°_{(750)}/-2.303\ R\ (750)$$

$$+25.2\ kJ/2.303\ (0.008314\ kJ/K)\ (750\ K) = -1.75$$

$$K_{(750)} = \text{antilog}\ (-1.75) = 10^{-1.75} = \mathbf{1.76 \times 10^{-2}}$$

i) $H_2C{=}CH{-}CH_3$ (H–C(H)(H)–C(H)=C(H)–H) + H–O–H ⟶ H–C(H)(H)–C(H)(H)–C(H)(H)–O–H

Bonds broken: C==C and one H—O
Bonds formed: one C—C, one C—O and one C—H
From the table in Chapter 18:

Bonds Broken	Bonds Formed
1 ×O—H = 1 × 460 = 460 kJ	1 ×C—C = 1 × 345 = 345 kJ
1 ×C=C = 1 × 615 = 615 kJ	1 ×C—H = 1 × 415 = 415 kJ
1075 kJ	1 ×C—O = 1 × 360 = 360 kJ
	1120 kJ

ΔH (reaction) = [Sum of BDE for bonds broken) − [Sum BDE for bonds formed]

ΔH (reaction) = [1075] − [1120] = **−45 kJ**

IX PRACTICE PROBLEMS

1. For the following processes, predict the sign of the entropy for the forward direction. Then state whether the process as written is spontaneous or non-spontaneous based **only** on the **entropy** criterion.

   a) $C_6H_{6\,(l)} + 3\,H_{2\,(g)} \longrightarrow C_6H_{12\,(l)}$
   b) $CH_3OH_{\,(l)} + H_2O_{\,(l)} \longrightarrow CH_3OH_{\,(aq)}$
   c) $C_4H_6O_{3\,(l)} \longrightarrow C_3H_6O_{\,(l)} + CO_{2\,(g)}$
   d) $CH_3CH_2OH_{\,(aq)}$ at 1.0 M $\longrightarrow CH_3CH_2OH_{\,(aq)}$ at 5.0 M

	ΔH°f kJ/mole	ΔS°J/K
($N_{2(g)}$)	0	191.5
($H_{2(g)}$)	0	130.7
($N_2H_{4(l)}$)	50.6	121.2
($H_2O_{(l)}$)	−285.8	69.9
($H_2O_{(g)}$)	−241.8	188.7
($H_2O_{2(l)}$)	−187.3	109.6
($HCO_2H_{(l)}$)	−424.7	129.0
($CH_{4(g)}$)	−74.9	186.3
($CO_{2(g)}$)	−393.5	213.7
($CO_{(g)}$)	−110.5	197.67
($CH_3OH_{(l)}$)	−238.7	126.8
($O_{2(g)}$)	0	205.1

(Continued)

	ΔH°f kJ/mole	ΔS°J/K
($HNO_{3(g)}$)	−135.1	266.4
($NO_{(g)}$)	90.25	210.8
($NO_{2(g)}$)	33.2	240.1
($C_{(s)}$)	0	5.7

2. For the reaction: $N_2H_{4\,(l)} + 2\,H_2O_{2(l)} \longrightarrow N_{2\,(g)} + 4\,H_2O_{(g)}$
   a) Predict the sign of the entropy change for the forward reaction.
   b) Calculate the numerical values for standard enthalpy and standard entropy from enthalpies of formation and absolute entropies.
   c) Based on the signs of ΔH° and ΔS°, state the spontaneity conclusion: always spontaneous; never spontaneous; spontaneous at sufficiently low temperatures; spontaneous at sufficiently high temperatures.
   d) Calculate the numerical values for standard free energy, $\Delta G°_T$ at temperatures of 200 K and 500 K and determine the spontaneity.
   e) Assume for this part that the reaction is reversible. Calculate the hypothetical numerical values for K_{200} and K_{500} for the forward reaction.

For problems 3 through 8, complete the following parts:

a) Predict the sign of the entropy change for the forward reaction.
b) Calculate the numerical values for standard enthalpy and standard entropy from enthalpies of formation and absolute entropies.
c) Based on the signs of ΔH° and ΔS°, state the spontaneity conclusion: always spontaneous; never spontaneous; spontaneous at sufficiently low temperatures; spontaneous at sufficiently high temperatures.
d) Calculate the numerical values for standard free energy, $\Delta G°_T$ at temperatures the temperatures stated in the problem and determine the spontaneity for each.
e) Calculate the temperature at which $\Delta G°_T$ changes sign, (T at which $\Delta G°_T = 0$).
f) Assume each reaction is reversible and the temperatures are practical. Calculate the numerical value for K_T for the forward reaction for the temperatures stated in the problem.
g) For problems 7 and 8, calculate the value for the non-standard free energy (ΔG_T) under the concentration conditions stated.

3. $HCO_2H_{\,(l)} \longrightarrow CO_{2\,(g)} + H_{2\,(g)}$ Analysis at 100 K and 500 K

4. $CO_{\,(g)} + 2\,H_{2\,(g)} \longrightarrow CH_3OH_{\,(l)}$ Analysis at 200 K and 500 K

5. $N_{2\,(g)} + 5O_{2\,(g)} + 2\,H_2O_{\,(g)} \longrightarrow 4\,HNO_{3\,(g)}$ Analysis at 298 K and 60 K

6. $2\,NO_{\,(g)} + O_{2\,(g)} \longrightarrow 2\,NO_{2\,(g)}$ Analysis at 298 K and 1000 K

7. $CH_{4\,(g)} + H_2O_{\,(g)} \longrightarrow CO_{\,(g)} + 3\,H_{2\,(g)}$ Analysis at 298 K and 1200 K

   Calculate (ΔG_T) at 298 K for partial pressures of CO = 0.20 atm.; H_2 = 0.30 atm.; CH_4 = 2.0 atm.; H_2O = 1.5 atm.

8. $C_{\,(s)} + 2\,H_{2(g)} \longrightarrow CH_{4(g)}$ Analysis at 298 K and 1000 K

   Calculate (ΔG_T) at 298 K for partial pressures of H_2 = 0.20 atm.; CH_4 = 4.0 atm.

X ANSWERS TO PRACTICE PROBLEMS

1. a) $C_6H_{6\,(l)} + 3\,H_{2\,(g)} \longrightarrow C_6H_{12\,(l)}$

Entropy change is negative (decreases): a greater number of smaller, simpler molecules produces a lesser number of larger, more complex molecules; total number of gas molecules decreases. Based on entropy, the reaction as written is non-spontaneous.

b) $CH_3OH_{\,(l)} + H_2O_{\,(l)} \longrightarrow CH_3OH_{\,(aq)}$

Entropy change is positive (increases): formation of a solution (mixture) from pure components. Based on entropy, the reaction as written is spontaneous.

c) $C_4H_6O_{3\,(l)} \longrightarrow C_3H_6O_{\,(l)} + CO_{2\,(g)}$

Entropy change is positive (increases): a lesser number of larger, more complex molecules produces a greater number of smaller, simpler molecules; total number of gas molecules increases. Based on entropy, the reaction as written is spontaneous.

d) $CH_3CH_2OH_{\,(aq)}$ at 1.0 M $\longrightarrow$ $CH_3CH_2OH_{\,(aq)}$ at 5.0 M

Entropy change is negative (decreases): solution becomes more concentrated: reverse of dilution. Based on entropy, the reaction as written is non-spontaneous.

2.

$N_2H_{4\,(l)}$	+	$2\,H_2O_{2\,(l)}$	$\longrightarrow$	$N_{2\,(g)}$	+	$4\,H_2O_{\,(g)}$
ΔH°f: 50.6 kJ		−187.3 kJ		0 kJ		−241.8 kJ
S° : 121.2 J/K		109.6 J/K		191.5 J/K		188.7 J/K

a) The balanced equation shows 3 moles of liquid reactants breaking down to form 5 moles of gas products. The standard entropy change is predicted to be **positive**: a fewer number of larger reactant molecules break down to form a greater number of product molecules; the system becomes less ordered (more disordered).

b) $\Delta H^\circ_{(reaction)} = [0 + 4(-241.8\text{ kJ})] - [(50.6\text{ kJ}) + 2(-187.8\text{kJ})]$

$$\Delta H^\circ_{(reaction)} = \mathbf{-642.2\ kJ/mole\text{-}reaction}$$

$$\Delta S^\circ_{(reaction)} = [(191.5\text{ J/K}) + 4(188.7\text{ J/K})] - [(121.2\text{J/K}) + 2(109.6\text{J/K})]$$

$$\Delta S^\circ_{(reaction)} = \mathbf{+\ 605.9\ J/K} \text{ or } \mathbf{+0.6059\ kJ/K}$$

c) Enthalpy is negative: contributes to spontaneity; entropy is positive: contributes to spontaneity. Both enthalpy and entropy show the same trend toward spontaneous: reaction is **always spontaneous**.

d) $\Delta G^\circ_T = \Delta H^\circ - T\,\Delta S^\circ$

$$\Delta G^\circ_{(200)} = -642.2\text{ kJ} - [(200\text{ K})(0.6059\text{ kJ/K})]$$

$$= -642.2\text{ kJ} - 121.18\text{ kJ} = \mathbf{-763.38\ kJ/mole}$$

Forward reaction is **spontaneous** at 200 K.

$$\Delta G^\circ_{(500)} = -642.2 \text{ kJ} - [(500 \text{ K})(0.6059 \text{ kJ/K})]$$

$$= -642.2 \text{ kJ} - 302.95 \text{ kJ} = \mathbf{-945.15 \text{ kJ/mole}}$$

Forward reaction is **spontaneous** at 500 K.

e) $\log_{10} K = \Delta G^\circ_T / 2.303\ RT$

$\log_{10} K_{200} = (-763.38 \text{ kJ})/(-2.303)(0.008314 \text{ kJ/mole-K})(200 \text{ K})$

$\log_{10} K_{200} = 199.35$; $K = \text{antilog}(199.35) = 10^{199.35} = 2.24 \times 10^{199}$

$\log_{10} K_{500} = (-945.15 \text{ kJ})/(-2.303)(0.008314 \text{ kJ/mole-K})(500 \text{ K})$

$\log_{10} K_{500} = 98.72$; $K = \text{antilog}(98.72) = 10^{98.72} = 5.25 \times 10^{98}$

3. $HCO_2H_{(l)} \longrightarrow CO_{2\,(g)} + H_{2\,(g)}$

	$HCO_2H_{(l)}$		$CO_{2\,(g)}$	+	$H_{2\,(g)}$
ΔH°f:	−424.7 kJ		−393.5 kJ		0 kJ
S° :	129.0 J/K		213.7 J/K		130.7 J/K

a) The balanced equation shows 1 mole of liquid reactants breaking down to form 2 moles of gas products. The standard entropy change is predicted to be **positive**: a fewer number of larger reactant molecules break down to form a greater number of product molecules; the system becomes less ordered (more disordered).

b) $\Delta H^\circ_{rxn} = [(-393.5 \text{ kJ}) + (0 \text{ kJ})] - [(-424.7 \text{ kJ})]$

$\Delta H^\circ_{(reaction)} = \mathbf{+31.2 \text{ kJ/mole-reaction}}$

$\Delta S^\circ_{rxn} = [(213.7 \text{ J/K}) + (130.7 \text{ J/K})] - [(129.0 \text{ J/K})]$

$\Delta S^\circ_{(reaction)} = \mathbf{+215.4 \text{ J/K}}$ or $\mathbf{+0.2154 \text{ kJ/K}}$

c) Enthalpy is positive: contributes to non-spontaneity; entropy is positive: contributes to spontaneity. Enthalpy and entropy trends oppose each other. High temperature favors the direction indicated by the entropy: reaction is **spontaneous** at sufficiently **high** temperatures.

d) $\Delta G^\circ_T = \Delta H^\circ - T\,\Delta S^\circ$

$$\Delta G^\circ_{(100)} = +31.2 \text{ kJ} - [(100 \text{ K})(+\,0.2154 \text{ kJ/K})]$$

$$= +31.2\text{kJ} - 21.54 \text{ kJ} = \mathbf{+9.7 \text{ kJ/mole}}$$

Forward reaction is **non-spontaneous** at 100 K.

$$\Delta G^\circ_{(500)} = +\,31.2 \text{ kJ} - [(500 \text{ K})(+0.2154 \text{ kJ/K})]$$

$$= +31.2 \text{ kJ} - 107.7 \text{ kJ} = \mathbf{-76.5 \text{ kJ/mole}}$$

Forward reaction is **spontaneous** at 500 K.

e) $T_{\Delta G} = {}_0 = \Delta H°/\Delta S° = (31.2\ kJ)/(0.2154\ kJ/K) =$ **144.8 K**; $\Delta G°_{(145)} = 0$

f) $\log_{10} K = \Delta G°_T/-2.303\ RT$

$\log_{10} K_{100} = (+9.7\ kJ)/(-2.303)(0.008314\ kJ/mole\text{-}K)(100\ K)$

$\log_{10} K_{100} = -5.07; K = antilog(-5.07) = 10^{-5.07} = 8.5 \times 10^{-6}$

$\log_{10} K_{500} = (-76.5\ kJ)/(-2.303)(0.008314\ kJ/mole\text{-}K)(500\ K)$

$\log_{10} K_{500} = 7.99; K = antilog(7.99) = 10^{7.99} = 9.77 \times 10^{7}$

4. $CO_{(g)}$ + $2\,H_{2(g)}$ $\longrightarrow$ $CH_3OH_{(l)}$

	$CO_{(g)}$	$2\,H_{2(g)}$	$CH_3OH_{(l)}$
$\Delta H°f$:	−110.5 kJ	0 kJ	−238.7 kJ
S° :	197.67 J/K	130.7 J/K	126.8 J/K

a) The balanced equation shows 3 moles of gas reactants combining to form 1 mole of liquid products. The standard entropy change is predicted to be **negative**: greater number of smaller molecules combine to form a fewer number of product molecules; the system becomes more ordered (less disordered).

b) $\Delta H°_{rxn} = [(-238.7\ kJ)] - [(-110.5\ kJ) + 2(0\ kJ)]$

$\mathbf{\Delta H°_{(reaction)} = -128.2\ kJ/mole\text{-}reaction}$

$\Delta S°_{rxn} = [(126.8\ J/K)] - [(197.67\ J/K) + 2(130.7\ J/K)]$

$\Delta S°_{(reaction)} =$ **−332.27 J/K** or **−0.3323 kJ/K**

c) Enthalpy is negative: contributes to spontaneity; entropy is negative: contributes to non-spontaneity. Enthalpy and entropy trends oppose each other. Low temperature favors the direction indicated by the enthalpy: reaction is **spontaneous** at sufficiently **low** temperatures.

d) $\Delta G°_T = \Delta H° - T\,\Delta S°$

$\Delta G°_{(200)} = -128.2\ kJ - [(200\ K)(-0.3323\ kJ/K)]$

$= -128.2\ kJ + 66.46\ kJ =$ **−61.7 kJ/mole**

Forward reaction is **spontaneous** at 200 K.

$\Delta G°_{(500)} = -128.2\ kJ - [(500\ K)(-0.3323\ kJ/K)]$

$= -128.2\ kJ + 166.15\ kJ =$ **+ 38.0 kJ/mole**

Forward reaction is **non-spontaneous** at 500 K.

e) $T\Delta_G = {}_0 = \Delta H°/\Delta S° = (-128.2\ kJ)/(-0.3323\ kJ/K) =$ **386 K**; $\Delta G°_{(386)} = 0$

f) $\log_{10} K = \Delta G°_T/-2.303\ RT$

$\log_{10} K_{200} = (-61.7\ kJ)/(-2.303)(0.008314\ kJ/mole\text{-}K)(200\ K)$

$\log_{10} K_{200} = 16.11;\ K = antilog(16.11) = 10^{16.11} = 1.29 \times 10^{16}$

$\log_{10} K_{500} = (+38.0\ kJ)/(-2.303)(0.008314\ kJ/mole\text{-}K)(500\ K)$

$\log_{10} K_{500} = -3.97;\ K = antilog(-3.97) = 10^{-3.97} = 1.07 \times 10^{-4}$

5.

	$2\ N_{2\,(g)}$	+	$5O_{2\,(g)}$	+	$2\ H_2O_{(g)}$	$\longrightarrow$	$4\ HNO_{3\,(g)}$
ΔH°f:	0 kJ		0 kJ		−241.8 kJ		−135.1 kJ
S° :	191.7 J/K		205.1 J/K		188.8 J/K		266.4 J/K

a) The balanced equation shows 9 moles of gas reactants forming 4 moles of gas products. The standard entropy change is predicted to be **negative**: greater number of smaller reactant molecules forming a fewer number of product molecules; the system becomes more ordered (less disordered).

b) $\Delta H°_{(reaction)} = [4(-135.1\ kJ)] - [2(0\ kJ) + 5(0\ kJ) + 2(-241.8\ kJ)]$

$\Delta H°_{(reaction)} = \mathbf{-\ 56.8\ kJ/mole\text{-}reaction}$

$\Delta S°_{(reaction)} = [4(266.4\ J/K)] - [2(191.7\ J/K) + 5(205.1\ J/K) + 2(188.8\ J/K)]$

$\Delta S°_{(reaction)} = \mathbf{-720.9\ J/K}$ or $\mathbf{-0.721\ kJ/K}$

c) Enthalpy is negative: contributes to spontaneity; entropy is negative: contributes to non-spontaneity. Enthalpy and entropy trends oppose each other. Low temperature favors the direction indicated by the enthalpy: reaction is spontaneous at sufficiently low temperatures.

d) $\Delta G°_T = \Delta H° - T\ \Delta S°$

$\Delta G°_{(298)} = -56.8\ kJ - [(298\ K)(-0.721\ kJ/K)]$

$= -56.8\ kJ + 214.9\ kJ = +158.1\ kJ/mole$

Forward reaction is **non-spontaneous** at 298 K.

$\Delta G°_{(60)} = -56.1\ kJ - [(60\ K)(-0.721\ kJ/K)]$

$= -56.1\ kJ + 43.3\ kJ = \mathbf{-12.8\ kJ/mole}$

Forward reaction is spontaneous at 60 K. (hypothetical)

e) $T_{\Delta G\ =\ 0} = \Delta H°/\Delta S^0 = (-56.8\ kJ)/(-0.721\ kJ/K) = 79\ K;\ \Delta G°_{(79)} = 0$

f) $\log_{10} K = \Delta G°_T/-2.303\ RT$

$\log_{10} K_{298} = (+158.1\ kJ)/(-2.303)(0.008314\ kJ/mole\text{-}K)(298\ K)$

$\log_{10} K_{298} = -27.7;\ K = antilog\ (-27.7) = 10^{-27.7} = 5.1 \times 10^{-28}$

$\log_{10} K_{60} = (-12.8\ kJ)/(-2.303)(0.008314\ kJ/mole\text{-}K)(60\ K)$

$\log_{10} K_{60} = 11.1$; $K = \text{antilog}\ (11.1) = 10^{11.1} = 1.3 \times 10^{11}$

6. | | $2\ NO_{(g)}$ | + | $O_{2(g)}$ | $\longrightarrow$ | $2\ NO_{2(g)}$ |
|---|---|---|---|---|---|
| $\Delta H°f$: | 90.25 kJ | | 0 kJ | | 33.2 kJ |
| S° : | 210.8 J/K | | 205.1 J/K | | 240.1 J/K |

a) The balanced equation shows 3 moles of gas reactants forming 2 moles of gas products. The standard entropy change is predicted to be **negative**: greater number of smaller reactant molecules forming a fewer number of product molecules; the system becomes more ordered (less disordered).

b) $\Delta H°_{(reaction)} = [2(33.2\ kJ)] - [2(90.25\ kJ) + (0\ kJ)J$

$\Delta H°_{(reaction)} = \mathbf{-114.1\ kJ/mole\text{-}reaction}$

$\Delta S°_{(reaction)} = [2(240.1\ J/K)] - [2\ (210.8\ J/K) + (205.1\ J/K)]$

$\Delta S°_{(reaction)} = -146.5\ J/K$ or $\mathbf{-0.1465\ kJ/K}$

c) Enthalpy is negative: contributes to spontaneity; entropy is negative: contributes to non-spontaneity. Enthalpy and entropy trends oppose each other. Low temperature favors the direction indicated by the enthalpy: reaction is spontaneous at sufficiently low temperatures.

d) $\Delta G°_T = \Delta H° - T\ \Delta S°$

$\Delta G°_{(298)} = -114.1\ kJ - [(298\ K)(-0.1465\ kJ/K)]$

$= -114.1\ kJ + 43.7\ kJ = -70.4\ kJ/mole$

Forward reaction is spontaneous at 298 K.

$\Delta G°_{(1000)} = -114.1\ kJ - [(1000\ K)(-0.1465\ kJ/K)]$

$= -114.1\ kJ + 146.5\ kJ = +\ 32.4\ kJ/mole$

Forward reaction is non-spontaneous at 1000 K.

e) $T\Delta_G =_0 = \Delta H°/\Delta S° = (-114.1\ kJ)/(-0.1465\ kJ/K) = \mathbf{779\ K}$; $\Delta G°_{(779)} = 0$

f) $\log_{10} K = \Delta G°_T/-2.303\ RT$

$\log_{10} K_{298} = (-70.4\ kJ)/(-2.303)(0.008314\ kJ/mole\text{-}K)(298\ K)$

$\log_{10} K_{298} = 12.3$; $K = \text{antilog}\ (12.3) = 10^{12.3} = 2.0 \times 10^{12}$

$\log_{10} K_{1000} = (32.4\ kJ)/(-2.303)(0.008314\ kJ/mole\text{-}K)(1000\ K)$

$\log_{10} K_{1000} = -1.69$; $K = \text{antilog}\ (-1.69) = 10^{-169} = 2.0 \times 10^{-2}$

7. $CH_{4\,(g)}$ + $H_2O_{\,(g)}$ $\longrightarrow$ $CO_{\,(g)}$ + $3\,H_{2\,(g)}$

	$CH_{4\,(g)}$	$H_2O_{\,(g)}$	$CO_{\,(g)}$	$3\,H_{2\,(g)}$
$\Delta H°f$:	–74.8 kJ	–241.8 kJ	–110.5 kJ	0 kJ
S° :	186.3 J/K	188.8 J/K	197.7 J/K	130.7 J/K

a) The balanced equation shows 2 moles of gas reactants breaking down to form 4 moles of gas products. The standard entropy change is predicted to be positive: a fewer number of larger reactant molecules break down to form a greater number of product molecules; the system becomes less ordered (more disordered).

b) $\Delta H°_{rxn} = [(-110.5\text{ kJ}) + 3(0\text{ kJ})] - [(-74.8\text{ kJ}) + (-241.8\text{ kJ})]$

$\Delta H°_{(reaction)} = $ **+206.1 kJ/mole-reaction**

$\Delta S°_{rxn} = [(197.7\text{ J/K}) + 3(130.7\text{ J/K})] - [(186.3\text{ J/K}) + (188.8\text{ J/K})]$

$\Delta S°_{(reaction)} = +\,214.7\text{ J/K or } +\,0.215\text{ kJ/K}$

c) Enthalpy is positive: contributes to non-spontaneity; entropy is positive: contributes to spontaneity. Enthalpy and entropy trends oppose each other. High temperature favors the direction indicated by the entropy: reaction is spontaneous at sufficiently high temperatures.

d) $\Delta G°_T = \Delta H° - T\Delta S°$

$\Delta G°_{(298)} = +206.1\text{ kJ} - [(298\text{ K})(+0.215\text{ kJ/K})]$

$= +206.1\text{ kJ} - 64.1\text{ kJ} = $ **+142 kJ/mole**

Forward reaction is **non-spontaneous** at 298 K.

$\Delta G°_{(1200)} = +206.1\text{ kJ} - [(1200\text{ K})(+0.215\text{ kJ/K})]$

$= +206.1\text{ kJ} - 258\text{ kJ} = $ **–51.9 kJ/mole**

Forward reaction is **spontaneous** at 1200 K.

e) $T_{\Delta G\,=\,0} = \Delta H°/\Delta S° = (206.1\text{ kJ})/(0.215\text{ kJ/K}) = $ **958 K**; $\Delta G°_{(958)} = 0$

f) $\log_{10} K = \Delta G°_T/-2.303\,RT$

$\log_{10} K_{298} = (+142\text{ kJ})/(-2.303)(0.008314\text{ kJ/mole-K})(298\text{ K})$

$\log_{10} K_{298} = -24.9;\ K = \text{antilog}(-24.9) = 10^{-24.9} = 1.3 \times 10^{-25}$

$\log_{10} K_{1200} = (-51.9\text{ kJ})/(-2.303)(0.008314\text{ kJ/mole-K})(1200\text{ K})$

$\log_{10} K_{1200} = 2.26;\ K = \text{antilog}(2.26) = 10^{2.26} = 1.8 \times 10^{2}$

g) $$Q = \frac{(P_{CO})(P_{H2})^3}{(P_{CH4})(P_{H2O})} = \frac{(0.20)(0.30)^3}{(2.0)(1.5)} = 1.8\times10^{-3}$$

$\Delta G_{298} = (+142\text{ kJ/mole-rxn}) + 2.303\,(0.008314\text{ kJ/mole-K})(298\text{ K})\log_{10}(1.8 \times 10^{-3})$

$\Delta G_{298} = (+142\text{ kJ/mole-rxn}) + 2.303\,(0.008314\text{ kJ/mole-K})(298\text{ K})(-2.745)$

$\Delta G_{298} = (+142\text{ kJ/mole-rxn}) + (-15.7\text{ kJ/mole}) = +126.3\text{ kJ/mole-rxn}$

8. $C_{(s)}$ + $2\,H_{2\,(g)}$ ⟶ $CH_{4\,(g)}$

	$C_{(s)}$	$2\,H_{2\,(g)}$	$CH_{4\,(g)}$
ΔH°f :	0 kJ	0 kJ	–74.9 kJ
S° :	5.7 J/K	130.7 J/K	186.3 J/K

a) The balanced equation shows 3 moles (2 moles gas) of reactants forming 1 mole of gas products. The standard entropy change is predicted to be negative: greater number of smaller reactant molecules forming a fewer number of product molecules; the system becomes more ordered (less disordered).

b) $\Delta H°_{(reaction)} = [(-74.9\text{ kJ})] - [(0\text{ kJ}) + 2(0\text{ kJ})]$

$\Delta H°_{(reaction)} = \mathbf{-74.8\ kJ/mole\text{-}reaction}$

$\Delta S°_{(reaction)} = [(186.3\text{ J/K})] - [(5.7\text{ J/K}) + 2(130.7\text{ J/K})]$

$\Delta S°_{(reaction)} = \mathbf{-80.8\ J/K}$ or $\mathbf{-0.0808\ kJ/K}$

c) Enthalpy is negative: contributes to spontaneity; entropy is negative: contributes to non-spontaneity. Enthalpy and entropy trends oppose each other. Low temperature favors the direction indicated by the enthalpy: reaction is **spontaneous** at sufficiently **low** temperatures.

d) $\Delta G°_T = \Delta H° - T\,\Delta S°$

$\Delta G°_{(298)} = -\ 74.8\text{ kJ} - [(298\text{ K})(-0.0808\text{ kJ/K})]$

$= -74.8\text{ kJ} + 24.1\text{ kJ} = -50.8\text{ kJ/mole}$

Forward reaction is **spontaneous** at 298 K.

$\Delta G°_{(1000)} = -\ 74.8\text{ kJ} - [(1000\text{ K})(-0.0808\text{ kJ/K})]$

$= -74.8\text{kJ} + 80.8\text{kJ} = \mathbf{+6.0kJ/mole}$

Forward reaction is **non-spontaneous** at 1000 K.

e) $T\Delta_G{=}_0 = \Delta H°/\Delta S° = (-74.8\text{ kJ})/(-0.0808\text{ kJ/K}) = \mathbf{927\ K}$; $\Delta G°_{(927)} = 0$

f) $\log_{10} K = \Delta G°_T/-2.303\ RT$

$\log_{10} K_{298} = (-50.8\text{ kJ})\,/(-\ 2.303)(0.008314\text{ kJ/mole-K})(298\text{ K})$

$\log_{10} K_{298} = 8.90$; $K = \text{antilog}(8.90) = 10^{8.90} = 7.9 \times 10^8$

$\log_{10} K_{1000} = (+6.0\text{ kJ})/(-2.303)(0.008314\text{ kJ/mole-K})(1000\text{ K})$

$\log_{10} K_{1000} = -0.313$; $K = \text{antilog}(-0.313) = 10^{-\ 0.313} = 0.49$

g) $$Q = \frac{(P_{CH4})}{(P_{H2})^2 \chi C\text{-solid}} = \frac{(4.0)}{(0.20)^2(1)} = 100$$

$\Delta G_{298} = (-\ 50.8\text{ kJ/mole-rxn}) + 2.303\ (0.008314\text{ kJ/mole-K})(298\text{ K})\log_{10}(100)$

$\Delta G_{298} = (-\ 50.8\text{ kJ/mole-rxn}) + 2.303\ (0.008314\text{ kJ/mole-K})(298\text{ K})(2)$

$\Delta G_{298} = (-\ 50.8\text{ kJ/mole-rxn}) + (+11.4\text{ kJ/mole}) = -39.4\text{ kJ/mole-rxn}$

26 Acid/Base Equilibrium, pH, and Buffers

The general concepts of acids, bases, and acid/base reactions were described in Chapter 9 and Chapter 23 described techniques for solving acid/base equilibria. In this chapter, these skills are combined for a complete analysis of all reaction combinations of strong and weak acids and bases.

I ACID AND BASE DISSOCIATION REACTIONS: ACID AND BASE STRENGTH

Acid molecules react with water in an acid/base reaction with water molecules acting as the base; the result is acid dissociation in water or acid ionization. Note that the format showing the base role of water in the reaction indicates that H_3O^+ is the conjugate acid of water as a base.

$$\text{H—••Acid}_{(aq)} + H_2O•• \rightleftarrows ••\text{Acid}^-_{(aq)} + (H_2O••\text{—H})^+_{(aq)}$$

Since H_3O^+ ($H_2O••—H^+_{(aq)}$) is considered to be identical to $H^+_{(aq)}$ for this reaction, the alternative notation is used for equilibrium calculations of acid dissociation (Chapter 23).

$$\text{H—Acid}_{(aq)} \xrightarrow{H_2O\ \text{solvent}} H^+_{(aq)} + (\text{Acid})^-_{(aq)}$$

The dissociation (ionization) reaction of acids with water as a base is the reference reaction for classification of acid strength. Acids can very generally be defined as either weak acids or strong acids based on the equilibrium constant for the acid dissociation reaction with water. Classification through the value of the equilibrium constant allows more precision than the previous approximation (Chapter 9) of ≈ 100% dissociation for strong acids and ≈ < 1% dissociation for weak acids.

Acids that have an acid dissociation equilibrium constant with a numerical value greater than one (and usually greater than 100) can be considered **strong** acids:

$$\underset{\textbf{(strong)}}{\text{H—Acid}} \underset{\longleftarrow}{\xrightarrow{H_2O}} H^+_{(aq)} + (\text{Acid})^-_{(aq)} \quad \mathbf{K_a} = \frac{[H^+][(\text{Acid})^-]}{[\text{H—Acid}]} > \mathbf{1}(\text{usually} > 10^2)$$

Example:

$$\underset{\textbf{(strong)}}{\text{H—Cl}} \underset{\longleftarrow}{\xrightarrow{H_2O}} H^+_{(aq)} + Cl^-_{(aq)} \quad \mathbf{Ka = 1 \times 10^7} \quad (\cong \mathbf{100\%}\,\text{reaction})$$

Acids that have an acid dissociation equilibrium constant with a numerical value less than one (and usually less than 0.01) can be considered **weak** acids:

$$\underset{\textbf{(weak)}}{\text{H—Acid}} \underset{\longleftarrow}{\xrightarrow{H_2O}} H^+_{(aq)} + (\text{Acid})^-_{(aq)} \quad \mathbf{K_a} = \frac{[H^+][(\text{Acid})^-]}{[\text{H Acid}]} < 1\ (\text{usually} < 10^{-2})$$

Example:

$$\underset{\textbf{(weak)}}{CH_3COO\ \ H} \underset{\longleftarrow}{\xrightarrow{H2O}} H^+_{(aq)} + CH_3COO^-_{(aq)} \quad \mathbf{K_a = 1.8 \times 10^{-5}} \quad (\cong \mathbf{0.5\%}\ \text{reaction})$$

Certain acids are polyprotic; in reactions with a base they can transfer more than one proton (H^+ ion); each individual proton transfer has a separate equilibrium constant.

Example:

$$H_2SO_{4(aq)} \underset{\longleftarrow}{\xrightarrow{H_2O}} H^+_{(aq)} + HSO_4^-{}_{(aq)} \quad \mathbf{K_a} = \frac{[H^+][HSO_4^-]}{[H_2SO_4]} = \mathbf{6.3 \times 10^4}$$

$$HSO_4^-{}_{(aq)} \underset{\longleftarrow}{\xrightarrow{H_2O}} H^+_{(aq)} + SO_4^{-2}{}_{(aq)} \ \mathbf{K_a} = \frac{[H^+][SO_4^{-2}]}{[HSO_4^-]} = \mathbf{1.2 \times 10^{-2}}$$

The equilibrium constant for acid dissociation is often listed as a derived numerical value termed $\mathbf{pK_a}$ (the "p" refers to "power", signifying "exponent"). The pK_a is defined as the **negative** logarithm of the numerical value of the K_a: $\mathbf{pK_a = -log_{10}(K_a)}$

Example:

$$HCl \text{ has a } K_a \text{ of } 1 \times 10^7;\ pK_a = -\log_{10}(1 \times 10^7) = -7$$

$$CH_3COOH \text{ has a } K_a \text{ of } 1.8 \times 10^{-5};\ pK_a = -\log_{10}(1.8 \times 10^{-5}) = +4.75$$

$$H_2SO_4 \text{ has a } K_a \text{ of } 6.3 \times 10^4;\ pK_a = -\log_{10}(6.3 \times 10^4) = -4.8$$

$$HSO_4^- \text{ has a } K_a \text{ of } 1.2 \times 10^{-2};\ pK_a = -\log_{10}(1.2 \times 10^{-2}) = +1.9$$

Base strength in aqueous solution is measured as the concentration of aqueous hydroxide ion, $[OH^-{}_{(aq)}]$. Many strong bases are hydroxide containing ionic compounds, which have high solubility in water; thus, complete separation of dissolved ions produces high concentrations of hydroxide ion. In these cases, water acts only as the solvent and does not participate in an acid/base reaction.

$$\text{Metal(OH)} \xrightarrow{H_2O\ \text{solvent}} \text{Metal}^+_{(aq)} + OH^-_{(aq)} \quad \text{(variable solubility)}$$

Non-hydroxide bases undergo an acid/base reaction with the role of water as an acid; this reaction with water is the reference reaction for determining base strength. Note that (OH^-) is the conjugate base of water as an acid.

$$H\text{—}{\bullet\bullet}OH + {\bullet\bullet}Base^-_{(aq)} \rightleftarrows {\bullet\bullet}OH^-_{(aq)} + H\text{—}{\bullet\bullet}Base_{(aq)} \quad K_b = \frac{[OH^-][H\text{—}Base]}{[Base^-](\chi H_2O \cong 1)}$$

Non-hydroxide bases that have a base equilibrium constant ($\mathbf{K_b}$) with a numerical value greater than one can be considered **strong** bases. With some adjustment for the concentration of the water solvent, this approximately means that strong bases are those with a base strength equal to or greater than hydroxide ion.

Example:

$$H\text{—}{\bullet\bullet}OH + {\bullet\bullet}NH_2^-{}_{(aq)} \rightleftarrows {\bullet\bullet}OH^-_{(aq)} + (H\text{—}{\bullet\bullet}NH_2)_{(aq)};\ \mathbf{K_b} = \frac{[OH^-][NH_3]}{[NH_2^-]} = \mathbf{10^{22}}$$

$$= (NH_3)$$

Non-hydroxide bases that have a base equilibrium constant (**K_b**) with a numerical value less than one can be considered **weak** bases.

Example:

$$H{-}{\bullet\bullet}OH + {\bullet\bullet}NH_{3\,(aq)} \rightleftharpoons {\bullet\bullet}OH^-_{(aq)} + (H{-}{\bullet\bullet}NH_3)^+_{(aq)};\ \mathbf{K_b} = \frac{[OH^-][NH_4^+]}{[NH_3]} = \mathbf{1.8 \times 10^{-5}}$$

$$H{-}{\bullet\bullet}OH + {\bullet\bullet}^-OOCCH_{3\,(aq)} \rightleftharpoons {\bullet\bullet}OH^-_{(aq)} + H{-}{\bullet\bullet}OOCCH_{3(aq)};\ \mathbf{K_b = 2.0 \times 10^{-9}}$$

The equilibrium constant for the reaction of a non-hydroxide base has a similarly derived numerical value termed **pK_b**. The pK_b is defined as the **negative** logarithm of the numerical value of the K_b: **$pK_b = -\log_{10}(K_b)$**.

Example:

NH_2^- has a K_b of 1×10^{22}; $pK_b = -\log_{10}(1 \times 10^{22}) = -22$

NH_3 has a K_b of 1.8×10^{-5}; $pK_b = -\log_{10}(1.8 \times 10^{-5}) = +4.75$

CH_3COO^- has a K_b of 2.0×10^{-9}; $pK_b = -\log_{10}(2.0 \times 10^{-9}) = +9.25$

An important class of negative ions that act as bases are conjugate bases of acids: an acid molecule that has lost its H^+ ion is a possible base in a complementary reaction.

$$\underset{\text{acid}}{H{-}{\bullet\bullet}Acid} + {\bullet\bullet}Base^- \longrightarrow \underset{\text{conjugate base}}{\mathbf{{\bullet\bullet}Acid^-}} + H{-}{\bullet\bullet}Base$$

A base molecule, which has accepted an H^+ ion, is a possible acid in the complementary reverse reaction and is the conjugate acid of the base.

$$H{-}{\bullet\bullet}Acid + \underset{\text{base}}{{\bullet\bullet}Base} \longrightarrow {\bullet\bullet}Acid^- + \underset{\text{conjugate acid}}{(H{-}{\bullet\bullet}Base)^+}$$

Any general reversible acid/base reaction can be identified based on the acid, base, conjugate acid and conjugate base. The acid and base react in the forward direction; the conjugate acid and conjugate base react in the reverse direction.

Example:

acid		base		conjugate base		conjugate acid
H Acid	+	:Base	$\rightleftharpoons$	$(Acid)^-$	+	$(H\ \ Base)^+$
acid		**base**		**conjugate base**		**conjugate acid**
CH_3COOH	+	NH_3	$\rightleftharpoons$	CH_3COO^-		NH_4^+

The term "conjugate" signifies a relationship between the acid and base forms of a specific molecule; the term does not indicate that one form is necessarily less reactive or less important than the other. All combinations are designated as conjugate acid/base pairs; standard practice for such pairs generally lists information such equilibrium constants through the K_a, or pK_a, of the acid form.

Autoionization of Water

Water molecules in **neutral** water will self-ionize to form H^+ aqueous (or equivalently H_3O^+ aqueous) and OH^- aqueous; the reaction is an equilibrium:

$$H_2O_{(l)} + H_2O_{(l)} \rightleftarrows H_3O^+_{(aq)} + OH^-_{(aq)} \quad K_w = 1\times10^{-14}$$

or, equivalently:

$$H_2O_{(l)} \rightleftarrows H^+_{(aq)} + OH^-_{(aq)} \quad K_w = 1\times10^{-14}$$

Under neutral water conditions, only a very small percentage of water molecules will ionize. The molar concentrations of $[H^+_{(aq)}]$ and $[OH^-_{(aq)}]$ are determined from the equilibrium constant for the reaction, termed K_w. Neutral water ($\chi H_2O \cong 1$) does not appear in the equilibrium expression.

$$\mathbf{K_w = [H^+][OH^-] = 1 \times 10^{s14}}$$

The balanced equation for the ionization indicates a one-to-one ratio for the formation of ($H^+_{(aq)}$) and ($OH^-_{(aq)}$); **neutral** water has $[H^+]$ equal to $[OH^-]$. Setting **(x)** equal to the concentration changes for each of the two ions solves for the neutral concentrations: $\Delta[H^+] = \Delta[OH^-] =$ **(x)**.

	$H_2O_{(aq)}$ →	$H^+_{(aq)}$ +	$OH^-_{(aq)}$
starting conc. :	excess	0 M	0 M
change, Δ :	– x M	+ x M	+ x M
equilibrium conc. :	excess	x M	x M

$$K_w = [H^+][OH^-];\ 1 \times 10^{-14} = x^2;\ \mathbf{x} = (1 \times 10^{-14})^{1/2} = 1 \times 10^{-7}$$

$[H^+] = 1 \times 10^{-7}$ M for neutral water; **$[OH^-] = 1 \times 10^{-7}$ M** for neutral water.

The value of the constant K_w (1×10^{-14}) applies to all acid or base aqueous conditions. As $[H^+]$ increases through the addition of an acid, the $[OH^-]$ must decrease; as the $[OH^-]$ increases through the addition of a base, the $[H^+]$ must decrease.

Example: A certain solution has a $[H^+_{(aq)}]$ of 0.015 M. Calculate $[OH^-_{(aq)}]$.

$$K_w = [H^+][OH^-];\ 1\times10^{-14} = (0.015\text{M})[OH^-];\ \mathbf{[OH^-]} = \frac{(1\times10^{-14})}{(0.015)} = \mathbf{6.7\times10^{-13}\ M}$$

Example: A certain solution has a $[OH^-_{(aq)}] = 4.0 \times 10^{-5}$ M. Calculate $[H^+_{(aq)}]$.

$$K_w = [H^+][OH^-];\ 1\times10^{-14} = [H^+](4.0\times10^{-5}\text{M})\ \mathbf{[H^+]} = \frac{(1\times10^{-14})}{(4.0\times10^{-5})} = \mathbf{2.5\times10^{-10}\ M}$$

II Relationship between Acid (pK_a) and Conjugate Base Strength

The acid strength of an acid is inversely proportional to the base strength of its conjugate base. Very strong acids have very weak conjugate bases; very weak acids have strong conjugate bases. This conclusion is based on the spontaneity of the forward and reverse reactions. The degree of spontaneity (size of the negative value for ΔG) for the forward acid reaction correlates to the strength of the acid. The stronger the acid, the less spontaneous (size of the positive value for ΔG) the reverse conjugate base reaction will be, correlating to the weaker base. Conversely, the weaker the acid (unfavorable ΔG), the stronger conjugate base (favorable ΔG). In anthropomorphic terms, a strong acid that desperately wants to get rid of its proton has no desire to get it back through its conjugate base. A weak acid that desperately wants to keep its proton has a great desire to get it back through its conjugate base if it loses that proton.

A property of equilibrium constants states that if two component equations add to produce a specific required equation, then the equilibrium constant for the required equation is equal to the multiplication of the equilibrium constants for the two component equations.

$$A+B \underset{}{\overset{K_{ABCD}}{\rightleftharpoons}} C+D \underset{}{\overset{K_{CDEF}}{\rightleftharpoons}} E+F; \quad A+B \underset{}{\overset{K_{ABEF}=(K_{ABCD})\times(K_{CDEF})}{\rightleftharpoons}} E+F$$

The relationship between K_a of an acid and K_b of its conjugate base can be expressed through the related reference reactions with water as the base or the acid. HA represents the acid and $(A)^-$ represents the conjugate base.

$$HA_{(aq)} + H_2O_{(l)} \rightleftharpoons H_3O^+_{(aq)} + (A)^-_{(aq)} \qquad \text{reference for } K_a$$

$$(A)^-_{(aq)} + H_2O_{(l)} \rightleftharpoons HA_{(aq)} + OH^-_{(aq)} \qquad \text{reference for } K_b$$

To derive the reference reaction for K_b in terms of the reference reaction for K_a, first reverse the reference reaction for K_a. Recall that $K_{(reverse)} = 1/K_{(forward)}$.

$$(A)^-_{(aq)} + H_3O^+_{(aq)} \rightleftharpoons HA_{(aq)} + H_2O_{(l)} \quad K = \frac{1}{K_a}$$

This equilibrium equation is not the K_b reference reaction, but can be converted to the required equation by adding the equation for the autoionization of water; the compounds shown in boxes cancel:

$$(A)^-_{(aq)} + \boxed{H_3O^+_{(aq)}} \rightleftharpoons HA_{(aq)} + \boxed{H_2O_{(l)}} \quad K = \frac{1}{K_a}$$

$$\boxed{H_2O_{(l)}} + H_2O_{(l)} \rightleftharpoons \boxed{H_3O^+_{(aq)}} + OH^-_{(aq)} \quad K_w = 1\times10^{-14}$$

$$(A)^-_{(aq)} + H_2O_{(l)} \rightleftharpoons HA_{(aq)} + OH^-_{(aq)} \quad K_b$$

$$K_{ABEF} = (K_{ABCD}) \times (K_{CDEF}); \quad K_b = \frac{(1)}{K_a} \times (K_w); \quad K_b = \frac{K_w}{K_a};$$

$$\mathbf{K_b = \frac{1 \times 10^{-14}}{K_a}} \quad \text{or} \quad \mathbf{1 \times 10^{-14} = (K_a)(K_b)}$$

A useful form of the inverse relationship for pK_a and pK_b can be expressed by taking the negative $\log_{10}$ of both sides of the equation:

$$-\log_{10}(1 \times 10^{-14}) = -\log_{10}\{(K_a)(K_b)\}; \quad \text{Recall that } \log(a \times b) = \log(a) + \log(b).$$

$$-\log_{10}(1 \times 10^{-14}) = (-\log_{10}K_a) + (-\log_{10}K_b)$$

$$\mathbf{14 = pK_a + pK_b}$$

This equation can be used directly to find pK_b from pK_a or pK_a from pK_b.

Example: The K_a for HCl is 1×10^7; calculate the K_b and pK_b of (Cl^-) as a base.

$$K_b = \frac{1 \times 10^{-14}}{K_a}; \quad \mathbf{K_b} = \frac{1 \times 10^{-14}}{1 \times 10^7} = \mathbf{1 \times 10^{-21}}; \quad pK_b = -\log_{10}(1 \times 10^{-21}) = \mathbf{+21}$$

$$\text{or } \mathbf{pK_a} \text{ of HCl} = -\log_{10}K_a = -\log_{10}(1 \times 10^7) = \mathbf{-7}$$

$$14 = pK_a + pK_b;\ pK_b = (14 - pK_a);\ pK_b = 14 - (-7) = +21$$

Example: The K_b for NH_3 is 1.8×10^{-5}; calculate the K_a and pK_a of (NH_4^+) as an acid.

$$K_a = \frac{1 \times 10^{-14}}{K_b}; \quad \mathbf{K_a} = \frac{1 \times 10^{-14}}{1.8 \times 10^{-5}} = \mathbf{5 \times 10^{-10}}; \quad pK_a = -\log_{10}(5 \times 10^{-10}) = \mathbf{+9.25}$$

$$\text{or } \mathbf{pK_b} \text{ of } NH_3 = -\log_{10}K_b = -\log_{10}(1.8 \times 10^{-5}) = \mathbf{+4.75}$$

$$14 = pK_a + pK_b;\ pK_a = (14 - pK_b);\ pK_a = 14 - (4.75) = +9.25$$

The included "**Table of pKa Values for Common Acids and Bases**" provides the required data for equilibrium calculations and emphasizes the inverse relationship between the strength of the acid and the strength of the conjugate base.

III PROCESS FOR DETERMINING THE EQUILIBRIUM POSITION FOR ACID/BASE REACTIONS

The equilibrium position of an acid/base reaction describes which reaction, forward or reverse, is favored under standard concentration conditions:

acid		**base**		**conjugate base**		**conjugate acid**
H—Acid	+	Base or $(Base^-)$	$\rightleftarrows$	$(Acid)^-$	+	$(H\text{—}Base)^+$ or (H—Base)

The favored (spontaneous) direction is determined by the value of $\Delta G°_{(forward)}$ and is reflected in the corresponding equilibrium constant:

If the forward reaction is favored: $K_{(forward)} > 1$ and $K_{(reverse)} < 1$; the equilibrium favors product formation from reactants.

If the reverse reaction is favored: $K_{(forward)} < 1$ and $K_{(reverse)} > 1$; the equilibrium favors reactant formation from products.

Table of pKa Values for Common Acids and Bases

Acid	K_a	pK_a	Conjugate Base
Strongest Acid ↑			Weakest Base ↑
HI	$K_a = 1 \times 10^{10}$	$pK_a = -10$	I^-
HBr	$K_a = 1 \times 10^{9}$	$pK_a = -9$	Br^-
HCl	$K_a = 1 \times 10^{7}$	$pK_a = -7$	Cl^-
H_2SO_4 (first)	$K_a = 6 \times 10^{4}$	$pK_a = -4.8$	HSO_4^-
$CH_3OH_2^+$	$K_a = 1 \times 10^{2}$	$pK_a = -2$	CH_3OH
H_3O^+	$K_a = 50$	$pK_a = -1.7$	H_2O
HNO_3	$K_a = 25$	$pK_a = -1.4$	NO_3^-
H_2SO_3 (first)	$K_a = 1.7 \times 10^{-2}$	$pK_a = +1.8$	HSO_3^-
HSO_4^-	$K_a = 1.2 \times 10^{-2}$	$pK_a = +1.9$	SO_4^{-2}
$HClO_2$	$K_a = 1.0 \times 10^{-2}$	$pK_a = +2$	ClO_2^-
H_3PO_4 (first)	$K_a = 7.5 \times 10^{-3}$	$pK_a = +2.1$	$H_2PO_4^-$
HF	$K_a = 6.8 \times 10^{-4}$	$pK_a = +3.2$	F^-
HNO_2	$K_a = 4.5 \times 10^{-4}$	$pK_a = +3.3$	NO_2^-
CH_3COOH	$K_a = 1.8 \times 10^{-5}$	$pK_a = +4.75$	CH_3COO^-
H_2CO_3 (first)	$K_a = 4.3 \times 10^{-7}$	$pK_a = +6.4$	HCO_3^-
HOCl	$K_a = 3.0 \times 10^{-8}$	$pK_a = +7.5$	OCl^-
NH_4^+	$K_a = 5 \times 10^{-10}$	$pK_a = +9.25$	NH_3
HCN	$K_a = 4.9 \times 10^{-10}$	$pK_a = +9.4$	CN^-
C_6H_5OH	$K_a = 1.3 \times 10^{-10}$	$pK_a = +9.9$	$C_6H_5O^-$
H_2O	$K_a = 2 \times 10^{-16}$	$pK_a = +15.7$	OH^-
CH_3OH	$K_a = 1 \times 10^{-16}$	$pK_a = +16$	CH_3O^-
NH_3	$K_a = 1 \times 10^{-36}$	$pK_a = +36$	NH_2^-
CH_3CH_3	$K_a = 1 \times 10^{-62}$	$pK_a = +62$	$CH_3CH_2^-$
↓ Weakest Acid			↓ Strongest Base

The equilibrium position of an acid/base reaction is determined by comparing the values of K_a or pK_a for the acid vs. the K_a or pK_a for the conjugate acid formed from the reacting base in the equation. The pK_b of the base is not specifically required.

Step (**1**): Write the correct balanced proton transfer equation if required; use the techniques described in Chapter 9.

Step (**2**): Identify the acid in the forward reaction, then use the table of pK_a values to determine the pK_a of this acid.

Step (**3**): Identify the base in the forward reaction, then use the table of pK_a values to determine the pK_a of the conjugate acid of this base. The pK_a of the conjugate acid is derived from the pK_b of the base.

Step (**4**): Select the favored equilibrium direction by following the **rule**: The favored direction of any acid/base reaction ($K > 1$) always shows H^+ (proton) transfer from the stronger acid (lower pK_a value including sign) to produce the weaker acid (higher pK_a value).

Example: Determine the position of the following equilibrium:

$$CH_3COOH + NH_3 \rightleftarrows CH_3COO^- + NH_4^+$$

Step (**1**): The balanced equation is provided.

Step (**2**): In the forward direction, CH_3COOH transfers the proton; it is the acid in the forward direction. CH_3COOH has a pK_a of 4.7.

Step (**3**): In the forward direction, NH_3 accepts the proton; it is the base in the forward direction. NH_4^+ is the conjugate acid of NH_3 as a base; NH_4^+ has a pK_a of 9.25.

Step (**4**): The equilibrium is favored in the direction of H^+ (proton) transfer from the stronger acid (CH_3COOH; $pK_a = 4.7$) to produce the weaker acid (NH_4^+; $pK_a = 9.25$).

$$\begin{array}{lcll} CH_3COOH \;+\; NH_3 & \longrightarrow & CH_3COO^- \;+\; NH_4^+ & \mathbf{K>1} \\ pK_a = 4.7\ \textbf{(stronger)} & \leftarrow & \quad\quad pK_a = 9.3 & \textbf{(weaker)} \end{array}$$

Example: Determine the equilibrium position for the reaction of aqueous hydrochloric acid with aqueous sodium nitrate.

Step (**1**): $HCl_{(aq)} + \mathbf{NaNO_{4\,(aq)}} \rightleftarrows Cl^-_{(aq)}\ (\mathbf{Na^+_{(aq)}}) + \mathbf{HNO_{3\,(aq)}}$

$Na^+_{(aq)}$ is a spectator ion and is not involved in the acid/base reaction.

Step (**2**): In the forward direction, HCl transfers the proton; it is the acid in the forward direction. HCl has a pK_a of −7.

Step (**3**): In the forward direction, NO_3^- accepts the proton; it is the base in the forward direction. HNO_3 is the conjugate acid of NO_3^- as a base; HNO_3 has a pK_a of −2.

Step (**4**): The equilibrium is favored in the direction of H^+ (proton) transfer from the stronger acid (HCl; $pK_a = -7$) to produce the weaker acid (HNO_3; $pK_a = -2$).

$$\begin{array}{lcll} HCl_{(aq)} \;+\; NaNO_{3\,(aq)} & \longrightarrow & Cl^-_{(aq)}(Na^+_{(aq)}) \;+\; HNO_{3\,(aq)} & \mathbf{K>1} \\ pK_a = -7\ \textbf{(stronger)} & \leftarrow & \quad\quad pK_a = -2 & \textbf{(weaker)} \end{array}$$

Example: Determine the equilibrium position for the reaction of aqueous hydrofluoric acid with aqueous sodium hydrogen sulfate.

Step (**1**): $HF_{(aq)} + NaHSO_{4\,(aq)} \rightleftarrows F^-_{(aq)}\ (Na^+_{(aq)}) + H_2SO_{4\,(aq)}$

$Na^+_{(aq)}$ is a spectator ion and is not involved in the acid/base reaction.

Step (**2**): In the forward direction, HF transfers the proton; it is the acid in the forward direction. HF has a pK_a of +3.2.

Step (**3**): In the forward direction, HSO_4^- accepts the proton; it is the base in the forward direction. H_2SO_4 is the conjugate acid of HSO_4^- as a base; H_2SO_4 has a pK_a Of −4.8.

Step (**4**): The equilibrium is favored in the direction of H^+(proton) transfer from the stronger acid (H_2SO_4; $pK_a = -4.8$) to produce the weaker acid (HF; $pK_a = +3.2$).

$$\underset{pK_a = 3.2 \textbf{ (weaker)}}{HF_{(aq)}} + HSO_4^-(Na^+_{(aq)}) \underset{\longleftarrow}{\longrightarrow} F^-_{(aq)} + \underset{pK_a = -4.8 \textbf{ (stronger)}}{H_2SO_{4(aq)}} \qquad \mathbf{K < 1}$$

IV pH CALCULATIONS IN AQUEOUS SOLUTIONS: GENERAL CONCEPT OF pH

The concentration of $[H^+_{(aq)}]$ in a solution measures the general degree of acidity of the solution. This value is often listed as the derived numerical value termed **pH**, in which the "p" for power, or exponent, is defined in the same way as for pK_a. The pH of an aqueous solution is defined as the negative logarithm of the numerical value of the concentration of $(H^+_{(aq)})$, $[H^+_{(aq)}]$ in molarity: $\mathbf{pH = -log_{10}[H^+_{(aq)}]}$.

The basicity of a solution is inversely proportional to the acidity and can be expressed through the pH value, but a further unit can be employed; the **pOH** of an aqueous solution is defined as the negative logarithm of the numerical value of the $[OH^-_{(aq)}]$ in molarity: $\mathbf{pOH = -log_{10}[OH^-_{(aq)}]}$.

Example: Calculate the pH and pOH of neutral water.

From the previous calculation for neutral water: $[H^+] = 1 \times 10^{-7}$ M; $[OH^-] = 1 \times 10^{-7}$ M

$$\textbf{pH} \text{ of this solution} = -\log [H^+] = -\log(1 \times 10^{-7}) = -\log (10^{-7}) = -(-7) = 7;\ \mathbf{pH = 7}$$

$$\textbf{pOH} \text{ of this solution} = -\log [OH^-] = -\log (1 \times 10^{-7}) = -(-7) = 7;\ \mathbf{pOH = 7}$$

A useful form of the inverse relationship for pH and pOH is similar to the one expressed for pK_a and pK_b. Take the negative $\log_{10}$ of both sides of $K_w = 1 \times 10^{-14} = [H^+][OH^-]$.

$$-\log_{10}(1 \times 10^{-14}) = -\log_{10}\{[H^+_{(aq)}][OH^-_{(aq)}]\}$$

$$-\log_{10}(1\times10^{-14}) = (-\log_{10}[^+_{(aq)}]) + (-\log_{10}[OH^-_{(aq)}])$$

$$\mathbf{14} = \mathbf{pH} + \mathbf{pOH}$$

This equation can be used directly to find pH from pOH or pOH from pH.

Example: An aqueous solution shows a molar concentration of $H^+_{(aq)} = 0.001$ M; calculate pH and pOH.

$$pH = -\log [H^+] = -\log (0.001) = -\log(10^{-3}) = -(-3) = 3;\ \mathbf{pH = 3}$$

$$\mathbf{pOH = (14-pH) = 14-3 = 11}$$

Example: An aqueous solution shows a molar concentration of OH^- (aq) = 0.01 M; calculate pH and pOH.

$$pOH = -\log [OH^-] = -\log (0.01) = -\log(10^{-2}) = -(-2) = 2;\ \mathbf{pOH = 2}$$

$$\mathbf{pH} = (14-pOH) = 14-2 = \mathbf{12}$$

Example: An aqueous solution is 5.0×10^{-4} M $[OH^-_{(aq)}]$; calculate pH and pOH.

$$pOH = -\log [OH^-] = -\log (5 \times 10^{-4})$$
$$= -\{\log (5) + \log(10^{-4})\} = -\{(+0.70) + (-4)\} = -(-3.3) = \mathbf{3.3;\ pOH = 3.3}$$
$$\mathbf{pH} = (14-pOH) = 14-(3.3) = \mathbf{10.7}$$

The pH scale measures the acidity or basicity of aqueous solutions, usually between the values 0 and 14; negative numbers are also possible.

$$\mathbf{pH = 0} \longleftrightarrow \mathbf{pH = 7} \longleftrightarrow \mathbf{pH = 14}$$

most acidic ⟷ **neutral** ⟷ **most basic**

V pH CALCULATIONS IN AQUEOUS SOLUTIONS: REACTIONS OF ONE ACID OR BASE WITH WATER

A single specific acid in aqueous solution produces $H^+_{(aq)}$ through acid dissociation with water acting as the base. A non-hydroxide base in aqueous solution produces $OH^-_{(aq)}$ through the reaction of the base with water as an acid. A hydroxide base produces $OH^-_{(aq)}$ through a simple solubility process in water. Calculations of pH and pOH employ the techniques of simple stoichiometry (Chapter 8) and equilibrium calculations (Chapter 23). Interconversion of pH and pOH is accomplished with the use of the following equations:

$$\mathbf{pH = -\log_{10}[H^+_{(aq)}]} \qquad \mathbf{pOH = -\log_{10}[OH^-_{(aq)}]}$$

$$\mathbf{[H^+_{(aq)}] = antilog(-pH) = 10^{-pH}} \qquad \mathbf{[OH^-_{(aq)}] = antilog(-pOH) = 10^{-pOH}}$$

$$\mathbf{pH + pOH = 14}$$

Most of the possible reactions of acids or bases with water can be organized into four categories to aid in selection of the correct solution techniques.

Category **(1): Strong acids in water.** All strong acids can be considered nearly 100% dissociated in water due to the large values of the equilibrium constant. In these cases, the concentration of $H^+_{(aq)}$ is determined by solving the simple solution stoichiometry of the reaction; for a monoprotic acid, $[H^+_{(aq)}]$ in a solution of the acid is approximately equal to the initial concentration of the strong acid.

$$\underset{\textbf{(strong)}}{\text{H–Acid}_{(aq)}} \underset{\longleftarrow}{\overset{H_2O}{\longrightarrow}} H^+_{(aq)} + (\text{Acid})^-_{(aq)} \qquad \mathbf{K_a >> 1\ (\cong 100\%\ reaction)}$$

Category **(2): Strong hydroxide bases in water.** The concentration of $OH^-_{(aq)}$ is determined by simple solution stoichiometry based on the number of hydroxide ions in the formula of the base. Although not common for topics in this chapter, **strong** non-hydroxide bases are analyzed though a stoichiometric reaction with water as the acid.

Category **(3): Weak acids in water.** The concentration of $H^+_{(aq)}$ is determined by solving the acid dissociation equilibrium from a given value for K_a. An alternative problem is the calculation of the value of K_a from a given value for pH or pOH.

Category **(4): Weak bases in water.** The concentration of $OH^-_{(aq)}$ is determined by solving the equilibrium reaction of the weak base with water as the acid, given the value for K_b. An alternative problem is the calculation of the value of K_b from a given value for pH or pOH.

Example: Calculate the pH and pOH of a 2.5 M solution of HCl (strong acid).

The reaction of a strong acid in water is an example of category (1). $[H^+_{(aq)}]$ is determined by solving the simple stoichiometry that assumes that HCl is ≅ 100% ionized. Complete reaction of the limiting reagent (water is in excess) means that $[H^+_{(aq)}] \cong [H\text{—Acid}]_{initial}$. Use any necessary techniques described in Chapter 8. The first requirement is to write the correct balanced equation (Chapter 9).

	$HCl_{(aq)}$ $\xrightarrow{H_2O \text{ solvent}}$	$H^+_{(aq)}$ +	$Cl^-_{(aq)}$	$K_a = 1\times10^7$ (≅ 100% ionized)
starting conc.	: 2.5 M	≈ 0 M	0 M	
change, Δ	: −2.5 M	+2.5 M	+2.5 M	
finish	: 0 M	2.5 M	2.5 M	

$$\Delta[H^+_{(aq)}] = [2.5\text{ M }\Delta\text{ }HCl_{(aq)}] \times \frac{1\text{ }H^+}{1\text{ HCl}} = 2.5\text{ M}$$

$$\Delta[Cl^-_{(aq)}] = [2.5\text{ M }\Delta\text{ }HCl_{(aq)}] \times \frac{1\text{ }Cl^-}{1\text{ HCl}} = 2.5\text{ M}$$

The table entry for starting $[H^+_{(aq)}]$ as ≈ 0 M is due to the small amount of $H^+_{(aq)}$ present in neutral water formed form the autoionization equilibrium (K_w).

$$\mathbf{pH = -log_{10}[H^+_{(aq)}] = -log_{10}(2.5) = -0.40}$$

$$\mathbf{pH + pOH = 14;\ pOH = (14 - pH) = 14 - (-0.40) = 14.4}$$

Example: Calculate the pH and pOH of a solution of 0.050 M $Ba(OH)_2$.

A strong hydroxide base in water is an example of category (2); $[OH^-_{(aq)}]$ is determined by simple solution stoichiometry that assumes 100% solubility.

	$Ba(OH)_2$ $\xrightarrow{H_2O \text{ solvent}}$	$Ba^{+2}_{(aq)}$ +	$2\text{ }OH^-_{(aq)}$
starting conc.	: 0.050 M	0 M	≈ 0 M
change, Δ	: −0.050 M	+0.050 M	+0.10 M
finish	: 0 M	0.050 M	0.10 M

$$\Delta[Ba^{+2}_{(aq)}] = [0.050\text{ M }\Delta\text{ }Ba(OH)_{2(aq)}] \times \frac{1\text{ }Ba^{+2}}{1\text{ }Ba(OH)_2} = 0.050\text{ M}$$

$$\Delta[OH^-_{(aq)}] = [0.050\text{ M }\Delta\text{ }Ba(OH)_{2(aq)}] \times \frac{2\text{ }OH^-}{1\text{ }Ba(OH)_2} = 0.10\text{ M}$$

The table entry for starting $[OH^+_{(aq)}]$ as ≈ 0 M is due to the small amount of $OH^-_{(aq)}$ present in neutral water formed form the autoionization equilibrium (K_w).

$$pOH = -log\ [OH^-_{(aq)}] = -log\ (0.10) = -(-1) = \mathbf{1}$$

$$pH = (14 - pOH) = 14 - (1) = \mathbf{13}$$

Example: Calculate the pH and pOH of a solution of 0.0030 M $HBr_{(aq)}$. Use the table of pK_a values to determine the acid strength.

The pK_a of HBr is listed as 10^9; this is a reaction of a strong acid in water and falls into category **(1)**.

		$HBr_{(aq)}$	$\xrightarrow{H_2O\ solvent}$	$H^+_{(aq)}$	+	$Br^-_{(aq)}$	$K_a = 1 \times 10^9$ ($\cong$ 100% ionized)
starting conc.	:	0.0030 M		$\approx$ 0 M		0 M	
change, Δ	:	−0.0030 M		+0.0030 M		+0.0030 M	
finish	:	0 M		0.0030 M		0.0030 M	

$$\Delta[H^+_{(aq)}] = [0.0030\ M\ \Delta\ HBr_{(aq)}] \times \frac{1\ H^+}{1\ HBr} = 0.0030\ M$$

$$\Delta[Br^-_{(aq)}] = [0.0030\ M\ \Delta\ HBr_{(aq)}] \times \frac{1\ Br^-}{1\ HBr} = 0.0030\ M$$

$$\mathbf{pH} = -\log_{10}[H^+_{(aq)}] = -\log_{10}[(0.0030) = -\log(3 \times 10^{-3}) = \mathbf{2.5}$$
$$\mathbf{pOH} = (14 - pH) = 14 - (2.5) = \mathbf{11.5}$$

An alternate process for determining logs is to note that $[-\log(3 \times 10^{-3})]$ is the same as: $-\{\log(3) + \log(10^{-3})\} = -\{(+0.477) + (-3)\} = -\{(-2.523)\} = 2.523 = 2.5$.

Example: An unknown acid (symbolized as HA) is a weak acid. A 0.10 M starting concentration of this acid shows a pH of 2.97 after equilibrium is established. Calculate the K_a of this unknown acid.

A weak acid in water is an example of category **(3)**; in this case, an equilibrium must be solved to determine the value of K_a. The general steps described in Chapter 23 are employed.

(1) $\mathbf{HA_{(aq)} \rightleftharpoons H^+_{(aq)} + A^-_{(aq)}}$

(2) $\mathbf{K_a = \frac{[H^+][A^-]}{[HA]}}$

(3) The equilibrium $[H^+_{(aq)}]$ is determined from the pH of 2.97:
$pH = -\log[H^+] = 2.97$; $\mathbf{[H^+]}$ = antilog $(-2.97) = 10^{-2.97} = \mathbf{1.07 \times 10^{-3}\ M}$
Starting $[A^-]$ is assumed to be zero; starting $[H^+]$ is actually 1×10^{-7} M for pure water, but this is assumed to be approximately zero.

		$HA_{(aq)}$	$\longrightarrow$	$H^+_{(aq)}$	+	$A^-_{(aq)}$
starting conc.	:	0.10 M		$\cong$0 M		0 M
change, Δ	:	−? M		+? M		+? M
equilibrium conc.	:	? M		1.07×10^{-3} M		? M

$$\Delta[H^+] = (1.07 \times 10^{-3}\,M) - (0\,M) = +1.07 \times 10^{-3}\,M$$

$HA_{(aq)}$ is a reactant consumed in the reaction; change is negative:

$$\Delta[HA] = (1.07\times10^{-3}\ M\ \Delta\ H^+)\times\frac{1\,HA}{1\,H^+} = -1.07\times10^{-3}\ M$$

$A^-_{(aq)}$ is a product formed with the H^+; change is positive:

$$\Delta[A^-] = (1.07\times10^{-3}\ M\ \Delta\ H^+)\times\frac{1\,A^-}{1\,H^+} = +1.07\times10^{-3}\ M$$

		$HA_{(aq)}$	$\longrightarrow$	$H^+_{(aq)}$	+	$A^-_{(aq)}$
starting conc.	:	0.10 M		≅0 M		0 M
change, Δ	:	-1.07×10^{-3} M		$+1.07\times10^{-3}$ M		$+1.07\times10^{-3}$ M
equilibrium conc.	:	0.0989 M		1.07×10^{-3} M		1.07×10^{-3} M

$$K_a = \frac{[H^+][A^-]}{[HA]} = \frac{(1.07\times10^{-3}\ M)(1.07\times10^{-3}\ M)}{(0.0989\ M)} = \mathbf{1.16\times10^{-5}}$$

Example: The weak acid HOBr is dissolved in an aqueous solution; the starting concentration of this add is 2.50 M. Calculate the pH of the solution once equilibrium is established. The K_a of HOBr is 2.0×10^{-9}.

A weak acid in water is an example of category (3); in this case, an equilibrium must be solved to determine equilibrium concentrations from a given value of K_a.

(1) $\mathbf{HOBr_{(aq)} \rightleftarrows H^+_{(aq)} + OBr^-_{(aq)}}$

(2) $\mathbf{K_a = \dfrac{[H^+][OBr^-]}{[HOBr]}}$

(3) Set (**x**) equal to the molarity change of HOBr; the change must be negative since HOBr reacts to form the aqueous ions. The starting concentration of $[OBr^-]$ is assumed to be zero; starting $[H^+]$ is approximately zero. The concentration changes of the aqueous ions must be positive.

$$\Delta[H^+_{(aq)}] = (x\ M\ \Delta\ HOBr_{(aq)})\times\frac{1\,H^+}{1\,HOBr} = \mathbf{+x\ M}$$

$$\Delta[OBr^-_{(aq)}] = (x\ M\ \Delta\ HOBr_{(aq)})\times\frac{1\,OBr^-}{1\,HOBr} = \mathbf{+x\ M}$$

		$HOBr_{(aq)}$	$\longrightarrow$	$H^+_{(aq)}$	+	$OBr^-_{(aq)}$
starting conc.	:	2.50 M		@0 M		0 M
change, Δ	:	−x M		+x M		+x M
equilibrium conc.	:	(2.50 − x)M		x M		x M

(**4**) $K_a = \frac{[H^+][OBr^-]}{[HOBr]}$; $2.0\times10^{-9} = \frac{(x)(x)}{(2.50-x)}$; $2.0\times10^{-9} = \frac{x^2}{(2.50-x)}$

Solve the equation using the simplification technique; assume that x is much smaller than 2.50 for simplification: $(2.50-x) \cong (2.5)$.

$$2.0\times10^{-9} = \frac{x^2}{(2.5)} \quad ; \quad (2.0\times10^{-9})(2.5) = x^2; \quad 5.0\times10^{-9} = x^2$$

$$x = (5.0\times10^{-9})^{1/2} = \mathbf{7.1\times10^{-5}\ M}$$

$$\mathbf{[H^+] = \cong x \cong 7.1 \times 10^{-5}\ M; [OBr^-] = 7.1 \times 10^{-5}\ M}$$

$$\mathbf{[HOBr]} = 2.50-x = (2.50)-(7.1 \times 10^{-5}\ M) \cong \mathbf{2.50\ M}$$

(**5**) $K_a = \frac{[H^+][OBr^-]}{[HOBr]}$; $2.0\times10^{-9}\ ?=?\ \frac{(7.1\times10^{-5})(7.1\times10^{-5})}{(2.50)} = 2.0\times10^{-9}$✓

(**6**) **pH** $= -\log[H^+] = -\log(7.1 \times 10^{-5}) = \mathbf{4.2}$

Example: The compound CH_3NH_2 acts as a base in a manner similar to NH_3. Calculate the K_b of CH_3NH_2 if a 0.12 M starting concentration of this base shows a pH of 11.78 after equilibrium is established.

A weak base in water is an example of category (**4**); in this case, an equilibrium must be solved to determine the value of K_b.

The reaction of a weak base with water as the acid is used; the K_b expression does not include the mole fraction of water ($\chi \cong 1$).

(**1**) $CH_3NH_{2(aq)} + H_2O_{(l)} \rightleftharpoons CH_3NH_3^+{}_{(aq)} + OH^-{}_{(aq)}$ (**2**) $K_b = \frac{[CH_3NH_3^+][OH^-]}{[CH_3NH_2]}$

(**3**) The equilibrium $[OH^-_{(aq)}]$ is determined from the pOH calculated from the pH of 11.78: pOH = (14–pH) = 14–11.78 = 2.22

$$-\log[OH^-] = 2.22; \mathbf{[OH^-]} = \text{antilog}(-2.22) = 10^{-2.22} = \mathbf{6.0 \times 10^{-3}\ M}$$

		$\mathbf{CH_3NH_{2(aq)}}$	+ $\mathbf{H_2O_{(l)}} \longrightarrow$	$\mathbf{CH_3NH_3^+{}_{(aq)}}$	+	$\mathbf{OH^-{}_{(aq)}}$
starting conc.	:	0.12 M		0 M		≅0 M
change, Δ	:	– ? M		+ ? M		+ ? M
equilibrium conc.:		? M		? M		6.0×10^{-3} M

$$\Delta[OH^-] = (6.0 \times 10^{-3}M)-(0\ M) = +6.0 \times 10^{-3}M$$

$CH_3NH_{2(aq)}$ is a reactant consumed in the reaction; change is negative:

$$\Delta[CH_3NH_2] = (6.0\times10^{-3}M\ \Delta\ OH^-)\times\frac{1\ CH_3NH_2}{1\ OH^-} = -6.0\times10^{-3}\ M$$

$CH_3NH_3{}^+{}_{(aq)}$ is a product formed with the OH^-; change is positive:

$$\Delta[CH_3NH_3{}^+] = (6.0\times10^{-3}M\ \Delta\ OH^-)\times\frac{1\,[CH_3NH_3{}^+]}{1\,OH^-} = +6.0\times10^{-3}\,M$$

		$CH_3NH_{2(aq)}$	+ $H_2O_{(l)}$ ⟶	$CH_3NH_3{}^+{}_{(aq)}$	+ $OH^-{}_{(aq)}$
starting conc.	:	0.12 M		0 M	≅0 M
change, Δ	:	-6.0×10^{-3}M		$+6.0\times10^{-3}$ M	$+6.0\times10^{-3}$ M
equilibrium conc.:		0.114 M		6.0×10^{-3} M	6.0×10^{-3} M

$$\textbf{(4) } \mathbf{K_b} = \frac{[CH_3NH_3{}^+][OH^-]}{[CH_3NH_2]} = \frac{(6.0\times10^{-3}M)(6.0\times10^{-3}M)}{(0.114\ M)} = \mathbf{3.2\times10^{-4}}$$

Example: Calculate the pH of a solution of a 0.025 M initial concentration of the base ammonia (NH_3) after equilibrium is established. Ammonia is acting as a base with water as the acid; use the pK_a table to determine the category and required constant.

The conjugate acid of ammonia as a base is $NH_4{}^+$; its pK_a is 9.25. The pK_b and K_b of NH_3 is required: $pK_b = (14-pK_a) = 14-9.25 = 4.75$.

$$-\log[pK_b] = 4.75;\ K_b = \text{antilog}\ (-4.75) = 10^{-4.75} = \mathbf{1.8\times10^{-5}}$$

The K_b for ammonia classifies this as a weak base equilibrium in water, category **(4)**; solve for equilibrium concentrations from the value for K_b.

$$\textbf{(1) } \mathbf{NH_{3(aq)} + H_2O_{(l)} \rightleftarrows NH_4{}^+{}_{(aq)} + OH^-{}_{(aq)}}$$

$$\text{(2) } \mathbf{K_b} = \frac{\mathbf{[NH_4{}^+][OH^-]}}{\mathbf{[NH_3]}}$$

(**3**) Set (**x**) equal to the molarity change of NH_3; the change must be negative since NH_3 reacts with water to form the aqueous ions; starting $[NH_4{}^+]$ is zero; starting $[OH^-]$ is approximately zero. The concentration changes of the aqueous ions must be positive.

$$\Delta[NH_4{}^+{}_{(aq)}] = (x\,M\Delta NH_{3(aq)})\times\frac{1NH_4{}^\pm}{1NH_3} = +\ \mathbf{x\,M}$$

$$\Delta[OH^-{}_{(aq)}] = (x\,M\Delta NH_{3(aq)})\times\frac{1OH^-}{1NH_3} = +\ \mathbf{x\,M}$$

		$NH_{3(aq)}$	+ $H_2O_{(l)}$ ⟶	$NH_4{}^+{}_{(aq)}$	+ $OH^-{}_{(aq)}$
starting conc.	:	0.025 M		0 M	≅ 0 M
change, Δ	:	−xM		+xM	+x M
equilibrium conc. :		0.025 − x M		xM	xM

$$\text{(4) } K_b = \frac{[NH_4{}^+][OH^-]}{[NH_3]};\ 1.8\times10^{-5} = \frac{(x)(x)}{(0.025-x)};\ 1.8\times10^{-5} = \frac{x^2}{(0.025-x)}$$

Solve the equation using the simplification technique; in this case, assume that x is much smaller than 0.025: (0.025−x) ≅ (0.025).

$$1.8\times10^{-5} = \frac{x^2}{(0.025)};\ (1.8\times10^{-5})(0.025) = x^2;\ \ 4.5\times10^{-7} = x^2$$

$$x = (4.5\times10^{-7})^{1/2} = \mathbf{6.7\times10^{-4}\,M}$$

$$[NH_4^+] = x = 6.7\times10^{-4}\,M;\ [OH^-] = x = 6.7\times10^{-4}\,M$$

$$[NH_3] = (0.025-x) = (0.025)-(0.00067) = \mathbf{0.02433\,M}$$

(The simplification is only approximate.)

(5) $K_b = \frac{[NH_4^+][OH^-]}{[NH_3]};\ 1.8\times10^{-5}\ ?=?\ \frac{(6.7\times10^{-4})(6.7\times10^{-4})}{(0.02433)} = 1.845\times10^{-5}$ ✓

(6) pOH = −log[OH⁻] = −log (6.7 × 10^{-4}) = **3.2**

pH = (14−pOH) = 14−3.2 = **10.8**

VI pH CALCULATIONS IN AQUEOUS SOLUTIONS: SOLVING GENERAL ACID/BASE REACTIONS

Three types of complete problems are considered in this section. For general reactions of any acid with any base, if either the acid or the base (or both) is strong, the initial acid/base reaction is considered to go to ≅ 100% completion; this reaction is then solved through simple stoichiometry based on complete consumption of the limiting reagent. Reaction of a weak acid with a weak base is not covered in this section.

Problem Type (A): Strong acid reacts with a strong base; general process steps:

GenStep **(1)**: Write the balanced equation for the initial acid/base reaction (Chapter 9).

GenStep **(2)**: Solve the simple stoichiometry of the initial acid/base reaction based on complete consumption of the limiting reagent (Chapter 8).

GenStep **(3)**: Solve a second simple stoichiometry described by the reaction of the excess strong component with water according to the previous categories (1) or (2) from the previous section. If the initial reaction reaches stoichiometric equivalence, the solution is neutral.

Problem Type (B): Strong acid reacts with a weak base or a weak acid reacts with a strong base and the **strong** component is specifically in **excess**; general process steps:

GenStep **(1)**: Write the balanced equation for the initial acid/base reaction (Chapter 9).

GenStep **(2)**: Solve the simple stoichiometry of the initial acid/base reaction based on complete consumption of the limiting reagent (Chapter 8).

GenStep (**3**): Solve a second simple stoichiometry described by the reaction of the excess strong component with water according to the previous categories (1) or (2) from the previous section.

Problem Type (**C**): Strong acid reacts with a weak base or a weak acid reacts with a strong base and the **weak** component is specifically in **excess**, or the acid and base reach stoichiometric **equivalence**; general process steps:

GenStep (**1**): Write the balanced equation for the initial acid/base reaction (Chapter 9).

GenStep (**2**): Solve the simple stoichiometry of the initial acid/base reaction based on complete consumption of the limiting reagent (Chapter 8).

GenStep (**3**): Solve an **equilibrium** based on all weak aqueous acid and weak aqueous base components remaining and/or produced by the initial stoichiometric acid/base reaction. Follow the equilibrium procedures described for categories (**3**) or (**4**) from the previous section.

An important consideration for GenStep (**3**) is that conjugate bases of strong acids are very weak and do not measurably contribute to the basicity of neutral water. The conjugate acids of strong bases are very weak and do not measurably contribute to the acidity of neutral water. In the absence of a strong component, the acidity and basicity of aqueous solutions are determined by moderately weak acids with moderately weak conjugate bases. The moderately weak conjugate acid/base pairs for which equilibria must be solved generally range between a pK_a of ≈ 2 to pK_a of ≈ 13 for the acid component.

Example: Calculate the pH of a solution formed after reaction of 0.100 L of 0.10 M aqueous HBr with 0.100 L of 0.060 M aqueous NaOH.

The pK_a table indicates that HBr is a strong acid (pKa = −9); NaOH is a strong hydroxide base. The problem type can be identified as strong acid with strong base; follow the general steps for problem **Type** (**A**).

$$\text{GenStep (1): } HBr_{(aq)} + NaOH_{(aq)} \longrightarrow NaBr_{(aq)} + H_2O_{(l)}$$

GenStep (**2**): A general equation can calculate the concentration of any component [**A**] for a solution from provided individual concentrations and volumes.

$$[\mathbf{A}] = \frac{\text{moles of } \mathbf{A}}{\text{V(L) solution total}}; \qquad \text{moles } \mathbf{A} = \{\text{V(L) of } \mathbf{A}\} \times \{\mathbf{M} \text{ of } \mathbf{A}\}$$

$$\text{General equation for } [\mathbf{A}] = \frac{\{\text{V(L) of } \mathbf{A}\} \times \{\text{M of } \mathbf{A}\}}{\text{V(L) solution total}}$$

V(L) total solution = (0.100 L HBr $_{(aq)}$) + (0.100 L NaOH $_{(aq)}$) = 0.200 L total

$$\text{initial } [\mathbf{HBr}] = \frac{(0.100\,\text{L})(0.10\,\text{M})}{0.200\,\text{L total}} = \mathbf{0.050\,M}$$

$$\text{initial } [\mathbf{NaOH}] = \frac{(0.100\,\text{L})(0.060\,\text{M})}{0.200\,\text{L total}} = \mathbf{0.030\,M}$$

	$HBr_{(aq)}$	+ $NaOH_{(aq)}$	$\longrightarrow$ $NaBr_{(aq)}$	+ $H_2O_{(l)}$
starting conc.:	0.050 M	0.030 M	0 M	excess
change, Δ :	−0.030 M	−0.030 M	+0.03 M	
finish :	0.020 M	0 M	0.030 M	

GenStep (**3**): NaOH is limiting; HBr is in excess. $Na^+_{(aq)}$ has no effect on acidity; $Br^-_{(aq)}$ is the very weak conjugate base of a strong acid and does not contribute to basicity.

	$HBr_{(aq)}$	$\longrightarrow$ $H^+_{(aq)}$	+ $Br^-_{(aq)}$ (≅ **100% ionized**)
starting conc.:	0.020 M	≅ 0 M	0.030 M
change, Δ :	−0.020 M	+0.020 M	+0.020 M
finish :	0 M	0.020 M	0.050 M (no effect on pH)

$$[H^+_{(aq)}] \text{ from the stoichiometric calculation} = \mathbf{0.020\ M}$$
$$\mathbf{pH} = -\log_{10}[H^+_{(aq)}] = -\log_{10}[(0.020) = -\log(2.0 \times 10^{-2}) = \mathbf{1.7}$$

Example: Calculate the pH of a solution formed after reaction of 0.100 L of 0.10 M aqueous HBr with 0.100 L of 0.060 M aqueous NaCN.

The pK_a table indicates that HBr is a strong acid ($pK_a = -9$). The aqueous Na^+ ion is a spectator ion; the base in the reaction is aqueous CN^-. The conjugate acid of the base is HCN with a pK_a of 9.4 ($K_a = 4.0 \times 10^{-10}$); the pK_b of CN^- is (14−9.4) = 4.6. This is a moderately weak base. The problem type can be identified as strong acid with weak base; the problem type is **Type (B)** or **Type (C)** depending on which component is in excess.

GenStep (**1**): $HBr_{(aq)} + NaCN_{(aq)} \longrightarrow NaBr_{(aq)} + HCN_{(aq)}$

GenStep (**2**): Use the general equation: $[\mathbf{A}] = \dfrac{\{V(L) \text{ of } \mathbf{A}\} \times \{\mathbf{M} \text{ of } \mathbf{A}\}}{V(L) \text{ solution total}}$

$$V(L) \text{ total solution} = (0.100 \text{ L HBr}_{(aq)}) + (0.100 \text{ L NaCN}_{(aq)}) = 0.200 \text{ L total}$$

$$\text{initial}\,[\mathbf{HBr}] = \frac{(0.100\,L)(0.10\,M)}{0.200\,L\ \text{total}} = \mathbf{0.050\,M}$$

$$\text{initial}\,[\mathbf{NaCN}] = \frac{(0.100\,L)(0.060\,M)}{0.200\,L\ \text{total}} = \mathbf{0.030\,M}$$

	$HBr_{(aq)}$	+ $NaCN_{(aq)}$	$\longrightarrow$ $NaBr_{(aq)}$	+ $HCN_{(aq)}$
starting conc.:	0.050 M	0.030 M	0 M	0 M
change, Δ :	−0.030 M	−0.030 M	+0.03 M	+0.03 M
finish :	0.020 M	0 M	0.030 M	0.030 M

GenStep **(3)**: NaCN is limiting; HBr is in excess. This identifies the reaction as problem **Type(B)** with the strong component in excess. As in the previous example, $Na^+_{(aq)}$ has no effect on acidity and very weak base $Br^-_{(aq)}$ does not contribute to basicity. However, the stoichiometric reaction has formed the weak acid HCN (pKa = 9.4).

GenStep **(3)** for problem **Type (B)** is to solve a second simple stoichiometry described by the reaction of the excess strong component with water. This procedure is based on the fact that the weak acid will make no measureable contribution to acidity when in the presence of a measurable amount of strong acid. The resulting starting concentrations are derived from the first stoichiometric table.

	$HBr_{(aq)}$	$\longrightarrow$	$H^+_{(aq)}$	+	$Br^-_{(aq)}$ (≅100% ionized)
starting conc.:	0.020 M		≅0 M		0.030 M
change, Δ :	−0.020 M		+0.020 M		+0.020 M
finish :	0 M		0.020 M		0.050 M (no effect on pH)

$$[H^+] = 0.20\text{ M};$$

$$pH = -\log_{10}[H^+_{(aq)}] = -\log_{10}[(0.020) = -\log(2.0 \times 10^{-2}) = 1.7$$

A calculation of the acid equilibrium for HCN can demonstrate the validity of ignoring the weak acid in the previous example. The starting concentration of $H^+_{(aq)}$ is 0.20 M derived from the excess strong acid HBr described by the stoichiometric table.

	$HCN_{(aq)}$	$\longrightarrow$	$H^+_{(aq)}$	+	$CN^-_{(aq)}$
starting conc. :	0.030 M		0.020 M		0 M
change, Δ :	−x M		+x M		+x M
equilibrium conc.:	(0.030 − x) M		(0.020 + x) M		x M

$$\text{(4) } K_a = \frac{[H^+][CN^-]}{[HCN]};\quad 4.0\times10^{-10} = \frac{(0.020+x)(x)}{(0.030-x)}$$

The simplification rule is applicable and would state that the $H^+_{(aq)}$ contribution from HCN, (x), can be ignored: (0.020 + x) ≅ (0.020). If the equation is solved, the result is **x** = 6.0×10^{-10} M. **[H⁺]** = 0.20 M + (6.0×10^{-10} M) ≅ 0.20 **M**

Example: Calculate the pH of a solution formed after reaction of 0.100 L of 0.60 M aqueous HBr with 0.100 L of 0.90 M aqueous NaCN.

HBr is a strong acid ($pK_a = -9$); the weak base in the reaction is aqueous CN^-. The conjugate acid of the base is HCN with a pK_a of 9.4 ($K_a = 4.0 \times 10^{-10}$). The problem type can be identified as **Type (B)** or **Type (C)** depending on which component is in excess.

$$\text{GenStep (1): } HBr_{(aq)} + NaCN_{(aq)} \longrightarrow NaBr_{(aq)} + HCN_{(aq)}$$

$$\text{GenStep (2): } [\mathbf{A}] = \frac{\{V(L)\text{ of }\mathbf{A}\}\times\{M\text{ of }\mathbf{A}\}}{V(L)\text{ solution total}}$$

V(L) total solution = (0.100 L $HBr_{(aq)}$) + (0.100 L $NaCN_{(aq)}$) = 0.200 L total

$$\text{initial}\,[\mathbf{HBr}] = \frac{(0.100\,\text{L})(0.60\,\text{M})}{0.200\,\text{L total}} = \mathbf{0.30\,M}$$

$$\text{initial}\,[\mathbf{NaCN}] = \frac{(0.100\,\text{L})(0.90\,\text{M})}{0.200\,\text{L total}} = \mathbf{0.45\,M}$$

	$HBr_{(aq)}$ +	$NaCN_{(aq)}$ ⟶	$NaBr_{(aq)}$ +	$HCN_{(aq)}$
starting conc.:	0.30 M	0.45 M	0 M	0 M
change, Δ :	−0.30 M	−0.30 M	+0.30 M	+0.30M
finish :	0 M	0.15 M	0.30 M	0.30 M

GenStep (**3**): HBr is limiting; NaCN is in excess. This identifies the reaction as problem **Type (C)** with the weak component in excess. $Na^+_{(aq)}$ and the very weak base $Br^-_{(aq)}$ do not contribute to basicity. The reaction has produced a solution of a combination of a weak acid with its conjugate weak base. Solving this equilibrium will determine the acidity/basicity of the aqueous solution. The starting concentrations for the equilibrium calculations are those that remain or are produced from the initial stoichiometric reaction.

	$HCN_{(aq)}$ ⟶	$H^+_{(aq)}$ +	$CN^-_{(aq)}$
starting conc. :	0.30 M	≅ 0 M	0.15 M
change, Δ :	−x M	+ x M	+x M
equilibrium conc. :	(0.30 − x) M	x M	(0.15 + x) M

$$\textbf{(4)}\; K_a = \frac{[H^+][CN^-]}{[HCN]};\quad 4.0\times10^{-10} = \frac{(x)(0.15+x)}{(0.30-x)}$$

The simplification rule is applicable: assume (0.15 + x) ≅ (0.15); (0.30 − x) ≅ (0.30).

$$4.0\times10^{-10} = \frac{x(0.15)}{(0.30)};\ (4.0\times10^{-10})(0.30) = x(0.15);\ \frac{(4.0\times10^{-10})(0.30)}{(0.15)} = x$$

$$\mathbf{x = 8.0\times10^{-10}\,M;\ [H^+] = x = 8.0\times10^{-10}\,M};\ [CN^-] \cong 0.15\,M;\ [HCN] \cong 0.30\,M$$

$$\mathbf{pH} = -\log_{10}[H^+_{(aq)}] = -\log(8.0\times10^{-10}) = \mathbf{9.1}$$

Example: Calculate the pH of a solution formed after reaction of 0.050 L of 0.020 M aqueous acetic acid (CH_3COOH) with 0.025 L of 0.040 M aqueous NaOH.

NaOH is a strong hydroxide base. Acetic acid is a weak acid with a pK_a of 4.75 ($K_a = 1.8 \times 10^{-5}$). The problem type is a weak acid reacting with a strong base and can be identified as **Type (B)** or **Type (C)** depending on which component if either is in excess.

GenStep (**1**): $CH_3COOH_{(aq)} + NaOH_{(aq)} \longrightarrow CH_3COO^-_{(aq)}\ (Na^+_{(aq)}) + H_2O_{(l)}$

$$\text{GenStep } (\mathbf{2}):\ [\mathbf{A}] = \frac{\{V(L)\text{ of }\mathbf{A}\}\times\{M\text{ of }\mathbf{A}\}}{V(L)\text{ solution total}}$$

V(L) total solution = (0.050 L $CH_3COOH_{(aq)}$) + (0.025 L $NaOH_{(aq)}$) = 0.075 L total

$$\text{initial}\,[\mathbf{CH_3COOH}] = \frac{(0.050\,L)(0.020\,M)}{0.075\,L\,\text{total}} = \mathbf{0.0133\,M}$$

$$\text{initial}\,[\mathbf{NaOH}] = \frac{(0.025\,L)(0.040\,M)}{0.075\,L\,\text{total}} = \mathbf{0.0133\,M}$$

	CH_3COOH	+	$NaOH$	$\longrightarrow$	$CH_3COO^-\,(Na^+)$	+	H_2O
starting conc.:	0.0133 M		0.0133 M		0 M		excess
change, Δ :	−0.0133 M		−0.0133 M		+0.0133 M		
finish :	0 M		0 M		0.0133 M		

GenStep **(3)**: The reaction is at stoichiometric equivalence. This identifies the reaction as the other result for problem **Type (C)**. Unlike reactions of strong acid with strong base at equivalence, the solution is not neutral since the conjugate base of a weak acid was produced. Solving this weak base equilibrium will determine the acidity/basicity of the aqueous solution. The pK_b of (CH_3COO^-) is (14−4.75) = 9.25; 5.6×10^{-10}. The starting concentrations for the equilibrium calculations are those that remain or are produced from the initial stoichiometric reaction.

	CH_3COO^-	+	$H_2O_{(l)}$	$\longrightarrow$	$CH_3COOH_{(aq)}$	+	$OH^-_{(aq)}$
starting conc. :	0.0133 M				0 M		≅0 M
change, Δ :	− x M				+x M		+x M
equilibrium conc.:	(0.0133 − x) M				x M		x M

$$\textbf{(4)}\; K_b = \frac{[CH_3COOH][OH^-]}{[CH_3COO^-]};\; 5.6\times10^{-10} = \frac{(x)(x)}{(0.0133-x)};\quad 5.6\times10^{-10} = \frac{x^2}{(0.0133-x)}$$

Assume that x is much smaller than 0.0133; (0.0133−x) ≅ (0.0133)

$$5.6\times10^{-10} = \frac{x^2}{(0.0133)};\quad (5.6\times10^{-10})(0.0133) = x^2;\quad 7.475\times10^{-12} = x^2$$

$$x = (7.475 \times 10^{-12})^{1/2} = \mathbf{2.73 \times 10^{-6}\,M};\; \mathbf{[OH^-]} = x = \mathbf{2.73 \times 10^{-6}\,M}$$

$$[CH_3COOH] = x = 2.73 \times 10^{-6} M$$

$$[CH_3COO^-] = 0.0133 - x = (0.0133) - (0.00000273) = 0.0133\,M$$

$$\mathbf{pOH} = -\log_{10}[OH^-_{(aq)}] = -\log\,(2.73 \times 10^{-6}) = \mathbf{5.56}$$

$$\mathbf{pH} = (14 - pOH) = 14 - 5.56 = \mathbf{8.44 = 8.4}$$

Note that at the stoichiometric equivalence point for reaction of any weak acid with strong base, the resulting aqueous solution is not neutral (pH of 7). The solution is basic because the initial stoichiometric reaction forms a weak conjugate base, while the conjugate acid of the strong base does not contribute to acidity; in this case, the conjugate acid of (OH^-) is neutral water.

The converse is true for reaction of any strong acid with a weak base at stoichiometric equivalence. The resulting aqueous solution is acidic because the initial stoichiometric reaction forms a weak conjugate acid, while the conjugate base of the strong acid does not contribute to basicity.

VII THE COMMON ION EFFECT AND ACID/BASE BUFFERS

An acid/base buffer is a solution of a weak acid with its corresponding weak conjugate base; the combination of weak acid plus weak base stabilizes the solution against large pH changes upon the addition of extra strong base or strong acid.

The buffering process works through the neutralization of the added strong acid or base by the complementary weak base or acid. An added strong acid will be neutralized by reaction with the weak conjugate base; this will increase the concentration of the weak acid. An added strong base will be neutralized by reaction with the weak acid; this will increase the concentration of the weak conjugate base. The net result is that a buffered solution "exchanges" (converts) an added strong component to the corresponding weak component by reaction. The buffering process is effective approximately until one of the weak components is exhausted (completely reacted).

The most effective buffers have nearly equal concentrations of the conjugate acid and base forms and an acid pK_a very close to the desired pH of the buffer solution.

Example: Calculate the pH of a solution containing only initial concentration 1.0 M acetic acid. Then calculate the pH of the solution of 1.0 M acetic acid if 0.10 M $H^+_{(aq)}$ from the strong acid HCl is added. K_a of acetic acid = 1.8×10^{-5}. Use the simplification technique.

$$\textbf{(1)}\ CH_3COOH_{(aq)} \rightleftarrows H^+_{(aq)} + CH_3COO^-_{(aq)} \qquad \textbf{(2)}\ K_a = \frac{[H^+][CH_3COO^-]}{[CH_3COOH]}$$

(3)		$CH_3COOH_{(aq)}$	$\longrightarrow$	$H^+_{(aq)}$	+	$CH_3COO^-_{(aq)}$
starting conc.	:	1.0 M		≅ 0 M		0 M
change, Δ	:	– x M		+ x M		+x M
equilibrium conc.:		(1.0 – x) M		x M		x M

$$\textbf{(4)}\ K_a = \frac{[H^+][CH_3COO^-]}{[CH_3COOH]};\quad 1.8\times10^{-5} = \frac{(x)(x)}{(1.0-x)};\ \text{simplification: } (1.0-x) \cong (1.0)$$

$$1.8\times10^{-5} = \frac{x^2}{1.0};\quad x = (1.8\times10^{-5})^{1/2} = \mathbf{4.2\times10^{-3}\,M}$$

$$\mathbf{[H^+] = x = 4.2 \times 10^{-3}\,M};\ [CH_3COO^-] = x = 4.2 \times 10^{-3}\,M$$

$$[CH_3COOH] = 1.0\,M - x = (1.0) - (4.2 \times 10^{-3}\,M) = 0.996\,M$$

$$\mathbf{pH} = -\log [H^+] = -\log (4.2 \times 10^{-3}) = \mathbf{2.4}$$

If 0.10 M $H^+_{(aq)}$ from HCl is added to the previous solution, the same equilibrium is analyzed but with the starting concentration of $[H^+_{(aq)}]$ equal to 0.10 M instead of approximately zero. Since this solution contains no weak acetate base, the added $H^+_{(aq)}$ is not neutralized; the complete additional 0.10 M must be included with the starting $H^+_{(aq)}$ in the equilibrium calculation.

(3)		$CH_3COOH_{(aq)}$	$\longrightarrow$	$H^+_{(aq)}$	+	$CH_3COO^-_{(aq)}$
starting conc.	:	1.0 M		0.10 M		0 M
change, Δ	:	–x M		+x M		+x M
equilibrium conc.:		(1.0 – x) M		(0.10 + x) M		x M

$$\textbf{(4) } K_a = \frac{[H^+][CH_3COO^-]}{[CH_3COOH]};\quad 1.8\times10^{-5} = \frac{(x)(0.10+x)}{(1.0-x)};\ (1.0-x) \cong (1.0)$$

$$(0.10+x) \cong (0.10);\quad 1.8\times10^{-5} = \frac{(x)(0.1)}{1.0};\quad x = \frac{(1.8\times10^{-5})}{0.10} = \mathbf{1.8\times10^{-6}\ M}$$

$$[H^+] = x = 0.10 + (1.8 \times 10^{-6} M) = \cong \mathbf{0.10\ M}$$

$$[CH_3COO^-] = x = 1.8 \times 10^{-6} M$$

$$[CH_3COOH] = 1.0\ M - x = (1.0) - (1.8 \times 10^{-6} M) = \cong 1.0\ M$$

$$\mathbf{pH} = -\log [H^+] = -\log (0.10\ M) = \mathbf{1.0}$$

The pH of the solution is determined by the added strong acid; the acetic acid and small amount of acetate have no measurable effect of the solution pH.

Example: Calculate the pH of a solution containing 1.0 M acetic acid plus 0.50 M sodium acetate. Then calculate the pH of the solution containing 1.0 M acetic acid plus 0.50 M sodium acetate if 0.10 M $H^+_{(aq)}$ from the strong acid HCl is added. K_a of acetic acid = 1.8×10^{-5}. Use the simplification technique.

$$\textbf{(1) } CH_3COOH_{(aq)} \rightleftarrows H^+_{(aq)} + CH_3COO^-_{(aq)} \quad \textbf{(2) } K_a = \frac{[H^+][CH_3COO^-]}{[CH_3COOH]}$$

(3) In this problem, the starting concentration of $[CH_3COO^-]$ is **not** zero but is 0.50 M from the sodium acetate (the Na^+ ion is a spectator ion).

	$CH_3COO_{(aq)}$	$\longrightarrow$	$H^+_{(aq)}$	+	$CH_3COO^-_{(aq)}$
starting conc. :	1.0 M		≅ 0 M		0.50 M
change, Δ :	−x M		+x M		+x M
equilibrium conc.:	(1.0 − x) M		x M		(0.50 + x) M

$$\textbf{(4) } K_a = \frac{[H^+][CH_3COO^-]}{[CH_3COOH]};\ 1.8\times10^{-5} = \frac{(x)(0.50+x)}{(1.0-x)}$$

$$(1.0-x) \cong (1.0) \text{ and } (0.50 + x) \cong (0.50)$$

$$1.8\times10^{-5} = \frac{(0.5)(x)}{(1.0)};\ x = \frac{(1.0)(1.8\times10^{-5})}{(0.5)} = \mathbf{3.6\times10^{-5}\ M}$$

$$[H^+] = x = \mathbf{3.6 \times 10^{-5} M}$$

$$[CH_3COO^-] = 0.50\ M + x = 0.50 + (3.6 \times 10^{-5} M) = \cong 0.50\ M$$

$$[CH_3COOH] = 1.0\ M - x = (1.0) - (3.6 \times 10^{-5} M) = \cong 1.0\ M$$

$$\mathbf{pH} = -\log [H^+] = -\log (3.6 \times 10^{-5}) = \mathbf{4.45}$$

The higher pH for the buffer is due to added weak base.

Addition of 0.10 M $H^+_{(aq)}$ to the buffer solution results in an initial stoichiometric reaction of the strong acid with the weak conjugate base. The $H^+_{(aq)}$ is the limiting reagent and is completely consumed; $\mathbf{CH_3COO^-_{(aq)}}$ is in excess.

	$CH_3COO^-_{(aq)}$ +	$H^+_{(aq)}$ ⟶	$CH_3COOH_{(aq)}$ (≅ 100% reaction)
starting conc.:	0.50 M	0.10 M	1.0 M
change, Δ :	−0.10 M	−0.10 M	+0.10 M
final conc. :	0.40 M	≅0 M	1.10 M

The initial stoichiometric reaction has altered the relative concentrations of the weak acid and the weak base. Solving the acid equilibrium determines the pH.

	$CH_3COOH_{(aq)}$ ⟶	$H^+_{(aq)}$ +	$CH_3COO^-_{(aq)}$
starting conc. :	1.1 M	≅0 M	0.40 M
change, Δ :	−x M	+x M	+x M
equilibrium conc.:	(1.1−x) M	x M	(0.40+x) M

$$\text{(4) } K_a = \frac{[H^+][CH_3COO^-]}{[CH_3COOH]}; \quad 1.8\times10^{-5} = \frac{(x)(0.40+x)}{1.1-x}$$

$$(1.1-x) \cong (1.1) \text{ and } (0.40+x) \cong (0.40)$$

$$1.8\times10^{-5} = \frac{(0.4)(x)}{(1.1)}; \quad \mathbf{x} = \frac{(1.1)(1.8\times10^{-5})}{(0.4)} = \mathbf{4.95\times10^{-5}\ M}$$

$$\mathbf{[H^+] = x = 4.95\times10^{-5}\ M}$$

$$[CH_3COO^-] = 0.40\text{ M} + x = 0.40 + (4.95\times10^{-5}\text{ M}) = \cong 0.40\text{ M}$$

$$[CH_3COOH] = 1.1\text{ M} - x = (1.1) - (4.95\times10^{-5}\text{ M}) = \cong 1.1\text{ M}$$

$$\text{pH} = -\log[H^+] = -\log(4.95\times10^{-5}) = \mathbf{4.31}$$

The 1.0 M non-buffer solution of acetic acid changed from a pH of **2.4** to a pH of 1.0 after addition of 0.10 M $[H^+_{(aq)}]$. The 1.0 M acetic acid/0.50 M acetate **buffer** solution changed from a pH of **4.45** to a pH of **4.31** after addition of 0.10 M $[H^+_{(aq)}]$, a change of a little more **0.1** pH unit.

VIII THE HENDERSEN–HASSELBACH EQUATION

The last example demonstrates a feature that allows a general equation to be derived for pH calculations under a specific set of conditions:

(1) The solution is a mixture of a weak acid and its weak conjugate base.
(2) The solution must contain a measurable concentration of each of these components.
(3) The combination of K_a value and concentration must allow the simplification technique to be applicable to both the weak base and weak acid.

Under these conditions, a general solution to a general acid dissociation equilibrium can be produced. **HA** is the weak acid; **A⁻** is the weak conjugate base.

$$\textbf{(1) } HA_{(aq)} \rightleftharpoons H^+_{(aq)} + A^-_{(aq)} \quad \textbf{(2) } K_a = \frac{[H^+][A^-]}{[HA]}$$

(3) $[HA]_0$ is initial concentration of the weak acid HA.
$[A^-]_0$ is the initial concentration of the weak conjugate base.

	$HA_{(aq)}$	$\longrightarrow$	$H^+_{(aq)}$	+	$A^-_{(aq)}$
starting conc. :	$[HA]_0$ M		$\cong 0$ M		$[A^-]_0$ M
change, Δ :	−x M		+x M		+x M
equilibrium conc.:	$([HA]_0 - x)$ M		x M		$([A^-]_0 + x)$ M

(4) $K_a = \dfrac{[H^+][A^-]}{[HA]}$; $K_a = \dfrac{(x)([A^-]_0 + x)}{([HA]_0 - x)}$

Simplification: $([HA]_0 - x) \cong ([HA]_0)$; $([A^-]_0 + x) \cong ([A^-]_0)$

$[H^+] \cong x$; $K_a = \dfrac{[H^+][A^-]}{[HA]}$ becomes $K_a \cong \dfrac{[H^+][A^-]_0}{[HA]_0}$

Take the negative log of both sides:

$$-\log_{10}(K_a) = -\log_{10}([H^+]) + \frac{(-\log_{10}\{[A^-]_0\})}{\{[HA]_0\}}$$

Solve for $\{-\log_{10}([H^+])\}$: $-\log_{10}([H^+]) = -\log_{10}(K_a) - \dfrac{(-\log_{10}\{[A^-]_0\})}{\{[HA]_0\}}$

$$\text{or } -\log_{10}([H^+]) = -\log_{10}(K_a) + \log_{10}\frac{\{[A^-]_0\}}{\{[HA]_0\}}$$

$$\mathbf{pH = pK_a + \log_{10}\frac{\{[A^-]_0\}}{\{[HA]_0\}} = \text{Hendersen–Hasselbach equation}}$$

or

$$\mathbf{pH_{(buffered\ solution)} = pK_{a(buffering\ acid)} + \log_{10}\{[conjugate\ base]/[acid]\}}$$

Example: Use the Hendersen–Hasselbach equation to calculate the pH for the two resulting buffer solutions from the previous example.

a) The initial buffer concentrations are: $[CH_3COO^-] = 0.50$ M; $[CH_3COOH] = 1.0$ M.
b) The reacted buffer concentrations are: $[CH_3COO^-] = 0.40$ M; $[CH_3COOH] = 1.1$ M.

The pK_a of acetic acid $= -\log_{10}(1.8 \times 10^{-5}) = 4.75$

(a) $pH = pK_a + \log_{10}\dfrac{\{[A^-]_0\}}{\{[HA]_0\}}$; $pH = 4.75 + \log_{10}\dfrac{\{(0.50)\}}{\{(1.0)\}}$

$$\mathbf{pH} = 4.75 + \log_{10}(0.5) = (4.75) + (-0.3) = \mathbf{4.45}$$

(b) $\mathbf{pH} = 4.75 + \log_{10}\dfrac{\{(0.40)\}}{\{(1.1)\}} = 4.75 + \log_{10}(0.364) = 4.75 + (-0.44) = \mathbf{4.31}$

IX PRACTICE PROBLEMS

1. Calculate the K_a for an unknown acid HA if a 0.150 M starting concentration of the acid has a pH of 4.0 after equilibrium is reached.

2. Calculate the K_a for nicotinic acid. The starting amount of nicotinic acid ($C_6H_4NO_2H$) is 1.00 gram in 0.0600 L; the solution has a pH of 2.70 after equilibrium is reached.

$$C_6H_4NO_2H_{(aq)} \underset{}{\overset{H_2O}{\rightleftarrows}} H^+_{(aq)} + C_6H_4NO_2^-{}_{(aq)}$$

3. Calculate the equilibrium pH for a solution of the weak acid HCN. The starting concentration of the acid is 0.150 M. The K_a of HCN = 4.0×10^{-10}.

4. Calculate the equilibrium pH of 4 tablets (total mass = 1.30 grams) of aspirin (acetylsalicylic acid, $C_8H_7O_2COOH$) dissolved in 50.0 mL of water; this is the starting amount of aspirin. The K_a of $C_8H_7O_2COOH = 3.27 \times 10^{-4}$. (Use the simplification technique; however, this will only give an approximate solution in this case.)

$$C_8H_7O_2COOH_{(aq)} \overset{H2O}{\rightleftarrows} H^+_{(aq)} + C_8H_7O_2COO^-{}_{(aq)}$$

5. Calculate the pH of a solution formed by addition of 150 mL of 0.80 M NH_3 with 80 mL of 1.0 M HCl; pK_b of NH_3 = 4.75. Use the simplification technique.

6. Carbonic acid (H_2CO_3) with the conjugate base hydrogen carbonate (HCO_3^-) is the major buffering system in the blood; this acid can be used to model the blood buffer. Write the correct acid dissociation equilibrium; then complete the following parts dealing with this equilibrium.
   a) Calculate the pH of a solution containing only 0.020 M (starting) carbonic acid.
   b) Calculate the pH of the solution containing only 0.020 M carbonic acid if 0.001 M $H^+_{(aq)}$ is added.
   c) Calculate the pH of a solution containing 0.020 M carbonic acid plus 0.100 M hydrogen carbonate from sodium hydrogen carbonate.
   d) Calculate the pH of the solution containing 0.020 M carbonic acid plus 0.100 M sodium hydrogen carbonate if 0.001 M $H^+_{(aq)}$ is added.
   Use the simplification technique in all cases; in some cases, the calculations will only be approximate. K_a of carbonic acid = 4.3×10^{-7}

7. a) Calculate the pH of a buffer solution formed by adding 0.20 mole of $NH_{3(aq)}$ and 0.30 mole of $NH_4Cl_{(aq)}$ to form 1.00 L of solution.
   b) Calculate the pH of this buffer solution if 0.15 mole of $NaOH_{(aq)}$ is added to the 1.00 L of buffer solution. Use the Hendersen–Hasselbach equation for this problem; pK_b of NH_3 = 4.75.

X ANSWERS TO PRACTICE PROBLEMS

1. **(1)** $HA_{(aq)} \rightleftarrows H^+_{(aq)} + A^-_{(aq)}$ **(2)** $K_a = \dfrac{[H^+][A^-]}{[HA]}$

(3) pH= $-\log[H^+] = 4.0$; equilibrium $[H^+]$ = antilog $(-4.0) = 10^{-4.0} = 1.0 \times 10^{-4}$

		$HA_{(aq)}$	$\longrightarrow$	$H^+_{(aq)}$	+	$A^-_{(aq)}$
starting conc.	:	0.150 M		≅ 0 M		0 M
change, Δ	:	– ? M		+ ? M		+? M
equilibrium conc.:		? M		1.0×10^{-4} M		? M

$$\Delta[H^+] = (1.0 \times 10^{-4} M) - (0\ M) = +1.0 \times 10^{-4} M$$

$$\Delta[HA] = (1.0\times10^{-4}\ M\ \Delta\ H^{+})\times\frac{1\ HA}{1\ H^{+}} = -1.0\times10^{-4}\ M$$

$$\Delta[A^{-}] = (1.0\times10^{-4}\ M\ \Delta\ H^{+})\times\frac{1\ A^{-}}{1\ H^{+}} = +1.0\times10^{-4}\ M$$

		$HA_{(aq)}$ $\longrightarrow$	$H^{+}_{(aq)}$ +	$A^{-}_{(aq)}$
starting conc.	:	0.150 M	≅ 0 M	0 M
change, Δ	:	-1.0×10^{-4} M	$+1.0\times10^{-4}$ M	$+1.0\times10^{-4}$ M
equilibrium conc.	:	0.1499 M	≅ 1.0×10^{-4} M	1.0×10^{-4} M

(4) $$K_a = \frac{[H^{+}][A^{-}]}{[HA]} = \frac{(1.0\times10^{-4}\ M)(1.0\times10^{-4}\ M)}{(0.1499\ M)} = \mathbf{6.67\times10^{-8}}$$

2. **(1)** $C_6H_4NO_2H_{(aq)} \rightleftarrows H^{+}_{(aq)} + C_6H_4NO_2^{-}_{(aq)}$ **(2)** $K_a = \frac{[H^{+}][C_6H_4NO_2^{-}]}{[C_6H_4NO_2H]}$

(3) pH = −log $[H^{+}]$ = 2.70; equilibrium $[H^{+}]$ = antilog (−2.70) = $10^{-2.70} = 2.0\times10^{-3}$

$$\text{moles } C_6H_4NO_2H = \frac{1.00\,g}{123.1\,g/mole} = 0.00812 \text{ mole}$$

$$\text{Starting concentration of } C_6H_4NO_2H = \frac{0.00812 \text{ mole}}{0.0600\ L} = 0.135\ M$$

		$C_6H_4NO_2H_{(aq)}$ $\longrightarrow$	$H^{+}_{(aq)}$ +	$C_6H_4NO_2^{-}{}_{(aq)}$
starting conc.	:	0.135 M	≅0 M	0 M
change, Δ	:	− ? M	+ ? M	+ ? M
equilibrium conc.	:	? M	2.0×10^{-3} M	? M

$$\Delta[H^{+}] = (2.0\times10^{-3}\,M) - (0\ M) = +\,2.0\times10^{-3}\,M$$

$$\Delta[C_6H_4NO_2H] = (2.0\times10^{-3} M\,\Delta H^{+})\times\frac{1\ C_6H_4NO_2H}{1\ H^{+}} = -2.0\times10^{-3} M$$

$$\Delta[C_6H_4NO_2^{-}] = (2.0\times10^{-3} M\,\Delta H^{+})\times\frac{1\ C_6H_4NO_2^{-}}{1\ H^{+}} = +2.0\times10^{-3}\ M$$

		$C_6H_4NO_2H_{(aq)}$ $\longrightarrow$	$H^{+}_{(aq)}$ +	$C_6H_4NO_2^{-}{}_{(aq)}$
starting conc.	:	0.135 M	≅ 0 M	0 M
change, Δ	:	$-\,2.0\times10^{-3}$ M	$+2.0\times10^{-3}$ M	$+2.0\times10^{-3}$ M
equilibrium conc.	:	0.133 M	≅ 2.0×10^{-3} M	2.0×10^{-3} M

(4) $$K_a = \frac{[H^{+}][C_6H_4NO_2^{-}]}{[C_6H_4NO_2H]} = \frac{(2.0\times10^{-3}\ M)(2.0\times10^{-3}\ M)}{(0.133 M)} = 3.0\times10^{-5}$$

3. **(1)** $HCN_{(aq)} \rightleftharpoons H^+_{(aq)} + CN^-_{(aq)}$ **(2)** $K_a = \frac{[H^+][CN^-]}{[HCN]}$

(3)

	$HCN_{(aq)}$	$\longrightarrow$	$H^+_{(aq)}$	+	$CN^-_{(aq)}$
starting conc. :	0.150 M		≅ 0 M		0 M
change, Δ :	– x M		+ ? M		+? M
equilibrium conc.:	? M		? M		? M

$$\Delta[H^+_{(aq)}] = (x\,M\,\Delta\,HCN_{(aq)}) \times \frac{1H^+}{1HCN} = +x\,M$$

$$\Delta[CN^-_{(aq)}] = (x\,M\,\Delta\,HCN_{(aq)}) \times \frac{1CN^-}{1HCN} = +\,x\,M$$

	$HCN_{(aq)}$	$\longrightarrow$	$H^+_{(aq)}$	+	$CN^-_{(aq)}$
starting conc. :	0.150 M		≅ 0 M		0 M
change, Δ :	– x M		+ x M		+ x M
equilibrium conc.:	(0.150 – x) M		x M		x M

(4) $K_a = \frac{[H^+][CN^-]}{[HCN]}$; $4.0 \times 10^{-10} = \frac{(x)(x)}{(0.150 - x)}$

Simplification: (0.150–x) ≅ (0.150)

$$4.0 \times 10^{-10} = \frac{(x^2)}{(0.15)}; (4.0 \times 10^{-10})(0.15) = x^2;\ 0.60 \times 10^{-10} = x^2$$

x = (0.60 × 10–10)1/2 = 7.7 × 10⁻⁶ M

[H+] = x = 7.7 × 10⁻⁶ M; [CN–] = x = 7.7 × 10⁻⁶ M

[HCN] = 0.150 – x = (0.150) – (7.7 × 10⁻⁶ M) ≅ 0.150 M

pH = – log[H+] = –log (7.7 × 10⁻⁶) = 5.11

(5) $K_a = \frac{[H^+][CN^-]}{[HCN]}$; $4.0 \times 10^{-10}\ ? = ?\ \frac{(7.7 \times 10^{-6})(7.7 \times 10^{-6})}{(0.150)} = 4.0 \times 10^{-10}$ ✓

4. **(1)** $C_8H_7O_2COOH_{(aq)} \rightleftharpoons H^+_{(aq)} + C_8H_7O_2COO^-_{(aq)}$ **(2)** $K_a = \frac{[H^+][C_8H_7O_2COO^-]}{[C_8H_7O_2COOH]}$

(3) moles $[C_8H_7O_2COOH] = \frac{1.30\,g}{180.2\,g/mole} = 0.00721\,mole$

$$\text{Starting concentration} = \frac{0.00721\,mole}{0.0500\,L} = 0.144\,M.$$

	$C_8H_7O_2COO_{(aq)}$	$\longrightarrow$	$H^+_{(aq)}$	+	$C_8H_7O_2COO^-_{(aq)}$
starting conc. :	0.144 M		≅ 0 M		0 M
change, Δ :	– x M		+ ? M		+? M
equilibrium conc.:	? M		? M		? M

$$\Delta[H^+{}_{(aq)}] = (x\,M\,\Delta\,C_8H_7O_2COOH_{\,(aq)}) \times \frac{1H^+}{1C_8H_7O_2COOH} = +\ x\,M$$

$$\Delta[C_8H_7O_2COO^-{}_{(aq)}] = (x\,M\,\Delta\,C_8H_7O_2COOH_{\,(aq)}) \times \frac{1C_8H_7O_2COO^-}{1C_8H_7O_2COOH} = +\ x\,M$$

		$C_8H_7O_2COO_{(aq)}$ $\longrightarrow$	$H^+_{(aq)}$ +	$C_8H_7O_2COO^-_{(aq)}$
starting conc.	:	0.144 M	≅ 0 M	0 M
change, Δ	:	− x M	+ x M	+ x M
equilibrium conc.:		(0.144 − x) M	x M	x M

(4) $K_a = \dfrac{[H^+][C_8H_7O_2COO^-]}{[C_8H_7O_2COOH]};\ 3.27\times10^{-4} = \dfrac{(x)(x)}{(0.144-x)}$

Simplification: (0.144 − x) ≅ (0.144)

$$3.27\times10^{-4} = \frac{x^2}{(0.144)};\ (3.27\times10^{-4})(0.144) = x^2\ ;\ 0.471\times10^{-4} = x^2$$

$$x = (0.471\times10^{-4})^{1/2} = \mathbf{6.86\times10^{-3}\ M}$$

$$[H^+] = x = 6.86\times10^{-3}\ M;\ [C_8H_7O_2COO^-] = x = 6.86\times10^{-3}\ M$$

$$[C_8H_7O_2COOH] = 0.144 - x = (0.144) - (6.86\times10^{-3}M) = 0.137\ M$$

$$\mathbf{pH} = -\log[H^+] = -\log(6.86\times10^{-3}) = \mathbf{2.16}$$

(5) $K_a = \dfrac{[H^+][C_8H_7O_2COO^-]}{[C_8H_7O_2COOH]};\ 3.27\times10^{-4}\ ?=?\ \dfrac{(6.86\times10^{-3})(6.86\times10^{-3})}{(0.137)} = 3.43\times10^{-4}$ approximate ✓

5. HCl is a strong acid ($pK_a = -7$); the weak base in the reaction is aqueous NH_3. The conjugate acid of the base is NH_4^+ with a pK_a of 9.25 ($K_a = 5.6\times10^{-10}$). The problem type can be identified as **Type (B)** or **Type (C)** depending on which component is in excess.

GenStep (**1**): $HCl_{(aq)} + NH_{3\,(aq)} \longrightarrow NH_4^+{}_{(aq)} + Cl^-{}_{(aq)}$

GenStep (**2**): $[\mathbf{A}] = \dfrac{\{V(L)\ of\ \mathbf{A}\}\times\{\mathbf{M}\ of\ \mathbf{A}\}}{V(L)\ solution\ total}$

V(L) total solution = (0.080 L $HCl_{(aq)}$) + (0.150 L $NH_{3(aq)}$) = 0.230 L total

$$\text{initial}\,[\mathbf{HCl}] = \frac{(0.080\,L)(1.0\,M)}{0.230\,L\ total} = \mathbf{0.348\,M}$$

$$\text{initial}\,[\mathbf{NH_3}] = \frac{(0.150\,L)(0.80\,M)}{0.200\,L\ total} = \mathbf{0.522\,M}$$

	$HCl_{(aq)}$	+	$NH_{3\,(aq)}$	$\longrightarrow$	$NH_4{}^+{}_{(aq)}$	+	$Cl^-{}_{(aq)}$
starting conc.:	0.348 M		0.522 M		0 M		0 M
change, Δ :	−0.348 M		−0.348 M		+ 0.348 M		+0.348 M
finish :	0 M		0.174 M		0.348M		0.348M

GenStep (**3**): HCl is limiting; NH_3 is in excess. This identifies the reaction as problem Type (C) with the weak component in excess. Solve the equilibrium of the ammonia base with water as the acid; the (Cl^-) very weak base does not contribute to basicity.

(1) $NH_{3(aq)} + H_2O_{(l)} \rightleftharpoons NH_4{}^+{}_{(aq)} + OH^-{}_{(aq)}$ **(2)** $K_b = \dfrac{[NH_4{}^+][OH^-]}{[NH_3]}$

(3) pK_b of NH_3 = 4.75; K_b = antilog (−4.75) = $10^{-4.75}$ = $\mathbf{1.8 \times 10^{-5}}$

(x) = molarity change of NH_3; the starting concentrations for the equilibrium calculations are those that remain or are produced from the initial stoichiometric reaction.

$$\Delta[NH_4{}^+{}_{(aq)}] = (x\ M\,\Delta\,NH_{3(aq)}) \times \frac{1\,NH_4{}^+}{1\,NH_3} = +\,\mathbf{x\,M}$$

$$\Delta[OH^-{}_{(aq)}] = (x\ M\,\Delta\,NH_{3(aq)}) \times \frac{1\,OH^-}{1\,NH_3} = +\,\mathbf{x\,M}$$

	$NH_{3\,(aq)}$	+ $H_2O_{(l)}$ $\longrightarrow$	$NH_4{}^+{}_{(aq)}$	+	$OH^-{}_{(aq)}$
starting conc. :	0.174 M		0.348 M		≅0 M
change, Δ :	− x M		+ x M		+x M
equilibrium conc.:	(0.174 − x) M		(0.348 + x) M		x M

(4) $K_b = \dfrac{[NH_4{}^+][OH^-]}{[NH_3]}$; $1.8\times10^{-5} = \dfrac{(0.348+x)(x)}{(0.174-x)}$

Simplification: (0.174−x) ≅ (0.174) and (0.348 + x) ≅ (0.348)

$$1.8\times10^{-5} = \frac{(0.348)(x)}{(0.174)};\ (1.8\times10^{-5})(0.174) = (0.348)(x)$$

$$\mathbf{x} = \frac{(3.13\times10^{-6})}{(0.348)} = \mathbf{9.0\times10^{-6}\,M}$$

$$\mathbf{[OH^-] = x = 9.0\times10^{-6}\,M;}$$

$[NH_4{}^+]$ = (0.348 + x) = (0.348) + (9.0 × 10^{-6}) ≅ 0.348 M (0.35 M; 2 significantly figures)

$[NH_3]$ = (0.174−x) = (0.174)−(9.0 × 10^{-6}) ≅ 0.174 M (0.17 M)

pOH = −log $[OH^-]$ = −log (9.0 × 10^{-6}) = **5.05** (5.0)

pH = (14 − pOH) = 14 − 4.05 = **8.95** (10.0)

(5) $K_b = \frac{[NH_4^+][OH^-]}{[NH_3]}$; 1.8×10^{-5} ?=? $\frac{(0.348)(9.0\times10^{-6})}{(0.174)} = 1.8\times10^{-5}$ ✓

6. **(1)** $H_2CO_{3(aq)} \rightleftarrows H^+_{(aq)} + HCO_3^-{}_{(aq)}$ **(2)** $K_a = \frac{[H^+][HCO_3^-]}{[H_2CO_3]}$

a)(3) $\Delta[H^+_{(aq)}] = (\mathbf{x\,M}\,\Delta H_2CO_{3(aq)}) \times \frac{1\,H^+}{1\,H_2CO_3} = \mathbf{+\,x\,M}$

$\Delta[HCO_3^-{}_{(aq)}] = (\mathbf{x\,M}\,\Delta H_2CO_{3\,(aq)}) \times \frac{1\,HCO_3^-}{1\,H_2CO_3} = \mathbf{+\,x\,M}$

		$\mathbf{H_2CO_{3(aq)}}$ $\longrightarrow$	$\mathbf{H^+_{(aq)}}$ +	$\mathbf{HCO_3^-{}_{(aq)}}$
starting conc.	:	0.020 M	≅ 0 M	0 M
change, Δ	:	−x M	+ x M	+ x M
equilibrium conc.:		(0.020 − x) M	x M	x M

(4) $K_a = \frac{[H^+][HCO_3^-]}{[H_2CO_3]}$; $4.3\times10^{-7} = \frac{x^2}{(0.020 - x)}$; $(0.020 - x) \cong (0.020)$

$4.3\times10^{-7} = \frac{x^2}{(0.020)}$; $(4.3\times10^{-7})(0.020) = x^2$; $\mathbf{x} = (8.6\times10^{-9})^{1/2} = \mathbf{9.3\times10^{-5}\ M}$

$\mathbf{[H^+] = x = 9.3 \times 10^{-5}\,M}$; $[HCO_3^-] = x = 9.3 \times 10^{-5}$ M

$[H_2CO_3] = 0.020\text{ M} - x = (0.020) - (9.3 \times 10^{-5}\text{ M}) = 0.0199$ M

$\mathbf{pH} = -\log[H^+] = -\log(9.3 \times 10^{-5}) = \mathbf{4.03}$

b) **(3)** In this problem, the starting concentration of $[HCO_3^-]$ is 0.100 M from the sodium hydrogen carbonate.

		$\mathbf{H_2CO_{3(aq)}}$ $\longrightarrow$	$\mathbf{H^+_{(aq)}}$ +	$\mathbf{HCO_3^-{}_{(aq)}}$
starting conc.	:	0.020 M	≅ 0 M	0.100 M
change, Δ	:	−x M	+ x M	+ x M
equilibrium conc.:		(0.020 − x) M	x M	(0.100 + x) M

(4) $K_a = \frac{[H^+][HCO_3^-]}{[H_2CO_3]}$; $4.3\times10^{-7} = \frac{(x)(0.1+x)}{(0.020-x)}$; $(0.02 - x) \cong (0.02)$; $(0.1 + x) \cong (0.1)$

$4.3\times10^{-7} = \frac{(x)(0.1)}{(0.020)}$; $\mathbf{x} = \frac{(4.3\times10^{-7})(0.02)}{(0.1)} = \mathbf{8.6\times10^{-8}\,M}$

$\mathbf{[H^+] = x = 8.6 \times 10^{-8}\,M;}$

$[HCO_3^-] = 0.100\text{ M} + x = 0.100 + 8.6 \times 10^{-8}\text{ M} \cong 0.100$ M

$[H_5CO_3] = 0.020\text{ M} - x = (0.020) - (8.6 \times 10^{-8}\text{ M}) \cong 0.020$ M

$\mathbf{pH} = -\log[H^+] = -\log(8.6 \times 10^{-8}) = \mathbf{7.07}$

(3) The starting concentration of $[H^+_{(aq)}]$ is 0.001 M from the added acid. Since this solution contains no weak base, the added $H^+_{(aq)}$ is not neutralized; the starting $H^+_{(aq)}$ is used in the equilibrium calculation.

		$\mathbf{H_2CO_{3\,(aq)}}$ $\longrightarrow$	$\mathbf{H^+_{(aq)}}$	+	$\mathbf{HCO_3^-{}_{(aq)}}$
starting conc.	:	0.020 M	0.001 M		0 M
change, Δ	:	−x M	+ x M		+x M
equilibrium conc.:		(0.020 − x) M	(0.001 + x) M		x M

$$\textbf{(4)}\ K_a = \frac{[H^+][HCO_3^-]}{[H_2CO_3]};\ 4.3\times10^{-7} = \frac{(0.001+x)(x)}{(0.020-x)};\ (0.02-x) \cong (0.02)$$

$[H^+] = (0.001 + x) \cong (0.001)$; Proof that x can be ignored:

$$4.3\times10^{-7} = \frac{0.001(x)}{(0.02)};\ \mathbf{x} = \frac{(4.3\times10^{-7})(0.02)}{(0.001)} = \mathbf{8.6\times10^{-6}\ M}$$

$$[H^+] = \cong 0.001\ M + x = 0.001 + 8.6\times10^{-6}\ M \cong \mathbf{0.001\ M}$$

$$[HCO_3^-] = x = 8.6\times10^{-6}\ M;$$

$$[H_2CO_3^-] = 0.020\ M - x = (0.020) - (8.6\times10^{-6}\ M) \cong 0.020\ M$$

$$\mathbf{pH} = -\log[H^+] = -\log(0.001) = \mathbf{3}$$

(d) **(3)** The stoichiometric reaction between the added strong acid and the conjugate weak base shows complete initial neutralization of the added strong acid as the limiting reagent; the weak base is in excess.

		$\mathbf{HCO_3^-{}_{(aq)}}$	+	$\mathbf{H^+_{(aq)}}$ $\longrightarrow$	$\mathbf{H_2CO_{3\,(aq)}}$ (≅ **100% reaction)**
starting conc.:		0.100 M		0.001M	0.020M
change, Δ	:	−0.001 M		−0.001M	+0.001M
final conc.	:	0.099 M		≅0M	0.021M

The weak acid/weak base equilibrium is solved with the changed concentrations.

		$\mathbf{H_2CO_{3\,(aq)}}$ $\longrightarrow$	$\mathbf{H^+_{(aq)}}$	+	$\mathbf{HCO_3^-{}_{(aq)}}$
starting conc.	:	0.021 M	≅0M		0.099M
change, Δ	:	− xM	+ xM		+xM
equilibrium conc.:		0.021 − xM	xM		0.099 + xM

$$\textbf{(4)}\ K_a = \frac{[H^+][HCO_3^-]}{[H_2CO_3]};\ 4.3\times10^{-7} = \frac{(x)(0.099+x)}{(0.021-x)}$$

$(0.021 - x) \cong (0.021)$ and $(0.099 + x) \cong (0.099)$

$$4.3\times10^{-7} = \frac{(x)\,(0.099)}{(0.021)};\ \mathbf{x} = \frac{(4.3\times10^{-7})(0.021)}{(0.099)} = \mathbf{9.1\times10^{-8}\ M}$$

$$[H^+] = x = \mathbf{9.1\times10^{-8}\,M}$$

$$[HCO_3^-] = 0.099\ M + x = 0.099 + 9.1\times10^{-8}M \cong 0.099\ M$$

$$[H_2CO_3] = 0.021\ M - x = (0.021) - (9.1\times10^{-8}\,M) \cong 0.021\ M$$

$$\mathbf{pH} = -\log\,[H^+] = -\log\,(9.1\times10^{-8}) = \mathbf{7.04}$$

7. a) **(1)** $NH_{3(aq)} + H_2O_{(l)} \rightleftarrows NH_4^+{}_{(aq)} + OH^-{}_{(aq)}$ **(2)** $K_b = \frac{[NH_4^+][OH^-]}{[NH_3]}$

(3) pK_b of $NH_3 = 4.75$; $K_b = \text{antilog}\,(-4.75) = 10^{-4.75} = 1.8\times10^{-5}$

$$\text{initial } [NH_3]_0 = \frac{0.20\,\text{mole}}{1.00\,L} = 0.20\,M;\ \text{initial}\,[NH_4^+]_0 = \frac{0.30\,\text{mole}}{1.00\,L} = 0.30\,M$$

		$\mathbf{NH_{3(aq)} + H_2O_{(l)}}$	$\longrightarrow$	$\mathbf{NH_4^+{}_{(aq)}}$	+	$\mathbf{OH^-{}_{(aq)}}$
starting conc.	:	0.20 M		0.30 M		≅ 0 M
change, Δ	:	−x M		+x M		+x M
equilibrium conc.:		(0.20 − x) M		(0.30 + x) M		x M

(4) The Hendersen–Hasselbach equation assumes that simplification applies to all values of (x) added to or subtracted from the starting concentrations; the starting concentrations are used directly. The equilibrium applied can be acid dissociation or the base reaction with water; the conjugate acid/base pair must be identified.

$NH_{3(aq)}$ acts as the base

$NH_4^+{}_{(aq)}$ acts as the acid; $\mathbf{pK_a} = (14 - PK_b) = 14 - 4.75 = \mathbf{9.25}$

$$pH = pK_a + \log_{10}\frac{\{[A^-]_0\}}{\{[HA]_0\}};\ pH = pK_a + \log_{10}\frac{\{[NH_3]_0\}}{\{[NH_4^+]_0\}}$$

$$pH = 9.25 + \log_{10}\frac{\{(0.20)\}}{\{(0.30)\}};\ \mathbf{pH} = 9.25 + \log_{10}(0.667) = 9.25 + (-0.18) = \mathbf{9.07}$$

b) Solve the stoichiometric reaction of the strong base with the weak conjugate acid in the buffer system; the partial ionic equation is:

$$NH_4^+{}_{(aq)} + NaOH_{(aq)} \longrightarrow NH_{3\,(aq)} + H_2O_{(l)}\ (+\,Na^+{}_{(aq)})$$

The starting buffer concentrations are the same as for part (a). 0.15 M NaOH was added and is the limiting reagent.

	$NH_4^+{}_{(aq)}$	+	$NaOH_{(aq)}$	$\longrightarrow$	$NH_{3\,(aq)}$	+	$H_2O_{(l)}$
starting conc.:	0.30 M		0.15 M		0.20 M		excess
change, Δ :	−0.15 M		−0.15 M		+0.15 M		
finish :	0.15 M		0 M		0.35 M		

Apply the Hendersen–Hasselbach equation with the altered concentrations directly:

$$\text{pH} = \text{pK}_a + \log_{10}\frac{\{[A^-]_0\}}{\{[HA]_0\}}\ ;\ \text{pH} = \text{pK}_a + \log_{10}\frac{\{[NH_3]_0\}}{\{[NH_4^+]_0\}}$$

$$\text{pH} = 9.25 + \log_{10}\frac{\{(0.35)\}}{\{(0.15)\}}\ ;\ \mathbf{pH} = 9.25 + \log_{10}(2.33) = 9.25 + (0.37) = \mathbf{9.62}$$

Index

Printed in the United States
by Baker & Taylor Publisher Services